CHEMISTRY FOR MEDICAL TECHNOLOGISTS

Chemistry

FOR MEDICAL TECHNOLOGISTS

WILMA L. WHITE, B.A.
Supervisor, Clinical Chemistry Laboratory,
Barnes Hospital, St. Louis, Mo.

MARILYN M. ERICKSON, B.S.
Supervisor, Clinical Chemistry Laboratory,
Barnes Hospital, St. Louis, Mo.

SUE C. STEVENS, B.A., M.A., Ph.D., F.A.I.C.
Director, Division of Endocrine Chemistry,
Jewish Hospital; Assistant Professor of Pathology,
Washington University School of Medicine,
St. Louis, Mo.

THIRD EDITION

With 275 illustrations and 1 color plate

The C. V. Mosby Company

Saint Louis 1970

THIRD EDITION

Second printing

Previous editions copyrighted 1958, 1965

Standard Book Number 8016-5431-9

Library of Congress Catalog Card Number 77-116590

Distributed in Great Britain by Henry Kimpton, London

• CONTRIBUTORS

MARY E. COMFORT, B.S.

Research Assistant, Metabolism Division, Washington University School of Medicine, St. Louis, Mo.

CHARLEY F. GUTCH, M.D.

Director, Dialysis Training Center, University of Utah College of Medicine, Salt Lake City, Utah; formerly Medical Director, Chronic Dialysis and Home Training Center; formerly Assistant Professor of Medicine (Renal Division), University of Colorado Medical Center, Denver, Colo.

EUGENE G. HAMILTON, M.D.

Director, Obstetrics and Gynecology, St. Mary's Hospital; Professor of Clinical Obstetrics and Gynecology, St. Louis University School of Medicine, St. Louis, Mo.

MARGIE ELLEN HAUSE, B.A., MT(ASCP)

Research Instructor in Anesthesia, Department of Surgery, Washington University School of Medicine, St. Louis, Mo.

JAMES N. HEINS, M.D.

Instructor in Medicine, Metabolism Division, Washington University School of Medicine, St. Louis, Mo.

CARL R. JOLLIFF, B.S.

Clinical Biochemist and Microbiologist, Laboratory Section, Lincoln Clinic, and Lincoln Medical Research Foundation; Visiting Lecturer, Department of Microbiology, University of Nebraska; Consulting Lecturer, Nebraska Wesleyan University, Lincoln, Neb.

NANCY KAY McADAM, A.B., MT(ASCP)

Assistant Supervisor, Clinical Chemistry Laboratory, Barnes Hospital, St. Louis, Mo.

JOSEPH R. SIMMLER, A.S., B.S., M.A.

Simmler & Son, Inc., St. Louis, Mo.

SISTER ANN MARIE SKINNER, R.S.M., M.A., MT(ASCP)

Research Instructor of Medicine, Washington University School of Medicine, St. Louis, Mo.

• CONSULTANTS

Hormones

WILLIAM H. DAUGHADAY, M.D.

Professor of Medicine, Washington University School of Medicine (Director, Metabolism Division); Associate Physician and Consultant for Clinical Chemistry, Barnes Hospital, St. Louis, Mo.

Rapid screening procedures

ALFRED H. FREE, Ph.D.

Vice President, Technical Services, Ames Company, Division of Miles Laboratories, Inc., Elkhart, Ind.

The kidney

CHARLEY F. GUTCH, M.D.

Director, Dialysis Training Center, University of Utah College of Medicine, Salt Lake City, Utah; formerly Medical Director, Chronic Dialysis and Home Training Center; formerly Assistant Professor of Medicine (Renal Division), University of Colorado Medical Center, Denver, Colo.

Acid-base metabolism

ALBERT ROOS, M.D.

Research Professor in Anesthesia, Department of Surgery, and Professor of Physiology and Biophysics, Washington University School of Medicine, St. Louis, Mo.

Normal pediatric values

VIRGINIA V. WELDON, M.D.

Instructor, Department of Pediatrics, Washington University School of Medicine, St. Louis, Mo.

• PREFACE

The revised edition has been changed to update methodology, including the addition of automated procedures. Manual methods have been retained or added, for use in the small laboratories and in doctors' offices.

We have taken into consideration the constructive criticisms offered and have attempted to move this revision toward the newer concepts of laboratory chemistry. It is impossible to include all methods that are serving a laboratory well. Automated methods are being revised constantly. Those for the Technicon system and for the Robot Chemist are given. Even though the Robot Chemist is no longer available, inclusion was considered of value to the study of present methods because of the number of instruments being used successfully, including the one in our laboratories.

The chapter on quality control has been enlarged, with emphasis on the industrial approach, the basis for the technique as applied in the medical laboratory. Recognizing that normal values are influenced by methodology, age, sex, and environmental conditions, we have included tables that are offered only as guides.

In the march forward we use guidelines for values and methods. Presented here are methods that a large group of scientists have found satisfactory for their laboratories. The methods of choice depend upon laboratory needs and equipment capabilities. New chapters on acid-base metabolism, atomic absorption, hormones, radioimmunoassay, chromatography, and electrophoresis and immunoelectrophoresis have been introduced, consistent with the progress that is occurring in these areas. The chapter on the kidney has been expanded to include physiology and dialysis because of the large dialysis and transplant programs in which the laboratories participate.

As your constructive recommendations arrive, a greater cross section of opinion will be evaluated and the book will become still more useful for your needs and ours.

Wilma L. White
Marilyn M. Erickson
Sue C. Stevens

• CONTENTS

CHEMISTRY FOR MEDICAL TECHNOLOGISTS

• 1

• Blood collection and sample preparation

The composition of body fluids in health is maintained within narrow physiologic limits. The human body consists of 50% to 70% fluid by weight—the fluids being classified as intracellular (within the cell) or extracellular (outside the cells). A balance is maintained between the fluids as well as their constituents, in the various compartments.

Blood, one of the extracellular fluids, makes up approximately 7.5% and 7% by weight of normal men and women, respectively. The concentrations of the wide variety of substances in the blood remain constant within narrow limits for the normal individual and these levels are known. When some part or parts of the body become involved in physiologic disorders or pathologic states, there may be a decrease or increase in the amount of the regular constituents or some abnormal factor may appear in the blood. From such changes as these, either a diagnosis or the extent of an abnormal condition may be determined.

With children and adults, the blood is usually secured from finger or ear and veins of the forearm, wrist, or ankle. With infants, the blood is generally taken from heel or toe and veins of the forearm, neck, or head. Since the usual practice is to obtain blood from the finger or from veins of the forearm, these two methods will be considered here in detail. Before collection of the blood is started, however, any apprehension displayed by the patient should be dispelled.

After a meal, food substances are absorbed into the bloodstream. However, other than glucose and inorganic phosphorus, the concentrations of the majority of blood chemical constituents are not significantly changed after a light breakfast. Therefore patients need not be in a fasting state prior to blood specimen collection for most of the chemical tests performed by the diagnostic laboratory.

BLOOD COLLECTION

Procedure for capillary puncture of finger or ear

An extremely old technique for obtaining blood is the puncture wound of the capillaries. Although the finger puncture is the most used, the blood drawer should be familiar also with the ear puncture and its good points and drawbacks. While the differences in chemistry values between the capillary blood of the ear and that of the finger are for most purposes insignificant, the following notations must be taken into consideration if blood from the ear is used for hematologic work.[1]

Unless the patient has been in the cold or under undue stress, the red cell counts and hemoglobin values are essentially the same as values obtained from a finger stick or from venous blood.

It has been found that the white cell count from the ear is slightly higher than the count from the finger. Also, larger white cells may be present in greater numbers.

There is a striking difference in results obtained from earlobe blood, versus blood from finger or vein, in certain pathologic conditions. The difference is seen in the white cell counts and the differential counts of infectious mononucleosis, leukemias, and bacterial endocarditis. The increase of white cells from the earlobe involves all types of white cells, especially the large cells such as monocytes and histiocytes that are seen in large numbers in bacterial endocarditis. This has been explained largely by the greater filtering capacity of the vascular bed of the earlobe in comparison to that in the tip of the finger.

In preparation of blood films for the study of leukemias, leukopenias, and certain infections, the earlobe is the most desirable source of blood. In blood from an earlobe puncture, the first two drops may contain more large cells. After several drops, the cells from this source are similar to those from the finger stick or the venipuncture.

The finger or ear is cleansed with an alcohol sponge, incision is made with a disposable sterile lancet, and drops of blood are available for examination. Detailed instructions for each step follow.

Preparing finger or earlobe

The fingertip is cleansed with a sterile gauze pad or sponge, moistened with 70% alcohol, leaving no moisture. The blood will form full rounded drops. If the patient has been outdoors in cold weather, the earlobe is first gently warmed to enhance circulation and then cleansed by an alcohol sponge leaving no moisture.

Puncturing finger or earlobe

By means of a loose wrist motion the finger is punctured with a sterile disposable lancet. The puncture is made by firmly grasping the finger and making a quick deliberate stab to a depth of about 3 mm into the finger. It is better to induce a good puncture rather than make the patient apprehensive by having to do a repeat. A deep puncture is no more painful than a superficial one.

Eliminating the first drop

The first drop of blood, containing tissue juices and foreign matter, is always wiped off. The diluted and contaminated drop of blood would give unreliable results.

Withdrawing blood

Free-flowing blood should be used for examination, although gentle pressure may be applied if needed to enhance the flow of blood. *Never* exert heavy pressure, because that may cause the flow of tissue juices in the blood.

Preventing further bleeding

After the required amount of blood has been taken, an antiseptic pad (70% alcohol pad) is placed on the puncture. The patient may be instructed to apply pressure to the wound until the bleeding has ceased. (See Fig. 1-1.)

Procedure for venipuncture

Larger volumes of blood needed for chemical analyses are obtained from the veins of the forearm, wrist, or ankle. Because of convenience and the larger

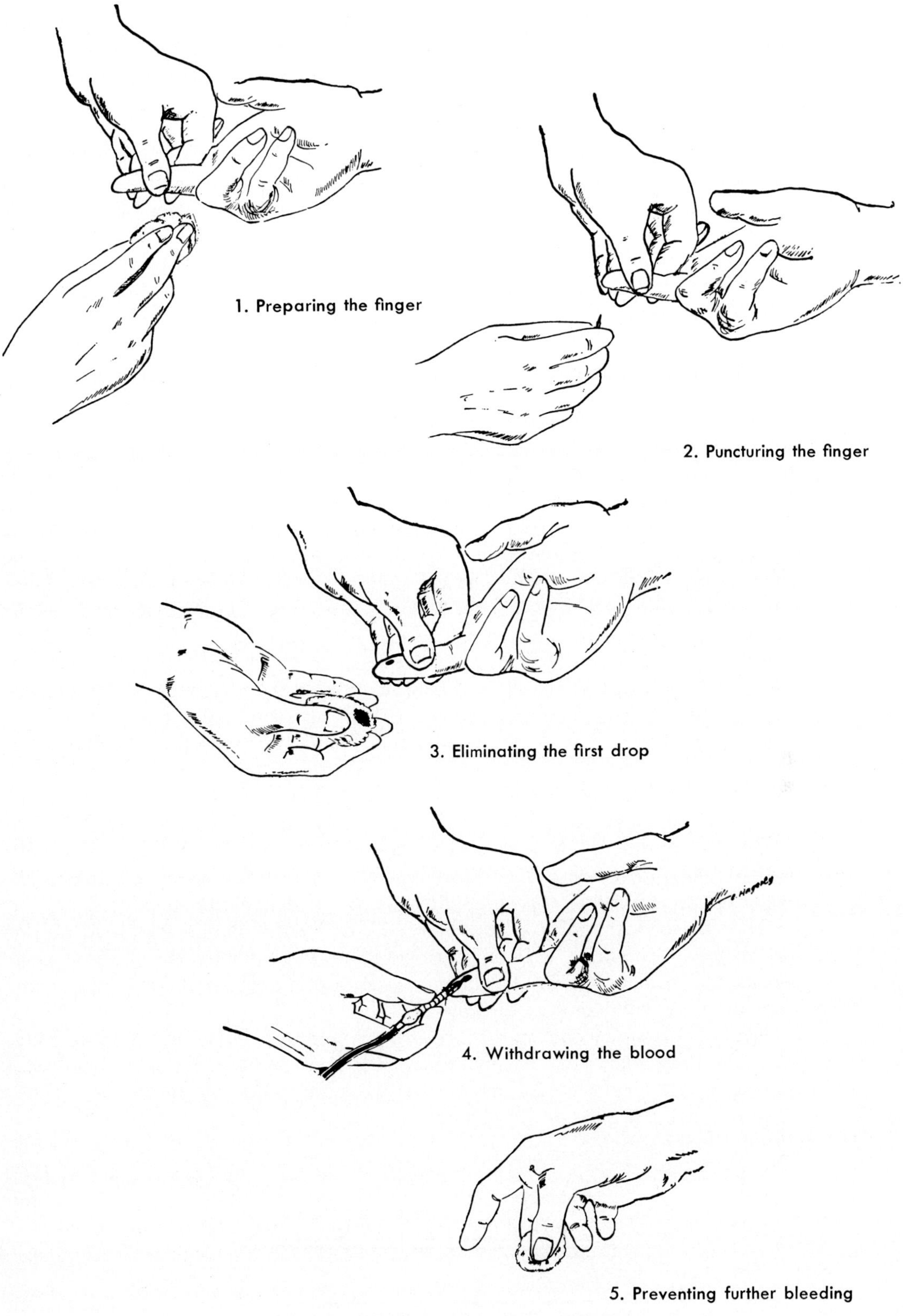

Fig. 1-1. Steps in the finger puncture.

size of the veins in the forearm, this area is usually chosen for a venipuncture. With good technique and sharp needles, the same vein may be used many times with little discomfort to the patient or damage to the vein. However, bandages, soreness, obesity, or scar tissue formed from poor techniques may sometimes make it necessary to use the veins of the wrist or ankle.

Obtaining a venipuncture is an art which demands compassion, knowledge, and skill to minimize discomfort to the patient. If the patient is not confined to a bed, have him seated comfortably in a chair with his arm placed upon a table, counter, or some other solid form. The chair and table should be on a level such that the arm will be extended downward at an angle and not raised above the shoulder level. Avoid low chairs and high tables. If the patient has a tendency to bend his arm, place a support under the elbow. Use books, a package of paper towels, or anything solid enough to hold the arm straight. A patient may feel faint from apprehension. Keep a fresh bottle of smelling salts available or moisten a piece of cotton with *dilute* ammonium hydroxide. Wave this under his nose and if you do not get immediate reponse, *call a physician.*

If a patient is in bed, place the blood-collecting tray where it will not be upset by either yourself or the patient. Try not to disturb the patient any more than is absolutely necessary. Take the arm available and gently place it toward you. If the patient shows apprehension, calmly explain what you are going to do and what you would like him to do. If there is still a slight pulling away of the arm, there are two things you may do. Fold a towel about four times and place it under his elbow to keep it from bending; then very gently place your elbow upon the patient's hand and wrist. The second suggestion is to hold the patient's elbow gently with your left hand and introduce the needle with the right—or vice versa if you are left-handed. A combination of the two steps is also successful. The object of this is to obtain blood without upsetting the patient.

The actual technique of obtaining venous blood is given in the following steps.

Preparing needle and syringe

The diameter or bore of the needle is indicated by its gauge number. The smaller the number, the greater the diameter. Gauges 20 and 21 are usually used for the forearm and 25 is used for the veins of the wrist and ankle. The size of the needle chosen is regulated by the size of the vein to be entered.

The choice in size of a syringe is regulated by the amount of blood needed for analysis. Ordinarily, a 10 ml syringe is most frequently used. The syringe and needle *must be sterile.* Do *not* use a syringe in which there is any dampness. Moisture will hemolyze the blood.

When the needle and syringe have been removed from their sterile packages, carefully place the needle firmly onto the syringe. Do not handle the cannula of the needle or the plunger of the syringe. Use the hub of the needle (Fig. 1-2) and the head of the plunger.

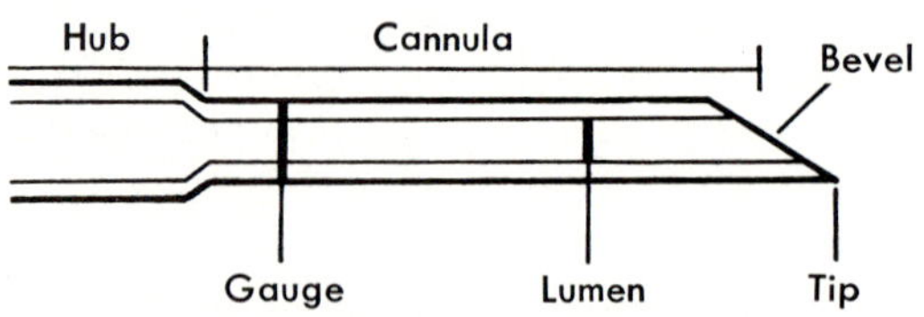

Fig. 1-2. Diagram of the needle.

Using Vacutainer tubes

Vacutainer tubes are widely accepted for blood letting. Rubber stoppers are color coded to identify the anticoagulant present, or the tube may be left plain and be used for blood that can be allowed to clot. The vacuum within the tube is sufficient to draw a predetermined volume of blood. The blood-drawing set contains a disposable double-pointed needle that screws into the holder. When the needle is in place, the Vacutainer tube is inserted into the holder, with the tube's rubber stopper touching the needle. It is then pushed forward until the top of the stopper reaches the guide line. Without breaking the vacuum the needle is partially embedded in the stopper in order to prevent leakage of blood at the time of venipuncture. Once the blood set has been assembled, the venipuncture needle is inserted into the arm in the routine manner and the needle in the Vacutainer stopper is pushed through the diaphragm, breaking the vacuum and thus allowing blood to flow into the tube. When more than one tube of blood is needed, the first tube may be removed from the needle and another be inserted while the needle is still in the vein.

Applying tourniquet

It is most desirable to enlarge the veins of the forearm so that they may become more prominent. With the area for puncture located, utmost care must be used for the proper distention of the vein. Let the arm hang down for a while. The veins will become more apparent.

A piece of soft rubber tubing is most commonly used for a tourniquet although types of tourniquets may range from pieces of cloth to elaborate mechanisms. A blood pressure cuff is an excellent tourniquet.

The tourniquet is placed above the elbow. The patient then continually opens and closes his hand to aid in filling the vein. During the time that the blood is being drawn the hand is kept closed.

Selecting a vein

After the arm has been hanging down until the veins are prominent, the most visible one is selected. If the veins are poorly dilated, even after pumping the hand, release the tourniquet, place a hot wet towel over the arm and cover with a dry towel. Keep this heat pack on the arm for 10 or 15 minutes. Electric blankets, heating pads, and hair driers have been used successfully to dilate the veins. A word of caution: do not use the hair dryer around any explosives. When the circulation is improved, reapply the tourniquet.

Applying antiseptic

Many antiseptic solutions are available, but the one most commonly used is 70% ethyl alcohol.

Lundy[2] has expressed the necessity of warming the sponge and antiseptic solution, at least to body temperature, before applying to the area of injection. Because the arm is warm it is more sensitive to the cold, and contact with a cold antiseptic will cause the vein to contract.[2,3]

Inserting needle

Normally, when the size of the vein to be entered is sufficiently large in relation to the needle size, place the bevel of the needle upward. If a large needle is all that is available for a small vein, however, place the needle into the vein with the bevel facing down. This will prevent a hematoma that might occur from puncturing the wall of the vein. Occasionally a small vein will collapse, covering the opening of the needle, and the blood drawer in attempting to adjust the needle to clear the passage may puncture the wall and cause a hematoma. Many veins have a tendency to "roll" when stuck with a needle. The blood drawer may hold or "fix" the vein either by holding the vein with his thumb or by grasping the patient's arm below the elbow and pulling the skin taut. The needle is held at an angle of approximately 25 degrees and no higher, for there is the possibility of going through the vein.

Withdrawing blood

When the arm has been prepared with antiseptic and the tourniquet tightened, the needle is aimed in line with the long axis of the vein. Never approach the vein from the side. Holding the syringe by both the plunger and the barrel, pierce the skin. After the blood has entered the syringe, the plunger of the syringe is slowly pulled back. Should the blood start to enter the syringe and then cease, the needle has either slipped out of the vein or gone through it. If it has slipped out, carefully reinsert it. If it has gone through the vein, slowly withdraw the needle until it is once again in the vein. This ability comes with experience. In remedying the above problems, be sure to avoid pushing the plunger forward in the syringe. In the case of a collapsed vein, if little or no blood has been obtained but there is a large air space within the syringe, obtain a new syringe and connect it to the needle. This, of course, is unpleasant for the patient, but it must be done.

Releasing tourniquet

Before the needle is withdrawn from the vein, the tourniquet must be released. Have the patient relax his fist before you release the tourniquet. Otherwise, you will still obtain a blood flow through the puncture.

Withdrawing needle

Withdraw the needle after the tourniquet has been removed or released. Place a dry, sterile pad or adhesive bandage over the puncture wound immediately.

Preventing further bleeding

The patient may gently apply pressure to the sterile pad covering the wound. Folding the arm up toward the shoulder, combined with the finger pressure, will aid in stopping the flow of blood. A hematoma is sometimes formed by the blood seeping into subcutaneous tissue, leaving a large black and blue mark painful to the patient.

• • •

Hematomas may be caused by several factors: (1) using too large a needle for the vein, therefore puncturing the wall; (2) failing to have the needle completely in the vein, thus allowing blood to enter into the tissues; (3) forgetting to release the tourniquet before withdrawing the needle, with the result that the pressure expels blood from the vein puncture into the tissues; (4) failing to apply finger pressure to the wound for a sufficient time; and last, but not least,

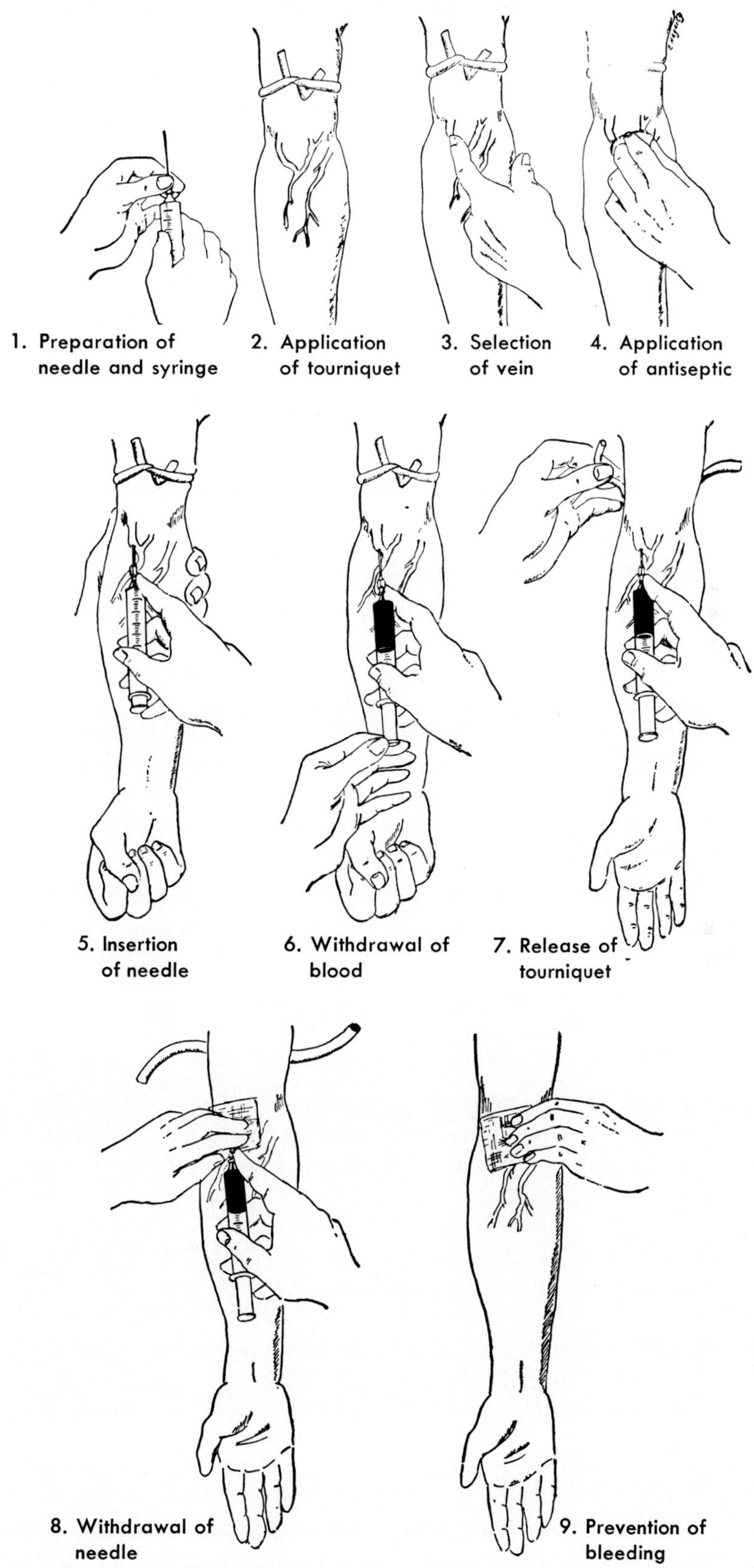

Fig. 1-3. Steps in the venipuncture. (From Seiverd, C. E.: Hematology for medical technologists, ed. 2, Philadelphia, 1958, Lea & Febiger.)

(5) not taking the time and care to do a good job the first time and then having to do repeated punctures of the vein.

The steps in the venipuncture are illustrated in Fig. 1-3.

SAMPLE PREPARATION

The blood usually requires some preparation before analysis. Serum is preferred to plasma or whole blood for most chemical analyses. However, in general, heparinized plasma may be used in place of serum, since the basic difference between them is the fibrinogen in plasma. Serum or plasma must be separated from the blood cells by centrifugation within an hour after collection. A few procedures require a protein-free filtrate. This can be prepared by using either whole blood (anticoagulated), serum, or plasma.

Serum

The reaction in blood clotting is a conversion of the soluble protein fibrinogen to the insoluble fibrin. This may be simply described as progressing in steps or phases. In the first phase, ionized calcium and thromboplastic factors (cephalin, tryptases, thromboplastins) activate prothrombin, forming thrombin. In the second phase, the thrombin converts fibrinogen into fibrin. This is the actual formation of the clot.

Plasma

In order to obtain plasma, the blood must be kept from clotting. Substances that prevent clotting are called anticoagulants. Some chemical anticoagulants remove the ionized calcium from the blood. Heparin prevents clotting by inactivating prothrombin. The more popular anticoagulants and the amounts required per milliliter of blood are given below:

Anticoagulant	*Amount per milliliter of blood*
Sodium oxalate	2 mg
Potassium oxalate	2 mg
Sodium fluoride	2 mg
Heparin	0.2 mg
Lithium oxalate	1 mg

A mixture of potassium oxalate and sodium fluoride is used as an anticoagulant and preservative for some glucose methods. However, since sodium fluoride inhibits enzyme activity, it cannot be used as an anticoagulant for any enzyme determination. It can be used in the urease method for blood urea nitrogen (BUN) if the amount does not exceed 2 mg/ml. Heparin, a physiologic anticoagulant, and EDTA (ethylenediamine tetra-acetate), a chelating agent binding calcium, are very popular anticoagulants used in chemistry and hematology laboratories, since there is no destruction of the cells.

To prepare test tubes containing the proper amount of potassium oxalate (10 mg), place 1 ml of a 1% solution of potassium oxalate (1 g/100 ml of distilled water) in each test tube. Evaporate to dryness in an oven that does not exceed 105° C. When blood is added, invert the tube until the dried oxalate is dissolved.

Protein-free filtrate

If an acid is added to whole blood, plasma, or serum, it precipitates the proteins. The proteins can then be removed by filtration or centrifugation. The clear fluid that remains is known as a protein-free filtrate.

A protein-free filtrate can be prepared in a number of ways. Procedures for three methods are given below. If it is necessary to prepare any of the reagents, refer to Appendix D and the number that follows the reagent. (Note: If only 1 ml of blood is available, divide all the quantities mentioned by 2.) Serum and plasma contain less protein than whole blood. Therefore, when using the Folin-Wu method for precipitating proteins from serum or plasma, mix 2 ml of serum or plasma with 16 ml of water and add 1 ml of 0.67 N sulfuric acid and 1 ml of 10% sodium tungstate. For the Somogyi method the concentration of zinc sulfate and of barium hydroxide in solution is not critical, but they must exactly neutralize each other.

Folin-Wu method

Reagents needed

Sulfuric acid, 0.67 N	(D-1)
Sodium tungstate, 10%	(D-2)

Procedure

1. Place 14 ml of distilled water in a 125 ml Erlenmeyer flask.
2. Invert the tube containing the oxalated blood several times in order to mix.
3. Using an Ostwald-Folin volumetric pipette, add 2 ml of the oxalated blood to the Erlenmeyer flask. Gently swirl the flask to hemolyze the cells.
4. While swirling the flask, slowly add 2 ml of 0.67 N sulfuric acid.
5. While swirling the flask, slowly add 2 ml of 10% sodium tungstate.
6. Let stand for a few minutes. Mix and then filter.

Haden method

Reagents needed

Sulfuric acid, 0.08 N	(D-3)
Sodium tungstate, 10%	(D-2)

Procedure

1. Place 16 ml of 0.08 N sulfuric acid in a 125 ml Erlenmeyer flask.
2. Invert the tube containing the oxalated blood several times in order to mix.
3. Using an Ostwald-Folin volumetric pipette, add 2 ml of the oxalated blood to the Erlenmeyer flask. Gently swirl the flask to hemolyze the cells.
4. While swirling the flask, slowly add 2 ml of 10% sodium tungstate.
5. Let stand for a few minutes. Mix and then filter.

Somogyi method

Reagents needed

Zinc sulfate, 5%	(D-4)
Barium hydroxide, 0.3 N	(D-5)

Procedure

1. Invert the tube containing the oxalated blood several times to thoroughly mix the specimen. With an Ostwald-Folin pipette, add 1.0 ml of blood into a 25 ml Erlenmeyer flask containing 5.0 ml of distilled water. The water dilutes the blood and lakes (hemolyzes) the cells.
2. Add slowly with constant agitation 2.0 ml of 5% zinc sulfate. Mix well.
3. Add slowly 2.0 ml of 0.3 N barium hydroxide.
4. Insert a rubber stopper and shake vigorously until a metallic click is heard and the absence of foam indicates complete precipitation of the

protein. This is called the isoelectric point. Exacting technique must be used to obtain this point of precipitation.

5. **Let the flask stand for 5 minutes and then filter or centrifuge.**

If the protein-free filtrate is not used the day it is prepared, it should be kept in a refrigerator to prevent decomposition. If it is to be kept several days or sent through the mail, a few drops of toluene or xylene should be added as a preservative.

In the final calculations for the above tests, a correction is always made for the dilution of the blood that was caused by preparing the filtrate. For example, in preparing the filtrate, the blood is diluted 1 in 10. This means that 10 ml of filtrate represents 1 ml of blood.

REFERENCES

1. Page, L. B., and Culver, P. J.: Syllabus of laboratory examinations in clinical diagnosis, rev. ed., Cambridge, Mass., 1961, Harvard University Press.
2. Lundy, J. S.: Suggestions to facilitate venipuncture in blood transfusion, intravenous therapy and intravenous anesthesia, Mayo Clin. Proc. **12**:122, 1937.
3. Abbott Laboratories, Medical Department: Parenteral administration, Chicago, 1959.

• 2

• Quality control

THE HISTORY AND FUNCTION OF QUALITY CONTROL

The history of quality control dates back to the Middle Ages. When man learned to build and to sell his talents as a craftsman, the quality of his efforts determined his success. Long periods of apprenticeship were required for learning a skill, and evidence of this training had to meet certain standards before an apprentice could become a master craftsman. In an effort to maintain quality of output, various methods for quality control have been introduced into the manufacturing market. Today, production quality in industry can be checked by on-line inspection.

STATISTICS

Statistics, a means of grouping and interpreting facts by use of numbers, was introduced about three centuries ago.[12] Statistical application was first employed in astronomy and physics and later utilized in biologic and social sciences. In the eighteenth and early nineteenth century three mathematicians—de Moivre, Laplace, and Gauss—independently devised a mathematical displacement distribution resembling a bell-shaped curve (Fig. 2-1), now commonly referred to as the gaussian curve. Such a distribution of values from the average or true value has been called the *law of errors.*

Statistical quality control

Statistical quality control (SQC) is a relatively new application. The first sketch of a quality control chart was made by Walter A. Shewhart, Bell Telephone Laboratories, in 1924.[8] Development of this new technique of utilizing charts to evaluate quality continued, and in 1931 Shewhart[14] published a book on statistical quality control. This eventually set the pattern for statistical methods as applied to process control. Dodge and Romig, also of the Bell Telephone Laboratories, adopted the statistical method for sampling inspection. The conclusion of their work resulted in the Dodge-Romig sampling inspection tables. The investigations carried out by Shewhart, Dodge, and Romig constitute the basis for the major part of statistical control.[3-5,14]

Industrial development

British industry readily accepted statistical quality control and by 1937 its use was widespread. In the United States, however, enthusiasm was lacking. American industry considered that the quality control measures already being applied were adequate and that introducing unnecessary mathematics would be an excessive burden.

With the outbreak of World War II the armed forces became a large consumer of industrial products and influenced the adoption of statistical quality control. The Bell Telephone Laboratories were invited to join the government

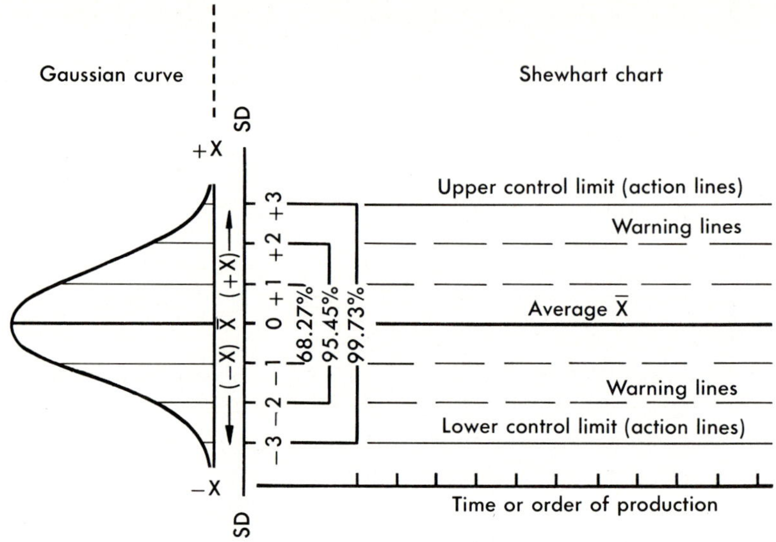

Fig. 2-1. Normal distribution: curve and chart.

in developing a sampling inspection program for Army Ordnance. The combined efforts resulted in sampling inspection tables for the Army Ordnance and the Army Service Forces.

The demand for this new method of evaluating both production and products and the shortage of personnel trained to use it led to the establishment of educational programs for teaching statistical quality control. In July, 1942, an intensive 10-day course was given at Stanford University for representatives of war industries and procurement agencies of the armed forces. The training sessions were sponsored by the Engineering Science and Management War Training Program of the U. S. Office of Education. A shorter (8-day) training period was given in Los Angeles. Courses were then provided in various centralized areas throughout the country by the Office of Production Research and Development of the War Production Board.

Clinical development

Compared with industry's application of statistical control, the clinical diagnostic laboratories' adoption of this tool for routine analysis control has been slow. The need for quality control in the medical diagnostic laboratory has long been recognized. Belk and Sunderman[1] in 1947 and Wootton and King[18] in 1953 reported results obtained from surveys in which blood specimens containing constituents that had been determined by reference laboratories were sent to a number of routine hospital laboratories. Wide variation in the analytic data obtained by the different laboratories confirmed the need for control procedures. In 1953 the first commercial control serum of known value was made available. It was in liquid form and was marketed under the name of Lab-Trol. Currently there are lyophilized sera and control urines and spinal fluids commercially prepared to enable the analyst to maintain a monitoring program of laboratory assay results.

Automation will not eliminate analytic problems in chemical determinations.

If a procedure is poor for measuring a specified constituent, improvement will come only with a better chemical method. Statistical quality control (SQC) will include variables inherent within a chemical determination. The following conditions are monitored by SQC: (1) reagent purity, (2) standard reliability, (3) accuracy of pipettes, autodilutors, and analysts, (4) cleanliness of all glassware and equipment, (5) calibration of the readout apparatus, (6) mechanical condition of the analytic hardware, and (7) proper calculation of the data obtained.

In recent years there has been a worldwide realization within medical laboratories that the needed degree of accuracy is not being achieved. Surveys carried out in routine hospital laboratories have revealed wide discrepancies between the various laboratory areas. With the introduction of statistical quality control into the laboratories the whole application has assumed the appearance of a "whodonit?" between the supervisory or managerial staff and the technical staff. For development of a productive, efficient laboratory, members of the technical staff must be oriented in the importance of instrument care, methodology limitations, analysis or quantitation demands, and purpose of a quality control program. They must be trained to utilize quality control rather than be policed by it. There certainly is a need for unknown constituents to be placed in the runs and for active participation in surveys or multiple-hospital control programs.

Purpose of quality control in laboratories

Many factors contribute to the quality of work produced in the laboratory. Management; professional and technical personnel; collection, delivery, and storage of specimens; analytic procedures; and transcription of records are a few of the variables related to the final result of an analysis. Team effort with understanding communications between all areas is needed for a proper program. Quality control is not a gathering of data for display or policing of the various work areas. It is a method of obtaining technical control by stimulating a team objective, with mutual assistance toward success in achieving a goal.

Starting a program

In setting up a program or revamping an existing one, it is important to establish simple rules that are easily understood. Consideration must also be given to the following:

1. Determine whether the program will be sectional (chemistry, hematology, microbiology, serology, etc.), with individual quality control supervisors, or will function as one overall program, including each laboratory and using one quality control team.
2. Choose the individual(s) who will assume the responsibility of inspector and define to all employees the range of authority invested in the inspector. If the laboratory is small, each person may assume the quality control responsibility with his duties on a rotation basis for a month.
3. Train all personnel in (a) proper methods of specimen collection, delivery, preparation, and storage and (b) correct performance of analytic methods and transcription of results. Doing this will include use of teaching sessions, clearly written directions, and the sending of individuals to certified workshops.
4. Determine the acceptable limits of each method.

5. For each procedure add the volumes of control needed in the normal and the abnormal ranges and choose the material to be used as the control.
6. Determine the type of tabulation to be carried out and the charts to be used in each case.

LABORATORY APPLICATION OF QUALITY CONTROL

Specimen collection

Collecting a specimen correctly is a vital step for an analytic method; yet it is practically impossible to eliminate all the errors of collection. A list of the tests and the right methods of specimen collection, including also the proper preservatives or anticoagulants, should be posted in every nursing and blood-drawing station. The instructions should be carried out as written. Physiologic sources, such as medication and foods, and improper collection of specimens may cause misleading laboratory results to be given the physician. If the error is subtle—in that it cannot be detected by the laboratory—such as incorrect labeling of a tube or combination of the urine of two patients in the same bottle, the effect could be harmful to the patient.

A few basic rules for specimen collection follow:

1. Read the directions and follow them.
2. Properly label all timed collections.
3. Do not draw blood from the arm receiving an intravenous fluid or transfusion of blood.
4. Collect a 24-hour urine during a period of exactly 24 hours and add to the designated bottle as soon as a specimen is voided. Each urine bottle may have a color-coded tag to identify the preservative and the group of tests for which its contents may be used.

A moment of care and observation may save the patient—and all concerned—grief, money, or time.

Specimen delivery and storage

Another area that can limit laboratory efficiency and patient care is the transport of specimens to the laboratory. Assuming that the specimens are drawn correctly and are contained in the proper tubes or bottles, there is still the problem of delivery. Smaller hospitals or medical centers have the advantage of utilizing laboratory employees for blood drawing and ward personnel for delivery of other specimens. The use of employees such as pages or dispatch messengers, who have no other contact with the specimen than its delivery, introduces a source of error difficult to control. Specimens that require processing within a limited time are adversely affected by delayed delivery. Thus errors may be compounded by prolonged delays in delivery, improper storage, and deterioration due to incorrect preservative before the analytic procedure is initiated.

Laboratory errors

The specimen's arrival in the laboratory does not guarantee infallibility in handling. Specimen identification must be carefully controlled throughout the analysis and during transcription of the results to the proper requisitions. Manual manipulations of the specimens and work sheets allow many possibilities for error.

Automation does not prevent erroneous handling of specimens. For example, sample plates may be loaded in a sequence different from that recorded on the work sheet, stats may be inserted during a run in the place of removed specimens and the latter not re-entered in the series later, or charts may be incorrectly numbered.

Controls

Quality control includes more than tabulating the results from a series of standards and controls. Factors to be taken into consideration include specificity of method, time required for analysis, the available and the needed equipment, cost of analysis, and proficiency of personnel. There must be a commonsense approach in choosing a reliable method and in establishing a continuing program for monitoring both the precision and the accuracy of test results obtained. The latter should be kept as consistent as possible, for an accurate monitoring of the area.

Various controls may be used:

1. Solutions of pure chemicals. Depending upon the method these controls may or may not be processed through an entire analysis. In prolonged procedures a solution containing a known concentration of the constituent is frequently analyzed along with the unknown specimen.

2. Patient controls. A specimen from one patient may be carried over in each batch run of a given test. This is an excellent method for checking multiple series of analyses throughout the day. The standard deviation may be calculated by dividing the average difference of the carryover specimen by 1.128, or on a long-term evaluation, by using this formula[16]

$$SD = (\Sigma d^2/2\,N)^{\frac{1}{2}}$$

in which d = the difference between duplicates and N = the number of duplicates performed.

3. Controls of pooled sera, cerebrospinal fluid (CSF), or urine. Sera and CSF with values in the normal range may be pooled separately. From this range, dilutions can be made for lower concentrations. For abnormal levels of constituents, in which the collection of material is not so prevalent, a group of laboratories could share in the pooling and in a monitoring program. A 24-hour collection of urine will supply the needs for most programs.

4. Commercially prepared sera and CSF. This is a more expensive program. Controls can be purchased as unassayed material or as analyzed products with target values. Quality control programs are now available commercially. With the wide selection of material, it is recommended that dependence not be placed on any one control and that after selection of the primary control other products be maintained for cross-checking.

5. Statistical evaluation of daily results. Such an application of statistics can be extremely useful in monitoring analytic precision. This method will detect a subtle shift in the total operation. It is a rapid technique for observing a curve.

The shelf-life of a control depends largely upon refrigeration. There is difficulty in maintaining pooled controls free from bacterial contamination. Frozen material should be held at a temperature of −20° C. Lyophilized material must be kept refrigerated. Constituents in biologic fluids vary in their stability, and pools should be cross-checked to validate the assay results.

Controls for monitoring hormone analyses are usually prepared from pooled collections of plasma and urine. Each new lot of pooled specimen is analyzed along with the previous batch and with any available commercially assayed controls. After such standardizing, the pooled material is divided into small aliquots and stored at −20° C.

Standards

A standard is a criterion for comparison that has been established by custom or authority. It may be a rule governing quality, quantity, or conduct. Standards for weights and measures such as the gram, pound, liter, gallon, meter, and yard are fixed by the National Bureau of Standards (NBS). Since its establishment in 1901, NBS has made available standard reference materials having certified properties and has provided specifications for weights and other units of measurement. The clinical chemist may need either to obtain from the Bureau a certified weight for calibrating weights available in the laboratory or send the weights to the Bureau for calibration, to request volumetric glassware that has been calibrated by the Bureau, and to purchase various chemicals for standardization. NBS has available salts for buffers, such as potassium hydrogen phthalate and phosphates, and chemicals for standards such as glucose, cholesterol, urea, uric acid, and creatinine.

Manufacturers who provide drugs or substances used in clinical practice must conform to the standards for these products as set forth in the *United States Pharmacopeia.* Included are assay procedures for determining compliance with pharmacopeial standards of purity and strength. Some professional organizations also publish standard or official procedures of analysis. *Official Methods of Analysis* is published by the Association of Official Agricultural Chemists; *Standard Methods for the Examination of Dairy Products,* by the American Public Health Association; and *Standard Methods for the Examination of Water and Sewage,* by the American Public Health Association and the American Water Works Association. A number of other organizations are active in programs for control and standardization of laboratory methods.

In an analytical chemistry laboratory a primary standard is defined as a chemically pure compound of known composition that can be accurately weighed and used in the preparation of a solution of known concentration. A secondary standard is one that can be prepared in approximate concentration, standardized against a primary standard, and then used to determine the concentration of another compound for which a primary standard is not available. For example, potassium hydrogen phthalate, a primary standard, may be used to determine the exact concentration of a solution of sodium hydroxide. The latter may then be used to determine the exact concentration of a solution of hydrochloric acid. In this case, the sodium hydroxide acts as a secondary standard.

The clinical chemistry laboratory is concerned with analyzing biologic fluids for various constituents, some of which are not available in pure form. When these compounds are to be used as standards, a known weight of the highest purity obtainable dissolved in a suitable medium is accepted. The solution with compounds of known concentration or primary standard is used as a basis for comparison with the constituents being measured in the unknown sample. Depending on the procedure, the standard may or may not be processed through all the steps of the method in the same manner as the unknown and controls. Regardless of the stage at which the standard is introduced into a procedure, the purpose is to make possible a comparison of the reaction of the unknown and that of the standard.

As a result of claims that standards in an aqueous system react differently from those in serum, commercial preparations have been made available that contain weighed amounts of various constituents added in a manner to duplicate human serum. The known levels of the added chemicals are indicated with each lot purchased. Since known absolute concentrations of desired constituents have been weighed into these products, they are considered, by the manufacturers, primary standards. However, the presence of unknown factors as a part of the medium negates this claim for such a product.

A standard solution must not be confused with a control sample. The latter is a specimen (serum, plasma, spinal fluid, urine) having approximately the same composition as the unknown that is to be analyzed. It is carried through the entire procedure in the same manner as the unknown and serves to monitor the precision of the determination. Concentrations of the constituents should approximate the levels of those in the unknown samples being assayed. Control samples are usually taken from pooled fluids that have been purchased commercially or prepared in the laboratory. They are preserved by lyophilizing, freezing, or sterilizing. The commercially available product is accompanied by a certificate listing the assayed values determined by reference laboratories and also the methods employed in the testing. A control sample can never be used as a standard; but a standard may be employed as a control.

Another type of standard used in clinical laboratories is called an internal standard. This term refers to the addition of a known amount of a pure compound to a routine sample being analyzed.

Data

Collecting

Data accumulated for control purposes should be kept in a special notebook. A serial listing of the daily monitoring makes possible a rapid comparison with previous values. At each work area a daily register of blanks, standards, working control, and baseline readings of the inverse colorimetric methods (AutoAnalyzer) should be maintained. This presents a continual check on chemistry, standards, and instrument conditions. The settings for the AutoAnalyzer colorimeter will indicate to the analyst if the photocells need to be replaced. With the lists available, control calculations can be performed at any time.

The unknown controls are tabulated and charted for visual display. At least 30 analyses should be tabulated for a consistent monitoring of a method. A large quality control program will exceed 30 analyses each month. Therefore the standard deviation is calculated after the desired population (total test) is obtained.

A small laboratory cannot maintain a large quality control program. The processing of control samples along with tests that are prolonged and time-consuming is limited. When there are fewer controls (less than 10 samples), it is more practical to compute the mean deviation than the standard deviation. The relative efficiency of mean deviation is high when measuring dispersion of the small group of samples, 91% for samples up to 10. However, with normal variations, the standard deviation when corrected for bias is more efficient.

Tabulating and studying

Preparing or reviewing data requires understanding of some basic terminology. When symbols or unfamiliar terms are used, the meaning must be explained.

Rates, ratios, and percentages

Rates, ratios, and percentages are used daily in the routine work. The AutoAnalyzer may be employed in an example for defining rate and ratio. *Rate* refers to a numerical proportion between two variables, such as the different sampling speeds per hour (20, 40, or 60/hr). *Ratio* denotes a quantitative relation between similar measurements such as the length of time a sample probe remains in the sample cup and its stay in the wash cup. A comparison of the time interval for each sampling step is the ratio—2:1, 1:1, or 1:2. Symbols, shown here in parentheses and commonly used to compare numbers, are (:) meaning *is to* and (::) meaning *as.* Therefore the statement may be made that 4 is to 2 as 2 is to 1 or 4:2::2:1. *Percentage* usually designates the frequency or quantity in proportion to a sum or total amount of 100 units. Therefore, a percentage is no more than a ratio that employs a base of 100 units.

A rule for computing a ratio is to *always* divide by the base. The base is the reference figure to which the other number is to be compared. For computing percentages, remember that the base should be 100 or more. The percentage of increase may be infinitely large but the percentage of decrease, unless nega-

Table 2-1. Sample work sheet for tabulating test results

A Number of tests	B Test results	C Difference from the mean	D Difference from the mean squared
1	5.8	0.0	0.00
2	5.6	0.2	0.04
3	5.9	0.1	0.01
4	5.8	0.0	0.00
5	5.7	0.1	0.01
6	5.8	0.0	0.00
7	5.8	0.0	0.00
8	5.8	0.0	0.00
9	5.4	0.4	0.16
10	5.7	0.1	0.01
11	5.7	0.1	0.01
12	5.6	0.2	0.04
13	5.8	0.0	0.00
14	5.7	0.1	0.01
15	5.8	0.0	0.00
16	5.8	0.0	0.00
17	5.8	0.0	0.00
18	5.9	0.1	0.01
19	5.7	0.1	0.01
20	5.8	0.0	0.00
21	5.9	0.0	0.00
22	5.7	0.1	0.01
23	5.7	0.1	0.01
24	5.8	0.1	0.01
25	(9.3)	(Specimen contaminated by cigarette ashes)	
Total	138.0		0.34

tive values are possible, does not exceed 100. The negative values usually occur in business where profit and loss data are involved.

In any computation, the decimal point is a source of potential error. A 48.9 numerical value, for example, might be recorded as 489 (ten times the actual figure) or as 4.89 (one tenth the designated value).

To assist in calculating the standard deviation and making a control chart for the daily plotting of tests, the following example is given. The tests being calculated are potassium results that were run routinely in the laboratory. The specimens used were obtained from pooled sera prepared and previously assayed. The potassium value is above the normal range for potassium. Specimens of the order listed last are called *outliers.* Quality control is needed for elevated and abnormal ranges, as well as for checks within the normal range. Table 2-1 illustrates also the results of careless and inferior technical application in analysis.

Standard deviation

Referring to Table 2-1, calculate the standard deviation.

1. **Total column B and divide by the number of tests done. Test 25 is eliminated from the averages because of a technical error.**

$$\text{Average or mean value} = \frac{138.0}{24} = 5.8$$

2. **Determine the absolute difference of each test from the average value (5.8). The difference is recorded in column C, which is *not* added. Do not bother placing plus or minus signs with the differences.**
3. **In column D, the difference from the mean is squared (multiplied by itself: e.g., 0.4 × 0.4 = 0.16). This column is totaled, giving a result of 0.34.**
4. **With this data, calculate the standard deviation. With n = number of tests done, the formula for calculating SD in this example is as follows:**

$$\text{Standard deviation} = \sqrt{\frac{\text{Sum of squared differences (column D)}}{n - 1}}$$

$$SD = \sqrt{\frac{0.34}{24 - 1}}$$

$$SD = \sqrt{\frac{0.34}{23}}$$

$$SD = \sqrt{0.0148}$$

$$SD = 0.12$$

The divisor (n − 1) is called the degrees of freedom in the sample. This term as applied to statistics refers to the number of independent differences possible with a series of observations. For example, given one sample with one result there is no variation possible. If there are ten results from a sample, then there are nine possible differences.

The symbol "$\sqrt{\quad}$" means taking the square root of a number, 0.0148 in the example shown. Values may be obtained from Table 2-2 or from Table A-8, in Appendix A, for the square roots of small numbers. In order to check the accuracy, multiply the square root by itself. A slide rule may be used to obtain either a square of a number or a square root.

Table 2-2. Squares and square roots

n	n^2	$\sqrt{n}$	n	n^2	$\sqrt{n}$	n	n^2	$\sqrt{n}$	n	n^2	$\sqrt{n}$
1	1	1.000	13	169	3.605	25	625	5.000	37	1369	6.082
2	4	1.414	14	196	3.741	26	676	5.099	38	1444	6.164
3	9	1.732	15	225	3.872	27	729	5.196	39	1521	6.244
4	16	2.000	16	256	4.000	28	784	5.291	40	1600	6.325
5	25	2.236	17	289	4.123	29	841	5.385	41	1681	6.403
6	36	2.449	18	324	4.242	30	900	5.477	42	1764	6.480
7	49	2.645	19	361	4.358	31	961	5.567	43	1849	6.557
8	64	2.828	20	400	4.472	32	1024	5.656	44	1936	6.634
9	81	3.000	21	441	4.582	33	1089	5.744	45	2025	6.708
10	100	3.162	22	484	4.690	34	1156	5.830	46	2116	6.782
11	121	3.316	23	529	4.795	35	1225	5.916	47	2209	6.856
12	144	3.464	24	576	4.898	36	1296	6.000	48	2304	6.928

n = the number.
n^2 = the number squared.
$\sqrt{n}$ = the square root of the number.

To find the square of a number

Set the hairline to the number on the D scale and read its square under the line on the A scale. The square may also be found by setting the hairline on the C scale and reading the square on the B scale.

To find the square root of a number

Set the hairline on the number of the A scale and read the square root under the line on the D scale. Reading from the B scale to the C scale may also be used.

Note: **Use the left half of the slide rule A scale for odd numbers left of the decimal point (1.30). Use the right half of the A scale for the even numbers before the decimal point (24.00).**

Whether it is found by means of chart, slide rule, or calculation, the standard deviation of 0.12 in this example represents ± 1 standard deviation. In other words, 68% of the tests will vary 0.12 from the average or mean value of 5.8. In view of the fact that controlling a test this closely is highly improbable, 3 deviations have been set as the "confidence limits." With some methods, it is desirable to hold to 2 deviations. A simple multiplying by 2 or by 3 will give the control limits desired:

	Lower	*Average*	*Upper*
± 1 SD = 0.12 mEq/L	5.7	5.8	5.9
± 2 SD = 0.24 mEq/L	5.6	5.8	6.0 (Warning lines)
± 3 SD = 0.36* mEq/L	5.4	5.8	6.2 (Action lines)

*0.36 is rounded off to 0.4.

Percent of the mean

It is now understood that standard deviation is the unit variation from an average or mean value. In the case of the example given, the unit variation is ± 0.12 mEq/L. The percent of the mean or average value called "coefficient of variation" is the percentage variation. To obtain this coefficient of variation (CV), the standard deviation value is divided by the mean or average value and then multiplied by 100. The following results are obtained by this formula:

$$\frac{\text{SD}}{\text{Mean value}} \times 100 = \text{Coefficient of variation}$$

±1 Standard deviation of the potassium = 0.12
Average (mean) value of the potassium = 5.8

$$\frac{0.12}{5.8} \times 100 = 2.06\% \quad 1\ \text{SD}$$

$$\frac{0.24}{5.8} \times 100 = 4.13\% \quad 2\ \text{SD}$$

$$\frac{0.36}{5.8} \times 100 = 6.20\% \quad 3\ \text{SD}$$

Control charts

There are at least four ways to present a chart for daily plotting of results:

1. The Shewhart control chart, with the mean and the deviations' boundaries drawn
2. The bell curve, inverted 90 degrees
3. The target control chart, with the rings representing the various deviations (not discussed here)
4. The cusum chart, discussed on p. 23

A convenient method of keeping the chart reusable from one month to another is placing the chart with its numerical values under a plastic cover. The average or mean value and the confidence limits are drawn on the plastic cover with a wax pencil. Marks can be wiped off with an alcohol sponge. An example of the Shewhart chart, with Table 2-1 data, is given in Fig. 2-2.

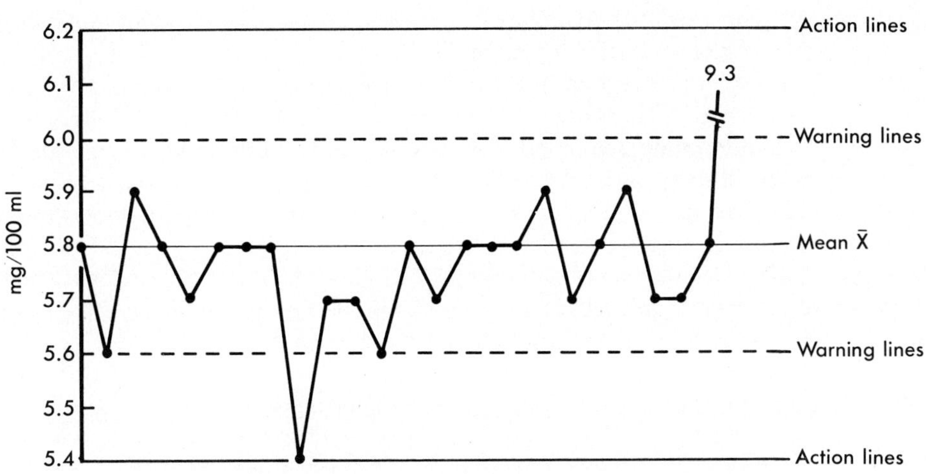

Fig. 2-2. Shewhart chart, using Table 2-1 data on potassium.

STATISTICAL METHODS

Control chart techniques

Shewhart chart

Shewhart[15] has suggested that control charts may have these three functions:

1. Specification—define the goal that management wishes to attain.
2. Production—act as an instrument to attain that goal.
3. Inspection—serve as a judge to verify whether the goal has been obtained.

His test on quality control set the pattern for later procedures and for application of statistical methods to process control. In this scheme of process control, samples of a fixed size are used and the statistics of the sample are plotted on a chart. When the point falls outside the chart, corrective action is initiated. The standard deviation (SD) lines may be called action lines.

Dudding and Jennett[6] in 1942 proposed that warning lines be drawn in less extreme positions than the action lines. Therefore the occurrence of data appearing between the warning and the action lines should be regarded as sufficient evidence of need for corrective measures. If the chart range is ± 3 SD, the warning lines would be at ± 2 SD; if the test limits are held at ± 2 SD, the warning lines should be at ± 1.5 SD. With a normal distribution of errors (gaussian), the warning limit will be exceeded for only 5% of the control results and the action limit for 0.3%. The control limits may be looked upon as follows:

SD	*% of observations*	*Beyond limits (approximate)*
± 1	68.27	1 out of 3
± 2	95.45	1 out of 20
± 3	99.73	1 out of 300

Levy and Jennings[11] in 1950 recommended the use of control charts in the clinical laboratory. Therefore some investigators who are medically oriented refer to the control chart as the Levy-Jennings chart.

When enough control samples have been analyzed, it is possible to determine the mean and the far points of distribution for the group. Sample fluctuations in a large group are distributed in a definite statistical pattern known as a normal distribution, illustrated by the bell or gaussian distribution curve. The bell curve is turned on its side, and the confidence limits are shown with SD lines of the control chart in Fig. 2-1.

The proper preparation of a control chart includes more than the grouping of a series of numbers. Methodology must be understood to prevent stringent demands being made on one method and overtolerance permitted for another. Establishment of warning lines requires that consideration be given to work performance under the influence of regular stress and the limitations of a routine day. Other important factors are listed below.

1. Accuracy and repeatability of the test must be reviewed with respect to low, normal, and elevated ranges that the method will have to cover, to identify the physiologic and pathologic limits.
2. Warning and action lines should be adjusted only after the method is under control and any large discrepancies have been corrected.
3. For a new method, duplicate analyses should be carried out and the charting include both the difference between and the average of the pairs (this is an excellent method for checking variations in multiple batch work).

4. Limits of individual test noise need to be determined and the difference between this noise and trouble signals recognized.

Allowable limits of error in methods form an area of debate. Tonks' formula[17] has set a maximum limit of ± 10% for determinations:

$$\text{Allowable limits of error (\%)} = \frac{\frac{1}{4}\text{ of normal range}}{\text{Mean of normal range}} \times 100$$

Some investigators insist on closer limits whereas others allow more than ± 10%. Without standardized methodology, agreements on limits of error for test procedures is not likely. It is important to recognize that close control must be maintained. A different confidence limit is important at the normal and at elevated levels.

Hoffman and Waid[7] have presented a method for using daily values of the normal range and calculating the mean and coefficient of variation. Each daily tally represents a controlled group or a shift in the method. This approach may be used as a supplement to the regular program.

Cumulative sum chart

The cumulative sum or cusum chart (CSC) is normally used in industry to maintain current control of a process.[2] It was proposed in 1954 by a British statistician, E. S. Page.[13] Recently the medical laboratories have started utilizing the chart to supplement the Shewhart chart.

The cumulative sum chart has an immediate advantage over other charts in that the work quality can be observed earlier by visual inspection.[10,13] The other

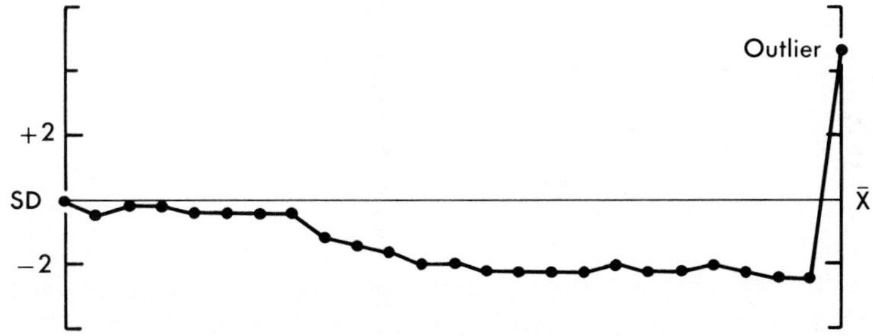

Fig. 2-3. Cusum chart of Table 2-1 data on potassium. Test out of control, as result of inferior technical application.

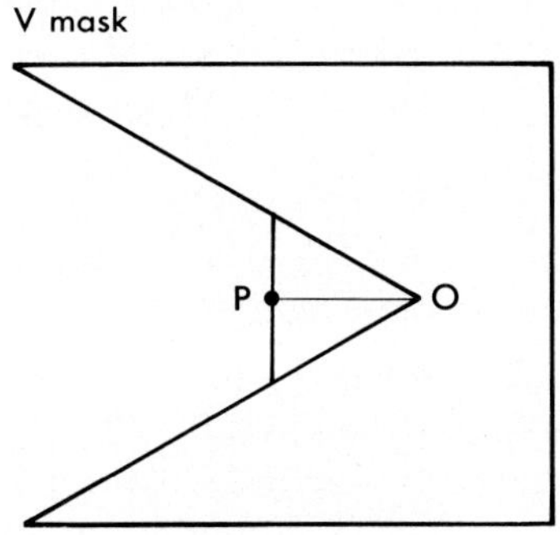

Fig. 2-4. V-mask.

Table 2-3. Tabulation of data for cusum chart

	Results	Difference	Plot		Results	Difference	Plot
1.	5.8	0	0	14.	5.7	−0.1	−1.1
2.	5.6	−0.2	−0.2	15.	5.8	0	−1.1
3.	5.9	+0.1	−0.1	16.	5.8	0	−1.1
4.	5.8	0	−0.1	17.	5.8	0	−1.1
5.	5.7	−0.1	−0.2	18.	5.9	+0.1	−1.0
6.	5.8	0	−0.2	19.	5.7	−0.1	−1.1
7.	5.8	0	−0.2	20.	5.8	0	−1.1
8.	5.8	0	−0.2	21.	5.9	+0.1	−1.0
9.	5.4	−0.4	−0.6	22.	5.7	−0.1	−1.1
10.	5.7	−0.1	−0.7	23.	5.7	−0.1	−1.2
11.	5.7	−0.1	−0.8	24.	5.8	0	−1.2
12.	5.6	−0.2	−1.0	25.	9.3	+3.5	+2.3
13.	5.8	0	−1.0				

charts display results plotted independently of each other. As observed in Fig. 2-3, the CSC has no warning or action lines. An accumulative variation on the CSC will travel in a horizontal pattern until it is out of control; then there will be a trend away from the previous pattern. The CSC can also be used to check blanks.

A V-shaped mask (Fig. 2-4) can be used to determine any process changes.[9] It also can be utilized to estimate the change in the process average. As each point is plotted on the CSC, it is evaluated to observe any shift from the $\overline{X}$ value. The point *P* of the mask is placed over the last point plotted on the chart, and the mask is leveled so that the *PO* of the mask is horizontal. If any of the previously plotted points fall under the mask, there has been a process change. If the points stay within the angle of the V, the process is considered in control. Dade has a cusum control chart with the mask lines on the chart. A ruler also may be used as a diagonal guide.

Tabulation of data on the cusum chart. A point of reference must be established for plotting the CSC. In clinical application the best reference to use is the mean ($\overline{X}''$) of the previous pool if the control material is still the same. With $\overline{X}''$ as the target, the data on the CSC will be the cumulative sums of $\overline{X} - \overline{X}''$. For example, if $\overline{X}''$ is 9.2 and $\overline{X}_1 = 9.2$, $\overline{X}_2 = 9.4$, and $\overline{X}_3 = 9.3$, the chart would be as follows: $\overline{X}(9.2) - \overline{X}''(9.2) = 0$, $9.4 - 9.2 = +0.2$, $9.3 - 9.2 = +0.1$. The cusum points would be 0, +0.2, and +0.3. Refer to Table 2-3 for the tabulation scheme.

Interpretation of the charts

Normal. With a normal distribution there is an almost even distribution on both sides of the mean line (Fig. 2-5, *A*). The cusum chart in Fig. 2-5, *B*, shows a normal horizontal path within the 95% confidence limits or ± 2 SD.

Systematic trend. Problems within the analytic procedure may show a systematic trend either up or down (Fig. 2-6). A trend upward may be caused by

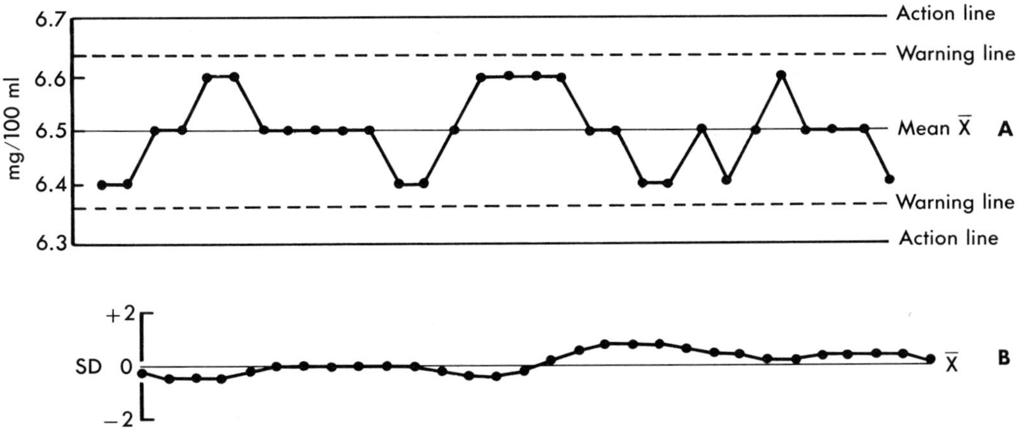

Fig. 2-5. A, Normal distribution. Shewhart chart. Total protein. **B,** Normal pattern. Cusum chart of **A.** Total protein.

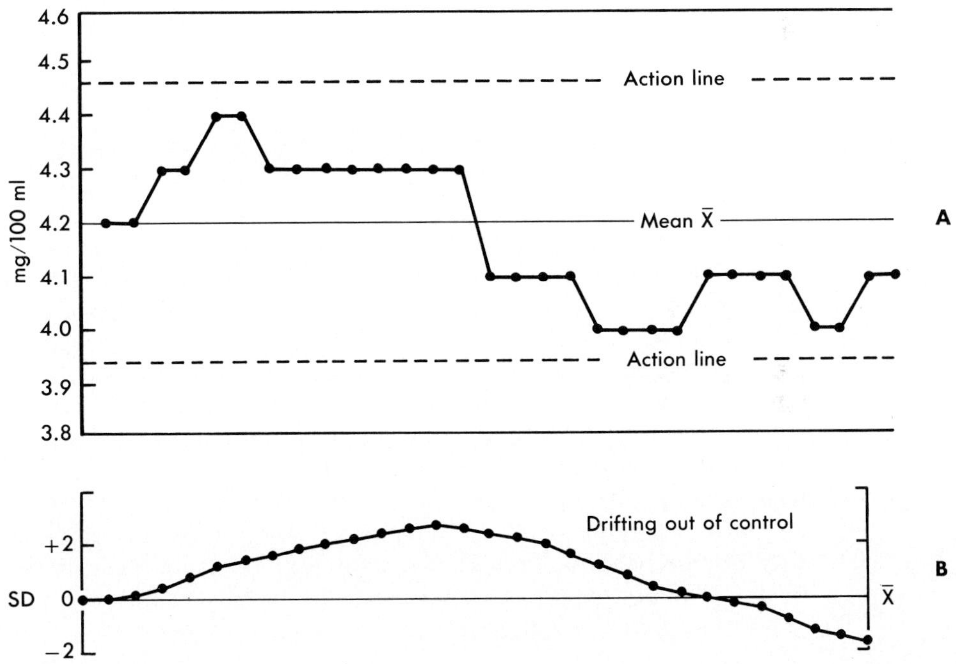

Fig. 2-6. A, Systematic trend resembling a shift. Shewhart chart. Albumin. **B,** Systematic trend more obvious. Cusum chart of **A.** Albumin.

(1) photocells or lamps that need replacing, (2) a deteriorating standard, (3) deterioration of one or more reagents, or (4) contamination in reagents or within the system. A trend downward may be caused by (1) need for rebuilding an AutoAnalyzer manifold, (2) standards that are too concentrated, and (3) contamination within reagents or systems.

Systematic shift. With the systematic shift, the points do not rise gradually, as with the trend. There is a distinct distribution accumulation on either the plus side or the negative side of the mean (Fig. 2-7). Upward shifts may be

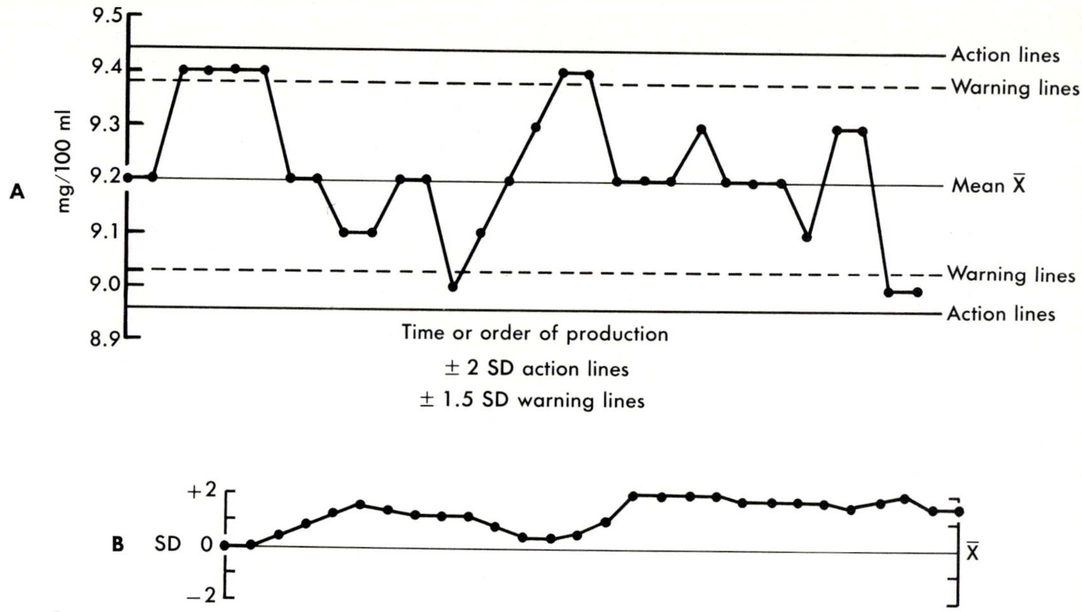

Fig. 2-7. A, Systematic shift. Shewhart chart. Calcium. **B,** Systematic shift. Cusum chart of **A.** Calcium.

caused by any of the following: (1) partial electrical failure in the instrument, (2) overdiluted new standard, (3) deteriorated reagent that has stabilized, (4) improperly prepared reagent, too dilute, (5) contaminated reagent, (6) inaccurate timer or one in need of repair, (7) deteriorated indicator, (8) dirty glassware, and (9) overheating in a temperature-sensitive analysis. A downward trend may result from (1) increased concentration of standards and reagents by evaporation of water (keep all bottles covered or stoppered), (2) contaminated reagents, (3) inaccurate timer, too rapid, (4) underheating in analysis, (5) new indicator or new standard that is too concentrated, or (6) contaminated glassware.

Outlier. A single point suddenly out of range (Figs. 2-2 and 2-3). The outlier may be due to an individual condition: (1) contamination of the single specimen, (2) faulty pipette, (3) incorrect dilution of the test, or (4) contamination of a tube, pipette, or AutoAnalyzer cup.

Frequency distribution

No two units are ever alike even though a product may be said to be uniform. The variability of the units may be tabulated in order to observe the characteristics of the total data (universe) from which the sample data came.

Two types of data may be separated into the categories of discrete data and continuous data.

Discrete data

Discrete data differs from cumulative data in that it is an integral unit—it either exists or it does not. There is no variable, or *half-wrong.* Of course, if averages of the data are taken into consideration, it is feasible to derive results in terms of fractions. The data are expressed in the form of a whole

Table 2-4. Number of outliers in 29 groups of calciums (30 tests/group, 870 total)

Group number	Number of outliers	Group number	Number of outliers
1	0	16	1
2	0	17	1
3	5	18	1
4	4	19	1
5	3	20	1
6	0	21	0
7	0	22	0
8	2	23	1
9	1	24	1
10	2	25	2
11	2	26	1
12	1	27	1
13	0	28	0
14	2	29	1
15	2	30 in process	

Table 2-5. Distribution of the number of outliers in 870 calciums

	Groups with designated number of outliers	
Number of outliers	Total	Fraction
0	8	0.2759
1	12	0.4137
2	6	0.2069
3	1	0.0345
4	1	0.0345
5	1	0.0345
Grand total	29	1.0000

$$\frac{\text{Total}}{\text{Grand total}} = \text{Fraction}$$

$$\frac{8}{29} = 0.2759$$

number, and variation in this number from result to result is in multiples of one. This is demonstrated in Tables 2-4 and 2-5. The example given is one method of tabulating results that are out of the acceptable range. It may represent a total day's run of all analyses, one group of analyses, or a report taken from the standard deviation charts. The frequencies may be represented by a histogram. However, with discrete data, remember that although the bars touch, the frequencies apply only to points on the X axis, not to a range of values. If a histogram is used, separate the bars from each other to emphasize the discrete nature of the variable. If histograms are used for the continuous

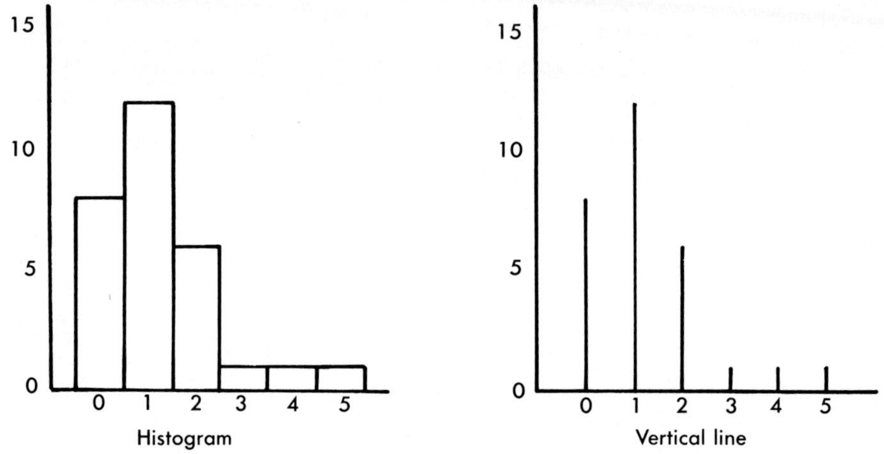

Fig. 2-8. Comparative graphs of discrete frequency distribution.

and discrete data, there will be less confusion if the discrete data are plotted using vertical lines at the abscissa values. Refer to Fig. 2-8 for the comparison.

Graphic forms

The frequency distribution may be displayed graphically by a histogram or curve. The average (mid-value) of each interval is plotted on the graph.

When the frequencies of the distribution are cumulated from the low end to the other end of the scale (low $\longrightarrow$ high), it is called a cumulative distribution. When the distribution is continuous, it can be best represented with a smooth-line curve called an orgive, rather than by a broken line (Fig. 2-9).

Purpose of the graph

The frequency distribution gives an overall picture of variation in a set of data. The characteristic pattern will give several bits of information.

1. Location of the data concentration will show if there is a dispersion (central tendency). An area with a single concentration is called a unimodal distribution. However, not all distributions are unimodal. There may be more than one concentration in a distribution, which is then called bimodal (Fig. 2-10).

2. Another answer that is needed concerns how much variability there is. Refer to Fig. 2-11. If the two figures are drawn on the same scale, then one obviously demonstrates a greater variability. Variations should be judged on the scale of the graph rather than the visual picture of the graph.

3. The graph shows the symmetry of the distribution. When it is lopsided and departs from symmetry, it is said to be skewed. With concentration of the data on the right side, the distribution is negatively skewed, and a positively skewed graph shows concentration on the left side. When there is no skewness, the graph is symmetric (Fig. 2-12).

4. The next characteristic to observe is the kurtosis or peakedness of the distribution. The graph may be symmetric yet differ from the normal graph in the following way. The smooth-line drawing demonstrates variations from the normal curve better than the histogram. Fig. 2-13 illustrates the normal curve (dotted

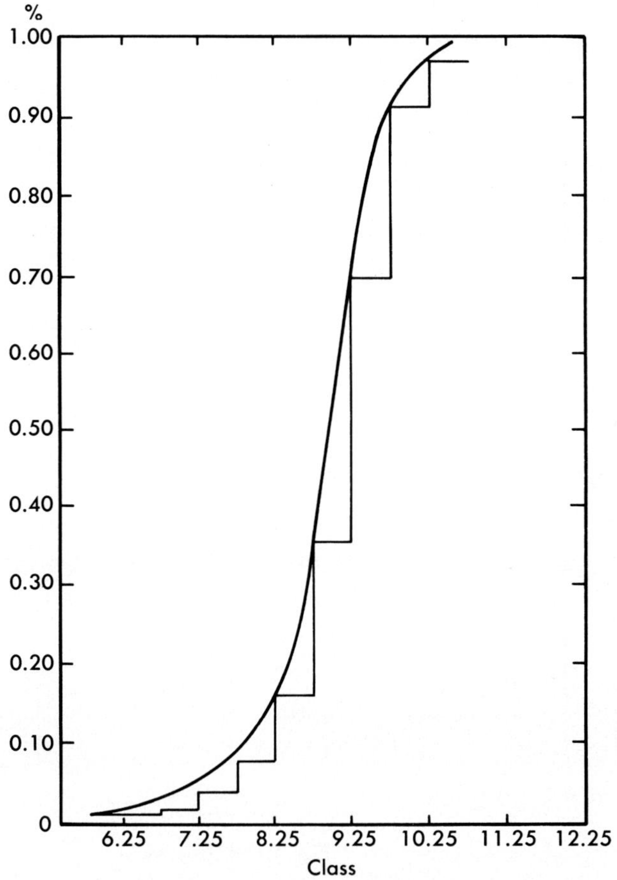

Fig. 2-9. Orgive curve and broken line showing cumulative frequency of calcium data in Fig. 2-16.

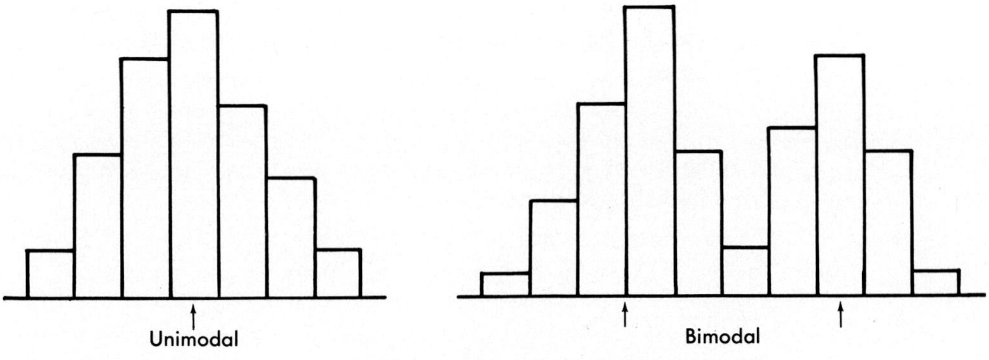

Fig. 2-10. Central tendency.

line) and the leptokurtic curve (solid line). The leptokurtic curve has a slender central distribution (modal) and the tails are higher. This is said to have a large kurtosis. The platykurtic curve (dotted line) is shown in Fig. 2-14. The central distribution area is broader and the tails appear to be bobbed or thin. This is said to have little kurtosis.

In some problems or evaluations, all that is needed is the graphic plot of the

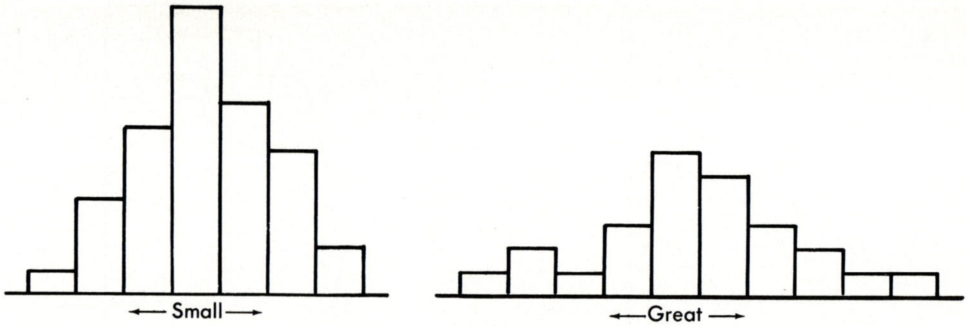

Fig. 2-11. Variability.

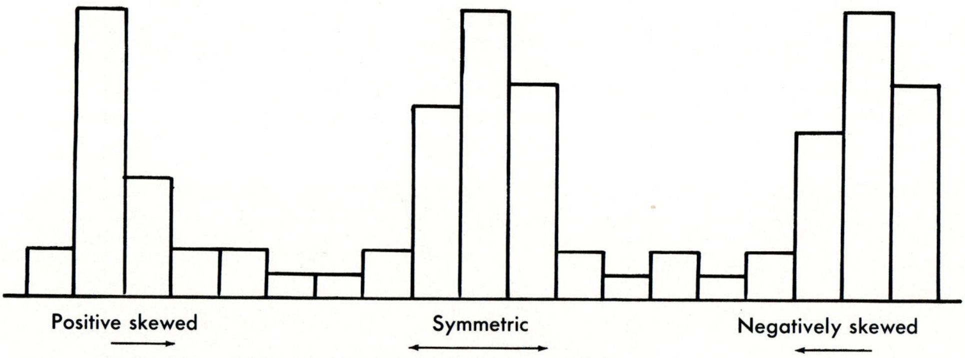

Fig. 2-12. Skewness.

frequency distribution. For those who wish to measure the skewness and kurtosis, the following formulas are offered:

1. A measure of skewness as coefficient of skewness, based on all data. This can be handled mathematically. However, it is affected by extreme variations.

$$\gamma_1 = \frac{\mu^3}{\sigma^3}$$

2. A measure of kurtosis as coefficient of kurtosis. It is approximately zero for a bell-shaped distribution.

$$\gamma_2 = \frac{\mu^4}{\mu_2^2} - 3 \text{ or } \beta_2 = \frac{\mu^4}{\mu_2^2}$$

Measures of central tendency

Three different measures of central tendency are the mean, the median, and the mode.

The arithmetic mean $(\overline{X})$ is used so often that it is sometimes just called "average" or "mean." It is considered the center of gravity, easy to compute, with a relatively small sampling error. It is based on all cases and is affected by extreme cases. When this is used, it should be designated as the arithmetic mean, since there are other means: the harmonic mean, which is defined as "the reciprocal of the arithmetic mean of the reciprocal of a series of values"; the geometric mean, which is "the Nth root of the product of the values"; and the

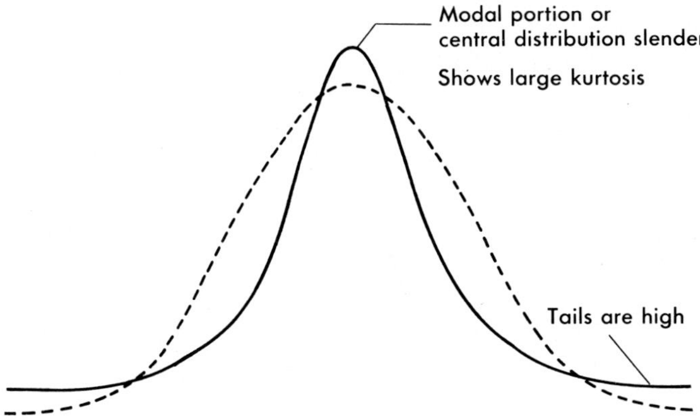

Fig. 2-13. Leptokurtic curve against a mesokurtic (normal) curve.

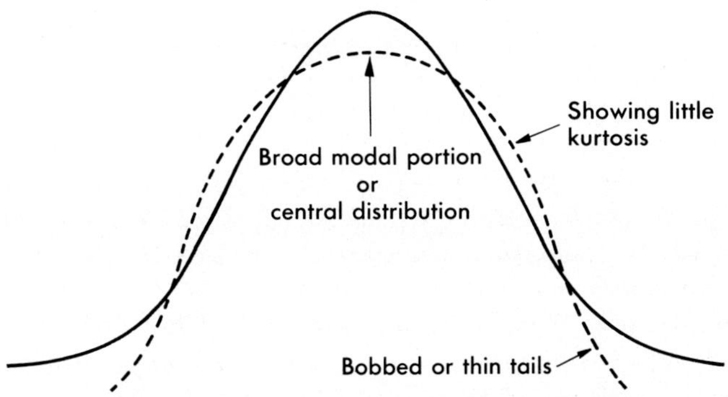

Fig. 2-14. Platykurtic curve against a mesokurtic (normal) curve.

quadratic mean, which refers to "the square root of the arithmetic mean of the square of the values." So! Still further details may be found in Croxton and Cowden.* Now, see the importance of specifying *arithmetic* mean?

The median (Mi) divides the total area in half. It is difficult to handle mathematically although it is not affected by extreme cases. If the universe is normal, there will be a greater sampling error than that of the arithmetic mean.

The mode (Mo) is the value in a series that appears most frequently. The mode is difficult to find in the universe.

Continuous data

Continuous data (collectively) refers to information that can be of value within an acceptable range. Gathering this type of data requires grouping into different intervals. When raw data are accumulated, there is no order in their arrangement. Therefore, to do anything statistical with the data, you must place them in order (array). The data are listed then according to numerical magnitude. In grouping the data into intervals or classes, it is simpler to keep the intervals a constant size. Two factors to take into consideration are (1) what size interval is to be used and (2) how the intervals are to be located.

*Croxton, F. E., and Cowden, O. J.: Applied general statistics, New York, 1939, Prentice-Hall, Inc., pp. 176-180.

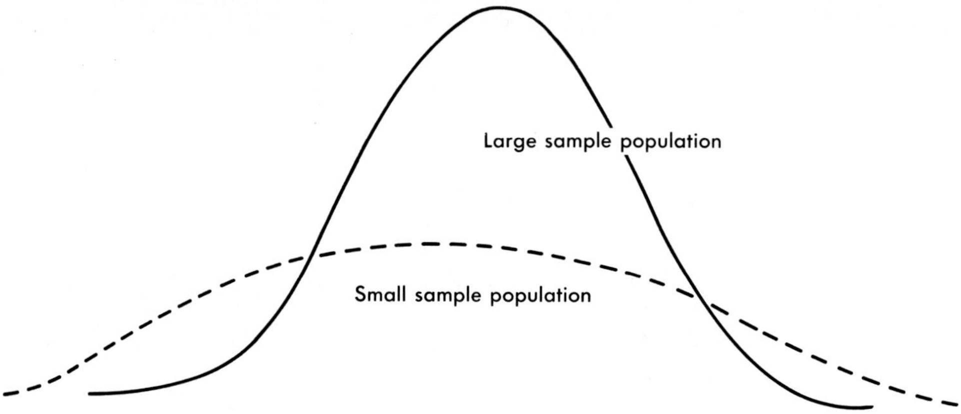

Fig. 2-15. Frequency distribution of large and small population sampling.

If there are too few intervals, the data will be bunched together and information lost. However, if the intervals are too spread, the distribution will be irregular and also of no value in studying the characteristics of the distribution.

In determining the format of the distribution, keep a few ideas in mind. Do not make a frequency distribution tally on less than 30 bits of data. See example of distribution shown in Fig. 2-15. This group may be studied in array. Five or six classes can be utilized for data of 100 and increased according to the volume of material to be tabulated. It is unusual to have as many as twenty intervals, unless the data are extensive. The analytic situation will determine the number of intervals used. The best compromise is to take ten to fifteen intervals with sampling over 100 bits of data. Unless a specific range is required for the subgroup, use the following as a guide:

$$\frac{\text{High value} - \text{Low value}}{\text{Number of intervals}} = \text{Subgroup size}$$

Tabulation of data

When the format has been decided, a tally of the frequency distribution can be made from the raw data. Data may be placed on index cards and sorted manually, or be keypunched and placed on the computer. Use of the computer is an efficient method if the data processing staff has been forewarned and a program has been written for frequency distribution.

An entry form designed for column listing of numerical values according to class is an effective means of checking incorrect entries. Now, a tally form (Fig. 2-16) is drawn, and the values falling within the class interval are recorded with tally marks. The frequency column is a final count; the cumulative frequency gives the total, and the percent column is obvious. If the tally marks are kept in neat order, the general picture of the distribution can be observed.

The percentages and the class intervals are plotted on probability paper (Fig. 2-17). Intersecting lines are drawn at the 2.5% and at the 97.5% acceptance range. This gives the lower and upper limits of distribution.

The t test

The comparing of lots of material—such as test kits, reagents, or controls—or two different methods has to be accomplished with individual measurements,

Class interval	Data range	Class midpoint	5 15 25 35 45 55 65 75 85 95 105 115 125 135 145 155	Frequency total	Cumulative total	Fraction
6.0- 6.5	6.06- 6.55	6.25	||	2	2	0.0040
6.5- 7.0	6.56- 6.95	6.75	|	1	3	0.0060
7.0- 7.5	6.96- 7.55	7.25	𝍸	5	8	0.0170
7.5- 8.0	7.56- 8.05	7.75	𝍸 ||||	9	17	0.0370
8.0- 8.5	8.06- 8.55	8.25	𝍸 𝍸 𝍸 𝍸 ||	22	39	0.0850
8.5- 9.0	8.56- 9.05	8.75	𝍸 𝍸 𝍸 𝍸 𝍸 𝍸 𝍸 |||	38	77	0.1688
9.0- 9.5	9.06- 9.55	9.25	𝍸 𝍸 𝍸 𝍸 𝍸 𝍸 𝍸 𝍸 𝍸 𝍸 𝍸 𝍸 𝍸 𝍸 𝍸 𝍸 𝍸 𝍸 |	91	168	0.3688
9.5-10.0	9.56-10.05	9.75	𝍸 ||||	154	322	0.7072
10.0-10.5	10.06-10.55	10.25	𝍸 |||	103	425	0.9335
10.5-11.0	10.56-11.05	10.75	𝍸 𝍸 𝍸 𝍸 𝍸 |||	28	453	0.9950
11.0-11.5	11.06-11.55	11.25	|	1	454	0.9971
11.5-12.0	11.56-12.05	11.75	|	1	455	0.9992
12.0-12.5	12.06-12.55	12.25				
12.5-13.0	12.56-13.05	12.75				
13.0-13.5	13.06-13.55	13.25				
13.5-14.0	13.56-14.05	13.75				
14.0-14.5	14.06-14.55	14.25	|			

Fig. 2-16. Frequency distribution tally (calcium).

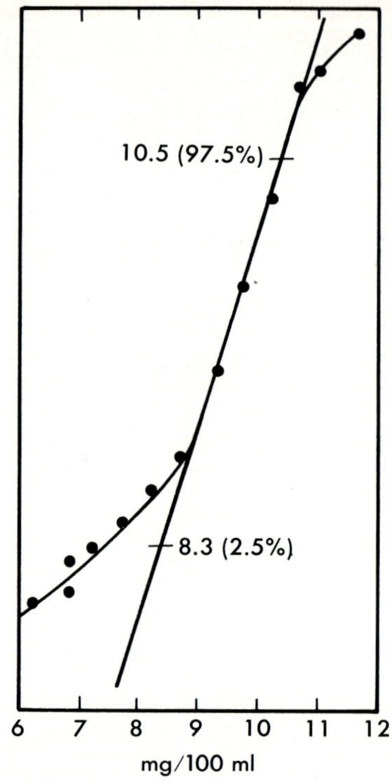

Fig. 2-17. Distribution curve from Fig. 2-16.

and this next question raised. Is the difference between the measured items significant? The average or the standard deviation may be within acceptable limits of a single evaluation, when one lot or test is not being compared to another.

For example, however, the PBI method of test kit A can be compared to an automated method, B. Refer to Table 2-6. The results of the first column are squared in the second. Both columns are totaled. To obtain the sum of the x^2 (Σx^2), sums of A and B are totaled (130.15 + 137.30 = 267.45). The sum of the test columns (x_1 and x_2) are squared and divided by the number of tests performed (5).

$$\text{A column } (25.5)^2 = 650.25 \div 5 = 130.05$$

$$\text{B column } (26.2)^2 = 686.44 \div 5 = 137.28$$

Now, $\Sigma \frac{(\Sigma x)^2}{N}$ is equal to 130.05 + 137.28, or 267.33. The degrees of freedom for A and B are 5 − 1 = 4 (n − 1) individually and 4 + 4 = 8 total. The standard deviation is as follows:

$$\sqrt{\frac{267.45 - 267.33}{8}} = \sqrt{0.015} = 0.123 \text{ SD}$$

Is the difference between the methods due to chance or are they the same?

The average of the test for A = 5.10 and for B = 5.24, and the difference between these two averages is: 5.24 − 5.10 = 0.14.

The difference of 0.14 is not out of precision range for the test methods. Now, to determine if 0.14 is significant or not. The ratio of the 0.14 (difference between the two averages) to the standard deviation of 0.123 will be the determining factor.

Table 2-6. Comparison of PBI test results from two different methods

	Method A		Method B	
	x_1 (μg/100 ml)	x_1^2	x_2 (μg/100 ml)	x_2^2
	5.0	25.00	5.3	28.09
	5.3	28.09	5.2	27.04
	4.9	24.01	5.3	28.09
	5.2	27.04	5.2	27.04
	5.1	26.01	5.2	27.04
Total	25.5	130.15	26.2	137.30
Average	5.1		5.24	

This ratio to be calculated is called the t value.

$$t = \frac{\text{Difference between averages}}{\text{SD of difference between averages}}$$

$$= \frac{\overline{x} - \overline{y}}{\sigma}\sqrt{\frac{N_1 \cdot N_2}{N_1 + N_2}}$$

$$= \frac{0.14}{0.123}\sqrt{\frac{5 \cdot 5}{5+5}} = 1.138\sqrt{\frac{25}{10}} = 1.14\sqrt{2.5}$$

$$= 1.14 \cdot 1.58 = 1.8 \text{ t value}$$

If the t value is large, the story is told that the difference between averages is much greater than chance variation. Table A-10 in Appendix lists the t values, which must be expected in order to forewarn that the method differences are significant.

The degrees of freedom of A + B was 8. Refer to Table A-10 in Appendix. For this degree of freedom, the t value of 2.31 must be exceeded at the 5% probability level. The t value holds good for the 10% level. Therefore the tests are assumed to be identical. If the 10% level had been exceeded, then at least 10 more tests should have been performed and evaluated.

CONCLUSION

Analytic conditions and quality controls must be under continuous review. The need for more specific and improved methodology is evident.

A few rules that will serve advantageously in a good quality control program are (1) keep instruments in good working order with a scheduled preventive maintenance program, (2) properly ground electric appliances, (3) do not use equipment beyond its capabilities, (4) keep all equipment routinely calibrated, (5) maintain clean glassware and use disposable materials when economically feasible, (6) train personnel to utilize good technique, (7) rotate workers weekly or every two weeks to prevent boredom and lethargic reactions, and (8) maintain an active program for quality control as a successful team activity.

GLOSSARY

Symbols | **Definitions**

accuracy Deviation of an estimate from the true value. The word used in this sense does not mean precision. It may mean bias.

AD $\frac{\Sigma F X'}{\eta}$ ***average deviation*** Deviation divided by the number of determinations run. The method is rapid for obtaining the magnitude of the standard deviation and is a rough check on calculations. Is valid only on long runs of data. The relationship between the average and the standard deviation should show the SD greater than the AD by a factor of $\sqrt{\frac{\pi}{2}} = 1.25$.

bias Tendency of sample statistic to deviate in one direction from a true value, as by reason of nonrandom sampling.

BIPP Binominal probability paper.

block Group of test or experiments.

column mean square Column sum of squares divided by $\eta - 1$ (degrees of freedom).

P (A/B) ***confidence coefficient*** Chance that a confidence interval has of including the universal value.

confidence limits High and low limits of a confidence interval.

continuous sampling plan Continuous production without a change of lots or groups.

±1 SD 68.27%
±2 SD 95.45%
±3 SD 99.73% ***control limits*** Standard deviations based on the normal distribution (68.27%, 95.45%, and 99.73%) of the total observation or universe.

correction factor A constant or adjustment factor applied to observed values for small sample size to obtain a better fit of data to the theoretical χ^2 distribution.

$\eta - 1$ ***degrees of freedom*** Number of independent differences or variations possible among individual observations. If only one observation is made, there is zero degree of freedom, no other observation being available for comparison.

duplicates Unit or divided specimens produced or analyzed under the same conditions. This term is used whether there are two or more duplicates.

expected value Arithmetic mean value of a variable.

experimental error Deviation from the expected value, due to the limitations of methodology or production.

f ***fixed factors, variables, or effects*** Factors, variables, or effects that are constant or the same for all samples.

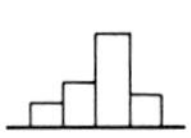

frequency distribution Grouping of the values of a variable in order of size and plotting of the frequencies according to size.

histogram Bar diagram representing a frequency distribution.

$\bar{X}$ 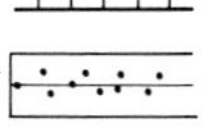 ***homoscedastic*** Having a uniform scatter.

interaction Tendency of two variables to combine, giving a result that differs from the sum of their independent activities.

γ_2 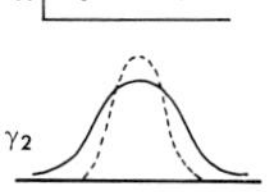 ***kurtosis*** Degree of peakedness in the curve. It is approximately zero for a bell-shaped distribution. $\frac{\mu}{\mu_2^2} - 3$

$\bar{X}$ ***mean (arithmetic)*** Sum of a series of tests divided by the total number.

Me ***mean error*** Difference between the true result and the mean obtained.

Mi ***median*** Line of equality where there are as many cases above the line as there are below.

50% of cases > Mi
50% of cases < Mi

Mo ***mode*** Class of data with the greatest frequency, indicating the point of central tendency.

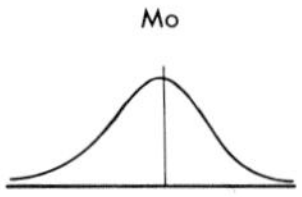

normal distribution Distribution of data in a frequency tabulation resembling a a bell-shaped curve.

$\bar{X}$ 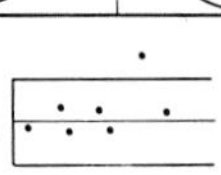 ***outlier*** Result in an extreme position, beyond acceptable limits. An obviously wild error not relating to analytic variations; therefore it is not considered in tabulating data.

parameter Constant of a universe, describing an individual characteristic.

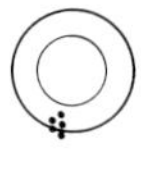

precision Exactness in the repeated determination of a test. The smaller the standard deviation, the higher the precision. Precision does not refer to accuracy; it denotes reproducibility.

P ***probability*** Relative frequency of chance. Varies from 0 to 1.

random Condition of disorder or disarray without prediction of individual results.

R

range Simplest measure of general variability describing the difference between the highest value and the lowest value.

$x_H - x_L$

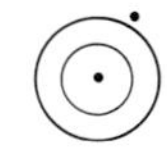

relative error Mean error of a series of tests, representing a percentage of the true result.

relative standard deviation Reference to coefficient of variation.

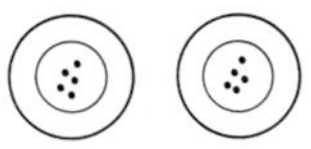

reproducibility Ability to repeat or reproduce the same data on different occasions.

η sample Group of items or data from a larger group (universe). Unless specified, samples are considered to be in a random state.

sampling plan Plan that designates the sample size and acceptance criteria.

sampling scheme Coordinated group of sampling plans utilized toward a certain goal.

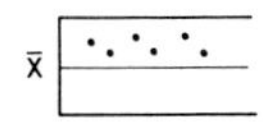

sensitivity Ability of a test to detect deviations from a target value or goal.

shift Congregating of well-distributed values on either side of the mean.

significant Test result deviating from the confidence limits.

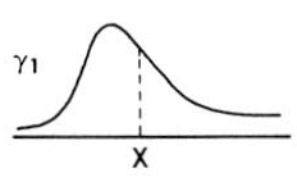

skewed Asymmetric curve based on total data. Affected by extreme variation. Can be handled mathematically. $\gamma_1 = \frac{\mu^2}{\sigma^3}$

SD or s $\sqrt{\frac{\Sigma(\bar{X} - X)^2}{\eta - 1}}$

standard deviation Square root of the average of the squared deviations from the arithmetic mean.

SE $\sqrt{\frac{SD}{n}}$

standard error Error around the mean for a series of tests.

statistics Collection and use of numerical facts to estimate a universe parameter.

subgrouping Dividing a group of items into subgroups such that, if assignable causes are present, these will become common to all elements of the subgroups.

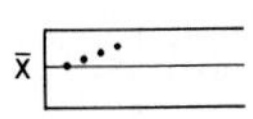

trend Values increase or decrease in succession from the mean.

universe Group (either infinitely or extremely large) from which samples are taken for tabulation.

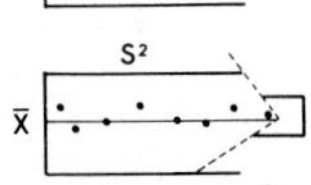

variance Square of the standard deviation.

V mask V-shaped mask or overlay used with the cusum chart.

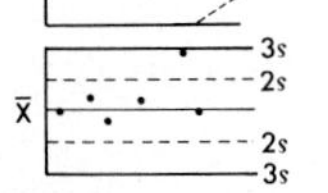

warning limits Limits on the control chart (usually at ± 2 SD) Data exceeding these limits alert for immediate attention.

REFERENCES

1. Belk, W. P., and Sunderman, F. W.: A survey of the accuracy of chemical analyses in clinical laboratories, Amer. J. Clin. Path. **17**:853, 1947.
2. van Dobben de Bruyn, C. S.: Cumulative sum tests: theory and practice, New York, 1968, Hafner Publishing Co., Inc.
3. Dodge, H. F.: Using inspection data to control quality, Manufacturing Industries **16**: 517, 613, 1928.
4. Dodge, H. F.: Statistical aspects of sampling inspection plans, Mechanical Engineering **57**:645, 1935.
5. Dodge, H. F., and Romig, H. G.: Sampling inspection tables, single and doubling sampling, New York, 1959, John Wiley & Sons.
6. Dudding, B. P., and Jennett, W. J.: Quality control charts (British Standard 600R: 1942), London, 1942, British Standards Institution.
7. Hoffman, R. G., and Waid, M. E.: The average of normals. Method of quality control, Amer. J. Clin. Path. **43**:134, 1965.
8. Industrial quality control **4**:23, July, 1947.
9. Kemp, K. W.: The average run length of the cumulative sum chart when a V-mask is used, J. Royal Statistical Society (s.b.) **23**:149, 1961.
10. Kemp, K. W.: The use of cumulative sums for sampling inspection schemes, Applied Statistics **11**:16, 1962.
11. Levy, S., and Jennings, E. R.: The use of control charts in the clinical laboratory, Amer. J. Clin. Path. **20**:1059, 1950.
12. Moroney, M. J.: Facts from figures, Harmondsworth, England, 1951, Penguin Books, Ltd.
13. Page, E. S.: Continuous inspection schemes, Biometrika **41**:100, 1954.
14. Shewhart, W. A.: Economic control of quality of manufactured product, New York, 1931, Van Nostrand Co., Inc.
15. Shewhart, W. A.: Statistical method from the viewpoint of quality control, Washington, D. C., 1939, Graduate School, Department of Agriculture, chap. 1.
16. Snedecor, G. W.: Queries, Biometrics **8**:85, 1952.
17. Tonks, D. B.: A study of the accuracy and precision of clinical chemistry determinations in 170 Canadian laboratories, Clin. Chem. **9**:217, 1963.
18. Wootton, I. D. P., and King, E. J.: Normal values for blood constituents: interhospitals differences, Lancet **1**:470, 1953.

• 3

• Normal values

NANCY KAY McADAM, A.B., MT(ASCP)

The "normal values" presented in this chapter are presented only as a guide to what one might expect to find in a normal, healthy individual.

The question "What is normal?" is a subject that has been under discussion for years. It seems that the answer to this question is just as elusive now as it was years ago—maybe more so.

Many investigators have divided the healthy population into groups according to sex and age in an attempt to refine normal ranges. However, there are many variables that have been found to affect "normal" ranges. Such things as diet, economic level, geographic location, farm or city life, climate, season, and time of day are just a few of the factors affecting such ranges. In determining what is normal for an individual, we should also consider whether he leads an active or a sedentary life. Is he fasting or in a postabsorptive state? Computer analyses of multiphasic screening programs are uncovering even more variables that affect normal values.

Added to these personal variations are the variations in methods of analysis. There are many methods, differing in specificity, and many modifications of these methods are being used to assay a single constituent. Units used in reporting enzyme activity vary almost as widely.

Obviously, no single normal range is going to include all normal, healthy people or exclude all people with abnormal or pathologic conditions. The only recourse in the problem of normal values is for each laboratory to establish its own normal ranges for each substance and for the methods used and to define the limits of a range. See, as examples, Tables 3-1 and 3-2.

Table 3-1. Normal values

Constituent	Material analyzed	Normal value: Adult	Normal value: Children	Comments	Page
Acetone	Serum[1]	0.3-2.0 mg/100 ml	0.3-2.0 mg/100 ml[3]	Elevated[4] BSP PSP	
	Urine[2]	Negative	About 50 mg/24 hr[3]		383
Acetylcholine	CSF[5]	None			
Acidity	Gastric content[6]	Free HCl: 0°-30° Total acid: 10°-50°		45 min after histamine Free HCl ♂: 30°-85° ♀: slightly less than ♂ Total acid ♂: 67°-100° ♀: 67°-82°	307
	Urine		20-40 mEq/24 hr[3]		403
Albumin	See Protein				
Alcohol, ethyl	Whole blood[7,8]	0%-0.11% (subclinical)		Stimulation: 0.09%-0.22% Confusion: 0.18%-0.33% Stupor: 0.27%-0.44% Coma: 0.36%-0.55% Death: over 0.45% Elevated[2] Isopropyl alcohol Methyl alcohol	496
Aldolase	Serum	3-10 units/ml[9] 1.8-4.9 IU/L[10]	Newborn: 4 times adult[10] Child: 2 times adult		
Aldosterone	Urine[1]	2-23 μg/24 hr			449
Allantoin	Urine[11,12]	25-45 mg/24 hr			
Aluminum	Whole blood[13] Serum[13] Urine[14,15] Feces[14,15] CSF[13]	21-94 μg/100 ml 46 μg/100 ml (av) 0.078 mg/L 41.92 mg/24 hr Trace			
Amino acid nitrogen	See Table 3-2				

Continued.

Table 3-1. Normal values—cont'd

Constituent	Material analyzed	Normal value: Adult	Normal value: Children	Comments	Page
Ammonia nitrogen	Whole blood	75-196 μg/100 ml[16]	45-80 μg/100 ml[10]	Elevated[2,4]	248
	Plasma[1]	20-150 μg/100 ml		Ammonium salts	
	CSF[17]	0-14.7 μg/100 ml		Exercise	
	Feces[18,19]	251-881 μg/100 ml		Hemolysis	
	Sweat[13]	5-9 mg/100 ml		Increased mineral acid intake	
	Tears[13]	0.005 g/100 ml			
	Urine	20-70 mEq/24 hr[20]	20-70 mEq/L[3]	Ion exchange resins	
	Saliva[13]	0.5-9.9 mg/100 ml		Low-carbohydrate diet	
				Meat diet	
				Normal pregnancy	
				Standing of specimen	
				Starvation	
				Decreased	
				Increased alkali intake	
				Sodium salts	
				Vegetarian diet	
Amylase	Whole blood[3]		60-180 Somogyi units %		279
	Serum[16]	40-140 Somogyi units			
	Urine[16]	♂ 38-263 units/hr			405
		♀ 47-310 units/hr		Elevated[4]	
				Chlorides	
				Codeine	
				Demerol	
				Ethanol (gross)	
				Fluoride	
				Lipemia	
				Methanol	
				Morphine	
				Saliva	
				Decreased	
				Citrate	
				Oxalate	
Arginase	Whole blood[21]	30-80 units %			
Arginine	See Table 3-2				
Arsenic	Blood[16]	<3 μg/100 ml			
	Urine[16]	Up to 100 μg/L			510
Ascorbic acid	Whole blood[22]	0.4-1.5 mg/100 ml	0.5-1.0 mg/100 ml		
	Plasma[1]	0.6-1.6 mg/100 ml			
	Serum[3]		0.4-1.5 mg/100 ml		
	Urine	15-50 mg/24 hr[23]	>5% of oral[10] 20 mg/kg dose in 24 hr		
	CSF[24]	1-8 mg/100 ml			
	Tears[13]	0.14 mg/100 ml			

Table 3-1. Normal values—cont'd

Constituent	Material analyzed	Normal value: Adult	Normal value: Children	Comments	Page
Barbiturate	Serum[25-27]	Negative	Therapeutic[10] Short-acting <0.3 mg/100 ml Long-acting <3.0 mg/100 ml	Phenobarbital 1.0-4.0 mg/100 ml—awake 4.0-5.0 mg/100 ml—sedated 5.0-8.0 mg/100 ml—comatose; reflexes 8.0-11.0 mg/100 ml—no reflexes 11 mg/100 ml— comatose; circulatory and/or respiratory difficulty	498
Barium	CSF[13]	Trace			
Bilirubin	Serum	Total: 0.5-1.4 mg/100 ml*	—	Elevated[2,4] Atabrine Carotene Lipochrome Methanol Vitamin A	256
				Decreased Aldomet	
		1 min[1] 0-0.4 mg/100 ml		Caffeine Chlorine Citrate	
	CSF	None found[28]	None found[3]	Hemolysis	
	Feces	None found[29]	None found[3]	Light	
	Urine[30]	Too small for detection		Protein Urea	392
Biotin	Urine[31,32]	3-31 μg/24 hr			
Bismuth	Urine[33-35]	Negative			
Blood (guaiac or benzidine)	Urine[2]	Negative	Negative	False positives[4] Bromides Copper Ferricyanide Formalin Iodides Nitric acid Permanganate Pus	399

*Cord	*Premature (mg/100 ml)*	*Full-term[10] (mg/100 ml)*
blood	—	2.5
24 hr	8.0	6.0
48 hr	12.0	7.5
3-5 days	24.0	12.0
1 mo		0.1-0.7

Continued.

Table 3-1. Normal values—cont'd

Constituent	Material analyzed	Normal value: Adult	Normal value: Children	Comments	Page
Blood—cont'd	Feces[36]	Negative	Negative	False positives Bromides Iodides Iron-containing medications Meat	318
Boron	Serum[16]	Up to 0.09 mg/100 ml			
	Urine[16]	Up to 0.3 mg/100 ml			
	CSF[13]	Trace			
Bromide	Serum[16]	0-1.5 mg/100 ml		Toxic level 136 mg/100 ml (17 mEq/L)[22] Elevated Halogens Hemolysis	502
	Urine[16]	<0.25 mg/100 ml			
	CSF[13]	0.14-0.38 mg/100 ml			
Bromsulphalein (BSP)	Serum[16]	5 mg/kg IV <5% retention at 45 min <2% at 60 min 2 mg/kg IV: <50% retention at 5 min None to trace at 30 min	Newborn: up to 20% Older: <5%	Elevated[2,4] Androgens Barbiturates Enovid Estrogens Halogens Hemolysis Lipemia Morphine Phenolphthalein Posture PSP Radiographic contrast media Telepaque Decreased Ascites Nonfasting state	259
Calcium	Serum	Total[1]: 8.5-10.5 mg/100 ml Ionized: 4.2-5.2 mg/100 ml	9.0-11.5 mg/100 ml[3]	Elevated[4] Ammonium oxalate Contaminated glassware Potassium Sodium	208 213

Table 3-1. Normal values—cont'd

Constituent	Material analyzed	Normal value: Adult	Normal value: Children	Comments	Page
Calcium—cont'd	CSF	3.9-5.1 mg/100 ml[13]	4.5-5.5 mg/100 ml[3]	Decreased Citrate Delayed separation from cells EDTA Heparin Insulin Magnesium Oxalate Phosphorus Sulfate	
	Urine	50-400 mg/24 hr[16]	<150 mg/24 hr[3]	Decreased pH increase	212
	Feces[37,38]	0.4-0.8 g/24 hr			
	Sweat[13]	1-8 mg/100 ml			
	Amniotic fluid[13]	7.2 mg/100 ml			
	Saliva[13]	5.2-9.7 mg/100 ml			
Carbon dioxide combining power	Serum[16]	22-30 mEq/L[16]	22-30 mEq/L[3]	Decreased[4] EDTA	184
	CSF[28]	24-29 mEq/L[28]	18-31 mEq/L[3]		
	Urine[39,40]	20-200 vol%			
	Amniotic fluid[13]	37 vol%			
	Tears[13]	26 mEq/L[13]			
	Saliva[13]	3.5-10.7 mEq/L			
Pressure (Pco_2)	Whole blood	35-45 mm Hg[1]	35-45 mm Hg[3]		
Carbon monoxide	Whole blood[40-42]	0-0.84 vol%		15-20 vol%—symptoms produced 30-75 vol%—usually fatal	502
Carbon monoxide hemoglobin	Whole blood[16]	<5% of total hemoglobin			206
Carotene	Serum	40-200 μg/100 ml[1]	Newborn: 25 μg/100 ml[3] Thereafter: 60-180 μg/100 ml		
Catecholamines	Urine	Up to 18 μg/100 ml (random sample)[16] Epinephrine: up to 20 μg/24 hr Norepinephrine: 10-80 μg/24 hr	0.4-2.0 μg/kg/24 hr[10] Epinephrine: < 10 μg/24 hr Norepinephrine: <100 μg/24 hr[3]	Elevated[2,4] Bananas (fluorometry) Erythromycin Exercise Methyldopa Quinidine Quinine Tetracycline	466

Continued.

Table 3-1. Normal values—cont'd

Constituent	Material analyzed	Normal value: Adult	Normal value: Children	Comments	Page
Catecholamines—cont'd	CSF[43]	Epinephrine: 0.90-0.93 μg/L Norepinephrine: 2.4-3.1 μg/L			
Cephalin-cholesterol flocculation (CCF)	Serum	24 hr: negative–1+ 48 hr: negative–2+[1]	Infants: erratic Thereafter: negative–1+[3]	Elevated[2,4] Aldomet (may give also false negative) Bacterial contamination Light Temperature	245
Ceruloplasmin	Serum	23-50 mg/100 ml[1]	28-56 mg/100 ml[3]		
Chloride	Serum CSF Feces[37] Sweat[13] Tears[13] Urine[16] Amniotic fluid[13] Saliva[13]	95-103 mEq/L[1] 119-130 mEq/L[13] 3 mEq/24 hr 10-132 mEq/L 118-138 mEq/L 170-254 mEq/24 hr 104 mEq/L 8.4-17.7 mEq/L	94-106 mEq/L[3] 118-127 mEq/L[3]	Elevated[4] Bromide Contaminated glassware Ion exchange resin therapy KCl Mercurial diuretics NH[4]Cl Protein Saline (IV) Steroids Decreased ACTH Hemolysis	180 329 181, 405
Cholesterol Esters	Serum	70%-78% of total cholesterol[16]	65%-70% of total cholesterol[3] 3 day-1 yr: 51-132 mg/100 ml 2-14 yr: 99-173 mg/100 ml[10]	Elevated[4] Bile salts Bilirubin Bromide Hemoglobin Lipemia Pregnancy Protein Tryptophan Vitamin A Decreased Androsterone Oxalate	250
Total	Serum	150-250 mg/100 ml[1]	Newborn: 50-100 mg/100 ml Thereafter: 150-275 mg/100 ml[3]		

Table 3-1. Normal values—cont'd

Constituent	Material analyzed	Normal value		Comments	Page
		Adult	Children		
Cholesterol—cont'd	CSF[13]	0.24-0.50 mg/100 ml			
	Amniotic fluid[13]	2 mg/100 ml			
	Urine[13]	0-0.07 mg/kg/24 hr			
	Saliva[13]	3-15 mg/100 ml			
Choline (free)	Serum[44]	0.3-1.5 mg/100 ml			
	Sweat[13]	7.1 μg/100 ml			
Cholinesterase	Plasma	0.5-1.3 pH units[16]	0.5 pH unit or more/hr[3]		
	CSF[45]	Present			
Citrate (citric acid)	Serum[13]	1.4-3.0 mg/100 ml			
	CSF[13]	0.04 mg/100 ml			
	Urine[13]	3-20 mg/kg/24 hr			
Colloidal gold	Serum[46]	Negative–1+			
	CSF[1]	0000000000– 0001111000			
Colloidal red	Serum[47,48]	Negative–1+			
Compound S		See Tetrahydro S			
Congo red	Serum[1] 10 ml of 1% solution IV	>60% remains at 1 hr		Hemolysis[3]	
Copper	Serum	♂ 70-140 μg/100 ml ♀ 85-155 μg/100 ml[16]	6 mo-5 yr: 27-153 μg/100 ml 5-17 yr: 94-234 μg/100 ml[10]	Elevated[4] Cobalt Contaminated needles Iron Nickel	270
	CSF[13]	6-20 μg/100 ml			
	Feces[14,15]	1.96 mg/24 hr			
	Urine	0-100 μg/24 hr[22]	0-25 μg/24 hr[3]		
	Sweat[13]	0.006 mg/100 ml			
	Saliva[13]	2-22 μg/100 ml			
Coproporphyrins	Urine	Random sample: 20 μg/100 ml 24 hr sample: 50-160 μg[1]	 50-250 μg[3]	Elevated[2] Alcohol Barbiturates Chlorpromazine Penicillin	549
	Feces	300-1100 μg/24 hr[49]	400-1000 μg/24 hr[3]	Phenazopyridine Sedatives Sulfonamides Terramycin Thora-Dex Thorazine	
Cortisol	See 17-Hydroxy-corticosteroids				450

Continued.

Table 3-1. Normal values—cont'd

Constituent	Material analyzed	Normal value: Adult	Normal value: Children	Comments	Page
Creatine	Serum[1]	♂ 0.2-0.6 mg/100 ml ♀ 0.6-1.0 mg/100 ml	Somewhat higher		151
	Urine	♂ 0-150 mg/24 hr ♀ 0-250 mg/24 hr[16] 6% of creatinine[16]	Newborn-1 yr: may equal creatinine Older: 30% of creatinine[22]	In pregnancy, may go up to 12% of creatinine[22]	153
Creatine phosphokinase (CPK)	Serum[1]	0-36 IU			299
Creatinine	Serum	0.6-1.3 mg/100 ml[2]	0.4-1.2 mg/100 ml[10]	Elevated[4]	148
	CSF[13]	0.54-1.91 mg/100 ml		Ascorbic acid	
	Sweat[13]	0.1-1.3 mg/100 ml		BSP	
	Urine	♂ 20-26 mg/kg/24 hr[6] ♀ 14-22 mg/kg/24 hr or ♂ 1.0-1.8 g/24 hr ♀ 0.7-1.5 g/24 hr	15-25 mg/kg/24 hr[3]	Diacetic acid Heat High glucose with acetone Levulose Protein PSP Pyruvate	152
	Amniotic fluid[13]	2 mg/100 ml			
	Saliva[13]	0.28-0.46 mg/100 ml			
Creatinine clearance					341
Cryoglobulins	Serum[1]	Negative			518
Cystine	See Table 3-2				
Dilantin	Serum[22]	None		Therapeutic level: 1-11 μg/ml	
Estrogens (total)	Urine[16]	♂ 4-25 μg/24 hr ♀ 4-60 μg/24 hr			476
Fat	Serum	30-70 mg/100 ml (neutral)[51,52]	10-350 mg/100 ml[10]		
	Feces	Total: 10%-25% of dry weight or <5 g/24 hr Neutral: 1%-5% of dry weight[1]	Total: <5g/24 hr[52]		320
Fatty acids	Serum	250-500 mg/100 ml[53]	380-465 mg/100 ml[3]		
	CSF[13]	1-3 mg/100 ml			
	Amniotic fluid[13]	23 mg/100 ml			

Table 3-1. Normal values—cont'd

Constituent	Material analyzed	Normal value: Adult	Normal value: Children	Comments	Page
Fatty acids—cont'd	Feces				
	Free	5%-13% of dry weight	1%-10% of dry weight[3]		
	Combined	5%-15% of dry weight[1]			
Fibrinogen	Plasma	200-400 mg/100 ml[16]	200-500 mg/100 ml[10]		246
	CSF[54]	None			
Fluoride	Serum[16]	<10 μg/100 ml		Fatal dose of sodium fluoride: 1-4 g[55]	
	Urine[16]	<800 μg/24 hr			
	Saliva[13]	0-0.005 mEq/L			
Folic acid	Serum[16]	5-21 ng/ml			
	Urine[56]	3.8-23.8 μg/24 hr			
	Sweat[13]	0.53-0.88 μg/100 ml			
Fructose	CSF[13]	2.0-7.5 mg/100 ml			
	Amniotic fluid[13]	1-2 mg/100 ml			
	Urine[22]	Negative			
Galactose	Urine	Negative[57]	Up to 15 mg/100 ml on milk diet[10]	Elevated[4] Ethanol ingestion	381
	Blood[57]		Up to 20 mg/100 ml		
Globulin	See Protein				
Globulin turbidity (acid-precipitable)	Serum[58]	4-8 units			
Glucose					138
Total reducing substances	Whole blood[59]	80-120 mg/100 ml			
	Plasma/serum[60]	70-100 mg/100 ml			329
	CSF[13]	45-93 mg/100 ml			
	Sweat[13]	2.8-40 mg/100 ml			
	Tears[13]	2.8-7.7 mg/100 ml			
	Urine[13]	7-21 mg/kg/24 hr		Elevated[4] Amino acids Ascorbic acid Aureomycin Beta-Chlor Chloral hydrate Chloromycetin Chlortetracycline Formaldehyde Oxytetracycline Penicillin (gross) Salicylates Streptomycin Terramycin Uric acid	377

Continued.

Table 3-1. Normal values—cont'd

Constituent	Material analyzed	Normal value		Comments	Page
		Adult	Children		
Glucose—cont'd	Saliva[13]	11.3-28.0 mg/100 ml			
True	Whole blood	60-100 mg/100 ml[16]	60-90 mg/100 ml	Elevated[4]	
	Serum/plasma	70-110 mg/100 ml[16]	Newborn:	ACTH	
			20-80 mg/100 ml	Decreased	
				Standing of sample	
	CSF	40-80 mg/100 ml[16]	40-80 mg/100 ml[3]		144
	Tears[13]	0-5 mg/100 ml			144
	Amniotic fluid[13]	4 mg/100 ml			
	Urine[3]		50-300 mg/24 hr		
	Sweat[13]	1-3 mg/100 ml			
Glucose tolerance				Decreased[4]	145
				Enovid	
Glucuronic acid	Urine[61]	0.3-1.0 g/24 hr		Elevated[4]	
				Aspirin	
				Camphor	
				Chloral hydrate	
				Menthol	
				Morphine	
				Sulfa	
Glucuronidase	CSF[45]	Present			
Glutathione	Whole blood[16]	24-37 mg/100 ml			
	Amniotic fluid[13]	13 mg/100 ml			
				Basal	413, 490
Growth hormone	Blood[16]	♂ <3 mμg/ml		Basal	
		♀ <5 mμg/ml			
Guanidine (imido-urea)	Serum[62]	0.032-0.48 mg/100 ml			
	Urine[62]	3-10 mg/24 hr			
Hexosamine	CSF[13]	5-18 mg/100 ml			
Hippuric acid	Urine[13]	1-10 mg/kg/24 hr			396
Histamine	Feces[63]	6-20 mg/500-600 g			
Hydrogen ion concentration (pH)	Serum[64]	7.35-7.45		Elevated[4]	
	Whole blood			Potassium oxalate	187
	Arterial	7.35-7.45[64]	7.35-7.45[3]	Sodium oxalate	
	Venous	7.33-7.43[64]	7.32-7.42[3]	Temperature	
	CSF	7.35-7.70[13]	7.33-7.42[3]	Decreased	
	Feces	6.9-7.7[19]	Breast-fed: 4.7-5.1	Ammonium oxalate	
			Cow's milk: >7.0[3]	Anaerobic storage at 36° C	
				EDTA	
	Gastric contents[45]	0.9-1.5		Exposure to air	
	Sweat[13]	3.8-6.5		Hemolysis	
	Tears[13]	7.3-7.7		Sodium citrate	

Table 3-1. Normal values—cont'd

Constituent	Material analyzed	Normal value: Adult	Normal value: Children	Comments	Page
	Urine[13]	4.8-7.8			368
	Saliva[13]	5.6-7.6			
α-Hydroxybutyric dehydrogenase	Serum[16]	140-350 units			
17-Hydroxycorticosteroids	Plasma	8-9 AM: 17-26 μg/100 ml 5 PM: approximately ½ the 8 AM value (fluorometric)[70]	8-20 μg/100 ml Newborn: lower (marked diurnal variation)[3]	After 25 USP units of ACTH, IM: 35-55 μg/100 ml[1]	450
	Urine	♂ 3-15 mg/24 hr ♀ 2-13 mg/24 hr or 2.0-4.2 mg/m^2/24 hr[94]		After 25 USP units of ACTH, IM: 200%-400% increase[1]	453
	CSF	<2 μg/100 ml[13]			
5-Hydroxyindoleacetic acid (5HIAA)	Urine	1-7 mg/24 hr[16]	33-75 μg/kg/24 hr[3]	Elevated[2] Acetanilid Amosene Bananas Banesin Causalin Halabar Mebralin Medigesic Mephenesin Methocarbamol Os-Cal-Gesic Robaxin Robaxisal Seconesin Tolseram Tolyd Decreased Chlorpromazine Phenothiazine-type drugs	407, 483
Hydroxyproline	Urine[10]		5-14 yr: 38-126 mg/24 hr		
Icterus index	Serum[65]	4-6 units		Elevated[4] Carotene Hemolysis Lipemia	
Immunoglobulins	Serum[1]	IgG 800-1500 mg/100 ml IgA 50-200 mg/100 ml IgM 40-120 mg/100 ml			591

Continued.

Table 3-1. Normal values—cont'd

Constituent	Material analyzed	Normal value: Adult	Normal value: Children	Comments	Page
Indican	Serum[66,67]	0.026-0.085 mg/100 ml			
	Urine[66,68]	4-20 mg/24 hr			397
Insulin	Serum[16]	4-24 μU/ml			435
Insulin tolerance	Serum[1]	50% or 60% of fasting glucose level			
Iodine					
Total	Serum[13]	2.5-16.9 μg/100 ml			
	CSF[13]	<0.001 mg/100 ml			
	Sweat[13]	0.5-1.2 μg/100 ml			
	Urine[13]	0.1-7.0 μg/kg/24 hr			508
	Saliva[13]	0-350 μg/100 ml			
Butanol-extractable (BEI)	Serum	3.5-6.5 μg/100 ml[1]	Newborn: 3-13 μg/100 ml 6 wk-16 yr: 3-7 mg/100 ml[10]	Elevated[1] X-ray contrast media	419
Protein-bound (PBI)	Serum	4.0-8.0 μg/100 ml (First trimester pregnancy: 6-12 μg/100 ml[16]	Birth: 4-8 μg/100 ml 2-14 days: 7-12 μg/100 ml 14-21 days: returns to adult level[3]	Elevated[4] Barium sulfate BSP Delvex Enovid Estrogens Exogenous iodine Cough medicine Dandruff control Gargles Suntan lotions Vitamins and cod-liver preparations Decreased ACTH BSP Chlorate Cortisone Cytomil Mercurial diuretics p-Aminobenzoate p-Aminosalicylic acid Salicylates Sulfonamides Testosterone Thiocyanate	421

Table 3-1. Normal values—cont'd

Constituent	Material analyzed	Normal value: Adult	Normal value: Children	Comments	Page
Uptake		20%-50%[2]		False depression[2] Iodine-containing drugs Some breads (iodine used as dough conditioner)	
See also Thyroxine					
Iron					
Total	Whole blood[13]	45.5-52.3 μg/100 ml		Elevated[4] ACTH Hemolysis Imferon Time of day (AM) Decreased ACTH Time of day (PM) Steroids	516
	Serum	♂ 90-150 μg/100 ml ♀ 70-130 μg/100 ml[2]	1 wk: 148 μg/100 ml 3 mo: 50 μg/100 ml 6-12 mo: 106 μg/100 ml 1-2 yr: 95 μg/100 ml 2-6 yr: 116 μg/μg/100 ml 6-12 yr:127 μg/100 ml[3]		
	CSF[13]	35 μg/100 ml			
	Sweat[13]	22-45 μg/100 ml			
	Urine[16]	Up to 0.15 mg/24 hr			
	Amniotic fluid[13]	0.07 mg/100 ml			
Binding capacity	Serum	250-410 μg/100 ml[2]	1 wk: 262 μg/100 ml 3 mo: 350 μg/100 ml 6-12 mo: 429 μg/100 ml 1-2 yr: 414 μg/100 ml 2-6 yr: 395 μg/100 ml 6-12 yr: 340 μg/100 ml[3]		
Isocitric dehydrogenase	Serum[16]	50-260 units			
17-Ketogenic steroids	Urine[16]	♂ 5-23 mg/24 hr ♀ 3-15 mg/24 hr			456
17-Ketosteroids	Plasma[16]	25-125 μg/100 ml			443
	Urine	♂ 6-18 mg/24 hr ♀ 5-15 mg/24 hr[95]	<12 yr: <5 mg/24 hr 12-15 yr: 5-12 mg/24 hr[1]	After 25 USP units of ACTH, IM: 50%-100% increase[1]	

Continued.

Table 3-1. Normal values—cont'd

Constituent	Material analyzed	Normal value: Adult	Normal value: Children	Comments	Page
17-Ketosteroids—cont'd	Amniotic fluid[13]	♂ fetus: 0.228 mg/100 ml ♀ fetus: 0.062 mg/100 ml			
Lactic acid	Whole blood	4.8-15.2 mg/100 ml[13]	6-16 mg/100 ml[3]		
	Serum[13]	36 mg/100 ml			
	CSF[13]	11-27 mg/100 ml			
	Sweat[13]	285-336 mg/100 ml			
	Urine[40]	50-200 mg/24 hr			
	Saliva[13]	0.17 mEq/L			
Lactic dehydrogenase	Serum	80-120 Wacker units[1] 87-186 IU/L[10]	2 hr-10 days: 308-1780 IU/L[10]	Elevated[2] Contact with clot Hemolysis Temperature Decreased Oxalate Temperature	296
	CSF[10]	6.3-30 IU/L	2 hr-10 days: 2.3-84 IU/L		
	Urine[1]	Up to 8300 Wroblewski units/8 hr			
Lactose	Urine[29]	Trace			380
Lead	Whole blood	18-49 μg/100 ml[13]	<50 μg/100 ml[10]		272
	Serum[13]	2.9 μg/100 ml			
	Feces[14,15]	0.32 mg/24 hr			
	Urine	0.16-1.10 μg/kg/24 hr[13]	<120 μg/24 hr[10]	Toxic level: >0.175 mg	
Leucine aminopeptidase	Serum[1]	5-20 Goldbarg units	5-24 Goldbarg units		
Lipase	Serum	0-1.5 Cherry-Crandall units[1]	20-136 IU/L[10]	Elevated[2,4] Codeine EDTA Morphine Temperature Decreased Hemolysis	283
	Duodenal juice		8-35 IU/L[10]		
Lipids					
Total	Serum	400-800 mg/100 ml[1]	Newborn: 170-440 mg/100 ml 3-10 days: 430-760 mg/100 ml 3 days-1 yr: 240-800 mg/100 ml 2-14 yr: 490-1090 mg/100 ml[10]		249

Table 3-1. Normal values—cont'd

Constituent	Material analyzed	Normal value: Adult	Normal value: Children	Comments	Page
Cholesterol		115-340 mg/100 ml			
Triglycerides		10-190 mg/100 ml			
Phospholipids		150-380 mg/100 ml	225 mg/100 ml		
Fatty acids		9.0-15.0 mM/L			
Neutral fat		0-200 mg/100 ml			
Phospholipid (P)		8.0-11.0 mg/100 ml			
	Urine[71-73]	0-50 mg/24 hr			320
	Feces[16]	0-7 g/24 hr (10%-27% dry weight)			
	Amniotic fluid[13]	48 mg/100 ml			
Lipoprotein		% of total			586
	Serum[16]	♂ ♀			
Lipalbumin		12-24 17-30			
β		36-59 29-53			
$\beta/(\alpha +$ lipalbumin)		0.89-2.5 0.6-1.8			
β/lipalbumin		1.6-4.5 1.1-2.8			
Lithium	Serum/plasma[1]	Negative		Therapeutic range: 0.5-1.0 mEq/L	511
Long-acting thyroid stimulator (LATS)	Serum[16]	None detectable			
Macroglobulins S19	Serum[16]	♂ 0.07-0.35g/100 ml ♀ 0.10-0.43 g/100 ml			517
Magnesium	Serum	1.5-2.0 mEq/L[16]	1.22-1.58 mEq/L (fluorometric)[10]		267
	Urine	0.42-2.40 mg/kg/24 hr[13]	0.1 mEq/kg/24 hr[10]		
	CSF	2.40-2.95 mg/100 ml[13]	2.8-3.3 mg/100 ml[3]		
	Feces[37]	17 mEq/L			
	Sweat[13]	4-286 μg/100 ml			
	Amniotic fluid[13]	2.2 mg/100 ml			
	Saliva[13]	0.16-1.06 mEq/L			
Manganese	Serum[16]	0.04-0.09 μg/100 ml			
	Urine[13]	0.1-1.4 μg/kg/24 hr			
	Sweat[13]	3-7 μg/100 ml			
	Feces[14,15]	3.69 mg/24 hr			
Melanin	Urine[1]	Negative			398

Continued.

Table 3-1. Normal values—cont'd

Constituent	Material analyzed	Normal value: Adult	Normal value: Children	Comments	Page
Mercury	Urine	<10 μg/24 hr[1]	<50 μg/24 hr[10]		511
Metanephrine, total	Urine[96]	0.3-0.9 mg/24 hr			472
Methanol	Whole blood[22]	Negative		May be fatal as low as 115 mg/100 ml[22] Fatal internal dose: 60-250 ml[55]	498
Mucoproteins	Serum	♂ 45-80 mg/100 ml ♀ 30-65 mg/100 ml[74,75]	45-105 mg/100 ml[3]		
	Urine[76]	♂ 120-170 mg/24 hr ♀ 85-125 mg/24 hr			
Myoglobin	Urine[16]	<1.6 mg/L			
Nickel	Urine[13]	2-4 μg/kg/24 hr			
Nicotinamide	Urine[13]	0.01-0.05 mg/kg/24 hr			
Nicotinic acid (niacin)	Urine[13]	0.002-0.020 mg/kg/24 hr			
	Sweat[13]	7-8.7 μg/100 ml			
Nitrogen (total)	Whole blood[13]	3000-4100 mg/100 ml			
	Urine[13]	130-300 mg/kg/24 hr			
	Feces[77]	1-2 g/24 hr			
	Sweat[13]	33.2 mg/100 ml			
	Tears[13]	0.158 g/100 ml			
	Plasma[13]	1200-1430 mg/100 ml			
Nonprotein nitrogen (NPN)	Whole blood	27-47 mg/100 ml[13]	22-40 mg/100 ml[3]		157
	Serum[13]	23-37 mg/100 ml		Elevated[4]	
	CSF[13]	12-28 mg/100 ml		EDTA	
	Sweat[13]	27-64 mg/100 ml		Temperature	
	Tears[13]	0.051 g/100 ml			
	Amniotic fluid[13]	24 mg/100 ml			
5'Nucleotidase	Serum[16]	Up to 1.6 units			
Osmolality	Serum	280-300 mOsm/kg[1]	285-295 mOsm/L[3]		216
	Urine	38-1400 mOsm/kg[1]			
Oxalic acid	Urine[13]	0.3-0.7 mg/kg/24 hr			

Table 3-1. Normal values—cont'd

Constituent	Material analyzed	Normal value: Adult	Normal value: Children	Comments	Page
Oxygen					
Capacity	Whole blood	17-22 vol%			
Content	Whole blood Arterial Venous	 15-22 vol% 11-16 vol%			
Percent saturation	Whole blood Arterial Venous	 94-96% 60-85%			224
Pressure (Po_2)	Whole blood Arterial	80-90 mm Hg[78,79]	85-100 mm Hg[3]		222
pH	See Hydrogen ion concentration				
Phenylalanine	See Amino acid				408
Phenylpyruvic acid	Urine[80]	Negligible		Elevated Aspirin Chlorpromazine Salicylates	
Phosphatase					
acid	Serum	0-1.1 Bodansky units 0.13-0.63 Bessey-Lowry unit 1.0-4.0 King-Armstrong units[1] ♂ 0.5-11.0 IU/L ♀ 0.2-9.5 IU/L[10]	1.0-5.0 King-Armstrong units[3] Newborn: 10.4-16.4 IU/L 2-13 yr: 8.6-13.0 IU/L[10]	Elevated[4] Androgens (in females) Hemolysis Prostatic massage Decreased Fluorides Oxalates Standing of sample	290
(tartrate labile)		0-0.5 King-Armstrong units[20]	5.0-14.0 Bodansky units	Elevated[2,4] Hemolysis	
alkaline	Serum	1.5-4.5 Bodansky units 0.8-2.3 Bessey-Lowry units 4.0-13.0 King-Armstrong units[1] 13.8-38.4 IU/L[10]	2.8-6.7 Bessey-Lowry units 15.0-30.0 King-Armstrong units[1] 1-3 mo: 73-226 IU/L 3-10 yr: 57-151 IU/L 10 yr-puberty: 57-258 IU/L 16 yr-adult: 13.8-38.4 IU/L[10]	Decreased Albumin Fluorides Oxalates Elevated/decreased Methyldopa (Aldomet, Aldoril)	285
Phospholipids	See Lipids				
Phosphorus (inorganic)	Serum	3.0-4.5 mg/100 ml[22]	First year: 4.0-7.0 mg/100 ml 1-12 yr: 5-6 mg/100 ml[3]	Elevated[4] Detergents Hemolysis Heparin	215

Continued.

Table 3-1. Normal values—cont'd

Constituent	Material analyzed	Normal value: Adult	Normal value: Children	Comments	Page
Phosphorus—cont'd	Urine	0.34-1.0 g/24 hr[16]	<1 g/24 hr[3]	Pituitrin Standing of sample Vitamin D	
	Feces[37]	0.51 g/24 hr			
	CSF[13]	1.25-2.10 mg/100 ml			
	Sweat[13]	9-43 μg/100 ml			
	Amniotic fluid[13]	2.7 mg/100 ml		Decreased Adrenalin Anesthesia Colloidal aluminum hydroxide Nonfasting state	
	Saliva[13]	7.4-21.1 mg/100 ml			
Pituitary gonadotropins (FSH)	Urine[98]	♂ 3-35 rat units ♀ 0-35 (premenopausal); 14-205 (postmenopausal)			
Porphobilinogen	Urine[16]	0-2 mg/24 hr		Elevated[4] Fresh warm urine Procaine Sulfonamides Urobilinogen (excessive)	396
Potassium	Serum/plasma	3.5-5.0 mEq/L[22]	Newborn: up to 7 mEq/L, falling rapidly to adult level[2]	Elevated[4] Calcium Cigarette ash and smoke Contact with cells Copper Hemolysis Iron Protein Venous stasis Decreased Acetazolamide ACTH *p*-Aminosalicylic acid Cortisone Phosphate Sulfate	178
	Urine[16]	26.0-133.0 mEq/L			176
	Feces[37]	12.0 mEq/24 hr			
	CSF[13]	2.2-2.9 mEq/L	2.1-3.4 mEq/L[3]		
	Sweat[13]	5.4-32.3 mEq/L			
	Tears[13]	7.7-22.1 mEq/L			
	Amniotic fluid[13]	4.8 mEq/L			
	Saliva[13]	14.3-37.9 mEq/L			
Pregnanediol	Urine	♂ 0-1.5 mg/24 hr ♀ Proliferative phase: 0.5-1.5 mg/24 hr	Negative[1]		481

Table 3-1. Normal values—cont'd

Constituent	Material analyzed	Normal value: Adult	Normal value: Children	Comments	Page
Pregnanediol—cont'd		Luteal phase: 2-7 mg/24 hr Postmenopausal phase: 0.2-1.0 mg/24 hr[16]			
Pregnanetriol	Urine[97]	< 6 yr: 0-0.2 mg/24 hr 7-16 yr: 0.3-1.1 mg/24 hr > 16 yr: 0.2-3.5 mg/24 hr			459
Protein					
Total	Serum	6.0-8.0 g/100 ml[22]	Premature: 4.6-6.4g/100 ml Newborn: 5-7.1 g/100 ml 1-3 mo: 4.7-7.4 g/100 ml 3-12 mo: 5.0-7.5 g/100 ml 1-15 yr: 6.5-8.6 g/100 ml[3]	Elevated[4] Bilirubin BSP Hemolysis Lipemia Decreased Ammonium ion Violent shaking	236
	CSF	12-43 mg/100 ml			327
	Lumbar	20-40 mg/100 ml	5-40 mg/100 ml		
	Cisternal	15-25 mg/100 ml	5-25 mg/100 ml		
	Ventricular	5-10 mg/100 ml[13]	1-15 mg/100 ml[3]		
	Urine	20-80 mg/24 hr[40]	<50 mg/24 hr[3]		370
	Tears[13]	0.67 g/100 ml			
	Feces	Negative[81]	Newborn: about 8% ingested protein[3]		
	Amniotic fluid[13]	260 mg/100 ml			
	Saliva[13]	156-630 mg/100 ml			
Albumin	Serum	4.0-5.0 g/100 ml[22]	Premature: 2.5-4.5 g/100 ml Newborn: 2.5-5.0 g/100 ml 1-3 mo: 3.0-4.2 g/100 ml 3-12 mo: 2.7-5.0 g/100 ml 1-15 yr: 3.2-5.0 g/100 ml[3]		237
	Tears[13]	0.39 g/100 ml			
	CSF[13]	40%-70% of total protein			
Globulin	Serum	2.0-3.0 g/100 ml[22]			237
	Tears[13]	0.28 g/100 ml			

Continued.

Table 3-1. Normal values—cont'd

Constituent	Material analyzed	Normal value: Adult	Normal value: Children	Comments	Page
Protein electrophoresis	Serum[16]				
	Albumin	55%-71%			
	Globulin				
	α_1	3%-6%			
	α_2	6%-11%			
	β	9%-15%			
	γ	10%-20%			
	CSF[1]				
	Albumin	52%			
	Globulin				
	α_1	5%			
	α_2	14%			
	β	10%			
	γ	19%			
Protein, Bence Jones	Urine[65]	Negative			377
PSP (based on 6 mg dye)	Urine[1]	*IV* — *Percent excreted*		Elevated[2]	347
		15 min — 20-50		BSP	
		30 min — 16-24		Ethoxazene	
		60 min — 9-17		Phenazopyridine drugs	
		120 min — 3-10		Phenolphthalein	
		IM		Serenium	
		1 hr — 40-60		Decreased	
		2 hr — 20-25		Penicillin	
				Salicylates	
				Sulfonamides	
Pyruvic acid	Plasma[13]	0.4-2.0 mg/100 ml			
	Urine[82]	0.5-2.5 mg/100 ml			
	CSF[13]	1.02 mg/100 ml			
	Amniotic fluid[13]	1.35 mg/100 ml			
Salicylates	Whole blood[22]	Negative		Therapeutic range: 20-25 mg/100 ml Toxic level: >30 mg/100 ml[22]	504
Serotonin	Blood[16]	0.05-0.20 μg/ml			
	Urine	See 5-HIAA			
Silver	Whole blood[13]	Trace			
	Feces[14,15]	0.058 mg/24 hr			
Sodium	Serum/plasma	136-148 mEq/L[13]	134-145 mEq/L[3]	Elevated[4]	169
				Calcium	
				Copper	
				Detergents	
				Excess saline	
				Iron	
				Potassium	
				Protein	
				Steroids	

Table 3-1. Normal values—cont'd

Constituent	Material analyzed	Normal value: Adult	Normal value: Children	Comments	Page
Sodium—cont'd				Decreased Hemolysis Mercurial diuretics Paracentesis Phosphate Sulfate	
	Urine[16]	43-217 mEq/24 hr			176
	Feces[20]	<10 mEq/24 hr			
	CSF[13]	131-152 mEq/L	130-165 mEq/L[3]		
	Sweat[13]	10-136 mEq/L			
	Tears[13]	108-176 mEq/L			
	Amniotic fluid[13]	127.6 mEq/L			
	Saliva[13]	3.5-24.4 mEq/L			
Solids	Urine[13]	55-70 g/24 hr			403
	Tears[13]	1.8 g/100 ml			
	CSF[13]	0.85%-1.70%			
	Sweat[13]	1.174%-1.597%			
	Saliva[13]	386-860 mg/100 ml			
Specific gravity	Whole blood[13]	1.052-1.064			
	Plasma[13]	1.024-1.030			
	Urine[13]	1.002-1.040		Elevated[2] Radiographic contrast media	369
	CSF	1.006-1.008[13]	1.005-1.009[3]		
	Sweat[13]	1.001-1.006			
	Amniotic fluid[13]	1.006-1.008			
	Saliva[13]	1.010-1.020			
Sulfate (inorganic)	Whole blood	0.45 mg/100 ml[13]	0.5-1.5 mg/100 ml[3]		
	Serum[13]	0.97-1.58 mg/100 ml			
Sulfonamides	Whole blood[1]	Negative		Therapeutic ranges[83]: Free sulfanilamide: 5-15 mg/100 ml Sulfadiazine: 5-13 mg/100 ml Sulfaguanidine: 2-4 mg/100 ml Sulfapyrazine: 4-10 mg/100 ml Sulfapyridine: 4-10 mg/100 ml Sulfathiazole: 5-13 mg/100 ml	505
	Urine[83]	Negative		Elevated[2] Phenacetin	
Sulfur	Whole blood[13]	0.28-0.65 mg/100 ml			
	Serum[13]	0.8-1.1 mg/100 ml			
	CSF[13]	0.6 mg/100 ml			

Continued.

Table 3-1. Normal values—cont'd

Constituent	Material analyzed	Normal value: Adult	Normal value: Children	Comments	Page
Sulfur—cont'd	Feces[37]	0.13 g/24 hr			
	Sweat[13]	0.7-7.4 mg/100 ml			
	Urine[13]	3.5-17.5 mg/kg/24 hr			
	Amniotic fluid[13]	3.7 mg/100 ml			
Takata-Ara reaction	Serum[84]	Negative			
	CSF[85]	Negative			
Tetrahydro S	Urine[16]	<1 mg/24 hr			461
Thiamine hydrochloride	Urine[86]	108-390 μg/24 hr			
Thiocyanate	Serum[87-89]	Negative		Therapeutic range: 8-12 mg/100 ml Toxic level: >15 mg/100 ml	
	Saliva[13]	3.1-27.5 mg/100 ml			
Thymol flocculation	Serum[1]	0-1+ in 24 hr			244
Thymol turbidity	Serum	0-5 units[1]	0-5 units[3]	Elevated[4] Bilirubin Hemoglobin Lipemia Decreased Albumin Oxalate Temperature increase	242
Thyroid-stimulating hormone (TSH)	Serum[16]	0-0.2 mU/ml (McKenzie)			414
Thyroxin-binding globulin (TBG)	Serum[1]	10-26 μg/100 ml		Elevated[4] Chlormadinone Enovid Ortho-Novum Provest	431
Thyroxin (T_4)					418, 419
By column as iodine	Serum[16]	2.9-6.4 μg/100 ml	Newborn-6 wk: about maternal level 6-12 wk: increases 12 wk-1 yr: drops to adult level		

Table 3-1. Normal values—cont'd

Constituent	Material analyzed	Normal value: Adult	Normal value: Children	Comments	Page
Radio-assay (Murphy-Pattee)	Serum[16]	4.0-11.0 μg/100 ml		Elevated[16] Dilantin Triiodothyronine (T_3) Decreased	419
Murphy-Pattee calculated as iodine[99]		2.6-7.1 μg/100 ml		Freezing Salicylates (gross)	
T^3 uptake	Serum[16]	10.0%-14.6%		Elevated Dicumarol	420
Free	Serum[16]	1.0-2.1 ng/100 ml		Dilantin Heparin Salicylates	420
Total	Serum[99]	4.0-11.0 μg/100 ml		Decreased Propylthiouracil therapy Triiodothyronine therapy	
Tin	Urine[90] Feces[14,15]	0.11 mg/L 22.88 mg/24 hr			
Transaminases glutamic oxalacetic (GOT)	Serum	10-40 Karmen units[22]	First week: 10-20 units Thereafter: 5-45 units[3]	Elevated[4] Hemolysis Leukocytes (CSF) Lipemia Opiates	293
	CSF		4-14 units[3]		
Glutamic pyruvic (GPT)	Serum	4-30 Karmen units[1]	First week: 10-90 units Thereafter: 5-45 units[3]	Elevated[4] Hemolysis Lipemia	293
Triglycerides	See Lipids				
Trypsin	Feces[1]	2+-4+			
Urea clearance	See p. 337				
Urea nitrogen	Whole blood	10.6-29.0 mg/100 ml[3]	Up to 40 mg/100 ml, falling rapidly to adult level[2]	Elevated[4] Acetone Chloral hydrate Chlorobutane Methyldopa Decreased Pregnancy Thymol	160

Continued.

Table 3-1. Normal values—cont'd

Constituent	Material analyzed	Normal value: Adult	Normal value: Children	Comments	Page
Urea nitrogen—cont'd	Serum/plasma[13]	13-26 mg/100 ml			164
	Urine[1]	6-17 g/24 hr			
	CSF	7.4-16.0 mg/100 ml[13]	6-28 mg/100 ml[3]		
	Sweat[13]	5-36 mg/100 ml			
	Tears[13]	0.03 g/100 ml			
	Amniotic fluid[13]	20 mg/100 ml			
	Saliva[13]	8.2-18.1 mg/100 ml			
Uric acid	Serum	♂ 2.5-7.0 mg/100ml[16]	2-6 mg/100 ml[3]	Elevated[4]	156
		♀ 1.5-6.0 mg/100 ml		Ascorbic acid	
	CSF[13]	0.5-2.8 mg/100 ml		Chlorothiazide	
	Sweat[13]	0.07-0.25 mg/100 ml		Nitrogen mustard	
	Urine[16]	250-750 mg/24 hr		Reducing substances	409
	Amniotic fluid[13]	4.4 mg/100 ml		Salicylates	
	Saliva[13]	0.5-2.9 mg/100 ml		Sodium oxalate	
				Whole blood	
				Decreased	
				Chlorine	
				Dicumarol	
				Potassium oxalate	
Urobilin	Urine[29]	10-130 mg/24 hr			395
Urobilinogen (stercobilinogen)	Feces	40-200 mg/24 hr[1]	40-280 mg/24 hr[3]	Elevated[2]	
				Acetone	
				Amidopyrine	
				p-Aminosalicylic acid	
	Urine	0.05-2.5 mg/24 hr[1]	1-3.5 mg/24 hr[3]	Antipyrine	393
				BSP	
				Chlorophyll	
				Chlorpromazine	
				Constipation	
				Hypaque	
				Light	
				Novocaine	
				Phenazopyridine	
				Pyridium	
				Thoradex	
				Thorazine	
				Decreased[4]	
				Antibiotics	
Uropepsin	Urine[16]	15-50 units/hr			
Uroporphyrins	Urine[1]	10-30 μg/24 hr			549

Table 3-1. Normal values—cont'd

Constituent	Material analyzed	Normal value: Adult	Normal value: Children	Comments	Page
Vanilmandelic acid (VMA)	Urine	0.5-7.0 mg/24 hr[100]	Infants: 83 μg/24 hr[1]	Elevated[4] Anileridine dihydrochloride Bananas Carbonate Chocolate Coffee Methocarbamol Tea Vanilla	470
Vitamin A	Whole blood[13]	60-70 IU/100 ml			
	Serum	69-158 IU/100 ml[13] 15-60 μg/100 ml[1]	15-100 μg/100 ml[3]		
Vitamin B_{12}	Whole blood[13]	0.06-0.14 μg/100 ml			
	Serum[13]	8-42 μμg/ml			
Vitamin C	Whole blood[13]	700-1900 μg/100 ml			
	Plasma	1000-2100 μg/100 ml[13]	500-1000 μg/100 ml[3]		
	CSF[13]	1.8 mg/100 ml			
	Urine[13]	0.1-0.4 mg/kg/24 hr			404
	Sweat[13]	70.5 μg/100 ml			
	Saliva[13]	0-0.372 mg/100 ml			
Vitamin E (tocopherol)	Serum[3]	0.6-1.1 mg/100 ml	Premature: 0.05-0.35 mg/100 ml Newborn: 0.1-0.35 mg/100 ml 2-5 mo: 0.2-0.6 mg/100 ml 6-24 mo: 0.35-0.8 mg/100 ml 2-12 yr: 0.55-0.9 mg/100 ml Breast-fed babies: 0.6-1.1 mg/100 ml		
Volume	Whole blood	♂ 70-85 ml/kg ♀ 59-73 ml/kg[91,92]	1 yr[10]: 69-112 ml/kg Older: 51-86 ml/kg		
	Plasma	♂ 40-48 ml/kg ♀ 37-46 ml/kg[91,92]	1 yr[10]: 66 ml/kg Older: 30-54 ml/kg		

Continued.

Table 3-1. Normal values—cont'd

Constituent	Material analyzed	Normal value: Adult	Normal value: Children	Comments	Page
Volume—cont'd	CSF[13]	90-150 ml			
	Gastric contents[6]	20-100 ml			
	Tears[13]	0.5-0.67 g/16 hr			
	Urine[20]	800-2000 ml/24 hr	1 wk: 50-300 ml/24 hr 6 mo: 350-550 ml/24 hr 3 yr: 500-700 ml/24 hr 10 yr: 700-1400 ml/24 hr		
	Amniotic fluid[13]	400-1200 ml			
Water					
	CSF[3]	99%			
	Urine[3]	7-42 ml/kg/24 hr			
	Tears[13]	98.2%			
	Sweat[13]	99.0%-99.5%			
	Feces[3]		Newborn: 80% Older: 65%		
	Amniotic fluid[13]	98.5%			
Weight	Feces[93]	Bulk: 100-200 g/24 hr Dry weight: 23-30 g/24 hr			
Xylose absorption	Serum	25-40 mg/100 ml (1-2 hr)[1]	30-40 mg/100 ml[10] (30-60 min)		
	Urine		*Percent of load*[10]		
	5 g/dose	1.1-2.4 g/5 hr	6 mo 11-30		
	25 g/dose	4.1-9.0 g/5 hr	6-12 mo 20-32 1-3 yr 20-42 3-10 yr 25-45 10 yr 25-50		
Zinc	Serum[16]	55-150 μg/100 ml			272
	Urine[16]	0.15-1.2 mg/24 hr			
Zinc sulfate turbidity	Serum	<12 units[1]	Newborn: <10 units 1-8 mo: <3 units 9-12 mo: <5 units Thereafter: <8 units[3]		

Table 3-2*

Amino acid[13]	Whole blood (mg/100 ml)	Plasma/serum (mg/100 ml)	CSF (mg/100 ml)	Urine (mg/kg/24 hr)		Sweat (mg/100 ml)	Saliva† (mg/100 ml)
				Total	Free		
Alanine	—	3.01-3.73	1.33	0.55	—	—	3.3-10.0
a-Alanine	2.79-5.11	—	—	—	—	—	—
Aminobutyric acid	—	0.22-0.35	—	—	—	—	—
Arginine	0.6-1.7	1.22-1.93	0.60	0.34-0.50	0.07-0.30	6.05-17.0	—
Asparagine	—	0.54-0.65	—	—	—	—	—
Aspartic acid	—	0.01-0.07	—	1.2-2.7	0.01-0.07	—	—
Citrulline	—	0.50	—	0-2.8	—	—	—
Cysteine plus cystine	—	1.08-1.30	—	—	—	—	—
Cystine	0.6-1.2	—	0.18	1.0-2.6	0.6-1.9	—	0.16-0.45
Glutamic acid	—	0.43-1.15	—	3.7-5.0	0-1.5	—	3.0-12.6
Glutamine	10	8.30	8.95	—	—	—	—
Glycine	1.8-2.5	1.34-1.73	0.17	6.5	2.2	—	1.9-15.5
Histidine	1.09-1.38	0.79-1.48	0.17	1.0-5.0	1.2-2.7	4.25-14.0	0.35-2.0
Hydroxyproline	—	—	—	0.02	—	—	—
Isoleucine	0.9-1.5	0.69-1.28	0.10	0.1-0.3	0.04-0.2	1.63-3.73	0.2-0.9
Leucine	1.4-2.0	1.42-2.30	0.14	0.22-0.45	0.05-0.17	1.98-3.75	0.025-0.300
Lysine	1.3-3.0	2.51-3.02	0.28	0.48-1.70	0.17-0.67	1.96-3.38	0.4-1.5
Methionine	0.392-0.457	0.33-0.43	0.04	0.10-0.17	0.03-0.10	—	0.005-0.010
1-Methylhistidine	—	0.04-0.17	—	—	—	—	—
3-Methylhistidine	—	0.04-0.13	—	—	—	—	—
Ornithine	—	0.62-0.80	—	0.15	—	—	—
Phenylalanine	0.8-1.2	0.69-0.95	0.19	0.21-0.54	0.09-0.23	1.70-3.47	0.6-2.5
Proline	—	2.01-3.34	—	0.3-0.9	0.03-0.20	—	0.35-1.50
Serine	—	1.01-1.25	—	0.5-0.7	0.2-0.5	—	1.0-1.8
Taurine	—	0.41-0.82	—	0.11-0.20	—	—	—
Threonine	1.3-2.0	1.21-1.72	0.28	0.36-2.60	0.11-0.35	2.13-8.18	0.4-5.6
Tryptophan	0.269-0.289	1.11	—	0.23-0.70	0.11-0.36	0.75-1.85	0.2-0.9
Tyrosine	0.8-1.4	0.81-1.45	0.20	0.44-0.82	0.15-0.30	1.32-5.45	0.2-1.0
Valine	2.0-2.9	2.37-3.71	0.21	0.25-0.42	0.04-0.18	2.40-4.35	0.7-2.2

*Based on data from Altman, P. L., and Dittmer, D. S.: Blood and other body fluids, Washington, D. C., Federation of American Societies for Experimental Biology.
†Paraffin-stimulated.

REFERENCES

1. Davidson, I., and Henry, J. B.: Todd-Sanford clinical diagnosis, ed. 14, Philadelphia, 1969, W. B. Saunders Co.
2. Garb, S.: Laboratory tests in common use, ed. 4, New York, 1966, Springer Publishing Co., Inc.
3. Cooke, R. E., and Levin, S.: The biologic basis of pediatric practice, New York, 1968, McGraw-Hill Book Co.
4. Forman, D. T.: Diagnostic interferences—materials affecting the diagnostic specificity of laboratory tests, Elmont, N. Y., Biochemex Laboratories, Inc.
5. MacDonald, R. P., Gerber, C., and Nielsen, A.: Amer. J. Clin. Path. **25**:1367, 1955.
6. Kolmer, J. A.: Clinical diagnosis by laboratory examinations, ed. 2, New York, 1949, Appleton-Century-Crofts.
7. Gettler, A. O., and Freireich, A. W.: J. Biol. Chem. **92**:199, 1931.
8. Turner, R. G.: Proc. Soc. Exp. Biol. Med. **32**: 1548, 1935.

9. Sibley, J. A., and Fleisher, G. A.: Mayo Clin. Proc. **29**:591, 1954.
10. O'Brien, D., Ibbott, F. A., and Rodgerson, D. O.: Laboratory manual of pediatric microbiochemical techniques, ed. 4, New York, 1968, Harper & Row, Publishers.
11. Wiechowski, W.: Beitr. Chem. Physiol. Path. **11**:109, 1908.
12. Young, E. G., and Conway, C. F.: J. Biol. Chem. **142**:839, 1942.
13. Altman, P. L., and Dittmer, D. S.: Blood and other body fluids, Washington, D. C., 1961, Federation of American Societies for Experimental Biology.
14. Kehoe, R. A., Cholak, J., and Story, R. V.: J. Nutr. **19**:579, 1940.
15. Kehoe, R. A., Cholak, J., and Story, R. V.: J. Nutr. **20**:85, 1940.
16. Handbook of specialized diagnostic laboratory tests, ed. 8, Van Nuys, Calif., 1968, Bio-Science Laboratories.
17. McDermott, W. V., Jr., Adams, R. D., and Riddell, A. G.: Ann. Surg. **140**:539, 1954.
18. Williams, J. L., and Dick, G. F.: J.A.M.A. **100**:484, 1930.
19. Robinson, C. S.: J. Biol. Chem. **52**:445, 1922; **66**:811, 1925; **108**:403, 1935.
20. Eastham, R. D.: Biochemical values in clinical medicine, ed. 3, Bristol, 1967, John Wright & Sons, Ltd.
21. Friedman, M. M., and Becker, E.: Clin. Chem. **1**:110, 1955.
22. Castleman, B., and McNeely, B. U.: New Eng. J. Med. **276**:167-174, 1967.
23. Roe, J. H., and Kuether, C. A.: J. Biol. Chem. **147**:399, 1943.
24. Wortis, I. S.: Arch. Intern. Med. **60**:264, 1937.
25. Goldbaum, L. R.: Anal. Chem. **24**:1604, 1952.
26. Koppanyi, T.: Clin. Chem. **5**:75, 1953.
27. Sunshine, I., and Hackett, E. R.: Amer. J. Clin. Path. **24**:1133, 1954.
28. Cantarow, A., and Trumper, M.: Clinical biochemistry, Philadelphia, 1955, W. B. Saunders Co.
29. Levinson, S. A., and MacFate, R. P.: Clinical laboratory diagnosis, Philadelphia, 1951, Lea & Febiger.
30. Golden, W. R. C., and Snavely, J. G.: J. Lab. Clin. Med. **33**:890, 1948.
31. Oppel, T. H.: Amer. J. Med. Sci. **205**:856, 1942.
32. Shull, G. M., Hutchings, B. L., and Peterson, W. H.: J. Biol. Chem. **142**:913, 1942.
33. Hanzlik, P. J., Lehman, A. J., Richardson, A. P., and Van Winkle, W., Jr.: Arch. Derm. Syph. **36**:725, 1937.
34. Kolmer, J. A., Brown, H., and Rule, A. M.: Amer. J. Syph. Gonorr. Ven. Dis. **24**:415, 1940.
35. Lehman, A. J., Richardson, A. P., and Hanzlik, P. J.: J. Lab. Clin. Med. **21**:95, 1935.
36. Johnson, B. C., Hamilton, T. S., and Mitchell, H. H.: J. Biol. Chem. **159**:231, 1945.
37. Shohl, A. T.: Mineral metabolism, New York, 1939, Reinhold Publishing Corp.
38. Shorr, E., Bernheim, A. R., and Taussky, H.: Science **95**:606, 1942.
39. Gamble, J. L.: J. Biol. Chem. **51**:295, 1922.
40. Peters, J. P., and Van Slyke, D. D.: Quantitative clinical chemistry, vol. 2, Baltimore, 1952, The Williams & Wilkins Co.
41. Christman, A. A., and Randall, E. L.: J. Biol. Chem. **102**:595, 1933.
42. Gettler, A. O., and Freimuth, H. C.: Amer. J. Clin. Path. Tech. (suppl.) **7**:79, 1943.
43. Davson, H.: Physiology of the cerebrospinal fluid, Boston, 1967, Little, Brown & Co.
44. Engel, R. W., Salmon, W. D., and Ackerman, C. J.: Methods of biochemical analysis, vol. 1, New York, 1954, Interscience Publishers, Inc.
45. Spector, Wm. S., editor: Handbook of biological data, Philadelphia, 1956, W. B. Saunders Co.
46. MacLagan, N. F.: Brit. J. Exp. Path. **25**:15, 1944.
47. Ducci, H.: J. Lab. Clin. Med. **32**:1273, 1947.
48. Strade, H. A., Dotti, L. B., and Ilka, S. J.: Gastroenterology **12**:934, 1949.
49. Sunderman, F. W., Jr., and Sunderman, F. W.: Amer. J. Clin. Path. **25**:1231, 1955.
50. Blix, G.: Acta Med. Scand. **64**:142, 175, 234, 1926.
51. Blix, G.: Biochem. Z. **305**:145, 1940.
52. Fowweather, F. S.: Brit. J. Exp. Path. **7**:15, 1926.
53. Man, E. B., and Gildea, E. G.: J. Biol. Chem. **90**:43, 1932.
54. Gornall, A. G., Bardawill, C. J., and David, M. M.: J. Biol. Chem. **177**:751, 1949.
55. Dreisbach, R. H.: Handbook of poisoning, ed. 4, Los Altos, Calif., 1963, Lange Medical Publications.
56. Wright, L. D., and Welch, A. D.: Science **98**:179, 1943.
57. Fister, H. J.: Manual of standardized procedures for spectrophotometric chemistry, New York, 1950, Standard Scientific Supply Corp.
58. Greenspan, E. M.: J. Mount Sinai Hosp. N. Y. **21**:270, 1955.
59. Folin, O., and Wu, H.: J. Biol. Chem. **41**:367, 1920.
60. Reiner, Miriam, editor: Standard methods of clinical chemistry, vol. 1, New York, 1953, Academic Press, Inc.
61. Freeman, S., Kanabrocki, E. L., and Inman, C. W.: J. Lab. Clin. Med. **47**:583, 1956.

62. Andes, J. E., and Myers, V. C.: J. Lab. Clin. Med. **22**:1147, 1937.
63. Hanke, M. T., and Koesler, K. K.: J. Biol. Chem. **59**:879, 1924.
64. Behrmann, V. G., and Fay, M.: Science **90**: 188, 1939.
65. White, W. L., and Frankel, S.: Seiverd's chemistry for medical technologists, ed. 2, St. Louis, 1965, The C. V. Mosby Co.
66. Meiklejohn, A. P., and Cohen, F. P.: J. Lab. Clin. Med. **27**:949, 1942.
67. Townsend, S. R.: J. Lab. Clin. Med. **23**:809, 1938.
68. Bradley, S. E.: Med. Clin. N. Amer. **29**:1314, 1945.
69. Hanzel, R. F.: Proc. Soc. Exp. Biol. Med. **30**:846, 1933.
70. DeMoor, P., Steeno, O., Raskin, M., and Hendrix, A.: Acta Endocr. **33**:297, 1960.
71. Brice, A. T., Jr.: J. Lab. Clin. Med. **15**:953, 1930.
72. Lazarus, J. A., and Marks, M. S., J. Urol. **56**: 246, 1946.
73. Page, I. H.: Amer. J. Med. Sci. **192**:217, 1936.
74. Rhees, M. C., Ellerbrook, L. D., and Lippincott, S. W.: Amer. J. Clin. Path. **24**:774, 1954.
75. Mandel, E. E., Gorsuch, T. L., and Cooper, G. E.: Clin. Chem. **1**:221, 1955.
76. Anderson, A. J., and MacLagan, N. F.: Biochem. J. **59**:638, 1955.
77. Wollaeger, E. E., Comfort, M. W., Weir, J. F., and Osterberg, A. E.: Gastroenterology **6**:83, 1946.
78. Gibbs, E. L., Lennox, W. G., Nima, L. F., and Gibbs, F. A.: J. Biol. Chem. **144**:325, 1942.
79. Van Slyke, D. D., and Neill, J. M.: J. Biol. Chem. **61**:523, 1924.
80. Folling, A.: Z. Physiol. Chem. **227**:169, 1934; Nord. Med. T. **8**:1054, 1934.
81. Fantus, B., Wozasek, O., and Steigmann, F.: Amer. J. Dig. Dis. **8**:296, 1941.
82. Friedemann, T. E., and Haugen, G. E.: J. Biol. Chem. **147**:415, 1943.
83. Bratton, A. C., and Marshall, E. K., Jr.: J. Biol. Chem. **128**:537, 1939.
84. Heath, C. W., and King, E. F.: New Eng. J. Med. **211**:1077, 1934.
85. Szanto, P. B., and Burack, S.: J. Lab. Clin. Med. **26**:1079, 1941.
86. Carden, G. A., Province, W. D., and Ferrebee, J. W.: Proc. Soc. Exp. Biol. Med. **45**:1, 1940.
87. Barker, M. H.: J.A.M.A. **106**:762, 1936.
88. Bowler, R. C.: Biochem. J. **38**:385, 1944.
89. Powell, W. N.: J. Lab. Clin. Med. **30**:1071, 1945.
90. Gonzales, T. A., Vance, M., Helpern, M., and Umberger, C. J.: Legal medicine, pathology and toxicology, ed. 2, New York, 1954, Appleton-Century-Crofts.
91. Gregerson, M. I.: J. Lab. Clin. Med. **29**: 1286, 1944.
92. Von Porat, B. T. D.: Acta Med. Scand. (supp.) **140**:256, 1951.
93. Wozasek, O., and Steigmann, F.: Amer. J. Dig. Dis. **9**:423, 1942.
94. Migeon, C. J., Green, O. C., and Eckert, J. P.: Metabolism **12**:718, 1963.
95. Smilo, R. P., and Forsham, P. H.: Postgrad. Med. **46**:146, 1969.
96. Pisano, J. J.: Clin. Chim. Acta **5**:406, 1960.
97. Bongiovanni, A. M., and Eberlein, W. R.: Anal. Chem. **30**:388, 1958.
98. Albert, A.: Recent Progr. Hormone Res. **12**: 227, 1956.
99. Murphy, B. P.: J. Lab. Clin. Med. **66**:161, 1965.
100. Pisano, J. J., Crout, J. R., and Abraham, D.: Clin. Chim. Acta **7**:285, 1962.

• 4

• AutoAnalyzer methods

ADAPTING MANUAL METHODS TO THE AUTOANALYZER

Most manual methods of analysis can be adapted to the AutoAnalyzer. The following steps are presented as a guide to automating a manual procedure.

1. Consider the kind of specimen required. Is it a simple dilution or a protein-free filtrate? Is the constituent to be measured dialyzable?

2. Consider the chemistry involved. What product of the reaction is measured? Can a different product be measured? Is a specimen blank correction necessary and, if so, can blanks be avoided by dialysis? Example:

$$p\text{-Nitrophenyl phosphate} \xrightarrow{\text{Alkaline phosphatase}} p\text{-Nitrophenol} + \text{Phosphate}$$

Either *p*-nitrophenol or phosphate could be measured. Both are dialyzable; therefore blank correction is not needed. In this case it is preferable to measure the *p*-nitrophenol because it has color and can be measured directly. The phosphate would have to be reacted with a reducing agent to produce a color that could be measured.

3. Plan a preliminary flow diagram. Decide what modules are needed. Can pump I handle the number of pump tubes necessary or will pump II be needed? Is a heating bath required? What temperature? Is a dialyzer needed? What will be used to measure the end product?

a. Colorimeter: Photocell or phototube? What filters?

b. Fluorometer: What filters?

c. Flame photometer: What filters?

Will any special glass fittings or coils be required? Solvaflex or Acidflex tubing?

4. Consider the proportions used in the manual method and match the proportions with the pump tubes (Table 4-1). Example:

	Sample	*Substrate*
Manual	0.1 ml	0.9 ml
Automatic	0.015-inch I.D.	0.045-inch I.D.

In systems using dialysis some adjustments of proportions are necessary. Figure on a 10% dialysis ratio (some constituents have a higher dialysis ratio) and adjust tubing sizes accordingly. Match the flow rate through the dialyzer so that it comes as close as possible to being the same on both sides of the membrane. This can be accomplished by adjusting the size of the air tubes.

5. Build and test the system.

6. Evaluate results. Has the desired sensitivity been achieved? If not, readjust the sample and reagent proportions, the incubation time, or the color development time.

7. If sensitivity is satisfactory, determine the optimum sampling rate. A proper rate is one in which a sample has an optical density (OD) or % T read-

ing equal to 95% of the steady state reading of the same sample. Also, a proper rate will keep the sample interaction at a minimum.

8. Check the automated method against the manual method. If results do not agree, investigate the cause.

9. Readjustment of reagent concentrations may be necessary, or advantageous, in some procedures.

NOTES ON GENERAL OPERATING PROCEDURE

1. Before turning a colorimeter on, be sure that proper filters are in place. Never allow the full intensity of the lamp to strike the photocells. The effect would be similar to that of a person's looking directly into the sun. Eyes require some time to refocus. If staring at the sun is done repeatedly, the eyes will suffer permanent damage and even blindness. The same type of damage is done to the photocells unless a set of filters is in place. If it is necessary to change filters for another procedure, turn the lamp off just long enough to change the filters.

In reading the following flow charts, assume that the standard 15 mm tubular flow cell is used, unless specified otherwise.

2. Before starting the pump, clean the platen and the manifold pump tubes with a cloth dampened with carbon tetrachloride. Wipe the rollers free of any grease at the same time. This will prolong the life of the pump tubes and contribute to a smooth flow through the tubes. When engaging the roller head assembly on pump I, be sure that the pump tubes are not overlapping. The following methods may be used with either pump I or pump II unless other directions are given.

3. Never release the roller head while reagents are being pumped. Mixed reagents will back up through the lines into the reagent bottles and cause contamination of the reagents. Clamp the lines with hemostats before releasing the roller head.

4. Allow sufficient time for a stable flow of reagents through the flow cell before setting the reagent baseline. This will lessen the chance of baseline drift during the test run.

5. Compensate for baseline drift by adjusting the standard curve to a standard placed in every tenth position in the sample plate.

6. When transferring reagent lines from one bottle to another (reagent or water), wipe the outside of the lines with a tissue to avoid contaminating the other bottle.

7. Each time a dialyzer membrane is changed, the bath water should be changed also. It is not enough to put an antibacterial agent in the water and forget about it. Invariably, something spills or leaks into the bath water. As acids or salts accumulate, corrosion of the metal parts takes place. Always use distilled or deionized water in the dialyzer bath. Be sure the temperature has stabilized at 37° C before running any procedure. If it becomes necessary to change a membrane in the middle of a run, the bath water may be changed when the run is completed. Always run fresh standards after the membrane has been changed before finishing the run.

8. Change the oil in the heating bath at least every 3 months. Remove all old oil from the coil, heating element, and thermoregulator by soaking the top part in the pot filled with a hot detergent solution. Before reassembling the heating bath, be sure there is no charred material coating the heating element. If necessary, scrape it off with a knife.

9. Be sure the circulating motors on the dialyzer and heating baths are working. If the liquid is not circulating, the temperature will vary considerably in different parts of the bath as the heating element cycles on and off. This can cause wide variation in test results.

10. If waste from the colorimeter is being collected in a container instead of draining into a sink, be sure that the debubbler line does not hang down below the surface of the liquid in the container. An all-liquid line would have no trouble in draining, but air bubbles would have to go down below the surface of the liquid to escape. Pressure is required to push the air down and out the end of the drain tube. The farther down below the surface the end of the tube is, the more pressure is required to push the air out. This buildup of pressure will affect the flow through the flow cell.

11. Check the recorder gain each day, or before starting a different procedure.

12. Always wait until the last specimen has recorded before removing lines from reagent bottles. If the lines are pulled out of reagents too soon, the last one or two specimens may be lost because there were no reagents with which to react. The sample has a longer distance to travel than do most reagents.

13. An "in-line" filter is now available that is very helpful. Each filter unit is self-contained and is complete with nipples for connection into the reagent

Table 4-1. Pump tube sizes and delivery volumes*

Tube I.D. (inch)	Shoulder colors	Clear standard delivery				Solvaflex delivery				Acidflex delivery	
		(ml/min)			(ml/hr)	(ml/min)			(ml/hr)	(ml/min approx)	(ml/hr)
		Min	Nom	Max	Max	Min	Nom	Max	Max		
0.005	Orange-black	0.005	0.015	0.029	1.74	0.005	0.015	0.029	1.74	n.a.†	
0.0075	Orange-red	0.016	0.03	0.048	2.88	0.016	0.03	0.048	2.88	n.a.†	
0.010	Orange-blue	0.032	0.05	0.072	4.32	0.032	0.05	0.072	4.32	n.a.†	
0.015	Orange-green	0.075	0.10	0.128	7.68	0.075	0.10	0.128	7.68	n.a.†	
0.020	Orange-yellow	0.13	0.16	0.19	11.4	0.13	0.16	0.19	11.4	n.a.†	
0.025	Orange-white	0.19	0.23	0.27	16.2	0.19	0.23	0.27	16.2	n.a.†	
0.030	Black	0.28	0.32	0.36	21.6	0.28	0.32	0.36	21.6	0.34	20.4
0.035	Orange	0.37	0.42	0.47	28.2	0.37	0.42	0.47	28.2	0.43	25.8
0.040	White	0.54	0.60	0.66	39.6	0.51	0.56	0.62	37.2	0.53	31.8
0.045	Red	0.73	0.80	0.87	52.2	0.64	0.70	0.76	45.6	0.64	38.4
0.051	Gray	0.92	1.00	1.08	64.8	0.81	0.88	0.93	55.8	0.78	46.8
0.056	Yellow	1.12	1.20	1.28	76.8	0.99	1.06	1.13	67.8	0.92	55.2
0.060	Yellow-blue	1.31	1.40	1.49	89.4	1.14	1.21	1.29	77.4	1.06	63.6
0.065	Blue	1.50	1.60	1.70	102.0	1.29	1.37	1.45	87.0	1.19	71.4
0.073	Green	1.90	2.00	2.10	126.0	1.60	1.69	1.78	106.8	1.44	86.4
0.081	Purple	2.37	2.50	2.63	157.8	1.92	2.02	2.12	127.2	1.71	102.6
0.090	Purple-black	2.77	2.90	3.03	181.8	2.31	2.42	2.53	151.8	2.03	121.8
0.100	Purple-orange	3.26	3.40	3.54	212.4	2.77	2.89	3.01	180.6	2.39	143.4
0.110	Purple-white	3.75	3.90	4.05	243.0	3.26	3.39	3.52	211.2	2.76	165.6

*Courtesy Technicon Corp., Tarrytown, N. Y.; from White, W. L., Erickson, M. M., and Stevens, S. C.: Practical automation for the clinical laboratory, St. Louis, 1968, The C. V. Mosby Co.

†Not available.

lines. It can be attached at any convenient point between the manifold tubing and the reagent bottle. The filter matrix is chemically inert and is unaffected by strong acids or bases. The manufacturer* states that the filter will remove completely from the reagents any debris (down to a 6-micron particle size).

14. One of the ways to adjust the sensitivity of a procedure that requires dialysis is to change the size of the sample, sample diluent, or recipient stream pump tube. When this is done, however, take care to see that the flow rates on each side of the dialyzer membrane are as close as possible to being equal. This can usually be accomplished by changing the air line to balance the change in a reagent or sample line. When calculating flow rates, the air lines must be included.

15. To determine the reagent consumption for a particular procedure, simply

*UniChem Corp.

Table 4-2. Wash ratios and volumes*

	Sample rate (analyses/hr)	Sample time (sec)	Wash time (sec)	Volume picked up by 0.030 sample line (ml)
Sampler I	20	120	60	0.64
	40	60	30	0.32
	60	40	20	0.21
Sampler II	10	240	120	1.28
(cams)	20	120	60	0.64
	30	80	40	0.43
	40	60	30	0.32
	50	48	24	0.26
	60	40	20	0.21
	70	34	17	0.18
	80	30	15	0.16
	90	27	13	0.14
	100	24	12	0.13
	110	22	11	0.12
	120	20	10	0.11

The following formulas are used to determine sample and wash times in seconds for cams employing the usual 2:1 sample-wash ratio:

$$\text{Sample time} = \frac{60\ \text{sec/min} \times 60\ \text{min/hr}}{\text{Sample rate/hr}} \times \frac{2}{3} = \frac{2400}{\text{Sample rate}}$$

$$\text{Wash time} = \frac{60\ \text{sec/min} \times 60\ \text{min/hr}}{\text{Sample rate/hr}} \times \frac{1}{3} = \frac{1200}{\text{Sample rate}}$$

The following formula may be used to determine the volume of sample picked up by various-sized pump tubes.

$$\text{Volume} = \frac{T}{60} \times F, \text{ where}$$

T = Sample time in seconds
F = Flow rate of pump tube in ml/min

*From White, W. L., Erickson, M. M., and Stevens, S. C.: Practical automation for the clinical laboratory, St. Louis, 1968, The C. V. Mosby Co.

note the size of the tube pumping the reagent and refer to Table 4-1, where the approximate ml/min and ml/hr values are given. To determine sample size, refer to Table 4-2.

16. Detailed information on the individual AutoAnalyzer modules has been published.[1]

WETTING AGENTS

A wide variety of wetting agents or surfactants are now being used in analytic chemical procedures. Their main purpose is to lower the surface tension of a liquid and thus permit a smoother flow through small-diameter tubing.

Sterox SE

One of the earliest wetting agents used in the clinical laboratory was Sterox SE,* the brand name for *p*-oxyethylene thioether. Sterox SE is a 100% nonionic type of surface-active agent and is also a wetting agent, emulsifier, and detergent. It has a specific gravity of 1.0495 at 25° C. This surfactant has been widely used in manual flame photometry. A concentration of 0.02% Sterox SE in standards and unknown solutions provides a smooth flow through the atomizer into the flame and helps to keep the burner clean.

Brij 35

The surfactant most commonly used in AutoAnalyzer methods is Brij 35,† a polyoxyethylene (23) lauryl ether. It is a nonionic white waxy solid with an approximate specific gravity of 1.05 at 25° C. Its pour point is approximately 33° C. The solution of Brij 35 supplied by the Technicon Corporation for use with their systems has a concentration of about 30%.

Triton X-100

Triton X-100‡ is a water-soluble isooctyl phenoxy-polyethoxy ethanol containing 10 moles of ethylene oxide. It is a biodegradable, anhydrous liquid, 100% active, nonionic surface-active agent. It is useful as a wetting agent, an emulsifier, and a detergent. The specific gravity of Triton X-100 is 1.065 at 25° C and its pour point is 7° C.

Triton X-405

Triton X-405‡ is a preparation of *p*-tertiary octylphenoxy-polyethoxy ethanols with about 40 moles of ethylene oxide. It is a liquid, 70% active, nonionic surface-active agent. The specific gravity of Triton X-405 is 1.11 at 25° C and its pour point is −4° C.

Both Triton X-100 and X-405 are ideal surfactants for enzyme determinations in continuous flow systems, since they do not interfere with enzyme activity.

Aerosol 22

Aerosol 22§ is a 35% aqueous solution of tetrasodium *N*-(1,2-dicarboxyethyl)-*N*-octadecyl-succinamate. It is an anionic surface-active agent with a specific

---

*Harleco.

†Atlas Chemical Industries, Inc.

‡Rohm & Haas Co.

§American Cyanamid Co.

gravity of 1.12 and a boiling point of 95° C (at 760 mm). Aerosol 22 is soluble in hot or cold water and saturated salt solutions and is insoluble on most organic solvents. Its surfactant properties are not as good as those of the other surface-active agents listed but it has proved of value in systems where other surfactants have been found to interfere in some way. One of these is the Technicon N-13b uric acid procedure.

FC-134

FC-134* is one of 3M's Fluorad group of surfactants. It is a cationic brown waxy solid. FC-134 is soluble in acetone and in ethyl, isopropyl, and methyl alcohols but is only slightly soluble in water. The solution supplied by the Technicon Corporation for use with the SMA 12/60 contains 0.2% FC-134.

ARW-7

ARW-7† is an anionic surfactant composed of a 40% by weight aqueous solution of sodium 2-ethylhexyl sulfate. It has a specific gravity of 1.144 at 20° C and is completely soluble in water. Commercially, it is used in caustic solutions for washing fruits and vegetables. In the laboratory, it is used in the biuret reagent for the SMA 12/60. Care should be used in handling ARW-7 because it will defat the skin and can damage the eyes in its undiluted form. It is corrosive to such materials as steel, aluminum, galvanized iron, and tin. ARW-7 should be stored in polyethylene containers in a place where the temperature does not go below 10° C or above 38° C (50° to 100° F); otherwise, phase separation will occur.

Ultrawet 60L

Ultrawet 60L surfactant,‡ also known as wetting agent A,§ is a 65.5% solution of triethanolamine linear alkylate sulfonate. It is an anionic biodegradable detergent with a specific gravity of 1.14 at 25° C. Commercially, it is used in formulas for shampoo, bubblebath, and specialty detergents. In the laboratory, it is used in the automated phosphorus procedure.

METHODOLOGY

Electrolytes (Na, K, CO_2, Cl)

Principle

The AutoAnalyzer electrolyte system determines sodium, potassium, carbon dioxide, and chloride simultaneously from a single specimen.[2]

Serum or heparinized plasma is mixed with a CO_2-free air–segmented stream of acid lithium nitrate solution. The diluted sample enters the dialyzer where the sodium, potassium, chloride, and lithium ions dialyze into a CO_2-free air–segmented stream of water. The lithium serves as an internal standard in the determination of sodium and potassium. The acid in the lithium nitrate solution liberates free CO_2 gas from the carbonate and dissolved CO_2 in the sample. As the sample stream leaves the dialyzer, it is joined and mixed with an acid antifoam

---

*3M Co.

†Technicon.

‡Arco Chemical Co.

§Technicon.

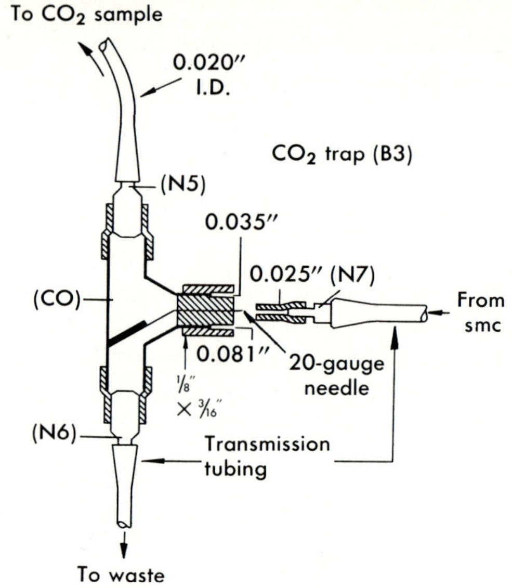

Fig. 4-1. CO_2 trap (B3).

reagent. The stream then enters a gas-liquid phase separator (CO_2 trap), where the liquid goes to waste and the gas phase is resampled (Fig. 4-1). The resampled CO_2 gas is mixed with a weakly buffered phenolphthalein solution. The CO_2 decolorizes the phenolphthalein solution and the loss of color is measured at 550 mμ in a 15 mm tubular flow cell.[3] This is called an inverse color reaction.

As the recipient stream leaves the dialyzer, a portion of the stream is resampled and mixed with chloride color reagent. The chloride ions (Cl^-) combine with the mercuric ions (Hg^{++}) in mercuric thiocyanate, $Hg(SCN)_2$, to form $HgCl_2$. The released $(SCN)^-$ reacts with Fe^{+++} to form the red complex—ferric thiocyanate, $Fe(SCN)_3$.[4] The intensity of this color is measured at 480 mμ in a 15 mm tubular flow cell.

The remainder of the recipient stream is pumped to the flame photometer, where the major portion is aspirated into the burner for the simultaneous determination of sodium and potassium. The excess liquid and air segments go to waste.

Modules needed

One sampler II
One proportioning pump II or two pump I's
One dialyzer with one set of dialyzer plates
Two colorimeters with 480 and 550 mμ filters
One flame photometer III or III A
Two double-pen recorders

Flow diagrams

See Figs. 4-2 and 4-3.

Reagents needed

Water recipient (0.5 ml Brij 35/L water)
Lithium nitrate stock, 1000 mEq Li/L in 2 N H_2SO_4 (D-6)

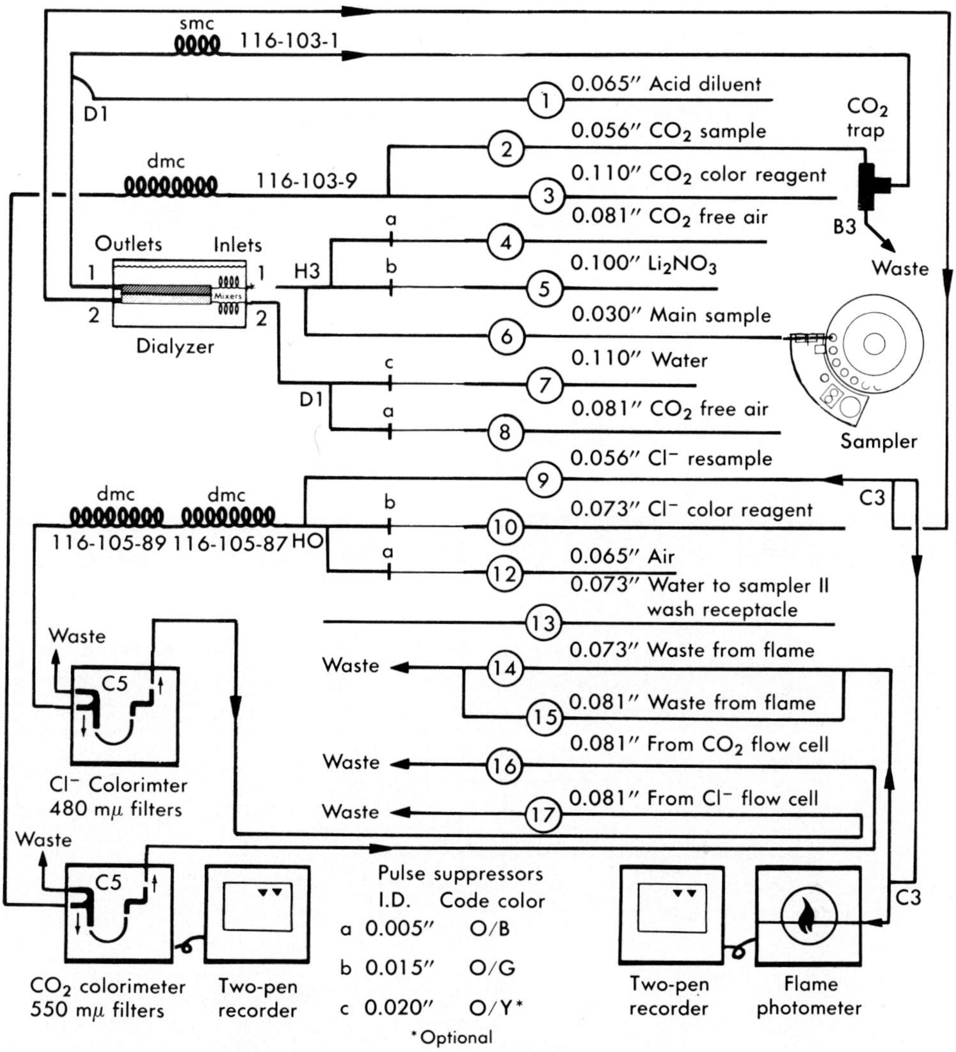

Fig. 4-2. Electrolytes (one pump II).

Lithium nitrate working solution, 125 mEq Li/L in 0.25 N H_2SO_4	(D-7)
Mercuric thiocyanate, saturated	(D-8)
Ferric nitrate, 0.5 M	(D-9)
Mercuric nitrate, 0.2 M in 0.2 N HNO_3	(D-10)
Chloride color reagent	(D-11)
Acid antifoam diluent	(D-12)
Sodium hydroxide wash solution	(D-13)
Phenolphthalein indicator, 1%	(D-14)
Sodium carbonate, 1 M	(D-15)
Sodium bicarbonate, 1 M	(D-16)
Carbonate-bicarbonate buffer	(D-17)
CO_2 color reagent	(D-18)
Sodium chloride stock standard, 1 M	(D-19)
Potassium chloride stock standard, 0.1 M	(D-20)
Ammonium chloride stock standard, 1 M	(D-21)

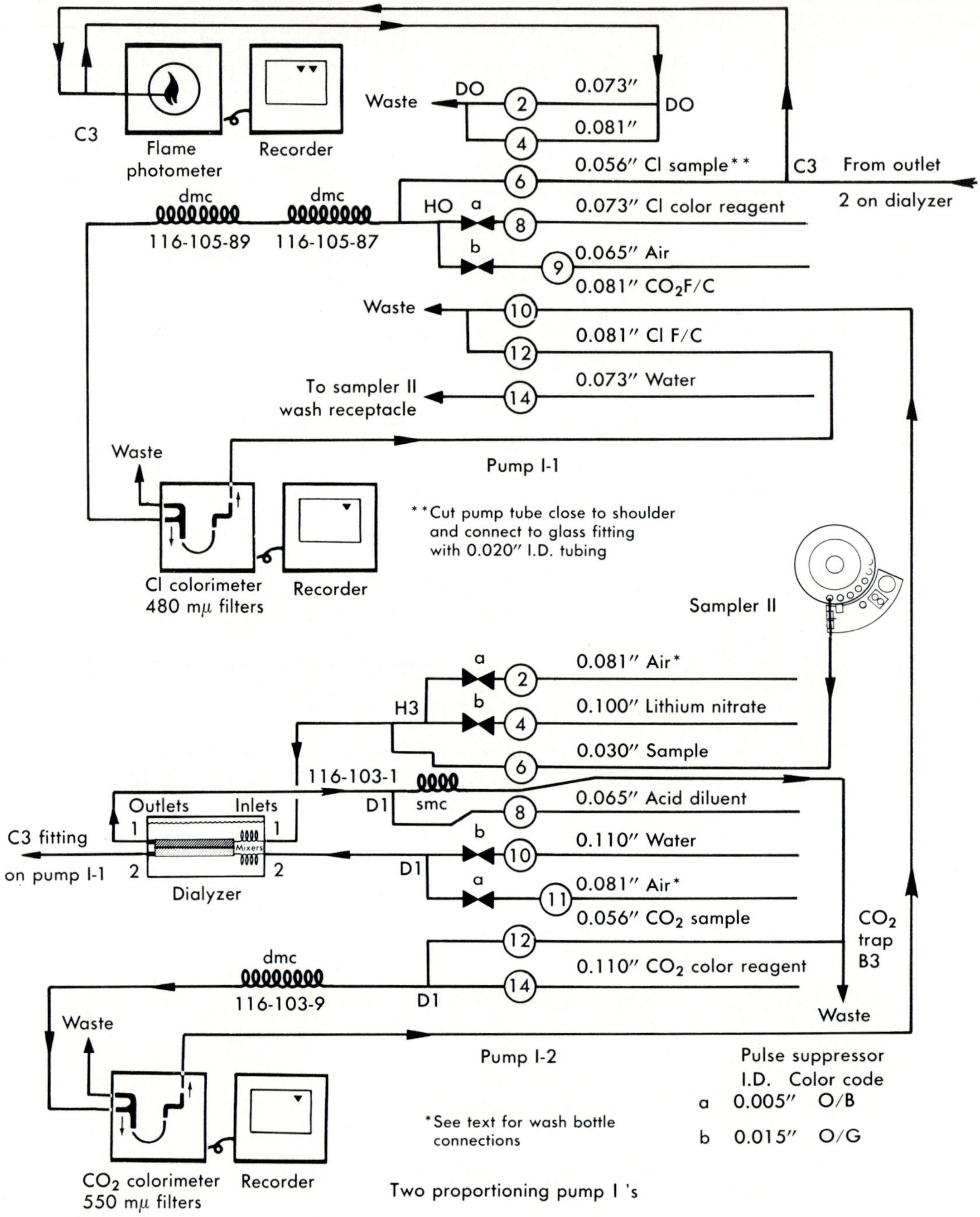

Fig. 4-3. Electrolytes (two pump I's).

Dilute stock standards

Prepare dilute stock standards as follows:

Standard A. Measure 100 ml of 1 M sodium carbonate stock, 200 ml of 0.1 M potassium chloride stock, and 100 ml of 1 M ammonium chloride stock standards and transfer to a 1 L volumetric flask. Dilute to volume with water and mix

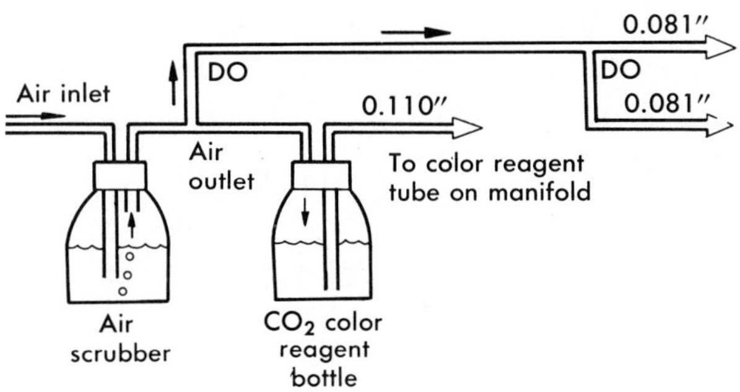

Fig. 4-4. CO_2 air scrubber (NaOH wash) connections.

thoroughly. Store in a Pyrex or polyethylene bottle. This standard contains 200 mEq of Na, 20 mEq of K, 120 mEq of Cl, and 100 mEq of CO_2/L.

Standard B. Dilute 800 ml of 1 M sodium chloride to 1000 ml. This standard contains 800 mEq of Na and 800 mEq of Cl/L.

Dilute working standards

Measure the following amounts of dilute standards A and B into 100 ml volumetric flasks. Dilute to volume and mix.

Working standard	Standard A (ml)	Standard B (ml)	Values in mEq/L of working standards Na	K	Cl	CO_2
1	5	10	90	1	86	5
2	10	10	100	2	92	10
3	15	10	110	3	98	15
4	20	10	120	4	104	20
5	25	10	130	5	110	25
6	30	10	140	6	116	30
7	35	10	150	7	122	35
8	40	10	160	8	128	40

Procedure notes

1. If flame III or III A is used, heparinized plasma or serum samples may be run at a rate of 60/hr (2:1 sample-wash ratio). This rate is possible only with flame III, III A, or IV, which operates on natural gas (or propane) and compressed air. The recommended rate for flame I (propane and oxygen) and for flame II (natural gas and oxygen) is 40 specimens/hr.

2. Because of the number of manifold pump tubes required it is recommended that either one pump II or *two* pump I's be used.

3. It is necessary for the air that segments the diluent and recipient streams to be bubbled through a sodium hydroxide wash solution to remove atmospheric CO_2 (Fig. 4-4). To accomplish this, the sodium hydroxide wash bottle must have an airtight cap with an air inlet tube that goes down below the level of the sodium hydroxide. The bottle should be no more than two thirds full. The outlet tube should be well above the level of the wash solution to avoid the possibility of sodium hydroxide being carried over into the CO_2 color reagent or to the sample stream. There is now available a carbon dioxide–absorbing cartridge (T11-0415-56)* that may be used in place of the sodium hydroxide wash

*Technicon Corp.

solution. The cartridge contains a color indicator that is pink while the absorbent is still active and turns tan/yellow when it is spent. The cap or stopper in the CO_2 color reagent should also be airtight. The tube carrying CO_2-free air into the bottle should just go through the stopper but not reach the level of the liquid. The outlet tube, connected to the color reagent pump tube, should be long enough to reach the bottom of the bottle. To avoid interference from atmospheric CO_2, the caps or stoppers of both the sodium hydroxide wash bottle and the CO_2 color reagent bottle *must* be airtight.

4. To maintain good bubble patterns and low noise levels, add 0.5 ml of Brij 35/L to the lithium nitrate solution, the water recipient, and the CO_2 color reagent.

Operating sequence

1. Turn on gas and air to the flame photometer and ignite the flame according to the instructions in the flame photometer manual. Allow 30 minutes for the instrument to warm up and stabilize. Do *not* allow any liquid to be aspirated into the burner until the flame is ignited. Turn on colorimeters.
2. While the flame photometer is warming up, start pumping water through all the lines. Connect the appropriate line to the flame photometer. Check to see that all parts of the manifold are connected properly and that both streams coming from the dialyzer are flowing properly.
3. Pump reagents through all lines except the CO_2 color reagent line. Leave it in water along with the sample line.
4. Adjust the CO_2 water baseline to 97% or 98% T. Check 0% T and, after the pen returns to the water baseline, check the recorder gain.
5. Connect the CO_2 color reagent line to the output tube on the color reagent bottle.
6. After the reagents have been pumping for about 10 minutes, adjust the chloride baseline to 97% or 98% T. Set the 0% T. Readjust the reagent baseline if necessary. Check the recorder gain.
7. Observe the reagent baseline (RBL) for CO_2. It should be about 20% T but not lower than 10% T nor higher than 30% T. If the RBL is lower than 10% T, dilute the color reagent with water. If the RBL is higher than 30% T, add a few drops more of phenolphthalein indicator and mix.
8. Place the sample line into a tube that contains working standard 8 (160 mEq of Na, 8 mEq of K, 40 mEq of CO_2, and 128 mEq of Cl/L).
9. Adjust the sodium pen to 95% T and the potassium pen to 90% T. Check the recorder gain on both pens.
10. Observe the steady-state position of the CO_2 pen on the chart while aspirating standard 8. It should be between 80% and 90% T—ideally, close to 90% T—for maximum sensitivity. If it reads less than 80% T, dilute the color reagent with water. If it reads greater than 90% T, add 2 to 3 drops of carbonate-bicarbonate buffer to the color reagent.
11. Place the sample line into a tube containing standard 2 (92 mEq of Cl/L). Note the position of the chloride pen on the chart when the steady state of this standard records. It should be between 60% and 65% T. Mercuric nitrate is added to the chloride color reagent to control the sensitivity. The 3.5 ml/L volume called for in the directions binds about 60 mEq/L of chloride ions in the sample. Consequently, only samples with chloride concentrations greater than 60 mEq/L will move the pen from the reagent baseline. This brings the physi-

ologic range of blood into a more readable and accurate portion of the chart. If the steady-state reading of the 92 mEq/L standard is lower (toward 0% T) than 60% T, add more mercuric nitrate in 0.1 ml steps. Excessive use of mercuric nitrate will result in higher noise levels. If the standard reads higher than 65% T, dilute the color reagent with a batch of chloride color reagent containing no mercuric nitrate.

12. Place the sample line in water and load standards on the sample plate in the following order: Three cups of standard 4 followed by standards 1 through 8. Place a cup containing standard 5 after standard 8 and in every tenth position on the plate. Fill the remaining positions with unknown samples to be analyzed. *Do not* use any water cups or leave any empty spaces. Fill any blank spaces with a standard cup.

NOTE: Do not pour standards into cups until the instrument is ready to start sampling. Standards exposed to air will absorb CO_2 from the atmosphere.

13. When the last sample has recorded, place all the reagent lines in water. After pumping water for 2 minutes, place the chloride color reagent line in a 25% bleach solution. Leave the other lines in water. Continue pumping for 10 minutes; then place the chloride color reagent line in water and flush for 10 minutes. Disconnect the line going to the flame photometer *before* turning the flame off.

Calculations

Values for the unknown samples may be determined by using Technicon comparators or by reading from a curve prepared by plotting standard readings versus concentration on graph paper. The percent transmission of CO_2 and chloride standards should be plotted versus concentration on semilog paper. The percent transmission of Na and K standards may be plotted versus concentration on linear graph paper.

System maintenance and troubleshooting

1. The dialyzer membrane should be changed at regular intervals, depending on use. Before changing the membrane, pump a 25% Clorox solution through the sample and reagent lines for 10 minutes. Follow with a 10-minute water wash and then change the membrane. This flushes out any deposits of precipitated protein in the tubing. Daily rinsing of the chloride color reagent line with bleach prevents a film buildup in the flow cell and keeps the mixing coils clean.

2. The acid antifoam diluent must be shaken each day before use because the Antifoam B has a tendency to settle out. If there is not enough antifoam in the solution, bubbling will occur in the CO_2 trap. (Refer to Fig. 4-1.) When this happens, a portion of the acid solution is resampled through the CO_2 line, mixes with the CO_2 color reagent and decolorizes the purple phenolphthalein. This is recorded on the CO_2 chart as a very erratic pattern, usually spiking toward 95% T.

3. The small capillary leading into the CO_2 trap or gas-liquid phase separator is the one spot on the manifold most likely to accumulate small particles of precipitated protein. (Refer to Fig. 4-1.) When this capillary becomes completely obstructed, pressure builds up back along the diluted sample line. Eventually one of the connections will "pop" but that connection may be far removed from the CO_2 trap. It just happened to be the "weakest link." In other words, if

a connection "pops" anywhere along the diluted sample line, look first for an obstruction in this small capillary.

4. Filter the chloride color reagent to avoid noisy recordings.

5. Clean the flame photometer glass chimney and mirror each day before lighting the flame.

6. Try to maintain a constant temperature in the room where the instrument is located. A changing room temperature will cause considerable drift in standardization, particularly for CO_2 and sodium.

Simultaneous glucose and urea nitrogen

Principle

The simultaneous AutoAnalyzer procedure for the determination of glucose and urea nitrogen[5] is based on the individual procedures for glucose and urea nitrogen. Two dialyzer plates are used, connected in series for the sample stream and in parallel for the recipient streams.

The automated glucose procedure was adapted from the manual procedure of Hoffman.[6] It is a method utilizing the potassium ferricyanide–potassium ferrocyanide oxidation reduction reaction. The yellow potassium ferr*i*cyanide solution is reduced by the glucose to the colorless ferr*o*cyanide. The loss of color is measured at 420 mμ using a flow cell with a 15 mm light path. The diffusion rate of glucose in biologic materials is similar to that of aqueous standards.

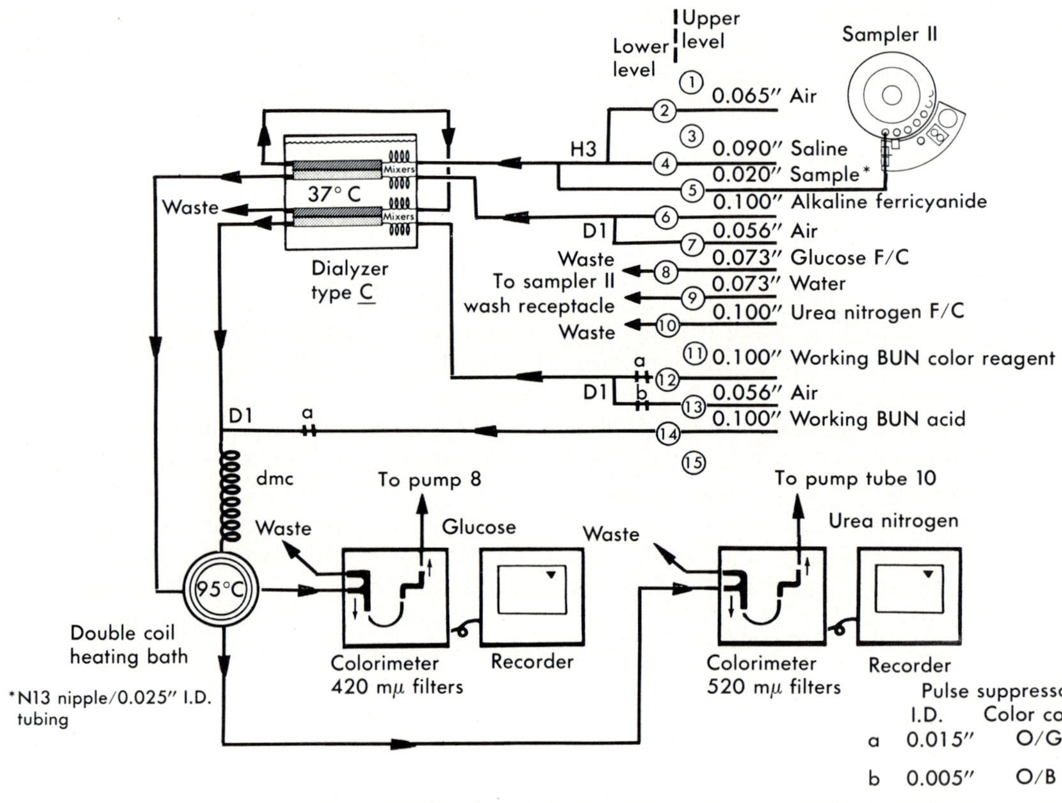

Fig. 4-5. Simultaneous glucose and urea nitrogen (two dialyzer plates).

The urea nitrogen method is a slightly modified version of the one described by Marsh and co-workers.[7] It is a modification of the carbamido-diacetyl reaction. In acid solution, diacetyl monoxime (2,3-butanedione-2-oxime) is hydrolyzed to diacetyl, which reacts directly with urea in the presence of acidic ferric ions. The pink color of the reaction product is intensified by the presence of thiosemicarbazide. This enables the determination to be run without the need of concentrated acid reagents. The colored reaction product is measured in a 15 mm flow cell at 520 mμ.

In the simultaneous procedure, the sample is diluted with saline, segmented with air, and pumped into a 37° C dialyzer bath where it goes through a mixing coil before entering the top side of a set of dialyzer plates. Glucose in the diluted sample dialyzes through a type C membrane into an air-segmented stream of alkaline potassium ferricyanide. Upon leaving the dialyzer, the recipient stream enters the 95° C heating bath (40-foot coil, 1.6 mm I.D.) for the color reaction to take place. The glucose reduces the ferricyanide (yellow) to ferrocyanide (colorless) in proportion to the amount of glucose present in the dialysate. The reaction stream then enters the colorimeter, where the loss of color is measured at 420 mμ. This is called an inverse color reaction.

The diluted sample stream emerging from the first dialyzer plate (which in a single glucose method would be pumped to waste) is pumped into the top of a second set of dialyzer plates. Urea is dialyzed into an air-segmented stream of working BUN color reagent. As the recipient stream leaves the dialyzer, it is joined by a stream of working BUN acid reagent, passed through a mixing coil, and then pumped into a second coil in the 95° C heating bath. As the acidified solution passes through the heating bath coil, the urea reacts with diacetyl and the color formation is intensified by thiosemicarbazide. The reaction stream then enters the colorimeter where the color intensity is measured at 520 mμ. This is a direct color reaction.

Modules needed

One sampler II
One proportioning pump (I or II)
One dialyzer bath with two sets of dialyzer plates
One double-coiled heating bath at 95° C
Two colorimeters with 420 mμ and 520 mμ filters
One double-pen recorder or two single-pen recorders

Flow diagram

See Fig. 4-5.

Reagents needed

Saline, 0.9%	**(D-22)**
Alkaline potassium ferricyanide	**(D-23)**
Diacetyl monoxime stock, 2.5%	**(D-24)**
Thiosemicarbazide stock, 0.5%	**(D-25)**
Working BUN color reagent	**(D-26)**
Ferric chloride–phosphoric acid stock	**(D-27)**
Sulfuric acid stock, 20%	**(D-28)**
Working BUN acid	**(D-29)**
Benzoic acid, saturated	**(D-30)**
Glucose stock standard, 10 mg/ml	**(D-31)**
Standard diluent, 0.01 N H_2SO_4 containing 40 mg/L PMA	**(D-32)**
Urea nitrogen stock standard, 10 mg urea N/ml	**(D-33)**

Working standards

Prepare the working standards in 100 ml volumetric flasks. Using volumetric pipettes, measure the amounts of the two stock standards indicated below into each of the flasks and dilute to volume with standard diluent (D-32).

Working standard	Glucose stock (ml)	Urea N stock (ml)	Glucose (mg/100 ml)	Urea N (mg/100 ml)
1		0.5	0	5
2	5	1	50	10
3		2	0	20
4	10	3	100	30
5		4	0	40
6	15	5	150	50
7		6	0	60
8	20	7	200	70
9		8	0	80
10	25	9	250	90
11		10	0	100
12	30	15	300	150

NOTE: If preferred, the glucose standards may be grouped together in either the first or last six working standards.

Procedure notes

1. The simultaneous glucose/BUN procedure may be run at a rate of 60 specimens/hr with a 2:1 sample-wash ratio. However, the amount of sample interaction or carryover can be reduced considerably by using a rate of 50/hr with a 1:1 sample-wash ratio.

2. Glucose and urea nitrogen may be determined in serum, plasma, spinal fluid, or anticoagulated whole blood. If whole blood is used, a mixer should be used on the sampler to maintain a homogeneous suspension of cells in the sample. Unless a fluoride preservative is used, all cells should be removed as soon as possible after the specimen is obtained. Glucose cannot be determined on urine because of the presence of high concentrations of uric acid, creatinine, and other reducing substances. Urea nitrogen can be determined on urine without prior treatment because the reaction is not affected by the presence of ammonia. The urine specimen need only be diluted with water to bring the urea nitrogen concentration within the range of the standards.

3. The bubble patterns of the system are improved and recorder noise is reduced by adding 0.5 ml/L of Brij 35 to the saline diluent, potassium ferricyanide, and working BUN color reagents. If surging occurs in the glucose heating bath coil, add more Brij 35 to the potassium ferricyanide (up to 2 ml/L).

4. Since glucose is determined by an inverse colorimetric procedure, it is necessary to establish a "reference" (colorless) baseline before getting a reagent baseline. This is accomplished by setting the pen to 97% or 98% T while pumping water through all the lines. The reagent baseline is then determined by putting the saline and potassium ferricyanide reagent lines in their respective bottles. This reagent baseline (RBL) should be at 13% T $\pm$ 3%. The RBL can be raised by adding a solution of 2% sodium carbonate in 0.9% sodium chloride or lowered by adding a solution of 5% potassium ferricyanide in 0.9% sodium chloride to the potassium ferricyanide solution. It would be advisable to calculate, at the time this adjustment is being made, the amount of potassium ferricyanide required in future batches of reagent to obtain the desired RBL.

A 300 mg/100 ml standard should read between 80% and 90% T. If it goes higher than 90% T, the RBL may be adjusted down toward 10% T but not below 10% T. If the 300 mg/100 ml standard still reads above 90% T, the ferricyanide reagent manifold pump tube should be increased to 0.110-inch I.D.

5. BUN color reagent should not be kept in a cold place, or precipitation may occur. However, a gentle warming should redissolve the precipitate.

If too much sensitivity is obtained, the concentration of sulfuric acid in the BUN acid may be reduced to 15%.

WARNING: Do not use the high speed of a two-speed pump while reagents are going through the lines. There is the danger of "popping" a line and spurting acid over everything in the near vicinity.

Operating sequence

1. Insert appropriate filters in each colorimeter *before* turning on the colorimeter lamps. Allow 20 minutes for optical systems to stabilize.
2. While colorimeters are stabilizing, start pumping water or saline with Brij 35 through sample and reagent lines. Check manifold connections and bubble patterns to see if everything is functioning properly.
3. When the colorimeters have stabilized, adjust the glucose water (or saline) baseline to 97% or 98% T. Check the gain of the glucose pen.
4. Place all the reagent lines in the appropriate reagent bottles and after about 8 minutes adjust the BUN baseline to 97% or 98% T. Check the gain of the BUN pen. Note the position of the glucose RBL.
5. Start sampler. Place a cup of water between the last standard and the first specimen.
6. When the last specimen has recorded, place all lines in water and rinse for 10 minutes.

System maintenance and troubleshooting

1. When using pump I, four helper springs at each end of the platen will facilitate adequate and even pumping.
2. When lowering the roller head assembly of pump I, be sure there is no overlapping of pump tubes of the manifold. If there is, raise the rollers, press the tubes down into proper position with one finger, and lower the rollers again.
3. Sample interaction can be reduced by using glass transmission tubing wherever possible, especially between the heating bath and the colorimeter.
4. If a noisy recorder pattern is obtained even though the reagents contain the proper amount of Brij 35, filter those reagents into clean bottles. One of the older urea nitrogen procedures called for the addition of urea to the diacetyl monoxime reagent. This was found to cause excessive noise on many systems and is no longer used.
5. The dialyzer membranes should be changed regularly according to use. Many problems can be avoided if a 25% Clorox solution is pumped through the entire system for about 15 minutes just before the membranes are changed. Follow the bleach with a 10-minute water wash before dismantling the dialyzer. The bleach will remove any protein buildup in the lines and any film that might have accumulated in the flow cell.
6. The hot alkaline potassium ferricyanide will gradually corrode the inside of the heating bath coil until it eventually perforates the wall of the coil. As the

inside diameter increases, the transit time through the coil increases. The longer dwell time in the heating bath results in higher glucose sensitivity and a shift in the calibration curves.[8,9]

Creatinine

Principle

The automated determination of creatinine is based on the classic manual method of Folin and Wu.[10,11] It was adapted for automation by Stevens and Skeggs[12] and by Chasson and colleagues.[13]

The sample is diluted with saline and pumped into the dialyzer where creatinine dialyzes into a water recipient stream. Picric acid and sodium hydroxide are mixed to form alkaline picrate and this stream joins with the recipient stream as it emerges from the dialyzer. They mix in a single mixing coil and then pass into a 40-foot time delay coil immersed in water at room temperature for color development. The reaction mixture then goes to the colorimeter, where the color intensity is measured at 505 mμ in a tubular flow cell.

Modules needed

One sampler II
One proportioning pump I
One dialyzer with one set of dialyzer plates
One colorimeter with 505 mμ filters
One single-pen recorder

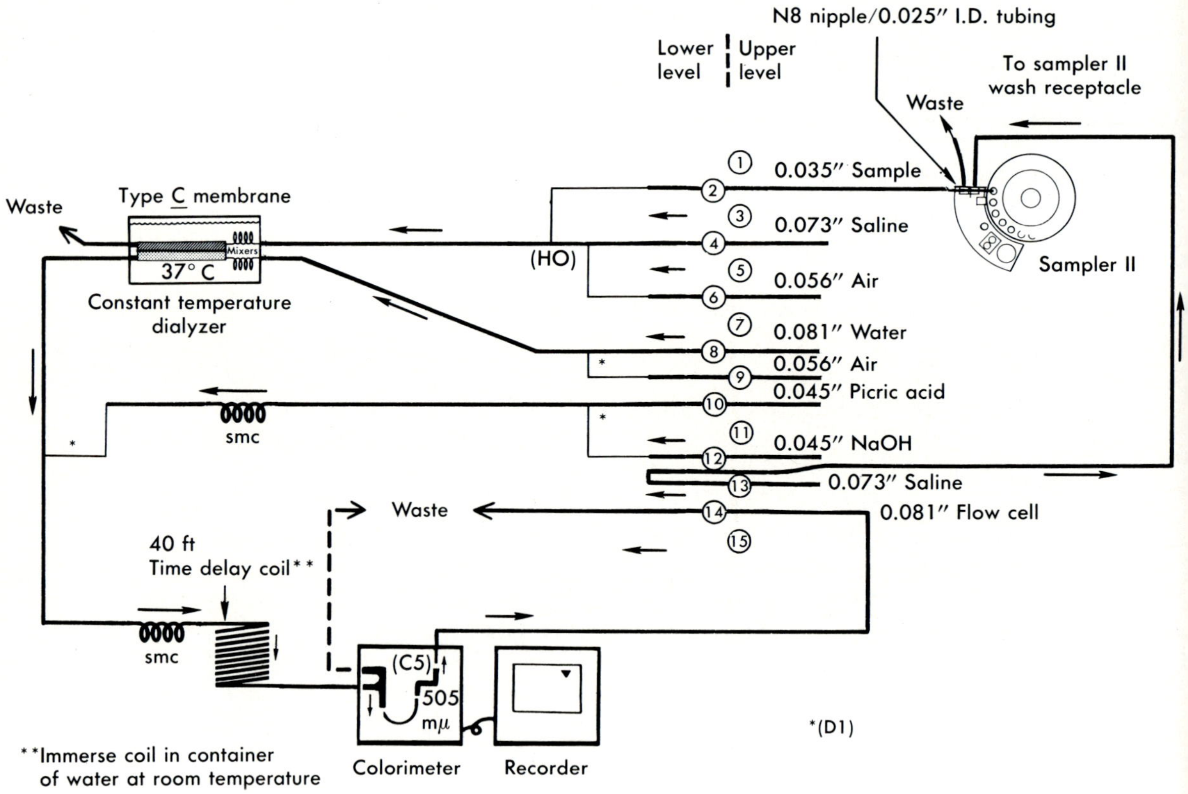

Fig. 4-6. Creatinine.

Flow diagram

See Fig. 4-6.

Reagents needed

Saline, 0.9%	(D-22)
Sodium hydroxide, 0.5 N	(D-34)
Picric acid, saturated	(D-35)
Creatinine stock standard, 1 mg/ml	(D-36)
Hydrochloric acid, 0.1 N	(D-37)
Hydrochloric acid, 0.02 N	(D-38)

Working standards

Measure the following amounts of creatinine stock standard, 1 mg/ml, into 100 ml volumetric flasks and dilute to volume with 0.02 N HCl.

Creatinine stock standard (ml)	Creatinine working standard (mg/100 ml)
1	1
2	2
3	3
5	5
7	7
10	10

Procedure notes

1. Samples may be run at a rate of 60/hr with a 2:1 sample-wash ratio. It may be advisable, however, to place a cup of saline between samples on the plate to provide better wash between samples.
2. Samples may include both serum and urine. Urines should be diluted 1:10 or 1:20 with water. Serum samples with creatinine values greater than 10 mg/100 ml should be diluted 1:2 or 1:3 with saline.
3. The bubble pattern is improved by using 0.5 ml of Brij 35/L of saline and water recipient.
4. To reduce noise level and baseline drift, filter all reagents and use saline instead of water in the sampler wash receptacle.
5. Wash the picric acid and sodium hydroxide lines, coils, and flow cell with 10% acetic acid for 5 minutes at the end of each day's run.

Operating sequence

1. Turn on colorimeter and allow for a 20-minute warmup.
2. Begin pumping water through all lines and check the system for leaks.
3. Place lines in reagent bottles. Leave the sample line in the sampler wash receptacle.
4. Pump reagents for 10 minutes; then adjust reagent baseline. Set the pen to 97% or 98% T and check 0% T. When the pen returns to the reagent baseline check the recorder again.
5. Begin sampling.
6. When the last sample has recorded, place the picric acid and sodium hy-

droxide lines in 10% acetic acid for 5 minutes. Then rinse with water for 10 minutes.

Calculations

The standard curve may be drawn on a Technicon plastic comparator or on graph paper by plotting % T versus concentration on semilog paper.

System maintenance and troubleshooting

1. The dialyzer membrane should be changed regularly according to use. Pump 25% Clorox through all lines for 5 minutes and rinse with water for 10 minutes before changing the membrane.

2. Since the picric acid is a saturated solution, some precipitation may occur. If this precipitate is pumped through the reagent line, it will cause excessive noise. If the precipitate will not go back into solution with agitation or gentle warming, the solution should be filtered.

Uric acid

Principle

The automated uric acid procedure[14] is adapted from the manual method of Brown[15] as described by Hawk and co-workers.[16]

The color reaction involves the reduction of the phosphotungstate complex to a phosphotungstite complex. Cyanide improves sensitivity and prevents turbidity.

The sample is diluted with saline and pumped into the dialyzer, where uric acid dialyzes into a saline recipient stream. As the dialysate leaves the dialyzer, it is joined and mixed with a cyanide-urea stream. This mixture is then joined by a phosphotungstic acid stream. The reaction mixture then passes through two double mixing coils for color development and on to the colorimeter where the intensity of the color is measured at 660 mμ in a 15 mm flow cell.

Modules needed

One sampler II
One proportioning pump (I or II)
One dialyzer with one set of dialyzer plates
One colorimeter with 660 mμ filters
One single-pen recorder

Flow diagram

See Fig. 4-7.

Reagents needed

Saline (with 0.5 ml of Aerosol 22/L)	(D-22)
Phosphotungstic acid	(D-39)
Sodium cyanide stock solution, 10%	(D-40)
Urea stock solution, 20%	(D-41)
Cyanide-urea working solution	(D-42)
Uric acid stock standard, 1 mg uric acid/ml	(D-43)

Working standards

Pipette the following volumes of uric acid stock standard into 100 ml volumetric flasks, dilute to volume with water, and mix thoroughly.

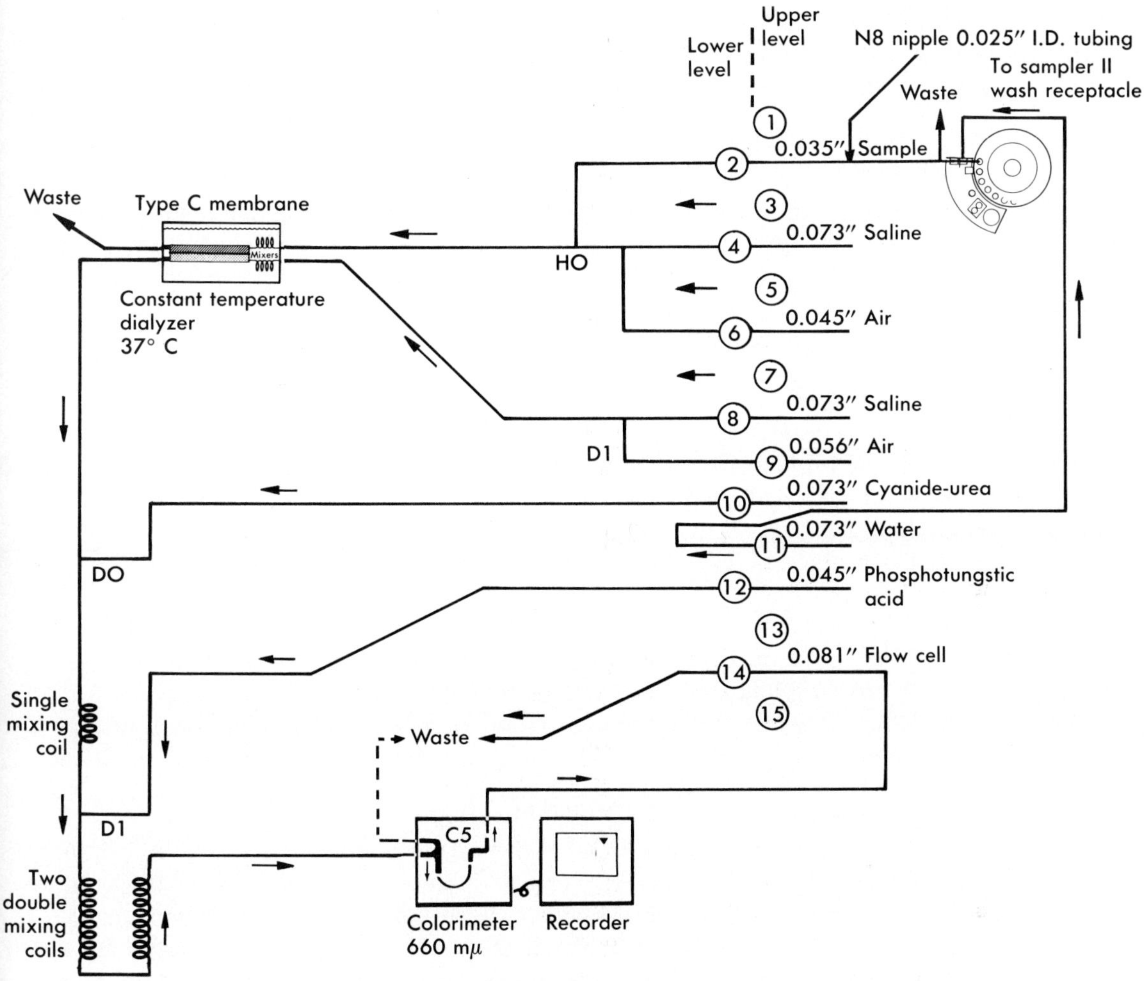

Fig. 4-7. Uric acid.

Uric acid stock standard (ml)	Uric acid working standard (mg/100 ml)
2	2
4	4
6	6
8	8
10	10
12	12

Store the working standards in the refrigerator.

Procedure notes

1. Samples may be run at a rate of 60/hr with a 2:1 sample-wash ratio. For better wash between samples, use a 50/hr cam with a 1:1 sample-wash ratio or place a saline cup between samples on the sample plate.

2. Samples may include nonhemolyzed serum or plasma and urine. Urines usually require a 1:5 or 1:10 dilution with water.

3. Serum or plasma samples with a uric acid concentration greater than 12 mg/100 ml should be diluted with saline.

4. Use 0.5 ml of Aerosol 22/L of saline to improve the bubble pattern and avoid noise.

5. Waste from the system should be flushed down a sink with running water. The mixing of an acid with the sodium cyanide will generate hydrogen cyanide gas which is extremely poisonous. If the waste liquid must be collected in a bottle, a neutralizing compound such as trisodium phosphate powder should be used in the bottle.

Operating sequence

1. Turn on colorimeter and allow a 20-minute warmup.

2. Begin pumping water through the sample and reagent lines. Check for leaks in the system and observe the nature of the bubble pattern coming from the dialyzer.

3. Place the reagent lines in their respective reagent bottles. Leave the sample line in water.

4. After about 8 minutes set the reagent baseline to 97% or 98% T. Check the 0% T and when the pen returns to the baseline, check the recorder gain.

5. Begin sampling standards and specimens.

6. When the last specimen has recorded, place the reagent lines in water and rinse for 10 minutes.

System maintenance and troubleshooting

1. The dialyzer membrane should be changed regularly according to use. Just prior to changing the membrane pump 25% Clorox solution through the sample and reagent lines for 5 minutes. Follow with a 5-minute water wash.

2. If turbidity is observed in the mixing coils it is probably caused by old sodium cyanide solution. Prepare a new sodium cyanide stock solution and a new cyanide-urea working solution.

Calcium

Principle

In this method[17] a metal-complexing dye, cresolphthalein complexone (phenolphthalein, 3′3″ bis-{[bis-(carboxymethyl)amino]methyl}-5′,5″-dimethyl; metalphalein; phthalein komplexon), is used to determine calcium. The use of this dye was introduced by Anderegg and co-workers.[18] The automated method was adapted by Kessler and Wolfman[19] from a spectrophotometric method of Pollard and Martin[20] and Stern and Lewis.[21] Gitelman[22] further modified the procedure by adding 8-hydroxyquinoline to minimize interference from magnesium without the need for a high base concentration.

The sample is mixed with 0.25 N hydrochloric acid and pumped into a dialyzer. The acid releases the protein-bound calcium and the calcium ions dialyze into a 0.25 N hydrochloric acid recipient stream. As the dialysate leaves the dialyzer it is joined by the dye and the base reagent streams. The mixture passes through two 3 mm double mixing coils for mixing and for color development. It then goes to the colorimeter, where the color intensity is measured at 580 mμ in a 15 mm flow cell.

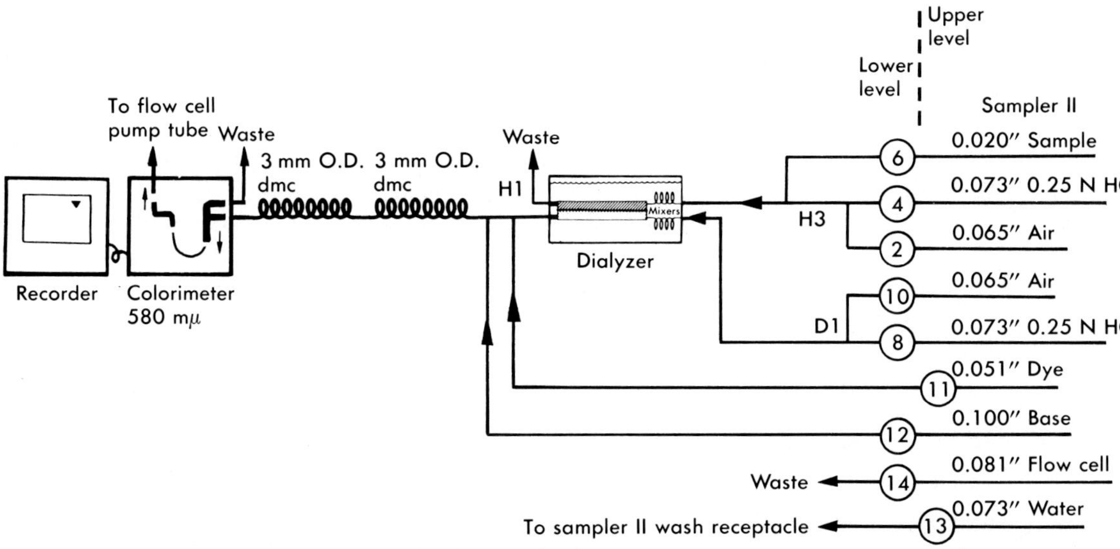

Fig. 4-8. Calcium.

Modules needed

One sampler II
One proportioning pump (I or II)
One dialyzer with one set of dialyzer plates
One colorimeter with 580 mμ filters
One single-pen recorder

Flow diagram

See Fig. 4-8.

Reagents needed

Hydrochloric acid, 0.25 N	(D-44)
Cresolphthalein complexone, 0.01% with 8-hydroxyquinoline	(D-45)
Calcium base	(D-46)
Calcium stock standard, 50 mg Ca/100 ml	(D-47)
Magnesium stock standard, 100 mg Mg/100 ml	(D-48)

Working standards

Prepare dilute working standards in 100 ml volumetric flasks. Pipette the following volumes of calcium and magnesium stock standards into the flasks,

Calcium stock standard (ml)	Magnesium stock standard (ml)	Calcium working standard (mg/100 ml)	Magnesium working standard (mg/100 ml)
5	2	2.5	2
10	2	5.0	2
15	2	7.5	2
20	2	10.0	2
25	2	12.5	2
30	2	15.0	2

add 1 drop of concentrated HCl to each flask, dilute to volume with water, and mix thoroughly.

Procedure notes

1. Calciums may be run at a rate of 60 samples/hr with a 2:1 sample-wash ratio. If a better wash between samples is desired, use a 50/hr cam with a 1:1 sample-wash ratio.

2. Determine calcium directly in serum, heparinized plasma (but no other anticoagulated plasma), urine, and in spinal, pleural or ascitic fluids. Collect urine specimens in bottles containing glacial acetic acid or 6 N hydrochloric acid to keep the calcium in solution. The amount of acid depends on the expected volume of urine. Use 10 ml of acid for a 24-hour collection. Adjust the urine to pH 2, using pH indicator paper.

3. Magnesium (2 mg/100 ml) is added to the working standards equivalent to the level normally present in serum.

4. Be sure that water rather than air is being aspirated through the sample line when the reagent baseline is being set. Sampling air will give a slightly higher % T reading.

5. Use 0.5 ml of Brij 35/L of 0.25 N hydrochloric acid to assure a good bubble pattern and low noise level.

Operating sequence

1. Turn on colorimeter and allow a 20-minute warmup.

2. Begin pumping water through reagent and sample lines. Check for leaks and irregular flow from dialyzer.

3. Place reagent lines in their respective reagent bottles. Leave the sample line in water.

4. After about 8 minutes, adjust the reagent baseline to 97% or 98% T. Check 0% T, and when the pen returns to the reagent baseline check the recorder gain.

5. Begin sampling standards and specimens.

6. When the last specimen has recorded, place the reagent lines in water and rinse for 10 minutes.

Calculations

The standard curve may be plotted either on a Technicon plastic comparator or on semilog paper.

System maintenance and troubleshooting

1. The dialyzer membrane should be changed regularly according to use. Pump 25% Clorox solution through sample and reagent lines for 10 minutes, followed by a 5-minute water rinse before changing the membrane.

2. Run the waste lines into a sink with running water. The mixture of acid with potassium cyanide in the base reagent will generate a small amount of highly poisonous hydrogen cyanide gas.

3. If it is not possible to set the reagent baseline up (toward 100% T) to 97% or 98% T, even with a No. 1 aperture, be sure first that the flow cell is clean. Disconnect it from the system and rinse with full-strength Clorox or chromic acid cleaning solution followed by water. Reconnect the flow cell and, if the baseline

still cannot be set to 98% T, add a neutral-density filter to the reference side or partially cover the aperture with black tape.

Inorganic phosphate

Principle

This procedure[23] is a modification of an earlier method[24] that used aminonaphthol sulfonic acid as a reducing agent. It is based on the formation of phosphomolybdic acid which is then reduced to the blue phosphomolybdous acid. This procedure uses the more stable reducing agent, stannous chloride–hydrazine sulfate. The use of this reducing agent in phosphate determinations was first reported by Hurst.[25] The automated procedure was adapted by Kraml.[26] No heating bath is required for this method as it was for the earlier method.

The sample is mixed with saline and pumped into the dialyzer where phosphorus ions dialyze into an air-segmented recipient stream. As the dialysate leaves the dialyzer, it is joined by an acid molybdate stream and mixes. This mixture is then joined by a stannous chloride–hydrazine sulfate stream and mixes again. The color produced is then measured at 660 mμ in a 15 mm tubular flow cell.

Modules needed

One sampler II
One proportioning pump (I or II)
One dialyzer with one set of dialyzer plates
One colorimeter with 660 mμ filters
One single-pen recorder

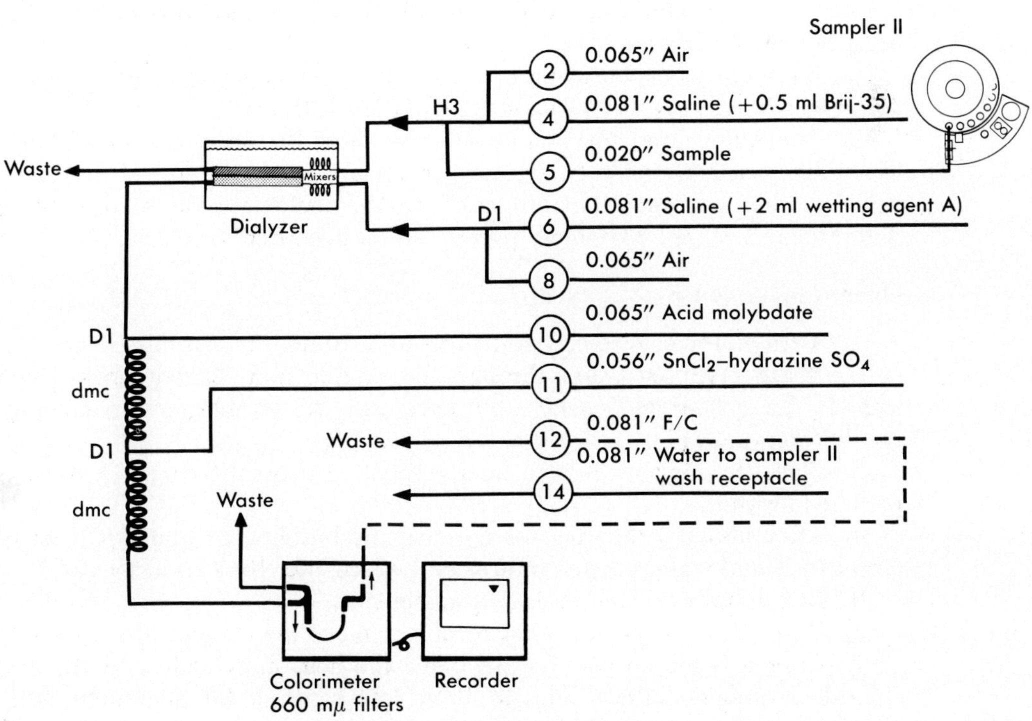

Fig. 4-9. Phosphate (via $SnCl_2$–hydrazine sulfate).

Flow diagram

See Fig. 4-9.

Reagents needed

Saline (with 0.5 ml Brij 35/L)	(D-22)
Saline (with 2 ml wetting agent A/L)	(D-22)
Hydrazine sulfate, 0.2%, in 1 N H_2SO_4	(D-49)
Stannous chloride–hydrazine sulfate working solution	(D-50)
Acid molybdate	(D-51)
Phosphorus stock standard, 1 mg P/ml	(D-52)

Working standards

Measure the following volumes of phosphorus stock standard into 100 ml volumetric flasks, dilute to volume with water and mix thoroughly.

Phosphorus stock standard (ml)	Phosphorus working standard (mg/100 ml)
1	1
3	3
5	5
7	7
10	10

Procedure notes

1. Phosphorus determinations may be run at a rate of 60 samples/hr with a 2:1 sample-wash ratio. For a better wash between samples, use a 50/hr cam with a 1:1 sample-wash ratio.
2. Inorganic phosphorus may be run on serum or heparinized plasma. Urine specimens usually require a 1:5 or 1:10 dilution with water.
3. The saline used for sample dilution should contain 0.5 ml of Brij 35/L. The saline used as the recipient stream should contain 2 ml of wetting agent A/L. If other wetting agents are used on the recipient side, a blue precipitate will form and cause a very noisy pattern on the recorder.

Operating sequence

1. Turn on the colorimeter and allow for a 20-minute warmup.
2. Begin pumping water through the sample and reagent lines. Check for leaks in the system and observe the nature of the bubble pattern coming from the dialyzer.
3. Place the reagent lines in their respective reagent bottles. Leave the sample line in water.
4. After about 8 minutes set the reagent baseline to about 97% or 98% T. Check 0% T and when the pen returns to the baseline check recorder gain.
5. Begin sampling standards and specimens.
6. When the last specimen has recorded, place the reagent lines in water.
7. After a 2-minute water wash place the acid molybdate and the stannous chloride–hydrazine sulfate lines in 1 N NaOH. Wash for 5 minutes with 1 N NaOH to remove traces of blue precipitate in the mixing coil and flow cell. Follow with a 10-minute water wash.

System maintenance and troubleshooting

1. The dialyzer membrane should be changed regularly according to use. Pump 25% Clorox solution through sample and reagent lines for 5 minutes followed by a 5-minute water wash before changing the membrane.

2. If the acid molybdate and stannous chloride–hydrazine sulfate lines are not washed with 1 N NaOH, a blue film will build up on the inside surface of the flow cell, resulting in a loss of sensitivity.

Serum glutamic oxalacetic transaminase (SGOT)

Principle

The basic reaction for this procedure is the same as it is in the manual method. However, the oxalacetic acid produced is dialyzed into a buffer and then coupled with the diazonium salt of *N*-butyl-4-methoxymetanilamide (azoene fast red [PDC] or fast ponceau L). The intensity of the color produced is measured at 460 mμ in a 15 mm tubular flow cell. This method[27] is based on the procedure of Morgenstern and associates.[28] By measuring dialyzed oxalacetic acid, interference from other diazo-reacting serum constituents is avoided and it is not necessary to run individual serum controls.

A serum sample is aspirated and mixed with an air-segmented substrate stream. It then enters one of two 40-foot coils in a 37° C heating bath for incubation. Upon leaving the heating bath, the substrate mixture enters a dialyzer, where the enzymatically generated oxalacetic acid is dialyzed into a citrate buffer. As the substrate stream leaves the dialyzer, it continues on to waste. As the

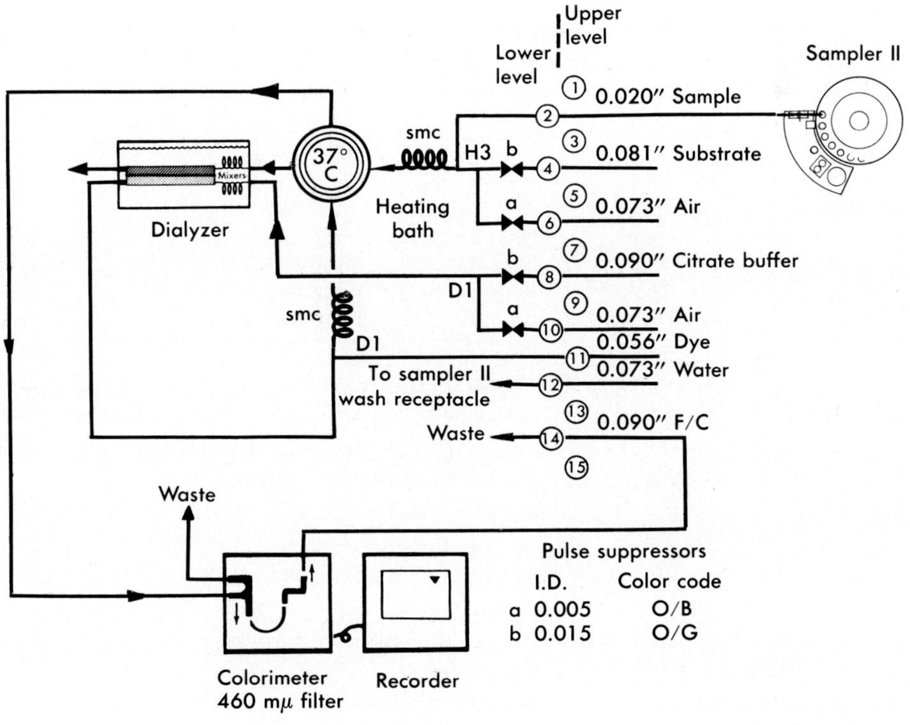

Fig. 4-10. SGOT by dialysis.

citrate buffer recipient stream leaves the dialyzer, it is joined and mixed with the diazonium salt solution. This mixture then returns to the 37° C heating bath to the second coil for color development. Upon leaving the heating bath, the reaction mixture goes to the colorimeter, where the intensity of color produced is measured at 460 mμ in a 15 mm flow cell.

Modules needed

One sampler II
One proportioning pump I
One heating bath, 37° C, with two coils
One dialyzer, one set of dialyzer plates
One colorimeter with 460 mμ filters
One single-pen recorder

Flow diagram

See Fig. 4-10.

Reagents needed

SGOT substrate	**(D-53)**
Phenyl mercuric acetate, 0.04%	**(D-54)**
Citrate buffer	**(D-55)**
Azoene fast red dye diluent	**(D-56)**
Azoene fast red dye solution	**(D-57)**
Alcoholic KOH wash solution, 2%	**(D-58)**

Reference controls

Enzyme "standards" are prepared from a frozen serum pool with high enzyme activity or from a lyophilized serum control such as Versatol E.* Dilute as follows:

Versatol E (ml)	+	Saline (ml)	=	Karmen units	Example Karmen units
0.1		0.9		V-E value ÷ 10	35
0.2		0.8		V-E value ÷ 5	70
0.3		0.7		V-E value × 3/10	105
0.4		0.6		V-E value × 2/5	140
0.5		0.5		V-E value ÷ 2	175
0.7		0.3		V-E value × 7/10	245
1.0		0.0		V-E value	350

These dilutions should be made fresh daily in small test tubes and mixed thoroughly before pouring into sample cups. CAUTION: Before accepting the given value of any lot of lyophilized control serum, the enzyme activity should be determined by a manual spectrophotometric method.

Procedure notes

1. Samples may be run at a rate of 60/hr with a 2:1 sample-wash ratio. If a number of elevated values are expected it is advisable to use a 1:1 sample-wash ratio to reduce the amount of sample interaction. Use a micro sample probe (0.016-inch I.D.) on the sampler.

---

*General Diagnostics Division, Warner-Chilcott.

2. Prepare the dye fresh each day, protect it from light by placing it in an amber glass bottle, and keep the bottle in ice during use. The use of premixed dye diluent is recommended. The inadvertent mixing of the dye with the undiluted alcohol or HCl will cause decomposition of the dye.

3. Wash the dye line, color development coil, and flow cell daily, at the end of the run, with 2% alcoholic KOH solution. Place the dye line in the KOH solution and allow the other reagent lines and the sample line to pump air. Wash for 2 minutes and then rinse all lines with water for 10 minutes.

Operating sequence

1. Turn on colorimeter and allow for a 20-minute warmup.
2. Start pumping water through all the lines and check for any leaks in the system.
3. Place the substrate and citrate buffer lines in their respective bottles. Leave the dye line in water.
4. After about 5 minutes place the dye line into the dye.
5. Wait about 8 minutes and then adjust the reagent baseline. Set the pen to 97% or 98% T and check the 0% T. When the pen returns to the reagent baseline, check the recorder gain.
6. Begin sampling.
7. At the end of the run place all lines in water for 5 minutes and then proceed to wash the dye line as directed in step 3 under Procedure notes.

Calculations

The standard curve may be plotted either on a Technicon plastic comparator or on semilog paper.

System maintenance and troubleshooting

1. The dialyzer membrane should be changed regularly according to use. A 25% Clorox wash for 5 minutes before changing the membrane will clear all the lines of any film buildup or protein particles in the sample line.
2. Unless the dye line is washed regularly with 2% alcoholic KOH the flow cell will become coated with the dye product and sensitivity will be decreased. Also, the coating built up in the lines and color development coil may flake off small particles that will cause a noisy pattern on the chart.
3. Filter all reagents to avoid noisy recordings.
4. Be sure the substrate and citrate buffer contain 1 ml of Triton X-405/L. Some commercially prepared reagents do not contain wetting agents.
5. Changing ambient temperature will cause considerable baseline drift.

Alkaline phosphatase

Two automated alkaline phosphatase methods are given. The procedure using a *p*-nitrophenyl phosphate substrate is the one used on the SMA 12/60. The procedure using a phenyl disodium phosphate substrate is the one used on the SMA 12/30.

Alkaline phosphatase (N-6b)

Principle

This automated method[29] is based on the manual procedure of Bessey and associates[30] using a *p*-nitrophenyl phosphate substrate, pH 10.25. It was modified for automation by Morgenstern and co-workers.[31] The phosphatase enzyme catalyzes the hydrolysis of the substrate while incubating at 37° C. The free *p*-nitrophenol is then dialyzed into a 2-amino-2-methyl-1-propanol (AMP) recipient buffer. Since *p*-nitrophenol is highly colored (yellow) in an alkaline solution, the color intensity can be measured directly at 410 mμ. Since dialyzed *p*-nitrophenol is measured, there is no need for specimen blank correction.

A sample is mixed with substrate and goes into a 37° C heating bath for incubation and hydrolysis. The reaction mixture then goes into a dialyzer where free *p*-nitrophenol dialyzes into the AMP recipient buffer. The yellow-colored dialysate then goes to the colorimeter, where the color intensity is measured at 410 mμ in a 15 mm tubular flow cell.

Modules needed

One sampler II
One proportioning pump (I or II)
One heating bath at 37° C
One dialyzer bath with one set of dialyzer plates
One colorimeter with 410 mμ filters
One single-pen recorder

Flow diagram

See Fig. 4-11.

Reagents needed

AMP stock solution, 50% w/v	**(D-59)**
AMP buffer, 0.625 M, pH 10.25	**(D-60)**

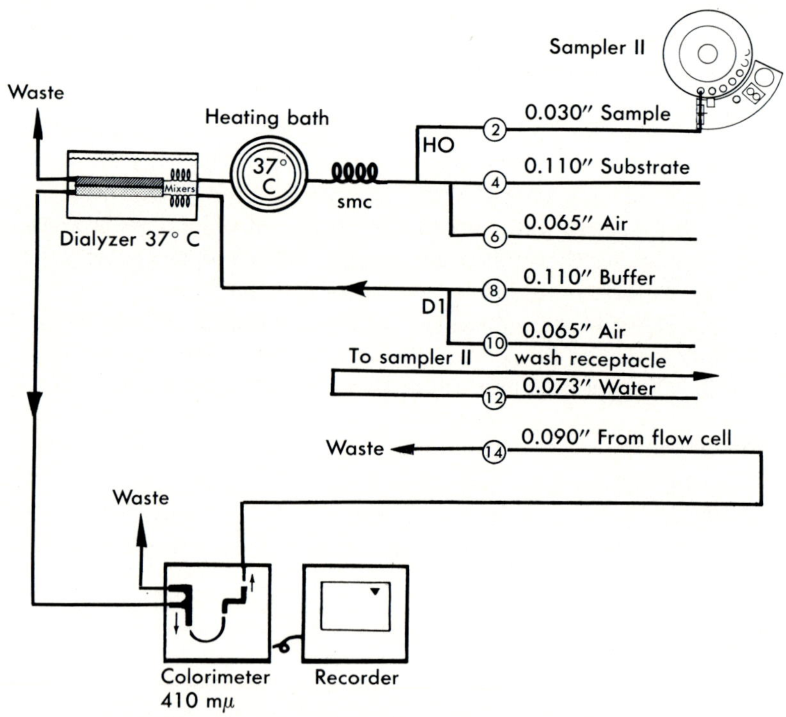

Fig. 4-11. Alkaline phosphatase (N-6b).

Magnesium chloride, 1.0 M (D-61)
p-Nitrophenyl phosphate working substrate (D-62)

Working standards

The International Unit, as proposed by the Enzyme Commission of the International Union of Biochemistry, is used as a measure of phosphatase activity by this method. One unit of enzyme activity is defined as the amount of enzyme that will catalyze the transformation of one micromole of substrate per minute under specified reaction conditions. It has been recommended that concentration be expressed as units per milliliter. Since values in the normal range are less than one International Unit, milli-Units (mU) are used. The suggested normal range for this method is 30 to 85 mU/ml.

Saline dilutions of a reference serum with a known alkaline phosphatase activity (e.g. Versatol E) may be used as working standards. These dilutions should be made just prior to use. Example:

Reference serum (ml)	Saline (ml)	Working standard (mU/ml)
0.2	0.8	29
0.4	0.6	58
0.6	0.4	87
0.8	0.2	116
1.0	0.0	145

Procedure notes

1. Samples may be run at a rate of 40/hr, using a 2:1 sample-wash ratio.
2. Serum specimens should not be hemolyzed.
3. Serum specimens with a phosphatase activity greater than the highest standard should be diluted with saline.

Operating sequence

1. Turn on colorimeter and allow a 20-minute warmup.
2. Begin pumping water through the sample and reagent lines. Check the system for leaks and observe the bubble pattern coming from the dialyzer.
3. Place the reagent lines in their respective reagent bottles. Leave the sample line in water.
4. After about 10 minutes set the reagent baseline to 97% or 98% T. Check 0% T and when the pen returns to the baseline check the recorder gain.
5. Begin sampling standards and serum specimens.
6. When the last specimen has recorded, place all lines in water and rinse for 10 minutes.

Calculations

The standard curve may be prepared by using a Technicon plastic comparator or by plotting % T versus mU/ml activity on semilog paper.

System maintenance and troubleshooting

1. The dialyzer membrane should be changed regularly according to use. Pump a 25% Clorox solution through all lines for 10 minutes followed by a 10-minute water rinse before changing the membrane.

Simultaneous alkaline phosphatase assay and control (N-28)

Principle

The automated procedure[32] for the determination of alkaline phosphatase activity is that of Marsh and associates,[33] which is based on modifications[34,35] of the King-Armstrong (K-A) method.

It is based on hydrolysis of the substrate phenyl disodium phosphate by action of the alkaline phosphatase enzyme in serum under specified conditions of pH, temperature, and time. The free phenol produced is measured by rapid condensation with 4-aminoantipyrine followed by oxidation with alkaline potassium ferricyanide to give a red product.

A sample is aspirated and divided into two streams—one for the assay and the other for the control. The assay sample is mixed with a buffer-substrate solution and goes into a 37° C heating bath coil for incubation. As the mixture leaves the heating bath, it is mixed with a 4-aminoantipyrine stream and a ferricyanide stream. The reaction mixture then goes to the colorimeter, where the color intensity is measured at 505 mμ in a 15 mm tubular flow cell. The control sample also is mixed with a buffer, 4-aminoantipyrine, and potassium ferricyanide. The mixture then goes to the colorimeter, where color intensity is measured at 505 mμ in a 15 mm tubular flow cell.

Modules needed

One sampler II
One proportioning pump (I or II—preferably, pump II)
One heating bath at 37° C
Two colorimeters with 505 mμ filters
One double-pen recorder with OD chart paper

Flow diagram

See Fig. 4-12.

Reagents needed

Magnesium chloride, 1% Mg^{++}	**(D-63)**
Alkaline buffer with magnesium	**(D-64)**
Phenyl disodium phosphate substrate	**(D-65)**
4-Aminoantipyrine	**(D-66)**
Potassium ferricyanide, 0.5%	**(D-67)**

Working standards

Saline dilutions of a reference serum of known alkaline phosphatase activity may be used as working standards. These dilutions should be made just prior to use. Example:

Reference serum (ml)	Saline (ml)	King-Armstrong units
0.2	0.8	7
0.4	0.6	14
0.6	0.4	21
0.8	0.2	28
1.0	0.0	35

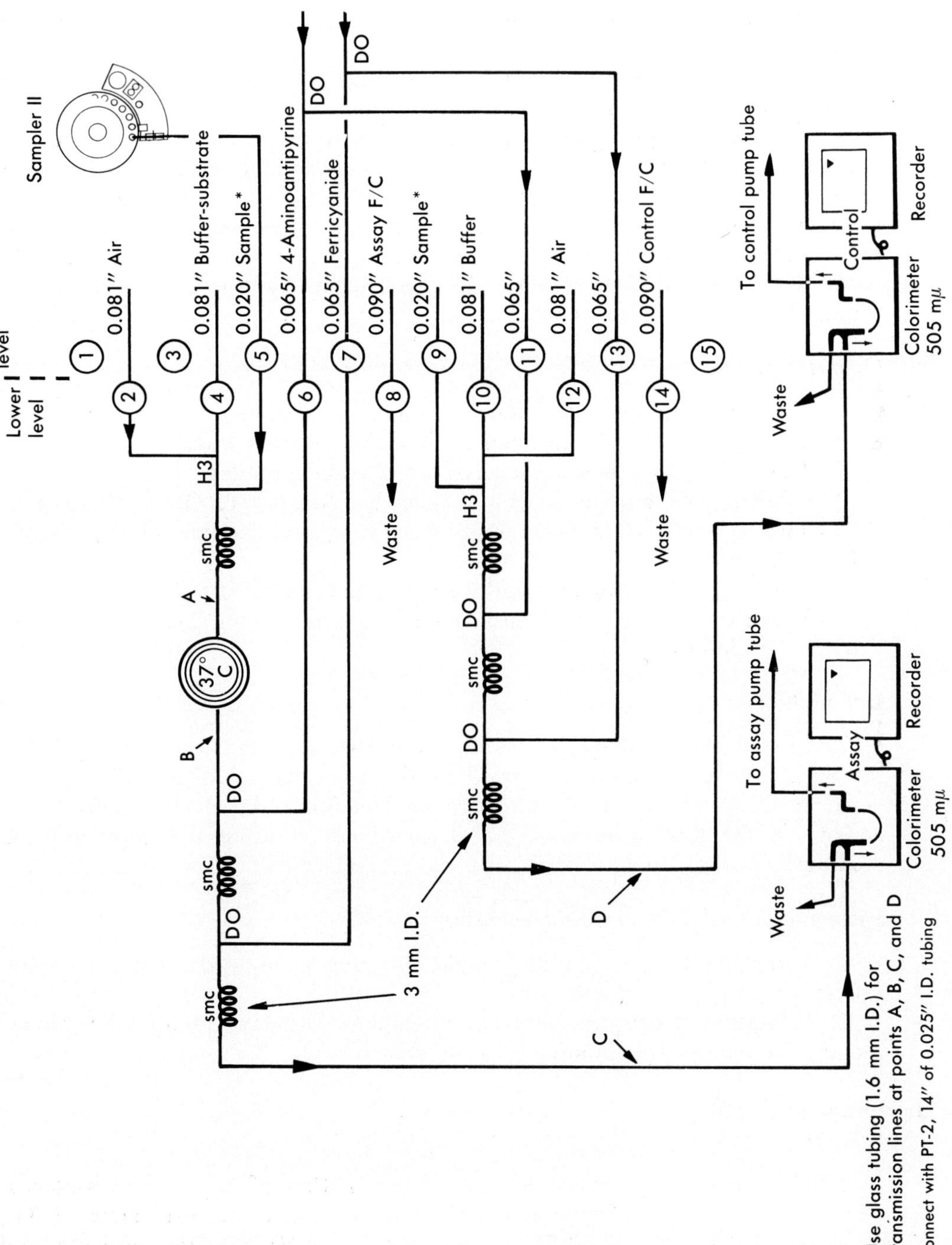

Fig. 4-12. Simultaneous alkaline phosphatase assay and control (N-28).

Procedure notes

1. Samples may be run at a rate of 60/hr, using a 2:1 sample-wash ratio.
2. Use glass tubing (1.6 mm I.D.) for transmission lines at points A, B, C, and D (Fig. 4-12) to prevent adsorption of phenol on the surface of the plastic tubing.
3. If pump I is used, add two helper springs under each end of the platen. There should be a total of four springs at each end.
4. Do not use a pump tube to supply water to the wash fluid receptacle on sampler II. In this case, an extra pump tube may inhibit the flow in the other tubes. Instead, place a water bottle above the level of the sampler and connect it directly to the wash receptacle. Start siphoning by aspiration and use an adjustable clamp on the line to regulate the flow.
5. Serum samples with alkaline phosphatase activity greater than the highest standard should be diluted with saline.

Operating sequence

1. Turn on colorimeters and allow a 20-minute warmup.
2. Begin pumping water through all lines and check for leaks.
3. Place all reagent lines in their respective reagent bottles.
4. After about 7 minutes, set the reagent baseline to 0.01 OD for both channels. Check 2.1 OD (0% T) and, when pens return to baseline, check recorder gain on both pens.
5. Begin sampling standards and unknown specimens.
6. When the last specimen has recorded, place reagent lines in water and rinse for 10 minutes.

Calculations

From each assay OD reading subtract the corresponding control OD reading for both standards and unknowns. Plot the corrected standard OD readings versus concentration on linear graph paper. Determine the alkaline phosphatase activity of the unknowns from the standard curve, using the corrected OD readings.

System maintenance and troubleshooting

1. Pump 25% Clorox through sample and reagent lines about once a week. Follow with a 10-minute water rinse.
2. Whenever it becomes necessary to replace a sample pump tube, always replace the other sample pump tube at the same time.

Lactic dehydrogenase (LDH)

Principle

This automated procedure for the determination of lactic dehydrogenase activity[36] is based on a method described by Hochella and Weinhouse.[37] The oxidation of *l*-lactate by NAD is catalyzed by LDH to give pyruvate and reduced NAD (NADH). The enzyme diaphorase catalyzes the coupling of NADH to an electron-acceptor dye (INT). The electron from NADH is transferred to the tetrazolium dye, resulting in the production of a colored formazan.

$$\text{Lactate} + \text{NAD} \xrightleftharpoons{\text{LDH}} \text{Pyruvate} + \text{NADH}$$

$$\text{NADH} + \text{INT} \xrightleftharpoons{\text{Diaphorase}} \text{NAD} + \text{Reduced dye (colored product)}$$

An air-segmented stream of substrate is joined by an INT dye stream and an NAD-diaphorase stream. This mixture is then joined by the sample stream and passes through a double mixing coil to a 37° C heating bath for color development. The reaction mixture then goes to the colorimeter, where the color intensity is measured at 505 mμ in a 15 mm tubular flow cell.

Modules needed

One sampler II
One proportioning pump (I or II)
One heating bath at 37° C
One colorimeter with 505 mμ filters
One single-pen recorder

Flow diagram

See Fig. 4-13.

Reagents needed

LDH substrate	**(D-68)**
Phosphate buffer, 0.067 M, pH 7.4	**(D-69)**
NAD-diaphorase	**(D-70)**
INT dye	**(D-71)**

Working standards

The Enzyme Commission of the International Union of Biochemistry defines one International Unit of enzyme activity as the amount of enzyme which will catalyze the transformation of one micromole of substrate per minute under specified reaction conditions. It is recommended that the concentration of enzyme solutions be expressed as Units per milliliter. Since the normal range of LDH activity in human serum is below one International Unit, activity is expressed

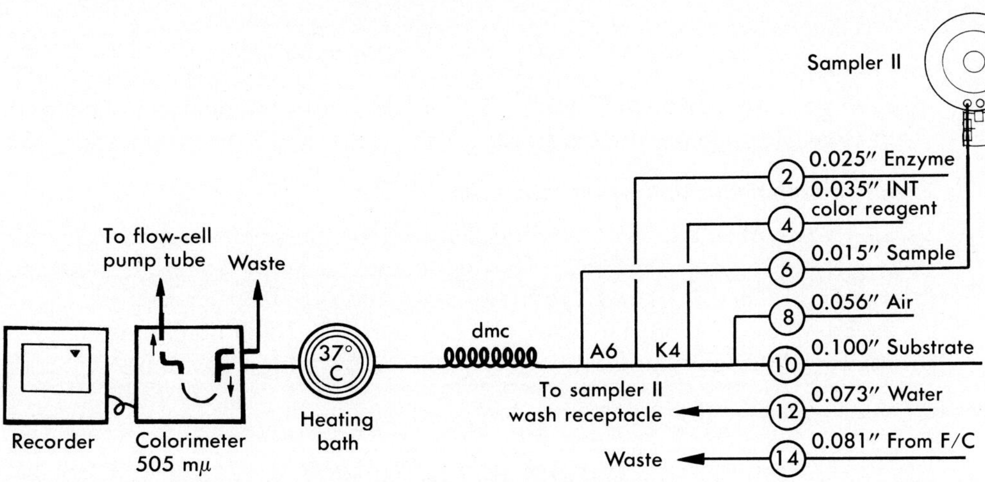

Fig. 4-13. LDH (N-60).

in milli-International Units per milliliter. The suggested normal range for this procedure is 90 to 200 mU/ml.

Dilutions of a reference serum of known LDH activity may be used as working standards. Example:

Reference serum (ml)	Saline (ml)	Working standard (mU/ml)
0.5	1.5	90
1.0	1.0	180
1.5	0.5	270
2.0	0.0	360

Procedure notes

1. Samples may be run at a rate of 60/hr, using a 2:1 sample-wash ratio.
2. Add Triton X-100 to the substrate as a wetting agent. Do not use any other wetting agents.
3. The tubing that carries the INT and NAD-diaphorase reagents from the bottles to the manifold should be of narrow bore. Use either 0.025- or 0.030-inch I.D. tubes.
4. A micro sample probe (0.016-inch I.D.) should be used and should be connected to the sample pump tube with 0.015-inch I.D. tubing.
5. Serum samples must be free of any fibrin particles that could plug the micro sample line.
6. Serum samples with LDH activity greater than the highest standard should be diluted with saline.

Operating sequence

1. Turn on the colorimeter and allow a 20-minute warmup.
2. Begin pumping water through all lines and check the system for leaks.
3. Place lines in reagents and leave sample line in water.
4. After about 7 minutes, set the reagent baseline to 97% or 98% T. Check 0% T and when the pen returns to baseline check the recorder gain.
5. Begin sampling standards and serum specimens.
6. When the last specimen has recorded, place all lines in water and rinse for 10 minutes.

Calculations

Prepare a standard curve on a Technicon plastic comparator or plot % T versus concentration on semilog paper.

System maintenance and troubleshooting

1. Eventually a red coating will deposit on the glass coils and flow cell. Clean these periodically by flushing with dichromate cleaning solution followed by a water rinse. The discolored Tygon tubing should then be replaced.
2. Pump a 50% Clorox solution through the INT, NAD-diaphorase, and sample lines for 10 minutes at least once a week. Follow with a 10-minute water rinse.

Total lactic dehydrogenase and isoenzyme index

Principle

This procedure is based on the Babson and Phillips manual method for determination of total lactic dehydrogenase activity[38] and Babson's method for LDH isoenzyme index.[39] (See Chapter 11.) The methods were adapted for use on the AutoAnalyzer by Rice.[40]

Lactic dehydrogenase catalyzes the oxidation of lactate to pyruvate. In this reaction NAD is reduced to NADH that is coupled to the reduction of a tetrazolium salt, 2-*p*-iodophenyl-3-*p*-nitrophenyl-5-phenyl tetrazolium chloride (INT), with phenazine methosulfate (PMS) serving as an intermediate electron carrier. The bright red formazan produced is measured at 520 mμ.

$$\text{Lactate} + \text{NAD} \underset{}{\overset{\text{LDH}}{\rightleftharpoons}} \text{NADH} + \text{Pyruvate}$$

$$\text{NADH} + \text{PMS} \longrightarrow \text{NAD} + \text{Reduced PMS}$$

$$\text{Reduced PMS} + \text{INT} \longrightarrow \text{PMS} + \text{Reduced INT (red color)}$$

To determine the isoenzyme index a 2 M *l*(+) lactate substrate is used as a heart LDH–inhibiting substrate and a solution containing 0.02 M *l*(+) lactate–2 M urea is used as a liver LDH–inhibiting substrate. The absorbance of a sample reaction with heart-inhibiting substrate divided by the absorbance of a sample reaction with liver-inhibiting substrate gives an isoenzyme index. A normal index range of 0.8 to 1.2 has been suggested for human serum. An index of less than 0.8 along with an elevated total LDH activity indicates the presence of an elevated level of LDH_1 (heart). An index greater than 1.2 along with an elevated total LDH activity indicates the presence of an elevated level of LDH_5 (liver).

In determining total LDH, the sample mixes with a buffer and then with the substrate and goes into a double mixing coil in the 37° C dialyzer bath for incubation. The reaction mixture emerges from the bath, mixes with INT dye, and returns to a single mixing coil in the dialyzer bath for color development. As this mixture emerges from the bath it is mixed with 0.1 N HCl to stop enzyme activity. The intensity of the red color produced is then measured at 520 mμ in a 15 mm tubular flow cell. A serum blank is run simultaneously, using a control reagent in place of the substrate.

When isoenzyme indices are desired, the LDH substrate is replaced with 2 M *l*(+) lactate and the LDH control reagent is replaced with 0.02 M *l*(+) lactate–2 M urea.

Modules needed

One sampler II
One proportioning pump (I or II)
One dialyzer bath or other 37° C water bath
Two colorimeters with 520 mμ filters
One two-pen recorder

Flow diagram

See Fig. 4-14.

Reagents needed

LDH color reagent	**(D-72)**
LDH buffer reagent	**(D-73)**
LDH substrate	**(D-74)**
LDH control reagent	**(D-75)**
Hydrochloric acid, 0.1 N	**(D-37)**
***l*(+) Lactate, 2 M**	**(D-76)**
***l*(+) Lactate, 0.02 M–2 M urea**	**(D-77)**

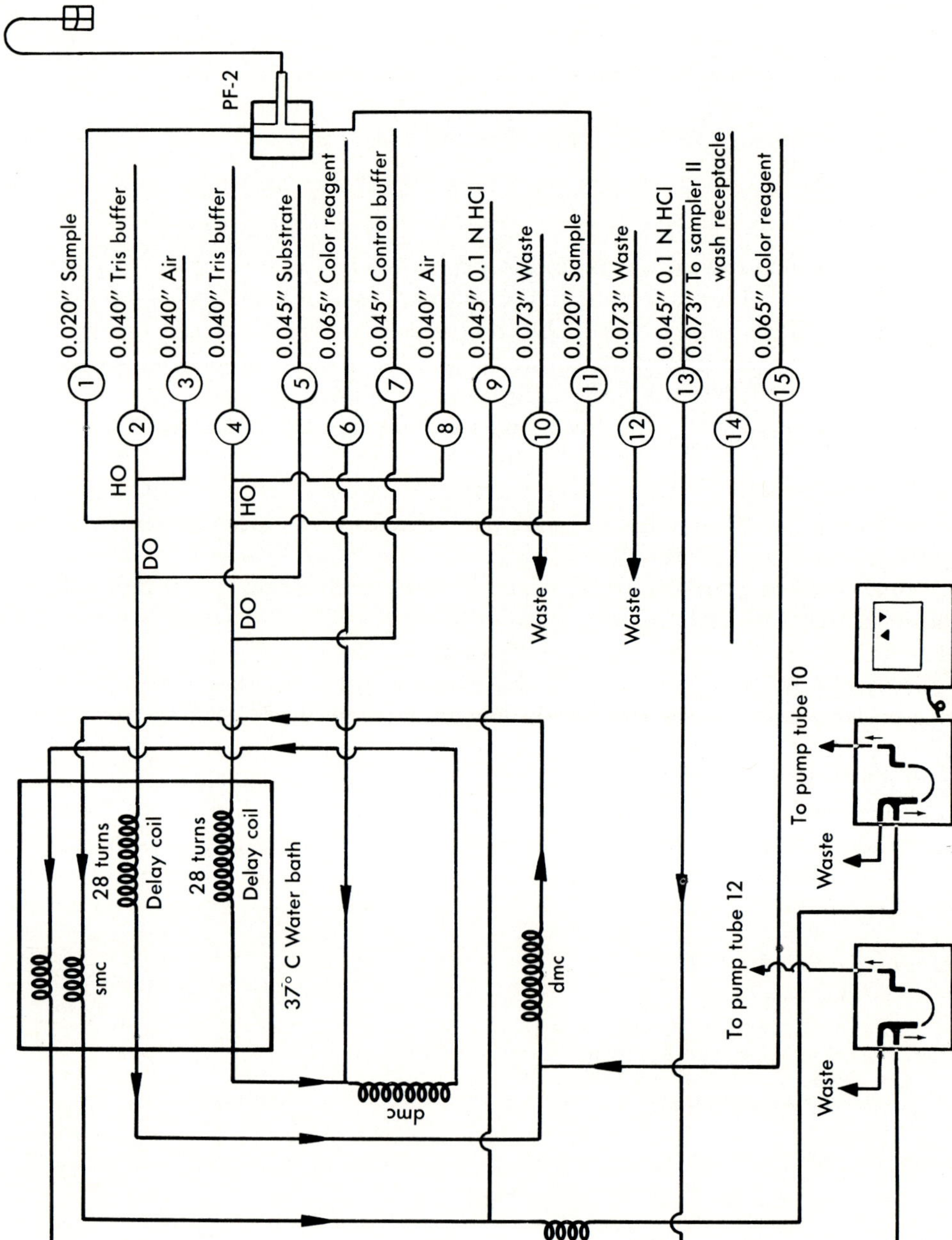

Fig. 4-14. Total LDH or LDH isoenzyme index.

Working standards

Commercial lyophilized enzyme standards such as Versatol-E and Versatol-E-N are used to prepare a standard curve. Example:

Tube	V-E-N (ml)	V-E (ml)	Units	Example
1	0.5	—	Given value	50
2	1.5	0.5	Values (1 + 3) ÷ 2	88
3	0.5	0.5	Values (1 + 5) ÷ 2	125
4	0.5	1.5	Values (3 + 5) ÷ 2	163
5	—	0.5	Given value	200

The suggested normal range for this procedure is 24 to 78 I.U.

Procedure notes

1. Samples may be run at a rate of 60/hr, using a 2:1 sample-wash ratio.
2. A dialyzer bath may be used to house the incubating coils without disturbing the connections to the dialyzer plates. Drill eight ⅛-inch holes in the center of the bath cover, insert N2 nipples, and connect the two double and two single coils to these nipples.
3. Triton X-100 may be added to the substrate (0.5 ml/L) as a wetting agent. Do not use any other wetting agents.
4. Serum samples with LDH activity greater than the highest standard should be diluted with saline.

Operating sequence

1. Turn on the colorimeter and allow a 20-minute warmup.
2. Begin pumping water through all lines and check the system for leaks.
3. Place lines in reagents and leave sample line in water.
4. After about 10 minutes, set the reagent baseline to 97% or 98% T. Check 0% T and when the pen returns to baseline check the recorder gain.
5. Begin sampling standards and serum specimens.
6. When the last specimen has recorded, place all lines in water and rinse for 10 minutes.

Calculations

For total LDH, prepare a standard curve on a Technicon plastic comparator or plot % T versus concentration on semilog paper.

To obtain the isoenzyme index, first convert all % T values to optical densities. Then divide the OD value of a specimen using the heart-inhibiting substrate by the OD value obtained using the liver-inhibiting substrate. To avoid converting % T to OD in this procedure, use optical density chart paper on the recorder.

System maintenance and troubleshooting

1. Pump a 50% Clorox solution through the sample and reagent lines for 10 minutes at least once a week. Follow with a 10-minute rinse with water.
2. Eventually a red coating will deposit on the glass coils and flow cell. This should be cleaned periodically by flushing with a dichromate cleaning solution followed by a water rinse. The discolored Tygon tubing should then be replaced.

Cholesterol (direct)

Principle

This procedure[41] employs a direct Liebermann-Burchard reaction for the determination of serum cholesterol. The manual procedure of Huang and associates[42] was adapted for automation by Levine and co-workers.[43]

Serum is mixed with the Liebermann-Burchard color reagent and pumped into a 40-foot glass coil in a 37° C heating bath for color development. When the reaction mixture leaves the heating bath it goes to the colorimeter where the color intensity is measured at 630 mμ in a 15 mm flow cell.

Modules needed

One sampler II
One proportioning pump (I or II)
One heating bath at 37° C
One colorimeter with 630 mμ filters
One single-pen recorder

Flow diagram

See Fig. 4-15.

Reagent needed

Liebermann-Burchard color reagent **(D-78)**

Working standards

Commercially prepared control sera in the elevated and normal ranges are used as standards. A third standard is prepared by mixing equal volumes of the normal and the elevated serum and calculating the value from the given values of the two. Example:

$$\text{Normal serum control} = 180 \text{ mg cholesterol/100 ml}$$

$$\text{Elevated serum control} = 420 \text{ mg cholesterol/100 ml}$$

$$1 \text{ part normal} + 1 \text{ part elevated} = \frac{180 + 420}{2} = 300 \text{ mg}$$

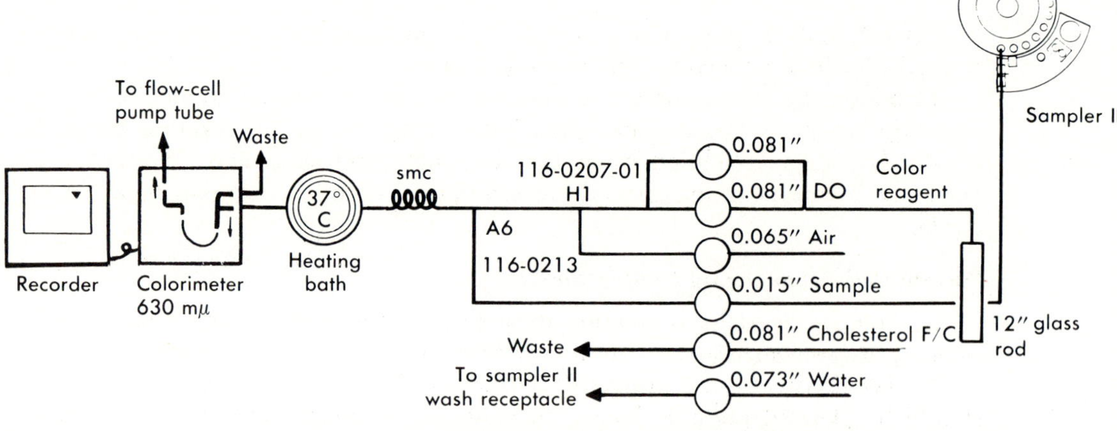

Fig. 4-15. Cholesterol via Liebermann-Burchard.

Procedure notes

1. The Liebermann-Burchard color reagent should not be exposed to atmospheric moisture any more than is absolutely necessary. The reagent is stable at least three months when stored in a refrigerator. Bring to room temperature before use. If it becomes discolored it should be discarded.

2. Precipitated protein may accumulate in the steel nipple in the A-6 fitting where the serum meets the color reagent. This should be cleaned daily with a fine steel probe.

3. The specific gravity of the color reagent should be 1.185 at room temperature (20° C). If the specific gravity is above 1.190, the reagent should be diluted with a solution made up of equal parts of acetic acid and acetic anhydride.

4. The reagent and flow-cell pump tubes should be changed daily.

Operating sequence

1. Turn on colorimeter and allow a 20-minute warmup.

2. Place the reagent line in the color reagent bottle and the sample line in the sampler wash receptacle. Begin pumping.

3. After about 7 minutes set the reagent baseline to 97% or 98% T. Check 0% T. When the pen returns to the baseline, check the recorder gain.

4. Begin sampling standards and specimens.

5. When the last sample has recorded, remove the reagent bottle. Wipe the glass rod with several layers of tissue. Allow the line to aspirate air for about 2 minutes and then place the line in water and wash for 10 minutes. Then remove all lines from water and continue pumping air until all water is pumped out.

System maintenance and troubleshooting

1. As stated previously, the steel nipple in the A-6 fitting should be cleaned daily.

2. Watch the entire system closely for leaks. The acids in the color reagent could cause serious damage if a leak should go undetected.

Simultaneous direct and total bilirubin

Principle

This procedure for the simultaneous determination of direct and total bilirubin[44] in serum is an adaptation by Gambino and Schreiber[45] of the manual method of Jendrassik and Grof.[46] The method is based on the formation of colored azobilirubin when bilirubin reacts with diazotized sulfanilic acid.

Total bilirubin (free and conjugated) is determined by reacting the serum sample with diazo reagent in the presence of a caffeine–sodium benzoate solution. Since free bilirubin is insoluble in water below pH 8, direct (conjugated) bilirubin is determined by reacting the serum sample with diazo reagent in an acidic aqueous medium.

The sample is aspirated and split into two streams—one for total bilirubin and the other for direct bilirubin. The total bilirubin sample stream is mixed with a caffeine solution, while the direct bilirubin sample stream is mixed with 0.05 N hydrochloric acid. Each stream is then mixed successively with diazo reagent, ascorbic acid, and tartrate buffer. The reaction mixtures then go to the colorimeters, where the color intensity of each is measured at 600 mμ in 15 mm flow cells. When running serum blanks, diazo I is used in place of the combined diazo reagent.

This method has high sensitivity and excellent linearity; it is insensitive to hemoglobin and to wide variations in protein concentration and in sample pH.

The caffeine–sodium benzoate accelerates the coupling of bilirubin to diazotized sulfanilic acid. The sodium acetate in the caffeine diluent buffers the pH of the diazo reaction. The ascorbic acid eliminates hemoglobin interference in the total reaction and stops additional color formation in the direct reaction. The strongly alkaline tartrate buffer brings the final pH to 13.4, which solubilizes the protein. It eliminates the effects of variation in sample pH, increases the molar absorptivity of azobilirubin, and increases specificity by shifting the absorbancy maximum to 600 mμ.

Modules needed

One sampler II
One proportioning pump II
Two colorimeters with 600 mμ filters
One double-pen recorder with OD chart paper

Flow diagram

See Fig. 4-16.

Reagents needed

Caffeine diluent	**(D-79)**
Sulfanilic acid (diazo I)	**(D-80)**
Sodium nitrite, 0.5% (diazo II)	**(D-81)**
Diazo reagent, combined	**(D-82)**
Tartrate buffer	**(D-83)**
Ascorbic acid, 4%	**(D-84)**
Hydrochloric acid, 0.05 N	**(D-85)**

Working standards

Working standards can be prepared by diluting a bilirubin-in-serum standard with water. A number of commercially prepared serum controls with elevated bilirubin values are available. A serum standard may be diluted up to fiftyfold without deleterious effects. However, some protein must be present. Aqueous bilirubin standards without protein are not reliable.[45]

Working standards should be prepared just prior to use and protected from light. The following is an example of how a bilirubin-in-serum standard with a value of 6.0 mg of bilirubin/100 ml would be diluted:

Serum standard (ml)	H_2O (ml)	Bilirubin working standard (mg/100 ml)
0.1	0.9	0.6
0.2	0.8	1.2
0.4	0.6	2.4
0.6	0.4	3.6
0.8	0.2	4.8
1.0	0.0	6.0

When running standards, use caffeine diluent in both channels. When the last standard has been aspirated, remove the 0.05 N HCl line from the caffeine

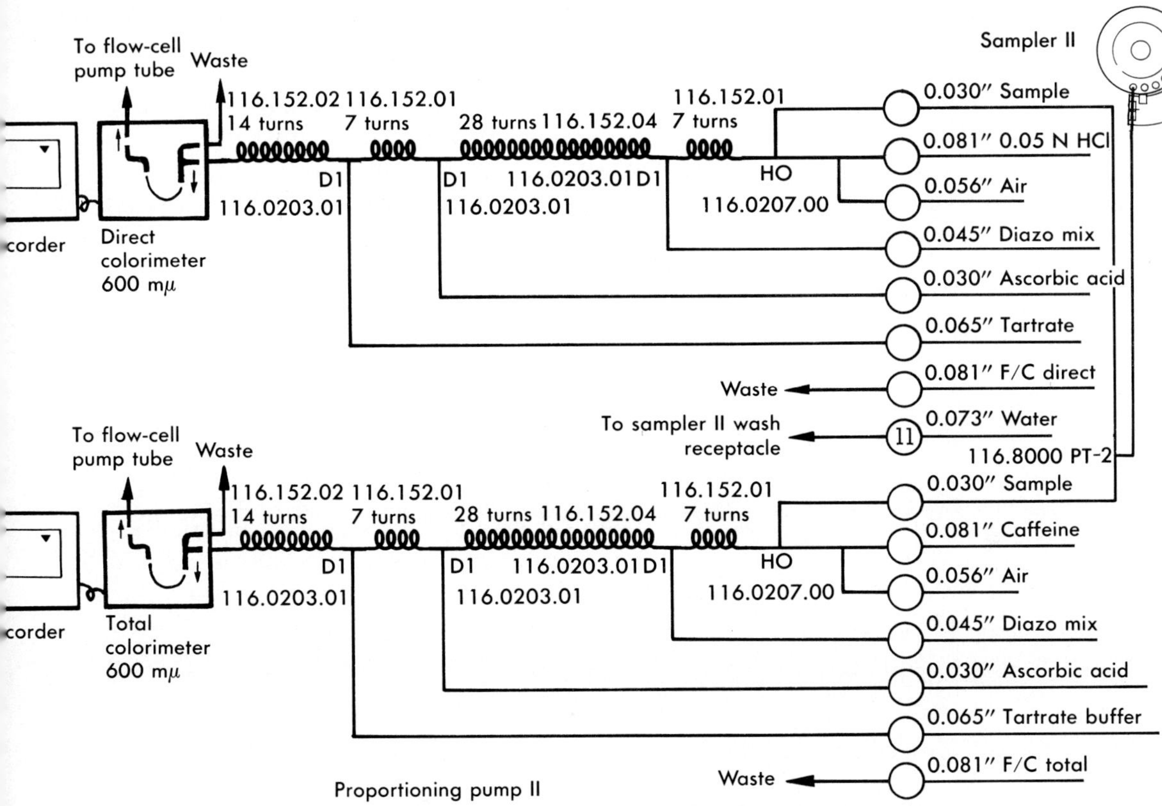

Fig. 4-16. Simultaneous direct and total bilirubin, pump II.

diluent and place in the HCl reagent bottle. Readjust reagent baseline and run unknown serum samples. All commercially prepared bilirubin standards are made with free (unconjugated) bilirubin. Conjugated (direct) bilirubin is not available as a stable standard.

Procedure notes

1. Samples may be run at a rate of 60/hr with a 2:1 sample-wash ratio.
2. Protect serum samples from light as much as possible. Bilirubin is altered by exposure to direct sunlight and fluorescent lights, giving lower results. An amber-colored sample plate cover is available to protect samples in the sampler tray.
3. Serum samples with bilirubin values above the highest standard should be diluted with water.

Operating sequence

1. Turn on colorimeters and allow a 20-minute warmup.
2. Begin pumping water through all the lines and check for leaks.
3. Place the reagent lines (except the 0.05 N HCl line) in their respective reagent bottles. Place the HCl line in the caffeine diluent bottle. Leave the sample line in water.
4. After about 5 minutes set the reagent baseline to 0.1 OD on both channels.

Check 2.1 OD (0% T). When the pens return to baseline, check the recorder gain on both pens.

5. Begin sampling standards.

6. One minute after the last standard has been aspirated, remove the 0.05 N HCl line from the caffeine diluent, wipe the outside of the line with tissue, and place the line in the 0.05 N HCl bottle.

7. Readjust the reagent baseline when the HCl has had time to reach the colorimeter.

8. Begin sampling unknown sera.

9. When the last sample has recorded, place the diazo reagent lines in a bottle containing diazo I (sulfanilic acid) and rerun the standards and unknown serum specimens.

10. When the last serum blank has recorded, place all reagent lines in water and rinse for 10 minutes.

Calculations

From each standard OD reading subtract the corresponding blank OD reading. Plot these OD readings versus bilirubin concentration on linear graph paper. Subtract blank OD from test OD for each sample and determine bilirubin value from standard curve. Each channel requires its own calibration curve.

System maintenance and troubleshooting

1. This system is relatively trouble-free. Problems can usually be traced to an improperly prepared reagent.

2. As the manifold ages, sensitivity decreases. After a month of daily use sensitivity may drop 12% to 15%.

Simultaneous total protein and albumin

Principle

Total protein is determined by the biuret reaction. This reaction produces a purple-colored complex of copper, in an alkaline solution, with two or more carbamyl groups (—CO—NR—) that are joined together directly or through a single atom of nitrogen or carbon.

The manual procedure of Weichselbaum[47] was adapted for the AutoAnalyzer by Stevens.[48]

The albumin procedure[49] is based on the quantitative binding of an anionic dye, 2-(4′-hydroxyazobenzene) benzoic acid (HABA), specifically to serum albumin. The manual procedure of Ness and co-workers[50] was adapted to the AutoAnalyzer by Nishi and Rhodes.[51]

The sample is aspirated from the sample cup and goes to a PT-2 fitting, where the sample is split into two streams. One sample stream joins an air-segmented biuret reagent stream and more biuret reagent is added. The reaction mixture then goes through three double mixing coils where color development takes place. It then goes to the colorimeter, where the color intensity is measured at 550 mμ in a 15 mm tubular flow cell.

The other sample stream joins an air-segmented HABA dye stream and more dye is added. The reaction mixture then goes through three double mixing coils where color development takes place. It then goes to the colorimeter, where the color intensity is measured at 505 mμ in a 15 mm tubular flow cell.

Serum blanks are obtained by rerunning the samples, using a tartrate-iodide solution in place of the biuret reagent and a phosphate buffer in place of the HABA dye.

Since the albumin portion of the manifold is identical to the total protein portion, one may run simultaneously either total protein and serum protein blanks or albumin and serum albumin blanks if two sets of filters of the same wavelength are available.

Modules needed

One sampler II
One proportioning pump (I or II)
Two colorimeters with 550 mμ and 505 mμ filters
One double-pen recorder

Flow diagram

See Fig. 4-17.

Reagents needed

Alkaline iodide	**(D-86)**
Biuret stock solution	**(D-87)**
Biuret working solution	**(D-88)**
Protein blank solution	**(D-89)**
HABA dye stock solution	**(D-90)**
Phosphate buffer	**(D-91)**
HABA dye working solution	**(D-92)**
Albumin stock standard, 10 g/100 ml	**(D-93)**

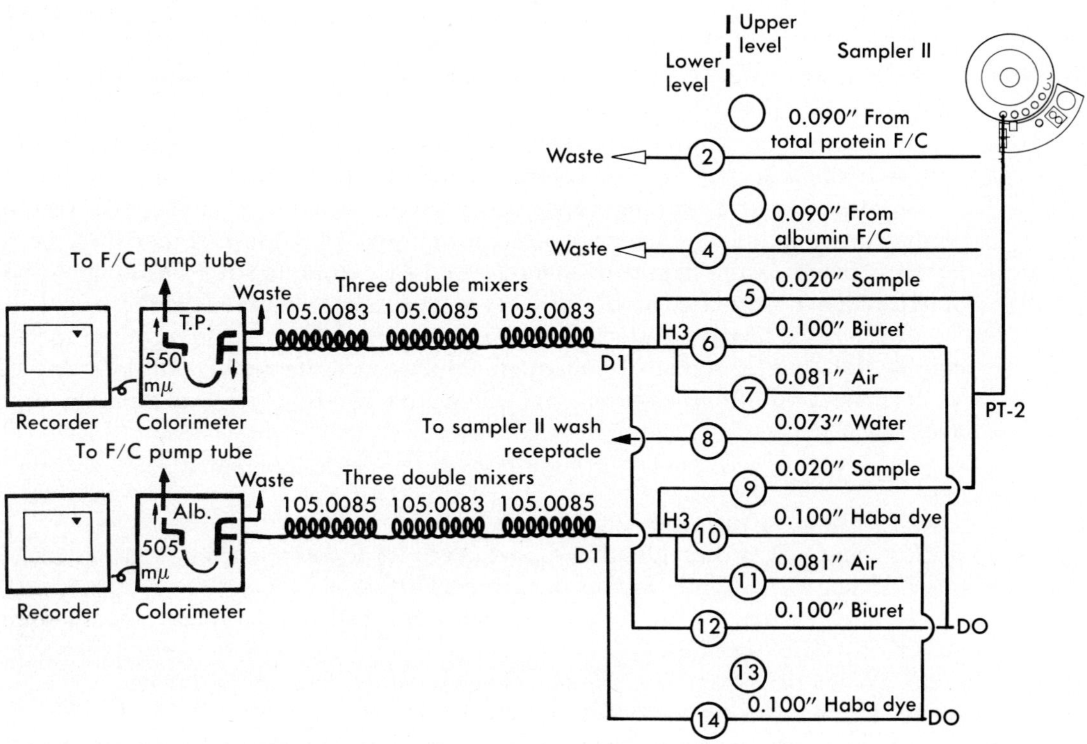

Fig. 4-17. Simultaneous total protein and albumin.

Working standards

Unless a laboratory has the facilities and the inclination to standardize their own total protein and human albumin standard it is strongly recommended that a stock standard be obtained from Technicon Corporation (T23-0204). Their standard is made from crystalline bovine albumin, fraction V. The total protein content and a factor with which to calculate the human albumin equivalent are given on the label. It is a good practice to verify the calibration curves with serum controls.

Pipette the following volumes of standardized albumin stock standard into 25 ml volumetric flasks and dilute to volume with saline containing 50 mg of sodium azide per 100 ml. Mix thoroughly.

Albumin stock (ml)	Total protein (g/100 ml)	X factor	Albumin (g/100 ml)
5	2	(e.g. 0.9)	1.8
10	4		3.6
15	6		5.4
20	8		7.2
Undiluted stock	10		9.0

These standards are stable for a year if kept refrigerated.

Procedure notes

1. This procedure can be run at a rate of 60 samples/hr, using a 2:1 sample-wash ratio.
2. Serum should be separated from the clot within 4 hours after blood is drawn. Care should be taken to avoid hemolysis. Refrigerated serum is stable for 2 days.
3. Serum blanks should be run on all specimens. This is especially important for turbid or icteric specimens.
4. The bovine albumin standard curves should be checked periodically with a human albumin standard or a commercially prepared control serum.
5. Since the dye-binding capacity of bovine albumin is about 90% that of human albumin, the stock bovine albumin solution (if a source other than Technicon is being used) should be standardized by comparing it to a human albumin fraction V standard or to a primary mercaptalbumin standard.[49,52]
6. The dye-binding capacity of serum albumin obtained from other species will be different from that of human albumin. Inaccurate results will be obtained if, for example, rabbit albumins are calculated by using human albumin standards.

Operating sequence

1. Turn on colorimeters and allow a 20-minute warmup.
2. Pump water through all lines and check for leaks.
3. Place reagent lines in their respective bottles.
4. After 5 minutes set reagent baselines for both total protein and albumin to 97% or 98% T. Check 0% T.
5. When pens return to baseline, check recorder gain on both pens.
6. Begin sampling standards and specimens.
7. When the last specimen has recorded, replace the biuret reagent with the protein blank solution and the HABA dye with the phosphate buffer.

8. After 5 minutes, reset the baselines and begin sampling the specimens again.

9. When the last blank has recorded, place all lines in water and wash for 10 minutes.

Calculations

Plot one curve for total protein and another for albumin—either on a Technicon plastic comparator or on graph paper (% T versus concentration on semilog paper). Determine from the curve the g/100 ml value for each test and subtract the g/100 ml value for each corresponding serum blank.

System maintenance and troubleshooting

1. The two sample pump tubes usually wear out before the other tubes. When this happens either the sample will not pass through the tube at all or the sampling is not reproducible. Always replace *both* sample pump tubes at the same time so that the sample flow rate will be about equal through both tubes.

2. After every 50 to 60 hours of operating time pump a 25% Clorox solution through the sample and reagent lines for 10 minutes to clean out the coils and the flow cells. Follow with a 10-minute water rinse.

REFERENCES

Notes on general operating procedure

1. White, W. L., Erickson, M. M., and Stevens, S. C.: Practical automation for the clinical laboratory, St. Louis, 1968, The C. V. Mosby Co.

Electrolytes

2. Electrolytes, Technicon Laboratory method file N-21b.
3. Skeggs, L. T.: An automatic method for the determination of carbon dioxide in blood plasma, Amer. J. Clin. Path. **33**:181, 1960.
4. Zall, D. M., Fisher, D., and Garner, M. O.: Photometric determination of chlorides in water, Anal. Chem. **28**:1665, 1956.

Glucose and urea nitrogen

5. Simultaneous glucose and urea nitrogen, Technicon Laboratory method file N-16b.
6. Hoffman, W. S.: A rapid photoelectric method for the determination of glucose in blood and urine, J. Biol. Chem. **120**:51, 1937.
7. Marsh, W. H., Fingerhut, B., and Miller, H.: Automated and manual direct methods for the determination of blood urea, Clin. Chem. **11**:624, 1965.
8. White, W. L., Erickson, M. M., and Stevens, S. C.: Practical automation for the clinical laboratory, St. Louis, 1968, The C. V. Mosby Co., p. 103.
9. Davis, H. A., Sterling, R. E., Wilcox, A. A., and Waters, W. E.: Investigation of a potential source of difficulty in the use of the AutoAnalyzer, Clin. Chem. **12**:428, 1966.

Creatinine

10. Folin, O.: On the determination of creatinine and creatine in urine, J. Biol. Chem. **17**:469, 1914.
11. Folin, O., and Wu, H.: A system of blood analysis, J. Biol. Chem. **38**:81, 1919.
12. Stevens, D. L., and Skeggs, L. T.: Creatinine, Technicon Laboratory method file N-11b.
13. Chasson, A. L., Grady, H. J., and Stanley, M. A.: Determination of creatinine by means of automatic chemical analysis, Amer. J. Clin. Path. **35**:83, 1961.

Uric acid

14. Uric acid, Technicon Laboratory method file N-13b.
15. Brown, H.: The determination of uric acid in human blood, J. Biol. Chem. **158**:601, 1945.
16. Hawk, P. B., Oser, B. L., and Summerson, W. H.: Practical physiological chemistry, ed. 13, New York, 1954, Blakiston Division, McGraw-Hill Book Co., pp. 564-565.

Calcium

17. Calcium, Technicon Laboratory method file N-3b.
18. Anderegg, G., Flaschka, H., Sallmann, R., and Schwarzenbach, G.: Ein auf erdalkaliionen ansprechendes Phthalein und seine analytische Verwendung, Helv. Chim. Acta **37**:113, 1954.
19. Kessler, G., and Wolfman, M.: An automated procedure for the simultaneous determination of calcium and phosphorus, Clin. Chem. **10**:686, 1964.

20. Pollard, F. H., and Martin, J.: The spectrophotometric determination of the alkaline-earth metals with murexide, Eriochrome Black T and with *o*-cresolphthalein Complexone, Analyst **81**:348, 1956.
21. Stern, J., and Lewis, W. H. P.: The colorimetric estimation of calcium in serum with *o*-cresolphthalein Complexone, Clin. Chim. Acta **2**:576, 1957.
22. Gitelman, H. J.: An improved automated procedure for the determination of calcium in biological specimens, Anal. Biochem. **18**:521, 1967.

Inorganic phosphate

23. Inorganic phosphate, Technicon Laboratory method file N-4c.
24. Inorganic phosphate, Technicon Laboratory method file N-4b.
25. Hurst, R. O.: The determination of nucleotide phosphorus with a stannous chloride–hydrazine sulfate reagent, Canad. J. Biochem. **42**: 287, 1964.
26. Kraml, M.: A semi-automated determination of phospholipids, Clin. Chim. Acta **13**:442, 1966.

Serum glutamic oxalacetic transaminase (SGOT)

27. Serum glutamic oxalacetic transaminase (SGOT), Technicon Laboratory method file N-25b.
28. Morgenstern, S., Oklander, M., Auerbach, J., Kaufman, J., and Klein, B.: Automated determination of serum glutamic oxaloacetic transaminase, Clin. Chem. **12**:95, 1966.

Alkaline phosphatase

29. Alkaline phosphatase, Technicon Laboratory method file N-6b.
30. Bessey, O. A., Lowry, O. H., and Brock, M. J.: Method for rapid determination of alkaline phosphatase with 5 cubic millimeters of serum, J. Biol. Chem. **164**:321, 1946.
31. Morgenstern, S., Kessler, G., Auerbach, J., Flor, R. V., and Klein, B.: An automated *p*-nitrophenyl phosphate serum alkaline phosphatase procedure for the AutoAnalyzer, Clin. Chem. **11**:876, 1965.
32. Simultaneous alkaline phosphatase assay and control, Technicon Laboratory method file N-28.
33. Marsh, W., Fingerhut, B., and Kirsch, E.: Adaptation of an alkaline phosphatase method for automatic colorimetric analysis, Clin. Chem. **5**:119, 1959.
34. Powell, M. E. A., and Smith, M. J. H.: The determination of serum acid and alkaline phosphatase activity with 4-amino-antipyrine (A.A.P.), J. Clin. Path. **7**:245, 1954.
35. Kind, P. R. N., and King, E. J.: Estimation of plasma phosphatase by determination of hydrolyzed phenol with aminoantipyrine, J. Clin. Path. **7**:322, 1954.

Lactic dehydrogenase

36. LDH, Technicon Laboratory method file N-60.
37. Hochella, N. J., and Weinhouse, S.: Automated assay of lactate dehydrogenase in urine, Anal. Biochem. **13**:322, 1965.

Lactic dehydrogenase and isoenzyme index

38. Babson, A. L., and Phillips, G. E.: A rapid colorimetric assay for serum lactic dehydrogenase, Clin. Chim. Acta **12**:210, 1965.
39. Babson, A. L.: The chemical differentiation of tissue lactic dehydrogenase, Clin. Chim. Acta **16**:121, 1967.
40. Rice, W.: Comparison of methods for LDH determination, presented at the 21st National Meeting, American Association of Clinical Chemists, Aug., 1969.

Cholesterol

41. Cholesterol (direct), Technicon Laboratory method file N-77.
42. Huang, T. C., Chen, C. P., Wefler, V., and Raftery, A.: A stable reagent for the Liebermann-Burchard reaction, Anal. Chem. **33**: 1405, 1961.
43. Levine, J., Morgenstern, S., and Vlastelica, D.: A direct Liebermann-Burchard method for serum cholesterol, Technicon Symposia 1967, New York, 1968, Mediad, Inc., pp. 25-28.

Bilirubin

44. Simultaneous direct and total bilirubin, Technicon Laboratory method file N-51a.
45. Gambino, S. R., and Schreiber, H.: The measurement and fractionation of bilirubin on the AutoAnalyzer by the method of Jendrassik and Grof, presented at Technicon International Symposium, Automation in Analytical Chemistry, no. 54, New York, 1964.
46. Jendrassik, L., and Grof, P.: Vereinfachte photometrische Methoden zur Bestimmung des Blutbilirubins, Z. Biochem. **297**:81, 1938.

Total protein and albumin

47. Weichselbaum, T. E.: Accurate and rapid method for the determination of proteins in small amounts of blood serum and plasma, Amer. J. Clin. Path. tech. sec. **10**:40, 1946.
48. Stevens, D. L.: Total protein, Technicon Laboratory method file N-14b.
49. Albumin, Technicon Laboratory method file N-15c.
50. Ness, A. T., Dickerson, H. C., and Pastewka,

J. V.: The determination of human serum albumin by its specific binding of the anionic dye, 2-(4′-hydroxybenzeneazo)-benzoic acid, Clin. Chim. Acta **12**:532, 1965.

51. Nishi, H. H., and Rhodes, A.: An automated procedure for the determination of albumin in human serum, Technicon Symposium, 1965, Automation in Analytical Chemistry, New York, 1966, Mediad Inc., pp. 321-323.

52. Pastewka, J. V., and Ness, A. T.: The suitability of various serum albumin products as standards for the quantitative analysis of total protein and albumin in human body fluids, Clin. Chim. Acta **12**:523, 1965.

• 5

• Robot Chemist methods

NOTES ON GENERAL OPERATING PROCEDURE

While preparing the initial setup of the instrument, open each pump to full capacity and cycle deionized water several times in order to flush out bubbles and to rinse the system. After the system has been primed, return the micrometer dial to the desired setting for analysis.

Check the operation of the pumps. If the instrument is not used daily, a pump will occasionally stick. The best prevention is to cycle water through the pumps daily. The pumps have a design of positive displacement that is controlled by the positive stops located in the micrometer barrel. When air pressure forces the aluminum piston up or down, a linkage attached to another piston in the cylinder moves that piston up or down within the limits of the stop. The metal stem attached to the linkage extends up behind the micrometer barrel. Taking a hemostat and carefully moving the stem up or down will usually loosen the pump's sticking or *frozen* condition. Detailed information on the operation of the Robot Chemist has been included in a recent book on automation for the clinical laboratory.[1]

Robot Chemist daily operational checklist

The following *start up* and *shutdown* procedures encompass those daily maintenance checks that will help ensure operational integrity of the Robot Chemist. These suggested procedures can be completed within less than 10 minutes of operator time each day.

Start up procedure

1. Turn on the *main power* circuit breaker.

2. Turn on air and water supplies with shutoff valves. Do not disturb regulator settings. Check both air and water pressure settings. Make sure that pilot lights indicate proper performance of water deionizer units.

3. Turn on the *operate* circuit breaker. The process pushbutton switch should be "off" with process step lights in the "0-1" configuration: the process step being 0 and the sample pickup, 1. If not, move to this configuration, using the *manual advance* switch.

4. Move all *reagent dispenser* pump switches to the "red" position. If a pump fails to actuate, gently lift the pump stem to free it. After pump has actuated, move switch to the *off* position. (NOTE: Micrometer settings may be left at the volume setting normally used. The volume setting must be greater than 0 prior to actuating pump.)

5. Press *process* pushbotton switch to on. Observe action of pipetter pumps. As above, pumps should be set at greater than 0 volume. If a pump fails to actuate at correct process step, lift the pump stem to free it.

6. Allow Robot Chemist to operate for several "process" cycles and check the following:

a. Water flow

(1) Wash wells for transfer pipette. Check that sufficient water flows in each well to fill the reservoir adequately to cover transfer pipette.

(2) Process turntable wash-drain. Check that flow is sufficient to fill reaction tubes to within ¼ inch of the top.

(3) Spectrophotometer probe well. Check that flow is sufficient to maintain water in the probe well with both the supply tubing for the pipetter and also the spectrophotometer probes immersed in the well. The spectrophotometer should be operating as well as the pipetter for this check. Be sure that the tubing coil for the transfer pipette is free of large bubbles.

(4) Process turntable bath. Check that water level of bath is being maintained.

b. Mechanical functioning

(1) Center-lift action. Observe that process turntable center-lift is not obstructed in its travel.

(2) Transfer arm action. Observe that transfer pipette is correctly positioned in both its horizontal and its vertical adjustments.

(3) Spectrophotometer cuvette "rinse-and-fill" action. Observe that spectrophotometer "rocker arms" are functioning smoothly (not necessarily in unison).

(4) Drain pump performance. Observe that drain pump is free of obstructions and is efficiently removing water from its multiple drain lines.

c. Miscellaneous

(1) Temperature control. When temperature of water bath is being controlled, note that heater pilot light remains on until desired temperature is achieved and then indicates, by intermittent *off-on* action, that control is maintained. Note control point on immersion thermometer.

(2) Paper tape supply. See that sufficient paper supply remains in readout printer to meet the usage requirement for the day.

7. Prepare instrument for performance of daily work load.

Shutdown procedure

8. Upon completion of the last test procedure scheduled for the Robot Chemist, place both spectrophotometer probes in the probe well and allow the spectrophotometer to continue to cycle. If a test procedure has been performed durthe day that is known to leave a chemical deposit or film on the walls of the cuvette, rinse cuvettes with a chemical appropriate for removal of the film before placing the probes in the probe well. Any unusual buildup of a chemical film on the cuvette walls should be removed for proper performance of the Robot Chemist. If chemical washing is not satisfactory, the cuvettes should be removed from the spectrophotometer and physically cleaned.

9. Allow the Robot Chemist to continue to operate in the "process" mode until the last "used" reaction tube in the turntable has passed under the wash-drain station.

10. After completion of steps 8 and 9, turn the process pushbutton off when the process step lights reach the "0-1" configuration.

11. Turn the operate circuit breaker off.

12. Turn air and water supplies off, using shutoff valves. Do not disturb regulator settings.

13. Turn the main power circuit breaker off.

14. Clean the top surfaces of the Robot Chemist to remove chemicals that may have been spilled.

Robot Chemist weekly maintenance schedule

1. Check reaction tubes in the process turntable for cleanliness. Certain test procedures will leave chemical films or stains that are not completely removed by the automatic wash-drain station. If the tubes appear to contain a chemical deposit, they should be removed and washed or replaced.

2. Drain the water from the process turntable and refill. Chemicals inadvertently spilled into the bath may initiate a corrosion process if not removed.

3. Remove front covers from reagent dispensers and pipetter and observe pumps in operation. If a leak is observed, it should be sealed by carefully tightening the appropriate tubing connector.

4. Check for proper water flow through the liquid valve "guard rings" of the pipetter, dispensers, and spectrophotometer. This supply should be checked at the appropriately identified drain tubing at the rear of the instrument. Adjust, if necessary, for slow, steady flow.

Robot Chemist extended-term shutdown

Occasionally, conditions necessitate shutdown of the instrument for periods of a week or longer. The following precautions will ensure an expeditious and orderly start up of the Robot Chemist.

1. Flush all reagent dispenser pumps with deionized water followed by a suitable antibacterial solution (such as 0.8% boric acid solution) to prevent bacterial growth. Leave both the reagent inlet and outlet lines immersed in a stock bottle of the antibacterial solution.

2. Remove all reaction tubes from the process turntable and clean thoroughly. Do not replace tubes in the turntable until ready to reactivate the instrument.

3. Drain the water from the process turntable.

4. Turn off water and air supplies at the main supply valves.

5. Disconnect electrical power from the instrument.

6. Inspect all surfaces of the instrument to be sure that no spilled chemicals remain.

7. Cover instrument with a large sheet of plastic film to prevent unnecessary exposure to dust and/or fumes. Allow sufficient venting to prevent moisture condensation.

METHODOLOGY

Urea nitrogen

Chaney-Marbach method[2]

Principle

Serum or plasma is incubated with urease, which hydrolyzes urea to form free ammonia. The ammonia is measured by the catalyzed indophenol reaction, which produces a stable blue color.

Reagents needed

Urease solution, buffered	(D-94)
Phenol color reagent	(D-95)
Dilute 400 ml of D-95 to 1000 ml with water.	
Alkali-hypochlorite reagent	(D-96)
Dilute 400 ml of D-96 to 1000 ml with water.	
Urea stock standard, 1 mg urea N/ml	(D-97)

Working standards

Dilute 1, 2, 3, 4, 6, and 8 ml of stock urea standard to 10 ml with water. These dilutions represent urea nitrogen concentrations of 10, 20, 30, 40, 60, and 80 mg of urea nitrogen/100 ml, respectively. Use the undiluted stock standard as a 100 mg/100 ml standard. Run as serum and prepare a standard curve by plotting absorbance versus concentration on linear graph paper.

Instrument setup

Timer (sec): 30

Pipetter: Probe wash: on Reagent pickup: left—on, right—off

Rack handler: Sampling 1—1

Dispensers:	Left	Right	Readout
Alternate	—	—	—
Every	1—2	3—4	Every

Pipetter	Spectromatic
Reagent, left: Buffered urease	Cuvette size: ½ inch
Reagent, right: None	Wavelength setting: 560 mμ
Sample pump (250 μl): 20 μl	Washes: 3
Reagent intake pump (1.25 ml): 0.30 ml	Reading: Sample vs water
Reagent delivery pump (5.00 ml): 1.30 ml	Zero to: Reagent blank

Turntable

Temperature: 37° C

No.	Slotted plate	Pump size	Volume delivered	Cycle	Pump	Reagent
0		250 μl	20 μl	Every		Sample
		1.25 ml	0.30 ml	Every		Buffered urease
		5.00 ml	1.00 ml	Every		Deionized water
21	Lower	5.00 ml	2.50 ml	Every	B	Phenol color reagent
21	Lower	5.00 ml	2.50 ml	Every	C	Alkali-hypochlorite
23	Upper					Stirrer
36				Every		Spectro sample pickup

Comments

The urea in the sample is hydrolyzed by urease and the ammonia liberated is determined by the Berthelot reaction. Ammonia reacts with an alkaline solution of sodium hypochlorite and phenol, in the presence of sodium nitroprusside as a catalyst, to produce a stable blue indophenol. The concentration of the ammonia is directly proportional to the color intensity of the indophenol.

The addition of EDTA to the buffered urease solution complexes some metals that might inhibit activity of the urease.

At the dilutions used it is not necessary to remove the proteins from the serum or plasma.

This is an extremely sensitive test. To increase the range, read at a wavelength away from the peak of the absorption curve.

Place the Spectromatic reference pickup in deionized water in the turntable wash well.

Reagent dispensers 2, 3, and 4 are not used; therefore, the switch positions are immaterial.

Cholesterol

Huang-Chen-Wefler-Raftery method[3]

Principle

Serum cholesterol is determined by a direct Liebermann-Burchard reaction, using a stable acidic reagent for color development. Reading the sample versus a serum blank corrects for possible interferences caused by the presence of free hemoglobin or bilirubin.

Reagents needed

Cholesterol color reagent	**(D-98)**
Acetic acid, 95%	**(D-99)**

Working standards

Prepare working standards by mixing commercially prepared serum controls in both the normal and elevated ranges. Example:

$$\text{1 part normal} = 180 \text{ mg/100 ml}$$

$$\text{3 parts normal} + \text{1 part elevated} = \frac{3(180) + 520}{4} = 265$$

$$\text{1 part normal} + \text{1 part elevated} = \frac{180 + 520}{2} = 350$$

$$\text{1 part normal} + \text{3 parts elevated} = \frac{180 + 3(520)}{4} = 435$$

$$\text{1 part elevated} = 520$$

Instrument setup

Timer (sec): 30

Pipetter: Probe wash: on Reagent pickup: left—on, right—off

Rack handler: Sampling 2—1

Dispensers:	Left	Right	Readout
Alternate	1	—	Alternate
Every	2	3—4	—

Pipetter	Spectromatic
Reagent, left: Glacial acetic acid	Cuvette size: ½ inch
Reagent, right: None	Wavelength setting: 610 mμ
Sample pump (250 μl): 200 μl	Washes: 3
Reagent intake pump (1.25 ml): 1.00 ml	Reading: Sample vs blank
Reagent delivery pump (5.00 ml): 0.50 ml	Zero to: Reagent blank

Turntable

Temperature: 37° C

No.	Slotted plate	Pump size	Volume delivered	Cycle	Pump	Reagent
0		250 μl	200 μl	Every		Sample
		1.25 ml	0.50 ml	Every		Glacial acetic acid
1	Lower	5.00 ml	4.50 ml	Alternate	B	Color reagent
2	Lower	5.00 ml	4.50 ml	Alternate	C	95% acetic acid
5, 6	Upper					Stirrers
11				Alternate		Spectro sample pickup
12				Alternate		Spectro reference pickup

Comments

A colored product is produced with serum cholesterol upon adding a mixture of glacial acetic acid, acetic anhydride, sulfuric acid, and anhydrous sodium sulfate.

Pumps B and C must be cleaned with glacial acetic acid prior to running the test procedure.

Since reagent dispensers 2, 3, and 4 are not used, the switch positions are immaterial.

Uric acid

Morgenstern-Flor-Kaufman-Klein method[4]

Principle

The difference in the amount of Cu^+ formed by the reaction of an alkanolamine-buffered cupric sulfate neocuproine reagent with uric acid before and after treatment of serum with uricase is used as a measure of uric acid content.

Reagents needed

Glycine buffer	(D-100)
Uricase solution, buffered	(D-101)
Cupric sulfate, 10%	(D-102)
2-Amino-2-methyl-1-propanol, 50%	(D-103)
Neocuproine solution	(D-104)
Cupric sulfate solution, buffered	(D-105)
Uric acid color reagent, combined	(D-106)
Uric acid stock standard, 2 mg/ml	(D-107)

Working standards

Prepare working standards by diluting 1, 2, 3, 4, 5, and 6 ml aliquots of stock standard to 100 ml with 0.9% saline. These dilute standards have concentrations of 2, 4, 6, 8, 10, and 12 mg/100 ml of uric acid. They are stable for about 1 week when stored in a refrigerator.

Instrument setup

Timer (sec): 30
Pipetter: Probe wash: on Reagent pickup: left—on, right—on
Rack handler: Sampling 2—1

Dispensers:	Left	Right	Readout
Alternate	—	—	Alternate
Every	1—2	3—4	

Pipetter	Spectromatic
Reagent, left: Buffered uricase solution	Cuvette size: ½ inch
Reagent, right: Glycine buffer	Wavelength setting: 455 mμ
Sample pump (250 μl): 250 μl	Washes: 3
Reagent intake pump (1.25 ml): 0.85 ml	Reading: Sample vs blank
Reagent delivery pump (5.00 ml): 1.70 ml	Zero to: Reagent blank

Turntable
Temperature: 37° C

No.	Slotted plate	Pump size	Volume delivered	Cycle	Pump	Reagent
0		250 μl	250 μl	Every		Sample
		1.25 ml	0.85 ml	Alternate		Buffered uricase solution
		5.00 ml	0.85 ml	Every		Deionized water
0		250 μl	250 μl	Every		Sample
		1.25 ml	0.85 ml	Alternate		Glycine buffer
		5.00 ml	0.85 ml	Every		Deionized water
34	Lower	5.00 ml	2.00 ml	Every	B	Deionized water
35	Lower	1.25 ml	1.20 ml	Every	A	Combined color reagent
37, 38	Upper					Stirrers
41				Alternate		Spectro sample pickup
42				Alternate		Spectro reference pickup

Comments

A smaller sample may be used but this will decrease the sensitivity of the test.

Reagent dispensers 2, 3, and 4 are not used; therefore, switch positions are immaterial.

Amylase

Caraway method[5]

Principle

Amylase catalyzes the hydrolysis of starch to polysaccharides and to glucose. Serum is incubated at 37° C with a starch substrate for a predetermined period of time. At the end of this time an iodine reagent is added. The formation of a blue-green color indicates the presence of unhydrolyzed starch. The intensity of the color is inversely proportional to the amylase activity.

Reagents needed

Starch substrate	**(D-108)**
Iodine reagent	**(D-109)**

Instrument setup

Timer (sec): 30
Pipetter: Probe wash: on Reagent pickup: left—on, right—off
Rack handler: Sampling 1—1

Dispensers:	Left	Right	Readout
Alternate	—	—	—
Every	1—2	3—4	Every

Pipetter	Spectromatic
Reagent, left: Starch reagent	Cuvette size: ½ inch
Reagent, right: None	Wavelength setting: 620 mμ*
Sample pump (250 μl): 20 μl	Washes: 3
Reagent intake pump (1.25 ml): 1.25 ml	Reading: Sample vs water
Reagent delivery pump (5.00 ml): 1.00 ml	Zero to: Water

Turntable
Temperature: 37° C

No.	Slotted plate	Pump size	Volume delivered	Cycle	Pump	Reagent
0		250 μl	20 μl	Every		Sample
		5.00 ml	1.00 ml	Every		Starch reagent
15	Lower	1.25 ml	1.00 ml	Every		Iodine reagent
		5.00 ml	4.00 ml	Every		Deionized water
17	Upper					Stirrer
20				Every		Spectro sample pickup

*If the 925 phototube is installed, set the wavelength at 660 mμ.

Comments

The amylase unit is defined as the amount of enzyme that will hydrolyze 10 mg of starch in 30 minutes to a stage at which no color is given by iodine.

The amylase units are calculated by using the following formula

$$800 \times \frac{\text{OD of control} - \text{OD of test}}{\text{OD of control}} = \text{Amylase units/100 ml}$$

where OD of control is obtained by using deionized water instead of serum sample.

The "800 factor" is obtained by applying the definition of amylase unit. In this procedure 1.0 ml of reagent containing 0.4 mg of starch is incubated with 20 μl of serum for 7½ minutes. Therefore:

$$\frac{0.4 \text{ mg starch}}{10 \text{ mg starch}} \times \frac{30 \text{ min}}{7.5 \text{ min}} \times \frac{100 \text{ ml}}{20\ \mu\text{l}} \times \frac{1000\ \mu\text{l}}{\text{ml}} = 800$$

For urine amylase, collect a 2-hour sample of urine and measure the volume. Run exactly as described for serum and report as units per hour, using the following:

$$\frac{\text{OD of control} - \text{OD of test}}{\text{OD of control}} \times 4\text{ v} = \text{Amylase units/hr}$$

where v is the volume of the 2-hour urine specimen.

Place the Spectromatic reference pickup in deionized water in the turntable wash well.

Creatine phosphokinase (CPK)

Nuttall-Wedin method[6]

Principle

In this method the forward CPK reaction is coupled to the pyruvate kinase reaction. The pyruvate produced is measured colorimetrically through the formation of the hydrazone with 2,4-dinitrophenylhydrazine.

$$\text{ATP} + \text{Creatine} \underset{}{\overset{\text{CPK}}{\rightleftharpoons}} \text{ADP} + \text{Creatine phosphate}$$

$$\text{ADP} + \text{Phospho-enol-pyruvate} \underset{}{\overset{\text{PK}}{\rightleftharpoons}} \text{Pyruvate} + \text{ATP}$$

Reagents needed

Sodium phosphate, dibasic, 0.1 M	**(D-110)**
Potassium phosphate, monobasic, 0.1 M	**(D-111)**
Phosphate buffer, pH 7.4	**(D-112)**
Pyruvic acid standard, 0.5 μmole/ml	**(D-113)**
CPK blank substrate	**(D-114)**
Creatine substrate	**(D-115)**
2,4-Dinitrophenylhydrazine	**(D-116)**
Mercaptoethanol, 0.25 M	**(D-117)**

Working standards

See manual procedure in Chapter 11.

Instrument setup

Timer (sec): 30

Pipetter: Probe wash: on Reagent pickup: left—on, right—on

Rack handler: Sampling 2—1

Dispensers:	Left	Right	Readout
Alternate	—	—	Alternate
Every	1—2	3—4	

Pipetter	Spectromatic
Reagent, left: CPK blank substrate	Cuvette size: ½ inch
Reagent, right: Creatine substrate	Wavelength setting: 440 mμ
Sample pump (250 μl): 50 μl	Washes: 3
Reagent intake pump (1.25 ml): 0.50 ml	Reading: Sample vs blank
Reagent delivery pump (5.00 ml): 0.50 ml	Zero to: Reagent blank

Turntable

Temperature: 37° C

No.	Slotted plate	Pump size	Volume delivered	Cycle	Pump	Reagent
99		250 μl	50 μl	Every		0.25 M Mercaptoethanol
0		250 μl	50 μl	Every		Sample
		1.25 ml	0.50 ml	Alternate		CPK blank substrate
0		250 μl	50 μl	Every		Sample
		1.25 ml	0.50 ml	Alternate		Creatine substrate
40		5.00 ml	0.50 ml	Every	A	2,4-Dinitrophenylhydrazine
50		5.00 ml	4.50 ml	Every	B	0.4 N NaOH
51, 52						Stirrers
55				Alternate		Spectro sample pickup
56				Alternate		Spectro reference pickup

Alkaline phosphatase

Bessey-Lowry-Brock method[7]

Principle

The substrate used in this procedure is *p*-nitrophenyl phosphate. Alkaline phosphatase catalyzes a reaction in which the phosphate group is split off, leaving free *p*-nitrophenol. The yellow color of the *p*-nitrophenol is measured at 410 mμ.

Reagents needed

Glycine buffer	(D-118)
p-Nitrophenyl phosphate stock substrate	(D-119)
p-Nitrophenyl phosphate working substrate	(D-120)
Hydrochloric acid, 6 N	(D-121)
Sodium hydroxide, 0.0075 N	(D-122)

Working standards

Saline dilutions of a reference serum of known alkaline phosphatase activity are used as standards. Run as serum and prepare a standard curve by plotting absorbance versus phosphatase activity on linear graph paper.

Instrument setup

Timer (sec): 30
Pipetter: Probe wash: on Reagent pickup: left—on, right—off
Rack handler: Sampling 2—1

Dispensers:	Left	Right	Readout
Alternate	1	—	Alternate
Every	2	3—4	—

Pipetter	Spectromatic
Reagent, left: Working substrate	Cuvette size: ½ inch
Reagent, right: None	Wavelength setting: 410 mμ
Sample pump (250 μl): 20 μl	Washes: 3
Reagent intake pump (1.25 ml): 0.20 ml	Reading: Sample vs blank
Reagent delivery pump (5.00 ml): 0.040 ml	Zero to: Reagent blank

Turntable
Temperature: 37° C

No.	Slotted plate	Pump size	Volume delivered	Cycle	Pump	Reagent
0		250 μl	20 μl	Every		Sample
		1.25 ml	0.20 ml	Every		Working substrate
		5.00 ml	0.20 ml	Every		Deionized water
49	Lower	5.00 ml	5.00 ml	Alternate	B	0.0075 N NaOH
50	Lower	5.00 ml	5.00 ml	Alternate	C	6 N HCl
53, 54	Upper					Stirrers
55				Alternate		Spectro sample pickup
56				Alternate		Spectro reference pickup

Comments

Reagent dispensers 2, 3, and 4 are not used and switch positions are immaterial.

Kind-King method[8]

Principle

The substrate disodium phenyl phosphate is hydrolyzed by the action of alkaline phosphatase in serum under specified conditions of pH, temperature, and time. The free phenol produced is measured by rapid condensation with 4-aminoantipyrine followed by oxidation with alkaline potassium ferricyanide to give a red product. The intensity of the red color is measured at 505 mμ.

Reagents needed

Carbonate buffer, pH 10.0	(D-123)
Disodium phenyl phosphate substrate	(D-124)
Sodium hydroxide, 0.5 N	(D-34)
Sodium bicarbonate, 0.5 M	(D-125)
4-Aminoantipyrine, 0.6%	(D-126)
Potassium ferricyanide, 2.4%	(D-127)

Working standards

Dilutions of a reference serum having a known alkaline phosphatase activity are used as standards. Plot absorbance of standards versus enzyme activity on linear graph paper.

Instrument setup

Timer (sec): 30

Pipetter: Probe wash: on Reagent pickup: left—on, right—on

Rack handler: Sampling 2—1

Dispensers:	Left	Right	Readout
Alternate	—	—	Alternate
Every	1—2	3—4	

Pipetter	Spectromatic
Reagent, left: Carbonate buffer	Cuvette size: ½ inch
Reagent, right: Substrate	Wavelength setting: 505 mμ
Sample pump (250 μl): 100 μl	Washes: 3
Reagent intake pump (1.25 ml): 1.20 ml	Reading: Sample vs blank
Reagent delivery pump (5.00 ml): 1.00 ml	Zero to: Reagent blank

Turntable

Temperature: 37° C

No.	Slotted plate	Pump size	Volume delivered	Cycle	Pump	Reagent
98	Lower	1.25 ml	1.00 ml	Every	A	Carbonate buffer
0		1.25 ml	1.00 ml	Alternate		Carbonate buffer
		250 μl	100 μl	Every		Sample
0		1.25 ml	1.00 ml	Alternate		Substrate
		250 μl	100 μl	Every		Sample
30	Lower	1.25 ml	0.80 ml	Every	D	0.5 N NaOH
31	Lower	5.00 ml	1.20 ml	Every	B	0.5 M $NaHCO_3$
32	Lower	1.25 ml	1.00 ml	Every	E	0.6% 4-Aminoantipyrine
33	Lower	5.00 ml	1.00 ml	Every	C	2.4% $K_3Fe(CN)_6$
35, 36	Upper					Stirrers
39				Alternate		Spectro sample pickup
40				Alternate		Spectro reference pickup

Comments

Removal of serum proteins is not necessary, since the aminoantipyrine reagent does not react with protein.

The NaOH and $NaHCO_3$ solutions may be combined into one reagent (2 parts NaOH + 3 parts $NaHCO_3$) and 2.00 ml be delivered from pump B.

Both left and right reagent pickup switches are on, enabling the pipetter to alternate at station *0*.

Since reagent dispensers 3 and 4 are not used, the switch positions are immaterial.

The concentration of ferricyanide may be increased if consistent color fading occurs.

Transaminase—SGOT

Reitman-Frankel method[9]

Principle

Serum glutamic oxalacetic transaminase (SGOT) catalyzes the transfer of an amino group from aspartic acid to α-ketoglutaric acid to form glutamic and oxalacetic acids. Oxalacetic acid reacts with 2,4-dinitrophenylhydrazine to form a brown dinitrophenylhydrazone. The hydrazone color, which is proportional to the oxalacetic acid formed, is measured at 505 mμ.

Reagents needed

SGOT substrate	(D-128)
2,4-Dinitrophenylhydrazine	(D-116)
Sodium hydroxide, 0.4 N	(D-129)

Working standards

Dilutions of a reference serum of known transaminase activity are used as standards.

Instrument setup

Timer (sec): 60
Pipetter: Probe wash: on Reagent pickup: left—off, right—on
Rack handler: Sampling 1—1

Dispensers:	Left	Right	Readout
Alternate	—	—	—
Every	1—2	3—4	Every

Pipetter	Spectromatic
Reagent, left: None	Cuvette size: ½ inch
Reagent, right: SGOT substrate	Wavelength setting: 505 mμ
Sample pump (250 μl): 200 μl	Washes: 3
Reagent intake pump (1.25 ml): 0.50 ml	Reading: Sample vs water
Reagent delivery pump (5.00 ml): 1.00 ml	Zero to: Reagent blank

Turntable
Temperature: 37° C

No.	Slotted plate	Pump size	Volume delivered	Cycle	Pump	Reagent
0		250 μl	200 μl	Every		Sample
		1.25 ml	0.50 ml	Every		SGOT substrate
		5.00 ml	0.50 ml	Every		Deionized water
30	Lower	1.25 ml	0.50 ml	Every	A	2,4-Dinitrophenylhydrazine
31	Upper					Stirrer
40	Lower	5.00 ml	5.00 ml	Every	B	Sodium hydroxide
41	Upper					Stirrer
49				Every		Spectro sample pickup

Comments

Place the Spectromatic reference pickup in deionized water in the turntable wash well.

Reagent dispensers 2, 3, and 4 are not used; hence, the switch positions are immaterial.

Transaminase—SGPT

Reitman-Frankel method[9]

Principle

Serum glutamic pyruvic transaminase catalyzes the transfer of an amino group from alanine to α-ketoglutaric acid to form glutamic and pyruvic acids. Pyruvic acid reacts with 2,4-dinitrophenylhydrazine to form its hydrazone. The hydrazone color, which is proportional to the pyruvic acid formed, is measured at 505 mμ.

Reagents needed

SGPT substrate	**(D-130)**
2,4-Dinitrophenylhydrazine	**(D-116)**
Sodium hydroxide, 0.4 N	**(D-129)**

Working standards

Dilutions of a reference serum of known transaminase activity are used as standards.

Instrument setup

Timer (sec): 60
Pipetter: Probe wash: on Reagent pickup: left—on, right—off
Rack handler: Sampling 1—1

Dispensers:	Left	Right	Readout
Alternate	—	—	—
Every	1—2	3—4	Every

Pipetter	Spectromatic
Reagent, left: SGPT substrate	Cuvette size: ½ inch
Reagent, right: None	Wavelength setting: 505 mμ
Sample pump (250 μl): 100 μl	Washes: 3
Reagent intake pump (1.25 ml): 0.50 ml	Reading: Sample vs water
Reagent delivery pump (5.00 ml): 0.50 ml	Zero to: Reagent blank

Turntable
Temperature: 37° C

No.	Slotted plate	Pump size	Volume delivered	Cycle	Pump	Reagent
0		250 μl	100 μl	Every		Sample
		1.25 ml	0.50 ml	Every		SGPT substrate
30	Lower	1.25 ml	0.50 ml	Every		2,4-Dinitrophenylhydrazine
31	Upper					Stirrer
40	Lower	5.00 ml	5.00 ml	Every		0.4 N NaOH
41	Upper					Stirrer
45				Every		Spectro sample pickup

Comments

Place the Spectromatic reference pickup in deionized water in the turntable wash well.

Reagent dispensers 2, 3, and 4 are not used; hence, the switch positions are immaterial.

Simultaneous SGOT and SGPT

Reitman-Frankel method[9]

In this procedure the same sample is used for simultaneous determinations of SGOT and SGPT. The method is a combination of the ones described for the individual measurement of the enzymes.

Instrument setup

Timer (sec): 60
Pipetter: Probe wash: on Reagent pickup: left—on, right—on
Rack handler: Sampling 2—1

Dispensers:	Left	Right	Readout
Alternate	—	—	—
Every	1—2	3—4	Every

Pipetter	Spectromatic
Reagent, left: SGOT substrate	Cuvette size: ½ inch
Reagent, right: SGPT substrate	Wavelength setting: 505 mμ
Sample pump (250 μl): 100 μl	Washes: 3
Reagent intake pump (1.25 ml): 0.50 ml	Reading: Sample vs water
Reagent delivery pump (5.00 ml): 0.50 ml	Zero to: Reagent blank

Turntable
Temperature: 37° C

No.	Slotted plate	Pump size	Volume delivered	Cycle	Pump	Reagent
0		250 μl	100 μl	Every		Sample
		1.25 ml	0.50 ml	Alternate		SGOT substrate
0		250 μl	100 μl	Every		Sample
		1.25 ml	0.50 ml	Alternate		SGPT substrate
30	Lower	1.25 ml	0.50 ml	Every	A	2,4–Dinitrophenylhydrazine
31	Upper					Stirrer
40	Lower	5.00 ml	5.00 ml	Every	B	0.4 N NaOH
41	Upper					Stirrer
45						Spectro sample pickup

Comments

Place the Spectromatic reference pickup in the turntable wash well.

Reagent dispensers 2, 3, and 4 are not used; hence, the switch positions are immaterial.

Lactic dehydrogenase

Babson-Phillips method[10]

Principle

Lactic dehydrogenase (LDH) catalyzes the oxidation of lactate to pyruvate. In the same reaction NAD is reduced to NADH. The reduction of NAD is coupled to the reduction of a tetrazolium salt, 2-*p*-iodophenyl-3-*p*-nitrophenyl-5-phenyl tetrazolium chloride, with phenazine methosulfate serving as an intermediate electron carrier.

$$\text{Lactate} + \text{NAD} \xrightleftharpoons{\text{LDH}} \text{NADH} + \text{Pyruvate}$$

$$\text{NADH} + \text{PMS} \longrightarrow \text{NAD} + \text{Reduced PMS}$$

$$\text{Reduced PMS} + \text{INT} \longrightarrow \text{PMS} + \text{Reduced INT (red color)}$$

Reduction of the INT dye produces a highly colored, bright red formazan.

Reagents needed

LDH color reagent	(D-72)
LDH buffer reagent	(D-73)
LDH substrate	(D-74)
LDH control reagent	(D-75)
Hydrochloric acid, 0.1 N	(D-37)

Working standards

Saline dilutions of a serum control with elevated LDH activity are used as working standards.

Instrument setup

Timer (sec): 30
Pipetter: Probe wash: on Reagent pickup: left—on, right—on
Rack handler: Sampling 2—1

Dispensers:	Left	Right	Readout
Alternate	—	—	Alternate
Every	1—2	3—4	

Pipetter	Spectromatic
Reagent, left: Control reagent	Cuvette size: ½ inch
Reagent, right: Substrate	Wavelength setting: 520 mμ
Sample pump (250 μl): 100 μl	Washes: 3
Reagent intake pump (1.25 ml): 0.50 ml	Reading: Sample vs blank
Reagent delivery pump (5.00 ml): 1.50 ml	Zero to: Reagent blank

Turntable

Temperature: 37° C

No.	Slotted plate	Pump size	Volume delivered	Cycle	Pump	Reagent
99	Lower	1.25 ml	0.20 ml	Every	D	Tris buffer
0		250 μl	100 μl	Every		Sample
		1.25 ml	0.50 ml	Alternate		Control reagent
		5.00 ml	1.00 ml	Every		Deionized water
0		250 μl	100 μl	Every		Sample
		1.25 ml	0.50 ml	Alternate		Substrate
		5.00 ml	1.00 ml	Every		Deionized water
8	Lower	1.25 ml	0.20 ml	Every	E	Color reagent
9	Upper					Stirrer
18	Lower	5.00 ml	4.00 ml	Every	F	0.1 N HCl
19	Upper					Stirrer
21				Alternate		Spectro sample pickup
22				Alternate		Spectro reference pickup

Comments

The color reagent is very light-sensitive and should be delivered through dark-colored reagent intake and delivery lines.

Both left and right reagent pickup switches are on, enabling the pipetter to alternate at station *0*.

Since reagent dispensers 1, 3, and 4 are not in use, the switch positions are immaterial.

LDH isoenzyme index

Babson method[11]

Principle

The principle involved in this test is the same as that described in chemical fractionation of lactate dehydrogenase in Chapter 11. A 2 M *l*(+) lactate substrate is used as a heart LDH–inhibiting substrate, and a solution containing 0.02 M *l*(+) lactate–2 M urea is used as a liver LDH–inhibiting substrate. The absorbance of a sample reaction with heart LDH–inhibiting substrate divided by the absorbance of a sample reaction with liver LDH–inhibiting substrate gives an isoenzyme index. A normal index range of 0.8 to 1.2 has been suggested for human serum. An index of less than 0.8 along with a elevated total LDH activity indicates the presence of an elevated level of LDH_1 (heart). An index greater than 1.2 along with an elevated total LDH activity indicates the presence of an elevated level of LDH_5 (liver).

Reagents needed

LDH color reagent	(D-72)
LDH buffer reagent (Tris buffer)	(D-73)
$l(+)$ Lactate, 2 M	(D-76)
$l(+)$ Lactate, 0.02 M–2 M urea	(D-77)
Hydrochloric acid, 0.1 N	(D-37)

Instrument setup

Timer (sec): 30
Pipetter: Probe wash: on Reagent pickup: left—on, right—on
Rack handler: Sampling

Dispensers:	Left	Right	Readout
Alternate	—	—	—
Every	1—2	3—4	Every

Pipetter	Spectromatic
Reagent, left: 2 M l (+)-lactate	Cuvette size: ½ inch
Reagent, right: 0.02 M l (+)-lactate–2 M urea	Wavelength setting: 520 mμ
Sample pump (250 μl): 100 μl	Washes: 3
Reagent intake pump (1.25 ml): 0.5 ml	Reading: Sample vs water
Reagent delivery pump (5.00 ml): 1.5 ml	Zero to: Water

Turntable
Temperature: 37° C

No.	Slotted plate	Pump size	Volume delivered	Cycle	Pump	Reagent
99	Lower	1.25 ml	0.20 ml	Every	D	Tris buffer
0		250 μl	100 μl	Every		Sample
		1.25 ml	0.50 ml	Alternate		2 M l (+) lactate
		5.00 ml	0.90 ml	Every		Deionized water
0		250 μl	100 μl	Every		Sample
		1.25 ml	0.50 ml	Alternate		0.02 M l (+) lactate–2 M urea
		5.00 ml	0.90 ml	Every		Deionized water
20		1.25 ml	0.20 ml	Every	E	Color reagent
22						Stirrer
30		5.00 ml	4.00 ml	Every	F	0.1 N HCl
32						Stirrer
36						Spectro probe

Bilirubin

Jendrassik-Grof method[12]

Principle

Serum is mixed with diazotized sulfanilic acid which, in conjunction with a coupling reagent, forms a red compound with bilirubin called azobilirubin. When an alkali is added, the red azobilirubin is converted to the alkaline or blue form. The method produces a stable color and has excellent sensitivity, linearity, and accuracy for conjugated bilirubin. It is insensitive to variations in sample pH, protein content, and hemoglobin concentration.

Reagents needed

Caffeine reagent	(D-79)
Sulfanilic acid, 1% (diazo I)	(D-80)
Sodium nitrite, 0.5% (diazo II)	(D-81)
Diazo reagent, combined	(D-82)
Tartrate buffer	(D-83)
Ascorbic acid, 4%	(D-84)

Working standards

Saline dilutions of a control serum with an elevated bilirubin concentration are used as working standards. Make dilutions just prior to use.

Instrument setup

Timer (sec): 30
Pipetter: Probe wash: on Reagent pickup: left—on, right—on
Rack handler: Sampling 2—1

Dispensers:	Left	Right	Readout
Alternate	—	—	Alternate
Every	1—2	3—4	—

Pipetter	Spectromatic
Reagent, left: Diazo blank	Cuvette size: ½ inch
Reagent, right: Diazo reagent	Wavelength setting: 600 mμ
Sample pump (250 μl): 250 μl	Washes: 3
Reagent intake pump (1.25 ml): 0.75 ml	Reading: Sample vs blank
Reagent delivery pump (5.00 ml): 1.25 ml	Zero to: Reagent blank

Turntable
Temperature: 37° C

No.	Slotted plate	Pump size	Volume delivered	Cycle	Pump	Reagent
0		250 μl	250 μl	Every		Sample
		1.25 ml	0.75 ml	Alternate		Diazo blank
		5.00 ml	0.50 ml	Every		Deionized water
0		250 μl	250 μl	Every		Sample
		1.25 ml	0.75 ml	Alternate		Diazo reagent
		5.00 ml	0.50 ml	Every		Deionized water
2	Lower	5.00 ml	2.00 ml	Every	F	Caffeine reagent
20	Lower	1.25 ml	0.25 ml	Every	D	Ascorbic acid
21	Lower	1.25 ml	1.25 ml	Every	E	Tartrate buffer
24, 25	Upper					Stirrers
29				Alternate		Spectro sample pickup
30				Alternate		Spectro reference pickup

Comments

Since bilirubin is light-sensitive, the samples should be kept from direct light.

Because reagent dispensers 1, 3, and 4 are not used, the switch positions are immaterial.

Malloy-Evelyn method[13]

Principle

Bilirubin in serum is reacted with diazotized sulfanilic acid to form a chromogenic compound called azobilirubin. The color intensity is measured at 540 mμ.

Reagents needed

Sulfanilic acid, 0.1%	**(D-131)**
Sodium nitrite, 0.5%	**(D-81)**
Diazo reagent	**(D-132)**
Diazo blank	**(D-133)**

Working standards

Dilutions of a reference serum (Versatol-A Alternate or Versatol Pediatric, see package insert) are used as working standards. Prepare a standard curve by plotting absorbance versus concentration on linear graph paper.

Instrument setup

Timer (sec): 30
Pipetter: Probe wash: on Reagent pickup: left—on, right—on
Rack handler: Sampling 2—1

Dispensers:	Left	Right	Readout
Alternate	—	—	Alternate
Every	1—2	3—4	—

Pipetter	Spectromatic
Reagent, left: Diazo blank	Cuvette size: ½ inch
Reagent, right: Diazo reagent	Wavelength setting: 540 mμ
Sample pump (250 μl): 200 μl	Washes: 3
Reagent intake pump (1.25 ml): 0.50 ml	Reading: Sample vs blank
Reagent delivery pump (5.00 ml): 2.30 ml	Zero to: Reagent blank

Turntable
Temperature: 37° C

No.	Slotted plate	Pump size	Volume delivered	Cycle	Pump	Reagent
0		250 μl	200 μl	Every		Sample
		1.25 ml	0.50 ml	Alternate		Diazo blank
		5.00 ml	1.80 ml	Every		Deionized water
0		250 μl	200 μl	Every		Sample
		1.25 ml	0.50 ml	Alternate		Diazo reagent
		5.00 ml	1.80 ml	Every		Deionized water
2	Lower	5.00 ml	2.50 ml	Every	B	Methanol, absolute*
4, 5	Upper					Stirrers
31				Alternate		Spectro sample pickup
32				Alternate		Spectro reference pickup

*If foaming becomes a problem, Dow Corning Antifoam B can be added to the methanol.

Comments

Since reagent dispensers 2, 3, and 4 are not used, the switch positions are immaterial.

Because of the known light-sensitivity of bilirubin the analysis should be performed without delay.

Total protein

Henry-Sobel-Berkman method[14]

Principle

This method uses Benedict's qualitative glucose reagent as a biuret reagent for the determination of serum proteins. A reddish blue complex is formed when serum or plasma is treated with biuret reagent. The intensity of the color is directly proportional to the concentration of proteins present.

Reagents needed

Biuret reagent	(D-134)
Sodium hydroxide, 15%	(D-135)
Protein stock standard, 10 g albumin/100 ml	(D-93)

Working standards

Dilute 2, 4, 6, and 8 ml aliquots of stock protein standard to 10 ml with 0.9% saline. These standards represent concentrations of 2, 4, 6, and 8 g of protein/100 ml. Use the undiluted stock as a standard for 10 g of protein/100 ml. Run as serum and prepare a standard curve by plotting absorbance versus concentration on linear graph paper.

Instrument setup

Timer (sec): 30

Pipetter: Probe wash: on Reagent pickup: left—off, right—on

Rack handler: Sampling 1—1

Dispensers:	Left	Right	Readout
Alternate	—	—	—
Every	1—2	3—4	Every

Pipetter	Spectromatic
Reagent, left: None	Cuvette size: ½ inch
Reagent, right: Sodium hydroxide	Wavelength setting: 545 mμ
Sample pump (250 μl): 100 μl	Washes: 3
Reagent intake pump (1.25 ml): 1.00 ml	Reading: Sample vs water
Reagent delivery pump (5.00 ml): 3.75 ml	Zero to: Reagent blank

Turntable

Temperature: 37° C

No.	Slotted plate	Pump size	Volume delivered	Cycle	Pump	Reagent
0		250 μl	100 μl	Every		Sample
		1.25 ml	1.00 ml	Every		NaOH
		5.00 ml	2.75 ml	Every		Deionized water
1	Lower	1.25 ml	1.25 ml	Every	D	Biuret reagent
3	Upper					Stirrer
32				Every		Spectro sample pickup

Comments

Place the Spectromatic reference pickup in deionized water in the turntable wash well.

Reagent dispensers 1, 3, and 4 are not used and the switch position for these is immaterial.

Albumin

HABA dye or Martinek method[15]

Principle

This method is based on the ability of HABA dye to be bound specifically by serum albumin. The color intensity of the reaction product is measured at 510 mμ. Although maximum sensitivity is obtained at about 485 mμ, there is interference from bile pigments at this wavelength. Reading at 510 mμ minimizes these effects.

Reagents needed

Phosphate buffer, pH 6.2 to 6.4	(D-136)
HABA dye stock solution	(D-137)
HABA dye working solution	(D-138)
Albumin stock standard, 10 g/100 ml	(D-93)

Working standards

Dilute 1, 2, 4, 5, and 7 ml aliquots of albumin stock standard to 10 ml with 0.9% saline. These standards represent concentrations of 1, 2, 4, 5, and 7 g of albumin/100 ml. Run as serum and prepare a standard curve by plotting absorbance versus concentration on linear graph paper.

Instrument setup

Timer (sec): 30
Pipetter: Probe wash: on Reagent pickup: left—on, right—off
Rack handler: Sampling 1—1

Dispensers:	Left	Right	Readout
Alternate	—	—	—
Every	1—2	3—4	Every

Pipetter	Spectromatic
Reagent, left: HABA dye	Cuvette size: ½ inch
Reagent, right: None	Wavelength setting: 510 mμ
Sample pump (250 μl): 100 μl	Washes: 3
Reagent intake pump (1.25 ml): 1.00 ml	Reading: Sample vs water
Reagent delivery pump (5.00 ml): 5.00 ml	Zero to: Reagent blank*

*Before determining concentrations of unknowns, subtract the reagent blank absorbance from the absorbance of the unknown.

Turntable
Temperature: 37° C

No.	Slotted plate	Pump size	Volume delivered	Cycle	Pump	Reagent
99	Lower	1.25 ml	1.00 ml	Every	A	HABA dye
0		250 μl	100 μl	Every		Sample
		5.00 ml	4.00 ml	Every		Deionized water
2	Upper					Stirrer
4	Upper			Every		Spectro sample pickup

Comments

Place Spectromatic reference pickup in deionized water in turntable wash well.

Since reagent dispensers 2, 3, and 4 are not used, the switch positions are immaterial.

REFERENCES

General operating procedures

1. White, W. L., Erickson, M. M., and Stevens, S. C.: Practical automation for the clinical laboratory, St. Louis, 1968, The C. V. Mosby Co.

Urea nitrogen

2. Chaney, A. L., and Marbach, E. P.: Modified reagents for determination of urea and ammonia, Clin. Chem. **8**:130, 1962.

Cholesterol

3. Huang, T. C., Chen, C. P., Wefler, V., and Raftery, A.: A stable reagent for the Liebermann-Burchard reaction, Anal. Chem. **33**:1405, 1961.

Uric acid

4. Morgenstern, S., Flor, R. V., Kaufman, J. H., and Klein, B.: The automated determination of serum uric acid, Clin. Chem. **12**:748, 1966.

Amylase

5. Caraway, W. T.: A stable starch substrate for the determination of amylase in serum and other body fluids, Amer. J. Clin. Path. **32**:97, 1959.

Creatine phosphokinase (CPK)

6. Nuttall, F. Q., and Wedin, D. S.: A simple rapid colorimetric method for determination of creatine kinase activity, J. Lab. Clin. Med. **68**:324, 1966.

Alkaline phosphatase

7. Bessey, O. A., Lowry, O. H., and Brock, M. J.: Method for rapid determination of alkaline phosphatase with 5 cubic millimeters of serum, J. Biol. Chem. **164**:321, 1946.
8. Kind, P. R. N., and King, E. J.: Estimation of plasma phosphatase by determination of hydrolysed phenol with amino-antipyrine, J. Clin. Path. **7**:322, 1954.

Transaminases

9. Reitman, S., and Frankel, S.: A colorimetric method for the determination of serum glutamic oxaloacetic and glutamic pyruvic transaminases, Amer. J. Clin. Path. **28**:56, 1957.

Lactic dehydrogenase

10. Babson, A. L., and Phillips, G. E.: A rapid colorimetric assay for serum lactic dehydrogenase, Clin. Chim. Acta **12**:210, 1965.

LDH isoenzyme index

11. Babson, A. L.: The chemical differentiation of tissue lactic dehydrogenase, Clin. Chim. Acta **16**:121, 1967.

Bilirubin

12. Jendrassik, L., and Grof, P.: Vereinfachte photometrische Methoden zur Bestimmung des Blut-bilirubins, Z. Biochem. **297**:81, 1938.
13. Malloy, H. T., and Evelyn, K. A.: The determination of bilirubin with the photoelectric colorimeter, J. Biol. Chem. **119**:481, 1937.

Total protein

14. Henry, R. J., Sobel, C., and Berkman, S.: Interferences with biuret methods for serum proteins; use of Benedict's qualitative glucose reagent as a biuret reagent, Anal. Chem. **29**:1491, 1957.

Albumin

15. Martinek, R. G.: Evaluation of a dye-binding method for determination of serum albumin, Clin. Chem. **11**:441, 1965.

• 6

• Glucose and nonprotein nitrogen compounds

Glucose

BLOOD GLUCOSE

Glucose, the Greek word for sweet, is one of the simple sugars that are the products of carbohydrate digestion. Upon leaving the small intestine, glucose enters the bloodstream and is carried to the liver. There it may be stored as glycogen, or it may be returned to the blood and oxidized as fuel for bodily activities.

The normal values for glucose depend upon the method of analysis. With the early methods, glucose and saccharoids (nonglucose substances) could not be separated. Consequently, the glucose values included saccharoids.

With the later methods of analysis, however, it was possible to separate glucose from saccharoids and thus determine glucose independently. It was then referred to as true glucose.

The normal values for the early methods of analysis—glucose plus saccharoids—are 80 to 120 mg/100 ml of blood. The normal values for the later methods of analysis—true glucose—are 65 to 95 mg/100 ml of blood.

When the glucose level rises above the normal range, the condition is known as hyperglycemia. When the glucose level falls below that range, the condition is referred to as hypoglycemia. Table 6-1 lists conditions in which an increased or a decreased glucose value may be expected. The diseases given in italics are of primary interest to the physician; the others are of somewhat lesser concern.

Insulin, a hormone produced by the pancreas, helps control the glucose level of blood. In diabetes, sufficient amounts of insulin are not produced, and the glucose level rises and some is excreted in the urine. The presence of excessive glucose in the urine is known as glycosuria.

The blood level at which glycosuria occurs is usually 160 to 180 mg/100 ml. This is known as the renal threshold value.

Some individuals have a low renal threshold value, and glycosuria occurs in the absence of any pathologic condition. This is referred to as renal glycosuria or "renal diabetes." And sometimes glycosuria will occur after a meal that is rich in carbohydrates. This is called alimentary glycosuria.

If glycosuria occurs and the physician suspects diabetes, he may request a glucose tolerance test. Glucose is given to the patient and glucose levels are measured at specific intervals. The diabetic takes longer to remove the sugar from the blood than a normal individual.

The more commonly used methods for the determination of glucose are discussed below.

Folin-Wu method

An alkaline copper solution is added to a protein-free filtrate. The glucose present reduces the copper to cuprous oxide. The cuprous oxide in turn reduces

Table 6-1. Conditions accompanied by abnormal glucose values

Increased glucose	Decreased glucose
Diabetes	*Hyperinsulinism*
Hyperthyroidism	Hypothyroidism
Hyperpituitarism	Hypopituitarism
Nephritis	Hepatic disease
Coronary thrombosis	Addison's disease
Infections	Pernicious vomiting
Pregnancy	Starvation
Uremia	

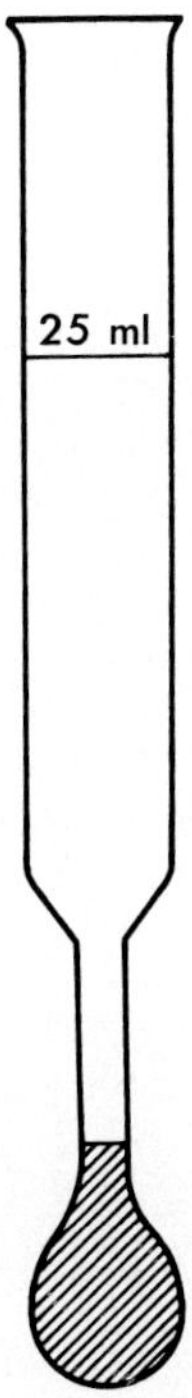

Fig. 6-1. Folin-Wu sugar tube.

a colorless phosphomolybd*ic* acid solution to blue phosphomolybd*ous* acid. The depth of color is measured in a colorimeter. The normal values for this method are 80 to 120 mg/100 ml of blood.

Benedict method

A special bisulfite copper solution is used that increases the specificity of the test for glucose at the expense of the saccharoids. This solution is added to a protein-free filtrate and the glucose reduces the copper to cuprous oxide. The cuprous oxide in turn reduces a colorless phosphomolybd*ic* acid solution to blue phosphomolybd*ous* acid. The depth of color is measured in a colorimeter. The normal values are 70 to 100 mg/100 ml of blood.

Somogyi-Nelson method

A Somogyi protein-free filtrate that removes reducing substances other than sugar is made. A copper tartrate reagent is added to the filtrate and the glucose reduces the copper to cuprous oxide. The cuprous oxide then reacts with an arsenomolybdate reagent to give a blue color. The depth of color is measured in a colorimeter. The normal values are 65 to 95 mg/100 ml of blood.

Glucose oxidase method (modified)

This is an enzymatic method that will respond to glucose only. No other physiologic constituent is measured. Only a Somogyi protein-free filtrate, which is free of metallic ions, can be used. The tungstic acid filtrate cannot be used with this method. The enzyme, glucose oxidase, catalyzes (or promotes) the oxidation of glucose with oxygen to gluconic acid and hydrogen peroxide. The hydrogen peroxide is detected by a chromogenic oxygen acceptor in the presence of peroxidase (horseradish). The normal range is 65 to 95 mg/100 ml.

Procedures for glucose determination

The following general considerations pertain to the procedure for glucose tests:

1. Specimen: whole blood, serum, or plasma.
2. Preservation of specimen: glucose disappears rapidly from whole blood on standing, but it may be stored for a few hours in a refrigerator; serum and plasma are more stable; filtrate may be stored overnight in a refrigerator if a few drops of toluene are added.
3. Some problems with glucose: contamination with cotton or paper cellulose (anthrone type method); delay in processing sample; nonspecific methods of analysis; interference by glutathione or ascorbic acid (glucose oxidase method).

Measurement of glucose

Somogyi-Nelson method[16,17]

Reagents needed

Alkaline copper reagent	(D-139)
Arsenomolybdate color reagent	(D-140)
Benzoic acid, 0.2%	(D-141)
Stock standard, 1.0 mg/ml	(D-142)

Working standard, 0.10 mg/ml

Pipette 10.0 ml of stock standard into a 100 ml volumetric flask and dilute to mark with 0.2% benzoic acid.

Procedure

Prepare a Somogyi protein-free filtrate. The filtrate *must* be crystal-clear.

Label and prepare Folin-Wu sugar tubes (Fig. 6-1) as described below:

Blank (1 tube)

2.0 ml of distilled water. Use volumetric pipette.

Standards (3 tubes)

1.0 ml of working standard and 1.0 ml of distilled water into three tubes. Use volumetric pipette.

Unknown (1 tube per test)

Pipette 1.0 ml of filtrate and 1.0 ml of distilled water into numbered sugar tube. Use volumetric pipette.

To all tubes:

1. Add 2.0 ml of alkaline copper reagent. Shake to mix.
2. Place into boiling water bath for 20 minutes. Time by clock.
3. At the end of 20 minutes, place the tubes into room-temperature water bath to cool. (CAUTION: Do not agitate the tubes when moving from the boiling water bath to the cooling bath. Agitation causes oxidation of cuprous oxide.)
4. When cool, add 2.0 ml of the arsenomolybdate reagent.
5. Mix well by grasping the top of the tube with the right hand and thumping the bottom of the tube against the palm of the left hand. Allow the tubes to stand for a few minutes and then mix again.
6. Add distilled water to the 25 ml mark *exactly*. Observe the meniscus! If over the mark, repeat the test!
7. Stopper and invert five times for a thorough mixing.
8. Transfer a portion of the solution to a cuvette and read against the blank at 540 millimicrons ($m\mu$). Read percent transmittance and convert to optical density or absorbance (A) (Table A-7, p. 621) or read directly optical density. Average the standards for the calculation of the glucose.

Calculations with an example

$$\frac{\text{A of unknown}}{\text{A of standard}} \times \text{mg standard} \times 1000 = \text{mg glucose/100 ml}$$

Standard readings (A) 0.310
0.310
0.309
0.929 ÷ 3 = 0.3096 or 0.310

Unknown reading (A) 0.365

$$\frac{0.365}{0.310} \times 0.1 \times 1000 = 117.7 \text{ or } 118 \text{ mg glucose/100 ml}$$

Somogyi-Nelson micromethod[16,17]

This method is the same as the preceding method of glucose determination by Somogyi-Nelson, with the exception that it is scaled down to small quantities so that *extreme accuracy* must be used. This method is mostly used on infants or patients whose veins are difficult to locate. The normal values are 70 to 100 mg/100 ml with capillary blood.

Reagents needed

Same as for the previous Somogyi-Nelson method

Working glucose standard (1.0 ml = 0.05 mg)

Pipette into a 100 ml volumetric flask 5.0 ml of stock standard. Dilute to mark with 0.2% benzoic acid.

Methods of collecting the blood

1. Finger, heel, or earlobe puncture: A pipette rinsed with heparin may be used and blood be pipetted directly from the puncture wound. The filtrate must be made immediately, since there is no fluoride protection. The glucose values run about 3% higher in the normal range than with the venipuncture. The elevated range may differ as much as 20 to 40 mg/100 ml.[15]
2. Venipuncture: A venipuncture may be made and the small amount of blood may be placed in a Kahn tube or small beaker containing small amounts of solid oxalate and fluoride.

Procedure

1. Pipette 0.2 ml of blood into a small test tube containing 1.0 ml of distilled water.
2. Add 0.4 ml of zinc sulfate. Mix thoroughly by tapping the tube with the finger.
3. Add 0.4 ml of barium hydroxide. Mix thoroughly by tapping the tube with the finger.
4. Let the tubes stand for 5 minutes.
5. Centrifuge for 10 minutes.
6. Pipette 0.5 ml of filtrate and 1.5 ml of distilled water into a Folin-Wu sugar tube.
7. Pipette 1.0 ml of standard (0.05 mg/ml) and 1.0 ml of distilled water into 3 Folin-Wu sugar tubes.
8. Pipette 2.0 ml of distilled water into a sugar tube for the blank.
9. The remainder of the test procedure is the same as that used in the Somogyi-Nelson method.

Calculations

$$\frac{\text{A of unknown}}{\text{A of standard}} \times \text{mg standard} \times 2000 = \text{mg glucose/100 ml}$$

Folin-Wu method[9]

An alkaline copper reagent is heated with a Folin-Wu filtrate. Cuprous oxide produced by the reaction of cupric hydroxide and glucose reduces phosphomolybdic acid to phosphomolybdous acid, which is blue. This blue color is compared to a standard solution of glucose. Values obtained using the Folin-Wu filtrate are about 20 to 30 mg/100 ml higher than those obtained using the Somogyi filtrate.

A modification of the Folin-Wu method enables measurement of "true glucose." The modification in this method is the use of the Somogyi filtrate to remove the nonglucose reducing substances. A change from the classic Folin-Wu method may be made by changing *only* the preparation of the protein-free filtrate.

Reagents needed

Alkaline copper reagent (D-143)
Phosphomolybdic acid (D-144)
Alternate phosphomolybdic acid (D-145)
This reagent may be used to reduce fading, instead of phosphomolybdic acid listed here as D-144; 3.0 ml should be used instead of 2.0 ml.
Stock standard, 1.0 mg/ml (D-142)
Benzoic acid, 0.2% (D-141)
Working standards (use volumetric flasks):

1. For 0.05 mg/ml dilute 5 ml of stock standard to 100 ml with benzoic acid.
2. For 0.10 mg/ml dilute 10 ml of stock standard to 100 ml with benzoic acid.
3. For 0.20 mg/ml dilute 20 ml of stock standard to 100 ml with benzoic acid.

Procedure

Prepare a Folin-Wu filtrate (or a Somogyi filtrate, if doing a true glucose). The filtrate *must* be crystal clear. Label and prepare Folin-Wu tubes as described below:

Blank (1 tube)
2.0 ml of distilled water

Standards (3 tubes)
2.0 ml of working standard (0.05 mg/ml)
2.0 ml of working standard (0.10 mg/ml)
2.0 ml of working standard (0.20 mg/ml)

Unknown (1 tube per test)
2.0 ml of filtrate into each tube.

To all tubes:

1. Add 2.0 ml of alkaline copper. Mix.
2. Place tubes into boiling water bath for 8 minutes. Time.
3. After 8 minutes, place tubes into room-temperature water bath to cool. Do not disturb the precipitate.
4. Add 2.0 ml of the phosphomolybdic acid. Mix. (*For alternate method:* Add 3.0 ml of alternate phosphomolybdic acid. Mix. This stabilizes the color, eliminating steps 5 and 6.)
5. To stabilize the color place the tubes into the boiling water bath again for 2 minutes. Time.
6. At the end of 2 minutes, return to the room-temperature water bath to cool.
7. Dilute to the 25 ml mark with distilled water. Stopper and mix thoroughly by inversion.
8. Pour into cuvette and read against the blank at 420 mμ.

Calculations

$$\frac{\text{A of unknown}}{\text{A of standard}} \times \text{mg standard} \times 500 = \text{mg glucose/100 ml}$$

NOTE: The standards are made so that a 2.0 ml volume will contain the following mg concentration:

1. (0.05 mg/ml) 0.1 mg equivalent to 50 mg/100 ml
2. (0.10 mg/ml) 0.2 mg equivalent to 100 mg/100 ml
3. (0.20 mg/ml) 0.4 mg equivalent to 200 mg/100 ml

Calculate the glucose, using the standard that reads closest to the unknown.

Dextrostix method

Dextrostix,* a rapid screening test, is a strip that may be used for measurement of glucose in a drop of whole blood. The test is based on the action of the enzyme glucose oxidase, which catalyzes the oxidation of glucose to gluconic acid and hydrogen peroxide. The hydrogen peroxide in the presence of peroxidase oxidizes the chromogen system, which is related to the amount of glucose present in the blood. The test is not intended to replace quantitative procedures.

Procedure

1. Place a large drop of whole blood (sufficient to cover the entire reagent area) on the reagent area of the printed side of the strip. (A thin film of blood tends to dry and is removed with difficulty during the washing.)
2. Allow the blood the remain on the strip for exactly 60 seconds; time accurately.
3. Quickly wash the blood from the reagent area, using a stream of water from a wash bottle and holding the strip vertically. This requires only 1 or 2 seconds.
4. Promptly (1 or 2 seconds) and decisively match the color with the color chart. Longer periods for matching allow the color of the strip to change giving unsatisfactory results. The system has been devised for whole blood. Do not use serum or plasma as the color shades will not correspond with the color chart.

Reaction

The values on the color chart are 40, 65, 90, 130, 150, 200, and 250 mg of glucose/100 ml of blood with color shades from gray to blue-purple.

*Ames Co., Division of Miles Laboratories, Inc.

SPINAL FLUID GLUCOSE

The normal values for glucose in spinal fluid are approximately two thirds that of the blood glucose or 45 to 93 mg/100 ml of fluid. Increased values may accompany brain tumors. Decreased values are seen in most types of meningitis. If the analysis cannot be done within the hour, either a protein-free filtrate must be made or the spinal fluid may be placed in a test tube containing sodium fluoride as a preservative.

Methods

Either the macromethod or the micromethod of Somogyi-Nelson is satisfactory. The spinal fluid glucose may be determined along with the blood glucose, using the same standards, and the calculations are the same. If tungstic acid is used to prepare a 1:10 filtrate, mix 1 ml of spinal fluid with 8 ml of water and add 0.5 ml of 0.67 N sulfuric acid and 0.5 ml of 10% sodium tungstate.

GLUCOSE IN URINE

Somogyi-Nelson method[1]

A quantitated glucose is occasionally requested for a 24-hour urine specimen. With the following modifications, this may be done by the Somogyi-Nelson method of glucose measurement. NOTE: The Benedict methods (quantitative and qualitative) are described on pp. 378 and 379.

Procedure

1. Measure the 24-hour urine volume.
2. In order to estimate the glucose concentration, do a Benedict semiquantitative test.
3. Differentiate the glucose from lactose as follows:
   a. Wash approximately a half cake of baker's yeast by placing it into approximately 15 ml of distilled water and centrifuging for 10 minutes. Decant the supernatant. Repeat the procedure.
   b. Add equal volumes of washed packed yeast and the urine to a test tube. Mix well either by shaking or by using a stirring rod.
   c. Incubate *unstoppered* for 1 hour at 37° C.
   d. Centrifuge the incubated tube for 10 minutes.
   e. Perform the Benedict test on the supernatant.
   f. Results of the second Benedict test:
      (1) If negative, all of the sugar present in the first test is probably glucose. However, in rare cases you may find fermentable reducing sugars such as maltose or fructose.
      (2) If as positive as the first test, all of the sugar is lactose or some other nonfermentable reducing substance such as various preservatives, galactose, or pentose.
      (3) If positive, but lower results than the original test, both glucose and lactose are present and the difference of the two tests represents the glucose
4. In the same manner used in the Somogyi-Nelson blood glucose method, prepare protein-free filtrates on both the untreated urine and the fermented urine.
5. Perform the glucose method on both filtrates. If the Benedict method shows results of the unfermented urine to be over 2 plus, then use a smaller amount of filtrate. Refer to the directions given for the Somogyi-Nelson method.

Calculations

The calculations are performed in the same way as for the blood glucose.

Since other reducing substances may be in the filtrate, in addition to glucose, the value of the fermented sample is subtracted from the value of the untreated urine to obtain the true concentration of glucose.

The 24-hour glucose excretion, expressed in milligrams, is obtained by the following formula:

$$\text{mg glucose/100 ml} \times \frac{\text{24-hour urine volume in ml}}{100} = \text{mg/24 hr}$$

GLUCOSE TOLERANCE

When glucose is given to a normal individual and to a diabetic, the normal individual removes the glucose at a faster rate than the diabetic. This is the principle of the glucose tolerance test. In the diabetic, the delay in reaction is due to lack of insulin. Criteria for a glucose tolerance test[5] are listed below:

FBS $<$ 100 mg/100 ml	1½ hr $<$ 140 mg/100 ml
1 hr $<$ 160 mg/100 ml	2 hr $<$ 120 mg/100 ml

Many methods for the glucose tolerance tests are used. Brief discussions of three methods follow.

Janney-Isaacson test

In the Janney-Isaacson test, fasting blood and urine specimens are taken. A known amount of glucose is given orally. Blood and urine specimens are then taken at ½, 1, 2, 3, and 4 hours after the ingestion of glucose. Glucose levels are

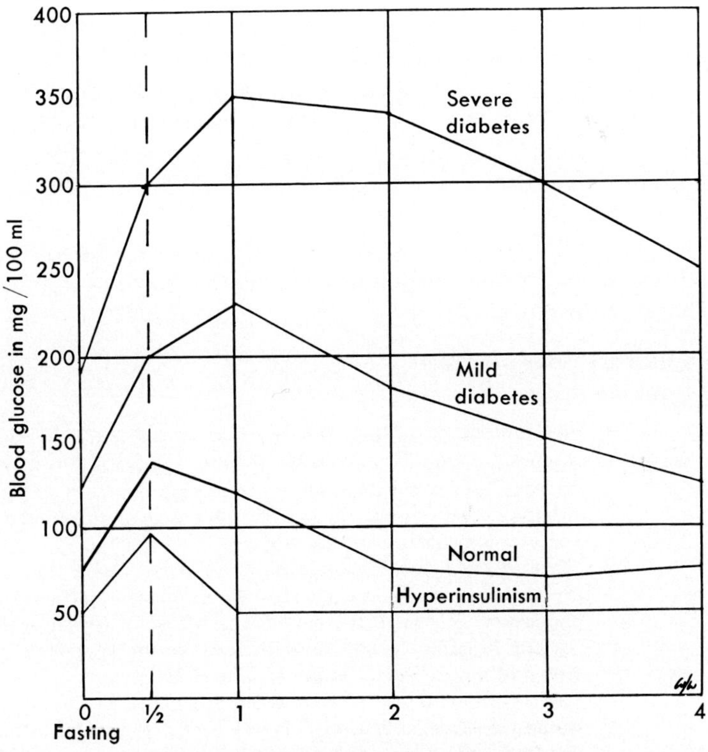

Fig. 6-2. Typical glucose tolerance curves in the Janney-Isaacson test and the intravenous test.

determined on all blood and urine specimens. Normal and abnormal results are indicated in Fig. 6-2.

Intravenous test

In the intravenous test, the following factors are taken into consideration: When glucose is taken by mouth, it is absorbed into the bloodstream from the small intestine. In health, the rate of absorption is fairly uniform. In diseases of the stomach and intestine, however, the rate of absorption may be altered. This would affect the results of the glucose tolerance test. The situation may be remedied by giving the glucose intravenously.

The intravenous test is performed as follows: Fasting blood and urine specimens are taken. Glucose is given intravenously by a physician. Blood and urine specimens are again taken at ½, 1, 2, and 3 hours after the injection. Glucose tests are run on all blood and urine specimens. Normal and abnormal results are indicated in Fig. 6-2.

Exton-Rose test

In the Exton-Rose test, fasting blood and urine specimens are taken. A known amount of glucose is given. One half hour later, blood and urine specimens are taken. A known amount of glucose is again given. Blood and urine specimens are again taken at ½ hour. Glucose levels are determined on all blood and urine specimens.

In a normal person, (1) the fasting blood specimen is normal, (2) the ½-hour blood specimen does not exceed the fasting blood specimen by more than 75 mg, (3) the 1-hour blood specimen does not exceed the ½-hour specimen by more than 10 mg, and (4) all urines are negative for glucose.

In the diabetic, the 1-hour blood specimen exceeds the ½-hour specimen by 10 mg or more, and the urines are usually positive for glucose.

In renal glycosuria, the blood glucose follows the normal curve, but the urine specimens are positive for glucose.

In alimentary glycosuria, the blood glucose follows the normal curve, the first and second urine specimens are negative for glucose, but the third specimen is positive for glucose.

Procedures for glucose tolerance tests

Janney-Isaacson test

1. Instructions to patient: The patient is instructed by the physician to maintain a normal carbohydrate diet (at least 300 g of carbohydrate per day) for three days prior to the test. The test is performed in the morning, and the patient is told not to eat or drink anything following the evening meal of the previous day.
2. Prepare the glucose solution as follows: Using the rough balance, weigh 200 g of glucose (dextrose) into a container and add 500 ml of tap water. Stir to dissolve. Place solution in a refrigerator for at least 1 hour in order to chill and thus make more palatable. Just prior to use, add a little lemon juice for flavoring.
3. Collect about 5 ml of venous blood, place in an oxalated tube containing fluoride and invert several times to dissolve the oxalate. Label the test tube: Fasting blood sugar. Collect a urine specimen and label: Fasting specimen. Store this blood specimen and all succeeding blood specimens in a refrigerator until you are ready to run the tests.
4. Obtain the patient's weight. Multiply this number by 2 and give the patient this number in milliliters of the glucose solution. This results

	Blood (mg / 100 ml)	Urine
Fasting specimen	75	Neg.
½-Hour specimen	135	Neg.
1-Hour specimen	120	Neg.
2-Hour specimen	75	Neg.
3-Hour specimen	70	Neg.

Fig. 6-3. Sample report for glucose tolerance test.

in 1.75 g of glucose per kilogram of weight. Example: A patient weighs 100 pounds. Multiplying 100 by 2, we get 200. The patient then receives 200 ml of the glucose solution. If a patient weighs over 250 pounds, simply give him all the glucose solution.

5. Immediately after giving the glucose, note and record the time.
6. Exactly ½ hour after the ingestion of glucose, take another blood and another urine specimen. Label each: ½-hour specimen.
7. Exactly 1 hour after the ingestion of glucose, take another blood and another urine specimen. Label each: 1-hour specimen.
8. Exactly 2 hours after the ingestion of glucose, take another blood and another urine specimen. Label each: 2-hour specimen.
9. Exactly 3 hours after the ingestion of glucose, take another blood and another urine specimen. Label each: 3-hour specimen. If the physician suspects hyperinsulinism, he may request 4- and 5-hour blood specimens. This gives him a more complete picture or curve.
10. Run glucose determinations on all blood specimens. Run *qualitative* glucose tests on the urine specimens.
11. Make the report as indicated in Fig. 6-3. Normal and abnormal curves are given in Fig. 6-2.

Intravenous test

1. Instructions to patient: The patient is instructed by the physician to maintain a normal carbohydrate diet (at least 300 g of carbohydrate per day) for 3 days prior to the test. The test is performed in the morning, and he is told not to eat or drink anything after the evening meal of the previous day.
2. Make a venipuncture on the patient and withdraw about 5 ml of blood. Place in a fluoride oxalate tube and invert several times to dissolve the oxalate. Label the test tube: Fasting blood sugar. Collect a urine specimen and label: Fasting specimen.
3. Obtain the patient's weight. Multiply this by 0.15 to get the dosage in grams (equivalent to 0.3 g/kg). Example: Patient weighs 100 pounds. And 100 times 0.15 equals 15 g.
4. Now the dosage in grams must be multiplied by a glucose concentration factor to get the number of milliliters to inject. If the sterile glucose solution on hand is a 50% solution, the factor is 100/50. If the glucose solution is a 33% solution, the factor is 100/33. Example: Patient is to get 15 g of glucose. The glucose on hand is a 50% solution. Therefore, the patient gets $15 \times 100/50 = 30$ ml of the 50% glucose solution.
5. Using *sterile* technique, the physician injects the proper amount of *sterile* glucose solution *very slowly* into a vein, taking about 5 minutes to complete the injection.
6. Immediately after giving the glucose, note and record the time.
7. Exactly ½ hour after the injection of glucose, collect another blood and another urine specimen. Label each of them: ½-hour specimen.
8. Exactly 1 hour after the injection of glucose, collect another blood and another urine specimen. Label each of them: 1-hour specimen.

9. Exactly 2 hours after the injection of glucose, collect another blood and another urine specimen. Label each of them: 2-hour specimen.
10. Exactly 3 hours after the injection of glucose, collect another blood and another urine specimen. Label each of them: 3-hour specimen.
11. Determine glucose levels of all blood specimens. Run *qualitative* glucose tests on all urine specimens.
12. Make the report as indicated in Fig. 6-3, noting that it was an intravenous test.

Exton-Rose test

1. Instructions to patient: The patient is instructed by the physician to maintain a normal carbohydrate diet (at least 300 g of carbohydrate per day) for 3 days prior to the test. The test is performed in the morning, and the patient is told not to eat or drink anything after the evening meal of the previous day.
2. Using the rough balance, weigh 100 g of glucose (dextrose) into a container. Dissolve in 650 ml of tap water. Divide the solution into two equal portions and place in a refrigerator for at least 30 minutes in order to chill and thus make more palatable. Just prior to use, add a little lemon juice for flavoring.
3. Make a venipuncture on the patient and withdraw about 5 ml of blood. Place in a fluoride oxalate tube and invert several times to dissolve the oxalate. Label the test tube: Fasting blood specimen. Collect a urine specimen and label: Fasting specimen.
4. Have the patient drink the first portion of glucose. Note and record the time. Exactly ½ hour later collect another blood and another urine specimen. Label each of them: ½-hour specimen.
5. Have the patient drink the second portion of glucose. Note and record the time. Exactly ½ hour later collect another blood and another urine specimen. Label each of them: 1-hour specimen.
6. Run glucose determinations on the three blood specimens. Run *qualitative* glucose tests on the three urine specimens.
7. Make the report as indicated by the accompanying chart, simply filling in the blood values and reporting the urine results as positive or negative.

Report of Exton-Rose test

	Blood (mg/100 ml)	Urine (pos. or neg.)
Fasting specimen	________	________
½-hour specimen	________	________
1-hour specimen	________	________

Nonprotein nitrogen compounds

Substances containing nitrogen may be divided into proteins and nonprotein nitrogens. At this point we will consider the nonprotein nitrogens. In general, they are the waste products of metabolism and are removed from the bloodstream by the kidneys. When they accumulate in the blood, they point to a flaw in the filtering system of the kidneys.

CREATININE

Creatinine is a waste product which is removed from the bloodstream by the kidneys. Its normal concentration is 1 to 2 mg/100 ml of blood and 0.6 to 1.3/100 ml of plasma. Increased values may be found in nephritis, urinary obstruction, and intestinal obstruction.

For the determination of creatinine, the method of Folin and Wu is commonly used. In this method, sodium picrate in alkaline solution is added to a blood or plasma filtrate. The ensuing reaction is called the Jaffé reaction. It consists of a reaction between creatinine and sodium picrate to form creatinine picrate. The creatinine picrate is red in color. The depth of color is measured in a colorimeter.

Procedure for creatinine[6-8]

The following general considerations pertain to the procedure for creatinine:

1. Specimen: serum or plasma.
2. Preservation of specimen: serum, plasma, or protein-free filtrate may be stored overnight in refrigerator.

Plasma creatinine

Reagents needed

Sulfuric acid, 0.67 N	(D-1)
Sodium tungstate, 10%	(D-2)
Saturated picric acid	(D-35)
Sodium hydroxide, 10%	(D-146)
Alkaline picrate solution	(See below)

Principle of plasma creatinine method

Creatinine yields a definite color reaction in the presence of picric acid in alkaline solution, due to the formation of a red tautomer of creatinine picrate (Jaffé reaction).

Procedure

1. With a 2.0 ml Ostwald-Folin pipette, pipette 2.0 ml of plasma into a 125 ml Erlenmeyer flask.
2. Add 14 ml of distilled water.
3. Add 2 ml of 0.67 N H_2SO_4. Mix by swirling.
4. Add 2 ml of 10% sodium tungstate. Stopper and shake.
5. Filter.

Alkaline picrate solution

The alkaline picrate solution is made up *just before using*. Set up the blank, standards, and unknowns first. Then make up the alkaline picrate solution and add to the tubes.

Determine the amount of solution needed for the test and mix in the following ratio:

5 volumes of picric acid (saturated)
1 volume of 10% NaOH
Mix well and use immediately.

Either in cuvettes or in test tubes add the following:

Blank

5.0 ml of distilled water
2.5 ml of alkaline picrate solution

Standards

Two standards of different concentrations are set up to check the calibration curve.

Standard 2 (from calibration curve)
1.0 ml of dilute standard (0.010 mg/ml)
4.0 ml of distilled water
2.5 ml of alkaline picrate solution
Mix.

Standard 4 (from calibration curve)
2.0 ml of dilute standard (0.010 mg/ml)
3.0 ml of distilled water
2.5 ml of alkaline picrate solution
Mix.

Table 6-2. Blood creatinine calibration chart

Tube	Working standard (ml)	H_2O (ml)	Alkaline picrate (ml)	A (520 mμ)	Creatinine equivalent (mg/100 ml)
Blank	0.0	5.0	2.5	0.00	0.0
1	0.5	4.5	2.5		1.0
2	1.0	4.0	2.5		2.0
3	1.5	3.5	2.5		3.0
4	2.0	3.0	2.5		4.0
5	2.5	2.5	2.5		5.0
6	3.0	2.0	2.5		6.0
7	3.5	1.5	2.5		7.0

Unknown

5.0 ml of protein-free filtrate
2.5 ml of alkaline picrate solution
Mix.

Allow the tubes to stand for 20 minutes. Read in a colorimeter or spectrophotometer at 520 mμ against the *blank* zero. Read absorbance (A) or read percent transmittance and convert to A from Table A-7. Calculate values from the calibration curve. The standards should coincide with their values on the curve.

Calibration curve for plasma creatinine

Stock standard (1 ml = 1 mg)

Weigh on analytical balance 1.0000 g of creatinine. If creatinine zinc chloride is used, weigh out exactly 1.6026 g of creatinine zinc chloride. Dissolve in 0.1 N HCl and make up to 1000 ml with 0.1 N HCl.

Working standard (1 ml = 0.01 mg)

Dilute 10.0 ml of stock creatinine standard to 1000 ml with 0.1 N hydrochloric acid.

Prepare a work sheet and set up the standards as shown in Table 6-2. Allow the color to develop for 20 minutes at room temperature. Read absorbance at 520 mμ and record readings in the A column on the work sheet. Plot A readings against the creatinine equivalent (mg/100 ml) on linear graph paper.

Creatinine is the least variable nitrogenous constituent of the blood, in which it exists to the extent of 1 to 2 mg/100 ml. In early nephritis, values of 2 to 4 mg/100 ml are noted, and in chronic hemorrhagic nephritis with uremia, 4 to 35 mg/100 ml may be noted. Creatinine is more readily excreted by the kidneys than urea or uric acid, and an increase of creatinine to 4 or 5 mg/100 ml or more in the blood is evidence of marked impairment of kidney function. Abnormally high creatinine values are accompanied by increased urea content of blood.

Serum creatinine and creatine[11]

Reagents needed

Picric acid, 0.040 M	(D-147)
Sodium hydroxide, 0.75 N	(D-148)
Sodium tungstate, 10%	(D-2)
Sulfuric acid, 0.67 N	(D-1)

Creatinine standard:

Stock standard: 1.5 mg/ml in 0.1 N HCl. Stable indefinitely in the refrigerator.

Working standard: 15 μg/ml. Dilute stock standard 1:100 with distilled water. Prepare fresh daily.

Creatine, working standard: 15 μg of creatine as creatinine per milliliter. No stock standard is stable, and working standard is made each day used.

Creatinine procedure

1. Into a 15 ml centrifuge tube pipette 2.0 ml of serum, 3.0 ml of distilled water, 1.0 ml of 10% sodium tungstate, and 2.0 ml of 0.67 N H_2SO_4. Mix and centrifuge.
2. Into test tubes pipette:
   Blank
   3.0 ml of distilled water
   Standard
   1.0 ml of working standard and 2.0 ml of distilled water
   Unknown
   3.0 ml of protein-free centrifugate
3. To each tube add 1.0 ml of picric acid and 1.0 ml of 0.75 N NaOH. Mix.
4. Allow to stand exactly 20 minutes and then read the absorbance against the reagent blank at 520 mμ.

Calculations

$$\frac{\text{A of unknown}}{\text{A of standard}} \times 0.015 \times \frac{100}{0.75} = \text{mg creatinine in 100 ml serum}$$

Creatine procedure

1. Determine creatinine (preformed) as described above.
2. Determine preformed creatinine and creatine as creatinine by pipetting into graduated centrifuge tubes:
   Blank
   6.0 ml of distilled water
   Standard
   1.0 ml of working standard and 5.0 ml of distilled water
   Unknown
   3.0 ml of protein-free centrifugate, 0.25 ml of 0.75 N NaOH, and 2.75 ml of distilled water
3. Add to each tube 1.0 ml of 0.04 M picric acid.
4. Immerse tubes in a boiling water bath so that the water level is above the liquid level in the tubes. Heat until the volume in each tube has decreased below 4.0 ml (requires between 1.5 and 2 hr).
5. Cool tubes to room temperature and add water to 4.0 ml.
6. To each tube add and mix 1.0 ml of 0.75 N NaOH.
7. At exactly 20 minutes read the absorbance against the reagent blank at 520 mμ.

Calculations

$$\frac{\text{A of unknown}}{\text{A of standard}} \times 0.015 \times \frac{100}{0.75} = \text{mg total creatinine/100 ml}$$

serum or preformed creatinine + creatine as creatinine

Total creatinine − preformed creatinine = mg creatine as creatinine/100 ml serum

For adults, the normal creatine/100 ml of serum is 0.2 to 0.6 mg in men and 0.35 to 0.9 mg in women. It is increased in hyperthyroidism and severe muscular disease.

Urine creatinine

Procedure

1. Suggested urine dilutions
   Dilution A (for volume less than 1000 ml/24 hr)
   Dilute 1 ml of urine to 3 ml with distilled water.
   Multiply the result by 3.
   Dilution B (for volume above 1000 to 1500 ml/24 hr)
   Dilute 1 ml of urine to 2 ml with distilled water.
   Multiply the result by 2.
2. Into 100 ml volumetric flasks pipette the following. (NOTE: Upon the addition of the 1.5 ml of 10% NaOH, start timing for 20 minutes.)
   Blank
   20.0 ml of saturated picric acid
   1.5 ml of 10% NaOH
   Standards (two)
   Standard 1
   0.5 ml of creatinine standard (1 mg/ml)
   20.0 ml of saturated picric acid
   1.5 ml of 10% NaOH
   Standard 2
   1.0 ml of creatinine standard (1 mg/ml)
   20.0 ml of saturated picric acid
   1.5 ml of 10% NaOH
   Unknown
   1.0 ml of diluted urine
   20.0 ml of saturated picric acid
   1.5 ml of 10% NaOH
3. Swirl flasks to mix.
4. Stand 20 minutes.
5. Dilute to mark with distilled water.
6. Stopper and mix thoroughly.
7. Read in spectrophotometer or colorimeter at 520 mμ against *blank*. Read OD or absorbance.

The purpose of the standards is to check the reagents and other conditions of the test. If the reading of the standard varies too much from the curve, the reagents should be checked. It may be necessary to construct a new curve.

Calibration curve for urine creatinine

With the stock standard (1 ml = 1 mg) make the following dilutions, using 0.1 N HCl:
Dilution A (1 ml = 0.25 mg)
Mix in a test tube 2.5 ml of stock standard and 7.5 ml of 0.1 N HCl.
Dilution B (1 ml = 0.1 mg)
Mix in a test tube 1.0 ml of stock standard and 9.0 ml of 0.1 N HCl.

In 100 ml volumetric flasks, set up the following concentrations according to the work sheet shown in Table 6-3. Allow the color to develop 20 minutes. Dilute to mark with distilled water and mix well. Pour into proper tubes and read absorbance at 520 mμ.

Plot the optical density readings on the vertical (ordinate) of linear graph paper, allowing each division of the graph paper to represent 0.005.

Plot the milligrams per milliliter on the horizontal (abscissa), allowing each division to represent 0.01 mg/ml.

The curve does not form a straight line.

The daily excretion of creatinine by an adult of medium weight averages about 1.25 g. The average adult normal range for a male is 20 to 26 mg/kg and for females is 14 to 22 mg/kg. The value is nearly constant from day to day for a given individual, being influenced by the diet very little unless the diet consists

Table 6-3. Urine creatinine calibration chart

Tube	Stock standard (1 ml = 1.0 mg)	1:4 dilution (1 ml = 0.25 mg)	1:10 dilution (1 ml = 0.1 mg)	Picric acid (ml)	NaOH (10%)	A (520 mμ)	Creatinine equivalent (mg/100 ml)
Blank	0.00	0.00	0.00	20	1.5	0.00	0.00
1	2.00	0.00	0.00	20	1.5		2.00
2	1.50	0.00	0.00	20	1.5		1.50
3	1.00	0.00	0.00	20	1.5		1.00
4		3.00	0.00	20	1.5		0.75
5		2.00	0.00	20	1.5		0.50
6		1.00	0.00	20	1.5		0.25
7			2.00	20	1.5		0.20
8			1.00	20	1.5		0.10

of a heavy meat diet which contains much preformed creatinine. The excretion of creatinine is, to a certain extent, a measure of muscular efficiency and of the amount of active muscle tissue in the body. Relative to body weight, therefore, less creatinine is excreted by obese persons.

Creatinine excretion is decreased in disorders associated with muscular atrophy and muscular weakness. It increases with increased tissue catabolism as in fever.

Urine creatine

Creatine occurs in only very small amounts in the urine of normal adults, but it is found in larger amounts in the urine of children. Creatine ingestion in adults has little effect on the urinary excretion. In a fasting state, the amount is markedly increased. The normal adult range is 0 to 200 mg/24 hr. Creatine also appears in the urine after high water ingestion. It is found in many pathologic conditions associated with malnutrition and disintegration of muscular tissue and in fevers. Large amounts are found in diseases of the muscles.

The method of determining urine creatine is based upon the Jaffé reaction. Creatine, if boiled with acid, is transformed into creatinine. By determining the content of creatinine before and after the acid treatment it is possible to calculate the amount of creatine originally present in the urine.

Procedure

1. **Pipette 10 ml of diluted urine into a Folin-Wu sugar tube.**
2. **Add 10 ml of 1 N HCl.**
3. **Cover top of tube tightly with aluminum foil and put into boiling water bath for 30 minutes. Use timer.**
4. **Cool in room-temperature water bath and dilute to 25 ml mark with distilled water.**
5. **Stopper and mix well by inverting tube several times.**
6. **In 100 ml volumetric flasks set up the following:**
   Total creatinine
   2.5 ml of the above acid-treated urine (2.5 ml = 1.0 ml of diluted urine)
   20.0 ml of saturated picric acid
   2.0 ml of 10% NaOH
   Swirl to mix.

Preformed creatinine
1.0 ml of diluted urine (*not* the acid-treated urine)
20.0 ml of saturated picric acid
1.5 ml of 10% NaOH
Swirl to mix.

Blank
20.0 ml of saturated picric acid
1.5 ml of 10% NaOH
Swirl to mix.

Standards
Standards are the same as for the urine creatinines; urine creatines may be run with urine creatinines.

7. Swirl flasks to mix.
8. Let stand 20 minutes. Use timer.
9. Dilute to mark with distilled water.
10. Stopper and mix thoroughly.
11. Read absorbance at 520 mμ against the *blank.*
12. Calculate the two unknowns (*performed creatinine and total creatinine*) from the urine creatinine standard curve.
13. The results of the preformed and total creatinine test are then multiplied by the dilution of urine.

Calculations of the creatine

Total creatinine − Preformed creatinine = mg creatine/ml

or

(T. creatinine − P. creatinine) × 100 = mg creatine/100 ml

or

$$\text{mg/100 ml creatine} \times \frac{\text{24 hr urine vol}}{100} = \text{mg creatine/24 hr}$$

URIC ACID

Uric acid is a waste product that is removed from the bloodstream by the kidneys. Conditions in which an increase or a decrease may be expected are listed in Table 6-4.

Procedures for uric acid

The following general considerations pertain to the procedures for uric acid:

1. Specimen: serum or plasma.
2. Precautions: chlorine in water causes oxidation; uricase is inhibited by formaldehyde in standards; results increase (colorimetric method) with the presence of ascorbic acid, methydopa, or glutathione.

Table 6-4. Conditions accompanied by abnormal uric acid values

When increased		When decreased
Gout	Acute infections	Salicylate therapy
Nephritis	Intestinal obstruction	Atophan therapy
Arthritis	Urinary obstruction	Yellow atrophy of liver
Eclampsia	Metallic poisoning	
Leukemia	Hypertension	
Polycythemia	Following exercise	
Diabetes		

3. Preservation of specimen: serum, plasma, or protein-free filtrate may be stored overnight in a refrigerator.
4. CAUTIONS:
   a. Cyanide solutions are *poisonous.*
   b. Cyanide solutions should be stored in a refrigerator.
   c. Cyanide solutions should be cold (about 12° C) when added to the test tube.

The more commonly used methods for the determination of uric acid are discussed briefly below.

Folin method

A blood filtrate is treated with urea-cyanide and phosphotungstic acid to form a phosphotungstate complex. The uric acid present reduces the phosphotungst*ate* complex to the blue phosphotungst*ite* complex. The depth of color is measured in a colorimeter and compared with a standard.

Brown method

A blood filtrate is treated with sodium cyanide, urea, and phosphotungstic acid to form a phosphotungstate complex. The uric acid present reduces the phosphotungst*ate* to the blue phosphotungst*ite* complex. The depth of color is measured in a colorimeter and compared with a standard.

Newton method

A special blood filtrate is made to remove interfering substances. The filtrate is treated with a urea-cyanide solution and lithium arsenotungstate to form an arsenotungstate complex. The uric acid present reduces the arsenotungst*ate* complex to the blue arsenotungst*ite* complex. The depth of color is measured in a colorimeter and compared with a standard.

Archibald method[2]

The uric acid in serum or plasma reduces phosphotungstate acid in an alkaline solution, thus producing a blue color. This method may also be used for urine without any modifications of the procedure. The phosphotungstate is not only a color-developing agent. It also precipitates the proteins. Plasma or serum is preferable to whole blood because the glutathione and ergothioneine from the red cells cause interference.[13]

Reagents needed

Uric acid stock standard, 1 mg/ml	(D-149)
Polyanethol sodium sulfonate	(D-150)
Glycerin-silicate reagent	(D-151)
Phosphotungstic acid	(D-152)
Sodium hydroxide, 0.5 N	(D-34)
Uric acid special reagent	(D-153)
Uric acid working standard, 0.005 mg/ml	

Pipette 0.5 ml of stock uric acid standard into a 100 ml volumetric flask and dilute to the mark with water. Prepare a fresh working standard daily.

Procedure

1. Pipette into a 125 ml Erlenmeyer flask 2.0 ml of plasma or serum. With urine make a 1:10 dilution and pipette 2.0 ml into the flask.
2. Add 16 ml of distilled water; swirl flask to mix.
3. Add 0.8 ml of 0.5 N NaOH, swirling the flask for at least 1 minute. For large volumes a mechanical shaker is most helpful.

4. Let the flask stand for 10 minutes.
5. Slowly add 1.2 ml of the phosphotungstic acid solution to the flask, with constant swirling of the flask to ensure a complete precipitation of protein.
6. Let the flask stand for 5 minutes.
7. Filter through Whatman No. 42 filter paper into large test tubes or flasks.
8. Set up the following tubes:
   Blank
   5.0 ml of water
   Standard
   5.0 ml of working standard
   Unknown
   5.0 ml of filtrate
9. Add 2.5 ml of glycerin-silicate reagent to each tube. Set timer for 15 minutes. Begin timing upon the addition of 2.0 ml of uric acid special reagent. Mix immediately. After 15 minutes the color is fully developed.
10. Set zero with blank and read absorbance or OD measurements on a spectrophotometer at 700 mμ or, with the aid of a No. 660 red filter, in a photoelectric colorimeter.

Calculations

$$\frac{\text{Absorbance of unknown}}{\text{Absorbance of standard}} \times 5 = \text{mg uric acid/100 ml}$$

e.g., Urine
Dilution 1:100
Filtrate 5 ml

$$\frac{\text{Absorbance of unknown}}{\text{Absorbance of standard}} \times 10 \times 5 = \text{mg uric acid/100 ml in urine}$$

Normal values[11]

	Men	Women
Serum or plasma	2.5 to 7.0 mg/100 ml	1.5 to 6.0 mg/100 ml
Urine, 24-hour	250 to 750 mg	250 to 750 mg

Henry method[12]

The uric acid in serum, plasma, or urine reduces an alkaline phosphotungstate solution to a tungsten blue. The alkali used is sodium carbonate. The depth of color is measured photometrically and compared with a standard.

Serum uric acid (sodium carbonate)

Reagents needed

Phosphotungstic acid (The directions given in the original publication were in error and were corrected by the authors in a later publication. For this reason the directions for preparing this reagent are given here instead of in Appendix D.)

Weigh 40.0 g of reagent grade sodium tungstate and transfer to a 1 L round-bottom boiling flask containing about 300 ml of distilled water. Swirl to dissolve. Add 32 ml of 85% *o*-phosphoric acid and mix. Attach a reflux condenser to the flask and reflux gently for 2 hours. Cool to room temperature and dilute to 1000 ml with distilled water. Weigh 32.0 g of lithium sulfate monohydrate and add to the above solution. Mix to dissolve. This reagent is stable indefinitely when stored in a refrigerator.

Sodium carbonate, 14% (D-154)
Uric acid, stock standard (D-149)
Working standard, 0.01 mg/ml
Dilute stock standard 1:100 in a volumetric flask.

Procedure

1. Prepare a Folin-Wu protein-free filtrate.
   Avoid: oxalated plasma, which may cause a crystalline precipitate.
2. Set up the following tubes:
   Blanks (2 tubes)
   3.0 ml of distilled water into each tube.
   Standards (3 tubes)
   3.0 ml of working standard into each tube.
   Unknown (1 tube per test)
   3.0 ml of unknown filtrate into its numbered tube.
3. *To all tubes add*
   1.0 ml of 14% Na_2CO_3 and mix.
   1.0 ml of phosphotungstic acid and mix.
4. Allow tubes to stand 15 minutes to develop color.
5. Read against the blank at 660 mμ within a 30-minute period. The blank must be colorless.

Calculations

$$\frac{\text{A of unknown}}{\text{A of standard}} \times 10 = \text{mg uric acid/100 ml}$$

The three standards are averaged unless they vary a great deal. In that case, repeat the test. The two blanks are used to check against each other.

Urine uric acid

Preparation of the urine

1. If the urine is cloudy, warm it to about 60° C to dissolve the urates and centrifuge.
2. Dilute 1:10 with distilled water.
3. Proceed with the method given for the serum uric acid. Everything is the same as in the preceding method with the exception of the calculation.

Calculation

$$\frac{\text{A of unknown}}{\text{A of standard}} \times 100 = \text{mg uric acid/100 ml}$$

NOTE: If the A reads less than 0.2 or more than 0.8, repeat the test, using the proper dilution.

TOTAL NONPROTEIN NITROGEN (NPN)

The components of total nonprotein nitrogen are the waste products of protein catabolism. These waste products include amino acids, creatine, creatinine, ergothioneine, glutathione, urea, uric acid, and many other components, unknown.

The tungstic acid protein-free filtrate is more commonly used for NPN than barium-zinc sulfate filtrates are. This is partly due to the fact that ergothioneine, glutathione, and uric acid are precipitated by the zinc.

The normal values for the total nonprotein nitrogen are 27 to 47 mg/100 ml of blood, using a tungstic acid filtrate, and 15 to 25 mg/100 ml of blood, using the zinc hydroxide filtrate. Higher values indicate faulty elimination and consequently point to a disorder of the kidneys. The conditions accompanied by abnormal NPN values are similar to those for urea nitrogen. These are given in Table 6-5.

A brief discussion of some of the more commonly used methods for the determination of total nonprotein nitrogen follows.

Folin-Wu method

A protein-free filtrate is made. To this is added a digestion mixture of sulfuric and phosphoric acids, its purpose being to aid in the liberation of ammonia. A few glass beads are added to prevent the solution from bumping, and the mixture is heated with a flame. This decomposes the nitrogen substances and forms an ammonium salt. Nessler's solution is then added to react with the ammonium salt and form the yellow dimercuric ammonium iodide. The depth of color is measured in a colorimeter and compared with a standard.

Koch-McMeekin method

A protein-free filtrate is made. To this is added a mixture of sulfuric acid and hydrogen peroxide, its purpose being to help liberate the ammonia. A few glass beads are added to prevent bumping, and the solution is heated with a flame. This decomposes the nitrogen substances and forms an ammonium salt. Nessler's solution is then added to react with the ammonium salt and form the yellow dimercuric ammonium iodide. The depth of color is measured in a colorimeter and compared with a standard.

The only difference between this method and the preceding one is in the digestion mixture. Folin and Wu use sulfuric and phosphoric acids for the digestion mixture, whereas Koch and McMeekin use sulfuric acid and hydrogen peroxide.

Berthelot color reaction (modified)

A protein-free filtrate (tungstic acid) is made. To this is added 30% sulfuric acid, its purpose being to aid in the liberation of ammonia by oxdizing the organic material. One or two glass beads or boiling stones are added to prevent the solution from bumping and the mixture is heated with flame (Bunsen burner). Phenol color reagent and alkali-hypochlorite reagent are then added. A blue color is formed by ammonia and phenol in the alkaline oxidizing medium, in the presence of sodium nitroprusside as catalyst. The depth of color is measured in a colorimeter and compared with a standard.

Procedure for total nonprotein nitrogen (NPN)

The following general considerations pertain to the procedure for total nonprotein nitrogen:

1. Specimen: whole blood or serum.
2. Preservation of specimen: whole blood, serum, or protein-free filtrate may be stored overnight in a refrigerator.
3. Precautions:
   a. Digestion must be done in a fume hood.
   b. Contents of tubes must be *cold* before the addition of Nessler's solution; otherwise cloudiness may develop.
   c. Overdiluting with water, excessive heating, causing frothing, or adding hydrogen peroxide (Superoxol) before the tube is cool, resulting in splattering of sample, produces low results.
   d. Extremely elevated sample will cloud. Repeat test, using a smaller amount of filtrate. Calculate accordingly.
   e. If distilled water is not free of nitrogen compounds, the blanks, standards, and unknowns will all read high.
   f. Improper precipitation of the protein-free filtrate or using too much sulfuric acid or sodium tungstate will cause a yellow precipitate to form during the digestion.

Koch-McMeekin method[8,14]

Reagents needed

Sulfuric acid, digestion mixture, 30%	(D-155)
Hydrogen peroxide, 30% (refrigerate)	(D-156)
Nessler's reagent	(D-157)
Standard solutions:	
Stock standard (1.0 ml = 0.5 mg nitrogen)	(D-158)
Working standard (1.0 ml = 0.025 mg nitrogen)	
1. Using a volumetric pipette, transfer 25.0 ml of stock standard to a 500 ml volumetric flask	
2. Dilute to volume with ammonia-free 0.1 N HCl	(D-37)

Procedure

1. Prepare either a Somogyi or a Folin-Wu protein-free filtrate, using 2.0 ml of oxalated whole blood.
2. Using a volumetric pipette, transfer 5.0 ml of filtrate to an NPN digestion tube graduated at 35 and 50 ml.
3. Add 1.0 ml of digestion mixture.
4. Add two or three boiling stones and gently heat the bottom of the tube with a gas flame until it begins to boil.
5. Boil vigorously until the solution begins to turn brown.
6. Continue heating until the tube is filled with dense white fumes.
7. Allow to cool; then add 3 drops of 30% hydrogen peroxide.
8. Cover with a watchglass and continue heating gently for exactly 3 minutes. (Use a timer.)
9. Allow to cool to room temperature and then rinse the underside of the watchglass into the tube, using a small stream of distilled water from a wash bottle.
10. Dilute to the 35 ml mark with distilled water.
11. Stopper and place the tube, along with a blank and standard tubes, into a cold water bath located in a refrigerator. These remain in the bath for a minimum of 15 minutes. NPN blanks and standards are not cooked with the unknowns. Set tubes up as follows:
    Blanks (2 tubes)
    1.0 ml of digestion mixture.
    Dilute to 35 ml mark with distilled water.
    Stopper and mix.
    Standards (3 tubes)
    5.0 ml of working standard.
    1.0 ml of digestion mixture.
    Dilute to 35 ml mark with distilled water.
    Stopper and mix.
12. *All tubes*
    After cooling for 15 minutes, add Nessler's reagent to the 50 ml mark.
    Stopper and mix by inverting tubes four times. If solution is not absolutely clear, repeat the test.
    Read A at 540 mμ or Klett units, using a No. 54 filter, against the reagent blank. The blank tubes must be clear and check against each other.

Calculations and dilutions

$$\frac{\text{Reading of unknown}}{\text{Reading of standard}} \times \text{Concentration of standard} \times \frac{100}{\text{Amount of blood used}} = \text{mg NPN/100 ml}$$

or

$$\frac{\text{RU}}{\text{RS}} \times 0.125 \times \frac{100}{0.5} = \text{mg NPN/100 ml}$$

and

$$\frac{\text{RU}}{\text{RS}} \times 25 = \text{mg NPN/100 ml}$$

Dilution, instead of using 5.0 ml of filtrate, use:

A. 3.0 ml of filtrate. Multiply mg NPN/100 ml by $\frac{5}{3}$.

B. 2.0 ml of filtrate. Multiply mg NPN/100 ml by $\frac{5}{2}$.

C. 1.0 ml of filtrate. Multiply mg NPN/100 ml by 5.

Normal range

15 to 25 mg/100 ml (Somogyi filtrate)
27 to 47 mg/100 ml (Folin-Wu filtrate)

UREA NITROGEN

Urea is the principal waste product of protein catabolism. Its formula is NH_2CONH_2. Since it is difficult to determine urea itself, and a relatively simple matter to analyze for the nitrogen in urea, it has become customary to determine the urea nitrogen. Most methods depend on the urea being converted to ammonium carbonate:

$$\underset{\text{Urea}}{C(=O)(NH_2)_2} + \underset{\text{Water}}{2H_2O} \xrightarrow[38°]{\text{Urease}} \underset{\text{Ammonium carbonate}}{(NH_4)_2CO_3}$$

The normal values for the blood urea nitrogen (BUN) are 6 to 29 mg/100 ml of blood. The corresponding values for urea may be found by multiplying these figures by the factor 2.14. Conditions in which an increased or a decreased urea nitrogen value may be expected are given in Table 6-5.

Karr method

A protein-free filtrate is treated with a buffer solution and the enzyme urease. The purpose of the buffer solution is to control the pH of the solution, since the urease reacts better at a certain pH (6.8). Upon incubation, the urease decomposes the urea and liberates ammonia, which forms ammonium carbonate. After incubation, gum ghatti is added to form a protective colloid and keep the solution from becoming cloudy. Nessler's solution is then added. This is an alkaline solution of the double iodide of mercury and potassium ($HgI_2 \cdot 2KI$). It reacts

Table 6-5. Conditions accompanied by abnormal urea nitrogen values

When increased		When decreased
Nephritis	Dehydration	Acute liver destruction
Intestinal obstruction	Malignancy	Amyloidosis
Urinary obstruction	Pneumonia	Pregnancy
Metallic poisoning	Surgical shock	Nephrosis
Cardiac failure	Addison's disease	
Peritonitis	Uremia	

with the ammonium carbonate to form yellow dimercuric ammonium iodide. The depth of color is measured in a colorimeter and compared with a standard.

Van Slyke-Cullen method

In the Van Slyke-Cullen method, oxalated blood is treated with the enzyme urease to form ammonium carbonate. Addition of potassium carbonate liberates the ammonia gas, which is collected in a boric acid solution. This is then titrated with a standard acid solution, and the urea nitrogen found by calculation.

Gentzkow method

Oxalated blood is diluted. To this is added the enzyme urease, which decomposes the urea to form ammonium carbonate. Next, the proteins are precipitated with sulfuric acid and sodium tungstate. The resulting filtrate containing the ammonium carbonate is treated with Nessler's solution to form the yellow dimercuric ammonium iodide. The depth of color is measured in a colorimeter and compared with a standard.

Folin-Svedberg method

A protein-free filtrate is made. To this is added a buffer solution to control the pH. Next, the enzyme urease is added to decompose the urea and form ammonium carbonate. The mixture is incubated to enhance the reaction. Sodium borate is added to the solution to liberate the ammonia gas, which is collected in a weak hydrochloric acid solution. Nessler's solution is then added to convert the ammonium ion to the yellow dimercuric ammonium iodide. The depth of color is measured in a colorimeter and compared with a standard.

Berthelot reaction

The urea in serum, plasma, or urine, is split, by the action of urease, into ammonia and CO_2. The ammonia is determined photometrically by the phenol-hypochlorite reaction of Berthelot, using sodium nitroprusside as a catalyst. Because of the very high dilution used, it is not necessary to precipitate and remove proteins.

Urograph* method

This method employs strips of chromatography paper (7 by 85 mm), which are banded with controlled amounts of reagents. As the serum or plasma travels up the Urograph chromatography paper, it first encounters a band of buffered urease. At this position the urea present is converted to an ammonia salt. Upon further migration the sample encounters a potassium carbonate band that produces free ammonia. The free ammonia produces a color change in the uppermost indicator band, which contains bromcresol green in tartaric acid. The height of the blue-green color developed is directly proportional to the amount of urea in the unknown sample.

Results obtained by this method compare well with those obtained by the AutoAnalyzer method when the directions are followed *exactly* as given in the package insert.[4]

---

*A preparation of General Diagnostics Division, Warner-Chilcott.

Procedure for urea nitrogen

The following general considerations pertain to the procedure for urea nitrogen:

1. Preparation of specimen: Whole blood, serum, plasma, or protein-free filtrate.
2. Preservation of specimen: Whole blood, serum, plasma, or filtrate may be stored overnight in a refrigerator.
3. Precautions: The following interfere with the urease reaction: excessive anticoagulant of oxalate or citrate; excessive buffer solution; glassware that has contained Nessler's solution or cyanide solution; traces of fluoride; loss of urease activity; turbid Nessler reagent.

Blood urea nitrogen[10]

Reagents needed

Urease solution	(D-159)
Sulfuric acid, 0.67 N	(D-1)
Sodium tungstate, 10%	(D-2)
Potassium persulfate, 2.5%	(D-160)
Potassium gluconate, 1%	(D-161)
Standard solution (same as for NPN)	(D-158)

Procedure

1. Pipette 2 ml of oxalated whole blood into a 125 ml Erlenmeyer flask.
2. Add 13.5 ml of water.
3. Add 0.5 ml of urease solution.
4. Mix well. Either stopper the flask or cover with Parafilm.
5. Place the flask in a 38° C water bath or incubator for 30 minutes.
6. Add 2 ml of 0.67 N sulfuric acid. Mix well and allow to stand for 5 minutes.
7. Add 2 ml of 10% sodium tungstate.
8. Stopper the flask and shake vigorously. There should be no frothing. Filter through a Whatman No. 1 filter paper or its equivalent.
9. Prepare the following NPN tubes:
   Blanks (2 tubes)
      1.0 ml of 2.5% potassium persulfate
      1.0 ml of 1% potassium gluconate
      Dilute to 35 ml mark with distilled water.
      Stopper and mix.
   Standards (3 tubes)
      5.0 ml of working standard (NPN standard)
      1.0 ml of 2.5% potassium persulfate
      1.0 ml of 1% potassium gluconate
      Dilute to 35 ml mark with distilled water.
      Stopper and mix.
   Unknowns (1 tube per test)
      5.0 ml of filtrate
      1.0 ml of 2.5% potassium persulfate
      1.0 ml of 1% potassium gluconate
      Dilute to 35 ml mark with distilled water.
      Stopper and mix.
10. *All tubes*
      Cool in ice bath for 15 minutes.
      Add Nessler's reagent to the 50 ml mark.
      Stopper and invert four times to mix. Solution must be clear.
      Read absorbance at 540 mμ (or Klett units, using a No. 54 filter) against the reagent blank. The blank tubes must be clear and check against each other.

Calculations and dilutions

The same as the NPN calculations and dilutions:

$$\frac{RU}{RS} \times 25 = \text{mg BUN/100 ml}$$

Normal range

8 to 20 mg/100 ml

Berthelot reaction[3]

Reagents needed

Buffered urease solution* (D-94)
Phenol color reagent* (D-95)
Alkali-hypochlorite reagent* (D-96)
Urea standard

Transfer 0.429 g of urea to a 1 L volumetric flask and dilute to volume with distilled water. Add a few drops of chloroform as a preservative and store in the refrigerator (1 ml = 0.2 mg of urea N).

Procedure for serum or plasma

1. Set up the following in test tubes (16 × 125):
   Blanks
   0.2 ml of buffered urease solution
   Standard
   0.2 ml of buffered urease solution
   0.02 ml of urea standard
   Unknown
   0.2 ml of buffered urease solution
   0.02 ml of serum or plasma
   The 0.02 ml aliquots of standard and unknown are added with TC micropipettes. Sahli hemoglobin pipettes are satisfactory.
2. Incubate the tubes in a water bath at 37° C for 15 minutes.
3. Then add 1.0 ml of phenol color reagent to each tube and mix.
4. Add 1.0 ml of alkali-hypochlorite reagent and mix promptly again.
5. Incubate tubes at 50° to 60° C for 5 minutes or at 37° C for 20 minutes.
6. Add 8 ml of water to all tubes and mix. If mixing must be accomplished by inversion, cover the tops of the tubes with clean Saran Wrap or Parafilm and hold with the thumb while inverting the tube.
7. Read A at 625 mμ (or Klett units, using a No. 54 filter). On a Klett photometer, Beer's law is obeyed with a No. 54 filter up to only about 250 Klett units and is not obeyed at all with a No. 62 filter. When specimens read higher than 0.800 A or 250 Klett units, dilute with water, read again, and calculate according to dilution.

Calculation

$$\frac{RU}{RS} \times 0.004 \times \frac{100}{0.02} = \text{mg urea N/100 ml}$$

$$\frac{RU}{RS} \times 20 = \text{mg urea N/100 ml}$$

Procedure for urine

1. Since urine normally contains some preformed ammonia, it is necessary to remove this ammonia before proceeding with the test.
2. Measure and record the volume of the urine.
3. Add 5 g of Permutit to a 10 ml aliquot of the urine and shake vigorously for 1 minute. Allow to stand for 10 minutes.

*These reagents are available commercially from Hyland Laboratories in the form of a kit with the trade name UN-Test.

4. Filter.
5. Transfer 1 ml of filtered urine to a 50 ml volumetric flask and dilute to the mark with water. Stopper and mix.
6. Using an 0.02 ml aliquot of diluted urine, follow the procedure as with serum or plasma.

Calculation

$$\frac{RU}{RS} \times 20 \times 50 = \text{mg urea N/100 ml urine}$$

$$\frac{RU}{RS} \times 1000 = \text{mg urea N/100 ml urine}$$

Azostix method

Azostix,* a screening procedure for blood urea, is a strip with a cellulose reagent area impregnated with urease and a semipermeable membrane over a yellow, buffered bromthymol blue indicator. Urea in a blood sample is hydrolyzed by urease to form ammonia or ammonium carbonate that is alkaline. The resulting alkalinity causes the bromthymol blue indicator to change from yellow to blue. The intensity of the greenish blue color is proportional to the urea nitrogen concentration.

Procedure

1. A large drop of blood is freely spread over the entire reagent area on the printed side of the strip. A thin film of blood tends to dry rapidly and is difficult to wash from the strip.
2. Wait exactly 60 seconds; time accurately.
3. Holding the strip vertically, wash the blood quickly from it with a sharp stream of water from a wash bottle. Use only one or two squirts from the wash bottle, as overwashing will remove some of the color. Shake off excess water.
4. Read immediately after washing. Allow no more than 1 or 2 seconds to read the color, which fades rapidly.

Reaction

The color chart represents four concentrations of urea nitrogen: 10, 20, 30, and 40 mg/100 ml of blood.

Urine urea nitrogen[10]

The formula for urea is $CO(NH_2)_2$. From the formula it can be seen that nitrogen is a definite fixed portion of urea, and it has become customary to determine the nitrogen present in urine and make the report as urea nitrogen. The normal values for urea nitrogen are 6 to 17 g/day. Decreased values may be found in acute nephritis, acidosis, and cirrhosis of the liver. Increased values may be seen in febrile conditions.

Reagents needed

Urease solution	(D-159)
Sulfuric acid 0.67 N	(D-1)
Sodium tungstate, 10%	(D-2)
Potassium persulfate, 2.5%	(D-160)
Potassium gluconate, 1%	(D-161)
Standard solution (same as for NPN)	(D-158)
Acetate buffer	(D-162)
Permutit†	

*Ames Co., Division of Miles Laboratories, Inc.
†May be purchased from Fisher Scientific Co.

Procedure

1. Measure and record exact volume of urine specimen.
2. Transfer 5.0 ml of urine to a 125 ml Erlenmeyer flask.
3. Add 5 g of Permutit, stopper, and shake vigorously for 1 minute. This removes any preformed ammonia.
4. Allow to stand for 10 minutes; then filter.
5. Dilute ten times by mixing 1.0 ml of Permutit-treated urine with 9 ml of distilled water in a large test tube or flask.
6. Transfer 2.0 ml of the diluted urine to a 125 ml Erlenmeyer flask.
7. Add 13.0 ml of distilled water.
8. Add 0.5 ml of urease solution.
9. Add 0.5 ml of acetate buffer.
10. Mix well and stopper or cover with Parafilm.
11. Incubate in a 38° C water bath for 30 minutes.
12. Add 2.0 ml of 0.67 N H_2SO_4.
13. Add 2.0 ml of 10% Na_2WO_4 (sodium tungstate).
14. Shake thoroughly to mix, and filter.
15. Transfer 5.0 ml of filtrate to an NPN digestion tube.
16. Add about 10 ml of water (distilled).
17. Add to the filtrate and water:
    1.0 ml of 2.5% potassium persulfate
    1.0 ml of 1% potassium gluconate
    Dilute to 35 ml mark with distilled water.
18. *Blanks* and *standards* are set up the same way as for the NPN.
19. Place all tubes in ice bath for 15 minutes.
20. Add Nessler's reagent to the 50 ml mark. Stopper and invert four times to mix.
21. Read absorbance against blank, using a 540 mμ or No. 54 filter for the Klett photometer.

Calculations

$$\frac{\text{RU}}{\text{RS}} \times 25 \times 10 \text{ (dilution)} = \text{mg urea nitrogen/100 ml}$$

or

$$\frac{\text{mg/100 ml}}{1000} \times \frac{\text{ml volume of 24-hour specimen}}{100} = \text{g urea nitrogen/24 hr}$$

Normal range

6 to 17 g/24 hr

Multiply mg urea nitrogen/100 ml urine by 2.14 to obtain mg urea/ 100 ml urine.

Care of equipment

Clean glassware is essential for accuracy in this procedure. It is highly recommended that all glassware used should be washed with concentrated nitric acid, thoroughly rinsed with tap water and finally with distilled water.

REFERENCES

1. Annino, J. S.: Clinical chemistry: principles and procedures, ed. 3, Boston, 1964, Little, Brown & Co.
2. Archibald, R. M.: Colorimetric measurement of uric acid, Clin. Chem. 3:102, 1957.
3. Chaney, A. L., and Marbach, E. P.: Modified reagents for determination of urea and ammonia, Clin. Chem. 8:131, 1962.
4. Erickson, M. M., Verdun, R. M., and White, W. L.: An evaluation of the Urograph method for the determination of blood urea nitrogen, Amer. J. Med. Tech. 31:127, Sept.-Oct., 1965.
5. Fajans, S. S., and Conn, J. W.: Approach to prediction of diabetes mellitus by modification of glucose tolerance test with cortisone, Diabetes 3:296, 1954.
6. Folin, O.: On the determination of creatinine and creatine in urine, J. Biol. Chem. 17:469, 1914.
7. Folin, O., and Doisy, E. A.: Impure picric acid as a source of error in creatine and cre-

atinine determinations, J. Biol. Chem. **28**:349, 1917.

8. Folin, O., and Wu, H.: A system of blood analysis, J. Biol. Chem. **38**:81, 1919.
9. Folin, O., and Wu, H.: A simplified and improved method for determination of sugar, J. Biol. Chem. **41**:367, 1920.
10. Gentzkow, C. J.: Accurate method for determination of blood urea nitrogen by direct nesslerization, J. Biol. Chem. **143**:531, 1942.
11. Henry, R. J.: Clinical chemistry; principles and technics, New York, 1964, Paul B. Hoeber, Inc., Medical Book Department of Harper & Row, Publishers.
12. Henry, R. J., Sobel, C., and Kim, J.: A modified carbonate-phosphotungstate method for the determination of uric acid and comparison with the spectrophotometric uricase method, Amer. J. Clin. Path. **28**:152, 645, 1957.
13. Jorgensen, S., and Nielsen, A. T.: Uric acid in human blood corpuscles and plasma, Scand. J. Clin. Lab. Invest. **8**:108, 1956.
14. Koch, F. C., and McMeekin, T. L.: A new direct nesslerization micro-Kjeldahl method and a modification of the Nessler-Folin reagent for ammonia, J. Amer. Chem. Soc. **46**: 2066, 1924.
15. Medical technologist's handbook for the determination of true glucose, Kalamazoo, Mich., 1960, The Upjohn Co.
16. Nelson, N.: Photometric adaptation of Somogyi method for determination of glucose, J. Biol. Chem. **153**:375, 1944.
17. Somogyi, M.: Notes on sugar determination, J. Biol. Chem. **195**:19, 1952.

• 7

• Electrolytes

This chapter considers various tests for electrolytes.

In good health, the body maintains electric neutrality. This means that there is a balance or equality between cation and anion groups. Sodium, potassium, calcium, and magnesium ions make up the cation group, sometimes called the total base. In the anion group (from inorganic and organic acids) are the electrolytes consisting of chloride, bicarbonate, protein, phosphate, sulfate, bromide, and iodide ions. These two groups in balance are shown in Table 7-1.

Since the complete set of the cation and anion groups is seldom run, a check may be made of the electrolyte balance as suggested in the following paragraphs. The sodium ion has the highest concentration of the cation group while the chloride and bicarbonate ions have the largest concentration of the anion group. Total the bicarbonate and chloride ions and subtract that total from the sodium ions. Normally, this difference averages 10 mEq/L with variation either way of 1 or 2 mEq. However, in a serious illness (pathologic condition) the difference may be appreciably greater than 10 or may be less than 5.

If there is a difference greater than 15 mEq/L, look at the other anions for an elevated value. Here are three of the most common causes of a great difference:

1. Kidney damage or uremia resulting in an elevated BUN or NPN. In addition there is a retention of endogenous acids that are either organic (amino or keto acids) or inorganic (phosphate or sulfate).
2. Diabetic acidosis with a retention of keto acids in the blood. There will be an elevation in the blood sugar, blood acetone, and a positive acetone result with the urine.
3. Various poisonings in which there are unmeasured acids in the bloodstream. An example would be poisoning from methanol where it is converted to formic acid.

A difference of less than 5 mEq/L is not as prevalent as the greater variation. Decreased albumin may be the cause of a close difference. The smaller amount of protein requires less sodium for electrical neutrality; therefore there will often be a low sodium accompanying a low albumin. However, look at the electrolyte results and repeat the test showing the most suspicious result first.

The total of the bicarbonate and chloride should *always be less* than the sodium. A quality control (QC) is most beneficial in eliminating doubts of a technical error. The electrolytes should always be added and compared with previous work on the patient. Other tests such as urea nitrogen, glucose, and protein are taken into consideration when totaling the electrolyte balance. Records of each abnormal test, with the patient's name and test results, are invaluable tools for the laboratory.

Table 7-1. Electrolyte composition of human plasma (approximate)

Cation	mEq/L	Anion	mEq/L
Na (sodium)	143.0	Cl (chloride)	104
K (potassium)	4.5	HCO_3 (bicarbonate)	29
Ca (calcium)	5.0	Protein	16
Mg (magnesium)	2.5	HPO_4 (phosphate)	2
		SO_4 (sulfate)	1
		Organic acids	3
Total	155.0	Total	155

Table 7-2. Approximate normal values for urine electrolytes

Electrolytes	Range
Sodium	43 to 217 mEq/24 hr
Potassium	26 to 133 mEq/24 hr
Chloride	170 to 254 mEq/24 hr
CO_2	None
Calcium	25 to 200 mEq/24 hr
Phosphorus	0.2 to 0.6 mEq/24 hr

There are times when results will not "check out" even after repetition of all tests and checking of the instruments. A new specimen should be requested because of two main problems that may arise before the blood reaches the laboratory:

1. A wet or contaminated syringe may have been used.
2. Blood may have been drawn during an intravenous infusion. Infusion may cause a great dilution factor or, if the infusion contains electrolytes, the sample will be contaminated.

Since urinary electrolytes are not excreted in a uniform concentration throughout the day, a 24-hour specimen should be collected. Electrolytes in the urine are dependent upon the following factors: (1) acid-base balance of the body, (2) intake of water and various ions, (3) state of body water: whether the patient is dehydrated or has an excess of fluids (edema), and (4) functioning ability of the kidneys.

Refer to Table 7-2 for approximate normal values obtained from normal individuals under an average diet and exercise program.

Table 7-3. Conditions accompanied by abnormal sodium values

When increased	When decreased
Nephritis	*Addison's disease*
Pyloric obstruction	*Alkali deficit*
Hypercorticoadrenalism	Dehydration
	Myxedema
	Sprue

SODIUM

Sodium is absorbed into the bloodstream from the small intestine. It plays a major role in controlling the water balance between blood and body cells. The normal concentration of sodium is 136 to 148 mEq/L or 313 to 340 mg/100 ml of serum.[9] A decreased value is of primary concern in Addison's disease—functional failure of the adrenal glands characterized by anemia, digestive disturbances, and bronzelike pigmentation of the skin. Conditions in which increased or decreased sodium values may be expected are given in Table 7-3.

Procedure for sodium

The following general considerations pertain to the procedure for sodium:

1. Specimen: serum or heparinized plasma.
2. Preservation of serum: may be stored overnight in a refrigerator.

The following pages describe a colorimetric procedure. For methods employing a flame photometer, refer to the manufacturer's manual. The Coleman and the IL methods are included here.

Use of Coleman flame photometer, model 21

Since the Coleman flame photometer is widely used, the method for this instrument is presented. The location of the instrument is important for the effectiveness of its performance. The following steps should be followed:

1. Place the instrument on a vibration-free table.
2. Keep the instrument from bright overhead lights or light coming through a window. Stray light will cause a fluctuation of readings.
3. Avoid direct drafts.
4. Keep area above the flame clear to avoid any dirt particles that might fall into the flame and result in erroneous values. Overhead obstacles would also be affected by the constant heat.
5. Ventilation requirements are the same as for a Bunsen burner.
6. Since the instrument is not water-cooled, there is no problem of water supply.

Oxygen

Installation of the oxygen tank should be within 6 feet of the instrument so that there will be easy access to the adjustments. A large tank with a two-stage regulator is preferable. The regulator should read in 1-pound intervals in the range between 10 and 15 pounds per square inch (psi). The oxygen requirement is about 10 to 12 ft^3/hr. A safety surge valve is connected to the output hose nipple of the oxygen regulator. This is a most important part of the instrument. The valve will instantly cut off the flow of oxygen to the flame photometer and remain closed until the differential pressure across it drops below 2 psi if any of the following conditions occur: (1) improper lighting of the flame, (2) accident, or (3) oxygen line rupture or detachment.

Gas

The burner of model 21 will work with either natural gas or any manufactured gas having a BTU (British Thermal Units) of 800 or over. If the gas pressure of the natural gas is low or irregular, a booster pump may be used. If no natural gas is available, use either propane or butane if the city fire laws permit. The propane requirement is 3 ft^3/hr and the butane is 2 ft^3/hr. (CAUTION: The tanks must stand upright to prevent the liquid from entering the regulator.) The regulator should have an output pressure gauge calibrated to allow a setting at 5 psi.

Gas and oxygen adjustment

Often an operator has hesitated to adjust a faulty flame and thus has not obtained full efficiency of the instrument. The procedure is as follows.

Natural or city gas. During his first experience of adjusting the flame the operator may be a bit apprehensive, but the method of adjustment is simple. Refer to Fig. 7-1, which shows the three valves involved in the regulation. Use a large screwdriver (not the pocket-size variety) for the valve adjusting. Better control is maintained for minute adjustments needed. During the adjustment, the machine will be noisy with a husky roar. This is a natural sound when too much gas and oxygen are coming through the mixing valve. If the flame "pops out" during the adjustment, the safety valve on the oxygen tank automatically cuts off the oxygen supply. Just turn the shutoff valve to the left. This cuts off both the gas and the oxygen supply to the burner. Close the oxygen regulator and start over again by lighting the flame.

Procedure

1. **Remove the two panels caps (chrome buttons) beside the shutoff valve. The top cap is the gas regulator and the cap in the five o'clock position is the mixer regulator for the gas and oxygen.**
2. **Remove the chimney for a better view of the flame.**

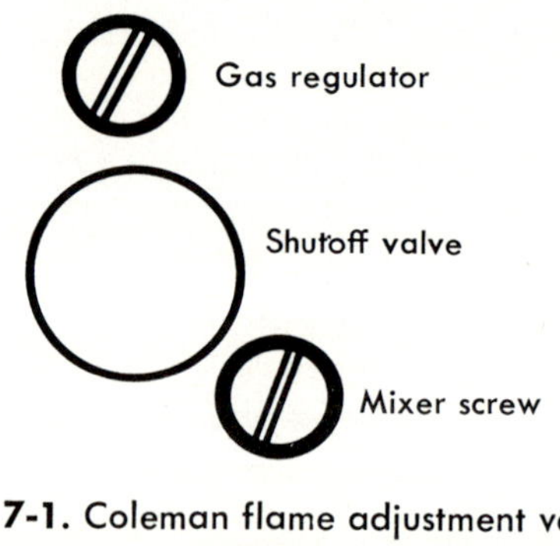

Fig. 7-1. Coleman flame adjustment valves.

3. Be sure the oxygen regulator valve on the oxygen tank is shut off.
4. With a screwdriver, close the mixer valve by turning the screw to the right. Just turn the valve until it stops—do *not* use force! When it stops, the valve is closed.
5. Open the shutoff valve by turning the large knob to the right.
6. Light the gas by holding the match flame inside the top edge of the screen.
7. If the height of the gas column is more than 8 to 10 inches, adjust it by turning the regulator valve to the right until the flame is decreased to the proper height.
8. Now adjust the oxygen regulator on the tank until the gauge reads 3 to 5 psi. The flame should be centered in the cylindrical (large) screen.
9. However, if the flame sounds rough or husky and starts to lift off the burner, slowly turn the mixer valve to the left until the sound is a steady hiss.
10. If a yellow color appears around the base of the flame or if dull yellow streaks appear at the bottom of the inner cone, slowly turn the mixer valve to the right until the yellow disappears. Sometimes just half a turn will correct this.
11. If the base of the flame turns yellow and then to a brilliant yellow-white flare with a blue column to the tip, too much oxygen is in the mixture. Reduce the oxygen by means of the regulator at the tank until the flame lifts from the atomizer tip; then adjust with the mixer valve again.
12. Increase the oxygen regulator until the pressure is 10 psi. *If the flame rumbles or lifts from the burner,* lessen the pressure and adjust the mixer valve by turning to the *right* slowly. *If the flame shrinks* to half its original height, adjust the gas flow by turning the regulator valve to the right until the correct height, of about 3½ inches, is obtained.
13. Increase the oxygen to 13 psi. Adjust with either the gas regulator valve or the mixer valve if necessary.
14. The fine adjustment is now made by slowly turning the mixer valve to the left until a yellow color just appears; then very carefully close the valve to the right until the yellow color disappears. The following conditions should develop:
    a. Slightly fuzzy blue inner cone, of about ¼ to ⅜ inch, burning steadily.
    b. Steady hissing sound, with no great change in sound when atomizing water.
15. If any of the following appear, instead, adjust as indicated:
    a. Unsteady flame—turn the mixer valve to the right until the flame is settled.
    b. Cone that is high and yellow-tipped—too much gas is supplied; turn gas regulator to the left.
    c. Husky roar when atomizing a sample—mixer valve is closed too far; turn the valve back to the left until a steady flame is obtained.
16. Replace the chrome caps into the panels and place the chimney back on the instrument.

When the instrument is operating normally, the burner body will remain around body temperature. However, the cylindrical and conical screens will be hot but not glowing. The unit is not water-cooled because of the "floating flame," resulting in a low burner temperature.

Propane or butane. The adjustment is the same as for natural or city gas, with the exception of using a built-in regulator for the gas tank. Then the valve of the regulator is adjusted for flame height. When operating on propane or butane the shutoff valve on the instrument panel must be left open *at all times* to prevent pressure building up in the gas hose should the tank regulator leak.

Procedure

1. Close the needle valve on the tank regulator by turning the valve to the right.
2. Turn the gas regulator handle to the left until the regulator is closed off and the handle turns freely.
3. Open the shutoff valve on top of the gas tank. Turn about one full turn.
4. Turn the gas regulator handle to the right until 5 psi shows on the gauge.
5. Light a match and hold by the cylindrical screen to light the burner.
6. Turn the needle valve until the flame height is 8 to 10 inches.
7. Adjust the oxygen and the mixer as given in the preceding directions.

Turning flame on

1. First, light the match. *Then* turn the gas on. You may either approach the burner through the door of the instrument or lift off the chimney.
2. As soon as the fluffy gas flame appears, turn the oxygen up to 13 psi slowly while you observe the flame for any abnormalities.
3. Once the flame is adjusted there is usually no problem. Watch to be sure there is plenty of oxygen and gas (if using tanks) before starting.

Turning flame off

1. Turn off the oxygen.
2. Then *immediately* turn off the gas. If using a tank, turn off the gas with the tank regulator or the tank shutoff valve. *Leave the needle valve alone.* If the machine is to be idle for any length of time, shut off both the tank shutoff valve and the regulator.

Cleaning the atomizer

For a trouble-free operation of the flame photometer, the burner system must be clean. During continued use of the instrument, the following units of the burner system may become ineffective due to accumulation of dirt and salts of various specimens. The problem may be solved as follows:

1. If during the operation of the instrument the flame suddently sits at a crooked angle or the atomizer does not take up the solution at a constant rate, close down the instrument. Allow a minute or so for cooling, then remove the atomizer and run a 0.012-inch cleaning wire through the *bottom end* of the capillary. A collection of salt may have accumulated at the atomizer orifice. Clean this by carefully scraping the crust off with your fingernail or a small brush. However, the atomizer should not be allowed to get in this condition.
2. If the wire fails to help the situation and the flame is still crooked, the oxygen line is clogged. With the cleaning tool, carefully thread the tubular shaft over the atomizer capillary. With a careful but firm motion, push straight up against the sliding plug of the atomizer. Do not bend the capillary. The flame should be burning while this is done. By pressure against the sliding plug the capillary will be raised, ejecting anything in the oxygen passage. This will cut off the oxygen supply temporarily and the large fluffy gas flame will appear. When the pressure is released, the oxygen will again flow into the burner and the flame will revert to the original mixture.

Cylindrical and conical flame screens

Another common problem is the accumulation of salts on the screens. At the end of every day, the screens and atomizer should be cleaned. If they are not clean, readings will drift or the flame will persist in showing a luminous light when distilled water is aspirated.

1. Turn the instrument off and allow the screens to cool.
2. With a brush wash the screens, in a solvent if needed; rinse well with tap water and then with distilled water.
3. Dry by shaking or by patting carefully with a towel. Reinstall on the burner.

Precautions

1. All glassware and distilled water must be cation-free. Wash the glassware in either Sterox SE or equivalent. All glassware used for calciums must be acid-washed.
2. It is best to check the new distilled water supply against the old. The water used for making the reagents should be used to dilute the specimens. The mixing of distilled water is one of the largest sources of error in flame photometry. To check the water, prepare a new reagent 1 and read against the old, using a sodium filter. The method is as follows:
   a. The pure reagent 1 (old) is poured into a sample cup or beaker.
   b. The sodium filter is inserted and the machine set at zero without aspirating a sample.
   c. While atomizing the old reagent 1, set the galvanometer on zero (the black scale).
   d. Pour the new reagent 1 into a beaker. If the reading lies within 0.2 division of zero, the water is satisfactory.

Sodium and potassium by flame photometry, using model 21 Coleman flame photometer

This procedure is done from a single dilution of plasma or serum. The volumes of specimen may be as small as 0.02 ml. For preparation of the reagents, reference is made to the manual of the instrument. This manual should be thoroughly studied and all parts of the instrument learned. The direct reading scale panel is recommended to eliminate calibrating the instrument, as well as to keep down a source of error in the final calculations.

Reagents needed

Reagent 5, Sterox SE, 0.02%	Refer to the manual
Reagent 15:	Refer to the manual
Sodium, 1.5 mEq/L	
Potassium, 0.05 mEq/L	
Sterox SE, 0.02%	

Procedure

1. Obtain heparinized blood and centrifuge. Do not allow the specimen to stand longer than necessary.
2. Pipette 0.5 ml of plasma into a 50 ml volumetric flask. Use an Ostwald-Folin pipette. If using a smaller volume of plasma, dilute as follows:
   0.10 ml plasma, use a 10.0 ml volumetric flask
   0.05 ml plasma, use a 5.0 ml volumetric flask
   0.02 ml plasma, use a 2.0 ml volumetric flask
3. Dilute to mark with working reagent 5.
4. Stopper and mix thoroughly.
5. If diluted sample is *over 5.0 ml,* use the regular 10 ml beaker or plastic cup and fill no more than ⅘ full.
   If the diluted sample is *5.0 ml or under,* use the Coleman micro sample container and fill to the top.
6. Turn the galvanometer coarse control to the left completely and the galvanometer fine adjustment halfway to the left. This will allow for adjustments needed during running of tests.
7. Light the flame.

Sodium

8. Place the filter (sodium) into the slot but do not press down until ready to use.
9. Attach the connecting cable in the back of the galvanometer. This connects the galvanometer to the flame photometer.
10. Now press the filter down to release the cutout switch. This switch protects the photometric system by disconnecting the circuit unless the filter is inserted completely into the filter slot.
11. On the flame photometer panel, you will find the *blk coarse* and the *blk fine* controls. Without aspirating any sample, and with the door closed, set the galvanometer index to zero on the left side of the scale.
12. Place a beaker ⅘ full of reagent 15 into the beaker holder attached to the door of the instrument. Close the door.
13. Turn the door latch to the right to raise the beaker to the atomizer.
14. Using the *galv coarse* adjustment and the *galv fine* adjustment, set working standard 15 at 150 mEq/L on the direct reading scale.
15. Clean the atomizer with reagent 5. Check the zero and reset if necessary.
16. Repeat the adjustment with standard reagent 15. It may take two or three adjustments to stabilize the instrument. If it needs these adjustments, make them. Take no shortcuts. This is another area in which gross errors are made.
17. After the machine is properly adjusted, read the plasma dilution.
18. Read the standard before *every* unknown and if necessary adjust the zero again.

Potassium

19. Aspirate distilled water and reagent 5 to cleanse the atomizer.
20. Replace the sodium filter with the potassium filter.
21. Again, zero the machine, with the door closed and nothing aspirating.
22. Set reagent 15 on 5.0 mEq/L on the direct reading scale.
23. Clean the atomizer with reagent 5.
24. Adjust the zero and standard until the machine is stable.
25. Read the plasma potassium, rinsing the machine after every reading *and* checking the standard before every unknown.

Use of Instrumentation Laboratory (IL) flame photometer, model 143

The IL flame photometer (Fig. 7-2) is rapidly gaining popularity because of its ease of operation, simultaneous digital readout of both sodium and potassium, linear range, and use of an internal lithium standard.

Oxidant

The IL flame uses compressed air as an oxidant. A pressure range of 25 to 50 psig of clean air, oil-free and moisture-free, must be maintained with a volume flow rate of 0.5 ft^3/min. Most laboratory compressed air supplies are not clean, and for this reason an air filter must be used. For optimum results the air filter and filter element supplied by IL should be used. An air compressor with filter tank and regulator assembly is also available for this purpose. To ensure adequate and constant delivery pressure, a pressure gauge should be installed on the inlet side of the air filter.

Fuel

The fuel used is 99% propane containing less than 50 ppm of sulfur by weight with a dew point of less than $-10°$ F at one atmosphere. It is supplied in 14-

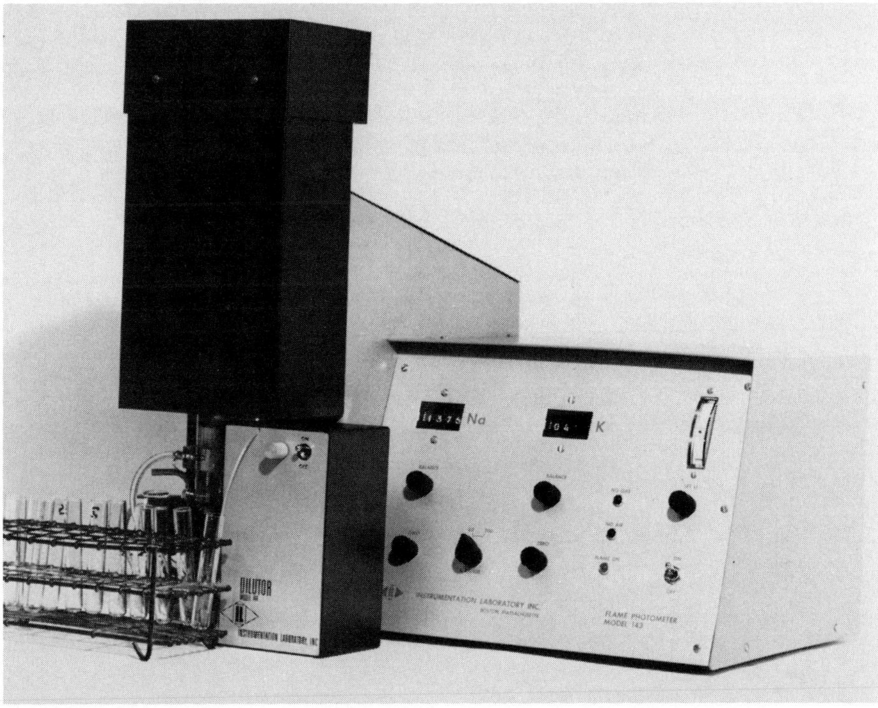

Fig. 7-2. Flame photometer, model 143. (Courtesy Instrumentation Laboratory, Inc., Lexington, Mass.)

ounce cylinders that have a service life of more than 8 hours of continuous operation. *Do not* substitute commercial grade propane for this reagent grade fuel.[14]

The gas and air pressures are regulated to give the proper gas to air ratio.

Flame ignition

Ignition of the flame is automatic and fail-safe. Turning the front panel toggle switch to the *on* position allows ignition to take place. If either air or fuel is not supplied at adequate pressures, all valves will close and a red light will appear on the front panel. A flame detector watches the flame. When a flame is not detected, all valves close and the red light comes on.

Procedure

1. **Open the propane shutoff valve at the top of the fuel cylinder one-half turn counterclockwise.**
2. **If the laboratory compressed air supply is being used, turn air on fully.**
3. **Turn the *on/off* switch to its upper position. The *flame on* indicator light should come on in a few seconds, indicating that the flame has ignited. When operating from an air compressor, this switch also turns the compressor on.**
4. **Place a lithium blank solution or distilled water on the sample stand, raise the stand, and aspirate for at least 5 minutes before calibrating.**

Instrument shutdown

1. **Aspirate lithium diluent for several minutes to thoroughly wash the atomizer and burner system.**

2. Turn the propane shutoff valve clockwise (closed). Finger tight is adequate. Do not overtighten this valve. Leave the lithium diluent in place. When the *flame on* indicator light goes out, the solenoid will shut off the gas supplies.
3. Allow the automatic safety valve at the bottom of the air filter to purge the condensed water vapor from the compressed air supply.
4. When dry, place the *on/off* switch in its lower position.

Sodium and potassium by flame photometry, using model 143 IL flame photometer

This assay is performed on a single 1 to 200 dilution of serum or plasma. Although the stock lithium solution and the sodium and potassium standards could be prepared in the laboratory, it is recommended that they be obtained from IL. Those formulations include a nonionic wetting agent as well as a mold inhibitor to ensure a long shelf-life.

Reagents needed

Lithium stock concentrate, 1500 mEq Li/L — Obtain from IL

Lithium diluent, 15 mEq Li/L

Dilute lithium stock concentrate 1 to 100 with distilled or deionized water.

140 Na/5 K mEq/L stock standard — Obtain from IL

100 Na/100 K mEq/L stock standard — Obtain from IL

Procedure

1. Dilute both standards and unknowns (serum or heparinized plasma and urine) 1 to 200 with lithium diluent. An automatic dilutor, one that will aspirate 0.025 ml of sample and dispense the sample with 4.975 ml of lithium diluent, is very handy for this purpose. However the dilution is accomplished, the standards must be diluted in the same way as the unknowns.
2. When the flame has stabilized, aspirate a fresh lithium diluent. Set the needle on the *lithium response meter* to the center of the reference triangle with the *lithium set control.*
3. Set both the Na and K digital concentration displays to 0000 with their respective zero controls. Lock each control after setting.
4. It is recommended that disposable plastic cups be used to hold standards and samples for analysis. Aspirate an appropriate working standard (140 Na/5 K for serum or 100 Na/100 K for urine, diluted) and set the potassium *range selector* switch to the required position (20 for serum or 200 for urine).
5. Set both the Na and K digital concentration displays to the corresponding values of the standard, using the respective *balance* controls. For example, the 140 Na/5 K working standard would be set to read 1400 and 0500, respectively.
6. Aspirate the diluted unknown specimens. Read and record each concentration directly from the digital concentration display.
7. It is advisable to rinse the atomizer with lithium diluent or distilled water between samples. Calibration should be checked after every five specimens.

System maintenance and troubleshooting

1. The air filter should be dismantled and cleaned at least every 2 months. Clean the plastic bowl and baffle with a mild detergent. Clean the felt filter element in toluene or other cleaner, and blow off with air. If the textile filter element is used, it should be replaced every 6 months. It cannot be cleaned.

2. If the aspirating tube becomes plugged, insert a 0.014-inch diameter stylet

to eliminate blockages. If the aspirator tube becomes bent, it should be replaced.

3. When inspection shows that the Lucite atomizer bowl is dirty and not wetting uniformly or when foreign materials have accumulated in this assembly, it should be removed and cleaned. Carefully disconnect the stainless steel gas line assembly and plastic air line at the front of the instrument. Disconnect the U-tube drain from the drain elbow assembly. The two long screws holding the atomizer assembly to the burners should be removed. Wash the atomizer bowl with a mild detergent and rinse well with distilled water. Carefully reassemble. Be sure the atomizer assembly forms an airtight seal with the burner assembly. Fill the drain tube with water. Aspirate a diluted serum sample for about 10 minutes to rewet the walls of the atomizer chamber.

4. The aspiration rate should be between 35 and 50 sec/ml. This can be adjusted for optimal signal strength by unlocking the setscrews on the side of the atomizer chamber and turning the large knurled nut on the aspirating tube. There are two positions on either side of the maximum that can give this rate. The lower one is preferred.

5. The burner head and ignition block should be cleaned regularly to prevent salt buildup. The burner is rather difficult to get to for a thorough cleaning but a major cleaning job may be delayed if the top of the burner is wiped clean with a damp cloth each day. This can be reached by removing the burner housing cover and the glass chimney. Take care not to damage the ceramic insulator around the spark electrode.

6. Clean the glass chimney with a gauze pad moistened with isopropyl alcohol. Do not use any liquids on the interference filters. Clean only with lens paper.

For more details on the care of the instrument consult the manufacturer's instruction handbook.[14]

Sodium by chemical analysis[3]

Although flame photometry is established in most laboratories, there may be instrument problems or no flame photometer be available. Therefore the method of Albanese and Lein for a chemical determination is given. This method gives a reasonable correlation with a flame photometer method.

With this method, the sodium is precipitated as sodium uranyl acetate. It is then dissolved in water and determined photometrically by the intensity of its yellow color.

Reagents needed

Uranyl zinc acetate reagent (D-163)
Ethanol, 95% (*do not use absolute ethyl alcohol*)
Trichloroacetic acid, 10% (D-164)
Sodium standard, 0.64 mg Na/ml (D-165)

Procedure

Do all tests (unknowns) in duplicate and average results. This will give a closer limit (± 3.6%) in agreement with the flame photometer.

1. **Pipette 0.50 ml of serum into a 15 × 125 mm test tube.**
2. **Pipette 2.0 ml of 10% trichloroacetic acid in the tube and mix.**
3. **Let stand for 5 minutes.**
4. **Set up the following tubes in duplicate:**
   Blank
   0.50 ml of distilled water.

Standard

0.50 ml of standard (0.32 mg of Na) into 3 test tubes.

Unknown

0.50 ml of supernatant into the test tubes.

5. Add 1.0 ml of uranyl zinc acetate reagent to all tubes and mix.
6. Refrigerate for 1 hour; then centrifuge.
7. Carefully pour off and discard supernatant. Drain tubes thoroughly.
8. Wash down the sides of the tube with 2 ml of 95% ethyl alcohol and resuspend the precipitate in the wash fluid. This method of mixing is sometimes called "spanking the tube."
9. Centrifuge the tubes and drain the supernatant as before.
10. Pipette 5.0 ml of distilled water into each tube and dissolve by spanking the tube. If turbid, centrifuge.
11. Transfer to spectrophotometer tubes and read against the reagent blank at 430 mμ.

Calculation

$$\frac{\text{A of unknown}}{\text{A of standard}} \times 0.32 \times \frac{100}{0.1} =$$

$$\frac{\text{A of unknown}}{\text{A of standard}} \times 320 = \text{mg sodium/100 ml}$$

or

$$\frac{\text{A of unknown}}{\text{A of standard}} \times 139 = \text{mEq sodium/L}$$

POTASSIUM

Potassium is absorbed into the bloodstream from the small intestine. A large portion enters the tissues of the body. Here it plays a major role in regulating the distribution of water between the tissues and blood.

The concentration of potassium is normally 3.6 to 5.0 mEq/L or 14 to 19.6 mg/100 ml of serum.[9] Conditions in which increased or decreased potassium values may be expected are given in Table 7-4.

Procedure for potassium

The following general considerations pertain to the procedures for potassium:

1. Specimen: serum or heparinized plasma. The serum or plasma must be obtained 20 minutes after blood is withdrawn from patient.
2. Preservation of serum: may be stored overnight in a refrigerator.

Table 7-4. Conditions accompanied by abnormal potassium values

When increased	When decreased	
Addison's disease	Hyperinsulinism	Hypercorticoadrenalism
Acute infections	Diabetes	Malignant growths
Pneumonia	Hereditary periodic paralysis	Chronic nephritis
Uremia	Overdosage of testosterone	Severe diarrhea
Acute bronchial asthma	Overdosage of desoxycorticosterone	Sprue

Potassium by chemical analysis[15]

Flame photometry is the preferred method for potassium determinations because of the accuracy and the speed of the instrument. However, if the flame photometer is not available, the method of Lochhead and Purcell can be used for the chemical determination. In the determination of potassium, regardless of the method used, there must be *no* hemolyzed cells in the serum. The specimen must be allowed to clot, be centrifuged as soon as possible, and the serum be removed from the clot. The use of heparinized plasma minimizes hemolysis and eliminates the waiting time for the clot to retract before centrifuging.

With the chemical method of analysis given, potassium is precipitated directly from the serum or plasma as potassium sodium cobaltinitrite. The cobalt in the precipitate is determined photometrically. The alkaline solutions of cobalt, in the presence of a trace of an amino acid such as glycine, reduce the Folin-Ciocalteu phenol reagent to a blue color. The intensity of the color is then measured against a standard.

Reagents needed

Sodium cobaltinitrite reagent	(D-166)
Sodium acetate, half-saturated	(D-167)
Wash solution, saturated with potassium sodium cobaltinitrite	(D-168)
Glycine, 7.5%	(D-169)
Sodium carbonate, 25%	(D-170)
Phenol reagent of Folin-Ciocalteu, stock	(D-171)
Potassium standard, 0.2 mg K/ml	(D-172)

Working phenol reagent

Prepare fresh. Mix 1 volume of the stock phenol reagent with 2 volumes of distilled water.

Procedure

Before starting the procedure, refilter the volume of sodium cobaltinitrite needed for the test and make certain that a centrifuge is available for the immediate centrifugation needed throughout the test.

1. **Using 12 ml or 15 ml conical-tipped graduated centrifuge tubes, set up the following in duplicate (the *blank* is set up later).**
   Standard
   0.20 ml of standard. This is a 0.04 mg concentration.
   Unknown
   0.20 ml of serum or plasma.
2. **Pipette 0.20 ml of half-saturated sodium acetate. Mix.**
3. **While constantly shaking the tubes, slowly add 0.5 ml of sodium cobaltinitrite.**
4. **Let the tubes stand at room temperature for 45 minutes. Use timer for exact timing.**
5. **Add 1.0 ml of distilled water. Mix.**
6. **Centrifuge *immediately* for *exactly* 15 minutes at 3000 rpm.**
7. **Carefully pour off supernatant and drain the tubes for about 15 minutes.**
8. **Add 1.0 ml of wash solution without disturbing the precipitate.**
9. ***Immediately* centrifuge for 15 minutes.**
10. **Pour off supernatant and drain the tubes for 5 or 10 minutes.**
11. **Pipette 1.0 ml of 70% ethanol, and mix thoroughly with a glass stirring rod. Add 3.0 ml of 70% ethanol, completely rinsing the precipitate off the stirring rod.**
12. ***Immediately* centrifuge for 5 minutes.**
13. **Pour off supernatant and drain the tubes for at least 5 minutes.**

NOTE: The test may be stopped at this point and continued the next day.

If it is to be held overnight, add 1 drop of distilled water to each tube to replace any lost by evaporation and stopper the tubes firmly.

14. **Label a *blank* tube and add 2.0 ml of distilled water to the blank and all tubes.**
15. **Place the tubes in a boiling water bath for 15 to 20 minutes, until the precipitate dissolves. If a minute amount of precipitate remains, this will not affect the analysis.**
16. **While the tubes are still hot, add 1.0 ml of glycine solution.**
17. **Add 1.0 ml of sodium carbonate solution. Mix.**
18. **Add 1.0 ml of working phenol reagent. Mix.**
19. **Place the tubes in a 37.5° C water bath for 15 minutes.**
20. **Cool to room temperature.**
21. **Add distilled water to the 6.0 ml mark. Stopper and mix.**
22. **Read against a water blank at 660 mμ. Convert the percent transmittance to absorbance.**

Calculation

$$\frac{\text{A of unknown} - \text{A of blank}}{\text{A of standard} - \text{A of blank}} \times 0.04 \times \frac{100}{0.2} = \text{mg K/100 ml}$$

$$\frac{\text{A of unknown} - \text{A of blank}}{\text{A of standard} - \text{A of blank}} \times 5.1 = \text{mEq K/100 ml}$$

CHLORIDE

The majority of methods used for the determination of chloride are based on the principle of precipitating the chloride as an insoluble salt. The analysis is done on either the precipitate or the excess precipitant. However, one of the most common methods of determining chloride concentration is the method of Schales and Schales, who in 1941 adapted the mercurimetric method for use in chloride determination. The chloride ion, in a tungstic acid filtrate of plasma or serum, is titrated with a standard solution of mercuric ions, with the formation of the soluble compound mercuric chloride ($HgCl_2$), which does not dissociate to form mercuric ions. The organic indicator 5-diphenylcarbazone yields a purple color when the excess mercury combines with the indicator. The color is stable although rather difficult to determine with urine, pleural fluid, and serum with elevated bilirubin.

Since the titration end point is dependent upon the individual's experience and eyesight, potentiometric methods have gained popularity by being simple, fast, accurate, and versatile. Icteric sera or turbid specimens pose no problem with these methods. In general, potentiometric methods utilize two electrodes, a voltmeter, and a galvanometer. The electrodes are immersed in a dilute nitric acid solution containing the specimen to be analyzed. By the means of adding silver ions with a microburette the chloride is precipitated as silver chloride. When the chloride has been precipitated, free silver ions change the potential, resulting in a shift in the reading on the galvanometer.

A recently manufactured instrument has eliminated the need for adding the silver ions with a microburette. The Buchler-Cotlove Chloridometer (Fig. 7-3) has a spool of silver wire that is immersed with the electrodes in a dilute nitric acid solution. The small stirrer, also attached to the electrode assembly, thoroughly mixes the specimen as it is being analyzed. With this instrument, the silver ions are released into the solution at a constant rate and when the chloride in the solution has been used completely, the machine automatically shuts off. Since the silver ions are being added at a constant rate, the amount of chloride

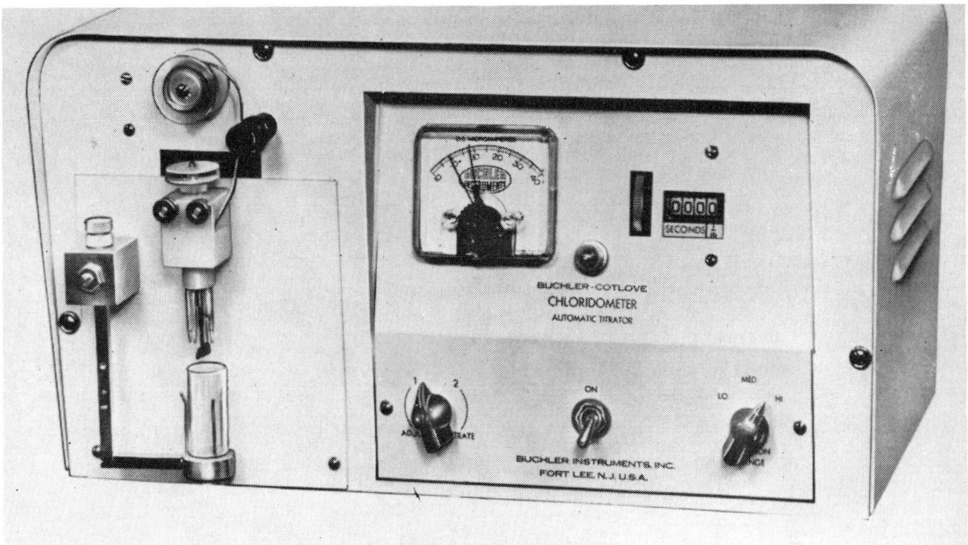

Fig. 7-3. Buchler-Cotlove Chloridometer. (Courtesy Buchler Instruments, Inc., Fort Lee, N. J.)

in the sample is proportional to the length of time in seconds necessary to complete the titration. A reagent blank and chloride standard are run and the reaction time of the unknown is calculated against the standard. A direct reader can be obtained with the instrument, which eliminates calculations. This potentiometric method is very satisfactory because it has greater sensitivity than the titration or colorimetric methods. The operation is simple, rapid, and automatic, eliminating technical errors and time-consuming precipitation of serum, tissue, urine, spinal fluid, or other biologic fluids.

Large quantities of bromide given to a patient will result in a falsely elevated chloride. This may be corrected by converting the bromide mg/100 ml to bromide mEq/L. The bromide value is then subtracted from the chloride value. If the bromide is calculated as sodium bromide, convert to bromide. Refer to Chapter 17. The calculation of an example is as follows:

$$\text{Cl(mEq/L)} - \frac{\text{Bromide (mg/100 ml} \times 10)}{80} = \text{True chloride value}$$

$$112 - \frac{120 \times 10}{80} =$$

$$112 - 15 = 97 \text{ mEq chloride/L (true chloride value)}$$

Chloride is absorbed into the bloodstream from the small intestine.

Note that whole blood has lower values than serum or plasma. Red cells, which make up almost half of whole blood, contain much less chloride than serum or plasma.

The normal concentration of chlorides is 95 to 103 mEq/L or 337 to 366 mg/100 ml of serum or plasma and 119 to 130 mEq/L or 422 to 461 mg/100 ml of spinal fluid.[9]

An increase in the chloride value is of primary concern in nephritis, prostatic obstruction, and eclampsia (convulsions). Conditions in which an increased or a decreased chloride level may be expected are given in Table 7-5.

Table 7-5. Conditions accompanied by abnormal chloride values

When increased	When decreased	
Nephritis	Addison's disease	Ether anesthesia
Prostatic obstruction	Burns	Typhus fever
Eclampsia	Diabetes	Anaphylactic shock
Anemia	Fevers	Uremia
Cardiac conditions	Intestinal obstruction	Vomiting
Hyperventilation	Metallic poisoning	Polycythemia
Hypoproteinemia	Pneumonia	Profuse sweating
Serum sickness	Heat cramps	Fasting
Urinary obstruction	Diarrhea	Hypercorticoadrenalism

The following methods are commonly used for the determination of chloride.

Whitehorn method

A protein-free filtrate is made. Nitric acid and a *measured excess* of standard silver nitrate solution are added to precipitate the chloride as silver chloride. An indicator is added and the excess silver nitrate is titrated with a standard thiocyanate solution. The amount used is an index of the chloride content. This is found by calculation.

Schales-Schales method

A protein-free filtrate is made. An indicator is added. The filtrate is then titrated with a standard mercuric nitrate solution. This reacts with the chloride to form undissociated mercuric chloride. When all the chloride has been removed, the addition of more mercuric nitrate gives a color reaction with the indicator. The amount of standard mercuric nitrate used in the titration is an index of the chloride content. This is then found by calculation.

Sendroy method

Serum or plasma is obtained. A phosphoric-tungstic acid solution is added to precipitate the proteins. Excess silver iodate is added to precipitate the chloride as silver chloride. This leaves free iodate which is proportional to the amount of silver chloride precipitated. The mixture is filtered. The free iodate is converted to free iodine by the addition of phosphoric acid and potassium iodide. The free iodine produces a color, the depth of which is measured in a colorimeter.

Procedures for chloride

The following general considerations pertain to the procedures for chloride:

1. **Specimen: serum or plasma. Do not use a tourniquet when withdrawing blood. If it is necessary to use a tourniquet to locate the vein, release it before withdrawing the blood. NOTE: the serum or plasma should be obtained 30 minutes after the blood is withdrawn.**
2. **Preservation of serum or plasma: may be stored overnight in a refrigerator.**

Whitehorn titration method*

The standard thiocyanate solution which is listed below may be made of sodium, potassium, or ammonium thiocyanate. Thiocyanate is often called sulfocyanate, and the ferric ammonium sulfate is often referred to as ferric alum.

Reagents needed

Concentrated nitric acid	
Ferric ammonium sulfate	
Standard silver nitrate solution	(D-173)
Standard thiocyanate solution	(D-174)

Procedure

1. Make a protein-free filtrate.
2. Pipette 10 ml of the filtrate into a porcelain dish. Add exactly 10 ml of the standard silver nitrate solution and *stir thoroughly* with a glass rod. Add 5 ml of the concentrated nitric acid and stir. Let stand 5 minutes to allow precipitation of the chloride as silver chloride.
3. Using a spatula, add about 0.3 g of the indicator, ferric ammonium sulfate.
4. Titrate with the standard thiocyanate solution until the salmon-red color of ferric thiocyanate persists for 15 seconds.

Calculation

1 ml of the standard thiocyanate solution is equivalent to 1 ml of standard silver nitrate which, in turn, is equivalent to 1 mg of sodium chloride. To find the number of milligrams of sodium chloride/100 ml of blood, proceed as follows. Subtract the milliliters of standard thiocyanate required from the milliliters of standard silver nitrate used (10 ml). Now multiple by 100. Example: Suppose 5.5 ml of the standard thiocyanate solution was required in the titration.

$$\text{Standard silver nitrate used} - \text{Standard thiocyanate required} \times 100 = \text{mg NaCl/100 ml blood}$$

$$(10) - (5.5) \times 100 = 450 \text{ mg NaCl/100 ml blood}$$

The figure 100 appears in the calculation because the 10 ml of filtrate used represents only 1 ml of blood. This 1 ml of blood must be multiplied by 100 to give the report in the usual form: mg/100 ml.

This titration is what is known as a back-titration. We know that we have added exactly 10 ml of the standard silver nitrate and we are titrating back to find out how much was not used to precipitate the chloride. In the above example, 5.5 ml of the silver nitrate was *not* used. Therefore, 4.5 ml was used (10.0 − 5.5 = 4.5), and each milliliter of this 4.5 represents 1 mg of chloride expressed as sodium chloride.

Schales-Schales titration method†

Reagents needed

Diphenylcarbazone indicator	(D-175)
Standard sodium chloride, 10 mEq/L or 58.5 mg/100 ml	(D-176)
Standard mercuric nitrate solution	(D-177)

Procedure

1. Make a tungstic acid filtrate.
2. Pipette 2 ml of the filtrate into a small flask.

*Adapted from Whitehorn, J. C.: System of blood analysis, J. Biol. Chem. **45**:449, 1921.

†Adapted from Schales, O., and Schales, S. S.: Simple and accurate method for determination of chloride in biological fluids, J. Biol. Chem. **140**:879, 1941.

3. Add 4 drops of the indicator diphenylcarbazone. Using a microburette graduated in hundredths, titrate with the standard mercuric nitrate solution. The end point is reached when the solution turns a pale violet or light purple color.

Calculation

Where E equals the number of milliliters of mercuric nitrate solution required for 2 ml of standard sodium chloride solution:

$$\text{ml standard mercuric nitrate used} \times 5.85 \times \frac{100}{E} = \text{mg sodium chloride/100 ml blood}$$

CO_2 CONTENT

*Method using the Van Slyke manometric apparatus**

Reagents needed

Caprylic alcohol
Lactic acid, 0.1 N (D-178)
Sodium hydroxide, 5 N (carbonate-free) (D-179)
Mercury

Procedure

1. Before beginning analysis for CO_2, remove metal jewelry such as rings and watches. Mercury forms amalgams with most metals, so do not dip coins or trinkets into the mercury. Although there is a temptation for the inexperienced operator to "play" with this element it must be stressed that continual carelessness may result in mercurial poisoning.
2. Expel water from the chamber by placing a vacuum hose into the cup and as the water is suctioned out of the cup, lift the mercury bulb carefully so the remaining water in the chamber and a few drops of mercury are expelled into the waste jar. Close stopcock 1 (Fig. 7-4). Place the leveling bulb in lower position.
3. Add 3 drops of caprylic alcohol to the cup and carefully turn stopcock 1 until the stopcock channel is filled. Caprylic alcohol prevents foaming.
4. Add 3.0 ml of 0.1 N lactic acid to the cup. The lactic acid liberates the carbon dioxide.
5. Using a Van Slyke pipette (or a 1 ml graduated measuring pipette but not a blow-out pipette), add 0.5 ml of serum or plasma to the cup. By placing the tip of the pipette into the lactic acid near the bottom of the cup before releasing the serum, contact with air is avoided. For the *blank*, use 3.5 ml of lactic acid instead of 0.5 ml of serum and 3.0 ml of lactic acid.
6. Slowly turn stopcock 1, releasing the serum and lactic acid into the chamber. Allow a drop of the caprylic alcohol to remain behind.
7. Seal the stopcock channel with mercury and lower the bulb to bring the mercury meniscus down to the 50 ml mark. With the bulb in the lower position, agitate the contents in the extraction chamber for 2 minutes.
8. Carefully turn stopcock 2 and let the solution rise (without oscillation) to the 2 ml mark.
9. Read the mercury manometer for P_1. The P_1 equals the millimoles of the initial gas pressure.
10. Add 0.3 ml of 5 N NaOH to the cup.
11. Turn stopcock 1 carefully, admitting the alkali to the reaction chamber. No air must enter the chamber.
12. Seal the stopcock with mercury, and by raising and lowering the bulb three times, raise and lower the contents of the chamber. This ensures complete absorption of carbon dioxide by the alkali.

*Arthur H. Thomas Co.

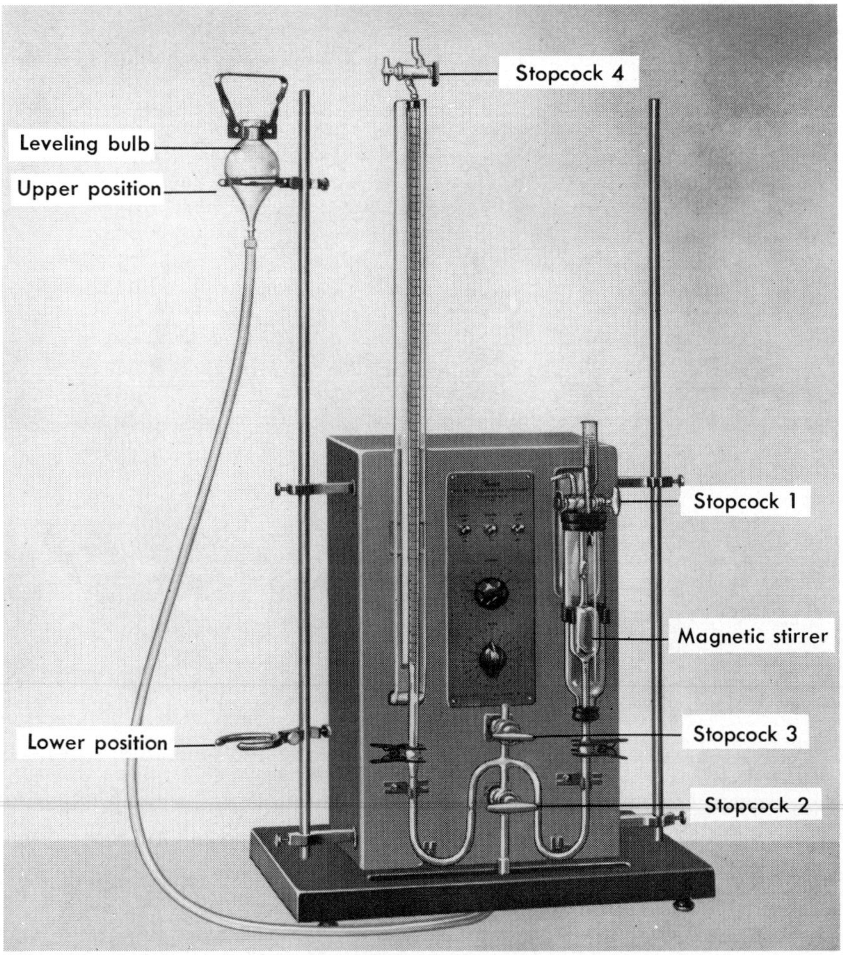

Fig. 7-4. Van Slyke manometric blood gas apparatus. (Courtesy Arthur H. Thomas Co., Philadelphia, Pa.)

Calculations

The *blank* value represents the amount of carbon dioxide in the reagents. The value usually runs between 1 and 4 mm. The P_1 value equals millimeters of the initial gas pressure. The P_2 value equals millimeters of pressure due to all gases except carbon dioxide:

$$P_1 - P_2 = \text{mm of carbon dioxide pressure}$$

The millimeter value of CO_2 pressure in the *blank* is subtracted from that of the serum, and the result is multiplied by the appropriate temperature factor (Table 7-6):

$$(P_1 - P_2) - \text{Blank} \times \text{Factor} = \text{mEq/L}$$

Care of the Van Slyke apparatus

Since the stopcocks are under pressure, they must be kept well greased. With the Arthur H. Thomas unit, whose method is described, the apparatus may be dismantled for cleaning. The ball-and-socket joint makes dismantling extremely

Table 7-6. Temperature factors for CO_2 determinations

Temperature (°C)	Factor
18	0.2416
19	0.2404
20	0.2392
21	0.2380
22	0.2366
23	0.2354
24	0.2342
25	0.2330
26	0.2320
27	0.2308
28	0.2298
29	0.2286
30	0.2276
31	0.2266
32	0.2256
33	0.2246
34	0.2236

easy. However, the unit may be cleaned without being dismantled. The following method may be used for the old and the new Van Slyke units.

1. Attach a vacuum line to the stopcock 4 at the top of the manometer column and suction out all the mercury.
2. With stopcock 1 open, and the vacuum line attached to the open stopcock 4, pour acetone into the cup until the complete system is filled.
3. Close stopcock 4 and allow the acetone to remain in the unit for 15 minutes to dissolve the grease particles.
4. Keeping stopcock 2 closed at all times, drain the system by vacuum through stopcock 3.
5. Repeat, filling the system with acetone, and let the acetone remain in the system for 15 minutes. After draining the acetone through stopcock 3, the system should be free of grease particles.
6. Fill the system with either chromic acid or 10% nitric acid. Let the cleaning solution remain in the system for at least 1 hour before draining.
7. Rinse the system at least three times with distilled water, then twice with acetone.
8. Leaving stopcocks 1 and 4 open, attach the vacuum to stopcock 3 to completely dry the system.
9. Carefully grease the stopcocks and fill the system with clean mercury.

Cleaning mercury

A simple way to clean mercury is to mix the mercury with sugar and then dissolve the sugar by washing with water. Place the dirty mercury in a large mortar. Pour the sugar into the mortar and grind the sugar and mercury with a pestle. The dirty sugar is then rinsed or dissolved off the mercury by running water. The water may be vacuumed off. The procedure is repeated until the sugar remains clean. All moisture is then vacuumed off the mercury, and it is ready to be filtered into the storage bottle. Place a small pinhole in the bottom of the filter paper. The mercury is then filtered to remove small traces of dirt not previously removed and to absorb any remaining moisture.

BLOOD pH, Pco_2, AND Po_2

The pH is a measure of the hydrogen ion concentration. The normal pH values for blood are 7.35 to 7.45. Increased pH values are found in uncompensated alkalosis. Decreased pH values are found in uncompensated acidosis.

Blood pH may be determined electrometrically, colorimetrically, or from calculations by applying the carbon dioxide tension and bicarbonate content of plasma in the Henderson-Hasselbalch equation. (Refer to Chapter 8.) The pH of whole blood and that of plasma are considered to be the same for practical application.

The colorimetric procedures employ indicator color changes that have been established by potentiometric methods and correlated with hydrogen ion concentrations. Unless the methodology is carefully controlled, errors of as much as 1.0 pH may occur. The measurements are affected by temperature, indicator characteristics, color of the sample, salt concentration, and colloidal material in the sample.

In the electrometric method the pH is measured by determining the voltage developed by two electrodes in a solution. In many of the earlier investigations of blood pH a hydrogen electrode was employed. However, a glass electrode was subsequently developed that was more practical for measuring the pH of biologic fluids. The electrode consists of a glass tube closed at the bottom by a thin-walled bulb of soft glass that has high electric conductance. The bulb contains a buffered chloride solution and a silver chloride–coated wire that is connected to a conductor cable which may be attached to a pH amplifier. A theoretical explanation of the response of the glass electrode to hydrogen ion concentration is lacking. A potential is developed at the glass-solution interface that depends upon the pH of the solution in which the electrode is immersed. For measurement of the potential a second or reference electrode is required—to complete the electric circuit and to furnish a constant potential with which the varying potential of the glass electrode can be compared. As the reference electrode, a calomel electrode, prepared from mercury and mercurous chloride in an aqueous solution of potassium chloride, may be used. It is usually connected to the unknown solution by a saturated potassium chloride bridge.

The pressure of carbon dioxide (Pco_2) in arterial blood may be determined by an indirect calculation of the pressure of CO_2 from a nomogram (Fig. 7-5) based on the Henderson-Hasselbalch equation. This equation is based on three unknowns: the blood pH, the plasma bicarbonate concentration, and the partial pressure of the CO_2 in the blood. If any two are known, the third may be calculated. The nomogram provides a means of obtaining a Pco_2 value without long calculations.

Great care must be used in doing a blood pH determination, for an error of 0.02 unit can mean an error of about 1.8 mm Hg in the Pco_2 value.

An example in using the nomogram is as follows:

A straight line drawn through the given points on any two scales crosses the other two scales at points indicating simultaneously occurring values. For example, a patient has the following results: an arterial pH of 7.40 and a total

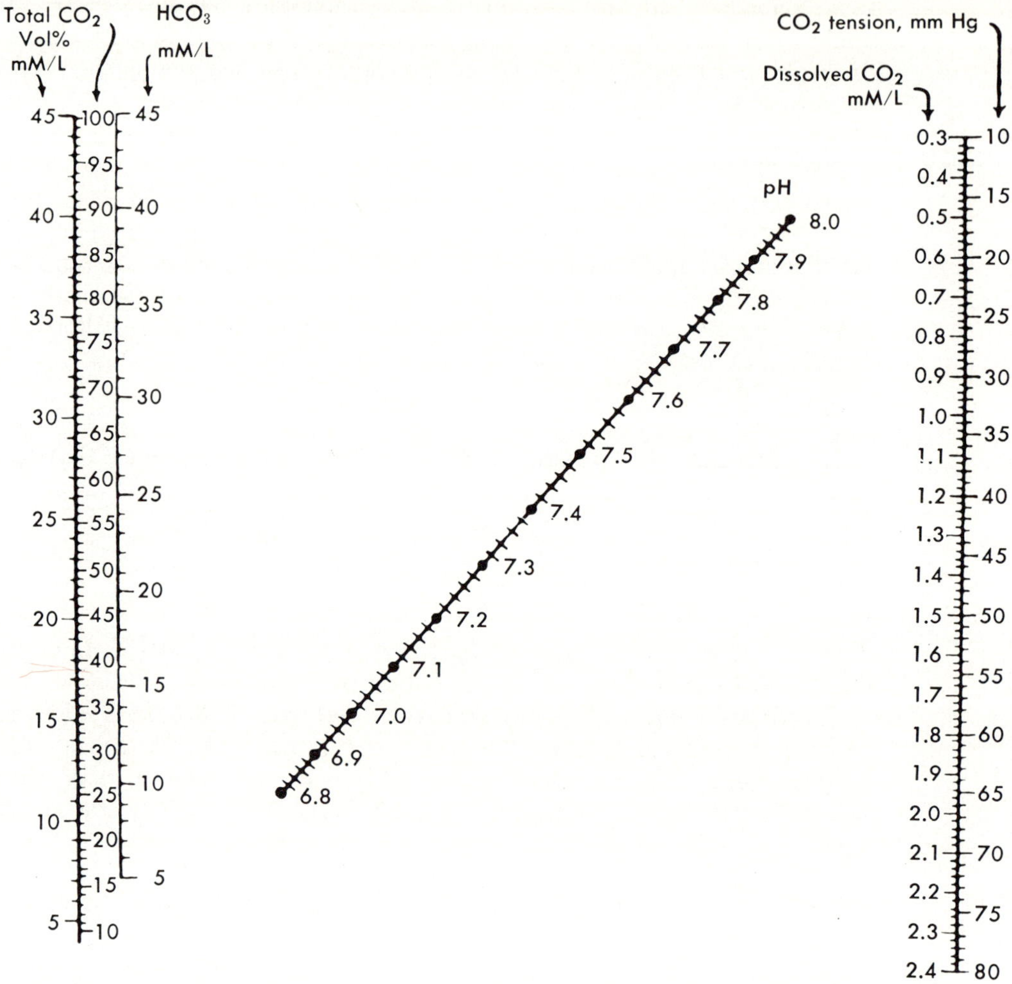

Fig. 7-5. Nomogram for calculating P_{CO_2}. (From McLean, F. C.: Physiol. Rev. **18**:495, 1938.)

CO_2 of 28 mEq/L (or millimoles per liter). The P_{CO_2} value is needed. Draw a line from the total CO_2 column reading 28 through the pH column, intersecting the 7.40 point. The line intersecting the mm Hg column gives a value of 45. Therefore, the P_{CO_2} is 45 mm Hg.

Astrup method

During the great Danish polio epidemic of 1952 simpler and faster methods of determining the acid-base status of patients were needed. At that time Astrup started measuring blood pH routinely; and gradually with Siggaard-Andersen and their associates,[1,2,19,20] a unique system was developed for evaluating acid-base metabolism, which included those blood constituents they considered necessary: the actual pH, carbon dioxide tension, standard bicarbonate, and base excess or deficit. The procedure involves measuring both the actual blood pH and the pH after equilibration of two portions of the blood with carbon dioxide–oxygen gas mixture of two known CO_2 tensions, one higher and one lower than normal. The values obtained are applied to a nomogram to derive other pertinent information.

Terminology

Astrup and associates[5] have defined the terms that they employ in describing acid-base metabolism.

actual pH A measure of the hydrogen ion concentration of anaerobically drawn blood at the patient's partial pressure of CO_2 and temperature. It reflects total acid-base changes caused by variation in the P_{CO_2} and/or base excess (respiratory and nonrespiratory disturbances). A decreased pH is termed total acidosis and an increased pH, total alkalosis.

actual P_{CO_2} (mm Hg) The partial pressure of CO_2 or the CO_2 concentration in a gas phase in the patient's blood drawn anaerobically. As a measure of the respiratory component, it is directly proportional to the CO_2 production and inversely proportional to the respiration. The P_{CO_2} multiplied by 0.03 will give the concentration of carbonic acid in plasma in millimoles per liter. When the P_{CO_2} is decreased, the condition is termed respiratory alkalosis and when increased, respiratory acidosis.

base excess (BE) (mEq/L) The nonrespiratory quantity, a measure of the excess or deficit of base. By definition it is zero for fully oxygenated blood with pH of 7.40 at P_{CO_2} of 40 mm Hg. When nonvolatile acid accumulates in the blood, the BE value decreases; and when base accumulates or nonvolatile acid is removed, the BE value increases. A negative BE value is termed nonrespiratory acidosis and a positive BE value, nonrespiratory alkalosis.

buffer base (BB) (mEq/L) A nonrespiratory quantity introduced by Singer and Hastings, considered to be the sum of all buffer anions that are available to accept hydrogen ions, mainly bicarbonate and proteinate ions. At a constant base excess it varies with the protein and hemoglobin concentrations. However the compounds that function as buffer bases depend upon the pH of the blood.

standard bicarbonate (mEq/L) The concentration of bicarbonate in plasma from blood that has been completely oxygenated to saturate the hemoglobin and equilibrated with a P_{CO_2} of 40 mm Hg.

actual bicarbonate (mEq/L) A measure of the bicarbonate of the plasma from anaerobically drawn blood.

total CO_2 (mEq/L) The CO_2 of plasma obtained from the bicarbonate and carbonic acid in anaerobically drawn blood.

acidosis A condition caused by the accumulation in the body of an excess of acid or the loss from the body of base.

alkalosis or basosis A state caused by accumulation in the body of an excess of base or by the loss of acid.

Apparatus

The system (Fig. 7-6) is modular and consists of the following:

A *pH meter* with an expanded scale. The pH scale is based upon a series of standard buffer solutions.

A *micro capillary electrode unit* with a thermostated capillary glass electrode and a calomel electrode. The glass electrode is a pH-sensitive glass capillary inside a glass tube containing a liquid of constant pH. It is inserted in a glass water jacket that is maintained at constant temperature by circulation of water from the circulating water thermostat. The electrode is connected by a cable to the pH meter. An exchangeable polyethylene capillary inserted in the polyvinyl chloride adapter provides the means for filling the glass capillary electrode. Over the end of the electrode is a cap, connected to thc suction pump. On closing, this provides light suction to draw solutions and air into the electrode. The reference or calomel electrode is inserted in a U-tube system with saturated potassium chloride solution. It is maintained at constant temperature by circulating in the water jacket water from the circulating water thermostat. A porous plate that

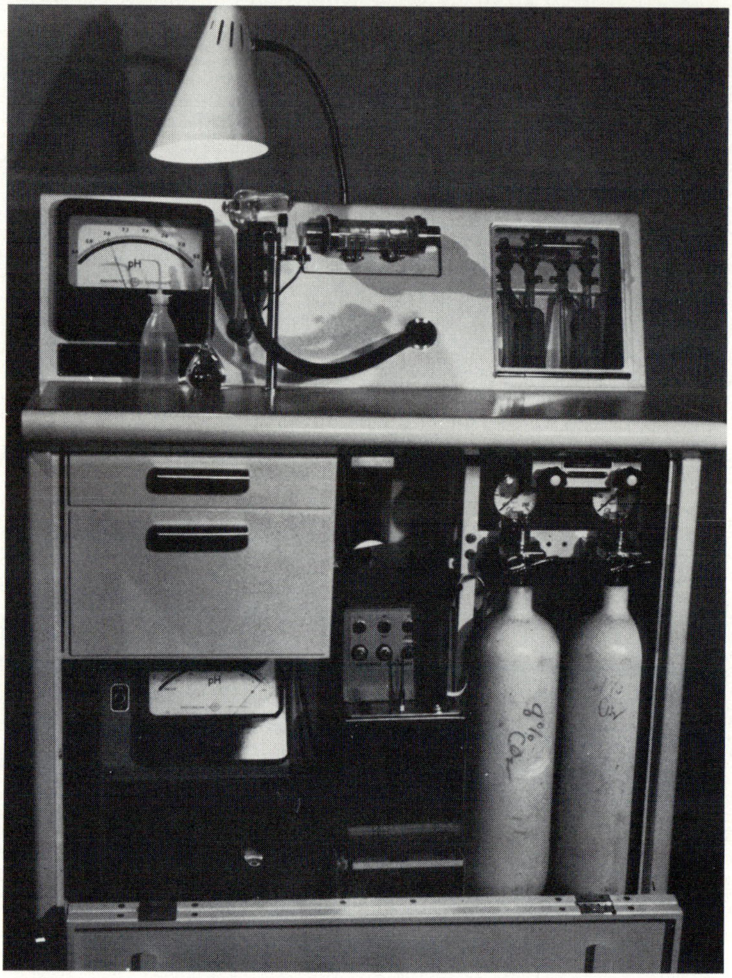

Fig. 7-6. Astrup blood micro system.

protects it from contamination forms the base of a compartment into which the capillary attached to the glass electrode and containing the sample is dipped. Thus a liquid junction in the electrode system is established.

An *equilibration chamber and humidifier unit* or microtonometer. This equipment equilibrates simultaneously two samples of blood in duplicate with gas mixtures at two different carbon dioxide tensions. The equilibration chamber consists of two pairs of glass tubes sealed into a water jacket maintained at 37° C and mounted on a shaft connected to a motor for vibration. The two gases are saturated with water vapor in the thermostated humidifiers and are allowed to flow to the equilibration apparatus through tubes maintained at 37° C. With the gases conditioned, no evaporation or condensation of water occurs in the equilibration tubes.

Water circulation thermostat. This thermostat with a built-in suction pump is a 7 L stainless steel tank that maintains water at any setting between 10° C and 60° C within $\pm 0.1°$ C. The pump circulates 16 L of water each minute and also acts as a suction device to draw the sample, buffers, and wash solutions through the glass electrode. The temperature should be the temperature of the patient, usually 37° C.

Two gas cylinders and two sets of reduction valves and gauges for oxygen–carbon dioxide gas mixtures having approximately 4% and 8% CO_2, respectively. The exact CO_2 content is known and the carbon dioxide tension is calculated daily. The gauges are set to permit in each tube a gas flow of approximately 50 ml/min.

All the units and accessories can be mounted on a portable cart.

Reagents needed

Gas mixtures, oxygen–carbon dioxide, in cylinders containing approximately 4.0% and 8.0% carbon dioxide, respectively, but accurately determined. Analyzed gases may be obtained commercially. The P_{CO_2} is calculated daily, using the following equation:

$$P_{CO_2} = \frac{B - 47}{100} \times CO_2 \text{ percent}$$

where B is the barometric pressure in mm Hg and 47 is the vapor pressure of water at 37° C.

Standard buffer solutions:

Two primary standards are prepared according to the National Bureau of Standards using salts obtained from the Bureau:

pH 7.386 at 37° C: dissolve 1.179 g KH_2PO_4 and 4.302 g Na_2HPO_4 in freshly boiled distilled water and dilute to 1 L at 25° C.

pH 6.840 at 37° C: dissolve 3.388 g KH_2PO_4 and 3.533 g Na_2HPO_4 in freshly boiled distilled water and dilute to 1 L at 25° C.

Secondary standard, commercial concentrated buffer standard, pH approximately 7.0. Dilute according to directions, usually 1:25, and standardize against the buffers prepared according to the National Bureau of Standards. Prepare fresh daily for checking the pH meter setting.

Potassium chloride solution, saturated.

Specimen collection (Fig. 7-7)

Warm a finger for 5 to 10 minutes in water at approximately 45° C or warm an earlobe, using an electric light bulb. The skin is punctured with a lancet and blood allowed to flow freely; squeezing must be avoided. After the first drop of blood is wiped away, 3 or 4 glass capillary tubes (50 to 75 µl) containing heparin are filled. One end of each tube is sealed with modelling wax, a short piece of steel wire inserted, and the other end sealed. The blood is mixed with the heparin by moving a magnet along the outside of the tube. If the analysis is not to be completed within 30 minutes, store the capillaries in a horizontal position at 4° C or in ice water. The values obtained for arterialized capillary blood are the same as for arterial blood.

Procedure

1. **Adjust the pH meter weekly, using the two primary buffers, and check repeatedly for drift between pH measurements with a secondary buffer that is prepared fresh daily.**
2. **Rinse the glass electrode with distilled water and dry by sucking air through it. The buffer and water should be prewarmed (37° C) as room-temperature solutions and prolonged air drying produce a cooling effect, resulting in lower pH values.**
3. **Remix the blood in one of the capillaries, cut off the seal from one end, and remove the stirring wire. Maintain the capillary in a horizontal position, break the seal from the closed end, and insert the polyethylene capillary of the glass electrode. Using suction, fill the polyethylene capillary and electrode capillary by pressing the suction cap against the plastic tip at the back of the glass electrode.**
4. **The glass electrode unit is replaced in the holder and the polyethylene**

Text continued on p. 196.

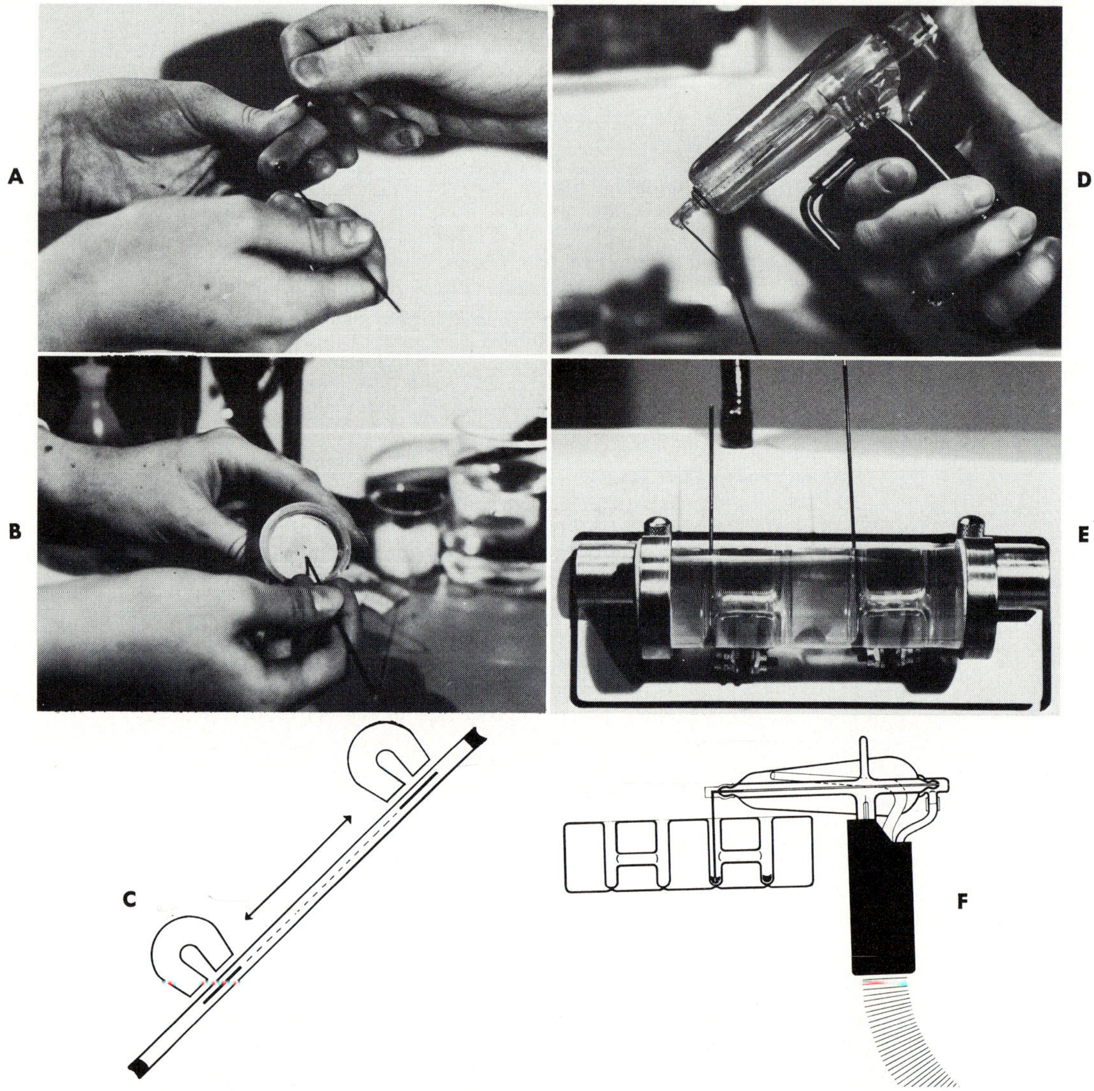

Fig. 7-7. Specimen collection procedure. **A,** Sample capillary with heparin is anaerobically filled with blood by placing tip of capillary in center of blood drop. **B,** One end of capillary is closed with sealing plasticine, a piece of iron (3 to 4 mm) is inserted, and finally, the other end is closed. **C,** Blood is mixed by moving a magnet back and forth along the capillary. Where plasticine and blood meet, ends of the capillary are cut with glass cutter. The two ends can then be broken off. Iron wire is removed with magnet **D,** Plastic tip of glass electrode is inserted in capillary, blood is sucked into electrode, and pH is measured. **E,** Blood from two other capillaries is introduced into the thermostated equilibration chambers and shaken for 3 minutes while O_2-CO_2 mixtures of known P_{CO_2} bubble through. Equilibration is made at a low P_{CO_2} (about 30 mm Hg) in the left-hand chamber, and at a high P_{CO_2} (about 60 mm Hg) in the right-hand chamber. **F,** After equilibration, the blood is sucked into the electrode and the pH is measured as before. Actual pH of the blood and pH at two different CO_2 tensions are now known. By use of the nomogram, all relevant acid-base values can be computed.

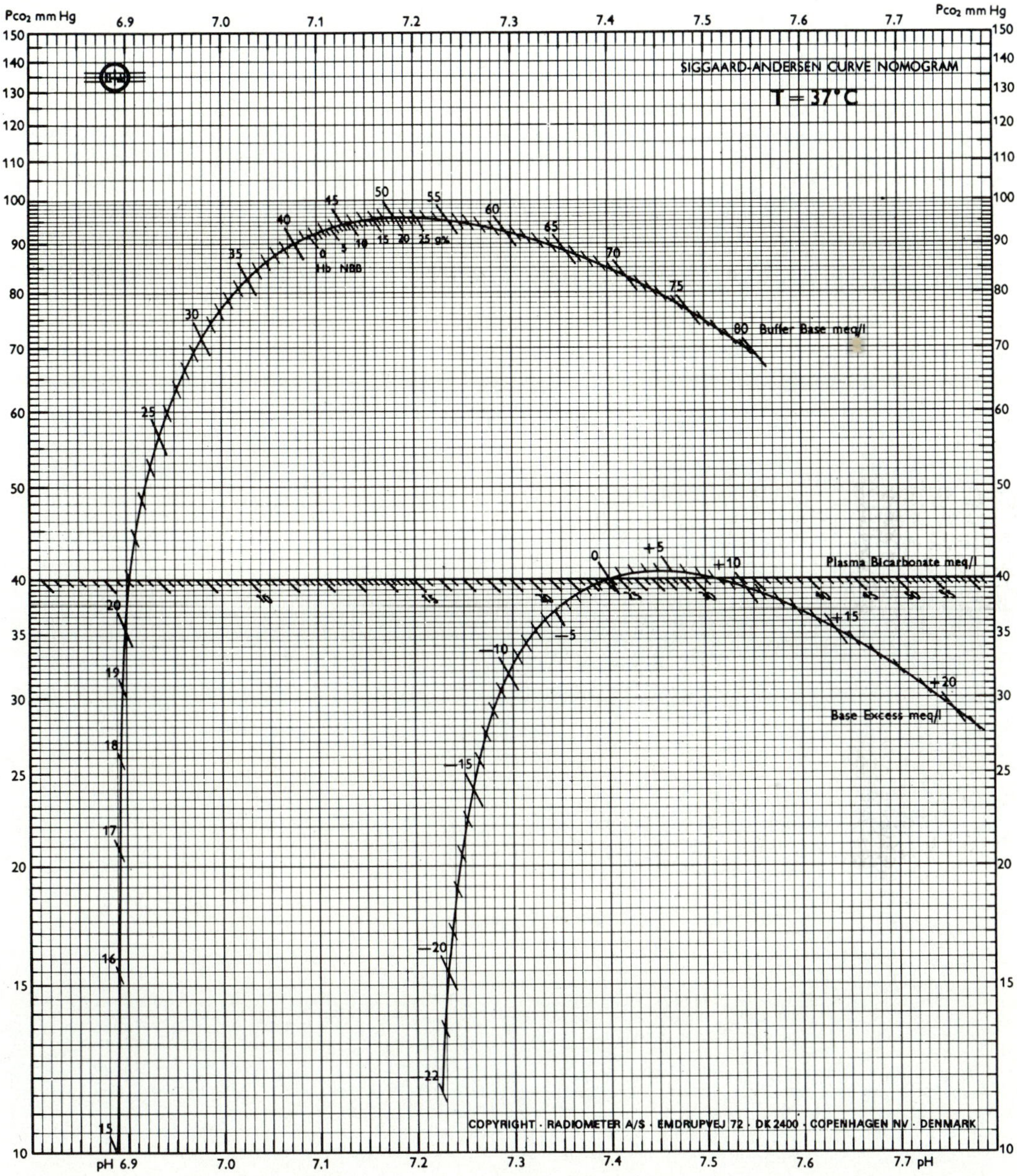

Patient's name:		Barometric pressure	mm Hg	READINGS			RESULTS	
		CO_2 percentage	Cylinder No 1: %	Before equilibration	Actual pH:		Actual Pco_2	mm Hg
Dept:	Sample No.:		Cylinder No 2: %				Base Excess	meq/l blood
Date:		CO_2 partial pressure	Cylinder No 1: mm Hg	After equilibration	high Pco_2	pH:	Buffer Base	meq/l blood
Hour of Sampling:			Cylinder No 2: mm Hg		low Pco_2	pH:	Plasma Bicarb. at Pco_2 = 40 mm Hg	meq/l plasma
Remarks:		Hemoglobin:	g/100 ml	Readings made by:			Actual Bicarb.	meq/l plasma
		Oxygen Saturation:	percent	Signature:			Total CO_2	meq/l plasma

Code: 984-201

References: Siggaard-Andersen, O. and Engel, K.: Scand. J. clin. Lab. Invest. 12, 177, 1960. Radiometer Reprint AS7
Siggaard-Andersen, O.: Scand. J. clin. Lab. Invest. 14, 598, 1962. Radiometer Reprint AS18

Fig. 7-8. Siggaard-Andersen curve nomogram. (Courtesy The London Co., Cleveland, Ohio.)

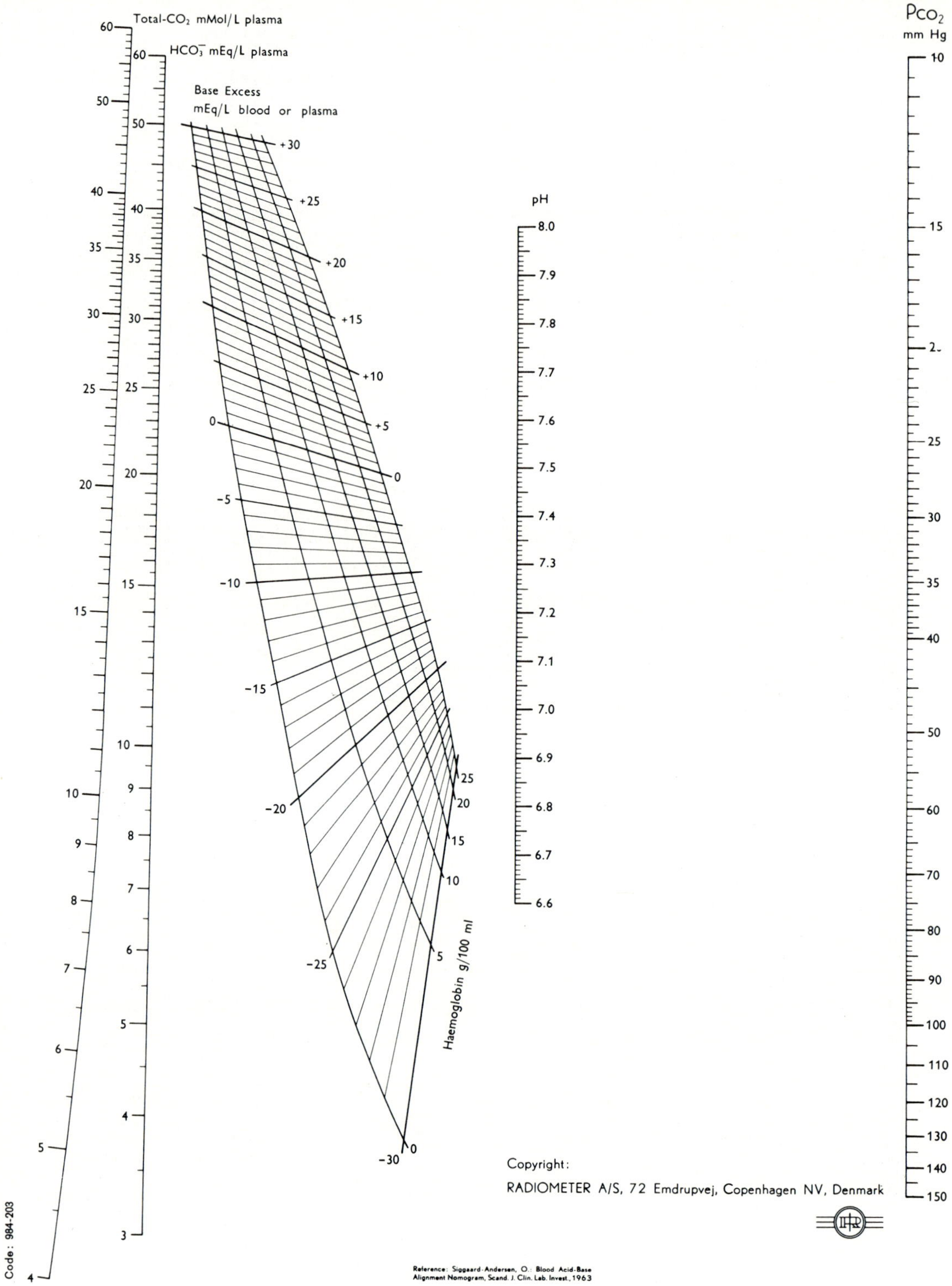

Fig. 7-9. Siggaard-Andersen alignment nomogram. (Courtesy The London Co., Cleveland, Ohio.)

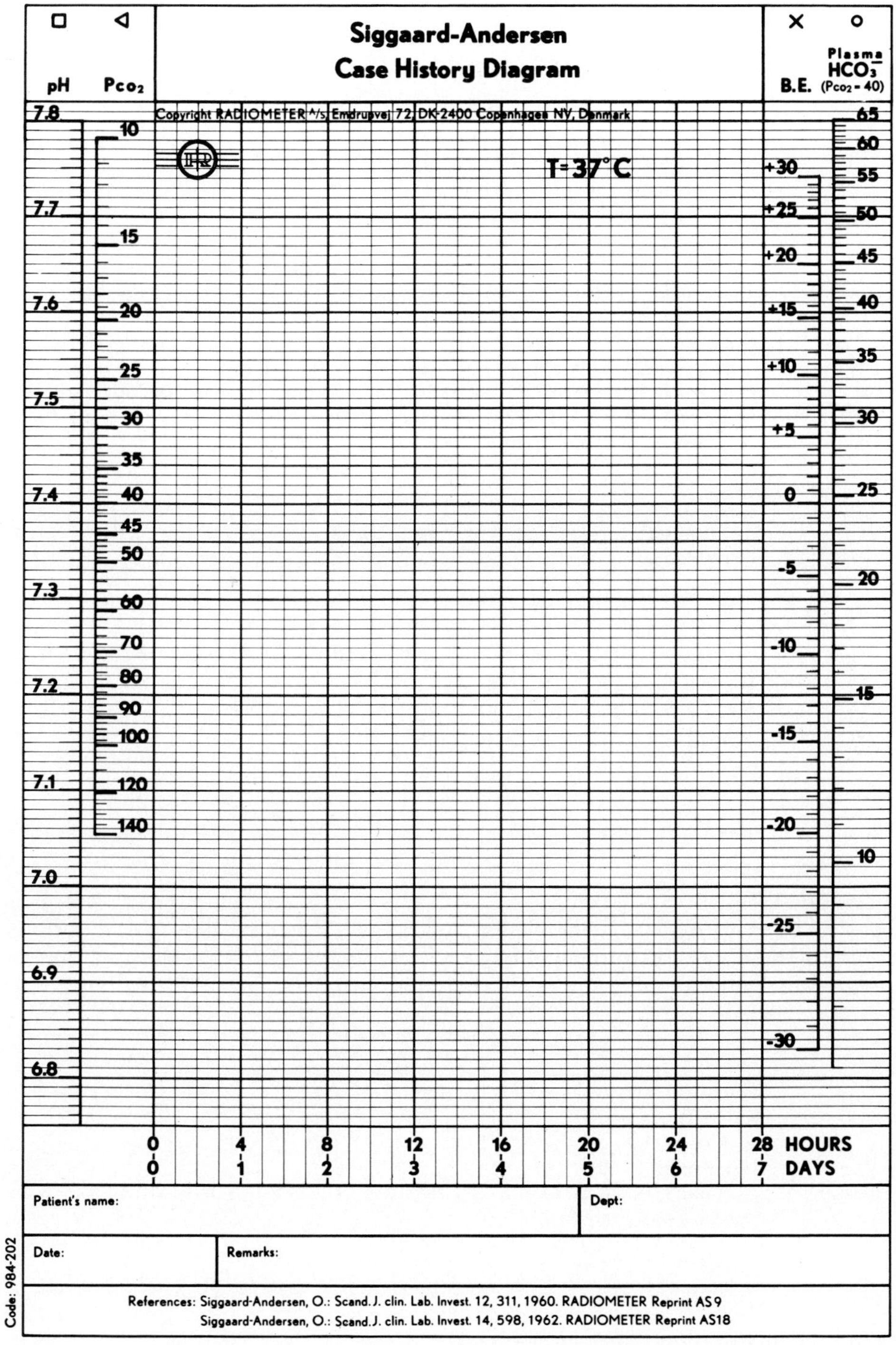

Fig. 7-10. Siggaard-Andersen case history diagram. (Courtesy The London Co., Cleveland, Ohio.)

capillary immersed in the KCl compartment of the calomel electrode. The pH is measured after 10 to 20 seconds.

5. **Into each equilibration receptacle introduce blood from the other capillary tubes. Shake by vibration of the chamber for 3 minutes while exposing the blood, spread as a film, to oxygen–carbon dioxide gas mixtures of known concentration at 37° C. The gases are passed through humidifiers before entering the equilibration chamber.**
6. **Measure the pH of the two equilibrated samples with the electrode assembly.**
7. **Between measurements the glass electrode is washed with buffer solution and distilled water and then air dried.**
8. **The relation between pH and log P_{CO_2} is a straight line. The pH values of the two equilibrated blood samples are plotted against the corresponding P_{CO_2} values for the gases, using the Siggaard-Andersen nomogram (Fig. 7-8), on which the pH is the abscissa and the log P_{CO_2} the ordinate. By drawing a line through the two points the P_{CO_2}, standard bicarbonate, base excess, and buffer base can be read from the nomogram. Fig. 7-9 is an alignment nomogram representing the same functions as the curve nomogram. In Fig. 7-10 is shown a case history diagram that permits the physician to maintain a record of the various blood values at a number of specified intervals and observe quickly the acid-base status of the patient.**

Comments

A mean normal temperature for man of 37° C is used in all measurements. If the patient's temperature varies by more than 2° C, the measurements should be corrected to actual body temperature; or the water circulation thermostat may be reset if a number of tests are to be run for a different body temperature. The pH of blood decreases by approximately 0.0147 pH unit (0.01 unit for plasma) for each degree C increase in temperature. It is also necessary that the temperature of the electrode and liquid junction be maintained within ±0.1° C of the body temperature, usually 37° C.

The pH meter must be well grounded; otherwise the instrument will drift.

Blood samples for actual pH must be introduced into the electrode anaerobically so that there will be no loss of CO_2 resulting in a simultaneous rise in pH. There must be no air bubbles in the polyethylene capillary or glass electrode capillary.

Before using new tanks of gas, check the pH values obtained by equilibration with these gas mixtures against results with the gas from the old tanks. If the results do not agree, request another analysis of the gas mixtures by the supplier.

Normal values (37° C)[5]

	Men		Women	
Actual pH, whole blood	**7.38 to 7.44**	**(7.40)**	**7.39 to 7.44**	**(7.41)**
Actual P_{CO_2} (whole blood), mm Hg	**34 to 45**	**(39.3)**	**31 to 42**	**(36.4)**
Base excess (whole blood), mEq/L	**−2.4 to +2.3**	**(−0.1)**	**−3.3 to +1.2**	**(−1.0)**
Standard bicarbonate (plasma), mEq/L	**23.0 to 25.4**		**22.4 to 25.8**	
Total CO_2 (plasma), mM/L	**23.5 to 27.9**		**22.2 to 27.4**	

Use of radiometer blood micro system, type BMS2

The blood micro system, type BMS2, is a compact thermostat unit containing a new micro pH electrode that requires less than 30 μl of sample. Contained in

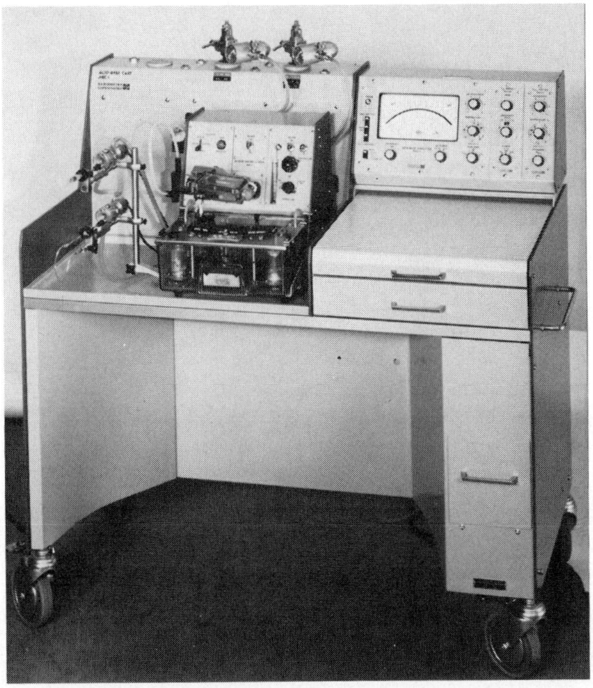

Fig. 7-11. Blood micro system, type BMS2, and acid-base analyzer, type PHM71. (Courtesy The London Co., Cleveland, Ohio.)

a single water bath controlled to $\pm 0.02°$ C are a tonometer, gas humidifier, and reservoirs for buffer and for rinsing solutions. There are outlets for external circulation of thermostated water, making possible the connection of separate P_{O_2} or P_{CO_2} electrodes or electrode assemblies. By combining this system with an acid-base analyzer, type PHM71, a complete system for acid-base analysis using the equilibration method (Astrup technique) is set up (Fig. 7-11).

A combination of a blood micro system, type BMS3, with a digital acid-base analyzer, type PHM72, provides a complete system for digital readout of pH, P_{CO_2}, and P_{O_2}.

Use of Beckman pH meter

The determination of blood pH is done by means of a closed glass electrode whose voltage output at a given temperature and pH is read directly from the scale of a pH meter (Fig. 7-12).

The pH of venous blood is usually 0.02 to 0.04 pH unit lower than that of arterial blood.

In obtaining the blood, the syringe should be rinsed with a sterile solution of heparin before the sample is drawn. Do not use oxalates or fluorides as anticoagulants, because they will alter the pH of the blood.

Some techniques given below will help obtain an accurate blood pH determination.

1. **Draw the venous or arterial blood in the heparinized syringe. Avoid getting any air bubbles in the syringe. Place a cork on the end of the needle to prevent air from getting into the syringe or the blood from leaking out.**

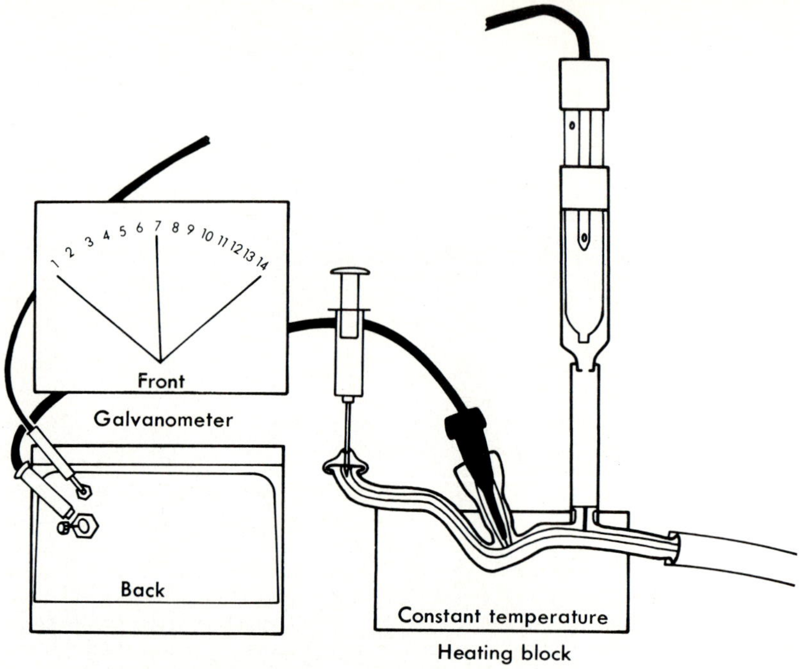

Fig. 7-12. Blood pH electrode unit.

2. ***Place the syringe in ice to prevent glycolysis.*** **The pH should be read within 10 minutes after the blood is drawn.**
3. **Most manufacturers of pH meters also have available an electrode system, specifically designed for blood pH, which allows the pH to be measured without exposing the blood to the air (Fig. 7-12). This type of system is essential for an accurate blood pH.**
4. **Adjust the meter with a buffer, pH 7.0, and rinse the electrode with saline before introducing the blood.**
5. **Allow the blood to come to the temperature of the system before taking a reading.**
6. **Introduce more of the blood and take another reading. Repeat this procedure until readings are reproducible. Usually the first reading is lower than the final one, due to dilution with saline.**

Use of IL model 113-SI pH-blood gas analyzer (Fig. 7-13)[13]

pH electrode assembly (Fig. 7-14)

Components. The three major components of the pH electrode assembly are the glass measuring electrode, the reference electrode, and the liquid junction assembly. The liquid junction assembly should be filled to the upper junction *(S)* with saturated KCl. There should be a ⅛- to ¼-inch layer of KCl crystals on the bottom of both sides of the assembly. The reference electrode should be filled to the rubber cap *(N)* with saturated KCl and have a thin layer of crystals at the bottom of the electrode covering the ceramic plug *(T)*. The measuring electrode contains a pH-sensitive glass capillary surrounded by an electrolyte solution of constant pH that is permanently sealed into the electrode assembly.

Slope standardization. The actual millivolt output of any glass electrode can vary over a period of time from a theoretical output. A one-point standardization is not adequate without a periodic second-point check. A one-point stan-

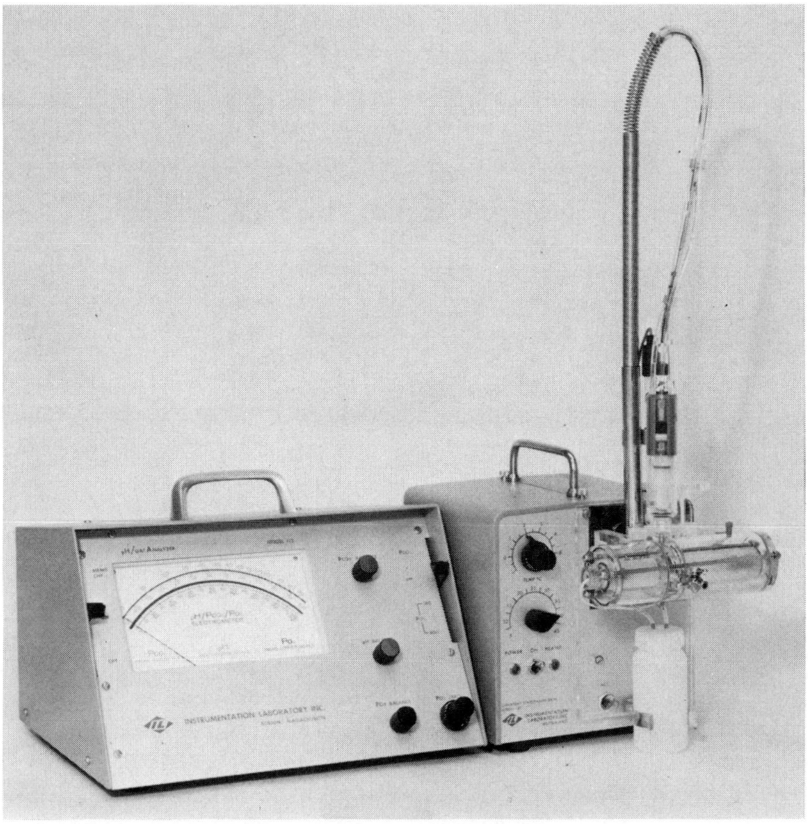

Fig. 7-13. Blood gas analyzer. (Courtesy Instrumentation Laboratory, Inc., Lexington, Mass.)

dardization may merely be a pivot point around which the standard curve rotates a few degrees. By use of two reference points a standard slope can be established that corresponds to the pH scale on the meter. After the slope has been set, only one standard buffer is needed to calibrate the instrument before determining a blood pH. The slope should be checked at least once a week.

1. **Turn the vacuum pump on.**
2. **Flush the capillary of the measuring electrode with saline and air by depressing the black vacuum button *(F)*. Avoid prolonged aspiration just prior to calibration. This would cool the measuring surface of the pH-sensitive glass and lead to a false calibration.**
3. **Pour a small amount of pH 6.84 and 7.384 buffers into labeled 10 ml beakers.**
4. **Wipe the external surface of the plastic sampling tube with a tissue and briefly flush the capillary with pH 6.84 buffer. Fill the capillary with the buffer. Avoid trapping air bubbles in the capillary.**
5. **Leave a small droplet of buffer on the end of the capillary and set the measuring electrode in the liquid junction assembly.**
6. **Set the mode selector switch on the meter to the pH position. Allow the meter needle to stabilize and then set it to pH 6.84, using the red pH *balance control* knob on the front panel.**
7. **Remove the electrode from the liquid junction assembly, wipe the plastic sampling tube, and flush the capillary with saline and air.**

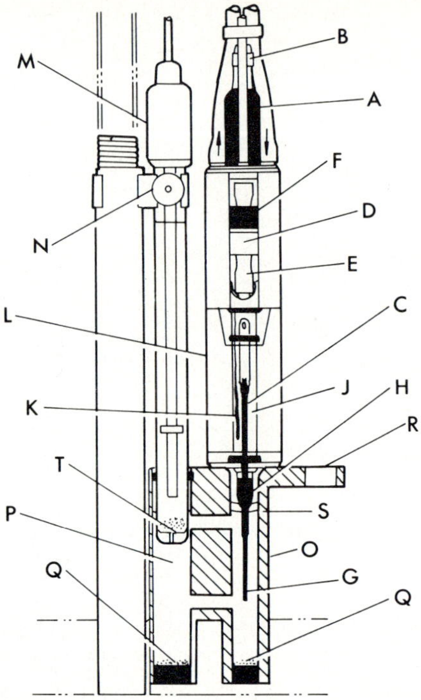

Fig. 7-14. Ultra-micro pH electrode assembly: **A,** pH glass electrode (11053); **B,** coaxial UM electrode lead (18091); **C,** active capillary glass; **D,** sample aspiration roller assembly; **E,** rubber tube; **F,** vacuum aspiration button; **G,** sample plastic tip (18083); **H,** sample tip holder (18088); **J,** electrolyte chamber; **K,** silver–silver chloride half-cell; **L,** water jacket; **M,** reference electrode (13077); **N,** rubber cap (18092); **O,** UM liquid junction assembly (18092); **P,** saturated KCl; **Q,** undissolved crystals; **R,** support to hold pH electrode; **S,** filling height of KCl in liquid junction assembly; and **T,** ceramic plug.

8. Wipe the sampling tube, briefly flush the capillary with pH 7.384 buffer, and then fill the capillary with the buffer.
9. Set the electrode in the liquid junction assembly.
10. Set the meter needle to pH 7.384, using the red pH *slope control* knob on the back of the meter.
11. Recheck with both buffers until no further adjustments of either the *balance control* or *slope control* knobs are necessary.
12. Lock the *slope control* knob by rotating the outer ring to the lock position.

Daily use. Once the slope has been standardized, only the pH 7.384 buffer is needed to calibrate the instrument for daily use. This is done by setting the meter needle to 7.384, using the red pH *balance control* knob.

1. Flush the capillary with saline and air. Set the roller to its lowest position.
2. Wipe the sampling tip. Fill the capillary of the measuring electrode with blood until the sample just appears in the larger glass chamber immediately above the active capillary *(C)*. Heparinized blood in a syringe or capillary tube is most commonly used although the sample may be aspirated directly from a finger, heel, or earlobe stick. In such a case, where the blood supply is limited, the operator should use the roller assembly to aspirate the sample. The assembly is gently rolled with the

thumb toward the top of the electrode. Care must be taken to avoid aspirating air bubbles.

3. **Leaving a small droplet on the end of the sampling tip, set the measuring electrode in the liquid junction assembly.**
4. **Read the pH of the sample from the pH scale on the meter.**
5. **Immediately flush the capillary with air and saline.**
6. **Flush the capillary with pH glass electrode cleaner. This is a 1% pepsin solution in 0.1 N HCl. Pepsin is an enzyme that catalyzes the digestion or breakdown of protein in the presence of a dilute acid. Unless this cleaner is used regularly, a protein film will build up on the wall of the capillary, which will affect the pH measurements. The electrode cleaner should be stored in a refrigerator when not in use. If this solution turns yellow, it should be discarded. An old solution can cause problems.**
7. **When the measuring electrode is not in use, leave either saline or buffer in the capillary.**
8. **Rinse and recalibrate with pH 7.384 buffer before each blood pH determination.**

System maintenance and troubleshooting

1. The life expectancy of both the reference and measuring electrodes can be extended with proper care. The reference electrode should be filled with saturated KCl and have a thin layer of crystals covering the ceramic plug in the bottom. With evaporation of water, crystallized KCl may form a heavy plug in the bottom of the electrode. This can cause problems. Remove the KCl solution by aspiration and fill with distilled water. Place the electrode in a beaker of moderately hot distilled water and gently invert the electrode periodically. When the crystal plug has dissolved, remove the fluid by aspiration and refill with fresh saturated KCl. Be sure there is a thin layer of loose crystals on the bottom.

Never aspirate saturated KCl into the capillary of the measuring electrode. The saturated solution will dehydrate the electrode and a KCl crystal may plug the capillary. Keep the pH-sensitive glass capillary free of film by cleaning regularly with the cleaning solution.

2. Avoid using blood specimens collected under mineral oil. If the oil is aspirated into the capillary, it will coat the pH-sensitive glass and interfere with pH measurement.

3. Bleach may also be used to clean the electrode, but the bleach must be *thoroughly* rinsed out afterward. If an electrode will have to be discarded because neither of these cleaners has been effective, try cleaning it with ether. Ether will dissolve oils and plexiglass, so avoid getting any ether on any of the plexiglass parts. Since ether will dehydrate the electrode, it will need to be hydrated for 12 to 24 hours or longer. Remove the pH electrode from its outer jacket and place it in a test tube of pH 6.84 buffer.

4. If electrodes are stored dry, they must be hydrated for 24 hours before they are used.

5. If the pH-sensitive glass capillary is broken, the meter may be set without difficulty but it will not respond to a change in pH.

6. Air bubbles in the capillary will cause the meter needle to peg off scale.

7. A small bubble on the side of the capillary will cause noise on the meter. Improper grounding of the system will also cause noise on the meter.

8. Avoid contaminating the buffers. Always wipe the sampling tip before dipping into the buffer solution. Never pour unused buffers back into the bottles. Keep the bottles tightly capped.

9. The KCl salt bridge in the liquid junction assembly *must* be saturated. An unsaturated solution increases electrical resistance. This liquid junction assembly should be cleaned regularly and refilled with fresh KCl solution and crystals.

***P_{CO_2}* electrode assembly** (Fig. 7-15)

Components. P_{CO_2} is measured as a change in pH when CO_2 gas diffuses across a membrane into an electrolyte solution surrounding a pH electrode. The P_{CO_2} electrode consists of a combination pH glass and reference electrode arranged behind a gas permeable membrane. CO_2 diffuses through the membrane in proportion to the P_{CO_2} difference across the membrane. As the CO_2 enters the electrolyte solution, it combines with water to form carbonic acid. The electrode measures the increase in carbonic acid as a change in pH and reports the signal logarithmically in terms of P_{CO_2}. A tenfold increase in P_{CO_2} is nearly equal to a decrease of one pH unit.

The gas-permeable P_{CO_2} membrane is a 1-mil silastic (ILastic) film that acts as a selective barrier. It is permeable to gases but not to ions. Therefore the ions in a sample cannot contribute to the pH change in the electrolyte solution. Several layers of nylon mesh are fastened to the end of the electrode to act as a spacer between the electrode and the silastic membrane. In this space is a thin film of aqueous P_{CO_2} electrolyte solution that must cover the glass measuring surface of the pH electrode, both to function as a layer to dissolve and react CO_2 and to act as a liquid junction between the glass and reference electrodes. The reference chamber portion of the combination pH-reference electrode is filled with P_{CO_2} electrolyte solution and is concentric around the measuring chamber.

***P_{CO_2}* slope standardization.** Two analyzed gas mixtures of known CO_2 content are used as calibration standards for the P_{CO_2} electrode assembly. One gas is usually about 5% CO_2, 12% O_2, the balance nitrogen and is referred to as the "low" gas. The other gas is usually about 10% CO_2 and the balance nitrogen and is referred to as the "high" gas. The gas cylinders are tagged with the analyzed content. The analysis should be accurate to within three hundredths of one percent (0.03%).

1. **To slope the P_{CO_2} electrode assembly it is necessary to know the CO_2 ratio of the two calibrating gases. This is calculated by dividing the percent composition of the low gas into the percent composition of the high gas and multiplying the quotient by 10. Example:**

$$\frac{10.57\%\ CO_2}{5.02\%\ CO_2} \times 10 = 21.06$$

2. **The low gas is bubbled through the humidifier on the back of the constant temperature control module and is introduced into the measuring chamber of the blood gas assembly through the stopcock at the bottom of this assembly. The gas is vented at the top fitting through a length of vinyl tubing (⅛-inch I.D. × 12 inches) down to a beaker of water. The end of the tube should be below the surface of the water so that**

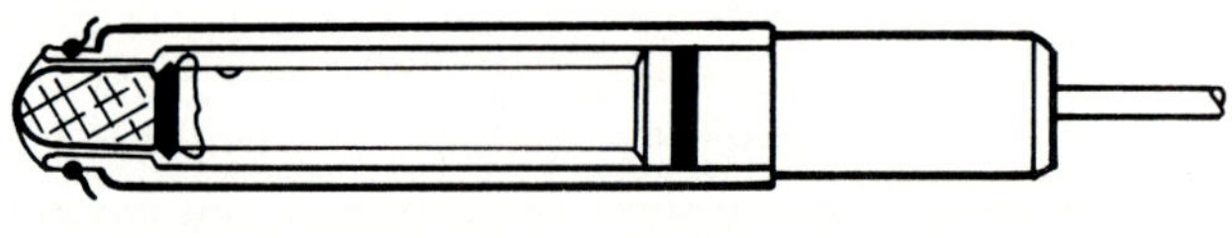

Fig. 7-15. P_{CO_2} combination pH glass and reference electrode with Lucite jacket.

the flow rate of gas may be observed. The flow rate should be adjusted so that the bubbles may be counted—about one bubble per second. Check to see that the water bath is at 37° C.

3. Move the mode selector switch to P_{CO_2}.
4. Set the meter needle to read 10 mm on the green P_{CO_2} scale, using the green *balance control* knob.
5. Introduce the high calibrating gas into the measuring chamber. Adjust the flow rate so that it is slightly slower than the low gas. Excessive gas flow will dry out the electrolyte.
6. Set the meter needle to the CO_2 ratio, using the P_{CO_2} *slope control* knob on the back of the meter.
7. Recheck setting with both the low and high gases until no further adjustments need be made. Lock the *slope control* knob by rotating the outer ring on the knob.
8. Introduce the low gas into the measuring chamber. Set the meter needle to the P_{CO_2} value of the calibrating gas corrected for the barometric pressure at the time of calibration.
9. The slope should be checked at least once a week.

Calculations of P_{CO_2} of analyzed gases. The pressure of a gas varies with a change in barometric pressure (BP). For this reason the barometric pressure, in millimeters of mercury (not corrected to sea level), must be known for calculating the P_{CO_2} of the calibrating gas at the time of standardization. The vapor pressure of water at 37° C, which is 47 mm, must be subtracted from the barometric pressure. The gases are saturated with water vapor at the measuring surface of the P_{CO_2}, but the composition of the analyzed gas is reported on a dry basis.

To calculate P_{CO_2}, subtract the pressure of water vapor (Pwv) at 37° C from the barometric pressure and multiply the difference by the $\%CO_2$ in the gas divided by 100.

$$(\text{BP} - \text{Pwv}) \times \frac{\%\,CO_2}{100} = P_{CO_2}$$

$$(755 - 47) \times 0.0502 = 35.5 \text{ mm Hg} = P_{CO_2}$$

• • •

Discussion of introduction of the blood sample and system maintenance and troubleshooting will follow the description and calibration of the P_{O_2} electrode assembly.

P_{O_2} electrode assembly (Fig. 7-16)

The P_{O_2} electrode produces a current, at a constant polarizing voltage (0.6 V), that is directly proportional to the partial pressure of oxygen (P_{O_2}) diffusing to the reactive surface of the electrode. The reduction of oxygen at the P_{O_2} cathode results in the generation of the current.

$$O_2 + 2H_2O + 2e^- \longrightarrow H_2O_2 + 2OH^- + 2e^- \longrightarrow 4OH^-$$

Each molecule of oxygen reacts with four electrons and it is this flow of electrons that is measured. This electrode is specific for oxygen because of the polypropylene membrane used that is permeable to gases (O_2) but not to reducible ions in the sample. A silver–silver chloride anode incorporated into the electrode provides the electrons for this cathode reaction. Silver is oxidized at this anode providing electrons as they are needed.

$$4Ag + 4Cl^- \longrightarrow 4AgCl + 4e^-$$

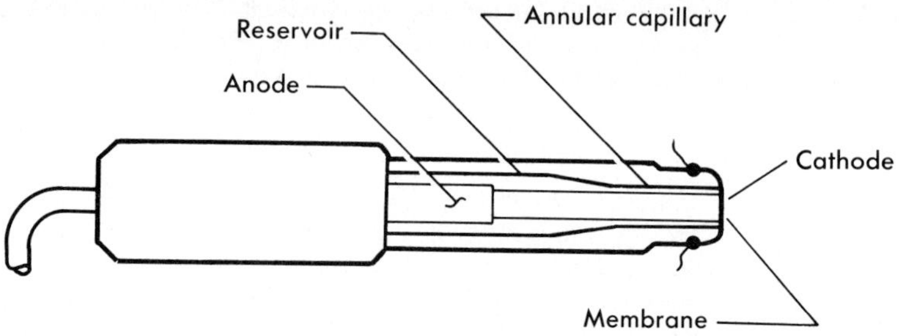

Fig. 7-16. Po_2 electrode assembly.

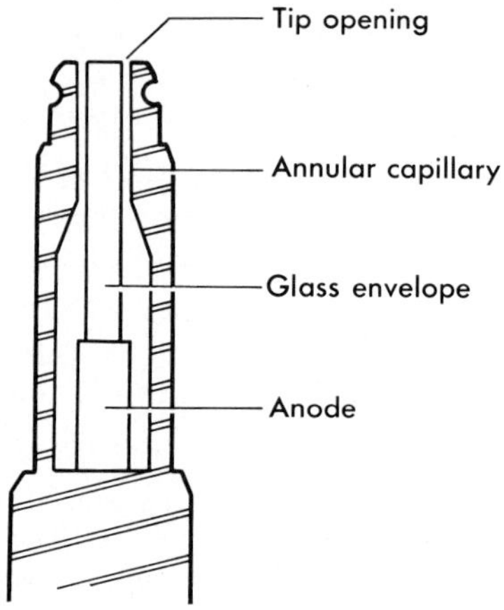

Fig. 7-17. Cross section of oxygen electrode.

A reserve chamber filled with Po_2 electrolyte solution surrounds the anode and supplies electrolyte to the tip of the cathode as needed through an annular capillary opening (Fig. 7-17).

It would seem that because of the consumptive type of cell the life-span of the Po_2 electrode would be relatively short. However, because of the micro design of the cathode and the current levels used, it would take nearly 8 years to consume all the oxygen in a normal blood contained in the ultramicro sample chamber.

Po_2 slope standardization

1. Introduce the oxygen-free gas (high CO_2 gas) into the measuring chamber. Adjust the flow rate to one bubble per second. Allow the gas to flow for 10 seconds; then close the stopcock and vent the gas to the atmosphere. Allow about 20 seconds for temperature stabilization. Move the mode selector switch to 160 Po_2.
2. Set the meter needle to zero on the black Po_2 scale, using the Po_2 *zero control* knob.
3. Introduce the 12% oxygen gas mixture (low CO_2 gas) into the measuring chamber. Adjust the flow rate to one bubble per second. After about a 10-second flow close the stopcock and allow the temperature to stabilize.
4. Set the meter needle to the calculated Po_2 for this gas, using the Po_2 *balance control* knob.

5. **Repeat the procedure with both gases until no further adjustments are needed. Lock the Po_2 *zero control* knob.**
6. **If a blood specimen has a Po_2 value above 160 mm, move the mode selector switch to 800 Po_2. Another gas mixture with a Po_2 in the range of 300 to 400 mm should be used to calibrate this scale. If such a gas is not immediately available, the results obtained on the higher scale can be assumed to be reasonably accurate for clinical purposes if the lower scale was carefully calibrated.**

Calculation of Po_2 of analyzed gases. The calculation of Po_2 is the same as for Pco_2:

$$(BP - Pwv) \times \frac{\% O_2}{100} = Po_2$$

$$(755 - 47) \times \frac{11.94}{100} = 84.5 \text{ mm Hg} = Po_2$$

Sample introduction. Blood samples are usually collected in a heparinized syringe and the syringe is inserted into the bottom stopcock. The blood is injected into the measuring chamber very slowly to avoid trapping any air bubbles in the chamber. Blood is added until it appears in the vinyl tubing at the top of the chamber. Allow for temperature stabilization. Read both Pco_2 and Po_2, moving the mode selector switch to the appropriate position. If sample volume permits, add more sample and take a second reading on each gas tension.

If the blood sample is collected in a heparinized capillary tube, the sealed tube must be carefully opened at both ends. The tube is then inserted through the open stopcock until it seats in the O-ring near the measuring chamber. Place the plastic tip of the pH electrode down into the fitting at the top of the measuring chamber until a tight fit is obtained. Using the manual roller assembly, slowly draw the sample into the measuring chamber. While the pH electrode is still in place, remove the capillary and close the stopcock. The pH electrode may now be removed. Allow for temperature stabilization and read gas tensions from the meter scales.

Cleaning the measuring chamber. Immediately after the gas measurements have been taken, flush the chamber with saline until all visible traces of blood have been removed. Then flush the chamber with a detergent/antifoam solution. Leave this solution in the chamber when not in use.

System maintenance and troubleshooting

1. It is very important that the gas-measuring chamber be kept as clean as possible. The detergent/antifoam solution removes any lipid film that is left by the blood. A lipid film has a tendency to hold air bubbles on the surface of the membranes. An air bubble will cause an unstable, low reading for either gas. The detergent/antifoam solution can be made up by adding 0.2 ml of any neutral liquid detergent and 0.2 ml of IL antifoam to a liter of water. NOTE: *Do not* use this solution to clean the pH electrode. The antifoam will coat and reduce the output of the pH-sensitive glass.

2. Dried blood will accumulate in and around the stopcock. Remove the stopcock and soak in water. Use a moist cotton-tipped applicator stick to clean around the O-ring that the stopcock fits up against.

3. Do not cool the tips of the electrodes and evaporate electrolyte by rapid or prolonged gas flows. Cooling and bubbles under the membranes can lead to false calibrations.

4. The power selector switch on the left side of the meter has a membrane check position that tests the electrical isolation of the electrode system from

ground. With the switch in this position and a conducting liquid in the measuring chamber the meter needle will go to 10 mm on the P_{CO_2} scale or 0 mm on the P_{O_2} scale if the membranes are intact. The mode selector switch must also be in the appropriate position. If either of the membranes is leaking, the electrolyte will provide an electrical path to ground and the meter needle will peg off to the extreme left. Only one membrane can be checked at a time.

5. If a bad membrane check has been obtained, blow the liquid out of the sample chamber and remove the electrode assembly from the Plexiglas cuvette. Look for moisture on the outside of the electrode or drops of electrolyte solution on the membrane. If there does not appear to be a puncture in the membrane, dry the cuvette with a cotton-tipped applicator stick, replace the electrode, and retest. If the membrane is punctured, change the membrane and dry the cuvette before replacing the electrode.

6. It is advisable to replace the electrolyte solution in the inner chambers of the electrodes at the time the membranes are being changed. The old electrolyte can be removed by aspiration with a 26-gauge needle and syringe. Refill with fresh electrolyte solution, again using the needle and syringe. Be sure to use the right electrolyte solution. They are not the same.

7. After the membranes are in place, check to see that there are no air bubbles trapped under the membranes.

Blood handling

1. Blood should be collected anaerobically in an airtight syringe or a capillary tube. A syringe should be sealed by sticking the needle into a cork or rubber stopper. Both ends of a capillary tube should be sealed with clay or wax.

2. Do not use mineral oil. Heparin should be used as an anticoagulant. Any other anticoagulant may have a significant effect on the pH.

3. Blood pH, P_{CO_2} and P_{O_2} should be determined within 5 minutes of drawing for optimal accuracy. If samples cannot be analyzed within 10 minutes of withdrawal, they should be stored in ice water.[7] Samples to be stored should be drawn in small (5 ml or less) syringes so that the sample will cool faster and slow down blood metabolism more rapidly.

Use of CO-Oximeter (IL model 182) (Fig. 7-18)[12,17,22]

The CO-Oximeter measures true saturations in relation to the total hemoglobin concentration. The measurement at three wavelengths has an advantage over a two-wavelength system. With the latter in the measurement of HbO_2 and Hb relatively small amounts of HbCO have an adverse effect on the methodology. The present system with the three wavelengths takes into account the effect of the HbCO and measures the concentration.

The simplicity of the instrument has reduced to a minimum technical errors that would normally interfere with accurate measurements. If it is calibrated improperly or is not given proper maintenance, then poor results will, naturally, be the product. Measurements are independent of the total Hb concentration of the sample and are not affected by diagnostic dyes such as methylene blue or Cardio-green.

Model 182 combines an absorption spectrophotometer with a special-purpose analog computer. Three separate interference filters are used as monochromators. At selected wavelengths they monitor the absorbance of HbO_2, HbCO, and the deoxygenated Hb in the sample. The three absorbance values obtained

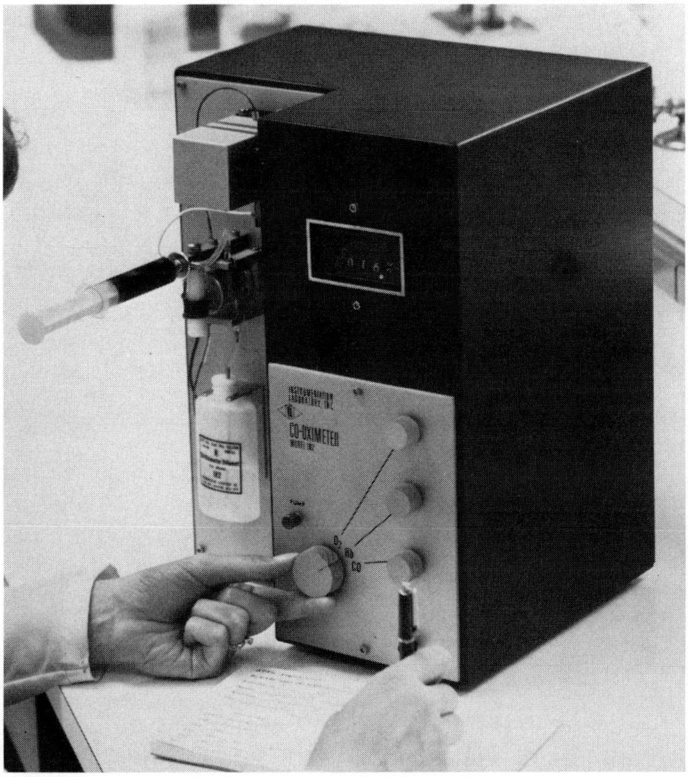

Fig. 7-18. CO-Oximeter, model 182. (Courtesy Instrumentation Laboratory, Inc., Lexington, Mass.)

are programmed in a matrix circuit board that instantly solves simultaneous equations for the three unknowns. These determinations may be achieved with a basic spectrophotometer but they are rather cumbersome.

Answers are presented on the digital counter in a sequence that the operator selects. The HbO_2 and HbCO values are linear over the entire 0 to 100% range; the total Hb concentration is reproducible to 0.1 g of Hb, from 6 to 20 g/100 ml of blood.

Procedure

Into the instrument, from the collection syringe, is introduced 400 μl of blood. In subsequent steps the sample is drawn up and diluted by a proportioning pump, passed through a mechanical hemolyzer, brought to a constant temperature, and pumped into the measuring cuvette. The absorbance values at the three selected wavelengths (548, 568, and 578 mμ) are monitored. The operator selects a mode, and the answer from the computer system is read on the digital counter. The control switch is turned to obtain the other two values. After the readings have been completed, the sample is expelled into a waste flask by depressing a pushbutton. A rinsing agent is introduced to clean the system for the next sample. The oxygen capacity and content are calculated from the known values.

The precision optics can be verified in the laboratory by introducing blood samples of known saturation values.

CALCIUM AND PHOSPHORUS

Calcium and phosphorus tend to maintain an equilibrium in the blood. They are usually considered together, since disturbances of one quite often result in a disturbance of the other. A low calcium is often accompanied by a high phosphorus. This does not always hold true, however, since during bone formation high phosphorus and sometimes high calcium levels are noted. This is found when fractures are healing and also explains the higher phosphorus level in children. Rickets is another exception, in that calcium is usually normal while phosphorus may be down to 2 mg/100 ml. In some cases, however, the results may be reversed, or both calcium and phosphorus levels may be low.

Patients with renal problems will have an elevated phosphorus level due to retention of phosphates, usually excreted in large quantities by normally functioning kidneys. An elevated BUN or NPN will generally accompany this situation.

Calcium exists in the body in an ionized form and in a protein-bound form, which is mainly attached to albumin. When a calcium determination is done, the *total* calcium is being determined, regardless of the proportion of the above two forms. The ionized fraction is normally about 4.2 to 5.2 mg/100 ml, or about 50% to 60% of the total calcium, while the protein-bound calcium is 40% to 50% of the total.

CALCIUM

Calcium is absorbed into the bloodstream from the small intestine. It is used in the formation of bone and the clotting of blood. The normal calcium values are 8.5 to 10.5 mg/100 ml of serum. In growing children, these values are slightly increased.

Decreased calcium values are of primary concern in tetany—a disease caused by faulty calcium metabolism and characterized by convulsions and muscular twitchings. Increased and decreased calcium values may be found in those conditions given in Table 7-7.

Clark-Collip modification of Kramer-Tisdall method

Ammonium oxalate is added to serum to precipitate the calcium as calcium oxalate. This is centrifuged. The precipitate is washed with dilute ammonium hydroxide to remove impurities and excess oxalate. Dilute sulfuric acid is added to convert the calcium oxalate to oxalic acid. After being heated to about 70° to 90° C (to enhance the following reaction) the oxalic acid is titrated with

Table 7-7. Conditions accompanied by abnormal calcium values

When increased	When decreased	
Carcinoma	*Tetany*	Osteomalacia
Hyperparathyroidism	Celiac disease	Parathyroidectomy
Hypervitaminosis	Hypoparathyroidism	Pregnancy
Multiple myeloma	Nephritis	Rickets
Polycythemia vera	Nephrosis	Sprue
		Vitamin D deficiency

standard potassium permanganate. The titration is an oxidation-reduction reaction, the oxalic acid being oxidized and the permanganate being reduced. The calcium content is found by calculation and reported in mg/100 ml of serum.

Roe-Kahn method

Trichloroacetic acid is added to precipitate the serum proteins. After filtering, sodium hydroxide and trisodium phosphate are added to precipitate the calcium as tricalcium phosphate. The mixture is centrifuged. The precipitate is washed with an alkaline alcohol solution to remove impurities and excess phosphate. Acid molybdate and aminonaphtholsulfonic acid reagent are added to form a color complex with the tricalcium phosphate. The depth of color is compared with a standard and measured in a colorimeter.

Bachra-Dauer-Sobel EDTA titration method, using cal-red indicator

Potassium hydroxide is added to make the serum alkaline. Cal-red indicator is added to the specimen, and titration with disodium ethylenediamine tetraacetate (EDTA) is done immediately. The calcium content is found by calculation and reported in mg/100 ml of serum. Icteric and hemolyzed serum may be used after preparation for analysis. Urine calcium, after preparation of the specimen, may also be determined by this method.

Procedures for calcium

The following general considerations pertain to the procedures for calcium:

1. Specimen: serum; the blood should be centrifuged and the serum obtained no later than 30 minutes after the blood is withdrawn from the patient. (If the blood stands, calcium diffuses into the cells and thus gives lower results.)
2. Preservation of serum: may be stored overnight in a refrigerator.
3. Precautions: Wash the precipitate exactly as directed in the procedure. Failure to remove interfering substances may give high results.

Clark-Collip modification of Kramer-Tisdall method for calcium titration

Reagents needed

Sodium oxalate, 0.100 N	(D-180)
Potassium permanganate, 0.100 N	(D-181)
Ammonium oxalate, 4%	(D-182)
Ammonium hydroxide, dilute (0.5%)	(D-183)
Sulfuric acid, 1 N	(D-184)

NOTE: Review for beginning student. To obtain the 0.01 N solution from the 0.1 N, dilute 10 ml of 0.1 N solution to 100 with distilled H_2O in a volumetric flask.

Procedure

1. About 30 minutes after the withdrawal of 8 ml of blood, centrifuge the blood and pour off the serum.
2. Prepare and standardize a 0.01 N potassium permanganate solution as follows:

   0.01 N sodium oxalate: This is a standard solution of exact normality to be used to determine the strength of approximately 0.01 N potassium permanganate, since this solution is unstable.

   Approximately 0.01 N potassium permanganate

   Standardization of 0.01 N potassium permanaganate: Pipette 5 ml of 0.01 N sodium oxalate into a 125 ml flask or similar container. Add 10 ml of 1 N (approximate) sulfuric acid. Heat to about 90° C by placing into a beaker containing boiling water. Using a burette, titrate with the approximately 0.01 N potassium permanganate. The titration will require about 5 ml so add 4 ml rather rapidly and then titrate drop by drop. The end point is reached when a single drop causes the solution to remain a faint pink color. Take the number of milliliters required and compute the factor for the 0.01 N potassium permanganate as follows.

Computation of factor. The factor tells the strength of the potassium permanganate solution in comparison to an exact 0.01 N solution. If the potassium permanganate solution is weaker than 0.01 N, the factor will be less than 1.0; if stronger than 0.01 N, the factor will be more than 1.0. Of course, if the potassium permanganate is exactly 0.01 N, the factor will be 1.0. Find the factor as follows: the number of milliliters of 0.01 N sodium oxalate used in the titration is divided by the number of milliliters of potassium permanganate required to reach the end point. Example: If 4.8 ml of potassium permanganate were needed, the factor is:

$$\frac{5.0 \text{ (ml of 0.01 N sodium oxalate used)}}{4.8 \text{ (ml of potassium permanganate required)}} = 1.04$$

3. Pipette 2 ml of serum into a 15 ml graduated centrifuge tube. Add 2 ml of distilled water and 1 ml of 4% ammonium oxalate. Mix thoroughly (Vortex mixer).
4. Let stand for 30 minutes in order to allow the ammonium oxalate to precipitate the calcium as calcium oxalate.
5. Centrifuge at 3000 rpm for at least 10 minutes. The precipitated calcium oxalate will then be seen as a small white precipitate in the bottom of the tube.
6. With a slow steady motion, invert the centrifuge tube, thus discarding the supernatant fluid and leaving the precipitate in the bottom of the tube. In this inverted position, place the tube on a piece of filter paper and allow to drain for a few minutes.
7. Remove any impurities from the precipitate by washing it *twice* in the following manner: Add 2 ml of dilute ammonium hydroxide to the centrifuge tube. Suspend all the precipitate by tapping the bottom of the tube. Wash down the sides of the tube with another 1 ml of the dilute ammonium hydroxide. Centrifuge for 10 minutes. Pour off the supernatant fluid with a slow steady motion and, with the centrifuge tube still in the inverted position, place it on a piece of filter paper to drain.
8. Add 1 ml of 1 N (approximate) sulfuric acid. (This converts the calcium oxalate to oxalic acid.) Dissolve the precipitate by tapping the bottom of the tube. Wash down the sides of the tube with another milliliter of the sulfuric acid.
9. Fill a 250 ml beaker, about half full with water, and bring water to a boil. Place the centrifuge tube in this for 1 minute. (This is done because the following oxidation-reduction reaction proceeds better at a high temperature.)
10. Remove from water bath. Using a burette graduated in hundredths, titrate with the 0.01 N potassium permanganate solution until a *faint* pink color persists for 1 minute.
11. Take the number of milliliters of 0.01 N potassium permanganate used and the factor for the 0.01 N potassium permanganate and make the calculation according to the following:

$$\text{ml used} \times \text{Factor} \times 0.2 \times \frac{100}{2.0} = \text{mg calcium/100 ml serum}$$

where ml used = ml of 0.01 N potassium permanganate used in titration

Factor = Factor for potassium permanganate

0.2 = Amount of calcium that 1 ml of 0.01 N potassium permanganate is equivalent to

100 = Amount of serum reported on (100 ml)
2.0 = Amount of serum used

Example: If the factor for the 0.01 N potassium permanganate is found to be 1.04, and 0.9 ml of the potassium permanganate is used in the titration, the calcium value is found as follows:

$$\text{ml used} \times \text{Factor} \times \frac{\text{Calcium}}{\text{equivalent}} \times \frac{\text{Volume}}{\text{correction}} = \text{mg/100 ml serum}$$

$$0.9 \times 1.04 \times 0.2 \times \frac{100}{2} = 9.4 \text{ mg/100 ml serum}$$

Bachra-Dauer-Sobel method for determination of calcium by EDTA titration[4]

Calcium is determined by titration with disodium ethylenediamine tetraacetate (EDTA), using cal-red* as the indicator. With urine, icteric serum, and hemolyzed serum, the calcium is first obtained as the oxalate. With clear serum (nonicteric or nonhemolyzed) the calcium is titrated directly.

Reagents needed

Potassium hydroxide, 1.25 N	(D-185)
EDTA solution	(D-186)
Cal-red indicator	(D-187)
Sodium citrate, 0.05 M	(D-188)
Calcium standard (1 ml = 100 μg Ca)	(D-189)
Caprylic alcohol	
Ammonium oxalate, 10%	(D-190)
Hydrochloric acid, 1 N	(D-191)
Ammonium hydroxide, 5%	(D-192)

Procedure

1. Set up the following in 15 ml conical centrifuge tubes:

   Standard (3 tubes)

   0.50 ml of working standard.

   Unknown

   0.50 ml of serum plus 1 small drop of caprylic alcohol.
2. Proceed with one tube at a time to obtain a sharp end point. If samples remain in an alkaline medium beyond 10 minutes, the end points are not sharp.
3. Pipette 2.5 ml of 1.25 N potassium hydroxide into the tube and mix.
4. Shake the indicator solution each time before using, to mix. Pipette 0.25 ml of indicator solution into the tube. Mix.
5. *Immediately* titrate with EDTA solution until the color changes from wine-red to blue. The color change may be observed by placing the tube against a bright light.

Calculation

$$\frac{\text{Titration volume of unknown}}{\text{Titration volume of standard}} \times 0.05 \times \frac{100}{0.5} =$$

$$\frac{\text{Titration volume of unknown}}{\text{Titration volume of standard}} \times 10 = \text{mg Ca/100 ml}$$

or

$$\frac{\text{Titration volume of unknown}}{\text{Titration volume of standard}} \times 5 = \text{mEq/L calcium}$$

---

*Cal-red, or 2-hydroxy-1-(2-hydroxy-4-sulfo-1-naphthylazo)-3-naphthoic acid, may be obtained from Scientific Service Laboratories, Inc.

Icteric and hemolyzed serum

There are instances when the only available specimen for analysis is either icteric or hemolyzed. Using the specimens that are either icteric or hemolyzed will give invalid results. The following treatment will enable the specimen to be used.

1. **Pipette 0.50 ml of serum into a conical tipped centrifuge tube.**
2. **Add 0.63 ml of distilled water and 0.13 ml of 10% ammonium oxalate. Mix.**
3. **Incubate in a 56° C water bath for 15 minutes.**
4. **Centrifuge for 10 minutes at 3000 rpm.**
5. **Pour off the supernatant and invert the tube on a filter paper or blotter to drain.**
6. **Dissolve the precipitate in 0.25 ml of 1 N HCl and add 0.25 ml of 0.05 M sodium citrate. Mix.**
7. **The tube is now ready for the addition of KOH (step 3 of the preceding method).**
8. **The procedure is carried out in the same way as with the clear serum.**

Urine calcium

Calcium on a 24-hour collection of urine may be done with the serum method after the following preparation of the urine sample. If calcium determination only is to be done, the specimen may be collected in a gallon bottle containing 10 ml of glacial acetic acid. If other tests are requested, where an acidified specimen would be undesirable, collect the specimen without the acid. Shake and mix the urine well, taking care that if there is any precipitate in the urine, none is left adhering to the walls of the bottle. Allow the urine to come to room temperature and measure the total volume in a graduated cylinder. The urine must be at room temperature, and the measurement must be accurate. Pipette 10 ml of the well-mixed urine into a test tube and acidify to pH 1 with glacial acetic acid. Use nitrazine paper to check the pH. Proceed as follows:

1. **Place the tube of acidified urine in a water bath (use a Pyrex beaker with a thermometer and beaker clamp).**
2. **Heat the specimen to 60° C for 15 minutes, mixing occasionally with a stirring rod.**
3. **Pipette 0.50 ml of the mixed urine into a conical tipped centrifuge tube.**
4. **Add 0.1 ml of 10% ammonium oxalate. Mix.**
5. **Add 1 drop of methyl red indicator (0.1%). Mix.**
6. **With a pipette or dropper add, dropwise, 5% ammonium hydroxide (NH_4OH) until an orange color is obtained.**
7. **Place the tube in a boiling water bath for 20 minutes. Use the Pyrex beaker again.**
8. **Cool the tube to room temperature and centrifuge for 10 minutes at 3000 rpm.**
9. **Decant the supernatant and invert the tube on filter paper or blotter to drain.**
10. **Dissolve the precipitate in 0.25 ml of 1 N HCl.**
11. **Add 0.25 ml of 0.05 M sodium citrate. Mix.**
12. **Proceed as in step 3 of the serum procedure. The standard is set up just as it was for the serum. (NOTE: If less than 0.5 ml of EDTA is used in the titration of the calcium in the urine, use a larger volume of urine—1.0 ml or more—instead of the 0.50 ml that was to be precipitated.) This is step 3 of the urine method.**

Calculation

$$\frac{\text{Titration volume of unknown}}{\text{Titration volume of standard}} \times 10 = \text{mg Ca/100 ml}$$

$$\frac{\text{Titration volume of unknown}}{\text{Titration volume of standard}} \times 5 = \text{mEq Ca/L}$$

To calculate the mg/100 ml or mEq/L for the 24-hour urine, the example is set, using the 0.50 ml of urine precipitated (step 3).

$$\frac{\text{Titration volume of unknown}}{\text{Titration volume of standard}} \times 0.05 \times \frac{1}{0.50} \times \frac{\text{Urine volume}}{(24\text{ hr})} =$$

$$\frac{\text{Titration volume of unknown}}{\text{Titration volume of standard}} \times 0.1 \times \frac{\text{Urine volume}}{(24\text{ hr})} = \text{mg Ca/24 hr}$$

$$\frac{\text{Titration volume of unknown}}{\text{Titration volume of standard}} \times 0.025 \times \frac{1}{0.50} \times \frac{\text{Urine volume}}{(24\text{ hr})} =$$

$$\frac{\text{Titration volume of unknown}}{\text{Titration volume of standard}} \times 0.05 \times \frac{\text{Urine volume}}{(24\text{ hr})} = \text{mEq Ca/24 hr}$$

Calcium by flame photometry

Although calcium in serum has been successfully determined by chemical analysis and atomic absorption spectroscopy (refer to Chapter 10), there has been more difficulty with flame photometry. However, if the instrument purchased is suitable for calcium determinations (Coleman model 21 flame photometer) and if care is used, accurate determinations can be produced. Glassware for the calcium determinations *must be acid-washed* in chromic acid each time before use. Glassware that is contaminated or nonchemically clean will give grossly elevated results. Standards in serum (available commercially) produce the most reliable results. Urine calcium determinations are more reliable when done by chemical analysis or atomic absorption.

Ionized (diffusible) calcium

Determining ionized calcium is not a practical procedure for the routine laboratory. A very low ionized calcium results in tetany, causing severe muscle spasms and tremors. Therefore, a rapid method for calcium is desirable, as well as a method for determining the ionized calcium content. The following method of estimating ionized (diffusible) serum calcium is given. This differs from the McLean and Hastings[16] nomogram by utilizing the albumin and globulin ratio instead of the total protein in order to give a closer estimation of the ionized calcium value.

An explanation of the K value is as follows. Originally, a nomogram was developed from data obtained by Hopkins and co-workers.[11] Total calcium, ionized (diffusible) calcium, and total serum protein with an A/G ratio of 1.8 were analyzed. The resultant factor (K) was 11.7.

Hanna and co-workers[8] assayed 100 thoroughly screened patients to obtain the K factor used in the following formula.

By using Hanna's findings, the following constant is derived. First, the chemical equation would be:

$$\text{Ca proteinate} \longrightarrow Ca^{++} + \text{Protein}$$

Following the law of mass action, "The speed of a chemical reaction is proportional to the active masses of the reacting substances," the following equation is:

$$\frac{\text{Ionized (diffusible) calcium} \times \text{Total protein}}{\text{Nonionized (nondiffusible) calcium}} = \text{K}$$

By assaying over 200 tests in duplicate the following values were obtained for the new K factor: Total calcium, 10.35 mg/100 ml; ionized calcium, 6.16 mg/100 ml; albumin, 4.40 g/100 ml, globulin, 3.60 g/100 ml, and A/G ratio, 1.22.

The following equation is set up to obtain the K factor:

$$\frac{6.16 \times 8.0}{10.35 - 6.16} = 11.8 \text{ (K value)}$$

Since it is understood how the K value is obtained, the following formula may be used to calculate an ionized calcium:

$$\frac{11.8 \times \text{Total calcium (mg/100 ml)}}{11.8 + \text{Total protein (g/100 ml)}} = \text{Ionized calcium (mg/100 ml)}$$

Using the nomogram

The nomogram shown in Fig. 7-19 is constructed with the K value being 11.8. The following is an example to show how the nomogram works. The patient has an albumin of 4.0 g/100 ml, a globulin of 3.0 g/100 ml, and a total calcium of 10.0

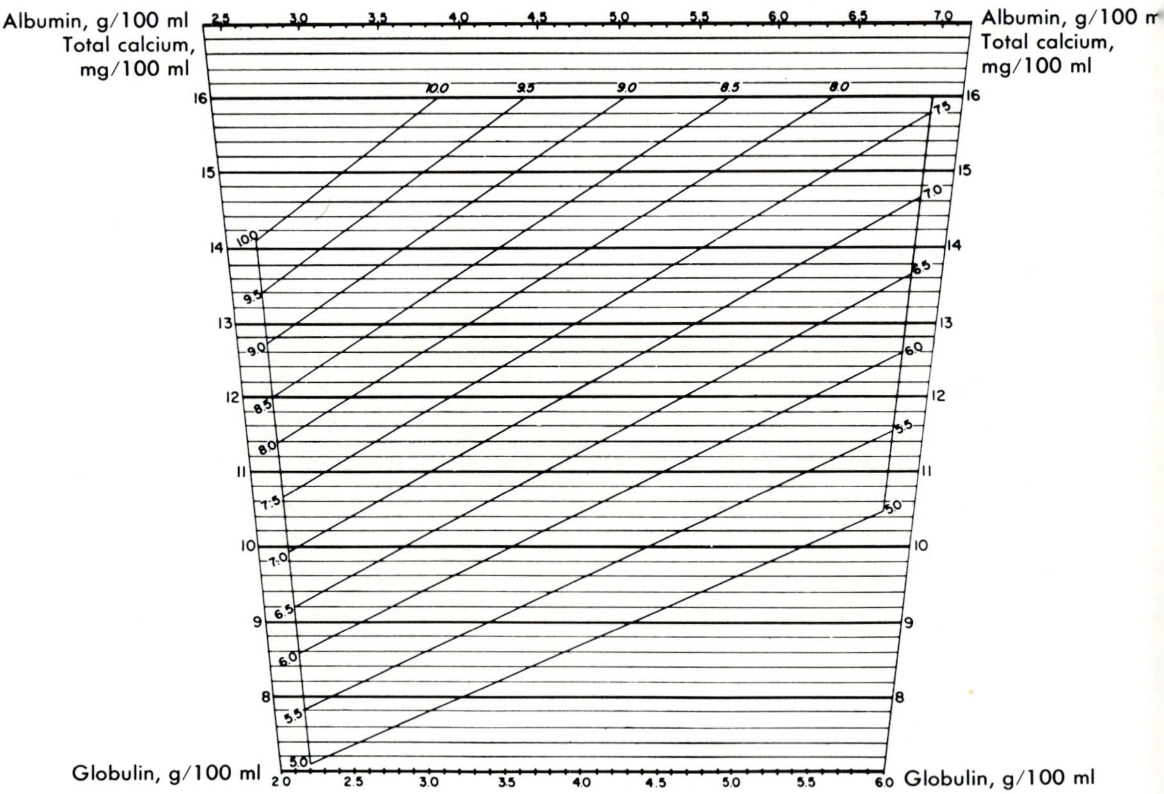

Fig. 7-19. Nomogram for estimating diffusible serum calcium. (From Hanna E. A., Nicholas, H. O., and Chamberlin, J. A.: Clin. Chem. **10**:235, 1964.)

mg/100 ml. Using the nomogram, draw a line from the top line (albumin, 4.0 g/100 ml) to the bottom line (globulin, 3.0 g/100 ml). The total calcium line at 10.0 mg/100 ml intersects the protein line, giving an ionized calcium result of 6.3 mg/100 ml.

PHOSPHORUS

Phosphorus is absorbed into the bloodstream from the small intestine. It is used in forming bone and in regulating the pH of blood. The normal phosphorus values are 3.0 to 4.5 mg/100 ml of serum, with growing children having slightly higher values. Changes from normal values are sometimes of special importance (Table 7-8).

For the determination of phosphorus, the method of Fiske and SubbaRow is commonly used. Trichloroacetic acid is added to serum to precipitate the proteins. After filtering, a molybdate solution is added to unite with the phosphorus to form phosphomolybdic acid. The phosphomolybdic acid is then reduced by an organic reagent to form a blue color. The intensity of blue color is proportional to the amount of phosphorus present. The depth of color is measured in a colorimeter.

Procedure for phosphorus

The following general considerations pertain to the procedure for phosphorus:

1. **Preparation of patient: must be fasting.**
2. **Preparation of blood: serum. The serum should be obtained 30 minutes after the blood is withdrawn.**
3. **Preservation of serum: refrigerate.**
4. **Precautions: follow the directions *exactly* as given in the procedure.**

Fiske-SubbaRow method[6]

Phosphates (inorganic) in serum

Reagents needed

Trichloroacetic acid, 10%	**(D-164)**
Molybdate reagent	**(D-193)**
Sulfuric acid, 10 N	**(D-194)**

Table 7-8. Conditions accompanied by abnormal phosphorus values

When increased	When decreased
Nephritis	*Rickets*
Healing bone fractures	Hyperparathyroidism
Hyperinsulinism	Myxedema
Hypervitaminosis	Osteomalacia
Hypoparathyroidism	Sprue
Uremia	Idiopathic steatorrhea
Pyloric obstruction	Neurofibromatosis
Starvation	Ether and chloroform anesthesia
Following Pituitrin administration	Following insulin and Adrenalin administration
	Lobar pneumonia

Reducing reagent (aminonaphtholsulfonic acid) (D-195)
Phosphorus stock standard, 0.1 mg/ml (D-196)
Phosphorus working standard (1 ml = 0.01 mg P)
Using a 100 ml volumetric flask, dilute 10 ml of stock standard to 100 ml with distilled water.

Procedure

Preparation of protein-free filtrate

1. **Pipette 1 ml of serum into a 15 ml test tube.**
2. **Add 9 ml of 10% TCA. Stopper and shake.**
3. **Allow the tube to stand for 15 minutes and then centrifuge or filter through Whatman No. 42 filter paper.**

Development of color

1. **Set up the following in 15 ml test tubes:**

   Blank

   5.0 ml of TCA into tube.
   3.6 ml of distilled water.

   ***Standards* (three tubes)**

   4.5 ml of TCA into each tube.
   1.6 ml of distilled water.
   2.5 ml of working standard.

   Unknown

   5.0 ml of filtrate into the test tube.
   3.6 ml of distilled water.
2. **Add 1.0 ml of molybdate reagent to all tubes. Mix.**
3. **Add 0.4 ml of cold reducing reagent. Mix thoroughly.**
4. **Allow the tubes to stand for 15 minutes (use timer) for the full development of color.**
5. **Read against reagent blank at 660 mμ or use a filter No. 66.**

Calculation

$$\frac{\text{A of unknown}}{\text{A of standard}} \times 0.025 \times \frac{100}{0.5} =$$

$$\frac{\text{A of unknown}}{\text{A of standard}} \times 5 = \text{mg phosphorus/100 ml}$$

Urine phosphorus

To determine urine phosphorus, it is necessary to make a 1 to 5 or a 1 to 10 dilution with distilled water. This dilution is treated in the same manner as the serum. The result obtained is then multiplied by the appropriate dilution factor, either 5 or 10.

OSMOLALITY

Measurement of solute (ionic and nonionic) concentration of biologic fluids is important in the care of certain medical and surgical patients. These solutes can be divided into three general categories. (1) electrolytes, (2) organic solutes of small molecular weight (urea, glucose, creatinine), and (3) colloids (chiefly protein).

When a solute is added to a solvent the colligative properties of the solvent change in a linear fashion with the increased amount of solute: the freezing point and vapor pressure decrease, and the osmotic pressure and boiling point increase. With the present commercial instruments, the determination of freezing point lowering is more satisfactory than other colligative properties for measuring solute concentration in biologic fluids. The freezing point of a solution is related to the osmotic concentration of that solution or to the concentration of particles of solute per unit amount of solvent (water). The molal freezing point lowering is the change in freezing point produced by one mole of a nonionic solute in 1000 g of solvent. When one mole of a nonionic solute is added to one kilogram of water, the freezing point is decreased 1.858° C. There is an interaction between the solute and solvent referred to as "activity," which includes among other factors the degree of dissociation of the solute. The addition of one mole of an ionic solute such as sodium chloride lowers the freezing point almost twice as much because of the dissociation. For example, one mole of NaCl plus one kilogram of water lowers the freezing point $\phi \times n \times 1.858°$ C, where n is the number of ions from the molecule and ϕ is the osmotic coefficient, or about 0.93 at this concentration. The latter varies with the sodium chloride concentration in solution.

Measurement of freezing point depression is carried out using an osmometer, the operation of which is based upon the comparison of the freezing point of an unknown with the freezing point of a standard solution of known molality. The unit for reporting is milliosmol per kilogram of water or simply milliosmol.

Pure water may be cooled or supercooled to −40° C without freezing. However, the addition of a crystal of ice or a dust particle will cause rapid ice formation and a temperature return to 0° C, the normal freezing point. Each gram of water at 0° C produces 80 calories of heat (latent heat of fusion). Crystallization of the ice does not occur instantaneously; the more dilute the solution, the more rapidly the equilibrium temperature is reached. This equilibrium is the plateau referred to in the procedure and is the end point for the reading.

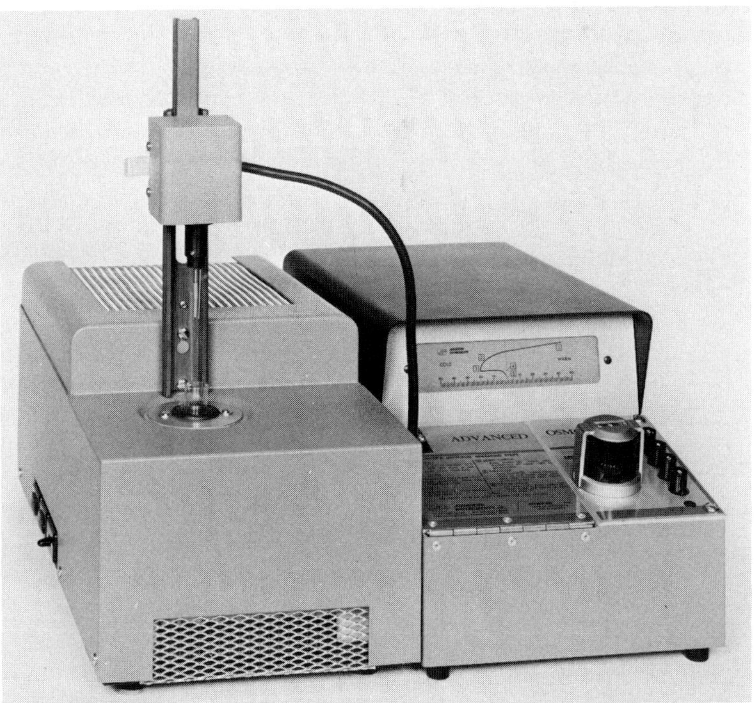

Fig. 7-20. Advanced osmometer. (Courtesy Advanced Instruments, Inc., Newton Highlands, Mass.)

For describing the operation of an osmometer, the one manufactured by Advanced Instruments[18] will be used as an example (Fig. 7-20). It is made in four modules:

1. Head—this module holds the thermistor probe and adjustable stir mechanism.
2. Control unit—located on this module are pushbuttons keyed to operating steps printed on the control panel, with some of the seldom-used controls under a hinged panel.
3. Air-cooled thermoelectric refrigerator—this unit has an adjustable thermostatic control and houses modular internal electronics, built-in stir motor, bath level control, and a 350 ml bath containing a mixture of ethylene glycol and water. The bath should be completely changed every 3 or 4 months because of the evaporation of water, which increases bath viscosity and decreases heat conductivity. The cooling is slowed and reproducibility decreased.
4. Galvanometer—this module has an illustrated scale showing the freezing curve, numbered to correspond with pushbuttons and operating steps. The spotlight of the galvanometer traces the path through the freezing cycle.

A sample (2.0 ml or 0.2 ml) is pipetted into a sample tube and the tube placed in the opening on the deck of the bath and immersed in the cooling bath ($-7° \pm 0.5°$ C). The head is lowered, and cooling takes place rapidly while the sample is being stirred with the measuring circuit turned on. The rate of cooling decreases as the temperature of the sample approaches that of the bath. At a fixed interval, which is designated on the galvanometer scale, the stir wire is activated and it vibrates violently for one second. The sample freezes and the temperature rises to a plateau. The galvanometer light is adjusted until it comes to rest before it starts drifting to the left. This is the reading.

A standardized procedure in processing standards and samples is essential for reproducible results. The thermister probe should be precisely in the center of the liquid sample, both laterally and vertically. All samples should start from the same temperature. Precooling will shorten the freezing cycle but may lead to prefreezing.

Normal values for osmolality of human serum from 104 blood donors had a mean value and standard deviation of 290 ± 3.8 mOsm/L.[21] A deviation of ± 10 mOsm from normal is considered significant.

For clinical situations the ratio between the serum sodium concentration and osmolality (mEq/mOsm) is often used.[10] For normal patients this should range from 0.43 to 0.50. The ratio will be elevated in hypernatremic patients without a change in other serum metabolites. In hyponatremia in the absence of an increase in other serum solutes, the ratio should be normal. A calculation of the milliosmolal concentration may be made from the serum concentrations of electrolytes, urea, and glucose, by applying the following:

$$\text{mOsm} = 1.86\ (\text{Na, mEq/L}) + \frac{\text{Glucose (mg/100 ml)}}{18} + \frac{\text{BUN (mg/100 ml)}}{2.8}$$

For normal patients the determined osmolality should not exceed the calculated by more than 5 to 8 mOsm/L.

For urine concentration studies the measurement of osmolality is a direct indication of the formation of concentrated urine by the kidney and is more reliable than specific gravity measurements. Osmolality determinations have definite application in following dialysis therapy, the course of uremia, hyperosmolal nonketonic coma in diabetes, and fluid administration in postsurgical patients.

Plasma hypotonicity resulting from low plasma sodium concentration may occur after surgical operations, with congestive cardiac failure and in hepatic failure with ascites. Trauma, acute illness, and many chronic wasting diseases produce accumulation of water and salt, with water in excess of salt resulting in hypotonicity of the body fluids. Patients with advanced liver and kidney disease or severe diabetes mellitus may have an increased plasma osmolality of 20 to 50 mOsm/L from nonelectrolyte solutes such as urea and other related products, glucose, bilirubin, and pathologic constituents.

Conversion of milligrams/100 ml to milliequivalents per liter

Milligrams per 100 ml is a term that deals with weight, whereas milliequivalents per liter deals with combining power. For example, the element calcium has a specific weight, which may be given as mg/100 ml. However, it also has combining power, determined by its valence, which may be expressed as mEq/L.

The following abbreviations are usually used:

mg = milligram
ml = milliliter
mEq (or meq) = milliequivalent

To convert mg/100 ml to mEq/L, use the formula:

$$\text{mEq/L} = \text{mg/100 ml} \times \frac{10}{\text{Equivalent weight}}$$ where 10 is used to convert 100 ml to 1 L.

Example: Convert 325 mg of sodium/100 ml to mEq/L. Equivalent weight of sodium is 23. Use the formula:

$$\begin{aligned}\text{mEq/L} &= \text{mg/100 ml} \times \frac{10}{\text{Equivalent weight}}\\ &= 325 \times \frac{10}{23}\\ &= 141\end{aligned}$$

A shortcut in the above formula may be taken by first dividing 10 by 23. The number 10 divided by 23 equals 0.435; this is known as the conversion factor for sodium. The formula then becomes:

$$\text{mEq/L} = \text{mg/100 ml} \times \text{Conversion factor}$$

Thus, to convert 325 mg of sodium/100 ml to mEq/L by using the conversion factor of 0.435:

$$\begin{aligned}\text{mEq/L} &= \text{mg/100 ml} \times \text{Conversion factor}\\ &= 325 \times 0.435\\ &= 141\end{aligned}$$

By multiplying the units in column two of Table 7-9, by the factor given, the desired unit (in column four) may be obtained.

Conversion of milliequivalents per liter to milligrams/100 ml

To convert milliequivalents per liter to milligrams/100 ml use the formula:

$$\text{mg/100 ml} = \text{mEq/L} \times \frac{\text{Equivalent weight}}{10}$$

where the figure 10 is used to convert 1 L to 100 ml

Table 7-9. Electrolyte conversion factors

Electrolyte	Unit	Factor	Unit
Na	mg/100 ml	0.435	mEq/L
Na	mEq/L	2.3	mg/100 ml
Na	mEq/L	0.0585	g/L as NaCl
K	mg/100 ml	0.256	mEq/L
K	mEq/L	3.91	mg/100 ml
K	mEq/L	0.0746	g/L as KCl
Cl	mg/100 ml	0.282	mEq/L
Cl	mEq/L	3.55	mg/100 ml
Cl	mEq/L	0.0585	g/L as NaCl
CO_2	vol %	0.45	mEq/L
CO_2	mEq/L	2.24	vol %
Ca	mg/100 ml	0.5	mEq/L
Ca	mEq/L	2.0	mg/100 ml
PO_4	mg/100 ml	0.581	mEq/L
PO_4	mEq/L	1.72	mg/100 ml

Example: Convert 4.5 mEq of calcium per liter to milligrams/100 ml. The equivalent weight of calcium is 20. Use the formula:

$$\begin{aligned} \text{mg/100 ml} &= \text{mEq/L} \times \frac{\text{Equivalent weight}}{10} \\ &= 4.5 \times \frac{20}{10} \\ &= 9.0 \end{aligned}$$

Conversion of volumes percent to millimoles per liter

Gases are sometimes expressed in millimoles per liter instead of volumes percent. A mole is the molecular weight expressed in grams and a millimole is one thousandth part of this.

To convert carbon dioxide in volumes percent to millimoles (mM) per liter:

$$\text{mM/L} = \text{vol\%} \times 0.446$$

Example: Convert 50 vol% of carbon dioxide to millimoles per liter,

$$\begin{aligned} \text{mM/L} &= \text{vol\%} \times 0.446 \\ &= 50 \times 0.446 \\ &= 22.3 \end{aligned}$$

Conversion of millimoles per liter to volumes percent

To convert carbon dioxide in millimoles per liter to volumes percent,

$$\text{vol\%} = \text{mM/L} \times 2.24$$

Example: Convert 22.3 millimoles per liter of carbon dioxide to volumes percent.

$$\begin{aligned} \text{vol\%} &= \text{mM/L} \times 2.24 \\ &= 22.3 \times 2.24 \\ &= 50 \end{aligned}$$

REFERENCES

1. Astrup, P., Jorgensen, K., Siggaard-Andersen, O., and Engel, K.: The acid-base metabolism. A new approach, Lancet **1**:1035, 1960.
2. Astrup, P., and Siggaard-Andersen, O.: Micromethods for measuring acid-base values of blood. In Sobotka, H., and Stewart, C. P., editors: Advances in clinical chemistry, vol. 6, New York, 1963, Academic Press, Inc.
3. Albanese, A. A., and Lein, M.: Microcolorimetric determination of sodium in human biologic fluids, J. Lab. Clin. Med. **33**:246, 1948.
4. Bachra, B. N., Dauer, A., and Sobel, A. E.: The complexometric titration of micro and ultramicro quantities of calcium in blood serum, urine and inorganic salt solutions, Clin. Chem. **4**:107, 1958.
5. Current concepts of acid-base measurements, Ann. N. Y. Acad. Sci. **133**:3, 1966.
6. Fiske, C. H., and SubbaRow, Y.: The colorimetric determination of phosphorus, J. Biol. Chem. **66**:375, 1925.
7. Gambino, S. R., Astrup, P., Bates, R. G., Campbell, E. J. M., Chinard, F. P., Nahas, G. G., Siggaard-Andersen, O., and Winters, R.: Report of the *ad hoc* committee on acid-base methodology, Amer. J. Clin. Path. **46**:376, 1966.
8. Hanna, E. A., Nicholas, H. O., and Chamberlin, J. A.: Nomogram for estimating diffusible serum calcium, Clin. Chem. **10**:235, 1964.
9. Henry, R. J.: Clinical chemistry, principles and technics, New York, 1964, Paul B. Hoeber, Inc., Medical Book Department of Harper & Row, Publishers.
10. Holmes, J. H.: Measurement of osmolality in serum, urine and other biologic fluids by the freezing point determination, pre-workshop manual on urinalysis and renal function studies, 1962, American Society of Clinical Pathologists, Advanced Instruments, O-2 Holmes.
11. Hopkins, T., Howard, J. E., and Eisenberg, H.: Ultrafiltration studies on calcium and phosphorus in human serum, Bull. Johns Hopkins Hosp. **91**:1, 1952.
12. Horvath, S. M., and Roughton, F. J. W.: Improvements in gasometric estimation of carbon monoxide in blood, J. Biol. Chem. **144**:747, 1942.
13. Instruction handbook, model 113-Sl pH-blood gas analyzer, Watertown, Mass., 1967, Instrumentation Laboratory, Inc.
14. Instruction handbook, model 143 flame photometer, Watertown, Mass., 1967, Instrumentation Laboratory, Inc.
15. Lochhead, H. B., and Purcell, M. K.: Rapid determination of serum potassium employing glycine-phenol reagent, Amer. J. Clin. Path. **21**:877, 1951.
16. McLean, F. C., and Hastings, A. B.: The state of calcium in the fluids of the body, J. Biol. Chem. **108**:285, 1935.
17. Mensh, A. H., and Zettner, A.: Standardization of hemoglobinometry by atomic absorption spectrophotometry, Washington, D. C., 1966, American Society of Clinical Pathology Annual Meeting.
18. Osmometry and cryoscopy school outline, Newton Highlands, Mass., Advanced Instruments, Inc.
19. Siggaard-Andersen, O.: The acid-base status of the blood, Scand. J. Clin. Lab. Invest. **15** (supp. 70): entire issue, 1963.
20. Siggaard-Andersen, O., Engel, K., Jorgensen, K., and Astrup, P.: A micro method for determination of pH, carbon dioxide tension, base excess and standard bicarbonate in capillary blood, Scand. J. Clin. Lab. Invest. **12**:172, 1960.
21. Stevens, S. C., Neumayer, F., and Gutch, C. F.: Serum osmolality as a routine test, Nebraska Med. J. **45**:447, 1960.
22. Van Slyke, D. D., and Neill, J. M.: The determination of gases in blood and other solutions by vacuum extraction and manometric measurement, J. Biol. Chem. **61**:523, 1924.

• 8

• Acid-base balance

MARGIE ELLEN HAUSE, B.A., MT(ASCP)

PARTIAL PRESSURE OF A GAS

The actions of a gas depend upon the pressure it exerts, and the pressure any gas exerts—whether it is alone or is mixed with other gases—is called the partial pressure of that gas. If the barometric pressure is known, the partial pressures of the gases can be calculated. If the barometric pressure of *dry air* is 760 mm Hg, the partial pressures of oxygen, carbon dioxide, and nitrogen can be calculated as follows:

$$P_{O_2} = (760 \text{ mm Hg})\ 20.93/100 = 159 \text{ mm Hg}$$

$$P_{CO_2} = (760 \text{ mm Hg})\ 0.03/100 = 0.2 \text{ mm Hg}$$

$$P_{N_2} = (760 \text{ mm Hg})\ 79.04/100 = 601 \text{ mm Hg}$$

20.93%, 0.03%, and 79.04% here denote the percentage composition of room air. Room air usually contains some water vapor. When air is inspired, it becomes saturated with water vapor at body temperature. Water vapor exerts a partial pressure of 47 mm Hg at body temperature (37° C). *The pressure is directly dependent on temperature, but independent of barometric pressure.* Since total gas pressure in the trachea must equal 760 mm Hg, only 760 − 47 = 713 mm Hg is available for the sum of the partial pressures of O_2, CO_2, and N_2. The P_{O_2} of moist inspired gas in the trachea is, therefore, 20.93% of 713, or 149 mm Hg, and not 20.93% of 760, or 159, as in dry air.

PARTIAL PRESSURE OF OXYGEN (P_{O_2})

The body tissues use oxygen and form carbon dioxide. They remove oxygen from arterial blood and add carbon dioxide to it. In order to maintain a steady level of P_{O_2} and P_{CO_2}, the lungs must supply to the alveoli as much oxygen as is removed from the blood by the tissues and must remove the amount of carbon dioxide added to the blood by these same tissues. This is done continuously by the process of ventilation. The actual carriage of oxygen in the blood may be best understood through the following explanation.

Since gas diffuses from a place in which its partial pressure is high to one in which its partial pressure is low, the mixed venous blood entering the lungs has a P_{O_2} lower than that of alveolar air. Resulting diffusion of oxygen from the alveolar air into venous blood converts it into arterial blood. Oxygen contained in arterial blood is carried in two ways: (1) as dissolved oxygen in physical solution and (2) in chemical combination with hemoglobin in the erythrocytes. The amount dissolved and the amount combined both depend on the P_{O_2} of arterial blood. Arterial blood is transported to the body tissues where P_{O_2} is lower than that of arterial blood. This P_{O_2} difference drives the oxygen from the blood into the tissues. The movements of carbon dioxide in lungs and tissues are in a direction opposite to that of oxygen. Thus venous blood contains less O_2 and more CO_2 than arterial blood.

OXYGEN SATURATION OF HEMOGLOBIN

As mentioned previously, oxygen contained in arterial blood is carried in two ways: as dissolved oxygen in physical solution and as an element combined with hemoglobin in erythrocytes. The oxygen and hemoglobin together form *oxyhemoglobin:*

$$O_2 + Hb \rightleftharpoons HbO_2$$

As Comroe[1] has illustrated (Fig. 8-1), the largest portion of oxygen in blood

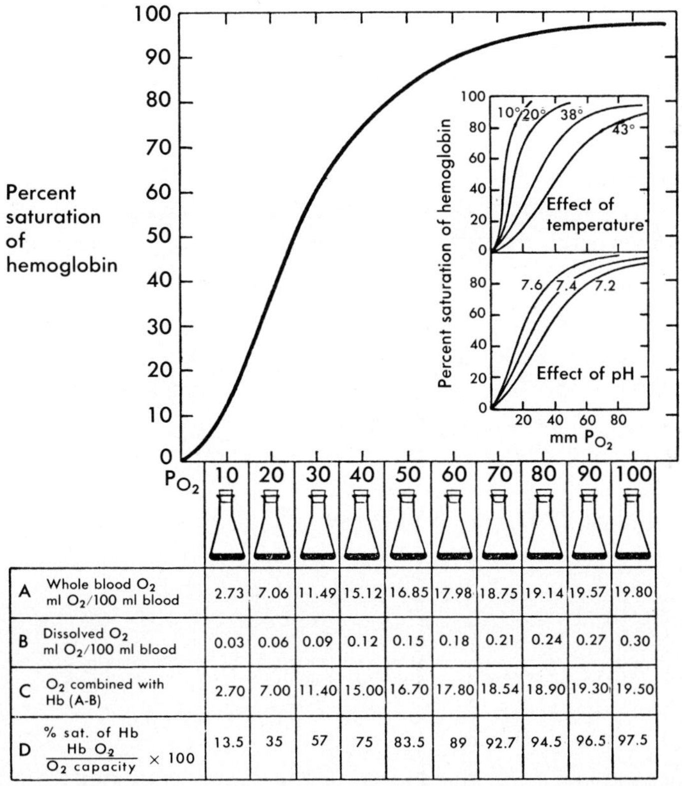

A	Whole blood O_2 ml O_2/100 ml blood	2.73	7.06	11.49	15.12	16.85	17.98	18.75	19.14	19.57	19.80
B	Dissolved O_2 ml O_2/100 ml blood	0.03	0.06	0.09	0.12	0.15	0.18	0.21	0.24	0.27	0.30
C	O_2 combined with Hb (A-B)	2.70	7.00	11.40	15.00	16.70	17.80	18.54	18.90	19.30	19.50
D	% sat. of Hb $\frac{Hb\ O_2}{O_2\ capacity} \times 100$	13.5	35	57	75	83.5	89	92.7	94.5	96.5	97.5

Fig. 8-1. HbO_2 dissociation curves. Large graph shows a single dissociation curve, applicable when the pH of the blood is 7.40 and temperature 38° C. Blood O_2 tension and saturation of patients with CO_2 retention, acidosis, alkalosis, fever, or hypothermia will not fit this curve because the curve shifts to the right or left when temperature, pH, or P_{CO_2} is changed. In the smaller graphs are shown effects on the HbO_2 dissociation curve, caused by change in temperature (upper right) and change in pH (lower right). A small change in blood pH occurs regularly in the body; for example, when mixed venous blood passes through the pulmonary capillaries, P_{CO_2} decreases from 46 to 40 mm Hg and pH rises from 7.37 to 7.40. During this time, blood changes from a pH of 7.37 dissociation curve to a pH of 7.40 curve. (From Physiology of respiration by J. H. Comroe, Jr. Copyright © 1965, Year Book Medical Publishers, Inc. Used by permission.)

is bound to hemoglobin. At P_{O_2} of 100 mm Hg, 100 ml of normal blood contains 19.8 ml of O_2 in the combined state, and only 0.3 ml physically dissolved in the blood water. It has been calculated that 1 g of hemoglobin combines with 1.34 ml of oxygen. Since the average amount of hemoglobin in 100 ml of blood is 15 g, fully (100%) saturated hemoglobin can carry 20.1 ml of O_2/100 ml of blood. The extent of the combination of hemoglobin with oxygen is termed *percent saturation:*

$$\frac{O_2 \text{ content}}{O_2 \text{ capacity}} \quad \frac{(O_2 \text{ combined with Hb})}{(\textit{maximum} \text{ amount of } O_2 \text{ that can be combined with Hb})} \times 100 = \% \text{ saturation}$$

OXYGEN DISSOCIATION CURVE

The relation between percent saturation and P_{O_2} is expressed in the dissociation curve of hemoglobin or oxygen dissociation curve. (See Fig. 8-1.) A knowledge of this curve is essential if one is to fully understand the remarkable properties of hemoglobin. The following points should be noted:

1. At a P_{O_2} of 100 mm Hg (normally found in arterial blood) Hb is about 97% saturated with O_2. At tensions greater than 100 mm Hg hemoglobin cannot combine with much more oxygen; it is probably completely saturated at 250 mm Hg.

2. There is very little change in the amount of oxygen held by hemoglobin between the oxygen tensions of 70 and 100 mm Hg. This is called the flat part of the curve. It enables a man living at a high altitude to receive enough oxygen

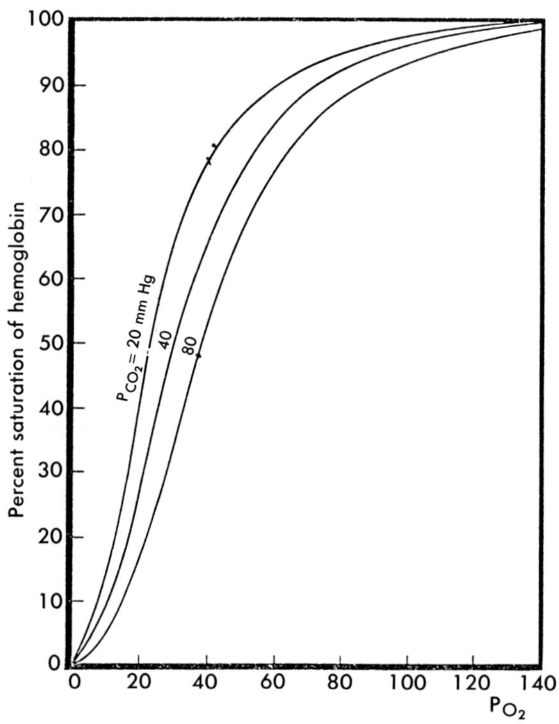

Fig. 8-2. Dissociation curve of hemoglobin at 38° C and at partial pressures of carbon dioxide equal to 20, 40, and 80 mm Hg.

even though his alveolar and arterial P_{O_2} will be lower than that at sea level. It also enables people with respiratory and pulmonary problems and an arterial tension between 70 and 100 mm Hg to be nearly fully saturated. The small degree of arterial oxygen deficiency in these patients can thus be more accurately estimated by the direct measurement of P_{O_2} (which may be considerably reduced), rather than O_2 content or saturation.

3. The curve is S shaped and steep between the tensions of 100 and 40 mm Hg; thus the change in O_2 content (saturation) is relatively great when P_{O_2} is slightly reduced. Since P_{O_2} of active tissues comes within this low range, oxyhemoglobin in blood supplying such tissues readily gives up its oxygen. In contrast, the flat part of the curve enables the body to receive enough oxygen even in the face of a large decrease in P_{O_2}.

4. The shape of the dissociation curve depends on the temperature and on pH or P_{CO_2} of the medium. A decrease in pH or an increase in P_{CO_2} shifts the curve downward and to the *right.* (See Fig. 8-2.) Conversely, a rise in pH or a decrease in P_{CO_2} shifts the curve upward and to the *left.* The shift in the dissociation curve that occurs when the P_{CO_2} rises reduces the oxygen saturation percentage at a constant P_{O_2}. This means that at the tissue level the increased P_{CO_2} by itself drives off oxygen from the hemoglobin. Conversely, the fall in P_{CO_2} occurring when blood passes through the lungs causes the curve to shift upward and to the left and enhances oxygen uptake by the hemoglobin.

CARRIAGE OF CARBON DIOXIDE IN THE BLOOD

The lungs and the kidneys are the organs that excrete acids. The lungs excrete the volatile carbonic acid as CO_2; the kidneys excrete the nonvolatile acids. Arterial blood comes to the tissues with a large amount of oxyhemoglobin (HbO_2) and a minimum amount of carbon dioxide. Oxyhemoglobin gives up oxygen to the tissues and the tissues release carbon dioxide into the blood. This carbon dioxide is hydrated and ionized into $H^+ + HCO_3^-$ (equations 1 and 2). Simultaneously, some of the oxyhemoglobin is reduced, as discussed above. Reduced hemoglobin is a weaker acid than oxyhemoglobin and thus capable of accepting more H^+ from the ionization of carbon dioxide. Davenport[2] has calculated that for every millimole of oxygen used by the tissues, 0.7 mM of carbon dioxide or H^+ is taken up by this reduced hemoglobin—with no change in blood pH.

As carbon dioxide diffuses into the plasma, three reactions can happen:

1. The largest fraction enters the red cell and is buffered by the hemoblogin.
2. Dissolved carbon dioxide forms some carbamino compounds with plasma proteins.
3. The rest of the amount dissolved remains in the plasma.

Both in plasma and in the cells, dissolved CO_2 reacts with water to form carbonic acid:

$$CO_2 + H_2O \rightleftharpoons H_2CO_3 \qquad (1)$$

The reaction is far to the left; thus the plasma concentration of dissolved carbon dioxide is about 1000 times greater than the concentration of carbonic acid. The small amount of carbonic acid formed is ionized according to the equation below:

$$H_2CO_3 \rightleftharpoons H^+ + HCO_3^- \qquad (2)$$

Hydrogen ions are buffered by the weak buffering system in the plasma; the bicarbonate ions remain in the plasma. As stated previously, the largest fraction of carbon dioxide enters the red cell. There a small amount remains dissolved; a significant amount forms carbamino-CO_2 compounds with hemoglobin. The largest fraction is hydrated into carbonic acid which in turn is ionized into $H^+ + HCO_3^-$, according to equations 1 and 2. Again:

$$CO_2 + H_2O \rightleftharpoons H_2CO_3 \rightleftharpoons H^+ + HCO_3^-$$

Hydrogen ions are buffered by the hemoglobin, and many of the formed bicarbonate ions diffuse into the plasma because the bicarbonate concentration in the cells is greater than that in the plasma. Electric neutrality demands that there be the same number of positive charges in solution as there are negative charges. This movement of bicarbonate ions causes a decrease in the negative charges within the red cell. To compensate for this deficit, negatively charged chloride ions move *into* the red cell from the plasma (the "chloride shift," first described by Hamburger). The hydration of CO_2 to H_2CO_3 is a slow process; however, an enzyme called carbonic anhydrase, present in the red cell, speeds up the reaction. The exchange of bicarbonate ions from the red cell with chloride ions from the plasma increases the total CO_2 content of the plasma. It is clear that this increase is mainly due to the strong buffering power of the hemoglobin; the plasma buffers are secondary.

CARBON DIOXIDE COMBINING POWER OF ALKALI RESERVE

Prior to development of the pH glass electrode, a method was introduced by Van Slyke and Cullen[3] for measuring gross changes in plasma bicarbonate concentration. Clinicians in 1917 (date that the method was first used) were primarily interested in acid-base disturbances associated with diabetes, vomiting, diarrhea, and renal failure rather than those associated with respiratory disorders. The method was, from their standpoint, a simple and practical laboratory procedure quite capable of giving them an approximation of the disturbances mentioned. It consisted of equilibrating plasma with carbon dioxide at a P_{CO_2} of 40 mm Hg and analyzing the sample for total CO_2 content. Bicarbonate concentration or "alkali reserve" was then determined from an analysis of the total carbon dioxide content less the amount dissolved (1.2 mM/L at 40 mm Hg). Because equilibration is done *after* the plasma is separated from the red cell, any changes observed in bicarbonate concentration are a reflection of the buffering power of plasma alone. Since it is known that the main buffering action is carried out by the hemoglobin within the red cell, and not in the plasma, this is a source of error. No information can be obtained concerning respiratory disorders, since the procedure involves titration from the original unknown P_{CO_2} to a P_{CO_2} of 40 mm Hg. If the blood initially had a high P_{CO_2}, it would lose carbon dioxide during equilibration; bicarbonate concentration would fall and pH rise as the blood moved toward a P_{CO_2} of 40 mm Hg. If alkali reserve is above the normal point, extra fixed base is present and metabolic alkalosis is the condition. If it is below the normal point, extra fixed acid is present and metabolic acidosis is the condition. In either case, there is no way of telling from which direction the blood approached a P_{CO_2} of 40 mm Hg; therefore, no information is available on the existence or extent of respiratory disturbances. This fact was pointed out by Peters and Van Slyke[4] in their book on methods. Today, with the easy availability of P_{CO_2} and pH electrodes, there is no reason to continue this method in the laboratory.

HENDERSON-HASSELBALCH EQUATION

This equation is used in the laboratory to calculate the partition of *carbon dioxide in plasma.* It applies only to weak acids and their salts. Since pH can be measured, as well as total CO_2 and P_{CO_2}, the plasma bicarbonate or dissolved carbon dioxide concentration can be calculated by employing one of the following equations:

$$pH_p = 6.11 + \log \frac{[HCO_3^-]_p}{S\ P_{CO_2}} \qquad S = 0.0301$$

or

$$pH_p = 6.11 + \log \frac{[\text{Total } CO_2]_p - S\ P_{CO_2}}{S\ P_{CO_2}} \qquad S = 0.0301$$

The pK value for this system is 6.11. This is the ionization or dissociation constant of the weak acid, H_2CO_3, in plasma. S = the solubility coefficient of CO_2 in plasma at body temperature—or millimoles of dissolved CO_2 per liter, per 1 mm Hg P_{CO_2}. The quantity $[\text{Total } CO_2]_p$ is similarly expressed in millimoles per liter.

If equipment such as the Instrumentation Laboratory blood gas system is used, the first equation will be applicable. From the measured pH and P_{CO_2} values, the bicarbonate concentration of the plasma at normal body temperature can be calculated. Although the IL system is offered whole blood in a syringe or capillary tube, the pH measured is actually that of the *plasma* component of the blood. The same syringe of whole blood is now offered to the P_{CO_2} electrode. Because equilibrium is rapidly established, the measured P_{CO_2} applies both to plasma and to red cells.

It is equally correct to determine the acid-base status of plasma by measuring the plasma pH (as in the first equation) and the concentration of total CO_2 of plasma. This total CO_2 concentration can be determined with the Van Slyke apparatus and includes the bicarbonate fraction, the physically dissolved CO_2, and the small amount of carbamino-CO_2 in the plasma. The plasma will have to be separated from the red cells and introduced into the Van Slyke apparatus under strict anaerobic conditions. The second of these equations can then be used and solved for P_{CO_2}. The carbamino-CO_2 introduces a small but unavoidable error. The relationship shown in both equations pertains only to plasma, not to whole blood. Venous blood is usually analyzed (although arterial blood could be used just as well, since bicarbonate concentrations in the two differ only slightly).

SINGER-HASTINGS NOMOGRAM AND BLOOD BUFFER BASE

A nomogram is a chart whose length and shape have values satisfying a given equation or relationship. In 1948, Singer and Hastings[5] described such a nomogram (Fig. 8-3) employing the Henderson-Hasselbalch equation. Their objective was to devise a nomogram for the calculation of all the factors necessary in the evaluation of disturbances in acid-base balance. This was done on the basis of two determinations, (1) the plasma pH and (2) either plasma or whole blood CO_2 content, together with the hematocrit. They used the term "*blood buffer base*" or BB^+ (Fig. 8-4) to denote the fraction of the total base

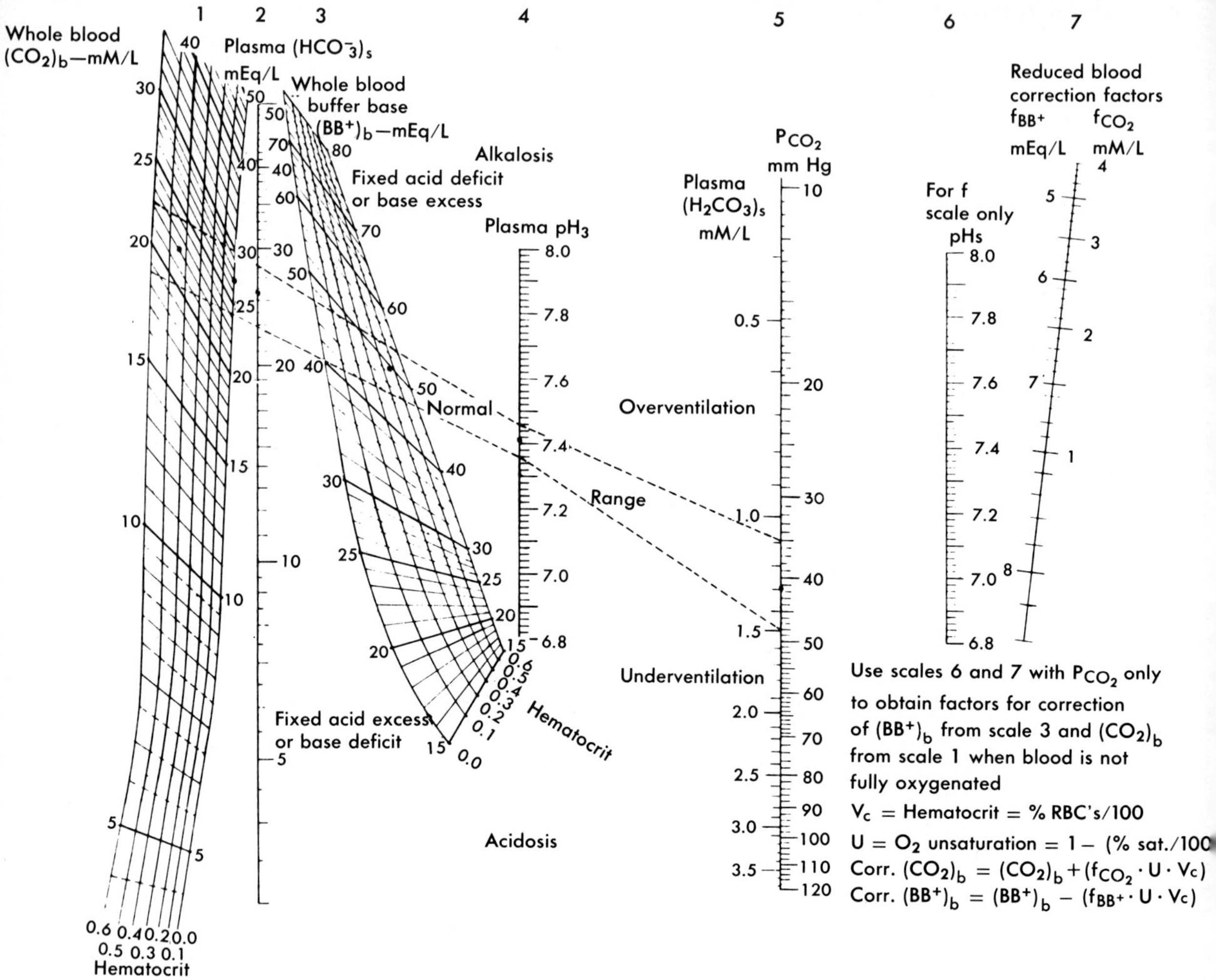

Fig. 8-3. Singer-Hastings nomogram for acid-base balance. Nomogram for the acid-base balance of human blood at 37° C. For oxygenated blood (scales 1 to 5) a straight line through points given on two of the scales intersects the remaining scales at simultaneously occurring values of the other variables. The position of this line indicates the kind and magnitude of any disturbance of the acid-base balance. When the oxygen saturation is less than 90%, the whole blood CO_2 and BB^+ figures should be corrected by factors derived from the use of scales 5, 6, and 7, for the most precise results. The normal range is for arterial, not venous, blood. (From Singer, R. B., and Hastings, A. B.: Medicine **27**:223, 1948.)

B^+, equivalent to the sum of $(HCO_3^-)_b$ and $(P^-)_b$. An equation for the term could be written as follows:

$$BB^+ = (P^-)_b + (HCO_3^-)_b + \text{Carbamino-}CO_2$$

HB + Plasma protein

A *decrease* in total base or an increase in fixed acid will produce a corre-

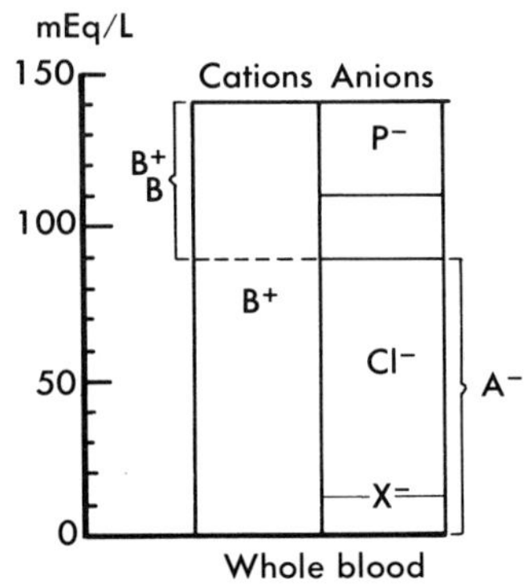

Fig. 8-4. Whole blood.

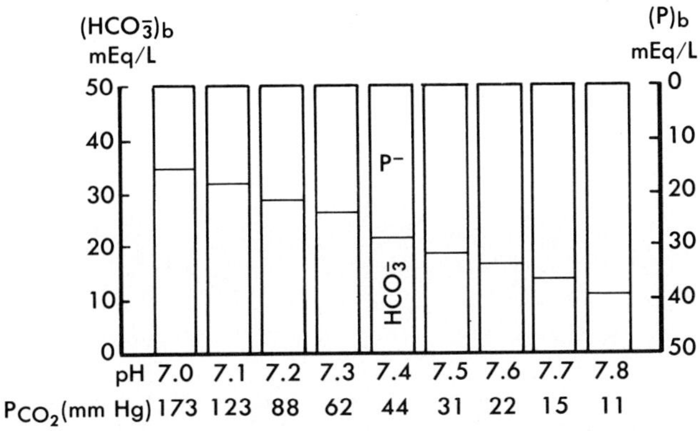

Fig. 8-5. Effect of buffer anions of varying CO_2 pressure (P_{CO_2}) and pH. Oxygenated whole blood. Hematocrit (Vc) = 0.45. Whole blood buffer base $(BB^+)_b$ = 50 mEq/L.

sponding *decrease* in buffer base. An increase in total base or a decrease in fixed acid will produce an *increase* in buffer base. These changes are called nonrespiratory or "metabolic" because they are altered *only* by body changes in absorption, secretion, or excretion. In contrast, *buffer base remains constant* with variations in P_{CO_2}. (See Fig. 8-5.) This illustration shows the $(HCO_3^-)_b$ and $(P^-)_b$ values vary with changes in P_{CO_2}, but the *buffer base remains the same.* Since P_{CO_2} is altered rapidly by changes in lung ventilation, it is considered the primary factor in respiratory disturbances. Because all the blood buffers are taken into account, *whole blood* buffer base is determined. The major role that hemoglobin plays in buffering the blood makes the packed cell volume or hematocrit an essential factor in the determination of blood buffer base. The Singer-Hastings nomogram assumes full oxygenation of the hemoglobin. If reduced blood were used, the value for $(P^-)_b$ would be about 4 milliequivalents lower at each pH unit. Singer and Hastings have therefore inserted a correction factor for use with blood less than 90% saturated, in order to more accurately determine buffer base.

The nomogram shown on p. 228 consists of seven scales, each with the proper designation. Scales 2, 4, and 5 constitute the Henderson-Hasselbalch equation applied to plasma. If any two variables are known, such as either CO_2 content (whole blood or plasma) and pH or P_{CO_2} and pH, a line can be drawn connecting the two points. This same line will intersect the other scales at their corresponding values, giving a clear picture of the acid-base status of that particular blood at the time it was drawn.

In the construction of the nomogram, normal values were assumed for plasma protein concentration (72 g/L of plasma) and for hemoglobin concentration (20 mM oxygen capacity/L of cells). Deviations in these concentrations will affect the accuracy of the nomogram. However, since the buffer value of plasma is small and red cell hemoglobin concentration has a narrow normal range (18.7 to 21.2 mM/L), the errors introduced would not be clinically significant.

In summary, by means of the nomogram *all the factors* in acid-base balance can be determined from blood of a given hematocrit and oxygen saturation if *two* parameters are known.

An error that often exists in clinical diagnosis is the assumption that an increase in CO_2 content represents alkalosis and a decrease in CO_2 content represents acidosis. Actually, total CO_2 content can be high or low in either acidosis or alkalosis. Also, normal values for CO_2 content may be obtained even in the

Table 8-1. Examples of acid-base disturbances (assume hematocrit, 0.45; saturation, 98%; normal plasma protein)

Etiology	Whole blood CO_2 (mM/L)	Plasma HCO_3 (mEq/L)	Whole blood BB^+ (mEq/L)	pH	P_{CO_2} (mm Hg)	Disturbance
1. Acute asthma	28.0	31.0	50.0	7.25	70.0	Uncompensated respiratory acidosis
2. Emphysema	35.0	40.0	60.0	7.37	70.0	Partially compensated respiratory acidosis
3. During anesthesia and with patient on respirator	15.0	18.9	50.0	7.71	15.0	Uncompensated respiratory alkalosis
4. Patient on respirator for prolonged period	10.5	12.5	40.0	7.53	15.0	Partially compensated respiratory alkalosis
5. Condition rarely seen; respirations are immediately stimulated as pH goes down	13.3	14.4	34.0	7.17	40.0	Uncompensated metabolic acidosis
6. Diabetic acidosis	8.2	10.2	34.0	7.32	20.0	Partially compensated metabolic acidosis
7. Condition rarely seen; respirations depressed as pH goes up	29.0	35.0	60.0	7.56	40.0	Uncompensated metabolic alkalosis
8. Condition commonly seen after prolonged vomiting from any cause and nasal gastric suction; HCl lost from stomach	33.0	37.8	60.0	7.46	55.0	Partially compensated metabolic alkalosis

presence of severe acidosis or alkalosis. *Thus a differential diagnosis cannot be made based solely on CO_2 content.*

Table 8-1 shows a number of examples of disturbed acid-base balance. The second, fifth, and sixth columns contain directly measurable values: whole blood CO_2, pH, and P_{CO_2}. Plasma HCO_3^- and whole blood buffer base (BB^+) have been derived from the nomogram. A few points might be made about the table.

1. In pure, uncompensated respiratory acid-base abnormalities (examples 1 and 3) blood buffer base is unchanged, as discussed previously; P_{CO_2}, pH, and CO_2 content are abnormal. Partial compensation (examples 2 and 4) leads to a movement of pH in the normal direction, while blood buffer base becomes abnormal, due to retention or removal of base.

2. In uncompensated metabolic disturbances (examples 5 and 7) all values, except P_{CO_2} are abnormal. Partial compensation is accomplished by superimposition of respiratory changes, again leading to a movement of pH in the normal direction.

3. Examples 2 and 8 have about the same values for whole blood CO_2 content and buffer base. Yet the disturbances which they represent are diametrically opposite. The pH in example 2 is moving in an acid direction and that in example 8 is moving in an alkaline direction.

APPARATUS

It is becoming increasingly important to know pH, P_{CO_2}, and P_{O_2} values both in research and in clinical medicine. The Instrumentation Laboratory blood pH/gas analyzer, briefly mentioned earlier in this chapter, measures all three parameters—pH, P_{CO_2}, and P_{O_2}—in one unit. From the direct measurement of pH and P_{CO_2}, buffer base, bicarbonate, and CO_2 content can be obtained by using the Singer-Hastings nomogram or similar nomograms. From the direct measurement of pH and P_{O_2}, blood oxygen saturation can be determined by using the oxygen dissociation curve or its nomogram. NOTE: Oxygen saturation determined indirectly from pH and P_{O_2}, although fairly accurate, is not the *true* oxygen saturation. This can only be obtained directly from a CO-Oximeter instrument (Chapter 7), which measures the carbon monoxide portion of total hemoglobin.

Ultra-Micro blood/gas system (113-S1) (Fig. 8-6)

The *electrometer* displays four color-coded scales. The top green scale gives P_{CO_2} directly in millimeters of mercury between 10 and 100. The center red scale gives the pH in the physiologic range of 6.80 to 8.00. The bottom two black scales indicate P_{O_2} in the ranges of 0 to 160 and 160 to 800 mm Hg, respectively.

A *constant-temperature control module* together with the circulating water bath keeps the temperature at 37° C. A Plexiglas cylinder provides a reservoir for the water circulating around both the P_{CO_2} and P_{O_2} electrodes. Another Plexiglas block acts as a water jacket around the pH electrode.

The *water pump* provides water circulation for the entire unit. A *vacuum pump* provides reduced pressure for drawing samples into the Ultra-Micro pH electrode.

Two *gas cylinders* are used in the calibration and slope standardization of the electrodes. The gas mixture used in routine work contains 5% CO_2, 12% O_2, and the balance nitrogen, closely matching normal arterial blood. A cylinder of

10% CO_2 and the balance nitrogen is used for the slope standardization. To ensure accuracy, all IL gas mixtures are carefully analyzed by the Scholander method before delivery.

The *IL Ultra-Micro blood pH electrode assembly* can be used to measure samples of blood from a capillary tube or syringe. The total sample size is 25 to 30 microliters. A roller assembly allows the blood to be withdrawn or ejected from the electrode assembly. All pH glass electrodes have a similar design. A solution of constant pH (usually HCl) is contained on one side of a glass membrane. In contact with the other side of this glass is the solution of unknown pH (your sample). A potential or voltage is established that is proportional to the pH difference of these solutions. Since the pH of one solution is constant, this developed potential is a measure of the pH of the other solution. To measure the potential, half-cells are introduced into the constant, internal solution and into the unknown, external solution. The internal half-cell sealed within the chamber of constant pH is a wire of silver–silver chloride. The external half-cell, or reference electrode, is usually a saturated mercury–calomel electrode. The function of this external half-cell is simply to provide a steady reference voltage against which voltage changes at the glass pH membrane can be compared.

The *IL P_{CO_2} electrode* is a modification of the electrode developed by Sever-

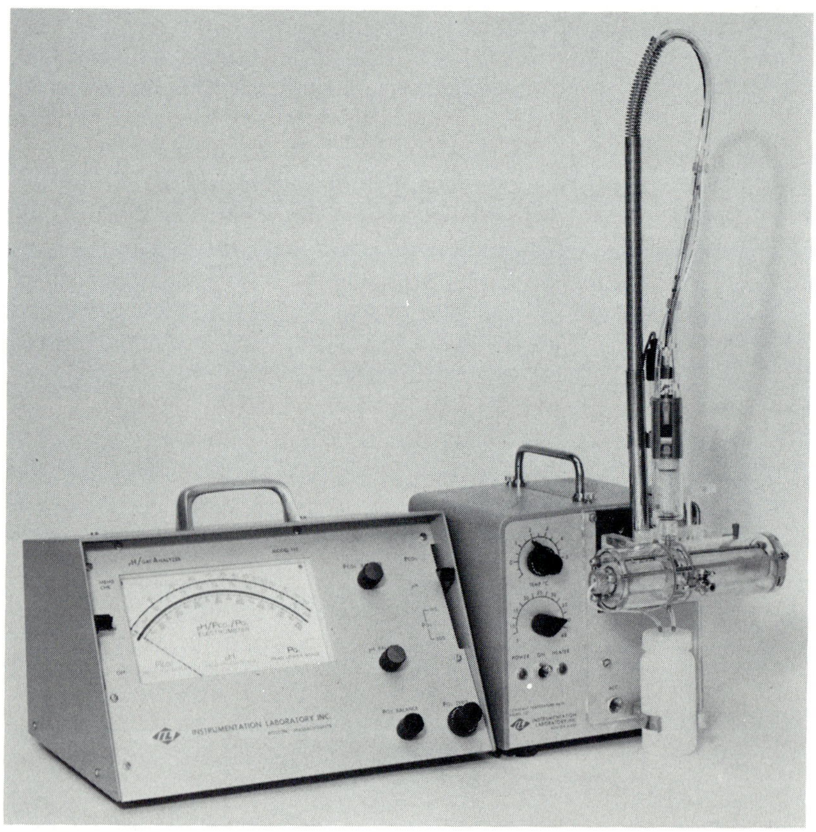

Fig. 8-6. Blood pH/gas analyzer. (Courtesy Instrumentation Laboratory, Inc., Lexington, Mass.)

inghaus and Bradley[6] in 1958. Both the pH glass electrode and the reference electrode are part of one common electrode arranged behind an outer gas-permeable ILastic membrane. A thin film of dilute $NaHCO_3$ covers the glass measuring surface of the pH electrode. This functions both as a layer in which CO_2 dissolves and reacts and as a liquid salt junction between the glass and the reference electrodes. Because diffusion rates are slow, the pH glass electrode must be placed close to the gas-permeable outer membrane. In order to provide close proximity and wetting of the electrode, a nylon mesh several layers thick is used. This nylon mesh inner membrane acts as a spacer between the outer membrane and the tip of the electrode. The outer membrane is permeable to gases (CO_2) but not to ions. CO_2 diffuses through this membrane in proportion to the P_{CO_2} difference across the membrane. The reaction produces carbonic acid. The pH electrode senses the change in carbonic acid as a change in pH and reports this signal logarithmically in terms of P_{CO_2}. A decrease of one pH unit will produce a tenfold increase in P_{CO_2}.

The *IL* P_{O_2} *electrode* is essentially a refined Clark[7] electrode measuring P_{O_2} by the polarographic technique. This technique is based on the fact that if a constant voltage (−0.6v) is maintained at a cathode, with respect to a reference anode, an electrolytic current is set up as dissolved oxygen is reduced at the cathode. The intensity of the current is directly related to the P_{O_2} being analyzed.

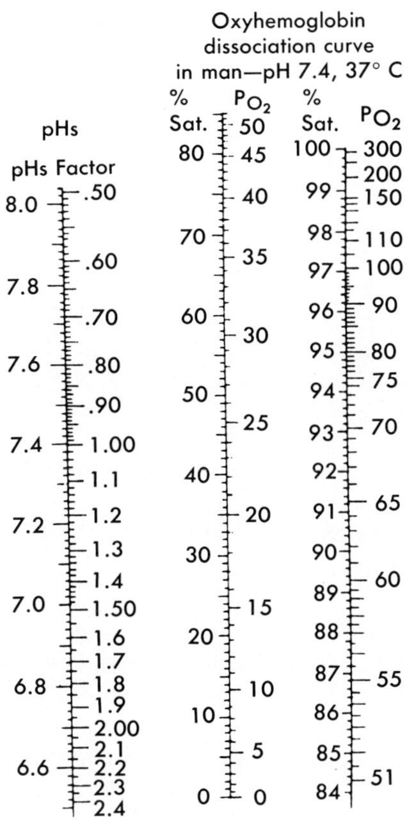

Fig. 8-7. Severinghaus oxyhemoglobin dissociation curve correction chart. (From Severinghaus, J. W.: J. Appl. Physiol. **12**:485, 1958.)

The electrode consists of a silver–silver chloride anode and a platinum cathode covered with an oxygen-permeable membrane. Each molecule of oxygen reacts with four electrons and it is this flow of electrons which is measured. The anode provides electrons as needed for the cathode reaction. Removed from the tip of the electrode, the anode is housed in a reserve chamber of electrolyte (usually buffered KCl). In addition to stabilizing the anode, the reserve chamber keeps the tip of the electrode supplied with electrolyte through a capillary opening. This eliminates frequent membrane changes due to electrolyte evaporation at the tip of the electrode. The small cathode (0.001 inch in cross section) is sealed in a glass envelope at the tip of the electrode. This micro cathode, together with the polypropylene membrane and the thin layer of electrolyte, minimizes changes in temperature, pressure, and flow of the sample.

Slide rule calculators are now available that enable the analyst to quickly determine the O_2 saturation from pH and Po_2 readings. This calculation is based on the Severinghaus[8] oxyhemoglobin dissociation curve correction chart shown in Fig. 8-7.

Example: Arterial blood analyzed at 37° C in the IL instrument had a Po_2 of 65 mm Hg and a pH of 7.28. What is the percentage oxygen saturation? Since the blood was analyzed anaerobically, its saturation was unchanged and the only correction needed in calculating saturation is this for pH:

$$\frac{65}{1.13} = 57.5 \text{ mm Hg } (Po_2 \text{ at } 37^\circ \text{ C and pH } 7.40)$$

From the chart the saturation = 88.5%.

GENERAL INFORMATION PERTAINING TO THE IL pH/BLOOD GAS ANALYZER

For the most accurate and reliable results, pH and blood gas analysis should be done immediately after the blood is drawn. In the range of 100 mm Hg Po_2, the rate of oxygen consumption is about 0.5 vol%/hr. The higher the Po_2 the faster the rate of consumption. From this fact it should be obvious that samples must be measured immediately, particularly where the Po_2 measurement is critical.

Use gas-tight heparinized syringes to collect the blood. Avoid trapping air bubbles; if any appear, quickly eject them and cap the syringe. Never use grease or mineral oil; it inhibits the efficiency of the electrode membranes and may even cause permanent damage to the electrode. Rotate the syringe so that the cells are mixed with the plasma before injecting a sample into the chamber. This mixing is particularly important if the blood has been allowed to cool (to room temperature or below). The rewarming to 37° C (bath temperature) of plasma *not* in the presence of red cells will result in erroneous values for all parameters.

If for some reason it is not possible to analyze the samples immediately, cap the syringes and store them in ice. This will rapidly slow the rate of blood metabolism, particularly if small syringes are used. However, even the use of ice will not forestall deterioration forever. Three or four hours of ice storage is maximum if valid results are to be obtained.

Scrupulous cleaning of equipment cannot be overemphasized. Electrodes last longer and perform better when washed with the proper solution immediately after each blood sample.

REFERENCES

1. Comroe, J. H., Jr.: Physiology of respiration; an introductory text, Chicago, 1965, Year Book Medical Publishers, Inc.
2. Davenport, H. W.: The ABC of acid-base chemistry; the elements of physiological blood-gas chemistry for medical students and physicians, ed. 4, Chicago, 1958, The University of Chicago Press.
3. Van Slyke, D. D., and Cullen, G. E.: Studies of acidosis. I. The bicarbonate concentration of the blood plasma: its significance and its determination as a measure of acidosis, J. Biol. Chem. **30**:289, 1917.
4. Peters, J. P., and Van Slyke, D. D.: Quantitative clinical chemistry. Vol. II. Methods, Baltimore, 1932, The Williams & Wilkins Co.
5. Singer, R. B., and Hastings, A. B.: Improved clinical method for estimation of disturbances of the acid-base balance of human blood, Medicine **27**:223, 1948.
6. Severinghaus, J. W., and Bradley, A. F.: Electrodes for blood PO_2 and PCO_2 determinations, J. Appl. Physiol. **13**:515, 1958.
7. Clark, L. C., Jr.: Monitor and control of blood and tissue oxygen tensions, Trans. Amer. Soc. Artif. Intern. Organs **2**:41, 1956.
8. Severinghaus, J. W.: Oxyhemoglobin dissociation curve correction for temperature and pH variation in human blood, J. Appl. Physiol. **12**:485, 1958.

9

Liver and biliary system

The liver is the largest organ in the body and performs many complex functions. Therefore, it is necessary to employ a group of tests for investigating suspected cases of hepatic disease. Test selections are governed by the nature and stage of the disease and by the information desired.

Some of the individual tests used for study of the liver are affected by conditions in which the liver is not involved such as proteins in nephrosis, phosphatase in bone disease, and bilirubin and urobilinogen in hemoglobin processes. The major functions in which the liver is involved are (1) metabolism of carbohydrates, proteins, lipids, and pigments; (2) detoxification and conjugation processes; and (3) specific storage and excretion mechanisms.

CARBOHYDRATE METABOLISM

Carbohydrate metabolism in the liver includes conversion of certain amino acids to glucose (gluconeogenesis), conversion of glucose to glycogen (glycogenesis), breakdown of glycogen to glucose (glycogenolysis), and storage of glycogen. The liver aids in the maintenance of normal blood glucose levels. After an overnight fast, blood glucose may indicate hypoglycemia when there is disturbance of carbohydrate metabolism, as in cirrhosis of the Laennec type. In such cases glucose tolerance tests offer little diagnostic information, but a galactose tolerance test is of some value in differential diagnosis of jaundice. A normally functioning liver converts galactose to glucose but for patients with hepatocellular jaundice abnormal blood and urinary levels of galactose occur. However, the test at present is rarely used in clinical investigations.

PROTEIN METABOLISM

Protein metabolism in the liver involves deamination and transamination of amino acids; production of urea; and synthesis of serum proteins, prothrombin, and enzymes. Abnormalities of liver function may be associated with abnormal plasma levels of amino acids, proteins, urea, enzymes, and ammonia. Serum levels of albumin are lowered in patients having cirrhosis, viral hepatitis, or nutritional liver disease. This may not be reflected in total serum protein levels, since there is frequently an equal and simultaneous increase in globulin. The serum albumin concentration may reflect the severity of a chronic hepatic disease.

Blood urea levels do not reflect liver impairment until the condition is far advanced. Production of ammonia occurs with deamination of amino acids in the liver and by bacterial action on urea in the intestine. Under normal conditions ammonia is synthesized in the liver to urea. In liver disease, urea synthesis is impaired and accumulation of ammonia may occur in the blood. Blood ammonia determination is used in clinical investigations of hepatic coma and as one of the tests for indicating gastrointestinal bleeding in cirrhotic patients.

For patients with liver disease, plasma fibrinogen may be within normal limits or low; prothrombin and other clotting factors may be decreased. The level of blood plasma prothrombin depends not only on the ability of the liver to produce it but also on the availability of vitamin K. In obstructive jaundice with blockage of the flow of bile to the duodenum, the prothrombin activity may be low because of deficient intestinal absorption of vitamin K.

Many enzymes occur in the liver but those which are routinely used in clinical studies of liver disease are serum alkaline phosphatase, oxalacetic glutamic transaminase (SGOT), and pyruvic glutamic transaminase (SGPT). Enzyme patterns of an affected organ as indicated by the enzymes secreted into the blood are indications of the severity of the cell damage. Alkaline phosphatase has been used for differential diagnosis of jaundice. Increased levels of SGPT and SGOT are observed in patients with hepatic disease, but the SGOT may be elevated in nonhepatic disturbances, also. An increase in the lactic dehydrogenase isoenzyme LDH_5 is also associated with liver disease. Methods for the determination of these enzymes are given in Chapter 11.

Laboratory examinations

The blood proteins are made up partially of antibodies, enzymes, and hormones. These substances have varied functions. As a group, one of their major functions is to maintain the water balance between the blood and tissues.

Under normal conditions, the water balance is at equilibrium. However, if blood proteins are lost, the density of the blood decreases, and water flows toward the tissues. Swelling, or edema, results. This is found in nephrosis, a disease of the kidneys characterized by appearance of albumin in the urine.

The proteins measured most frequently in the clinical laboratory are albumin, globulin, and total protein. These substances are the most abundant solid material in blood, and their values are given in grams per 100 ml (g/100 ml) of serum. The normal values are given below.

Albumin	4.0 to 5.0 g/100 ml of serum
Globulin	2.0 to 3.0 g/100 ml of serum
Total protein	6.0 to 8.0 g/100 ml of serum

Some significant diseases in which abnormal protein values may be expected are given in Table 9-1.

Some of the more commonly used methods for the determination of proteins are discussed briefly below.

Table 9-1. Conditions accompanied by abnormal protein values

Disease	Albumin	Globulin	Total protein
Nephrosis	Decreased	Normal	Decreased
Multiple myeloma	Normal	Increased	Increased
Infectious hepatitis	Decreased	Increased	Decreased
Cirrhosis of liver	Decreased	Normal	Decreased

Greenberg method

1. The albumin is determined as follows. A sodium sulfate solution is added to a portion of serum. This precipitates globulin, which is removed by filtering or by centrifuging. (If done by centrifuge, ether is added to aid in the separation.) To the filtrate—containing the albumin—is added the phenol reagent of Folin and Ciocalteu. A blue color is produced, which is measured in a colorimeter and compared with a standard tyrosine solution.

2. The total protein is determined as follows. Another portion of serum is diluted with sodium chloride solution. The same phenol reagent of Folin and Ciocalteu is added. A blue color is produced. The depth of color is measured in a colorimeter and compared with a standard tyrosine solution.

3. The globulin is found as follows: Globulin = Total protein − Albumin.

Kingsley biuret reaction

1. The albumin is determined as follows. A sodium sulfate solution is added to a portion of serum. This precipitates the globulin, which is removed by filtering or by centrifuging. (If the centrifuge is used, ether is added to aid in the separation.) To the filtrate—containing the albumin—is added a biuret reagent that forms a reddish purple complex (the biuret reaction). The depth of color is measured in a colorimeter and compared with a standard.

2. The total protein is determined as follows. Another portion of serum is diluted with sodium chloride solution. The same biuret reagent is added. This forms a reddish purple complex. The depth of color is measured in a colorimeter and compared with a standard.

3. The globulin is determined as follows: Globulin = Total protein − Albumin.

Micro-Kjeldahl method

1. The total protein is found as follows. (The test calls for two separate determinations and a calculation.)

In the first determination, the total nitrogen is determined on serum by a micro-Kjeldahl method. This entails adding a digestion mixture, heating with a flame to decompose nitrogenous substances, adding Nessler's solution to produce a yellow-orange color, and then measuring the depth of color in a colorimeter.

In the second determination, the nonprotein nitrogen is determined on a protein-free filtrate by the micro-Kjeldahl method. This again calls for digestion and nesslerizing.

In the calculation, the total protein is found from the formula: Total protein = (Total nitrogen − Nonprotein nitrogen) × 6.25, where the 6.25 is a factor converting nitrogen to protein.

Example: Total nitrogen value is found to be 1000 mg/100 ml and nonprotein nitrogen is found to be 30 mg/100 ml. Therefore, total protein = (1000 − 30) × 6.25 = 6062.5 mg/100 ml, or 6.1 g/100 ml.

2. The albumin is determined as follows. The globulin is removed from the serum by precipitating with sodium sulfate and filtering. The filtrate contains the albumin. This is digested and nesslerized to determine the total nitrogen. The albumin is found from the formula: Albumin = (Total nitrogen − nonprotein nitrogen) × 6.25, where 6.25 is a factor converting nitrogen to protein. The nonprotein nitrogen value used above is derived from the total protein determination.

Example: Total nitrogen value is found to be 700 mg/100 ml, and nonprotein nitrogen is 30 mg/100 ml. Therefore: Albumin = (700 − 30) × 6.25 = 4187.5 mg/100 ml, or 4.2 g/100 ml.

3. The globulin is found as follows: Globulin = Total protein − Albumin.

Example: Using the above figures for total protein and albumin, Globulin = 6.1 − 4.2 = 1.9 g/100 ml.

Looney-Walsh method

1. The globulin is determined as follows. Serum is diluted with saline solution. To this are added gum ghatti and saturated ammonium sulfate, their function being to precipitate the globulin in the form of a colloidal suspension. The turbidity of the solution is then compared in a colorimeter with a standard that has been similarly treated.

2. The total protein is determined as follows. Serum is diluted with saline solution. To this, gum ghatti and sulfosalicylic acid are added, their purpose being to precipitate the proteins in the form of a colloidal suspension. The turbidity of the solution is then compared in a colorimeter with a standard that has been similarly treated.

The albumin is found as follows: Albumin = Total protein − Globulin.

Kingsley serum protein biuret method, modified by Weichselbaum

The biuret method is a fast, simple, and accurate procedure for protein determination. However, lack of sensitivity of the method makes it not too satisfactory for cerebrospinal fluid or urine protein analysis.

Protein produces a violet color in the presence of cupric ions in an alkaline solution, reaching its maximum color within 15 minutes. The color is then stable for several hours.

Weichselbaum added sodium potassium tartrate as a complexing agent and potassium iodide to prevent autoreduction.[9,25]

Reagents needed

Biuret reagent	(D-197)
Saline, 0.9%	(D-22)
Sodium sulfate, 24.5%	(D-198)
Anhydrous ether	

Procedure for albumin

1. Pipette 0.5 ml of serum into a test tube approximately 16 × 125 mm.
2. Add 7.5 ml of 24.5% sodium sulfate, using a 10 ml serologic pipette. Stopper and shake vigorously. (NOTE: Sodium sulfate at this concentration has a tendency to crystallize when the room temperature falls below 25° C. If the room temperature is below 25° C place the rack of tubes and a flask of 24.5% sodium sulfate in a large pan of water and adjust the temperature to 25° to 30° C by adding warm water, or use a water bath if available.)
3. Allow the tubes to stand for 15 minutes and then add approximately 3 or 4 ml of ether by pouring directly from the ether can into the tube. *Do not pipette ether! Do not smoke or have an open flame near this area.* Ether vapors travel along the tabletops of work area.
4. Shake the tubes vigorously. Use finger cots on the thumbs. At the beginning of shaking the tubes, carefully release the thumbs to relieve the pressure built up in the tube. This will prevent any ether and solution splattering around the laboratory.
5. Centrifuge the tubes 3 to 4 minutes at about 3000 rpm. The precipitated globulin will be packed in a firm layer between the top ether layer and the bottom solution layer containing the albumin.

6. This next step requires a careful technique. Use a 5.0 ml volumetric pipette and place the forefinger over the mouthpiece opening of the pipette to keep any ether or globulin from entering the tip. Tilt the tube to a slanting postion to dislodge the globulin layer. If it still adheres to the tube, gently tap the tube over the globulin area with the pipette. Carefully lower the pipette under the globulin layer and down into the albumin layer. Slowly remove 5.0 ml of the albumin layer and transfer into a test tube (large) or a Coleman cuvette.
7. Add 5.0 ml of saline.
8. Add 10.0 ml of biuret reagent and mix.
9. Prepare a reference blank as follows:
   a. Pipette 5.0 ml of 24.5% sodium sulfate into a large test tube or a Coleman cuvette.
   b. Add 5.0 ml of saline.
   c. Add 10.0 ml of biuret reagent and mix well.
10. Place all tubes into a water bath at 30° to 32° C for 30 minutes.
11. Read absorbance (A) at 555 mμ, or read % T and convert to A.
12. Refer to the calibration curve to convert A readings to grams of albumin per 100 ml (g/100 ml).

Procedure for total protein

1. Pipette 0.5 ml of serum into a large test tube (18 × 150 mm).
2. Add 20 ml of saline. (Use burette.) *Mix well.*
3. Transfer 5.0 ml of diluted serum to another tube (e.g., a 19 × 150 mm Coleman cuvette).
4. Add 5.0 ml of biuret reagent and mix well.
5. Prepare a reference blank by mixing 5.0 ml of saline with 5.0 ml of biuret.
6. Place the tubes in a water bath set at 30° to 32° C for 30 minutes.
7. Read A at 555 mμ.
8. Refer to the calibration curve to convert A readings to grams of protein per 100 ml.

Procedure for globulin

The amount of globulin is obtained by subtracting the amount of albumin found from the amount of total protein found.

Example:

Total protein	7.0 g/100 ml (or g%)
Albumin	4.0 g/100 ml
Globulin	3.0 g/100 ml

NOTE: The presence of lipids may cause a turbidity in the solutions. This may be solved by the following method. Freeze the serum; this splits the protein-lipid bond. Leave the specimen in the freezer overnight. Then allow the frozen specimen to come to room temperature. Add an equal volume of ether to the tube and shake vigorously. Centrifuge for about 3 minutes. Suction off the top ether layer. If the specimen is not clear, extract with ether a second time. Proceed with the routine protein analysis.

CALIBRATION CURVE FOR TOTAL PROTEIN AND ALBUMIN

For each new bottle of biuret reagent, a new calibration curve must be made. This is accomplished by using the new biuret solution with a *standard* protein solution, taking readings on the spectrophotometer, and plotting these readings against the concentration of protein on linear graph paper.

Standard protein solution preparation

A 2.8 mg/ml solution of protein may be prepared from a serum of known protein value.

Example: The commercial serum chosen has a known protein value of 7.0 g%, and 7.0 g% = 70 mg/ml. Let x represent the amount of standard to be used. Then:

$$\frac{2.8}{x} = \frac{70}{100}$$

$$70\,x = 280$$

$$x = \frac{280}{70}$$

$x = 4.0$ ml of control serum needed

With a 4.0 ml volumetric pipette, pipette 4.0 ml of commercial serum into a 100 ml volumetric flask. Dilute to 100 ml with saline. Mix well. This solution now contains 2.8 mg/ml of protein.

Procedure

1. Set up the following tubes in duplicate, using volumetric pipettes.

Tube	Protein standard	Saline	Biuret
1	1.0 ml	4.0 ml	5.0 ml
2	2.0 ml	3.0 ml	5.0 ml
3	3.0 ml	2.0 ml	5.0 ml
4	4.0 ml	1.0 ml	5.0 ml

2. Mix well and place in 30° to 32° C water bath for 30 minutes. Obtain A readings at a 555 mμ wavelength. Plot the A readings against the protein concentration on linear graph paper. The above readings will be plotted on one sheet of graph paper for both the total protein and the albumin. The protein concentration chart, calculated and ready for the graph paper, is as follows:

Tube	Total protein (g/100 ml)	Albumin (g/100 ml)
1	2.3	1.79
2	4.6	3.58
3	6.9	5.38
4	9.2	7.17

Method of obtaining protein concentrations

1. *Total protein* Adding 0.5 ml of serum to 20.0 of saline gives a volume of 20.5 ml as the total volume of diluted serum. Therefore:

$$\frac{\text{0.5 ml serum used}}{\text{20.5 ml, or total volume diluted serum}} = \text{0.0244 ml serum/ml diluted serum}$$

Now 5.0 ml of this diluted serum is used. Therefore 0.0244 × 5 = 0.122 ml of serum actually used. This must be corrected to 100 ml so that the result may be expressed in g/100 ml:

$$\frac{100}{0.122} = \text{819.6 or 820, which is the dilution factor}$$

To determine what the readings of the four standards should be plotted against, in grams per 100 ml, multiply the known amount of protein (2.8 mg = 0.0028 g) by the dilution factor of 820. Therefore:

	Total protein (g/100 ml)
0.0028 g (1 ml) × 820 =	2.3
0.0056 g (2 ml) × 820 =	4.6
0.0084 g (3 ml) × 820 =	6.9
0.0112 g (4 ml) × 820 =	9.2

2. ***Albumin*** **By calculation we find:**

$$\frac{0.5 \text{ ml serum used}}{8.0 \text{ ml, or total volume diluted serum}} = 0.0625 \text{ ml serum/ml diluted serum}$$

Now 5.0 ml of this diluted serum (filtrate) is used. Therefore 0.0625 × 5 = 0.3125 ml of serum actually used. This must be corrected to 100 ml so that the result may be expressed in grams per 100 ml:

$$\frac{100}{0.3125} = 320$$

This 320 must be multiplied by 2 because the total volume of the final albumin dilution is 20 (it is 10 for the total protein). Therefore, 320 × 2 = 640, the dilution factor for albumin.

To determine what the readings of the four standards should be plotted against, in grams per 100 ml, multiply the known amounts of protein by the dilution factor of 640:

	Albumin (g/100 ml)
0.0028 g (1 ml) × 640 =	1.79
0.0056 g (2 ml) × 640 =	3.58
0.0084 g (3 ml) × 640 =	5.38
0.0112 g (4 ml) × 640 =	7.17

ELECTROPHORESIS

The application of electrophoresis to the separation of serum proteins and the methodology are given in Chapter 20.

Conversion of total protein in grams per 100 ml to milliequivalents per liter

To convert total protein in g/100 ml to mEq/L, use this formula:

$$\text{mEq/L} = \text{g/100 ml} \times 2.43$$

Example: Convert 5.0 g/100 ml of total protein to mEq/L, using the formula:

$$\begin{aligned}\text{mEq/L} &= \text{g/100 ml} \times 2.43\\ &= 5.0 \times 2.43\\ &= 12.15\end{aligned}$$

THYMOL TURBIDITY

When the serum of a normal person is added to a buffered thymol solution, no appreciable precipitation occurs. However, when the serum of a patient with liver disease is added to a buffered thymol solution, a definite precipitation is produced. Consequently, the procedure can be used as a test of liver function.

The precipitation (turbidity) reaction is not completley understood. It is activated with an increase in the gamma (γ) globulin and is inhibited with a quantitative or qualitative decrease or change in the albumin.

The method of Maclagan, modified by Shank and Hoagland, is commonly used. In this method, serum is obtained. To this is added a thymol solution which has been buffered to a pH of 7.55. After 30 minutes, any turbidity produced is measured in a colorimeter or compared with a standard.

The normal values for the thymol turbidity test are 0 to 5 S.H. units. These values are exceeded in the following diseases: infectious hepatitis, cirrhosis of the liver, cancer of the liver, diabetes, nephrosis, Weil's disease, and parenchymatous liver disease.

Procedure for thymol turbidity test

The following general considerations pertain to the procedure for the thymol turbidity test.

1. Specimen: serum.
2. Preservation of serum: may be stored overnight in a refrigerator.
3. Precaution: avoid using lipemic serum.

Shank-Hoagland modification of Maclagan method[15,23]

To obtain a visual comparison of turbidity, the original Maclagan method used Kingsbury's gelatin standards. These permanent standards were also used for protein determinations in urine with sulfosalicylic acid.[11]

Shank and Hoagland adapted the method for a photometer or spectrophotometer by using $BaSO_4$ suspensions for standardization and reading the turbidities.

If the pH of the thymol-barbital buffer used is changed from a pH 7.80 to a pH 7.55, the sensitivity of the test is greater for detecting hepatocellular dysfunction, without an introduction of false positives.

Temperature control is important because the turbidity of the test decreases with an increase in the temperature. The pH of the thymol buffer also decreases with a temperature increase, affecting the sensitivity of the test. A maximum error of 10% may be found if the temperature deviates beyond $\pm$ 3° C; therefore the safe range is between 22° and 28° C. If necessary, use a controlled-temperature water bath at 25° C.[23]

Reagent needed

Thymol-barbital buffer, pH 7.55 (D-199)

Normal values

Shank-Hoagland units (buffer pH 7.55)	Range 0 to 5 units
Shank-Hoagland units (buffer pH 7.80)	Range 0 to 4 units

Procedure

1. Pipette 6.0 ml of thymol-barbital buffer into two 19 $\times$ 150 mm cuvettes.
2. Pipette 0.1 ml of water into the "blank" cuvette containing the thymol buffer. Mix.
3. Pipette 0.1 ml of serum into the "unknown" cuvette containing the thymol buffer. Mix.
4. Let the tubes stand for 30 minutes at a temperature of 25° $\pm$ 3° C. Use a controlled-temperature water bath if necessary.
5. Set the blank at 100% T and read the unknown in percent transmission at 650 mμ.

Calculation

The calculations may be done by reading from a calibration curve or by using a constant (K), which is averaged from three independent batches of barium chloride reagent.

NOTE: Since this method uses a 0.0962 *normal* solution of barium chloride, the results are recorded in Shank-Hoagland units. If the method used a 0.0962 *molar* solution of barium chloride, the results would be recorded in Maclagan units. Two Shank-Hoagland units equal one Maclagan unit.

The emphasis on molarity and normality in relation to the barium chloride solution is made for the following reason. In 1946, Shank and Hoagland[23] published the method of mixing barium chloride with dilute sulfuric acid for the barium sulfate suspensions. A misprint using the symbol N (normality) instead of M (molarity) occurred, resulting in the values for the unknowns being twice the results of the original method and the standards being only half the value expected. Although the authors corrected the error in reprints that were requested, many established this now-popular method in their laboratories, using the 0.0962 normality value. The Commission on Liver Disease of the Armed Forces Epidemiological Board has recommended that Shank-Hoagland units be adopted.

Calibration curve

Make the following reagents:

1. Barium chloride standard, 0.0962 normal
   a. Weigh 1.173 g of barium chloride ($BaCl_2 \cdot 2H_2O$) and place into a 100 ml volumetric flask.
   b. Dissolve and dilute to mark with water (distilled).
2. Sulfuric acid, 0.2 N
   a. In a volumetric flask, dilute 6 ml of concentrated sulfuric acid to 1 L with distilled water. CAUTION: add the acid to the water and then dilute to 1 L.
   b. Standardize to 0.2 N.
3. Stock barium sulfate
   a. Pipette 5.0 ml of the barium chloride standard into a 100 ml volumetric flask. Chill to 10° C.
   b. Dilute to mark with cold (10° C) 0.2 N sulfuric acid.
   c. Mix well.

Set up the tubes as indicated in Table 9-2 and read each tube three times against the zero standard at a wavelength of 660 mμ. Average each set of readings and draw the curve. Use semilog paper when plotting % T, or linear paper for A.

Obtain the constant (K) by converting the % T readings to A (see Table A-6):

$$K = \frac{\text{Units}}{A}$$

Obtain an average K value from the readings and use this to calculate the thymol turbidity units of serum specimens:

$A \times K = \text{Thymol turbidity units}$

THYMOL FLOCCULATION TEST[1]

The physiologic significance of this test is essentially the same as that of the test referred to as the "c.c.floc."

Table 9-2. Chart for thymol turbidity calibration

Units	$BaSO_4$ (ml)	H_2SO_4 (ml)	% T (average of three readings)
0	0	10.00	
5	1.35	8.65	
10	2.70	7.30	
15	4.05	5.95	
20	5.40	4.60	
25	6.75	3.25	
30	8.10	1.90	
35	9.45	0.55	

A thymol flocculation test is done by extending the reading of the thymol turbidity test done quantitatively. After the thymol turbidity has been done, let the tubes sit for 24 hours. If any flocculation occurs, it is recorded as 1+, 2+, 3+, or 4+. Normally there is no flocculation.

CEPHALIN-CHOLESTEROL FLOCCULATION[2,8,13]

In health, there is a balance between the albumin and gamma globulin of the serum. In liver disease, however, this balance is disturbed. This disturbance of balance produces a precipitation in a solution of cephalin and cholesterol.

For the test, the procedure of Hanger is commonly used. Two test tubes are prepared. One tube contains the patient's serum plus saline solution and the emulsion of cephalin and cholesterol. The other tube contains only saline solution and the emulsion of cephalin and cholesterol. This tube serves as a control. Both tubes are put aside and inspected for flocculation at the end of 24 and 48 hours.

A normal value is zero to 1 plus. In liver disease, there are varying degrees of flocculation. These flocculations are reported as 1+, 2+, 3+, and 4+.

Procedure for cephalin-cholesterol flocculation test

The following general considerations pertain to the procedure for the cephalin-cholesterol flocculation test:

1. Specimen: serum.
2. Preservation of serum: fresh serum should be used, if possible; however, serum may be refrigerated overnight; frozen serum may be used if it is thawed quickly at 45° C to avoid alteration of the proteins.
3. Precaution: Unclean glassware may produce a false flocculation.

Hanger method

Reagents needed

Ether
Saline, 0.9% (D-22)

Procedure

1. Prepare a *stock* ether solution of cephalin-cholesterol as follows. Vials containing the dried cephalin and cholesterol may be purchased from the Wilson Laboratories or the Difco Laboratories. If the vial containing the dried cephalin and cholesterol is from the Wilson Laboratories, add 8 ml of ether and mix. If the vial containing the dried cephalin and cholesterol is from the Difco Laboratories, add 5 ml of ether and mix. This stock solution is stable if kept in a refrigerator.
2. Prepare the *emulsion* of cephalin and cholesterol as follows. Place 30 ml of distilled water in a 50 ml beaker. Note the height of the water in the beaker and mark the spot with a marking pencil. Add 5 more ml of distilled water. Heat to 65° to 70° C. While stirring the distilled water, slowly add 1 ml of the stock ether solution of cephalin and cholesterol. Heat slowly to boiling. Let simmer until the final volume is 30 ml as shown by the mark on the beaker. Allow to cool to room temperature. Examine the emulsion in a strong light. If there is a precipitate, centrifuge the emulsion at high speed for 20 to 25 minutes. If a button of lipid material has risen to the top, gently push it aside and remove the smooth, milky supernatant emulsion by aspiration. This emulsion, when stored in a refrigerator, is stable for 1 week.
3. Pipette 0.2 ml of serum into a 12 ml conical-tipped centrifuge tube. Add 4.0 ml of saline. Each series of tests should include an emulsion control (4.0 ml of saline) and a negative and positive control serum when possible. Add 1.0 ml of cephalin-cholesterol-emulsion to each tube, includ-

ing the emulsion control. Mix and stopper. Place the tubes in a dark cabinet at 25° ± 3° C and read the reaction at the end of 24 hours (and at 48 hours, if requested). The degree of flocculation usually increases slightly in the second 24-hour period; i.e., a 1+ may become a 2+, etc. Normal sera also show this shift. Since there is no concrete evidence that a 48-hour reading increases the accuracy of the test and since this would produce a 1-day delay in reporting results, a single reading at 24 hours is believed to be sufficient.[9,20]

The following alternate procedure may be used: place the tubes in the dark at 37° C for 3 hours, then centrifuge at about 3000 rpm for 5 minutes, and read flocculation. Results obtained after 3 hours at 37° C agree very closely with the results obtained after 24 hours at 25° C.[2,9]

4. Read the reaction as follows:
   Negative. No flocculation or precipitation
   1+ Trace of flocculation or precipitation
   2+ Definite flocculation or precipitation
   3+ Heavy flocculation or precipitation, but with the supernatant remaining turbid
   4+ Precipitation complete and supernatant fluid clear

The emulsion control should be negative.

Normal range

0 to 1+

FIBRINOGEN

Fibrinogen is used in the clotting of blood. The normal values range from 200 to 400 mg/100 ml of plasma. These values are increased in infections and decreased in cirrhosis of the liver, phosphorus poisoning, and typhoid fever.

Fibrinogen screening test[6]

This method is performed by using 1 drop of blood from a finger stick. The blood is diluted with glycine-saline buffer. With the addition of the latex-antihuman fibrinogen reagent, the agglutination is compared with a normal control. The latex-antihuman fibrinogen reagent is prepared from polystyrene latex and antibody to human plasma fibrinogen. This gives a rapid and clear reaction for plasma fibrinogen levels over 100 mg/100 ml.

*Reagents needed**

Glycine-saline buffer diluent
Normal control
Latex-antihuman fibrinogen reagent

Procedure

1. After obtaining a finger stick, transfer (use capillary pipette furnished) 1 drop of the patient's blood to the dropper bottle containing a measured volume of glycine-saline buffer diluent. Cap bottle and mix well.
2. Using capillary pipettes, transfer 1 drop of the diluted blood to one of the printed card slides.
3. Using another capillary pipette, transfer 1 drop of the already diluted control to another card slide.
4. To both specimens, add 2 drops of the latex-antihuman fibrinogen reagent.
5. Using separate toothpicks, mix each of the diluted specimens with the

*These reagents, pipettes, and card slides may be obtained in a kit manufactured by Hyland Laboratories.

fibrinogen reagent. Carefully spread the mixed specimen over the card area within the circle.

6. Tilt the card slowly from side to side for about 15 to 20 seconds and observe for clumping.

Results

Blood specimens with plasma fibinogen levels of 100 mg/100 ml or less will not show agglutination in the antigen-antibody reagent. However, the specimens having fibrinogen levels within the normal range will have "clumping" comparable to that of the normal control.

Normal range

250 to 400 mg/100 ml

Precaution

The reagent must be kept at a temperature of 2° to 10° C (35° to 50° F).

Fibrinogen, quantitative[14]

This method is similar to that of Reiner and Cheung[19] with the exception that fibrinogen is removed from the plasma by precipitating with calcium chloride instead of thrombin.

The method presented utilizes the materials, standards, reagents, and techniques used for the chemical determination of proteins.

Reagents needed

Saline, 0.9%	(D-22)
Calcium chloride, 2.5%	(D-200)
Sodium hydroxide, 0.2 N	(D-201)
Biuret reagent	(D-197)
Protein standard, 3 mg/ml	(D-202)

Procedure

1. Use either a 3204 QBD Vacutainer tube (EDTA) or a regular syringe and a test tube containing EDTA to collect 5 ml of venous blood.
2. Centrifuge for 15 minutes at 3000 rpm.
3. Add 25 ml of saline to an Erlenmeyer flask.
4. Pipette 1.0 ml of EDTA plasma into the flask. Swirl to mix.
5. Add 1.0 ml of 2.5% $CaCl_2$; thoroughly mix.
6. Let stand for 1 hour. (A firm fibrin clot will form.)
7. With a glass stirring rod, pick up the fibrin clot by twisting the clot around the glass rod, squeezing out all of the saline plasma solution. (NOTE: Small parts of the clot may have separated from the main clot. These may be found by placing the flask against a dark background.)
8. Place all of the clot in a 125 ml Erlenmeyer flask.
9. Add about 15 ml of saline and swirl the flask to wash the clot.
10. Carefully pour off the saline, taking care not to lose any of the clot.
11. Do three saline washings.
12. After the last washing, carefully twist the clot around a clean glass stirring rod and gently press against the side of the flask to eliminate excess saline.
13. Transfer the clot to a 15 ml centrifuge tube. (Place the stirring rod against the lip of the test tube and gently pull the rod back away from the clot, which will slip off quite easily.)
14. Add 2.0 ml of 0.2 N NaOH.
15. Place a marble on the tube and heat in a boiling water bath until the fibrin clot dissolves.
16. Set the tube in a room temperature water bath to cool.
17. Add 2.0 ml of biuret reagent.
18. Stopper and mix. Let stand for 30 minutes.

19. Set up a blank and standard as follows:
    Blank
    - 2.0 ml of 0.2 NaOH
    - 2.0 ml of biuret reagent. Mix.

    Standard
    - 1.0 ml of protein standard, 3 mg/ml.
    - 1.0 ml of 0.2 N NaOH
    - 2.0 ml of biuret reagent. Mix.
20. Read the standard and test against the blank at 550 mμ.

Calculation

$$\frac{\text{A of unknown}}{\text{A of standard}} \times 300 = \text{mg fibrinogen/100 ml}$$

Normal values

Men 237.6 to 348.4 mg fibrinogen/100 ml (293.0 $\pm$ 55.4)
Women 255.1 to 356.1 mg fibrinogen/100 ml (305.6 $\pm$ 50.6)

BLOOD AMMONIA[5,21]

Blood is alkalinized with a mixture of potassium carbonate and potassium bicarbonate that converts the ammonium ions to ammonia gas. Ammonia diffuses from the blood-reaction mixture, is collected in sulfuric acid on a receiving rod, and is measured colorimetrically.

Specimen collection

Venous blood is drawn with a minimum of stasis in a heparinized Vacutainer tube and analyzed within one-half hour of the time of collection. Longer delay will result in variable increases in ammonia concentration.

Equipment needed

Diffusion bottles, 50 ml capacity, 43 $\times$ 73 mm with 13 mm inlet*
Receiving bottles, 10 ml capacity, 25 $\times$ 53 mm with 13 mm inlet*
Rubber stoppers to fit the above bottles*
Receiving rods, 6 $\times$ 80 mm, glass rod with a constriction 12 mm from one end.†
    The short end below the constriction is round so that when dipped into the acid it retains an acid film over its surface. Rod is fitted into the bored rubber stopper. This combination when fitted into the 50 ml bottle will allow the rod to extend halfway into the bottle and at the same time will seal it.
Stainless steel mixing rods, approximately 8 mm in diameter and 40 mm in length*
    These rods blend the reaction mixture and increase the diffusion surface.
Rotator*
    This device is a vertically mounted wheel on which are placed 18 spring clips to hold the diffusion bottles in a horizontal position 6 to 12 cm from the center of the wheel. The wheel is rotated at 50 rpm.
Spoon to deliver approximately 3.0 g of alkali mixture

Reagents needed

Alkali mixture
    Mix two parts, by weight, potassium carbonate ($K_2CO_3 \cdot 1\frac{1}{2}H_2O$) and 1 part potassium bicarbonate ($KHCO_3$). Store in sealed containers.

---
*Obtainable from Wheaton Glass Co.
†Obtainable from Macalaster Bicknell of Conn., Inc.

Sulfuric acid, 1 N

Mix 1 volume concentrated H_2SO_4 and 35 volumes deionized water.

Nessler's reagent of Vanselow

Dissolve 45.5 HgI_2 and 34.9 g KI in as little water as possible (approximately 150 ml). Add 112 g of KOH in 140 ml water and dilute to 1 L. Allow to stand for 3 days. Dilute 1:20 before use with deionized water.

Ammonium sulfate standard

Stock: Dissolve 1.415 g ammonium sulfate in water and dilute to 1 L to give 0.30 mg NH_3-N/ml.

Working standard: Prepare on the day of use by diluting the stock standard 1:100 (3 μg NH_3-N/ml).

Procedure

1. Into each diffusion bottle place 3 g of alkali mixture and 2 steel mixing rods. Arrange bottles in a horizontal position serially and rotate gently by hand to align the steel rods.
2. Add to diffusion bottles in duplicate:

   Blank

   2.0 ml of deionized water

   Unknown

   2.0 ml of blood

   Standard

   2.0 ml of working standard
3. Cap each bottle with a solid rubber stopper immediately after the pipetting and place it on the rotator. When pipetting has been completed, rotate the bottles for 1 minute to facilitate mixing of the blood and alkali reagent.
4. Remove bottles from the rotator in the order of analysis, replace the rubber stoppers with glass receiving rods that have been dipped into 1 N sulfuric acid up to the constriction, and return the bottles to the rotator.
5. Rotate the bottles at 50 rpm for 20 minutes. The diffused ammonia is received by the acid on the glass rod.
6. Prepare receiving bottles by adding 3.0 ml of the diluted Nessler's solution.
7. At the end of the diffusion period carefully remove the receiving rods from the bottles and immerse them directly into the Nessler's solution in the receiving bottles. Color development is complete in 5 minutes.
8. Measure the optical density of the yellow-orange solution at 395 mμ in a spectrophotometer, using the Nessler reagent as the reference.

Calculation

Optical density readings are corrected for the reagent blank, and calculations are made from the standard as follows:

$$\frac{\text{A of unknown}}{\text{A of standard}} \times 3.0 \times 100 = \mu g\ NH_3\text{-N}/100\ ml\ blood$$

Clinical findings

The normal fasting ammonia nitrogen (NH_3-N) concentration of venous blood is between 60 and 150 μg/100 ml, with a mean of 102 ± 23 (the SD) μg/100 ml.

LIPID METABOLISM[12]

Lipid metabolism in the liver includes functions related to promotion of the digestion and absorption of fats, storage of fats, oxidation and conjugation of fatty acids, and metabolism and excretion of cholesterol. Serum lipid levels may be altered in disease of the liver and biliary tract.

The laboratory test most frequently performed is that for cholesterol levels in the serum. However, neutral fat and phospholipid concentrations may show marked changes. When there is damage to the parenchymal cells, lowered lipid concentrations are noted. At the onset of jaundice in viral hepatitis a decrease in serum cholesterol occurs, largely resulting from a decrease in the esterified fraction. If liver damage is severe, esterified cholesterol concentrations may be below detectable limits. In cirrhosis of the Laennec type, serum lipids and cholesterol levels are usually low.

Biliary obstruction, cholangiolitic hepatitis, and toxic hepatitis (from drugs and chemicals) are associated with elevated concentrations of serum lipids.

TOTAL CHOLESTEROL AND CHOLESTEROL ESTERS[3]

Cholesterol is a fatlike substance that is found in blood, bile, and brain tissue. It serves as a precursor of bile acids and various steroid hormones.

Cholesterol is a very important member of a group of substances called sterols. Sterols are complex hydroaromatic alcohols distributed throughout all living matter. After extensive research, some evidence has shown that the animal sterol (cholesterol) is a mixture of two or more sterols that differ in respect to their biologic and physiochemical properties.

The animal sterols are probably absorbed from the intestinal tract along with neutral fats and other lipids. They form esters with the fatty acids and thus act as a vehicle for their carriage into the bloodstream. Some cholesterol is formed in the disintegration of red blood cells, and some is synthesized by the body but just "where" it is synthesized is unknown.

The cholesterol is more or less evenly divided between the plasma and the erythrocytes. In the blood cells and the tissue cells the cholesterol is almost all free, while in the serum about 60% to 70% is present as cholesterol esters of fatty acids. The serum and plasma studies have more clinical value than those of whole blood.

Normally, the serum cholesterol is relatively constant. However, disorders such as pernicious anemia may lower the cholesterol level to 50 mg/100 ml. Hemolytic jaundice produces a low cholesterol level, while obstructive jaundice maintains an elevated level of cholesterol.

Total cholesterol consists of free cholesterol and cholesterol esters. In discussions, when the word "cholesterol" is used by itself, it implies total cholesterol.

Table 9-3. Conditions accompanied by abnormal total cholesterol values

When increased		When decreased
Obstructive jaundice	Lipemia	Pernicious anemia
Nephrosis	Celiac disease	Hyperthyroidism
Diabetes	Leukemia	Severe infections
Hypothyroidism	Multiple sclerosis	Epilepsy
Xanthomatosis	Pregnancy	Gaucher's disease
Eclampsia	Aplastic anemia	Inanition
		Intestinal obstruction

The normal values for total cholesterol are 150 to 250 mg/100 ml of serum. Increased or decreased values may be found in those conditions listed in Table 9-3. Abnormal values are of primary interest in the disorders given in italics.

The normal value for cholesterol esters is 70% to 78% of the total cholesterol. Decreased values are found in diseases of the liver.

In clinical work it is sometimes necessary to determine free cholesterol and cholesterol esters separately instead of total cholesterol only. This is made possible by the fact that free cholesterol unites with digitonin to form cholesterol digitonide. This is insoluble in acetone ether, in which the cholesterol esters are freely soluble. Three methods are discussed below.

Bloor method

A commonly used method for the determination of total cholesterol and cholesterol esters is the method of Bloor.

To determine the total cholesterol, serum is obtained. An alcohol-ether mixture is added to extract the cholesterol. The alcohol-ether is driven off by evaporating just to dryness on a hot plate. The cholesterol is then taken up with chloroform. Acetic anhydride and sulfuric acid are added. This produces a green color (the Liebermann-Burchard reaction). The depth of color is measured in a colorimeter.

To determine cholesterol esters, serum is obtained. An alcohol-ether mixture is added to extract both the free cholesterol and the cholesterol esters. This is then filtered. To the filtrate is added an alcohol solution of digitonin. This precipitates the free cholesterol as cholesterol digitonin. The alcohol-ether is driven off by evaporating just to dryness on a hot plate.

The cholesterol esters in the residue are taken up with petroleum ether, the cholesterol digitonin being insoluble in petroleum ether. The ether is driven off on a hot plate. The cholesterol esters are then taken up with chloroform. To this is added acetic anhydride and sulfuric acid. A green color (the Liebermann-Burchard reaction) is produced. The depth of color is measured in a colorimeter.

Schoenheimer-Sperry method

The cholesterol and proteins are extracted by boiling solvents. The esters are then saponified or split into the free form by incubation with potassium hydroxide. The excess KOH is then neutralized with acetic acid. Digitonin, a complex polysaccharide that combines with free cholesterol to form an insoluble compound (cholesterol digitonide), is added to the filtrate that precipitates only free cholesterol. It is also added to the saponified filtrate that precipitates the total cholesterol. The two are left standing overnight to allow complete precipitation of cholesterol digitonide. Any impurities are washed out by acetone-ether, which also takes the precipitate from an aqueous medium to a completely dry state. Since the cholesterol digitonide is sparingly soluble, the precipitates are preheated in order to facilitate its solution in acetic acid. Acetic anhydride–sulfuric acid reagent is added to develop the color. After 30 minutes in the dark, the depth of color is read in a spectrophotometer and compared with standards.

Saifer-Kammerer method

The cholesterol and proteins are extracted by heated acetic anhydride-dioxane. Sulfuric acid is added to an aliquot of the extract, producing a green color, the Liebermann-Burchard reaction. The color is developed in 18 minutes, and the concentration is read against a standard with a colorimeter.

Procedures for total cholesterol and cholesterol esters

The following general considerations pertain to the procedures for total cholesterol and cholesterol esters:

1. Specimen: serum.
2. Preservation of serum: may be stored for a few hours in a refrigerator.
3. Precautions:
   a. Before use, all pipettes and other glassware must be dry because moisture interferes with the color development.
   b. During evaporation of the alcohol-ether, scorching of the residue must not occur. It is better to remove the tube while the residue is slightly moist than to allow scorching to take place.

Schoenheimer-Sperry method for cholesterol and esters[24]

The total cholesterol method has approximately ± 5% accuracy, while an approximate ± 2% accuracy may be obtained with the percentage of esters. This accuracy is, of course, dependent upon the analyst's willingness to use good technique and to resist taking shortcuts. Although this is a very accurate method, it is also very exacting.

Reagents needed

Acetone-alcohol mixture (D-203)
Digitonin solution (D-204)
Acetic acid, 10% (v/v) (D-205)
Acetone-ether mixture (D-206)
Potassium hydroxide solution (D-207)
Phenolphthalein, 1% (D-208)
Acetic acid, glacial
Acetic anhydride–sulfuric acid reagent

Make just before using. Do *not* use reagent if over 1 hour old. Place 20 ml of acetic anhydride in a glass-stoppered cylinder and chill in an ice water bath. When thoroughly chilled, cautiously add 1.0 ml of concentrated sulfuric acid, mix gently, and cool. After cooling, stopper the cylinder, shake vigorously, and return to the ice bath. Keep in the ice bath during use. Larger quantities may be made if needed, using the same proportions.

Cholesterol stock standard, 1.25 mg/ml (D-343)
Cholesterol working standard, 0.125 mg/ml

Dilute 10 ml stock standard to 100 ml with glacial acetic acid.

Procedure

1. Into a 25 ml Pyrex graduated mixing cylinder, add 10 ml of acetone-alcohol solution.
2. Pipette, dropwise, 1.0 ml of serum.
3. Place tubes into boiling water bath and bring contents to boil; remove from bath.
4. Use glass stirring rods to break up protein clumps.
5. Cool to room temperature in a water bath at room temperature.
6. Dilute to 25 ml with acetone-alcohol.
7. Filter into 125 ml Erlenmeyer flask and stopper to eliminate evaporation.

Total cholesterol

1. Into a 12 ml heavy-duty centrifuge tube, pipette 4.0 ml of the above filtrate.

2. To each tube add 3 drops of 50% KOH. Stir with glass stirring rods until thoroughly mixed. (NOTE: The rods are bent at one end so they can hang onto the side of the tube while the test is incubating.)
3. Incubate the tubes at 37° C for 45 minutes. Mix the solutions once or twice during the incubation.
4. After the incubation, add 2 ml of acetone-alcohol solution and 1 drop of phenolphthalein to each tube.
5. Add 10% (v/v) acetic acid, dropwise, with mixing until the pink color completely disappears.
6. Add 1 drop of acetic acid in excess.

Free cholesterol

1. Into a 15 ml centrifuge tube, pipette 8.0 ml of filtrate.
2. Add 1 drop of 10% (v/v) acetic acid. Mix with glass stirring rods.

Total and free cholesterol

Both tests may now be continued together.

1. Pipette 3.0 ml of digitonin into the total cholesterol tubes and 4.0 ml of digitonin into the free cholesterol tubes. (This is 0.5 ml of digitonin for every 1.0 ml of acetone-alcohol.)
2. Mix well with stirring rods.
3. Cover tubes to eliminate light, and let stand overnight to allow complete precipitation of the cholesterol digitonide. (The digitonin forms a 1:1 complex of cholesterol digitonide with the cholesterol.)
4. The next day, remove the stirring rods and hang them on a numbered rod, with the number corresponding to the tube number. This equipment can be made by taping numbers onto a glass or metal rod attached to a ring stand or similar supporting object.
5. Centrifuge tubes for 15 minutes, at 3000 rpm.
6. Pour off (decant) the supernatant from the packed precipitate.
7. Pipette 4.0 ml of acetone-ether into each tube, replace stirring rods, and mix thoroughly. Remove stirring rods.
8. Centrifuge tubes for 15 minutes, at 3000 rpm.
9. Decant the supernatant.
10. Wash the total cholesterols once more with the acetone-ether, and the free cholesterols twice more. Total washings for the total cholesterols—three, and total washings for the free cholesterols—four.
11. To remove the last traces of the acetone-ether left after the final decanting of the supernatant, place the tubes in warm water for a few minutes.
12. Replace the rods and place the tubes in a preheated sand bath at 110° C for 30 minutes. (The sand bath may contain about 3 cm of sand.)
13. Remove the tubes one at a time from the sand bath and add, immediately, 2.0 ml of glacial acetic acid, mixing well with the stirring rod.
14. Prepare working standards as follows (19 × 105 mm cuvette):
    a. Pipette 1.0 ml of working standard and 1.0 ml of glacial acetic acid into a cuvette (125 mg/100 ml).
    b. Pipette 2.0 ml of working standard into a cuvette (250 mg/100 ml).
15. Prepare the *blank* by pipetting 2.0 ml of glacial acetic acid into another cuvette.
16. Prepare enough of the acetic anhydride–sulfuric acid reagent to allow 4.0 ml for each tube, including the blank and standards.
17. Set a timer for 30 minutes but do not start timing.
18. Add 4.0 ml of the anhydride–sulfuric reagent to the blank tube, mix well with a stirring rod, and start timer. Pour the mixed solution into the cuvette for reading. Continue with each tube at 30-second intervals. Timing must be exact.
19. Place the rack of cuvettes in a dark place (cabinet or drawer) where the temperature is constant.

20. *Exactly* 30 minutes after the timer was started, set the blank at zero and read at 625 mμ. The readings are made at 30-second intervals, so that each density is read exactly 30 minutes after the reagent was added to the tube.

Calculations

Total cholesterol

$$\frac{\text{A unknown}}{\text{A standard}} \times 0.250 \times \frac{100}{0.16} = \text{mg total cholesterol/100 ml}$$

Free cholesterol

$$\frac{\text{A unknown}}{\text{A standard}} \times 0.125 \times \frac{100}{0.32} \times 0.5 = \text{mg free cholesterol/100 ml}$$

Cholesterol esters

Subtract the free cholesterol from the total cholesterol to obtain the concentration of the cholesterol ester. To report the esters in terms of percent of the total cholesterol, calculate as follows:

$$\frac{\text{mg/100 ml esters}}{\text{mg/100 ml total}} \times 100 = \text{\% Esters}$$

Complete results should be reported as follows:

Total cholesterol, in mg/100 ml
Free cholesterol, in mg/100 ml
Cholesterol esters, in mg/100 ml
Cholesterol esters, in % of total

Normal range of various age groups

Age (years)	Mean (mg/100 ml)	Range (mg/100 ml)
20	174	99 - 248
30	195	101 - 289
40	219	127 - 312
50	248	142 - 354
60	253	173 - 333

The data shown here applies to about 98% of the population. It is noted that in the older age groups, the level of cholesterol is significantly higher.[1,10]

Saifer-Kammerer modified method for total cholesterol

Serum or plasma mixed with acetic anhydride–dioxane is heated in a boiling water bath for 30 minutes, where extraction of cholesterol and cholesterol esters, precipitation of proteins, and conversion of water present into acetic acid occur simultaneously.

The color development of the test is the Liebermann-Burchard reaction, producing a green color after the addition of sulfuric acid. Twelve to eighteen tests may be run in one group. It is better to develop the color in groups of twelve. This depends upon the technique and the speed an analyst can manage accurately within 10 to 15 minutes.

Reagents needed

Acetic anhydride–dioxane (known as A-D solution)	(D-209)
Concentrated sulfuric acid	
Standard cholesterol solution (0.2 mg/ml)	(D-210)

Procedure

1. Set up the following 15 ml graduated centrifuge tubes. (These should have been washed in chromic acid and thoroughly dried. Water affects the test by drastically lowering the values. Due to toxic and unpleasant fumes, do the test under a fume hood.)
   Blank
   0.45 ml of distilled water into the labelled tube.
   Standards (three tubes)
   0.45 ml of distilled water into each tube.
   5.00 ml of cholesterol standard into each tube
   Unknown
   0.50 ml of serum or plasma into the tube.
2. From a burette (attached to a reservoir jar) add the A-D solution at a fast rate up to the 10 ml mark on each tube.
3. Mix thoroughly with glass stirring rods, scraping the clinging precipitate from the sides of the tubes.
4. Stopper with small air condensers (corks with small glass tubes drawn to a fine tip) and place in a boiling water bath for 30 minutes.
5. Remove the tubes from the boiling water bath and let cool to room temperature. Let stand or use a room temperature water bath.
6. Check the solution volume of each tube and, if necessary, add A-D solution to the 10.0 ml mark.
7. Stopper the tubes and mix thoroughly by inverting.
8. Remove the stoppers and centrifuge for 10 minutes at 3000 rpm.
9. With volumetric pipettes, transfer 5.0 ml of the supernatant to a 19 mm cuvette.
10. Set timer for 18 minutes. Upon the addition of the sulfuric acid to the blank tube, start the timer.
11. Add to all tubes, 0.25 ml of concentrated sulfuric acid.
12. Mix *thoroughly*.
13. By the timer, let the tubes be at room temperature for 5 minutes. This includes the time used in pipetting the acid.
14. For the remaining 13 minutes, place the tubes in a water bath at 30° to 32° C. Read % T (and convert to A) at a wavelength of 625 mμ.

Calculations

$$\frac{\text{A unknown}}{\text{A standard}} \times 0.5 \times \frac{100}{0.25} = \text{mg total cholesterol/100 ml}$$

or

$$\frac{\text{A unknown}}{\text{A standard}} \times 200 = \text{mg total cholesterol/100 ml}$$

Dilutions

1. Pipette 0.25 ml of serum or plasma and 0.225 ml of distilled water into a 15 ml graduated centrifuge tube. Bring volume to 10 ml with A-D solution. Multiply the result by 2.
2. Pipette 0.125 ml of serum or plasma and 0.338 ml of distilled water into a 15 ml graduated centrifuge tube. Bring volume to 10 ml with A-D solution. Multiply the result by 4.

BILE PIGMENT METABOLISM

Jaundice, caused by an abnormal accumulation of bilirubin in the tissues of the body, is a physical sign of liver damage or bile duct damage. Bilirubin is formed from hemoglobin by the reticuloendothelial cells throughout the body. It is transported in the blood (linked to protein, mostly albumin) to the liver, removed by the parenchymal cells, and converted by conjugation to a water-soluble form. Most of the bilirubin is conjugated with glucuronic acid. Conjugated bilirubin, but not the free, is excreted from the liver cells into the bile canaliculi, becoming one of the constituents of the bile that flows into the duodenum. The conjugated bilirubin is reduced in the intestines by bacteria to a group of colorless chromogens, urobilinogen. Some of the urobilinogen is absorbed from the intestinal tract, returned to the liver, and re-excreted into the bile as urobilinogen. There is usually a small amount that is not removed by the liver cells, which passes into general circulation and is excreted by the kidneys as urine urobilinogen.

Most of the urobilinogen is excreted in the feces although some is oxidized to urobilin by bacteria in the colon. The stool contains urobilinogen and urobilin.

An increase in bilirubin concentration in the blood may occur from an increase in the rate of destruction of red cells (as in hemolytic anemia, transfusion reactions, massive infarcts of lungs or heart), which produces more bile pigment than the normal liver can excrete; or a damaged liver may fail to excrete bilirubin formed in normal amounts. Jaundice will also result if obstruction of the excretory ducts of the liver prevents the excretion of bilirubin. There is a genetic defect that limits the enzyme activity in glucuronide conjugation and causes hyperbilirubinemia.

BILIRUBIN

Bilirubin is a bile pigment that is normally present in blood. The normal values are 0.5 to 1.4 mg/100 ml of serum. Increased values are found in the following diseases: obstructive jaundice, hemolytic jaundice, hepatic jaundice, infective hepatitis, and pernicious anemia.

Bilirubin is usually determined now by the method of Malloy and Evelyn. This method is essentially the same as the outdated procedure of van den Bergh.

In the van den Bergh method, both a qualitative test and a quantitative test for bilirubin were made. These tests were based on a color reaction. The absence of color development was normal for the qualitative test. If color did develop, however, it was then measured quantitatively.

It was formerly thought that the time interval of color development had a definite correlation with the origin of the bilirubin present. If the reaction took place immediately, the bilirubin came from bile (obstructive jaundice). If the reaction was delayed, the bilirubin came from hemolyzed red cells (hemolytic jaundice). Recent investigations, however, have ruled out this theory.

The qualitative test of van den Bergh was known as the direct reaction, and the quantitative test was referred to as the indirect reaction.

Like the van den Bergh, the method of Malloy and Evelyn also calls for a direct and an indirect reaction.

In the direct reaction, serum that is free from hemolysis is obtained. To this is added Ehrlich's diazo reagent, which reacts with the bilirubin to form a pink to reddish purple color. The depth of color is measured in a colorimeter.

In the indirect reaction, the tube containing the serum and the diazo reagent is used. To this is added alcohol, which extracts the coloring matter. The depth of color is then measured in a colorimeter.

Procedure for bilirubin

The following general considerations pertain to the procedure for bilirubin.

1. **Specimen: serum.**
2. **Preservation of serum: may be stored overnight in a refrigerator.**

3. Precaution: Avoid excessive hemolysis of red cells, which may be caused (1) by failure to remove needle from syringe when transferring blood to test tube and (2) by allowing blood to stand for several hours in a warm area.

Malloy-Evelyn modification for determination of direct and total bilirubin by diazo reaction[9,16]

In 1883, Ehrlich described the diazotizing reaction of bilirubin with a mixture of sulfanilic acid, hydrochloric acid, and sodium nitrite, resulting in the final violet compound called "azobilirubin."

Thirty years later, 1913, van den Bergh and Snapper quantitated this reaction. This bilirubin determination is known as the van den Bergh procedure using Ehrlich's diazo reagent.[18]

Malloy and Evelyn in 1937 published their adaptation of the above procedure to the photoelectric cell colorimeter.[16]

The serum is treated with Ehrlich's reagent, and the bilirubin in the serum reacts with the reagent, forming a red-violet (or purple) compound known as "azobilirubin." The intensity of the color is proportional to the bilirubin present in the serum and is measured photometrically against a standard.

Reagents needed

Methanol, absolute, A.R. grade	
Diazo blank solution	(D-133)
Solution A (sulfanilic acid, 0.1%)	(D-131)
Solution B, stock (sodium nitrite, 5%)	(D-211)
Solution B, working (sodium nitrite, 0.5%)	(D-212)
Working diazo reagent *(prepared just before using)*	
Mix 0.3 ml of working solution B with 10.0 ml of solution A.	

Precautions

1. Dilute the serum sample with water just before analysis. This is done to minimize the effect of the proteins slowly precipitating from the serum. Since "salt" interferes with the diazo reaction, do not use saline as a diluent.
2. The reagents must be added with exact measurements and in the sequence given in the method, or a precipitate will form, invalidating the test.
3. Bilirubin is sensitive to light (artificial and sunlight). If the serum is allowed to stand in a well-lighted area, there may be as much as a 50% drop in bilirubin concentration within 1 hour. Keep the serum in *total darkness* until analysis. Bilirubin in serum is considered stable under the following conditions[9]:
   a. In total darkness at room temperature, 2 days
   b. In refrigerator, 4 to 7 days
   c. If frozen, up to 3 months

Normal range

Direct levels	Less than 0.4 mg/100 ml (or mg%)
Total levels	Less than 0.8 mg/100 ml (or mg%)

Procedure

1. Set up two colorimeter tubes for each test. Label one *blank* and the other *unknown*. Pipette into *each* tube:
   a. 0.4 ml of serum
   b. 3.6 ml of distilled water
2. To the *blanks*, add 1.0 ml of diazo blank solution. Mix.
3. To the *unknowns*, add 1.0 ml of diazo reagent. Mix.

4. *Exactly* 1 minute later (use timer), read the *unknown* (tube 2) against its *blank* (tube 1) at 540 mμ wavelength. If only the total bilirubin is requested, omit this reading and proceed to the next step.
5. Pipette 5 ml of fresh methanol to each tube and mix by inverting gently. Avoid bubbles. Old methanol has a tendency to cause decreased results.[17]
6. Let the tubes stand at room temperature for 15 minutes and read the unknowns against their blanks. This is the total bilirubin.
7. CAUTION: If the solutions are transferred to small cuvettes, bubbles may cling to the sides of the cuvettes. Clear or dislodge the bubbles by tapping the cuvette gently.

NOTE: This method may be converted to the micromethod by scaling down ten or twenty times. Since the final volume is 1/10 or 1/20 the volume in the macroprocedure, it is necessary to use an instrument adapted for microcuvettes.

Calculation

Example: The bilirubin standard has a value of 5 mg/100 ml. Since 0.4 ml of the serum standard is used in the analysis, there is 0.02 mg of bilirubin in the 0.4 ml of serum.

A_1 = Absorbance of sample at 1 minute
A_s = Absorbance of standard at 15 minutes
A_t = Absorbance of sample at 15 minutes

$$\frac{A_1}{2A_s} \times 0.02 \times \frac{100}{0.4} =$$

$$\frac{A_1}{A_s} \times 2.5 = \text{mg bilirubin/100 ml at 1 minute or } \textit{direct bilirubin}$$

$$\frac{A_t}{A_s} \times 0.02 \times \frac{100}{0.4} =$$

$$\frac{A_t}{A_s} \times 5 = \text{mg } \textit{total bilirubin} \text{ /100 ml}$$

Total bilirubin − *Direct bilirubin* = *Indirect bilirubin*

Recommended standardization[4,7]

The need for a reliable bilirubin standard has become most evident in recent years. There is a more knowledgeable understanding of the severe jaundice caused by Rh and ABO incompatibility between a mother and her infant. Exchange transfusions are used to keep the bilirubin below 20 mg/100 ml in full-term infants and below 18 mg/100 ml in premature babies. Determining the need for the exchange is very dependent upon the bilirubin result.

There has been great variation from one laboratory to another. In order to standardize the method and obtain a uniform bilirubin standard, a subcommittee of the American Academy of Pediatrics* has made a study of various types of standards in use. The standard recommended was a crystalline bilirubin, defined in terms of its solubility and color intensity in chloroform. The color intensity must be read under strictly controlled conditions of temperature, light, weight, volume, wavelength, optical density, and light path.

Obviously, most laboratories do not have all the ideal conditions along with the proper spectrophotometer for measuring the color intensity (molar absorptiv-

*The Joint Bilirubin Committee is composed of members from the American Academy of Pediatrics, the College of American Pathologists, the American Association of Clinical Chemists, and the National Institutes of Health.

ity). Various manufacturers* have bilirubin standards in serum that meet the requirements suggested by the committee. These are more stable, uniform, and accurate than the standards most hospital laboratories are able to prepare. The controls are made under rigid quality control conditions, are assayed, and are then lyophilized. Lyophilizing is the process of quickly freezing the serum under unusually low temperatures and then using a high vacuum to dehydrate the frozen mass quickly. The product is stable under refrigeration. When reconstituted, the serum should be kept in the dark and used within 2 hours.

A calibration curve may be made with the lyophilized serum, or standards may be included with each group of bilirubin determinations.

Tests for bilirubin and urobilinogen in urine are given in Chapter 14 and for fecal bile pigment and urobilin, in Chapter 12.

DETOXIFYING MECHANISM

Conjugation is a function of the liver essential for the excretion of bilirubin, bile acids, and steroid metabolites. The same mechanism is used to convert foreign substances and toxic constituents to nontoxic products. The most frequently used test of this protective function of the liver is based upon a conjugation reaction whereby detoxication of benzoic acid with glycine results in the formation of hippuric acid that is excreted in the urine. (This test is described in Chapter 14.) The test is also a measure of the metabolic function of the liver, since the ability to synthesize the hippuric acid depends not only on the conjugating enzyme system of the liver but also on the availability of glycine stores in the liver.

EXCRETORY CAPACITY—BROMSULPHALEIN (BSP) TEST

The ability of the liver to remove from the circulation and eliminate from the body certain dyes has been used as the basis for liver function tests. The Bromsulphalein test is a sensitive indicator of the functioning ability of the liver cells.

When Bromsulphalein, an organic dye, is injected into the bloodstream, it is removed by the liver and is excreted into the bile. A healthy liver removes the dye at a faster rate than a diseased liver. Consequently, the rate of removal can be used as a test of liver function.

A commonly used method for the Bromsulphalein test is the procedure of Rosenthal and White, introduced in 1925. In this method, the patient is given 2 mg of dye per kg of body weight. The sterile solution of dye is injected into the patient's vein by the physician, special care being taken to see that a tourniquet is *not* used while the injection is being made.

Two samples of blood are taken, one 5 minutes after the injection and the other 30 minutes after the injection. Serum is obtained. Sodium hydroxide is added to bring out the color of the dye. The concentration of dye is then found either from a colorimeter reading or from a comparison with standards representing known quantities of Bromsulphalein.

With a healthy liver, the 5-minute sample of serum contains 20% to 50% of the dye and the 30-minute sample contains no dye. With a diseased liver, the 30-minute sample contains from 5% to 100% of the dye.

In 1939, MacDonald increased the amount to 5 mg of dye per kg of body weight. This measure is used almost exclusively because it increases the sensitivity of the test in detecting liver impairment. With the larger amount of dye in-

---

*Dade; Hyland Laboratories; and General Diagnostics Division, Warner-Chilcott.

jected, usually only one blood specimen is taken, 45 minutes later. Normally, less than 5% of the dye is retained 45 minutes after the injection.

In 1957, Seligson and co-workers introduced the use of an alkaline buffer to develop the color and an acid buffer for the blank. This was found to minimize errors caused by icterus, hemolysis, and lipemia. They also showed that free BSP has a different absorptivity and absorption peak than when it is bound to albumin as in serum. This was corrected by adding large amounts of an anion, *p*-toluene-sulfonate, to the alkaline buffer.

Procedure for Bromsulphalein (BSP) test

The general considerations that follow pertain to the procedure with Bromsulphalein.

1. Specimen: serum.
2. Preservation of serum: may be stored overnight in a refrigerator.
3. Precautions:
   a. If the patient has obstructive jaundice, this test should not be run, because the obstruction prevents the removal of dye and thus renders the test inaccurate.
   b. In some patients, the injected dye may cause chills, faintness, or headache. On rare occasions there may be a severe reaction. For this reason, the dye should be injected into the patient only by a physician.
4. To calculate the amount of BSP dye needed, weigh the patient and determine the amount of dye to be injected (5 mg/kg) by the following formula:

$$\frac{\text{Weight in pounds}}{22} = \text{ml of dye needed}$$

Table 9-4. BSP dye dosages

Patient weight in lb	BSP dye in ml,* at 5 mg/kg
60	2.7
70	3.2
80	3.6
90	4.1
100	4.5
110	5.0
120	5.5
130	5.9
140	6.4
150	6.8
160	7.3
170	7.7
180	8.2
190	8.6
200	9.1
210	9.5
220	10.0

*The amount of dye is given to the nearest 0.1 ml.

5. Table 9-4 shows the amount of dye needed according to the patient's weight in pounds.
6. Exactly 45 minutes after the injection of the dye, obtain a blood specimen from the arm not injected. Allow blood to clot, centrifuge, and remove the serum.

Seligson-Marino-Dodson method for sulfobromophthalein (BSP)[22]

Serum is diluted with an alkaline buffer and the absorbance is measured at 580 mμ. Acid buffer is then added and the absorbance measured. The difference of absorbance represents the BSP present.*

Reagents needed

Alkaline buffer, pH 10.6 to 10.7	(D-213)
Acid reagent, 2 M NaH_2PO_4	(D-214)
BSP standard, 5 mg/100 ml†	(D-215)

Procedure

1. Prepare the following tubes:
   Standard (50% retention standard)
   1.0 ml of standard into a colorimeter or test tube.
   7.0 ml of alkaline buffer.
   Mix.
   Unknown (serum)
   1.0 ml of serum into a colorimeter or test tube.
   7.0 ml of alkaline buffer.
   Mix.
2. Read the absorbance of both tubes against a water blank at a wavelength of 580 mμ or a filter of similar wavelength.
3. To the *unknown* tube:
   0.2 ml of acid reagent.
   Mix.
   Read the absorbance against a water blank at 580 mμ.

Calculation

$$\frac{\text{Absorbance of alkaline buffer} - \text{Absorbance of acid buffer}}{\text{Absorbance of standard}} \times 50 = \%\text{BSP}$$

Make the following corrections for body weight[26]:
110 to 149 lb. Add 1% retention
150 to 169 lb. No correction
170 to 189 lb. Subtract 1% retention
190 to 279 lb. Subtract 3% retention

The color of the Seligson method is stable for at least 24 hours and obeys Beer's law up to 100% retention.

Normal values

For dosage 5 mg/kg of BSP:

Time	Retention
30 minutes	Less than 10%
45 minutes	Less than 6%
60 minutes	Less than 3%

BSP is eliminated faster in infants and children than in adults.

*Available from Hynson, Westcott & Dunning, Inc., or from Dade.
†Diluted standard is stable for 1 week.

REFERENCES

1. Annino, J. S.: Clinical chemistry: principles and procedures, ed. 3, Boston, 1964, Little, Brown & Co.
2. Bunch, L. D.: A rapid cephalin-cholesterol flocculation test, Amer. J. Clin. Path. **28**:111, 1957.
3. Cantarow, A. C., and Trumper, M.: Clinical biochemistry, ed. 3, Philadelphia, 1945, W. B. Saunders Co.
4. Committee report: Recommendation on a uniform bilirubin standard, Clin. Chem. **8**:405, 1962.
5. Conn, H. O.: Blood ammonia. In Meites, S., editor: Standard methods of clinical chemistry, vol. 5, New York, 1965, Academic Press, Inc.
6. FI test, rapid slide test for hypofibrinogenemia, Los Angeles, Hyland Laboratories.
7. Gambino, S. R.: Bilirubin measurement in the newborn: need for a common reference standard in serum, Hosp. Top., Feb., 1964.
8. Hanger, F. M.: The flocculation of cephalin-cholesterol emulsions by pathological sera, Trans. Ass. Amer. Physicians **53**:148, 1938.
9. Henry, R. J.: Clinical chemistry; principles and technics, New York, 1964, Paul B. Hoeber, Inc., Medical Book Department of Harper & Row, Publishers.
10. Keys, A., Mickelsen, O., Miller, E. V. O., Hayes, E. R., and Todd, R. L.: The concentration of cholesterol in the blood serum of normal man and its relation to age, J. Clin. Invest. **29**:1347, 1950.
11. Kingsbury, F. B., Clark, C. P., Williams, G., and Post, A. L.: The rapid determination of albumin in urine, J. Lab. Clin. Med. **2**:981, 1926.
12. Kleiner, I. S., and Orten, J. M.: Biochemistry, ed. 7, St. Louis, 1966, The C. V. Mosby Co.
13. Knowlton, M.: Cephalin-cholesterol flocculation test. In Seligson, D., editor: Standard methods of clinical chemistry, vol. 2, New York, 1958, Academic Press, Inc.
14. Lackland, H.: Quantitative fibrinogen, Lab. Digest **28**:3, March-April, 1965.
15. Maclagan, N. F.: The thymol turbidity test as an indicator of liver dysfunction, Brit. J. Exp. Path. **25**:234, 1944.
16. Malloy, H. T., and Evelyn, K. A.: The determination of bilirubin with the photoelectric colorimeter, J. Biol. Chem. **119**:481, 1937.
17. Michaelsson, M.: Bilirubin determination in serum and urine, Scand. J. Clin. Lab. Invest. **13** (supp. 56):1, 1961.
18. Page, L. B., and Culver, P. J., editors: A syllabus of laboratory examinations in clinical diagnosis, rev. ed., Cambridge, Mass., 1961, Harvard University Press.
19. Reiner, M., and Cheung, H. L.: A practical method for the determination of fibrinogen, Clin. Chem. **5**:414, 1959.
20. Reinhold, J. G.: Flocculation tests and their application to the study of liver disease. In Sobotka, H., and Stewart, C. P., editors: Advances in clinical chemistry, vol. 3, New York, 1960, Academic Press, Inc.
21. Seligson, D., and Hirbara, K.: The measurement of ammonia in whole blood, erythrocytes and plasma, J. Lab. Clin. Med. **49**:962, 1957.
22. Seligson, D., Marino, J., and Dodson, E.: The determination of sulfobromophthalein in serum, Clin. Chem. **3**:638, 1957.
23. Shank, R. E., and Hoagland, C. L.: A modified method for the quantitative determination of the thymol turbidity reaction of serum, J. Biol. Chem. **162**:133, 1946.
24. Sperry, W. M.: A micromethod for the determination of total and free cholesterol, J. Biol. Chem. **150**:315, 1943.
25. Weichselbaum, T. E.: An accurate and rapid method for the determination of proteins in small amounts of blood serum and plasma, Amer. J. Clin. Path. **16** (tech. sec. 10):40, 1946.
26. Zieve, L., and Hill, E.: An evaluation of factors influencing the discriminative effectiveness of a group of liver function tests, Gastroenterology **28**:766, 1955.

• 10

• Atomic absorption spectroscopy

Atomic absorption spectroscopy is based on the fact that atoms of an element in the ground or unexcited state will absorb light of the same wavelength that they emit in the excited state. When atoms of a metallic element are raised from the ground state to a higher energy level, or excited state, they absorb energy from a source such as heat or light. Since the excited state is extremely unstable, the atoms quickly return to the ground state. In doing so, they release the absorbed energy in the form of light. The wavelengths of that light or the resonance lines are characteristic for each element. No two elements have identical resonance lines. Atomic absorption spectroscopy involves the measurement of light absorbed by atoms in the ground state. Flame photometry or flame emission spectroscopy employs the measurement of light emitted by atoms in the excited state. Most elements have several resonance lines but usually one line is considerably stronger than the others. The wavelength of this line is generally the one selected for measurement. Occasionally, however, it may be necessary to select another resonance line, either to reduce sensitivity or because a resonance line of an interfering element is very close to the one of interest.

If a beam of monochromatic light, wavelength 3247 Å, is passed through a flame and monochromator to a detector and a treated serum specimen is aspirated into the flame, only the copper atoms in the flame absorb light from the beam. The amount of light absorbed is proportional to the amount of copper in the serum specimen.

The sensitivity of atomic absorption techniques has stimulated investigation into the role of trace metals in biologic processes. A summary of these investigations was made recently by Christian.[1]

HOLLOW CATHODE LAMPS

The source of the monochromatic light is usually a hollow cathode lamp. This consists of a sealed glass tube or envelope containing a rod-shaped anode and a cup-shaped cathode that is about 1 cm in diameter and 1 cm deep. The cathode cup contains the metal of interest. Surrounding the electrodes within the glass envelope is a filler gas, usually neon or argon, at a pressure of about 3 mm Hg. When current is applied to the electrodes, the filler gas within the cathode becomes partially ionized and causes a cloud of metallic ions to sputter off the cathode. These metallic atoms collide with the gas ions and with each other, causing the metal atoms to enter the excited state in which they emit radiation characteristic of the component metal.

The newest hollow cathode lamps have cathodes shielded to contain the glow within the cathode. Glow outside the cathode is not only wasted, so far as absorption measurements are concerned, but it also shortens the life of the lamp. A restrictive shield is used for volatile or molten metals and an open shield is used for metals difficult to sputter.

Multi-element lamps are now available, some containing as many as seven elements.

SENSITIVITY AND DETECTION LIMITS

Sensitivity refers to the concentration of an element in aqueous solution that will produce an absorption of 1%. Generally, it is expressed as μg/ml/1%. *Detection limit* is defined as the concentration of an element in aqueous solution that gives a signal twice the size of the background variability.

INTERFERENCES

Although it may be claimed that certain elements can be determined without interferences, there are certain effects related to the flame that must be considered.

Matrix interferences

A significant difference between the surface tension or viscosity of the sample and that of the standards will cause them to be aspirated or atomized differently. High acid concentration can alter the absorption signal. Standards should have the same acid content as the samples. A total solids content of greater than 2% in aqueous solution may cause interference. For these solutions it may be more accurate to use the *method of additions* to minimize bulk interference effects.

Background absorption

A high salt content in a sample will cause an erroneously high analysis due to the light-scattering effect of the salt. Light is bounced off or reflected from the small particles of salt that still survive in the flame. Since this is light that does not reach the detector, it is measured as absorption. To correct for this light-scattering effect the apparent absorption should be measured at a nearby wavelength that is known to be nonabsorbing for the metal of interest. The background absorption is then subtracted from the analytic line absorption before the metal concentration of the sample is calculated.[2]

Acetylene cylinders

Erratic and inconsistent results are obtained when the pressure of an acetylene tank falls below 75 psi. This is especially true in work with metals that require a very rich acetylene flame. It is a problem caused by chemical interference from acetone vapors. Acetylene is dissolved in acetone in the tank. As the tank pressure decreases, the rate at which the acetone comes out of the tank increases.[3,4]

NOTE: It is dangerous to leave acetylene trapped in a line. When shutting the instrument down for any extended period of time, close the main valve at the tank and bleed all lines. No pressure should be registered on any of the acetylene gauges.

STANDARDS

Stock standards may be purchased from several sources. Harleco supplies atomic absorption standards with a concentration of 1000 μg/ml. For optimal accuracy, dilute working standards should be prepared just prior to use. Very dilute standards will change with time, as a result of plating of the metal to the sides of the container. The presence of 1% HNO_3 will, in many solutions, retard this plating and lengthen the life of the standard.

PERKIN-ELMER ATOMIC ABSORPTION SPECTROPHOTOMETER, MODEL 303

The discussion to follow concerning instrument parts, settings, and methods refers specifically to the Perkin-Elmer instrument.

Burner heads

There are at least five different burner heads available for this instrument. The one most frequently used in clinical work is the three-slot (Boling) burner head. The wide cross section of the flame permits more precise and more sensitive determinations, particularly for metals whose compounds do not dissociate readily in the flame—for example, calcium, chromium, tin. It will also burn concentrated solutions without clogging. Both the fuel and air flows should be adjusted to a reading of 10 on the flowmeter. Use a standard solution and adjust the fuel for maximum absorption when running calcium, magnesium, chromium, strontium, or tin. If a very lean (clear blue) flame is used, the heat will warp the slots unless water or some other solution is aspirated continuously.[5]

The short path burner head may be used when less sensitivity is desired.

Organic solvents

Organic solvent extractions generally enhance sensitivity and will improve detection limits by a factor of about four. Many solvents have been used but the one found most satisfactory is methyl isobutyl ketone. The fuel-to-air ratio should be adjusted to the settings recommended for the analysis. Then the fuel flow should be reduced gradually until the flame is as blue as possible but does not lift off the burner. The nebulizer also must be readjusted for organic solvents. Select an element with a wavelength greater than 250 mμ that requires a blue flame (using air, not N_2O)—for example, copper. Loosen the capillary locknut by turning it counterclockwise. While aspirating a standard in the solvent, turn the capillary counterclockwise until bubbling ("blowback") occurs in the solvent at the sample tube intake. Then slowly turn the capillary knob clockwise until maximum absorption is indicated on the null meter. Adjust the null meter needle back to midscale if necessary. When organic solvents are to be burned for more than one hour, it is necessary to replace the silicone O-ring in the burner end cap assembly with a Corkprene O-ring. The silicone O-ring must be replaced when aqueous solutions are burned again.[5]

METHOD OF ADDITIONS

When a sample matrix has a very high solids content whose effect on absorption is hard to duplicate with aqueous standards, the method of additions should be used.

Take three equal aliquots of the sample. Add a known volume of diluent to the first aliquot. Add the same volume of known standards to each of the other two aliquots so that each will have a different known concentration of the metal. Determine the percent absorption on each of the three aliquots. Convert percent absorption readings to absorbance and plot against concentration. For example: If 1 μg/ml of the metal was added to the second aliquot and 2 μg/ml to the

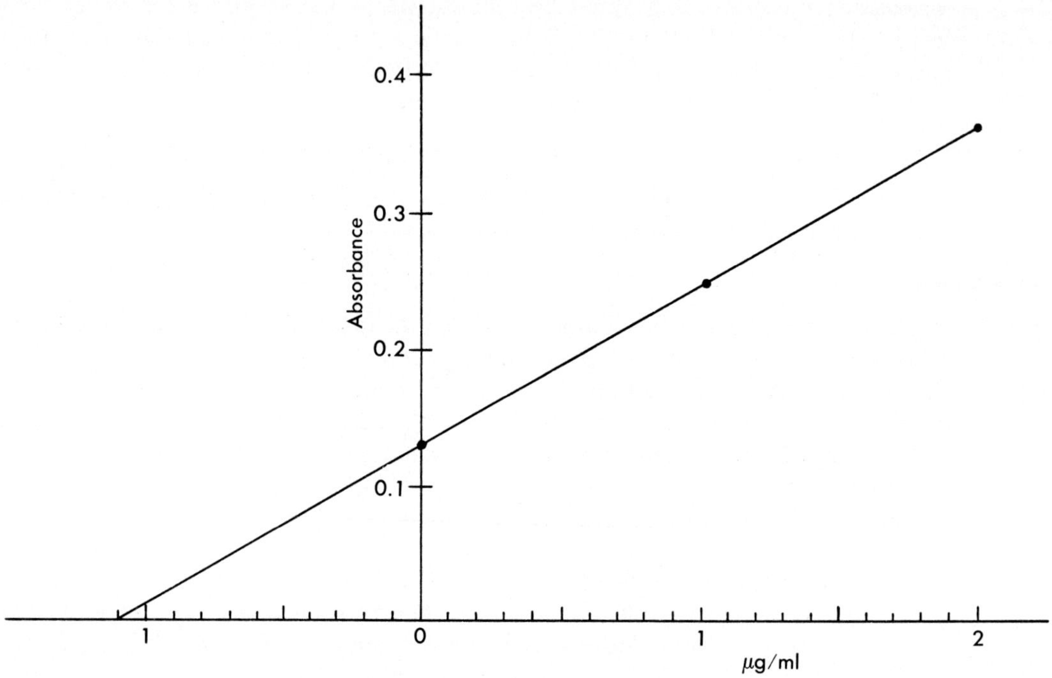

Fig. 10-1. Method of additions.

third, plot the absorbance of the first aliquot at zero concentration, the second at 1 μg/ml concentration, and the third at 2 μg/ml. Draw a line through the three points and extend it back until it intersects the concentration axis. This point will give the concentration of the metal in the original sample (Fig. 10-1).

METHODS

Calcium

Reagents needed

Lanthanum stock solution, 5% in 25% (v/v) HCl	(D-216)
Lanthanum working solution, 0.5% in 2.5% (v/v) HCl	(D-217)
Calcium stock standard, 50 mg Ca/100 ml	(D-47)

Intermediate standards

Pipette the following volumes of calcium stock standard into 100 ml volumetric flasks. Add one drop of concentrated HCl to each flask. Dilute to volume with water and mix thoroughly.

ml Ca stock	*mg Ca/100 ml*
5	2.5
10	5.0
15	7.5
20	10.0
25	12.5
30	15.0

Preparation of working standards and samples

If possible, use an automatic dilutor for these dilutions. Set it to aspirate 0.25 ml of sample (or intermediate standard) and to dispense the

Table 10-1. Instrument settings

Element	Source: Lamp	Source: ma†	Wavelength: Å	Wavelength: Dial	Range	Slit	Burner head	Acetylene pressure: Tank	Acetylene pressure: Flow‖	Air pressure: Regulator	Air pressure: Flow‖	Scale expansion
Calcium	H-C*		4227	211	Visible	4	3-slot	8	10	30	10	1X
Magnesium	H-C		2852	285	UV	5	Standard§	8	9¶	30	8	1X
Copper	H-C		3247	325	UV	4	3-slot	8	10	30	10	5X
Zinc	H-C		2138	214	UV	5‡	3-slot	8	10	30	10	1X
Lead	H-C		2833	283	UV	4	3-slot	8	10	30	10	5-10X

*Hollow cathode
†Use current given on lamp. Space was left for user to fill in.
‡Set to 4 when using a multi-element lamp containing copper.
§Use either standard or short path burner.
‖All flowmeter readings should be made at the center of the ball.
¶Use a standard solution of 0.5 μg Mg/ml and adjust the acetylene flow to obtain maximum absorption.

sample along with 4.75 ml of lanthanum working solution. Prepare working standards from the intermediate standards in the same manner. Using an automatic dilutor will ensure uniformity in the sample and standard dilutions.

Urine specimens should be collected in a jar containing 15 ml (for a 24-hour collection) of glacial acetic acid. If a specimen contains any sediment, a portion of the urine should be centrifuged. The clear portion may then be diluted in the same manner as the serum.

Procedure

1. Install the calcium lamp in the lamp compartment.
2. Turn the power switch on and adjust the source to the current given on the lamp.
3. Set the instrument to the proper wavelength, range, and slit (Table 10-1).
4. Install the 3-slot burner head.
5. Turn on compressed air supply and adjust the burner regulator to 30 psi and the flowmeter to 10.
6. Open the valve on the tank of acetylene and adjust the pressure coming from the tank to 8 psi. Adjust the fuel flowmeter to 10.
7. Be sure the **U** in the burner drain tube is filled with liquid and that the end of the tube is below the surface of a liquid. Light the flame.
8. Allow the flame to aspirate water for about 5 minutes; then adjust absorption to zero while aspirating lanthanum working solution.
9. Determine percent absorption on diluted standards and samples.
10. Convert percent absorption to absorbance (Table 10-2) and plot the absorbance of the standards versus concentration on linear graph paper.

Magnesium in serum and urine

Reagent needed

Magnesium stock standard, 100 mEq/L (D-218)

Table 10-2. Conversion of percent absorption (R) to absorbance

R	.0	.1	.2	.3	.4	.5	.6	.7	.8	.9
0.0	.0000	.0004	.0009	.0013	.0017	.0022	.0026	.0031	.0035	.0039
1.0	.0044	.0048	.0052	.0057	.0061	.0066	.0070	.0074	.0079	.0083
2.0	.0088	.0092	.0097	.0101	.0106	.0110	.0114	.0119	.0123	.0128
3.0	.0132	.0137	.0141	.0146	.0150	.0155	.0159	.0164	.0168	.0173
4.0	.0177	.0182	.0186	.0191	.0195	.0200	.0205	.0209	.0214	.0218
5.0	.0223	.0227	.0232	.0236	.0241	.0246	.0250	.0255	.0259	.0264
6.0	.0269	.0273	.0278	.0283	.0287	.0292	.0297	.0301	.0306	.0311
7.0	.0315	.0320	.0325	.0329	.0334	.0339	.0343	.0348	.0353	.0357
8.0	.0362	.0367	.0372	.0376	.0381	.0386	.0391	.0395	.0400	.0405
9.0	.0410	.0414	.0419	.0424	.0429	.0434	.0438	.0443	.0448	.0453
10.0	.0458	.0462	.0467	.0472	.0477	.0482	.0487	.0491	.0496	.0501
11.0	.0506	.0511	.0516	.0521	.0526	.0531	.0535	.0540	.0545	.0550
12.0	.0555	.0560	.0565	.0570	.0575	.0580	.0585	.0590	.0595	.0600
13.0	.0605	.0610	.0615	.0620	.0625	.0630	.0635	.0640	.0645	.0650
14.0	.0655	.0660	.0665	.0670	.0675	.0680	.0685	.0691	.0696	.0701
15.0	.0706	.0711	.0716	.0721	.0726	.0731	.0737	.0742	.0747	.0752
16.0	.0757	.0762	.0768	.0773	.0778	.0783	.0788	.0794	.0799	.0804
17.0	.0809	.0814	.0820	.0825	.0830	.0835	.0841	.0846	.0851	.0857
18.0	.0862	.0867	.0872	.0878	.0883	.0888	.0894	.0899	.0904	.0910
19.0	.0915	.0921	.0926	.0931	.0937	.0942	.0947	.0953	.0958	.0964
20.0	.0969	.0975	.0980	.0985	.0991	.0996	.1002	.1007	.1013	.1018
21.0	.1024	.1029	.1035	.1040	.1046	.1051	.1057	.1062	.1068	.1073
22.0	.1079	.1085	.1090	.1096	.1101	.1107	.1113	.1118	.1124	.1129
23.0	.1135	.1141	.1146	.1152	.1158	.1163	.1169	.1175	.1180	.1186
24.0	.1192	.1198	.1203	.1209	.1215	.1221	.1226	.1232	.1238	.1244
25.0	.1249	.1255	.1261	.1267	.1273	.1278	.1284	.1290	.1296	.1302
26.0	.1308	.1314	.1319	.1325	.1331	.1337	.1343	.1349	.1355	.1361
27.0	.1367	.1373	.1379	.1385	.1391	.1397	.1403	.1409	.1415	.1421
28.0	.1427	.1433	.1439	.1445	.1451	.1457	.1463	.1469	.1475	.1481
29.0	.1487	.1494	.1500	.1506	.1512	.1518	.1524	.1530	.1537	.1543
30.0	.1549	.1555	.1561	.1568	.1574	.1580	.1586	.1593	.1599	.1605
31.0	.1612	.1618	.1624	.1630	.1637	.1643	.1649	.1656	.1662	.1669
32.0	.1675	.1681	.1688	.1694	.1701	.1707	.1713	.1720	.1726	.1733
33.0	.1739	.1746	.1752	.1759	.1765	.1772	.1778	.1785	.1791	.1798
34.0	.1805	.1811	.1818	.1824	.1831	.1838	.1844	.1851	.1858	.1864
35.0	.1871	.1878	.1884	.1891	.1898	.1904	.1911	.1918	.1925	.1931
36.0	.1938	.1945	.1952	.1959	.1965	.1972	.1979	.1986	.1993	.2000
37.0	.2007	.2013	.2020	.2027	.2034	.2041	.2048	.2055	.2062	.2069
38.0	.2076	.2083	.2090	.2097	.2104	.2111	.2118	.2125	.2132	.2140
39.0	.2147	.2154	.2161	.2168	.2175	.2182	.2190	.2197	.2204	.2211
40.0	.2218	.2226	.2233	.2240	.2248	.2255	.2262	.2269	.2277	.2284
41.0	.2291	.2299	.2306	.2314	.2321	.2328	.2336	.2343	.2351	.2358
42.0	.2366	.2373	.2381	.2388	.2396	.2403	.2411	.2418	.2426	.2434
43.0	.2441	.2449	.2457	.2464	.2472	.2480	.2487	.2495	.2503	.2510
44.0	.2518	.2526	.2534	.2541	.2549	.2557	.2565	.2573	.2581	.2588
45.0	.2596	.2604	.2612	.2620	.2628	.2636	.2644	.2652	.2660	.2668
46.0	.2676	.2684	.2692	.2700	.2708	.2716	.2725	.2733	.2741	.2749
47.0	.2757	.2765	.2774	.2782	.2790	.2798	.2807	.2815	.2823	.2832

Table 10-2. Conversion of percent absorption (R) to absorbance—cont'd

R	.0	.1	.2	.3	.4	.5	.6	.7	.8	.9
48.0	.2840	.2848	.2857	.2865	.2874	.2882	.2890	.2899	.2907	.2916
49.0	.2924	.2933	.2941	.2950	.2958	.2967	.2976	.2984	.2993	.3002
50.0	.3010	.3019	.3028	.3036	.3045	.3054	.3063	.3072	.3080	.3089
51.0	.3098	.3107	.3116	.3125	.3134	.3143	.3152	.3161	.3170	.3179
52.0	.3188	.3197	.3206	.3215	.3224	.3233	.3242	.3251	.3261	.3270
53.0	.3279	.3288	.3298	.3307	.3316	.3325	.3335	.3344	.3354	.3363
54.0	.3372	.3382	.3391	.3401	.3410	.3420	.3429	.3439	.3449	.3458
55.0	.3468	.3478	.3487	.3497	.3507	.3516	.3526	.3536	.3546	.3556
56.0	.3565	.3575	.3585	.3595	.3605	.3615	.3625	.3635	.3645	.3655
57.0	.3665	.3675	.3686	.3696	.3706	.3716	.3726	.3737	.3747	.3757
58.0	.3768	.3778	.3788	.3799	.3809	.3820	.3830	.3840	.3851	.3862
59.0	.3872	.3883	.3893	.3904	.3915	.3925	.3936	.3947	.3958	.3969
60.0	.3979	.3990	.4001	.4012	.4023	.4034	.4045	.4056	.4067	.4078
61.0	.4089	.4101	.4112	.4123	.4134	.4145	.4157	.4168	.4179	.4191
62.0	.4202	.4214	.4225	.4237	.4248	.4260	.4271	.4283	.4295	.4306
63.0	.4318	.4330	.4342	.4353	.4365	.4377	.4389	.4401	.4413	.4425
64.0	.4437	.4449	.4461	.4473	.4485	.4498	.4510	.4522	.4535	.4547
65.0	.4559	.4572	.4584	.4597	.4609	.4622	.4634	.4647	.4660	.4672
66.0	.4685	.4698	.4711	.4724	.4737	.4750	.4763	.4776	.4789	.4802
67.0	.4815	.4828	.4841	.4855	.4868	.4881	.4895	.4908	.4921	.4935
68.0	.4948	.4962	.4976	.4989	.5003	.5017	.5031	.5045	.5058	.5072
69.0	.5086	.5100	.5114	.5129	.5143	.5157	.5171	.5186	.5200	.5214
70.0	.5229	.5243	.5258	.5272	.5287	.5302	.5317	.5331	.5346	.5361
71.0	.5376	.5391	.5406	.5421	.5436	.5452	.5467	.5482	.5498	.5513
72.0	.5528	.5544	.5560	.5575	.5591	.5607	.5622	.5638	.5654	.5670
73.0	.5686	.5702	.5719	.5735	.5751	.5768	.5784	.5800	.5817	.5834
74.0	.5850	.5867	.5884	.5901	.5918	.5935	.5952	.5969	.5986	.6003
75.0	.6021	.6038	.6055	.6073	.6091	.6108	.6126	.6144	.6162	.6180
76.0	.6198	.6216	.6234	.6253	.6271	.6289	.6308	.6326	.6345	.6364
77.0	.6383	.6402	.6421	.6440	.6459	.6478	.6498	.6517	.6536	.6556
78.0	.6576	.6596	.6615	.6635	.6655	.6676	.6696	.6716	.6737	.6757
79.0	.6778	.6799	.6819	.6840	.6861	.6882	.6904	.6925	.6946	.6968
80.0	.6990	.7011	.7033	.7055	.7077	.7100	.7122	.7144	.7167	.7190
81.0	.7212	.7235	.7258	.7282	.7305	.7328	.7352	.7375	.7399	.7423
82.0	.7447	.7471	.7496	.7520	.7545	.7570	.7595	.7620	.7645	.7670
83.0	.7696	.7721	.7747	.7773	.7799	.7825	.7852	.7878	.7905	.7932
84.0	.7959	.7986	.8013	.8041	.8069	.8097	.8125	.8153	.8182	.8210
85.0	.8239	.8268	.8297	.8327	.8356	.8386	.8416	.8447	.8477	.8508
86.0	.8539	.8570	.8601	.8633	.8665	.8697	.8729	.8761	.8794	.8827
87.0	.8861	.8894	.8928	.8962	.8996	.9031	.9066	.9101	.9136	.9172
88.0	.9208	.9245	.9281	.9318	.9355	.9393	.9431	.9469	.9508	.9547
89.0	.9586	.9626	.9666	.9706	.9747	.9788	.9830	.9872	.9914	.9957

$$A = -\log\left(1 - \frac{R}{100}\right)$$

A = absorbance

R = % absorption

Intermediate standards

Pipette 1, 2, 3, and 4 ml volumes of stock magnesium standard into 100 ml volumetric flasks and dilute to volume with distilled water. Mix. These standards equal 1, 2, 3, and 4 mEq mg/L.

Working standards

Dilute 1 ml of each of the intermediate standards to 50 ml with distilled water. Mix.

Sample preparation

Dilute 0.5 ml of serum or urine to 25 ml with distilled water and mix. Urine dilutions may need to be adjusted according to the magnesium concentration.

If an automatic dilutor is available, set it to aspirate 0.1 ml of sample and to deliver the sample plus 4.9 ml of distilled water. Dilute the intermediate standards in the same manner.

Procedure

1. Install the magnesium lamp in the lamp compartment.
2. Turn the power switch on and adjust the source to the current given on the lamp.
3. Set the instrument to the proper wavelength range, and slit (Table 10-1).
4. Install the standard burner head.
5. Turn on compressed air supply and adjust the burner regulator to 30 psi and the flowmeter to 8.
6. Open the valve on the acetylene tank and adjust the pressure coming from the tank to 8 psi. Adjust the fuel flowmeter to 9.
7. Be sure the **U** in the burner drain tube is filled with liquid and that the end of the tube is below the surface of a liquid. Light the flame.
8. Allow the flame to aspirate water for about 5 minutes; then adjust absorption to zero.
9. Determine percent absorption on diluted standards and samples.
10. Convert absorption to absorbance (Table 10-2) and plot the absorbance of the standards versus concentration on linear graph paper.

Automated calcium and magnesium

Calcium and magnesium can be automated by using a Technicon sampler and pump to dilute and deliver the specimen to the burner of the atomic absorption spectrophotometer. A manifold diagram is shown (Fig. 10-2) that may be used for either calcium or magnesium.[6] Lanthanum working solution (D-217) containing 0.5 ml Brij 35/L is used as the diluent in both procedures. The intermediate standards used in the manual procedures are poured into the sample cups and run along with the serum and urine specimens. Samples may be run at a rate of 50/hr with a 2:1 sample-wash ratio. An automatic null recorder readout and a 10 mv potentiometric recorder are required.

Copper in serum and urine

Reagents needed

Copper stock standard, 1000 μg/ml	(Purchase prepared standard)
Copper dilute stock standard, 100 μg/ml	(Dilute stock 10 to 100)
Trichloroacetic acid, 5% (TCA)	(D-219)
Sodium diethyldithiocarbamate, 1%	(D-220)
Sodium ethylenediamine tetraacetate (EDTA), 1%	(D-221)
Methyl isobutyl ketone (MIBK)	
Sulfuric acid, concentrated	
Nitric acid, concentrated	

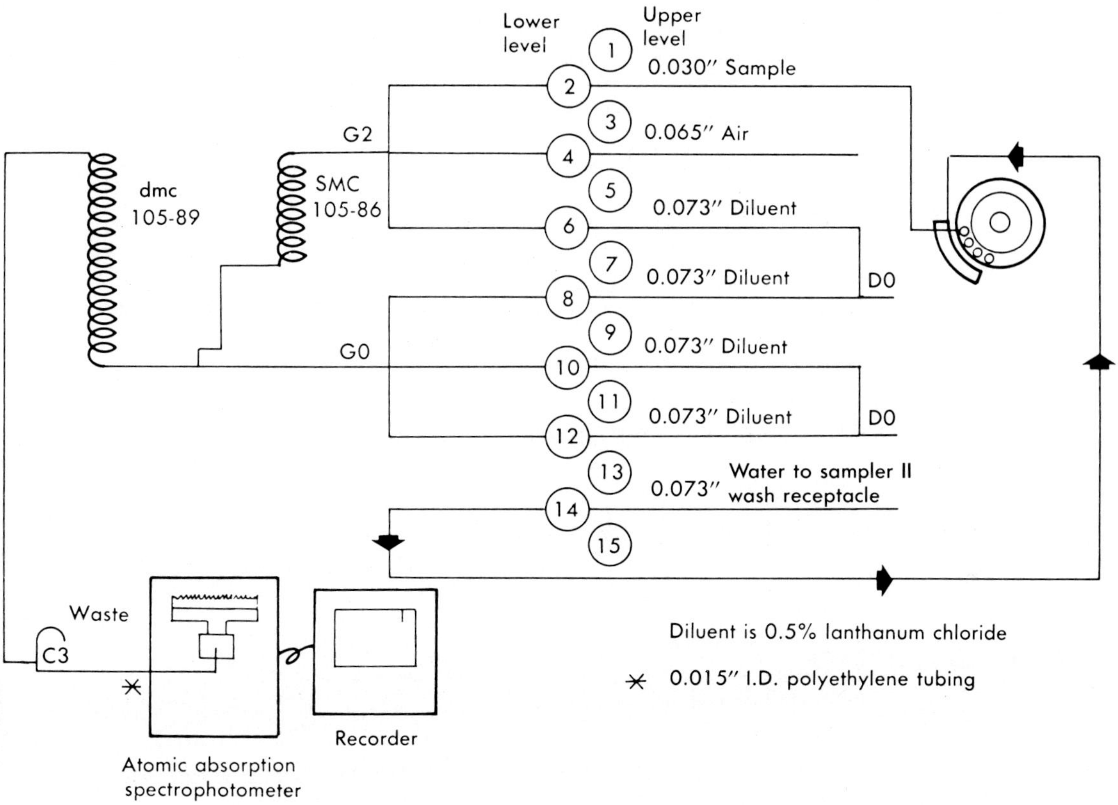

Fig. 10-2. Calcium or magnesium by atomic absorption. (Barnes Hospital, St. Louis.)

Working standards

Dilute 0.5, 1.0, 2.0, 3.0, and 5.0 ml volumes of dilute stock copper standard, 100 μg/ml, to 100 ml with water. These standards are equivalent to 50, 100, 200, 300, and 500 μg/100 ml. Extract these standards as described in the serum sample preparation. These organic standards may be used for both serum and urine.

Sample preparation—serum

1. Thoroughly mix 1.0 ml of serum with 5 ml of 5% TCA, allow to stand at room temperature for one hour and then centrifuge at 2000 rpm for 10 minutes.
2. Decant the supernatant into a 60 ml separatory funnel. Wash the precipitate with 5 ml of TCA and centrifuge. Add this supernatant to the separatory funnel.
3. Add 1 ml of 1% sodium diethyldithiocarbamate, 7 ml of 1% EDTA and 3 ml of MIBK to the funnel. Stopper and shake the funnel for 10 minutes.
4. Discard the lower aqueous layer and centrifuge the upper layer.
5. Adjust the flame and nebulizer as directed under Organic solvents. Determine the absorption on the extract and compare with the standards.

Sample preparation—urine

1. Transfer 25 ml of urine to a porcelain evaporating dish and add 2 ml of concentrated sulfuric acid. Boil gently. Add concentrated nitric acid repeatedly until the digestants are completely clear and colorless.
2. Transfer to a 25 ml volumetric flask. Rinse the evaporating dish with

a small amount of water and transfer the washings to the volumetric flask. Dilute to volume and mix.

3. Transfer a 10 ml aliquot to a 60 ml separatory funnel. Chelate and extract the copper as described under serum preparation beginning with step 3.
4. The calculated copper content is divided by 10 to obtain μg Cu/100 ml.

Zinc in plasma and urine[7]

Reagents needed

Zinc stock standard, 1000 μg/ml	(Purchase prepared standard)
Zinc dilute stock standard, 10 μg/ml	(Dilute stock 1 to 100)
Dextran, 6%*	
Dextran, 3% (1 part 6% dextran + 1 part water)	

Working standards

Dilute 5, 10, 15, and 20 ml volumes of dilute zinc standard, 10 μg/ml, to 100 ml with water and mix. Mix 2 ml of each standard with 2 ml of 3% dextran.

Sample preparation—plasma

1. Hackley and associates[7] found that serum zinc values were about 15% higher than plasma zinc. They recommend collecting blood specimens in tubes containing 0.1 ml of a 30% sodium citrate solution.
2. Mix 2 ml of plasma with 2 ml of water.
3. Determine percent absorption of standards and unknowns, using the 3-slot burner head.

Sample preparation—urine

1. Urine specimens should be acidified with a quantity of glacial acetic acid equal to about 1% of their volume.
2. Aspirate the undiluted urine directly into the flame along with undiluted working standards and determine percent absorption.

Lead in blood and urine

Reagents needed

Trichloroacetic acid, 5%	(D-219)
Ammonium pyrrolidine dithiocarbamate, 1%	(D-222)
Methyl isobutyl ketone (MIBK)	
Lead stock standard, 1000 μg/ml	(Purchase prepared standard)
Lead dilute stock standard, 10 μg/ml	(Dilute stock 1 to 100)

Working standards

Dilute 1, 3, 5, and 10 ml volumes of lead dilute stock standard to 100 ml with water to make standards containing 10, 30, 50, and 100 μg/100 ml lead. Put these standards through the same extraction procedure as the bloods.

Sample preparation—whole blood

1. Mix 5 ml of whole blood with 10 ml of 5% TCA. Allow to stand at room temperature for 1 hour, stirring occasionally with a glass rod.
2. Centrifuge for 10 minutes at 2000 rpm and decant the supernatant into a Babcock bottle. Wash the precipitate with 10 ml of distilled water, centrifuge, and add the supernatant to the same Babcock bottle.
3. Adjust the pH to 2.2 to 2.8 by adding 0.5 N NaOH (about 0.6 ml). pH paper may be used.
4. Add 1 ml of 1% ammonium pyrrolidine dithiocarbamate and 5 ml of MIBK saturated with distilled water.

*Abbott Laboratories.

5. Shake for 10 minutes mechanically or 2 minutes manually.
6. Add water to raise the organic phase into the neck of the bottle.
7. Using the 3-slot burner, adjust the flame and nebulizer in the manner described under Organic solvents, p. 265. Set the zero while burning water-saturated MIBK.
8. Aspirate the organic phase from the neck of each Babcock bottle containing standards and unknowns.

Sample preparation—urine

1. Transfer 30 ml of urine to a Babcock bottle and acidify to pH 2.2 to 2.8 with 5% TCA.
2. Follow the same procedure as with whole blood beginning with step 4.
3. Extract 30 ml portions of standards containing 2, 5, 10, and 30 μg/100 ml lead in the same manner.
4. Since different sample volumes are being extracted, it is necessary to prepare a separate standard curve for urines.

Interferences

Light-scattering effects may interfere with the analysis. This may be compensated for by subtracting the signal observed at the ion line at 2203 Å from the total signal measured at the absorption line.

REFERENCES

1. Christian, G. D.: Medicine, trace elements, and atomic absorption spectroscopy, Anal. Chem. 41:24A, 1969.
2. Slavin, W.: Atomic absorption spectroscopy, New York, 1968, Interscience Publishers, p. 72.
3. Manning, D. C.: The effect of acetone vapor in the air-acetylene flame, Atomic Absorption Newsletter 7:44, 1968.
4. Manning, D. C., and Chabot, H.: The effect of acetylene containing acetone vapor in atomic absorption analysis, Atomic Absorption Newsletter 7:94, 1968.
5. Analytical methods for atomic absorption spectroscopy, Norwalk, Conn., 1968, The Perkin-Elmer Corp.
6. White, W. L., Erickson, M. M., and Stevens, S. C.: Practical automation for the clinical laboratory, St. Louis, 1968, The C. V. Mosby Co., pp. 187-202.
7. Hackley, B. M., Smith, J. C., and Halsted, J. A.: A simplified method for plasma zinc determination by atomic absorption spectrophotometry, Clin. Chem. 14:1, 1968.

• 11

• Enzymes

An enzyme is an organic catalyst produced by a living cell, but it acts independently of the cell. (A catalyst is a substance that will excite or accelerate a reaction, but it is not itself involved or altered by the reaction.) Most of these organic enzymes are easily destroyed by heat, acids, or alkalies. They are active only within a limited pH range, are protein in nature, and will act on only a specific or particular substance or on closely related substances. The substrate is the substance upon which the enzyme acts. Most enzyme names are obtained by adding the suffix "ase" to the substance (substrate) acted upon. Therefore lipase, amylase, protease, sucrase, uricase, and lactase act upon fat, starch, protein, sucrose, uric acid, and lactose, respectively. Upon occasion, an enzyme is named after its function rather than its substrate. Two examples of this are dehydrogenases, which bring about the removal of hydrogen, and oxidases, which bring about oxidation. The enzymes that have been known for many years still maintain their original nondescriptive names—pepsin, ptyalin, trypsin, and papain.

The word "enzyme" has not always been used to describe the catalytic reaction, nor has the reaction always been understood. From early times man has known of the following chemical changes that occur spontaneously: (1) the fermentation of sugar with a final product of alcohol, (2) the souring of wine, and (3) the souring of milk.

The action in each case is a result of microorganisms, growing and changing the original substance. We know now that these changes are brought about by the action of enzymes. Years ago it was thought that the reactions were bound to living organisms and their life cycles and would occur only when those organisms were present in the substance in a living state.

Payen and Persoz (1833) precipitated a starch-digesting enzyme from a malt extract by means of alcohol. This was compared with the natural fermentation and the "ferments" that caused the souring of milk or wine. This enzyme or "ferment" was given the name *diastase*. During the same period of time that Payen and Persoz were working with the starch-digesting enzyme, Beaumont found the digestive action of the gastric juice to be due to a chemical substance. However, this latter was not isolated until 1836, when Schwann isolated the substance and called it *pepsin*. Other digestive substances such as saliva and pancreatic juice were also studied at that time. Although there was great activity during this period, the newly found substances were called ferments, not enzymes. The term "ferment" was used for any agent or substance bringing about a chemical reaction in biologic material. Later, Kuhne introduced the term "enzyme" as referring to a biologic catalyst. Buchner (1897) finally settled some heated debates about the action of the enzymes by proving that the microorganisms used intracellular enzymes to produce their effects.

The enzymes of the blood are present in very small quantities. Because of this, they cannot be measured in the usual terms of milligrams per 100 ml but must be measured in special units. The units vary with the method of determination.

In the determination of blood enzymes, the basic principle involved is this. The enzyme is allowed to decompose a substrate. After a definite period of time, the amount of decomposition is measured. The more the decomposition, the greater is the concentration of enzyme.

The fact that enzymes are dependent upon many physical and chemical factors presents a challenge to methodology and quality control. Although enzyme activity is specific, the range of the specificity is quite variable. For example, some enzymes will act on just one substrate, while another enzyme will act on differing substrates that are closely related. It is also possible to decompose a compound in various ways by different enzymes.

This type of analysis is based on measurable activity instead of recoverable weight. The weight concentration is more difficult to measure and is not as significant clinically as the measurement of the reaction rate on a substrate.

There has been confusion with recording units of the analysis. The Joint Sub-Commission on Clinical Enzyme Units of the International Union of Biochemistry and the International Union of Pure and Applied Chemistry recommends that the following be done whenever practicable. Define the enzyme units as moles (or micromoles) of substrate per liter (or milliliter) of serum, plasma, or urine transformed per minute by the enzyme under specified conditions. Therefore, the activity can be expressed in International Units.

Enzyme activity is very sensitive to variations in physical and chemical aspects of the method. Minute errors in one or more of the following factors may result in a gross error in the final determination.

FACTORS AFFECTING ENZYME ACTIVITY

Concentration of enyzme

Since the velocity of the enzyme reaction is directly proportional to the concentration of the enzyme, most procedures are chosen so the reaction is directly and linearly related to the concentration. The more enzyme, the faster the reaction; however, impurities may inhibit the reaction. This effect is shown in a linear curve early with a drastic drop-off of activity. In some cases there may be less than a 10% activity before the inhibitive reaction begins. If the concentration is high, the same type of activity results because the substrate may not be sufficient for maximum activity.

Concentration of substrate

Increased concentration of the substrate activates the velocity of the enzyme activity. The curve will be linear at the beginning and then curve to a plateau. The activity resulting in the linear part of the curve is called a first-order reaction. The plateau part of the curve demonstrates ineffectiveness of the increased concentration of substrate. This area is measured by the amount of product formed per unit time and is considered a zero-order reaction. With the substrate concentration placing activity in the area of the zero-order reaction, variations in the substrate will not invalidate the measurements of the enzyme concentration.

Hydrogen ion concentration

Each enzyme's optimum pH may be changed with the substrate or buffer. At its optimum pH, the enzyme will be most active, and any deviation from this level will result in lower reaction, inactivity, or destruction of the enzyme. The pH is dependent upon the source of enzyme, the substrate, and the buffer.

Temperature

Reaction rate increases with temperature. The temperature coefficient (Q_{10}) of a reaction is described as the increase in reaction rate with a 10° C temperature increase. The reaction rate or velocity is increased from 1.1 to 3 times for every 10° C rise in temperature. Optimum temperatures may reach 40° to 50° C for animal enzymes and 50° to 60° C for plant enzymes; above 60° C, enzyme destruction occurs. With a 10° C rise, the enzyme activity may increase from 50% to 100%. A small variation in the temperature, 1 or 2 degrees, may result in a 10% to 20% error in the test result. It is imperative to maintain a constant temperature.

Products of reaction

There is a retarding effect as the enzyme reaction progresses toward equilibrium. This is due to an accumulation of products of the reaction. Some of the reaction products are structurally similar to the substrate used. They therefore combine with the enzyme in a manner similar to that of the substrate—resulting in some inhibition of some of the enzyme.

Light rays

The specimen should be kept away from direct light, which may have an accelerating or an inhibiting effect. Ultraviolet light tends to inhibit the activity while blue or red light rays will enhance activity.

Time

Reaction time must be determined precisely. The specific reaction time must be accurate for quality results. A good timer must be used and periodically checked for accuracy.

Reagent purity and glassware

It is essential to use pure reagents. The glassware must be cleaned in chromic acid or some other agent that will make it chemically clean. Traces of heavy metals will inhibit reaction rate. For example, if a flask has contained Nessler's reagent, the mercury trace on the glass is likely to inhibit the reaction of urease used in urea determinations.

Opposite to this effect is the action of some divalent metal ions that activate certain enzymes.

Enzyme stability

Care must be taken in handling the enzyme solutions, vigorous shaking has a destructive effect, with denaturation of the enzyme.

The serum should not be allowed to stand at room temperature for a lengthy period. As an example, acid phosphatase has a very short stability period, even under refrigeration. It may be stabilized by adding disodium citrate.

Enzymes should be analyzed as soon as possible. The pH of fresh blood rises rapidly, due to the loss of CO_2, inactivating the enzyme.

DETERMINATION OF ENZYME REACTIONS

There are basically two methods of determining enzyme reactions.

Total change method

The total change method is the most familiar in the routine laboratory. Analysis is completed as rapidly as possible, using small amounts of substrate and larger amounts of enzyme. The result is calculated by means of a known constant of the substance.

Kinetic method

In the kinetic method the reaction of the substrate and enzyme is arranged so that the rate of reaction is slow enough to be measured accurately. Instead of a final reading to be compared to a standard curve, automatic recording (plotting) of the time against optical density is used to give information on the course of the reaction. Remember, the reaction curve is not always linear.

GENERAL INSTRUCTIONS

UV (ultraviolet) assay

Make optical density measurements at a wavelength of 366 or 340 nm. Not all photometers have the UV light source and filter. Now, most spectrophotometers have a dual source of lamps and filters, giving the laboratory greater versatility. Check with the instrument manual and note if the unit demands a reference cuvette. Measure the optical densities between 0.0 and 0.5.

Colorimetric assay

The colorimetric assay utilizes filters, and the test is read against a blank. The term "visible range" applies when readings are made at 400 to 700 nm. Occasionally a method cannot be read in the prescribed wavelength. Then, a standard curve is drawn with each run of that particular method. With enzyme work every batch run presents its own set of variances, which can develop into a full-blown problem with little effort. Watch conditions and run standards!

Cuvettes

Many times, excellent technique may be lost by reading the test in a cuvette of the wrong light path. This is true for both manual and automated techniques. When publishing a method or writing it out for laboratory use, be a "good guy" and give the wavelength. If the method is worth discussing, then give complete information so it will be a good contribution. Most enzyme assays may be read with a 1 cm light path; however, some enzymes such as α-oxoglutarate, acetoacetate, and β-hydroxybutyrate require a 2 cm light path. Before use, calibrate all the cuvettes and match those with closest transmittance readings (difference not to exceed $\pm$ 4 nm). Round cuvettes also are used and are matched. The round tube has variable light paths. To maintain a consistent reading in a tube, always place it into the well the same way, with the etched marker spot facing the operator or with it turned away from the operator. All tubes are read the same way. Handle all cuvettes with care—do not touch the outside. Always wipe the outside of the tube with a soft cloth before entering it into the well. Do not allow strong acids to remain in the tube for an extended period. When removing the fluid from the cuvette, place a small plastic tip on the vacuum line to prevent scratching the interior. When a tube is damaged by etching or chipping, throw it away!

Preparation of solutions

Reagents must be pure for all methodology, and their purity is extremely critical in enzyme analysis. Distilled water taken from metal equipment should be redistilled in an all-glass distilling apparatus. Deionized water although free from salts may have organic contaminants.

Reagent storage

All enzyme preparations and solutions should be stored at 2° to 4° C. Do not freeze, since freezing leads to loss of activity. Auxiliary reagents (perchloric acid, sodium hydroxide, etc.) are stable and are stored at room temperature in a safe area.

Glassware

Chromic acid is an excellent cleaning agent for glassware. If commercial detergents are used, check these for inhibiting action on the enzymes. All glassware must be thoroughly rinsed and then twice rinsed with distilled water. Haemo-Sol or a 30% sodium hydroxide solution will remove denatured protein.

Checking stability of DPNH solution

Add 0.05 ml of DPNH solution to 3.00 ml of redistilled water and read against a water blank in a 1 cm cuvette at 366 nm. The absorbance (A) should be over 0.350.

Probable sources of error

Dirty or contaminated glassware.
Distilled water purity not adequate.
Specimen unsatisfactory because of age of specimen, hemolysis, trace metals, inactivation by elevated temperature, or poor storage. (The specimen should be analyzed the same day it is collected.)
Incorrect pipetting.
Incorrect filter or wrong wavelength.
Solutions contaminated by microorganisms, appearing turbid. (Turbid or colored sample may be cleared after precipitation with perchloric acid.)
Solutions with expired stability date.
Poor reproducibility of test:
- Incorrect incubation temperature
- Poor analytic technique
- Refrigerated reagents not permitted to come to room temperature
- Improper mounting of the cuvette or use of dirty equipment
- Incorrect pH of sample for substrate assays

Initial absorbance too high. (Zero photometer with a solution of 0.005% to 0.01% aqueous potassium dichromate solution. If instrument has input sensitivity switch, utilize this for increasing the sensitivity.)

This chapter considers the following enzymes: amylase, lipase, alkaline and acid phosphatase, total acid and prostatic acid phosphatase, glutamic oxalacetic transaminase (SGOT), glutamic pyruvic transaminase (SGPT), lactic dehydrogenase (LDH), and creatine phosphokinase (CPK).

Table 11-1. Conditions accompanied by abnormal amylase values

When increased	When decreased
Acute pancreatitis	Cirrhosis of liver
Carcinoma of head of pancreas	Carcinoma of liver
Duodenal ulcer	Abscess of liver
Perforation of gastric ulcer	Hepatitis
Hyperthyroidism	Acute alcoholism
Mumps	Toxemias of pregnancy
Acute injury of spleen	
Renal disease with impaired renal function	
High intestinal obstruction	

Note that all the enzymes end in -ase. The enzyme amylase (Table 11-1) is sometimes referred to as diastase.

Methods

AMYLASE

A commonly used method for the determination of amylase is the procedure of Somogyi. In that method, three tubes are prepared.

The first tube contains serum, starch paste, and acid sodium chloride. The amylase in the serum will decompose the starch to reducing sugars. The acid sodium chloride is added to allow the reaction to take place at optimum conditions of pH and electrolytic content.

The second tube is a blank tube containing only the starch and acid sodium chloride. The purpose of this tube is to determine the amount of reducing substances in the starch, if any.

The third tube is a control tube containing only the serum and acid sodium chloride. The purpose of this tube is to determine the amount of reducing substances in the serum.

All three tubes are incubated for 30 minutes to enhance the reaction. Copper sulfate is then added to each tube to stop the reactions. Sodium tungstate is added to precipitate the proteins, and the contents of each tube are filtered.

A glucose determination is run on all three filtrates. The glucose values of the blank tube and control tube are added. This, of course, is the total glucose present in the starch and serum that is *not* due to the amylase activity. When this is subtracted from the glucose value of the tube containing the starch paste and serum, the amylase activity is obtained. This is reported as units of amylase. Thus:

Units of amylase = Glucose values of tube containing starch paste and serum *minus* glucose values of blank tube and control tube

Example:

$$\text{Units of amylase} = 200 \text{ mg/100 ml} - 100 \text{ mg/100 ml} = 100$$

Procedures for amylase

The following general considerations pertain to the procedure for amylase:

1. Specimen: serum or plasma.
2. Preservation of serum or plasma: may be stored in a refrigerator.
3. Precautions:
   a. Common errors result from saliva contamination, poor buffering, starch variability, and inhibition by oxalate and citrate.
   b. Temperature of water bath or incubator must not vary from figure given in procedure.

Leitz Photrometer method

Directions for preparing the reagents are given as follows:

Reagents needed

Purified cornstarch

1. In a tall 1000 ml cylinder place 100 g of cornstarch (U.S.P.)
2. Add 1000 ml of approximately 0.01 N hydrochloric acid.
3. Agitate vigorously and frequently for an hour.
4. Allow starch to settle (about an hour).
5. Pour off and discard the supernatant liquid.
6. Add 1000 ml of 0.05% sodium chloride.
7. Agitate vigorously and frequently for an hour.
8. Allow starch to settle.
9. Decant and discard the supernatant liquid.
10. Wash the starch once more with 0.05% sodium chloride.
11. Either decant the supernatant liquid, after allowing the starch to settle, or filter the suspension through a Buchner funnel.
12. Spread the starch on a clean plate and allow it to dry in air, protecting it from dust.

Cornstarch suspension, 1.5%

1. In a small beaker make a thin paste of 1.5 g of the purified cornstarch with 5 ml of distilled water.
2. Heat 90 ml of distilled water to boiling.
3. Pour the starch paste into the boiling water.
4. Rinse the beaker with 5 ml of distilled water and add the rinsings to the boiling liquid.
5. Boil for 1 minute with vigorous stirring.
6. Cover the beaker and allow to cool to room temperature.
7. Discard when starch settles out or mold appears.

Acid sodium chloride

1. Dissolve 10 g of reagent grade sodium chloride in about 800 ml of distilled water in a liter volumetric flask.
2. Add 3 ml of 0.01 N hydrochloric acid.
3. Dilute to volume with distilled water and mix well.

Copper sulfate, 5%

1. Dissolve 50 g of reagent grade crystalline copper sulfate ($CuSO_4 \cdot 5H_2O$) in distilled water in a liter volumetric flask.
2. Dilute to volume with distilled water and mix well.

Sodium tungstate, 6%

1. Dissolve 60 g of sodium tungstate ($Na_2WO_4 \cdot 2H_2O$—Folin special) in 850 ml of distilled water in a liter volumetric flask.
2. The solution should be neutral or slightly alkaline to phenolphthalein. If it is not, adjust by the addition of sulfuric acid or sodium hydroxide.
3. Dilute to volume with distilled water and mix well.

*Procedure**

Somogyi, M. Proc. Exp. Biol. Med. 32:538, 1934

In each of two test tubes labeled "A" and "B" place
5 ml of 1.5% cornstarch suspension and
2 ml of acid sodium chloride solution.

In a third test tube labelled "C" place
5 ml of distilled water and
2 ml of acid sodium chloride solution.

Place the tubes in a water bath at 40° C for 10 minutes.

Add 1 ml of serum or plasma to both tube "A" and tube "C."

Add 1 ml of distilled water to tube "B."

After mixing the contents of each tube, stopper it.

Place all three tubes in a water bath at 40° C and leave them there for exactly 30 minutes.

Remove the tubes from the bath and quickly add to each 1 ml of 5% copper sulfate solution.

Mix well.

Add to each tube 1 ml of 6% sodium tungstate solution.

Shake vigorously until foaming ceases.

Centrifuge or filter each, collecting the filtrates in correspondingly marked test tubes.

Refilter if filtrate is not clear.

Transfer 1 ml of each filtrate to correspondingly marked Folin sugar tubes.

Add 1 ml of distilled water to each tube.

Determine the amount of glucose in each according to the "glucose in blood" method, using the above diluted filtrates in place of the tungstic acid filtrates.

The number of amylase units per 100 ml of serum or plasma is obtained by subtracting the glucose values of tube "B" (starch blank) and that of tube "C" (serum blank) from the glucose value of tube "A" (total reducing substances), and then multiplying the result by 2.

NOTE: For the blood amylase and the micro glucose methods, the standards for blood glucose are used.

Somogyi method[1]

The Somogyi method is based on the measurement of time required for the amylase enzyme to convert (hydrolyze) starch to polysaccharides to glucose. The rate of speed of the conversion is proportional to the amount of amylase in the serum, urine, or fluid. Iodine is used as an indicator, yielding the following colors as the starch is converted to maltose:

Starch (deep blue) → Polysaccharides (red or violet) → Maltose (color of the iodine solution)

The end point of the reaction is an amber color. The color changes are specific enough for the test to be done visually. With experience, the analyst will be able to determine, by the color of the preceding check, how much time lapse can be allowed between color checks.

Reagents needed

Starch, sterile†

Iodine indicator (D-223)

*Procedure from Handbook for Leitz Photrometer and used by permission of E. Leitz, Inc., New York, N. Y.

†May be ordered from Sigma Chemical Co.

Procedure

Serum

1. For each serum to be analyzed, arrange the tubes in the following manner:

Test	a	b
4	0	oooooooo
3	0	oooooooo
2	0	oooooooo
1	0	oooooooo

   a. Test tubes containing serum, urine, or fluid to be analyzed.
   b. Kahn tubes (eight per test) to contain the iodine indicator.
2. Sterilize the cap of the starch substrate by swabbing with 70% alcohol. With a sterile syringe and needle, remove enough starch to allow 2.0 ml per test.
3. Place the tube containing the starch substrate into either a 37° or a 40° C water bath.
4. Pipette into each of the Kahn tubes, 0.25 ml of the iodine indicator.
5. Number tubes (regular size) for the tests and place in the water bath. Into each tube, pipette 2.0 ml of starch substrate.
6. Set timer for 40 minutes for a single test and add 1 minute for each additional test. For example, the timer would be set for 43 minutes for four tests.
7. Into the starch substrate of tube 1, pipette 0.5 ml of serum, mix thoroughly, and *start timer*. Exactly 1 minute later, pipette into tube 2 and at exactly 1-minute intervals add the test sera to their substrate tubes until all of the tests are started.
8. After 3 minutes, remove 0.25 ml of the substrate-serum mixture and place into the Kahn tube containing iodine indicator. Shake gently and note the color.
9. If the color is purple, wait for 3 minutes and withdraw another 0.25 ml of mixture and add to the second iodine tube. Record the elapsed time. Continue the testing and recording of time until the end point is reached. This is an amber or reddish brown color. Test 1 minute beyond the end point to verify that the reaction is completed.

Urine[2]

1. Urine amylase determinations should be performed on specimens collected during a timed interval. Measure and record the volume of urine and time of collection.
2. Adjust the pH to 7.0 to 7.4. If below a pH of 7.0 (acid), adjust with solid sodium carbonate. If the pH is above 7.4 (alkaline), adjust with solid potassium acid phosphate (KH_2PO_4).
3. Start with undiluted urine. If the reaction has gone past the end point in 3 minutes, dilute 1.0 ml of urine with 4.0 ml of 0.5% NaCl and begin again, using 0.5 ml of diluted urine.
4. The test for urine amylase is done in the same manner as for the serum.

Calculation

The factor to be used may be found on the bottle of starch substrate. However, if you incubate the substrate at 37° C, calculate by using a factor of 1800. If the incubation is done at a temperature of 40° C, calculate with a factor of 1600. The accompanying example contains three serum specimens and one urine specimen. One of the serum specimens was quite elevated and the end point was reached before the next spot test was done. The serum may be diluted and the test started again at a later time. For checking the progress of the reaction, the time of the spot test is recorded below the previous time check. At the end point, the time recorded (underlined) is subtracted from the starting time.

	Tube 1 (serum)	*Tube 2 (serum)*	*Tube 3 (serum)*	*Tube 4 (urine)*
Starting time	43 minutes	42 minutes	41 minutes	40 minutes
Spot checks	40 minutes	39 minutes	38 minutes	37 minutes
	37 minutes	Color beyond	35 minutes	34 minutes
	36 minutes	end point	20 minutes	31 minutes
		Dilute serum	18 minutes	28 minutes
		and start at	15 minutes	
		later time		
		Starting time for diluted tube 2:		
		24 minutes		
		23 minutes		
End point	35 minutes	22 minutes	11 minutes	25 minutes
	8 minutes	2 minutes	30 minutes	15 minutes

Subtracting the end point time from the starting time we have the following:

Test 1 8 minutes
Test 2 2 minutes (serum dilution: 1 ml of serum and 2 ml of 0.5% NaCl)
Test 3 30 minutes
Test 4 15 minutes (urine dilution: 1 ml of urine and 4 ml of 0.5% NaCl)

Serum calculation

$$\frac{\text{Factor of substrate}}{\text{Minutes required to obtain end point}} \times \text{Dilution (if any)} = \text{Units, percent amylase}$$

Test 1 $\frac{1600}{8} = 200$ Somogyi units/100 ml

Test 2 $\frac{1600}{2} \times 3 = 2400$ Somogyi units/100 ml

Test 3 $\frac{1600}{30} = 53$ Somogyi units/100 ml

Test 4 $\frac{1600}{15} \times 5 = 533$ Somogyi units/100 ml

Urine calculation

To obtain the amount of units excreted per hour the following calculation is necessary. The value of 533 Somogyi units/100 ml was obtained on 150 ml of urine collected in a 2-hour period:

$$\frac{\text{Somogyi units/100 ml}}{\text{Time of collection}} \times \frac{\text{Urine volume}}{\text{100 ml}} = \frac{533}{2} \times \frac{150}{100} = \text{399.75 or 400 Somogyi units excreted/hr}$$

Normal values

Normal serum	40 to 140 Somogyi units/100 ml
Borderline serum	Up to 300 Somogyi units/100 ml
Normal urine	50 to 400 Somogyi units/hr[2]
Urine in diseases other than pancreatitis	Up to 800 Somogyi units/hr
Urine in acute pancreatitis	Above 900 Somogyi units/hr

LIPASE

An elevated lipase value is of chief concern in acute pancreatitis. In this disease, both amylase and lipase are high. The amylase value, however, usually returns to normal after 3 days, whereas the lipase value generally takes 7 or more days to return to normal. Therefore, if the serum has not been obtained within

3 days after the attack, the lipase determination becomes the more significant diagnostic aid.

In addition to acute pancreatitis, elevated lipase values may be found in intestinal obstruction and carcinoma of the pancreas.

For the determination of lipase, the procedure of Cherry and Crandall is commonly used. In this method, serum is obtained. Two test tubes are secured. One tube is marked with the patient's test number and the other *control.* One milliliter of serum is added to each tube. The control tube is heated to destroy the enzyme. A buffer solution is added to both patient and control tubes, its function being to regulate the pH and electrolyte content to optimum conditions. The substrate, olive oil, is added to each tube. The tubes are then incubated at body temperature (37.5° C) for 24 hours. This enables the lipase to decompose the olive oil into fatty acids. Alcohol and an indicator are added. The amount of fatty acids produced in each tube is then found by titration with twentieth normal sodium hydroxide. The titration figure of the control is subtracted from the titration figure of the patient. The answer is the lipase value expressed in milliliters of 0.05 normal sodium hydroxide.

Lipase value = ml of 0.05 N NaOH used in titration of patient *minus* ml of 0.05 N NaOH used in titration of control

Example:

Lipase value = 2 ml − 1 ml = 1 ml of 0.05 N NaOH

Procedure for lipase

The following general considerations pertain to the procedure for lipase.

1. **Specimen: serum.**
2. **Preservation of serum: does *not* keep. Test must be run within 2 hours after the withdrawal of blood.**
3. **Precautions:**
   a. **Olive oil emulsion must not be rancid.**
   b. **Temperature of water bath or incubator must not vary from figure given in procedure.**
   c. **Timing of incubation must be exact.**

Cherry-Crandall method

The method of Cherry and Crandall is given below. The necessary reagents are listed and the procedure follows.

Reagents needed*

Sodium hydroxide, 0.05 N	**(D-224)**
Phosphate buffer, pH 7.0	**(D-225)**
Olive oil substrate, 50%	**(D-226)**
Phenolphthalein indicator, 1%	**(D-208)**
Ethyl alcohol, 95%	

Procedure

1. **The 50% olive oil emulsion should be prepared fresh every 2 weeks and stored in a refrigerator.**
2. **Make a venipuncture and withdraw about 7 ml of blood. Gently expel the blood into the test tube, let clot, and then centrifuge to obtain the serum. This must be used within a few hours, since the lipase activity is affected upon long standing.**

---

*May be purchased from Sigma Chemical Co.

3. Get two large test tubes. Label one with test number of the *patient* and label the other *control.*
4. Pipette 3.0 ml of distilled water into each tube. Add 1.0 ml of serum to each tube with an Ostwald-Folin pipette.
5. Destroy the enzyme in the control tube by heating in boiling water for 5 minutes. Place in room temperature water bath and cool to room temperature.
6. Add 0.5 ml of the phosphate buffer (pH 7.0) solution to both the patient's tube and the control tube.
7. Add 2.0 ml of the 50% olive oil emlusion to each tube.
8. Incubate both tubes for 24 hours at 37.5° C.
9. Add 3 ml of the 95% alcohol to each tube. Add 2 drops of the thymolphthalein indicator to each tube. Mix gently.
10. Using the 0.05 N sodium hydroxide, titrate each tube to a permanent *faint* pink color. Record the amount of sodium hydroxide used for each titration.
11. Find the lipase value as follows:

    Lipase value = ml of 0.05 N sodium hydroxide used in titration of patient's tube *minus* ml used in titration of control tube

    Example:

    Lipase value = 2.0 ml (used for patient) *minus* 1.1 ml (used for control)

    Lipase value = 0.9 unit

Normal values

0 to 1.5 units

ALKALINE AND ACID PHOSPHATASE[3,4]

The two phosphatases routinely analyzed are called alkaline and acid phosphatases according to the optimum pH at which they are measured.

The optimum pH of the alkaline phosphatase is 9.7 Alkaline phosphatase is found in the blood serum and plasma and in bone, kidney, liver, mammary gland, intestine, lung, spleen, leukocytes, seminiferous tubules, and the adrenal cortex. However, it is found chiefly in the bone and liver. Since this enzyme plays an important part in bone formation, there is an increase of activity in children. At birth, the alkaline phosphatase activity is low, rising rapidly in the first month of life, remaining highly elevated the first 2 years, and then decreasing to the normal child's range of activity.

The alkaline phosphatase increases in the blood plasma in bone disease. There have been various theories in regard to how this increased action takes place. One theory is that in the absence of normal bone synthesis, the bone's capacity for cellular activity is greater. With this greater activity there is an increased formation of phosphatase. Other explanations are given. The increased enzyme action may be one result of an overproduction of the enzyme in the bone attempting to compensate for the lesion, or there may be a forced discharge of the enzyme from the injured bone tissue. Diseases associated with this increase in the alkaline phosphatase are Paget's disease, rickets, bone atrophy, osteomalacia, osteoporosis, and bone malignancy. Moderately increased values have been found in hyperparathyroidism.

Alkaline phosphatase is also excreted into the bile by the normal liver. The serum values are usually increased in jaundice due to mechanical obstruction. Most liver diseases cause an elevation of results.

A decrease in the enzyme reaction, giving consistently low results, is found in individuals with hypophosphatasia (a hereditary bone disease), anemia, infectious hepatitis, or cretinism and in children who have an unusually early development.

The optimum pH of the acid phosphatase is 5.0. This enzyme was first discovered in male urine. Further investigation has shown the prostate gland to be extremely rich in this enzyme, thus leading to extensive investigation of tumors of the prostate. Normal blood plasma or serum contains small amounts of acid phosphatase. This may have its origin in the liver, spleen, bones, kidney, or prostate. It is present in women and children also, although the analysis for acid phosphatase is usually requested for diagnosis of prostate carcinoma.

Gutman and co-workers[5] (1936) first demonstrated that serum acid phosphatase levels increased significantly in many cases of metastatic prostatic carcinoma but seldom increased in the absence of metastases. (Metastasis is the transfer of diseased tissue from one part of the body to another.) However, there are three possibilities of obtaining normal results when metastasis has occurred:

1. Prostatic cells may produce very little enzyme because of very low androgen (hormone) level or because of anaplasia, which is the change of a normal cell structure to one similar to that of an embryo cell.
2. Cancer may have affected the lymph and blood channels so that the excessive amount of the enzyme cannot get into the bloodstream.
3. Castration, administration of estrogens, intensive irradiation, or radical prostatectomy inhibits the activity of the enzyme.

Women may also show elevations of acid phosphatase if they suffer from disease of any one of several organs (liver, spleen, kidney, or bone). There is an increase of the enzyme activity with carcinoma of the breast with extensive skeletal metastasis.

Since the red cells contain a high level of acid phosphatase, it is imperative that serum used for the acid phosphatase analysis contain no hemolysis.

Alkaline phosphatase

The normal values for alkaline phosphatase depend upon the method of determination. The values for three of the commonly used methods are listed here:

	Adults	*Children*
Bodansky units	1.5 to 4.5	5 to 14
King-Armstrong units	4 to 13.0	15.0 to 30.0
Bessey-Lowry-Brock units	0.8 to 2.3	2.8 to 6.7

Table 11-2. Conditions accompanied by abnormal alkaline phosphatase values

When increased		When decreased
Rickets	Neurofibromatosis	Scurvy
Hyperparathyroidism	Myositis ossificans	Hypoparathyroidism
Carcinoma of bone	Obstructive jaundice	Chronic nephritis
Paget's disease	Idiopathic steatorrhea	Celiac disease
Osteomalacia	Occlusion of pancreatic duct	Osteolytic sarcoma
Multiple myeloma	Abscess of liver	

Conditions in which an increased or decreased value may be expected are listed in Table 11-2. Abnormal values are of primary interest in those disorders given in italics.

For the determination of alkaline phosphatase, the three methods discussed below are commonly used.

Bodansky method

Serum is added to sodium glycerophosphate that has been buffered to pH 8.6 in order to ensure optimum conditions of pH and electrolyte content. The mixture is incubated at 37° C for exactly 1 hour. This allows the phosphatase enzyme in the serum to decompose the sodium glycerophosphate and liberate phosphorus.

A phosphorus determination is then run on this incubated serum and also on nonincubated serum. The phosphorus determination consists of adding trichloroacetic acid to both specimens to precipitate the proteins, adding molybdic acid to form a phosphate complex, adding a reducing agent to reduce the complex, and finally measuring the depth of color in a colorimeter.

The alkaline phosphatase units are then found by subtracting the phosphorus value of the nonincubated specimen from the phosphorus value of the incubated specimen. Thus:

Alkaline phosphatase units = Phosphorus value of incubated specimen *minus* phosphorus value of nonincubated specimen

Example:

Alkaline phosphatase units = 6 mg/100 ml − 3 mg/100 ml = 3

These units are usually reported as Bodansky units, a Bodansky unit being that amount of phosphatase which will liberate 1 mg of phosphorus/100 ml of serum under the conditions described above.

King-Armstrong method

Serum is added to disodium phenylphosphate that has been buffered to pH 9.0. The mixture is incubated at 37.5° C for 30 minutes to allow the phosphatase to decompose the disodium phenylphosphate and liberate phenol. A blank containing all the above ingredients is prepared. This is not incubated.

A phenol reagent is then added to the test and the blank. The reagent forms a color complex with the liberated phenol. The depth of color is compared with a standard and measured in a colorimeter. This determines the amount of phenol present.

The nonincubated specimen (blank) subtracted from the incubated specimen (test) gives the units of alkaline phosphatase.

Alkaline phosphatase units = Phenol value of incubated specimen *minus* phenol value of nonincubated specimen

The report is given in King-Armstrong units, one unit being the milligrams of phenol liberated by 100 ml of serum under the above conditions.

Bessey-Lowry-Brock method

Serum is added to *p*-nitrophenylphosphate that has been buffered to pH 10.3 to ensure optimum conditions for reaction. The mixture is incubated at 37° C for exactly 30 minutes. This allows the phosphatase to decompose the *p*-nitrophenylphosphate and liberate *p*-nitrophenol. The *p*-nitrophenol color is measured. Hydrochloric acid is added to decolorize the *p*-nitrophenol. The color due to serum is then measured. These units are reported as mM (millimoles) of substrate split per hour per liter of serum:

Alkaline phosphatase units or mM = *p*-Nitrophenol value of incubated serum *minus* decolorized value of the same specimen

Acid phosphatase

The normal values for acid phosphatase are 0 to 1 Bodansky unit or 0 to 2 King-Armstrong units. Increased values are found in the following conditions: carcinoma of the prostate, hyperparathyroidism, and acute granulocytic leukemia.

For the determination of acid phosphatase, the following three methods are commonly used.

Bodansky method

This is exactly the same procedure as the Bodansky alkaline phosphatase procedure described above except that the buffer solution is an acid buffer of pH 5.0 rather than the alkaline buffer of pH 8.6. The acid phosphatase units are found similarly:

Acid phosphatase units = Phosphorus value of incubated specimen *minus* phosphorus value of nonincubated specimen

King-Armstrong method

This is exactly the same procedure as the alkaline phosphatase procedure of King and Armstrong, which has been described above, except that the buffer solution is an acid buffer of pH 5.0 instead of the alkaline buffer of pH 9.0. The acid phosphatase units are found similarly:

Acid phosphatase units = Phenol value of incubated specimen *minus* phenol value of nonincubated specimen

Bessey-Lowry-Brock method, modified

This is the same procedure as for alkaline phosphatase except that an acid buffered substrate, pH 4.8, is used and serum blanks are not incubated. The *p*-nitrophenol liberated during the incubation period is measured. The nonincubated serum blanks are also measured. These units are reported either as Sigma units or as mM (millimoles):

Acid phosphatase units or mM = *p*-Nitrophenol value of incubated serum *minus* nonincubated serum

Procedures for alkaline and acid phosphatase

The following general considerations pertain to the procedures for alkaline and acid phosphatase.

1. **Specimen: serum. *Not* hemolyzed!**
2. **Preservation of serum: May be kept for 2 hours by adding a small drop of toluene and storing in a refrigerator.**
3. **Precautions:**
   a. **Temperature of water bath should not vary by more than ± 1° C.**
   b. **Timing of incubation must be exact.**

Serum alkaline phosphatase[6]

Bessey-Lowry-Brock method

Para-nitrophenyl phosphate is colorless in solution; but upon hydrolysis the phosphate group liberates *p*-nitrophenol, which is highly colored in an alkaline solution. The rate of hydrolysis of the *p*-nitrophenyl phosphate is proportional to the concentration of the enzyme present in the serum. The reaction may be shown as follows:

$$\underset{\text{(colorless in solution)}}{p\text{-Nitrophenyl phosphate} + H_2O} \xrightarrow{\text{Phosphatase}} p\text{-Nitrophenol} + H_3PO_4$$

(colorless in acid but highly colored—yellow—in alkaline solution)

Reagents needed*

Alkaline buffer, 0.1 M glycine, pH 10.5	**(D-227)**
Stock substrate, *p*-nitrophenyl phosphate	**(D-228)**
Alkaline buffered substrate, pH 10.3 to 10.4	**(D-229)**
Sodium hydroxide, 0.02	**(D-230)**
Hydrochloric acid, concentrated	

Procedure

1. **Use either 16 × 150 mm test tubes or spectrophotometer tubes for the test.**
2. **Into the *reagent blank* and *test* tubes, pipette 1.0 ml of alkaline buffered substrate.**
3. **Place the tubes (blank and tests) into a 37° C controlled water bath for 5 minutes, for the tubes and reagent to come to water bath temperature.**
4. **Set a timer for 30 minutes. Upon addition of the water to the reagent blank the timer is started, and every ½-minute thereafter the serum is added to its tube in the water bath.**
5. **Pipette 0.1 ml of distilled water to the blank and 0.1 ml of serum to each tube at ½-minute intervals.**
6. **Allow the tubes to incubate until *exactly* 30 minutes after the water was added to the blank.**
7. **Obtain an empty test tube rack in which to place the tubes as they are taken from the water bath.**
8. **Have in preparation a fast-flowing 10 ml serologic pipette (blowout) with a flask of 0.02 N NaOH. This must be added at timed intervals, also. The addition of the sodium hydroxide will stop the enzyme reaction.**
9. ***Exactly* 30 minutes by the timer, remove the blank from the water bath, add 10 ml of 0.02 N NaOH, and place tube into the test tube rack by the water bath. Continue at ½-minute intervals until the reaction has been stopped on all the tests.**
10. **Stopper the tubes with clean stoppers and mix by inverting.**
11. **Transfer the blank and tests to spectrophotometer tubes if these were not used during the incubation.**
12. **Read the tests against the reagent blank at 410 mμ wavelength.**
13. **Record the A of the serum samples.**
14. **Add 0.1 ml (2 drops) of concentrated HCl to each tube and mix by shaking tube. This removes the color due to the *p*-nitrophenol.**
15. **Read against the reagent blank at 410 mμ wavelength.**
16. **Record the A of the second readings. This is the A of the serum, since in an acid state the *p*-nitrophenol produces no color.**
17. **Subtract the second reading from the first reading where the *p*-nitro-**

*May be purchased already prepared from Sigma Chemical Co.

phenol produced color. This corrected optical density is then read from a curve or table made from the curve.

NOTE: When a value greater than 10 mM units is obtained, repeat the test with a smaller volume of serum or a shorter incubation time. Adjust the calculations accordingly: for example, use a 15-minute incubation and multiply the result by 2 for the correct values, or dilute the serum with saline and multiply by the dilution made. One millimolar (mM) unit of phosphatase activity liberates 1 millimole of *p*-nitrophenol/L of serum/hr (1 mM = 0.1391 g). If you are using the Sigma units, these are the same as the millimolar units.

Sources of error

1. Incorrect temperature of the water bath. The temperature must not vary more than ± 1° C. With an increase in temperature of 10° C the enzyme reaction will increase two to four times the initial value. The temperature limit if the animal enzyme is 40° to 50° C. Temperatures above this will cause a decrease in enzyme activity, and at 60° C the enzyme is destroyed. Plant enzymes can stand a higher temperature than the animal enzyme. The animal enzymes are less stable and have a tendency to become denatured when heated in the presence of water.[3]
2. Incorrect timing of the incubation period. The timing must be exact.
3. A substrate that is either too old or has an altered pH.
4. Incorrect pipetting.
5. Improper lighting. Light may either increase the enzyme rate or inhibit the rate of reaction. Ultraviolet rays or radium dispersion will inhibit enzyme reaction. Red or blue light will increase the reaction of the digestion enzymes such as pepsin, trypsin, and amylase.[3]
6. Glassware not chemically clean. If the glassware has been acid-washed and not thoroughly rinsed, the enzyme activity will be inhibited.
7. Violent shaking of the enzyme solutions, which will cause a denaturation of the protein enzyme, resulting in an inhibiting effect.
8. Separate storage, if possible, for glassware used for enzyme studies. Heavy metals such as mercury inhibit enzyme activity. Glassware used for Nessler's solution (containing mercury) should not be used for the BUN method utilizing urease. Aluminum ions inhibit peptic action, while toluene inhibits pepsin and trypsin.

TOTAL ACID PHOSPHATASE AND PROSTATIC ACID PHOSPHATASE[6]

Acid phosphatase is unstable and must be separated from the cells and refrigerated immediately. The serum must not be hemolyzed, since the red cells contain high levels of acid phosphatase. Since prostatic acid phosphatase may lose up to 50% of its activity within an hour at room temperature, 0.01 ml of 20% acetic acid/1.0 ml of serum is added to stabilize the enzyme.

Total acid phosphatase

The total acid phosphatase test uses the same stock substrate as the alkaline phosphatase.

Reagents needed*

Acid buffer, 0.09 M citric acid, pH 4.8	(D-231)
Stock substrate, *p*-nitrophenyl phosphate	(D-228)
Acid buffered substrate, pH 4.8 to 4.9	(D-232)
Sodium hydroxide, 0.1 N	(D-233)
Acetic acid, 20% (v/v)	(D-234)

*May be purchased from Sigma Chemical Co.

Procedure

1. Using either 16 × 150 mm test tubes or spectrophotometer tubes, set up the following test with a serum blank for each serum test to be analyzed.
2. The serum blanks are not incubated with the tests and may be read while the serum tests are incubating.
3. Into the *reagent blank* and *test* tubes, pipette 0.5 ml of acid buffer substrate.
4. Place the tubes (blank and tests) into a 37° C controlled water bath for 5 minutes, for the substrate to come to bath temperature.
5. Set the timer for 30 minutes. (NOTE: When placing the serum into the serum test tubes for incubation, it may also be added to the serum blank tube within the ½-minute interval between the tubes.)
6. Start timing at the addition of 0.2 ml of water into the *reagent blank*. Mix *gently* and replace the tube into the water bath.
7. At ½-minute intervals, pipette 0.2 ml of serum into the serum test tubes. Mix *gently* and replace into the water bath.
8. Incubate for 30 minutes.
9. *Serum blanks:* To the tubes containing 0.2 ml of serum, add 6.0 ml of 0.1 N NaOH. Stopper and mix by inversion or mix laterally by shaking the tube gently.
   a. Obtain the absorbance readings by reading the tests against a distilled water blank at wavelength of 410 mμ.
   b. Refer to the calibration chart for the acid phosphatase units.
10. *After incubating the serum tests for 30 minutes,* pipette 5.0 ml (use a fast-flowing serologic pipette) of 0.1 N NaOH into each tube at ½-minute intervals to stop the enzyme activity. The color is stable for several hours.
11. Obtain an absorbance reading by reading the tests against the *reagent blank* at a wavelength of 410 mμ.
12. Determine the acid phosphatase units from the calibration chart.
13. Subtract the "unit-results" of the serum blanks from the "unit-results" of the serum tests. This is the *corrected total acid phosphatase* of the serum sample.

NOTE: When a value greater than 2.8 mM units is obtained, repeat the test, with a smaller volume of serum or a shorter incubation time. Adjust the calculations accordingly: for example, use a 15-minute incubation period and multiply the result by 2 for the correct value, or dilute the serum with saline and multiply by the dilution made.

Normal values

Male total acid phosphatase	0.13 to 0.63 mM unit
Female total acid phosphatase	0.01 to 0.56 mM unit

NOTE: One millimolar (mM) unit of phosphatase activity liberates 1 millimole of *p*-nitrophenol per liter of serum per hour.

Serum prostatic acid phosphatase

Since the enzyme to be measured is highly unstable, refrigerate the blood immediately after drawing. After the clot has retracted (allow an hour) centrifuge and keep the serum at 0° to 5° C, or add 0.01 ml of 20% acetic acid per milliliter of serum to keep it stabilized.

*Reagents needed**

Tartrate acid buffer, pH 4.8	(D-235)
Acid buffer, 0.9 M citric acid, pH 4.8	(D-231)
Stock substrate, *p*-nitrophenyl phosphate	(D-228)
Acetic acid, 20% (v/v)	(D-234)

*May be purchased from Sigma Chemical Co.

Procedure

1. Using either 16 × 150 mm test tubes or spectrophotometer tubes, prepare two tubes as follows:
   Tube 1
   0.5 ml of stock substrate
   0.5 ml of tartrate acid buffer
   Tube 2
   0.5 ml of stock substrate
   0.5 ml of acid buffer
2. Place the tubes into the 37° C water bath for 5 minutes.
3. Set timer at 30 minutes. Start timing with the addition of 0.2 ml of serum into each tube. Mix *gently* and replace tube into water bath.
4. After the 30-minute incubation, add 5.0 ml of 0.1 N NaOH to each tube. Mix either by stoppering the tube and inverting or by lateral shaking of the tube.
5. Now, since the reaction has been stopped by the addition of NaOH, the tubes are read against a distilled water blank at a 410 mμ wavelength.
6. From the calibration chart, obtain the acid phosphatase units for tubes 1 and 2.
7. Subtract the "unit-results" of tube 1 from the "unit-results" of tube 2. This is the prostatic acid phosphatase of the serum.

Sources of error

The usual sources of error given earlier pertain to all enzymes, but two factors should be mentioned that frequently cause inaccurate results and thus misrepresent the correlation between the acid phosphatase and the prostatic acid phosphatase.

1. Care may not have been taken with the acid phosphatase determination. This enzyme is *unstable*, and correct precautions must be taken.
2. Several authors[7-9] have reported elevated total acid phosphatase and prostatic acid phosphatase after the physician has done a simple rectal palpitation of the prostate. The blood should be drawn for the phosphatase analysis before the examination, or a time lapse of 24 to 48 hours be allowed before drawing the blood for analysis.

Calibration of colorimeter for alkaline and acid phosphatase

The calibration may be done at room temperature. There is no incubation period in the calibration method.

Reagents needed

p-Nitrophenol, 10 mM/L* (D-236)
Sodium hydroxide, 0.02 N (D-230)
Working standard: Discard after use—stable 1 day

a. Into a 100 ml volumetric flask, pipette 0.5 ml of *p*-nitrophenol standard solution.
b. Dilute to 100 ml mark with 0.02 N NaOH.
c. Stopper and mix thoroughly.

Procedure

1. Pipette the following volumes into the tubes. Refer to Table 11-3.
2. Using a wavelength of 410 mμ or filter of equivalent wavelength, read the % T or A of the six tubes and record results in column 1. These are read against a reference blank containing 0.02 N NaOH.
3. Prepare the alkaline phosphatase curve by plotting the results in column 1 against the mM units of alkaline phosphatase in column 2.

*May be purchased from Sigma Chemical Co.

Table 11-3. Phosphatase calibration

Tube	Working standard (ml)	0.02 N NaOH (ml)	1 Column for scale reading % T or A	2 Equivalent to mM units, serum phosphatase	
				Alkaline	Acid
1	1.0	10.0		1.0	0.28
2	2.0	9.0		2.0	0.56
3	4.0	7.0		4.0	1.12
4	6.0	5.0		6.0	1.67
5	8.0	3.0		8.0	2.23
6	10.0	1.0		10.0	2.80

4. Prepare the acid phosphate curve by plotting the results in column 1 against the mM units of acid phosphatase in column 2.
5. If a Klett-Summerson colorimeter is used, the Klett readings may be converted to A readings by multiplying the Klett readings by $\frac{2}{1000}$. For example, a Klett reading of $310 \times \frac{2}{1000} = 310 \times 0.002 = 0.620$ A.
6. The calibration curve would be set up similar to this:

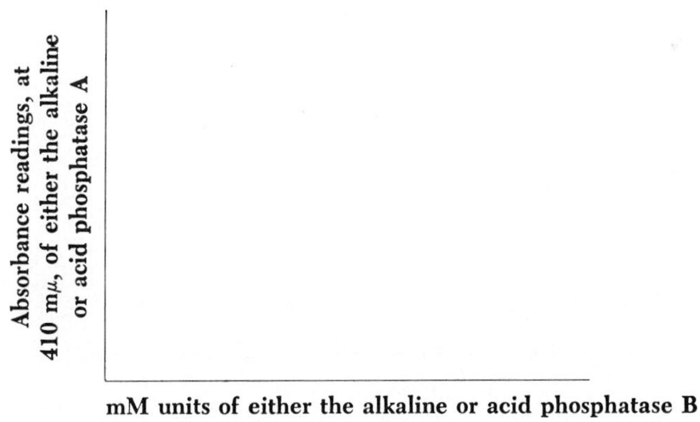

A chart made from the readings of the curves makes the calculations much easier and more convenient.

TRANSAMINASE

In 1940, an enzyme preparation was obtained from pig heart muscle and pigeon breast muscle. It was noted that in this preparation properties of both aspartic aminopherase and glutamic aminopherase were exhibited, thereby suggesting that only one enzyme existed. It had been noted previously that the differences were related primarily to activity rather than specificity. Therefore, the name suggested for the enzyme was *transaminase*. By utilizing tissue extracts, it was shown that the transaminase is largely limited to two reactions that can be demonstrated.[10]

Reaction 1 (SGOT)

Serum glutamic oxalacetic transaminase catalyzes the transfer of an amino group from aspartic acid to α-ketoglutaric acid to form glutamic and oxalacetic acids:

$$l(+)\text{ Glutamic acid} + \text{Oxalacetic acid} \leftrightharpoons \alpha\text{-Ketoglutaric acid} + l(+)\text{ Aspartic acid}$$

Reaction 2 (SGPT)

Serum glutamic pyruvic transaminase catalyzes the transfer of an amino group from alanine acid to α-ketoglutaric acid to form glutamic and pyruvic acids:

$$l(+)\text{ Glutamic acid} + \text{Pyruvic acid} \leftrightharpoons \alpha\text{-Ketoglutaric acid} + l(+)\text{ Alanine}$$

The abbreviations used to designate the enzymes that catalyze the reversible transaminase reactions serve also to monogram the main products of the equilibrium reactions:

Serum glutamic oxalacetic transaminase	SGOT
Serum glutamic pyruvic transaminase	SGPT

The transaminase enzymes are found in the tissues of many organs. There are large amounts of the enzymes present in the heart muscle, skeletal muscle, brain, liver, and kidneys, in descending order of concentration. Deterioration of the organs causes a release of abnormal amounts of the transaminase enzyme into the blood. An elevated serum glutamic oxalacetic transaminase may be found in cases of myocardial infarction, various liver diseases, and occasionally in renal diseases.

Normally, the ratio of $\frac{\text{SGOT}}{\text{SGPT}}$ is 1.0 or over, except in viral or infectious hepatitis, wherein the ratio is less than 1.0. Then the SGPT value is higher than the SGOT value, and both values are increased.[11,12]

SGOT and SGPT[13]

Reitman-Frankel method

The procedures for the SGOT and SGPT utilize similar conditions with the exceptions that alanine instead of aspartic acid is used in the SGPT and only a 30-minute incubation period is necessary. Since the two tests may be run simultaneously, either of the following incubation methods may be chosen:

1. The SGOT and SGPT may be started together. Then, at the end of the 30-minute incubation period, the SGPT tubes are removed from the water bath and the hydrazine (color reagent) is added immediately. This stops the reaction and starts the color development. The tubes may sit at room temperature until the SGOT tubes have finished incubating or they may be completed.
2. The SGPT may be started 30 minutes after the SGOT.

The following method is written so that both tests may be done either simultaneously or individually.

The instrument may be calibrated and tests read at 505 mμ. It is most important that both the calibration and the test readings be done on the same wavelength.

*Reagents needed**

Phosphate buffer, 0.1 M, pH 7.4	(D-342)
SGOT substrate, α-ketoglutarate, 2 mM/L; *dl*-aspartate, 200 mM/L	(D-128)
SGPT substrate, α-ketoglutarate, 2 mM/L; *dl*-alanine, 200 mM/L	(D-130)
Color reagent, 2,4-dinitrophenylhydrazine, 1 mM/L	(D-116)
Sodium hydroxide, 0.4 N	(D-129)

Procedure

1. Including a reagent blank for the SGOT and for the SGPT, pipette 1.0 ml of the proper substrate into each tube.
2. Place the tubes in a 37° C water bath for 5 minutes.
3. Set the timer for 60 minutes. Start timing when 0.2 ml of distilled water is added to the SGOT or SGPT reagent blank, whichever group is started first. Mix by swirling and replace tube in water bath.
4. At ½-minute intervals, pipette 0.2 ml of serum into each tube. Mix by swirling. Replace tube in water bath.
5. After incubating *exactly* 30 minutes, remove the SGPT tubes from the water bath at ½-minute intervals and *immediately* add 1.0 ml of color reagent to stop the reaction and start the color development.
6. After incubating *exactly* 60 minutes, remove the SGOT tubes from the water bath at ½-minute intervals and *immediately* add 1.0 ml of color reagent to stop the reaction and start the color development.
7. After a minimum of 20 minutes of standing with the color reagent (hydrazine), add 10.0 ml of 0.4 N NaOH. Stopper and mix by inverting the tubes. The final color develops in alkaline solution.
8. After standing 5 minutes, read at a wavelength of 505 mμ against the water blank. The blank tube should read very close to the same value on each run.
9. Calculate results from the calibration curve. If a value exceeds the upper limit of the curve, dilute serum with saline and multiply by the dilution made. For example, with a 1:10 dilution, multiply by 10.

Normal range

SGOT	8 to 40 units
Doubtful	40 to 50 units
SGPT	5 to 35 units
Doubtful	35 to 45 units

Calibration curve

The standards for the calibration curve are *not* incubated. The curve may be plotted either with optical density against the activity, on linear graph paper, or with the percent transmittance against the activity, on semilog paper. The units of the SGPT and SGOT are to be shown in the chart at 37° C and 40° C. The method was presented for a 37° C water bath, since the other enzyme tests in this chapter are calibrated for the use of a 37° C water bath.

Reagents needed

Pyruvate standard, 2 mM/L (D-237)

The remaining reagents are those used in the methods for the SGOT and SGPT.

Procedure

1. Pipette the following volumes into the tubes. Refer to Table 11-4.
2. Add 1.0 ml of color reagent to each tube. Mix.
3. Pipette 10.0 ml of 0.4 N NaOH to each tube. Stopper and invert to mix.

*May be purchased from Sigma Chemical Co.

Table 11-4. Transaminase calibration

Tube	Pyruvate (ml)	Substrate (ml)	H_2O (ml)	1 Absorbance % T	2 SGOT 37° C	2 SGOT 40° C	3 SGPT 37° C	3 SGPT 40° C
1	0	1.0	0.2		0	0	0	0
2	0.1	0.9	0.2		24	21	28	25
3	0.2	0.8	0.2		61	54	57	51
4	0.3	0.7	0.2		114	102	97	87
5	0.4	0.6	0.2		190	169	—	—

4. Wait 5 minutes; then read against a distilled water blank, using a wavelength of 505 mμ.
5. Record the readings of the five tubes in column 1.
6. Using the alkaline phosphatase as an example, plot the readings from column 1 against those of column 2 or column 3.
7. A chart made from the curves drawn allows for easier and consistent calculations.

LACTIC DEHYDROGENASE (LDH)[14]

Lactic dehydrogenase is an enzyme that is found in almost all tissues. For this reason, a number of pathologic conditions can cause an elevation of LDH activity. LDH activity is of particular interest as an aid in the diagnosis of myocardial infarction, hepatitis, and cirrhosis.

LDH catalyzes the oxidation of lactate to pyruvate. In this reaction NAD is reduced to NADH which is coupled to the reduction of a tetrazolium salt, 2-*p*-iodophenyl-3-*p*-nitrophenyl-5-phenyl tetrazolium chloride (INT), with phenazine methosulfate (PMS) serving as an intermediate electron carrier. The bright red formazan produced is measured at 520 mμ.

$$\text{Lactate} + \text{NAD} \xrightleftharpoons{\text{LDH}} \text{NADH} + \text{Pyruvate}$$

$$\text{NADH} + \text{PMS} \longrightarrow \text{NAD} + \text{Reduced PMS}$$

$$\text{Reduced PMS} + \text{INT} \longrightarrow \text{PMS} + \text{Reduced INT (red color)}$$

Reagents needed

LDH color reagent	(D-72a)
LDH buffer reagent	(D-73)
LDH substrate	(D-74)
LDH control reagent	(D-75)
Hydrochloric acid, 0.1 N	(D-37)

Working standards

Commercial lyophilized enzyme standards such as Versatol-E (V-E) and Versatol-E-N (V-E-N) are used to prepare a standard curve.

Example:

Tube	V-E-N (ml)	V-E (ml)	Units	Example
1	0.5	—	Given value	50
2	1.5	0.5	Values (1* + 3*) ÷ 2	88
3	0.5	0.5	Values (1* + 5*) ÷ 2	125
4	0.5	1.5	Values (3* + 5*) ÷ 2	163
5	—	0.5	Given value	200

*Tube number.

These standards are set up and read in exactly the same manner as the unknown serum specimens.

Procedure

1. Pipette 0.1 ml of serum and 0.2 ml of buffer reagent into each of two tubes, one labelled "test" and the other, "control."
2. Add 0.5 ml of substrate to the "test" and 0.5 ml of control reagent to the "control." Mix and place in a 37° C water bath for 5 minutes to warm.
3. At convenient but precisely timed intervals, pipette 0.2 ml of color reagent into each tube, mix, and return immediately to the water bath. Incubate for exactly 5 minutes.
4. Add 5 ml of 0.1 N HCl to each tube in the same sequence and time intervals as in step 3. Mix well.
5. Read absorbance at 520 mμ against distilled water within 20 minutes.

Calculations

Subtract the absorbance of the control from that of the test for standards and unknown sera. Plot the corrected absorbances of the standards versus LDH activity on linear graph paper. Determine the LDH activity of the unknowns from the standard curve, using the corrected absorbances.

Chemical fractionation of lactic dehydrogenase

Elevation in total LDH activity can be caused by a variety of pathologic conditions. Fractionation of the total activity into proportional amounts of isoenzyme activity has been demonstrated by a variety of techniques. However, it is important to remember how the isoenzyme activities make up the total LDH activities. In any procedure, the proportion of the total LDH activity is at best a measurement of the grouped activities of all five isoenzymes. The proportion of the total activity that is detected by the isoenzyme separation technique is a formation of the following three parameters:

1. Inherent chemical characteristics of each individual isoenzyme as related to temperature, pH, substrate concentration, buffer, coenzyme concentration, and so forth
2. Total reaction conditions for the assay
3. Relative amount of each isoenzyme in the total LDH activity

Combinations of the three parameters listed will regulate the proportional amount of isoenzyme determined.

The chemical inhibition of LDH heart and liver tissue isoenzymes has been studied by Babson.[15] Chemical inhibition has also been used to assess LDH isoenzymes in serum. This procedure separates the slow or hepatic type isoenzymes from the fast or cardiac type isoenzymes. To inhibit the activity of the electrophoretically fast migrating isoenzymes, 2 M lactate is used as a substrate.

The slow migrating LDH enzymes are similarly inhibited when the substrate consists of 0.02 M lactate and 2 M urea. The simultaneous use of these two substrates provides an *isoenzyme index* (I.I.), which can reflect an elevation of fast, slow, or intermediate LDH isoenzymes.

The chemical inhibition method indicates groupings of predominating isoenzymes that do correlate to the fast, slow, and intermediate electrophoretically migrating isoenzymes. The isoenzyme index may be divided into three groups of LDH isoenzyme information. First, an isoenzyme index (I.I.) of less than 0.80 indicates a predominance of the faster moving isoenzymes. Second, an isoenzyme index of greater than 1.20, indicates predominance of the slower moving isoenzymes. Finally, there may be a combination of both the fast and the slow moving isoenzymes, *or* a predominance of the intermediate moving isoenzymes. In either case, an isoenzyme index of 0.80 to 1.20 may be expected. Therefore, when the total serum LDH is elevated, an isoenzyme index in this mid range must be suspect for one of these latter two possibilities. An abnormal total LDH and an intermediate type isoenzyme index can reflect simultaneous cardiac and hepatic tissue damage, or diseases that produce isomorphic LDH_3 elevations.

The isoenzymes are divided into three groups and their occurrences are given below:

Group I (I.I. below 0.80). Tissues belonging to this group are distinguished by a high percentage of LDH_1 and LDH_2 and consequently a low percentage of LDH_4 and LDH_5. This fraction occurrence is seen in heart muscle but also is found in kidneys, brain, and erythrocytes.

Group II (I.I. above 1.20). Tissues in this group possess high percentages of LDH_4 and LDH_5, whereas LDH_1 shows the smallest percentage. This isoenzyme group is found in liver tissue, skeletal muscle, skin, and leukocytes.

Group III (I.I. 0.80 to 1.20). The vast majority of tissues belong to this group. As its distinctive feature, this group shows maximum activity in the middle fractions—LDH_2, LDH_3, and LDH_4. Tissues giving these middle elevations are lungs, spleen, ureter, uterus, pancreas, prostate, ovary, gallbladder, appendix, connective tissue, duodenum, thyroid gland, and lymph gland.

Reagents needed

Reagents listed for total lactic dehydrogenase	
LDH color reagent	**(D-72a)**
LDH buffer reagent	**(D-73)**
Hydrochloric acid, 0.1 N	**(D-37)**
***l*(+) Lactate, 2 M**	**(D-76)**
***l*(+) Lactate, 0.02 M to 2 M urea**	**(D-77)**

Procedure

1. **Into each of two tubes place 0.2 ml of LDH buffer reagent.**
2. **Add 0.5 ml of the heart-inhibiting substrate to one tube ("H" tube) and 0.5 ml of the liver-inhibiting substrate to the other ("L" tube).**
3. **Add 0.1 ml of serum to each tube, mix, and incubate 10 minutes at 37° C to provide the inhibitory effects.**
4. **Following this, add 0.2 ml of color reagent—that is, the NAD, the intermediate electron carrier, and the tetrazolium salt—to each tube and continue the incubation for exactly 5 minutes.**
5. **Terminate the reaction by adding 5 ml of 0.1 N HCl.**
6. **Read the absorbance of each tube at 520 nm with a distilled water blank.**

Calculation

$$\text{Isoenzyme index} = \frac{\text{A of "H" tube}}{\text{A of "L" tube}}$$

Interpretation

I.I.	*LDH isoenzyme fractions*
I. Below 0.80	Elevation of fast fractions, LDH_1 and LDH_2 —"cardiac"
II. Above 1.20	Elevation of slow fractions, LDH_4 and LDH_5 —"hepatic"
III. 0.80 to 1.20	a. Both fast and slow fractions elevated b. Intermediate fraction elevated LDH_3

CREATINE PHOSPHOKINASE (CPK)[16]

Because of the tissue specificity of creatine phosphokinase, there has been an increased interest in the determination of this enzyme as an aid in the diagnosis of acute myocardial infarction and primary muscle disease. High concentrations are found only in heart muscle, skeletal muscle, and brain. Only negligible activity is found in liver, pancreas, kidney, lung, or erythrocytes. Serum CPK activity elevates earlier, following myocardial infarctions, than either LDH or GOT.

In the method given, the forward CPK reaction is coupled to the pyruvate kinase reaction. The pyruvate formed is measured colorimetrically (at 440 mμ) through the formation of the hydrazone with 2,4-dinitrophenylhydrazine.

$$\text{ATP} + \text{Creatine} \xrightleftharpoons{\text{CPK}} \text{ADP} + \text{Creatine phosphate}$$

$$\text{ADP} + \text{Phosphoenolpyruvate} \xrightleftharpoons{\text{PK}} \text{Pyruvate} + \text{ATP}$$

Interferences are eliminated by the use of a serum blank.

Serum CPK activity is heat-labile. When serum is heated to 56° C for 15 minutes, there is a total loss of activity. Specimens will lose an average of 45% of their original activity over a period of 4 hours at room temperature. The addition of a sulfhydryl agent to the reaction mixture, however, will fully reactivate the enzyme. This has been found to be true of specimens kept up to 5 days at refrigerator temperatures. Several sulfhydryl agents have been used, such as cysteine and glutathione, but mercaptoethanol proved to be much more stable. The mercaptoethanol must be added separately and just before incubation since it is somewhat unstable at the pH of the reaction mixture.[16]

Hemoglobin concentration up to 1.25 g/100 ml and lipemia have no effect on the measurement of CPK activity with the use of a serum blank.

Nuttall and Wedin found that normal values ranged from about 6 to 25 IU for females and from 6 to 43 IU for males.

Results are expressed in international units (IU) per liter of serum; that is, μmoles of pyruvate (creatine phosphate) produced per liter of serum per minute at 37° C.

$$\frac{\mu\text{M pyruvate in tube}}{\text{30 minutes (incubation)}} \times 10{,}000^{*} = \mu\text{M/L/min (IU) CPK activity}$$

*0.1 ml of serum converted to 1 L.

Reagents needed

Reagent	
Sodium phosphate, dibasic, 0.1 M	(D-110)
Potassium phosphate, monobasic, 0.1 M	(D-111)
Phosphate buffer, pH 7.4	(D-112)
Pyruvic acid standard, 0.5 μmole/ml	(D-113)
CPK blank substrate	(D-114)
Creatine substrate	(D-115)
2,4-Dinitrophenylhydrazine, 0.001 M	(D-116)
Sodium hydroxide, 0.4 N	(D-129)
Mercaptoethanol, 0.25 M	(D-117)

Standard curve

1. To prepare a standard curve, set up the following tubes:

Tube	Standard (ml)	PO_4 buffer (ml)	μM/ml	IU
1	0.0	1.0	0.1	0
2	0.2	0.8	0.2	33
3	0.4	0.6	0.3	67
4	0.6	0.4	0.4	100
5	0.8	0.2	0.5	133
6	1.0	0.0	0.6	167

2. Add 1.0 ml of 2,4-dinitrophenylhydrazine to each tube, mix, and allow to stand at room temperature for 20 minutes.
3. Add 10 ml of 0.4 N NaOH and allow to stand for 5 minutes.
4. Read at 440 mμ using tube 1 as the reagent blank.
5. Plot optical density against concentration on linear graph paper.

Procedure

1. Set up two tubes for each unknown serum to be tested. Label one tube "test" and the other "blank."
2. Pipette 0.1 ml of 0.25 M mercaptoethanol into the bottom of each tube.
3. Pipette 1.0 ml of creatine substrate into each "test" tube and 1.0 ml of CPK blank substrate into each "blank" tube and one tube to be used as a reagent blank.
4. Place all tubes in a 37° C water bath for 5 minutes to warm.
5. At timed intervals, add 0.1 ml of serum to each of the two substrates and mix. Exactly 30 minutes later, add 1.0 ml of 2,4-dinitrophenylhydrazine to stop the reaction.
6. Mix and allow to stand at room temperature for 20 minutes.
7. Add 10 ml of 0.4 N NaOH, mix, and allow to stand for 5 minutes.
8. Read each tube against the reagent blank at 440 mμ.
9. Subtract the "blank" A from the "test" A and determine the CPK activity from the standard curve.

Hereditary nonspherocytic hemolytic anemia

SISTER ANN MARIE SKINNER, R.S.M., M.A., MT(ASCP)

The term "hereditary nonspherocytic hemolytic anemia" has been used as a catch-all category for a heterogeneous group of congenital hemolytic disorders that could not be made to fit into clearly defined syndromes. On the basis of response to the osmotic fragility, autohemolysis test, and morphologic characteristics, Selwyn and Dacie[17] classified these anemias into type I and type II.

A new classification of these disorders is based on the presence or absence of a demonstrable defect in glucose metabolism and its location in the pentose phosphate pathway or the Embden-Meyerhof pathway. The clinical severity of hemolysis varies from marked anemia requiring repeated transfusions to a complete compensation by moderately increased erythropoiesis. All have in common a hemolytic process on the basis of an intracorpuscular abnormality, absence of appreciable spherocytosis, and persistence of hemolysis after splenectomy.

G-6-PD deficiency

The defect most common in the pentose phosphate pathway is glucose-6-phosphate dehydrogenase deficiency (G-6-PD deficiency). Studies of red cell G-6-PD may be required in a variety of circumstances, including the following:

1. Hemolytic disease induced by oxidant drugs such as primaquine or sulfa compounds; incidence high in the American Negro
2. Differential diagnosis of chronic hemolytic disease
3. Search for etiology of acute hemolysis
4. Investigation of hemolytic disease of newborn
5. Population studies of frequency of a G-6-PD variant
6. Genetic investigations of G-6-PD inheritance

The most frequent clinical indication for study is a hemolytic episode suspected to be caused by G-6-PD deficiency. This deficiency corresponds to type I of Selwyn and Dacie.

Simple assay method for G-6-PD

Principle. There is an increase in absorbance at 340 mμ when triphosphopyridine nucleotide (TPN) is changed to reduced triphosphopyridine nucleotide (TPNH). Such reduction takes place when two electrons are transferred from glucose-6-phosphate (G-6-P) to TPN in the reaction catalyzed by G-6-PD. Mg ion is added to stabilize the enzyme:

$$\text{G-6-P} + \text{TPN} \xrightarrow[\text{Mg}^{++}]{\text{G-6-PD}} \text{6PG} + \text{TPNH}$$

Blood samples. Fresh blood samples drawn in EDTA are preferable. Blood drawn from a normal person at the time blood is drawn from the patient serves as a control.

Preparation of hemolysate. Blood is centrifuged and the plasma, buffy coat, and upper 10% of packed red cells are removed by aspiration. Cells are washed twice with 0.9% saline each time removing residual buffy coat and upper 10% of packed red cells. One-tenth milliliter (0.1 ml) of packed red cells is added to 9.9 ml of cold distilled water and allowed to hemolyze completely. This gives a 1:100 dilution of packed red cells. The hemoglobin content of this hemolysate is determined.

NOTE: It is expedient to use the method and hemoglobin chart set up in the hematology laboratory for determination of hemoglobin. For example, if hemoglobin is measured by the Drabkin method, in which 0.02 ml of blood is added to 10 ml of Drabkin solution, a 1:500 dilution is measured and the reading is charted as grams per 100 ml of blood. Therefore, if 1 ml of the 1:100 hemolysate is added to 10 ml of Drabkin solution, a 1:11 further dilution is obtained. Since 500 divided by 11 equals 45.5, one should divide the hemoglobin reading from the chart by 45.5 to obtain the hemoglobin in grams per 100 ml of 1:100 hemolysate.

Assay procedure. Each specimen is set up as follows:

Reagents	*Concentration*	*Volume (ml)*	*Final concentration (mM)*
G-6-P	6.25 mM	0.4	0.1
Tris buffer (pH 7.4)	250 mM	1.0	100
TPN	12.5 mM	0.1	0.5
$MgCl_2$	125 mM	0.2	10
H_2O	——	0.5	——
Hemolysate	1:100	0.3	
		2.5 total volume	

Water is substituted for hemolysate in the blank. The reaction is started by the addition of hemolysates containing the enzyme. The optical density (absorbance) is followed in cuvettes with a 1 cm light path at 340 mμ and at 25° C.

Calculation of results. After 2 to 3 minutes the rate of change becomes linear. Readings are taken each minute until this linear change per minute is established. One micromole per milliliter of TPNH has an absorbance of 6.22 in a light path of 1 cm at 340 mμ. Therefore

$$\Delta OD_{340}/\text{min} \div 6.22 \times 2.5 \text{ (vol)} \div \text{Hemoglobin in 2.5 ml}$$

equals micromoles of TPNH formed per minute per gram of hemoglobin. This is a measure of the amount of G-6-PD catalyzing the reaction.

Normal range. 4 to 7 μM of TPNH formed per minute per gram of hemoglobin.

NOTE: If the test cannot be run immediately, the blood should be refrigerated until time is available. Hemolysate should be made just prior to running the assay.

Pyruvate kinase deficiency

Pyruvate kinase deficiency is an enzymatic defect in the main pathway of glycolysis. Over a hundred cases in the United States have been reported; it is most common in people of Northern European origin and in Amish groups but has been found in Italians, Japanese, Mexicans, and American Negroes. Both sexes are equally affected. The majority of PK-deficient patients require transfusions until after splenectomy. The hemolytic anemia persists after splenectomy and the reticulocyte count rises to very high levels, but the hemoglobin is maintained at a comfortable level (7 to 9 g) without transfusions. Due to the high numbers of reticulocytes the red cells appear macrocytic. Clinical disease is usually limited to individuals with homozygous pyruvate kinase deficiency. Heterozygous persons usually have about half the normal PK activity but are usually free of clinical symptoms. The PK deficiency corresponds to the Dacie and Selwyn classification of type II.

Pyruvate kinase assay

Principle. Pyruvate kinase catalyzes the conversion of phosphoenolpyruvate (PEP) to pyruvate with regeneration of adenosine triphosphate (ATP). The assay procedure employs phosphoenolpyruvate as substrate, buffer, adenosine diphosphate (ADP), potassium chloride (KCl), magnesium sulfate ($MgSO_4$), lactic dehydrogenase (LDH), reduced diphosphopyridine nucleotide (DPNH), and hemolysate.

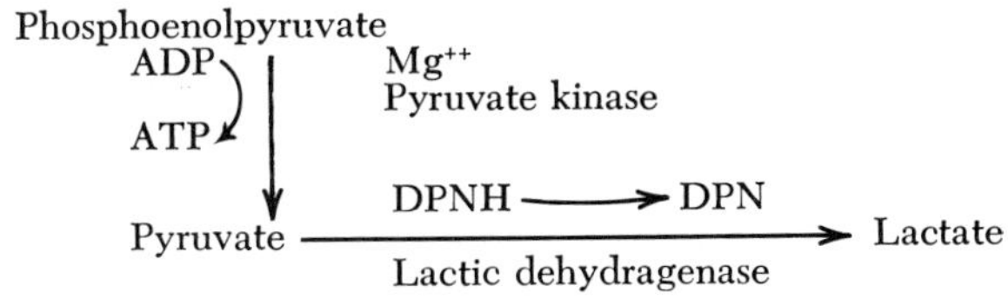

Blood sample. Same as for G-6-PD.
Hemolysate preparation. Same as for G-6-PD.
Assay procedure. Each specimen is set up as follows:

Reagents	*Concentration*	*Volume (ml)*	*Final concentration (mM)*
Triethanolamine buffer (pH 7.4)	125 mM	1	50
KCl	937.5 mM	0.2	75
$MgSO_4$	100 mM	0.2	8
ADP	5 mM	0.6	1.2
DPNH	5 mM	0.1	0.2
LDH	0.02 mg/0.1 ml	0.1	
PEP	75 mM	0.1	3
Hemolysate	1:100	0.2	
		2.5 total volume	

Water is substituted for hemolysate in the blank. The reaction is started by the addition of the hemolysate containing the enzyme. The absorbance is followed in cuvettes with a 1 cm light path at 340 mμ and at 25° C.

Calculation of results. Since DPNH has the same absorbance as TPNH, the results are calculated in the same way as G-6-PD assay. Therefore

$$\Delta OD_{340}/\text{min} \div 6.22 \times 2.5\ (\text{vol}) \div \text{Hemoglobin in 2.5 ml}$$

equals micromoles of DPN formed per minute per gram of hemoglobin. This is a measure of the amount of pyruvate kinase catalyzing the reaction.

Normal range. 4 to 8 μM of DPN formed per minute per gram of hemoglobin.

NOTE: Since pyruvate kinase is a labile enzyme, the test should be run as soon as possible after blood is collected. Blood and hemolysate should be kept in ice.

REFERENCES

1. Somogyi, M.: Micromethods for estimation of diastase, J. Biol. Chem. **125**:399, 1938.
2. Sachar, L.: Simplified laboratory test for acute pancreatitis, Amer. J. Clin. Path. **22**: 1117, 1952.
3. Kleiner, I. S., and Orten, J. M.: Biochemistry, ed. 6, St. Louis, 1962, The C. V. Mosby Co.
4. Woodard, H. Q.: The clinical significance of serum acid phosphatase, Amer. J. Med. **27**: 902, 1959.
5. Gutman, A. B., Tyson, T. L., and Gutman, E. B.: Serum calcium, inorganic phosphorus and phosphatase activity in hyperparathyroidism, Paget's disease, multiple myeloma, and neoplastic disease of bones, Arch. Intern. Med. **57**:379, 1936.
6. Bessey, O. A., Lowry, O. H., and Brock, M. J.: Method for rapid determination of alkaline phosphatase with 5 cubic millimeters of serum, J. Biol. Chem. **164**:321, 1946.
7. Ozar, M. B., Isaac, C. A., and Valk, W. L.: Methods for elimination of errors in serum acid phosphatase determinations, J. Urol. **74**: 150, 1955.
8. Sigma Technical Bulletin 104, St. Louis, 1963, Sigma Chemical Co.
9. Hock, E., and Tessier, R. N.: Elevation of serum acid phosphatase following prostatic massage, J. Urol. **62**:488, 1949.
10. Sobotka, H., and Steward, C. P.: Advances in clinical chemistry, vol. 1, New York, 1958, Academic Press, Inc.
11. DeRitis, F., Coltorti, M., and Giusti, G.: An enzyme test for the diagnosis of viral hepatitis; the transaminase serum activities, Clin. Chim. Acta **2**:70, 1957.

12. Sigma Technical Bulletin 505, St. Louis, 1964, Sigma Chemical Co.
13. Reitman, S., and Frankel, S.: A colorimetric method for the determination of serum glutamic oxalacetic and glutamic pyruvic transaminases, Amer. J. Clin. Path. **28**:36, 1957.
14. Babson, A. L., and Phillips, G. E.: A rapid colorimeter assay for serum lactic dehydrogenase, Clin. Chim. Acta **12**:210, 1965.
15. Babson, A. L.: The chemical differentiation of tissue lactic dehydrogenase, Clin. Chim. Acta **16**:121, 1967.
16. Nuttal, F. Q., and Wedin, D. S.: A simple rapid colorimetric method for determination of creatine kinase activity, J. Lab. Clin. Med. **68**:324, 1966.
17. Selwyn, J. G., and Dacie, J. V.: Autohemolysis and other changes resulting from the incubation in vitro of red cells from patients with congenital hemolytic anemia, Blood **9**:414, 1954.

• 12

• Digestion and absorption

The human digestive system performs three basic functions: ingestion, digestion, and absorption.

Ingestion is the basic mechanics of taking food into the mouth, grinding and breaking the material into smaller bits, and passing the material on by swallowing. Assisting the masticated material on its way are secretions of the parotid, submaxillary, sublingual, and buccal glands that produce about 1.5 L of saliva each day. There are two types of saliva: (1) a very fluid material that aids in diluting and dissolving dry food and (2) a more viscous fluid that contains mucus for lubricating food as it moves down the esophagus. The saliva acts also as a lubricant in the mouth, preventing the tongue from adhering to the roof and aiding in proper speech.

Saliva contains the digestive enzyme ptyalin, which attacks starch molecules and which would, if allowed to proceed with the complete reaction, convert the starch to maltose. Ptyalin functions well at the pH of the saliva (6.5 to 6.8) but is inactive at the acid condition of the stomach. However, food acts as a buffer in the stomach allowing ptyalin action until the acid of the stomach has worked through the mass of food.

Muscle contraction (peristalsis) in the wall of the esophagus and throughout the gastrointestinal tract moves food through the system. It requires about 6 seconds for solid food to reach the stomach from the mouth and by force of gravity liquid moves at a more rapid rate.

Digestion takes place in the stomach, a thick-walled muscular sac located on the left side of the body just under the lower ribs (Fig. 12-1). The size of the stomach changes as food is received and digested. The maximum capacity is about 2.5 L for the adult. The stomach lining (mucosa) contains millions of microscopic gastric glands that secrete gastric juice containing hydrochloric acid and the zymogen pepsinogen, which is converted to pepsin by the hydrochloric acid and autocatalytically. The gastric juice has a pH of approximately 1, but when the food contents are in the stomach, there is a pH of about 3. Gastric digestion is mainly concerned with the action of pepsin on proteins in the presence of hydrochloric acid, which is preparatory to the more complete hydrolyses that take place in the intestines. The rennin present causes milk to curdle, and the lipase of gastric juice shows little activity except on previously emulsified fats.

After the contracting action of the stomach, the food material has the consistency of a cream and the digestion is well advanced. The digested food from the stomach passes through the pylorus in regular spurts into the duodenum as chyme. This requires from 1 to 4 hours, dependent upon the volume and type of food intake. The sequence in which food types leave the stomach is carbohy-

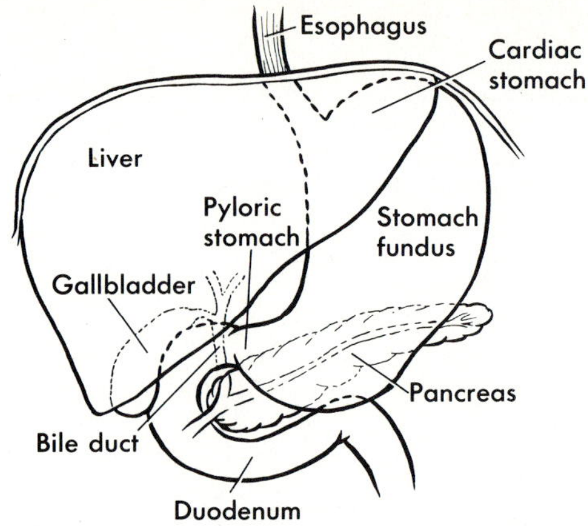

Fig. 12-1. Diagram of relations of stomach, liver, pancreas, and duodenum.

drates, proteins, fats. Hunger pangs result from the stomach contracting when empty.

The duodenum is the first segment (about 10 inches) of the small intestine. While the small intestine and most of the large intestine are mobile because of their mesentery connection to the dorsal wall, the duodenum is tightly secured in the abdominal cavity by ligaments that connect it to the liver, stomach, and dorsal wall. In the duodenum the chyme is mixed with digestive juices from the pancreas and with bile from the gallbladder. The principal digestive enzymes from the pancreas are (1) trypsin and chymotrypsin, which complete the breakdown of proteins to peptides, (2) peptidases that convert peptides to amino acids, (3) lipase, which breaks down fats, and (4) amylase, which finishes the conversion of starch to maltose. The alkaline (pH approximately 8.5) pancreatic secretions aid in adjusting the pH of the chyme to provide an optimum pH for action of the enzymes. About 500 ml of pancreatic juice is produced daily.

The liver is important to digestion, since it contributes bile to the digestive juice. The secretion excreted from the liver contains water, bile salts, bile pigments, inorganic salts, and a mixture of lipid materials including fats, cholesterol, and lecithin.

Bile is formed in all parts of the liver and is emptied into interlobular bile ducts which form larger ducts until two main ducts unite in the portal fossa and form the hepatic duct. The hepatic duct passes downward and joins the duct from the gallbladder, the cystic duct. The hepatic and cystic ducts form the common bile duct that passes downward and enters the duodenum below the pylorus. Bile is stored in the gallbladder until it is needed and there it is concentrated by removal of water and salts. Although the bile is secreted constantly, it passes to the duodenum only after food has been eaten. The muscular wall of the gallbladder contracts, passing the bile down the duct. Being alkaline, bile aids in the neutralization of the chyme.

The bile contributes to digestion by the action of the bile salts, since it is not a digestive enzyme. The bile salts act as emulsifying agent so that fat droplets within the chyme are formed during the churning motion of the intestine. With the fat broken into droplets greater surface area is provided for action of the lipase (fat-splitting enzyme) in effectively digesting the fat. When the bile duct is obstructed, the absence of bile salts in the intestine impairs digestion of the fat and causes much fat to be lost in the feces. The bile salts are reabsorbed from the lower part of the intestine and returned to the liver.

In the small intestine some amylase, lipase, and enzymes that break down disaccharides to monosaccharides are present. However, the main activity in the small intestine is the absorption of products of digestion. Food absorption is a complex process involving in part the diffusion of substances from the intestine through the lining cells into the blood and lymph capillaries.

Most of the food is absorbed in the small intestine and the residue is moved on into the large intestine. Since most of the water is absorbed, the waste material takes on a solid form. During this period bacterial activity occurs. By fermentation and putrefaction various gases (carbon dioxide, hydrogen, nitrogen, methane, hydrogen sulfide) and acids (acetic, butyric, lactic) are produced. The large intestine contains a significant amount of ammonia, presumed to be a product of the putrefaction of nitrogenous substances by bacteria. Bacteria decomposing lecithin may produce choline and related toxic amines such as neurine.

From 12 to 24 hours is required for the waste to pass through the colon and rectum. The resulting product, called feces, contains large quantities of bacteria (approximately one half of the mass), heavy metals, bile pigments and other substances secreted by the body, and indigestible remnants of food.

Gastric analysis

Gastric analysis consists of obtaining and examining the contents of the stomach after its activation by a suitable stimulant.

In health, the gastric contents are mostly water, salts, pepsin, and hydrochloric acid. In disease, the gastric contents may contain blood, bacteria, and lactic acid. The significant finding in disease, however, is a change in the hydrochloric acid content.

A decreased amount of hydrochloric acid is usually found in cancer of the stomach and in pernicious anemia. An increased amount of hydrochloric acid is generally seen in gastric ulcer.

COLLECTION OF SPECIMENS

The stomach is stimulated to activity by one of the methods given below. Directions for inserting the tube and removing the specimens follow.

Stimulation of stomach

Gastric secretion may be stimulated in a number of ways. The composition of each "test meal" discussed is given below. The alcohol consumption method is widely used and will be the one considered in the procedure that follows.

Ewald test meal

The Ewald test meal consists of either a roll or two slices of bread and 400 ml of water. Weak tea without cream or sugar may be substituted for the water. The meal should be well masticated. The gastric contents are removed 1 hour after the beginning of the meal.

Riegel test meal

The Riegel test meal consists of 150 g of mashed potatoes, 200 g of broiled beefsteak, and 400 ml of bouillon. The meal should be well masticated.

Alcohol consumption

The alcohol is generally given after insertion of the tube. The usual amount is 50 ml of a 7% solution of ethyl alcohol.

Histamine injection

Histamine is a powerful stimulant and is administered by a physician. It is usually given when the alcohol "test meal" fails to produce free hydrochloric acid. The dose depends upon the preference of the physician. Some physicians give 0.25 mg of histamine dihydrochloride or acid phosphate (0.25 ml of a 1:1000 dilution). Other physicians give 0.1 mg for every 22 pounds of body weight. The injection is given *subcutaneously*.

Insertion of tube

The stomach tubes most commonly used are Rehfuss, Kaslow, and Sawyer tubes (Fig. 12-2). The insertion of the tube requires a bit of technique. A suggested method follows.

Place an ordinary kitchen apron on the patient to prevent soiling his clothes.

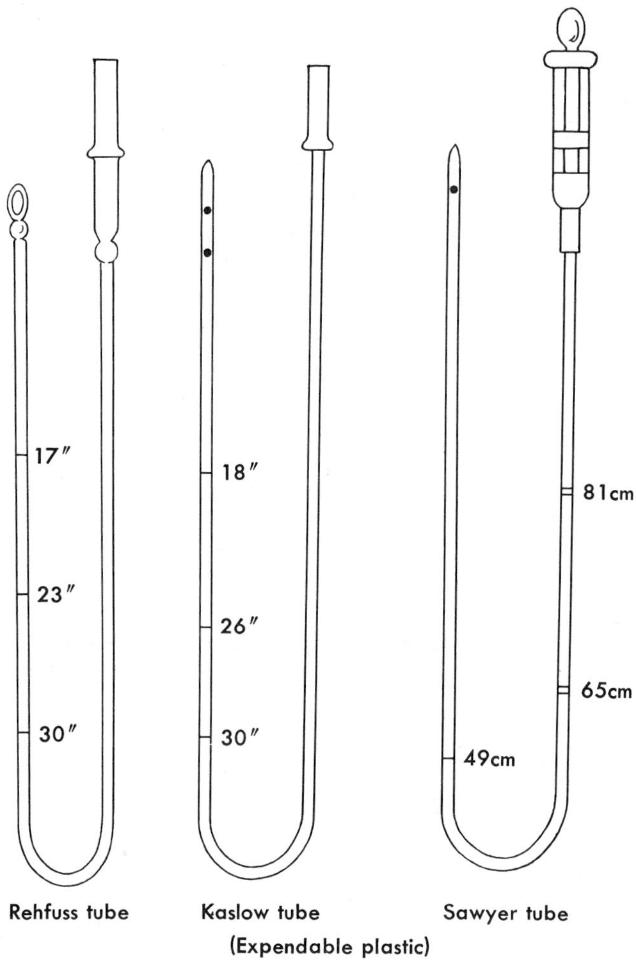

Fig. 12-2. Stomach tubes.

Give him a small emesis basin to expectorate into, if necessary. Explain that you have to insert the tube into his stomach and that his cooperation is essential. Also explain that there will be no pain but some discomfort in swallowing the tube.

Coil the tube in your left hand and take the tip of the tube in your right hand. Ask the patient to open his mouth and tilt his head back. Place the tip of the tube on the middle of the tongue and push slowly back until it is almost to the pharynx. Now tell the patient that you are going to insert it into the pharynx and that it may gag him a little but that he can help by swallowing. Thrust the tube gently but firmly into the pharynx. Tell the patient to swallow four to six times, and each time he swallows push the tube down the esophagus.

The tube should be inserted until the first mark on the tube is even with the teeth. At this point an attempt should be made to remove the stomach contents with a large syringe. If nothing can be removed, push the tube into the stomach until the second marking is even with the teeth and attempt to remove the contents. NOTE: If you think there is any danger that the tube has gone into the lungs rather than into the stomach, hold your end of the tube to your ear. If the inhalation and expiration of air can be heard, the tube is in the lungs and, of course, must be withdrawn.

Removal of specimens

The Sawyer tube is made with the syringe attached. However, with a Rehfuss or a Kaslow tube, a large syringe must be attached to your end of the tube. Now continue with the steps given below.

Procedure

1. **Withdraw all the fasting contents of the stomach by slowly pulling back the plunger of the syringe. Transfer to a container and label as fasting specimen.**
2. **Slowly inject 50 ml of a 7% solution of *ethyl* alcohol. Record the time.**
3. **Fifteen minutes after the injection of the alcohol, withdraw all the available contents and transfer to another container. Label: 15-Minute Specimen.**
4. **Thirty minutes after the injection of the alcohol, withdraw all the available contents and transfer to another container. Label: 30-Minute Specimen.**
5. **Put 2 drops of Töpfer's reagent in the 15-minute specimen. Also put 2 drops of Töpfer's reagent in the 30-minute specimen. If neither specimen turns red, it means that free hydrochloric acid is absent. In such cases inform the physician, as he may wish to give a histamine injection. This, of course, is a more powerful gastric stimulant than alcohol and will produce the greatest flow of gastric juice.**
6. **If the histamine injection is given, withdraw specimens 15 minutes and 30 minutes after the injection and label accordingly.**

FREE HYDROCHLORIC ACID

Reagents needed

Töpfer's reagent	**(D-238)**
Sodium hydroxide, 0.1 N	**(D-233)**

Procedure

1. **Treat each specimen as follows.**
2. **Measure and record the volume. Remove suspended material by centrifuging or by filtering through gauze or cotton.**
3. **Pipette 5 ml of the clear fluid into a small flask or beaker. Add a few**

drops of Töpfer's reagent. If the specimen does not turn red upon the addition of Töpfer's reagent, make a note that free hydrochloric acid is absent and titrate for total acidity as directed in the discussion of total acidity below.

4. If the gastric contents turn red when Töpfer's reagent is added, get a 2 ml burette, graduated in hundredths, and a bottle of 0.1 N sodium hydroxide. Titrate with the 0.1 N sodium hydroxide until the red color is replaced by a canary yellow color. Record the number of milliliters of 0.1 N sodium hydroxide used. *Save the specimen,* as it will be used later to determine the total acidity.

Calculation

The report is given in degrees of acidity. The degrees of acidity are the number of milliliters of 0.1 N sodium hydroxide that *would* be required to titrate 100 ml of gastric contents to the canary yellow color. Since only 5 ml of gastric contents is actually used, we must multiply by 100/5:

$$\text{Free HCl in degrees of acidity} = \text{ml of 0.1 N NaOH used} \times \frac{100}{5}$$

Example: If 0.8 ml of the 0.1 N sodium hydroxide was used to titrate the 5 ml of gastric contents, the degrees of acidity are found as follows:

$$\begin{aligned}\text{Free HCl in degrees of acidity} &= \text{ml of 0.1 N NaOH used} \times \frac{100}{5}\\ &= 0.8 \times \frac{100}{5}\\ &= 16\end{aligned}$$

Normal values

The normal values are 5 to 20 degrees for the fasting specimen and 20 to 70 degrees for specimens obtained after stimulation with alcohol or histamine. Decreased values may be found in cancer of the stomach and in pernicious anemia. Increased values may be seen in gastric ulcer. The total absence of free hydrochloric acid is referred to as achlorhydria.

TOTAL ACIDITY

Reagent needed

Phenolphthalein, 1% (D-208)

Procedure

After the free hydrochloric acid has been found in each specimen, treat each specimen as follows:

1. Add a few drops of a 1% alcoholic solution of phenolphthalein to the same specimen that was used for the free hydrochloric acid titration.
2. Continue to titrate with the 0.1 N sodium hydroxide, stopping when the mixture turns pink.
3. Record the number of milliliters of 0.1 N sodium hydroxide used.

Calculation

The report is given in degrees of acidity. The degrees of acidity are the *total* number of milliliters of 0.1 N sodium hydroxide that *would* be required to titrate 100 ml of gastric contents to the pink color. Since only 5 ml of gastric contents is actually used, we must multiply by 100/5.

$$\text{Total acidity in degrees of acidity} = \frac{\text{Total number of ml}}{\text{of 0.1 N NaOH used}} \times \frac{100}{5}$$

Example: If 0.8 ml is used to titrate for free hydrochloric acid and another

Name: John Smith Date: 6-4-70

Specimen	Quantity	Free HCl	Total Acidity	Occult Blood
Fasting	20 ml	5 degrees	15 degrees	Neg.
15 min	15 ml	20 degrees	50 degrees	Neg.
30 min	10 ml	24 degrees	70 degrees	Neg.

Fig. 12-3. Sample gastric analysis report.

1.0 ml is used to bring the specimen to the pink color of phenolphthalein, the total number of milliliters used is 1.8 (0.8 plus 1.0 equals 1.8).

$$\text{Total acidity in degrees of acidity} = \frac{\text{Total number of ml}}{\text{of 0.1 N NaOH used}} \times \frac{100}{5}$$

$$= 1.8 \times \frac{100}{5}$$

$$= 36$$

A sample gastric analysis report is given in Fig. 12-3.

Normal values

The normal values are 15 to 45 degrees for the fasting specimen and 50 to 100 degrees for specimens obtained after alcohol or histamine stimulation. Decreased values may be found in cancer of the stomach and in pernicious anemia. Increased values may be seen in gastric ulcer.

OCCULT BLOOD

In ulcers and cancer of the stomach, the gastric contents may contain occult blood. The procedures for the benzidine test and the guaiac test are given below. In interpreting the results of these tests, it must be remembered that small traces of blood may be due to injury sustained in passing the stomach tube.

Benzidine test

Reagents needed

Benzidine, saturated, in glacial acetic acid	(D-239)
Hydrogen peroxide, 3%	(May be purchased)
Ethyl alcohol, 95%	(May be purchased)

Procedure

1. Place about 1 ml of the filtered gastric fluid into a small test tube.
2. Add about 10 drops of a saturated solution of benzidine in glacial acetic acid. Mix.
3. Add about 10 drops of 3% hydrogen peroxide. Mix.

Reaction

A green to blue color indicates a positive reaction.

Guaiac test

Reagent needed

Guaiac reagent

Prepare just prior to use, by placing 2 ml of 95% ethyl alcohol and 0.2 g of powdered guaiac into a test tube, adding 2 ml of 3% hydrogen peroxide, and mixing.

Procedure

1. Pour about 5 ml of filtered gastric contents into a small test tube and add a few drops of glacial acetic acid. Mix.
2. Tilt the tube and overlay the contents with the guaiac reagent.

Reaction

A green to blue color at the junction of the two liquids indicates a positive reaction.

MISCELLANEOUS TESTS

Miscellaneous tests that are performed on gastric secretions are discussed below and some of the procedures follow.

Microscopic examination

Make a smear of the fasting specimen. Allow to dry and stain with Gram's stain. Examine for Boas-Oppler bacilli and sarcinae. Discussion follows.

Boas-Oppler bacilli

Boas-Oppler organisms are large gram-positive rods (Fig. 12-4). They are found in cancer of the stomach but must be present in large numbers to be of any significance. The presence of the bacilli indicates gastric stagnation.

Sarcinae

Sarcinae are large gram-positive cocci arranged in groups that look like bales of cotton. They may be found in ulcers but must be present in large numbers to be of any significance. Their presence also indicates gastric stagnation.

Bile

Small amounts of bile may be normally present in the gastric contents. Large amounts give the stomach secretions a yellow or green color. This usually indicates an obstruction at the beginning of the small intestine.

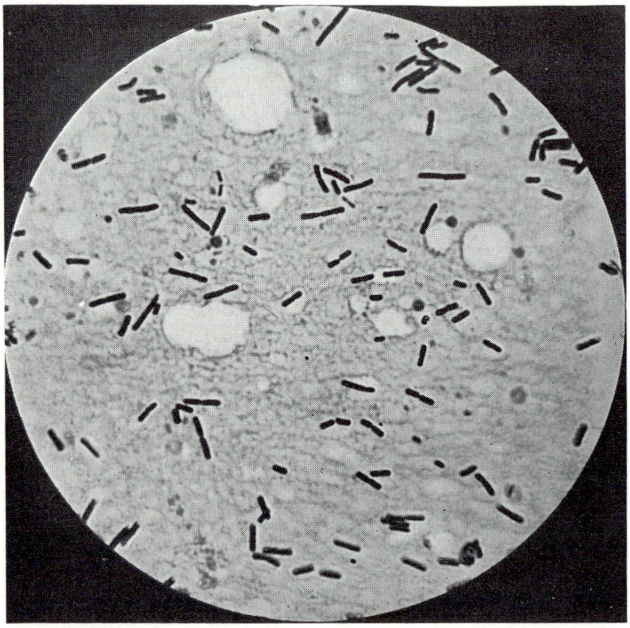

Fig. 12-4. Boas-Oppler bacilli. (From Levinson, S. A., and MacFate, R. P.: Clinical laboratory diagnosis, ed. 5, Philadelphia, 1956, Lea & Febiger.)

Procedure

1. Pour about 5 ml of concentrated nitric acid into a small test tube.
2. Tilt the tube to about a 30-degree angle.
3. Using a medicine dropper, *carefully* overlay the acid with the unfiltered gastric contents.

Reaction

If bile is present, a display of colors—yellow, green, red—will form at the junction of the two liquids.

Pepsin

Pepsin is an enzyme produced in the stomach and used in the digestion of protein. If the gastric contents show little or no pepsin, an organic disease of the stomach is indicated.

Reagent needed

Hydrochloric acid, concentrated

Procedure (Hammerschlag)

1. Prepare a 1% solution of egg albumin in the following manner. Take 5 ml of the white of an egg (use graduated cylinder) and place in a beaker. Add 57.6 ml of distilled water and 2.4 ml of concentrated hydrochloric acid. Mix and filter.
2. Get three large test tubes. Label them 1, 2, and 3. Pour 10 ml of the 1% solution of egg albumin into each tube.
3. Into tube 1, pour 5 ml of gastric secretion. Mix.
4. Into tube 2, pour 5 ml of distilled water. Add 0.5 g of pepsin. Mix.
5. Into tube 3, pour 5 ml of distilled water. Mix.
6. Place the three tubes in a 37° C water bath for 1 hour.
7. Calculate the amount of albumin in each tube by the method of Esbach that is used for the quantitative determination of protein in urine.

Interpretation

1. Tube 3 represents the amount of albumin in the 1% solution of egg albumin.
2. The difference between the amounts of albumin in tube 3 and tube 2 represents the amount of albumin that would be digested by the pepsin in normal gastric secretion.
3. The difference between the amounts of albumin in tube 3 and tube 1 represents the amount of albumin digested by the pepsin in the patient's gastric secretion.
4. The figure obtained from the patient's gastric secretion (point 3 above) should be equal to or exceed the figure obtained from normal gastric secretion (point 2 above).

Tubercle bacilli

The gastric washings are sometimes examined for tubercle bacilli. This is often done in suspected cases of tuberculosis when the organisms cannot be found in the sputum. The following test is performed in the morning, before the patient has had anything to eat or drink.

Reagent needed

Saline, 0.9% (D-22)

Procedure

1. Place a flask containing 50 ml of saline in a 37° C water bath.
2. Pass a stomach tube as directed in discussion on collection of specimens.
3. Remove the fasting stomach contents and place in a clean container.

4. Wash the stomach with the 37° C saline by injecting and removing this solution several times.
5. Finally withdraw the solution and place it in a clean container.
6. Centrifuge the fasting stomach contents and the saline washing at high speed for 15 minutes.
7. Examine the sediment for acid-fast bacilli; if found, a guinea pig inoculation should be performed.

Lactic acid

Lactic acid is an organic acid that is formed by the activity of bacteria on carbohydrates. Stomach contents containing large amounts of lactic acid are often found in cancer of the stomach.

Simon modification of Kelling test

Reagent needed

Ferric chloride, 10% (D-241)

Procedure

1. Obtain two large test tubes and label one Control and the other Test.
2. Pour about 20 ml of distilled water into the Control tube and add 4 drops of a 10% solution of ferric chloride. Mix.
3. Now pour half of this solution into the Test tube and add 1 ml of filtered gastric contents. Mix.

Reaction

Compare the Test tube against the Control tube. If lactic acid is present in large amounts the Test solution turns a deep yellow.

Combined acid

The combined acid is the total acid minus the free hydrochloric acid. Example: The total acid of a gastric specimen is 25 degrees and the free hydrochloric acid is 10 degrees. Find the combined acid:

Total acid	25 degrees
Free hydrochloric acid	10 degrees
Combined acid	15 degrees

TUBELESS GASTRIC ANALYSIS

Diagnex Blue

In the early 1950's, Segal and co-workers published several papers describing a new method for the evaluation of the stomach secretion of free HCl. It was found that after oral intake of a quininium-resin indicator, hydrogen ions from the HCl in the stomach can liberate quinine ions at a pH less than 3. After quinine hydrochloride is formed, it is absorbed into the intestine and is excreted into the urine approximately 15 minutes after ingestion. The quinine extracted from the urine is measured by ultraviolet light for fluorescence.

This work formed the basis of the Diagnex test marketed by Squibb & Sons. With further research Segal was able to develop a safer indicator material. The Diagnex Blue is prepared by reacting carbacrylic cation exchange resin with 3-amino-7-dimethyl-aminophenazathionium chloride (azure A). In the stomach with free HCl present, a reverse action takes place as the hydrogen ions of the resin are exchanged with azure A ions when the pH is less than 3. The released indicator is absorbed in the gastrointestinal tract and is excreted into the urine.

The gastric secretion is stimulated by taking two 250 mg tablets of caffeine sodium benzoate. For the test about 2 g of the Diagnex Blue granules (the equivalent of 100 mg of azure A carbacrylic resin) is administered.*

Specimen collection

Instruct the patient to be fasting and to discard the first morning urine specimen. Administer two caffeine sodium benzoate tablets with a full glass of water and collect urine one hour later for use as the control specimen. Give patient the Diagnex Blue granules suspended in one-fourth glass of water; any adhering to the glass are suspended in an additional small amount of water in order for the patient to consume the entire dose. After 2 hours have the patient empty his bladder and save this specimen. The urine may be blue, colorless, or a combination. The colorless urine may be converted to the blue form by acidifying, boiling, and standing for 2 hours for color development.

Procedure

1. **Dilute both control and test urines to 300 ml with distilled water.**
2. **Add 10 ml of control urine to two test tubes.**
3. **Add 10 ml of test urine to one test tube.**
4. **The visual color comparator block has two standards representing 0.3 and 0.6 mg of azure A. The control tubes are placed in front of the color standards and the test is placed in the middle. If the color of the test urine is the same as or more intense than the 0.6 mg standard, free HCl was present in the stomach.**
5. **If the test urine is of less intense color than the 0.6 mg standard, acidify it with 1 drop of Diagnex Blue reagent (Squibb acid-copper sulfate solution) or 2 drops of 10% HCl. Treat the control tubes in the same manner. Heat the three tubes in a boiling water bath for 10 minutes and allow them to cool for 2 hours. The color intensity is again compared.**

Comments

The procedure may be carried out using aliquots of the test specimen as controls by the additions of ascorbic acid which reduces the blue form of azure A to the colorless form. Two ascorbic acid capsules are supplied with each kit.

To confirm a negative test a check for any colorless azure A in the test aliquot is made by acidifying and boiling the urine to oxidize the indicator to the blue form. The blue color may fade during boiling but will reappear on cooling.

Interpretation

The presence of free HCl in the stomach is shown by a test urine color intensity equal to or greater than the 0.6 mg standard.

If the color intensity of the test urine is between the two standards, this is considered presumptive evidence for hypochlorhydria. An intensity less than the 0.3 mg standard is presumptive evidence for achlorhydria.

Since ascorbic acid decolorizes the azure blue, vitamin preparations should not be taken on the day preceding the test. The test may not be valid if there is malabsorption, renal disease, hepatic disease, or pyloric obstruction.

Duodenal analysis

The normal duodenal contents are colorless. Upon stimulation of the gallbladder, however, bile flows into the duodenum and gives the contents a brown,

*The tablets and granules are available as a kit from Squibb & Sons.

yellow, or green color. If bile fails to flow, the presence of gallstones may be suspected.

The samples of duodenal contents are usually obtained by the physician and brought to the laboratory. The method of obtaining the specimens is similar to the procedure in a gastric analysis except that the tube is allowed to pass into the duodenum. The gallbladder is then stimulated with a solution of magnesium sulfate.

An analysis of the duodenal contents consists of tests for amylase, trypsin, and bile and a microscopic examination. The procedures follow.

AMYLASE

The amylase of the duodenal contents is referred to as amylopsin or pancreatic amylase. The normal values depend upon the method of determination. For the procedure of Myers and Fine, which follows, the normal values are 5 to 200 units. Decreased values are found in chronic pancreatitis, pancreatic insufficiency, and fibrocystic disease of the pancreas.

Reagents needed

Sodium carbonate, 1%	(D-242)
Starch, 1%	(D-243)
Gram's iodine solution	(D-244)

Procedure (Myers-Fine)

1. Test the duodenal contents with nitrazine paper. If acid, add the 1% sodium carbonate dropwise until it is alkaline.
2. Label six large test tubes 1 to 6 and place in a test tube rack.
3. Make a serial dilution of the duodenal contents by doing the following. Place 1 ml of duodenal fluid in tubes 1 and 2. Pipette 1 ml of distilled water into all tubes except tube 1. Take tube 2 and mix the contents by drawing them into a 1 ml pipette and ejecting them several times. Take 1 ml of this mixture in tube 2 and transfer to tube 3. Mix the contents of tube 3 and transfer 1 ml to tube 4. Continue this process through the remaining tubes, discarding 1 ml from tube 6. The number of milliliters of duodenal fluid now in each tube is given below:

Tube	1	2	3	4	5	6
ml	1.0	0.5	0.25	0.125	0.062	0.031

4. Add 5 ml of the 1% starch solution to each tube. Mix. Incubate at 37° C for exactly 30 minutes.
5. Add 10 ml of cold water to each tube. Add 2 drops of Gram's iodine to each tube. Mix. Record the last tube in which there is a complete disappearance of blue color.

Calculation

By definition, the amylase activity equals the number of milliliters of starch solution that will be digested by 1 ml of duodenal contents under the conditions of the above test. Therefore:

$$\text{Amylase activity} = \frac{1.00}{\text{ml used}} \times 5$$

where 5 is the number of milliliters of starch solution digested, and ml used is the number of milliliters of duodenal contents in the last tube in which there is a complete disappearance of blue color.

Example: If tube 3 is the last tube in which there is a complete disappearance of blue color, the calculation is made as follows. As shown in step 3, tube 3 contains 0.25 ml of the duodenal contents. Therefore:

$$\text{Amylase activity} = \frac{1.00}{0.25} \times 5$$
$$= 20$$

Refer to Chapter 11 for Somogyi's amylase method.

TRYPSIN

Trypsin is an enzyme produced by the pancreas and used to digest protein. It is normally present in the duodenal fluid. The normal values depend upon the method of determination. For the procedure of Gross, which follows, the average tryptic activity is 2.5. Decreased values are found in chronic pancreatitis, pancreatic insufficiency, and fibrocystic disease of the pancreas.

Refer to Chapter 11 for discussion of enzymes.

Reagent needed

Acetic acid, 1.0% **(D-245)**

Procedure (Gross)

1. **Prepare a casein solution by dissolving 1 g of pure casein in 1 L of 0.1% sodium carbonate. Add 1 ml of chloroform to prevent bacterial decomposition.**
2. **Label ten large test tubes 1 to 10, and place in a test tube rack.**
3. **Pour 100 ml of the casein solution into a beaker and heat to 40° C. Pipette 10 ml into each tube.**
4. **Centrifuge the duodenal fluid. Pipette the clear fluid into each tube, using the amount called for below:**

Tube	1	2	3	4	5	6	7	8	9	10
ml	0.1	0.2	0.3	0.4	0.5	0.6	0.7	0.8	0.9	1.0

5. **Mix the contents of each tube. Place the tubes in a 37° C incubator for exactly 15 minutes.**
6. **Remove the tubes from the incubator. Add 3 drops of 1% acetic acid to each tube, noting whether the addition of the acid causes an increased turbidity. Record the first tube which shows *no* increased turbidity upon the addition of the acid. This is the first tube in which the casein has been completely digested by the trypsin. Calculate the tryptic activity as follows.**

Calculation

By definition, the tryptic activity is expressed as the effect of 1 ml of duodenal fluid on 10 ml of a 0.1% casein solution under the conditions of the above test. Thus:

$$\text{Tryptic activity} = \frac{1}{\text{ml used for digestion}}$$

Example: Tube 4 was the first tube which showed no increased turbidity upon addition of the 1% acetic acid. As noted in step 4, tube 4 contains 0.4 ml of duodenal fluid. Therefore:

$$\text{Tryptic activity} = \frac{1}{\text{ml used for digestion}}$$
$$= \frac{1}{0.4}$$
$$= 2.5$$

BILE

Bilirubin is the bile pigment normally present in duodenal fluid. Its absence often indicates an obstruction in the hepatic duct or common bile duct.

Procedure

1. Bile gives the duodenal fluid a yellow or green color. Note the color and report a 1 plus, 2 plus, 3 plus, or 4 plus, depending upon the depth of color.

MICROSCOPIC EXAMINATION

A microscopic examination of the duodenal fluid may give early warning of inflammation in the duodenum or biliary tract and also of the presence of parasites.

Procedure

1. Centrifuge a fresh sample of duodenal fluid.
2. Discard the supernatant (clear fluid) and place the sediment on a glass slide.
3. Look for an abnormal amount of pus cells, which indicates an inflammatory condition. A few cells are normally present.
4. In further observation of the slide, look for the cysts of amoebae, the eggs of hookworms, and for parasites such as *Giardia lamblia* and *Strongyloides stercoralis.*

Fecal analysis

Feces are waste products left in the wake of intestinal digestion and absorption. The average normal output of feces is 200 g/24 hr. In disease, deviations from the normal composition may occur. A few qualitative tests follow.

OCCULT BLOOD

Occult blood may be found in the stool of a person having ulcers or cancer. The patient should be on a meat-free diet for 4 days prior to the test to eliminate a false positive reaction. The following tests may be used to test for occult blood.

Reagents needed for occult blood tests

Benzidine, saturated, in glacial acetic acid (D-239)
Hydrogen peroxide, 3% (May be purchased)
Glacial acetic acid (May be purchased)
Ethyl alcohol, 95%
o-Tolidine, 4% (Make fresh monthly)
Using a volumetric flask, dissolve 4 g of *o*-tolidine in 100 ml of glacial acetic acid. Keep in brown bottle and refrigerate.
Hematest tablets

Benzidine test

Procedure

1. Take a piece of feces about the size of a pea and place on a spot plate.
2. Add a few drops of a saturated solution of benzidine in glacial acetic acid and mix with the feces, using an applicator stick.
3. Add a few drops of 3% hydrogen peroxide. Mix.

Reaction

A blue to green color represents a positive reaction. Disregard any other color. Report the test as positive or negative.

Guaiac test

Additional reagent needed

Guaiac reagent (Prepare just prior to use)
Into a test tube, place 2 ml of 95% ethyl alcohol and 0.2 g of powdered guaiac. Add 2 ml of 3% hydrogen peroxide. Mix.

Procedure

1. Take a piece of feces about the size of a pea and place on a spot plate.
2. Add 2 drops of glacial acetic acid and mix with the feces, using an applicator stick.
3. Add a few drops of the guaiac reagent. Mix.

Reaction

A blue to green color is a positive reaction. Disregard any other colors. Report the test as positive or negative.

Orthotolidine test

Procedure

1. Take a piece of feces the size of a pea and place on a spot plate.
2. Add a few drops of 3% hydrogen peroxide and mix with the feces, using an applicator stick.
3. Add a few drops of *o*-tolidine reagent. Mix.

Reaction

A blue to green color is a positive reaction. Disregard any other colors. Report the test as positive or negative.

Hematest*

Hematest, a tablet for detecting occult blood in stool, contains calcium acetate, tartaric acid, and strontium peroxide (which produce hydrogen peroxide when moistened), and *o*-tolidine as the chromogen.

Procedure

1. Place a small streak of solid feces on the special filter paper square supplied with the product. Do not use an emulsion of stool.
2. Place Hematest tablet on streak of stool.
3. Add 1 drop of water on tablet, wait 5 to 10 seconds, and flow a second drop on tablet allowing it to run down the sides onto the filter paper.

Reaction

Negative—If within 2 minutes color does not appear on the moistened filter paper square around the tablet, the test is considered negative.

Positive—A blue color appearing on the filter paper around the tablet within 2 minutes indicates a positive reaction. The amount of color and the time of its appearance provide a rough indication of the amount of blood in the stool.

BILE PIGMENT

Bile pigments are not normally present in feces but may be found in diarrhea and in obstructive jaundice.

Gmelin test

Reagents needed

Barium chloride, 10% (D-246)
Nitric acid, concentrated

Procedure

1. Using an evaporating dish or mortar, mix 7 ml of 10% barium chloride with a piece of feces about the size of a marble.
2. After an emulsion is made, filter the specimen.
3. When the funnel is completely drained, then add a few drops of concentrated nitric acid to the filter paper.

---

*Ames Co.

Reaction

A display of colors, one of which must be green, is a positive reaction. Report the test as positive or negative.

UROBILIN

Urobilin is normally present in feces. However, it is usually absent in obstructive jaundice.

Schmidt qualitative test

Reagent needed

Mercuric chloride, 10% (D-247)

Procedure

1. Place a piece of feces in an evaporating dish or mortar and mix thoroughly with 2 ml of 10% mercuric chloride.
2. Let the mixture stand overnight.

Reaction

If urobilin is present, a deep red color will be evident after 8 hours.

FECAL LIPIDS[1,2]

Normally the lipids of the feces have little to do with the lipids of the diet. They resemble the lipids of the blood and are undoubtedly secreted by the small intestine. Abnormal increase in the lipid is due to the food lipids, not those secreted by the liver or intestinal mucosa. Such increases may be due to blockage of the bile ducts or pancreatic ducts, failure of the pancreas to secrete pancreatic juice, or malabsorption. Malabsorption results when there is increased motility of the upper intestine and the food rushes through too rapidly. When the feces contain large amounts of fat, fatty acids, and soaps, the condition is called steatorrhea.

The term "malabsorption syndrome" is commonly used to describe a condition wherein absorption of the main nutrients is defective. This usually results in the passage of bulky, pale, fluid or semifluid, offensive, fatty stools. There are five main groups of disorders in which a malabsorption syndrome may develop. One group results from surgical operations on the gastrointestinal tract; a second is associated with constitutional diseases, such as diabetes; a third is due to granulomatous or neoplastic disease of the intestines; a fourth group is attributable to faulty preparation of food for absorption in the intestinal lumen; and a fifth group, which may be termed the "sprue group," includes such clinical conditions as celiac disease, idiopathic steatorrhea (nontropical sprue), and tropical sprue.

The estimation of fecal fat output is an important aid in diagnosing a malabsorption syndrome.

Determination of fat in feces

van de Kamer modified method[3,4]

An aliquot of a 72-hour collection of feces is saponified with concentrated potassium hydroxide in ethyl alcohol. This solution contains the soaps that were originally present in the stool plus those derived from the neutral fats and fatty acids. Hydrochloric acid is added to the solution to liberate the fatty acids. They are then extracted with petroleum ether. Ethyl alcohol is added to an aliquot of the petroleum ether layer, and the fatty acids are titrated with a standard solution of potassium hydroxide. Thymol blue is used as an indicator.

Preparation of patient and collection of feces

1. The patient should be on a known fat intake of at least 50 g of fat/day. If lower quantities are consumed, the fat in the feces originating from bacteria, intestinal cells, or bile (a total of 1 to 2 g is excreted every 24 hours) will influence the results.
2. The patient is given a carmine red dye capsule at the beginning of a 72-hour period and another at the end of 72 hours.
3. The collection of the stool should begin when the first red dye marker appears. All stool is then collected in the preweighed jar (2 or 4 quarts) until the second red dye marker appears. This may take more or less than 72 hours depending on intestinal motility.
4. The jar should be kept refrigerated during the collection.

Reagents needed

Ethyl alcohol, 95%, containing 0.4% capryl alcohol	(D-248)
Ethyl alcohol, 95%, neutral to thymol blue	
Potassium hydroxide, KOH, 33%	(D-249)
Hydrochloric acid, HCl, 25% (sp. gr. 1.13)	(D-250)
Petroleum ether, boiling point 90° to 100° C	
Sodium hydroxide, NaOH, 1.00 N	(D-251)
Thymol blue, 0.2% in 50% ethyl alcohol	(D-252)

Special equipment

1. Large mixer* to homogenize the entire 72-hour stool collection
2. Infrared lamp heater
3. Round-bottom centrifuge bottles, 250 ml capacity
4. Centrifuge carrier to hold the 250 ml bottles
5. Mason-type glass jars, 2- and 4-quart capacity
6. Microburette, 1.0 ml capacity

Procedure

1. Weigh the jar containing the stool collection and record the weight.
   Weight of stool = Total weight − Weight of empty jar
2. If the stool is formed (not watery), add enough water to equal about two-thirds the weight of the stool. Weigh the jar and record:
   Weight of blended stool = Total weight − Weight of empty jar
3. Attach the jar to the mixer and blend until the stool and water are thoroughly homogenized.
4. Weigh two 250 ml round-bottom centrifuge bottles. Do the test in duplicate.
5. Using a rubber bulb and a large-bore pipette (cut the tip off a 25 ml or 50 ml volumetric pipette), pipette about 10 g of the homogenized feces into each 250 ml centrifuge bottle. Weigh the bottles:
   Weight of blended sample = Total weight − Weight of empty bottle
6. To each bottle:
   a. Pipette 10 ml of 33% KOH.
   b. Add 40 ml of 95% ethyl alcohol containing 0.4% capryl alcohol.
   c. Add 10 to 12 glass beads.
7. Attach an air condenser to each bottle and set the bottles over an infrared lamp until the contents have boiled for 30 minutes. (A 25 ml volumetric pipette with the mouthpiece inserted through a cork serves well as a condenser.)
8. Cool in a room temperature water bath.
9. Add 17 ml of 25% HCl and cool again in the water bath.
10. Add 50 ml of petroleum ether (boiling point, 90° to 100° C).
11. Stopper the bottles and shake vigorously for 1 minute.

*May be obtained from Ivan Sorvall, Inc. This mixer has an adapter to fit Mason-type collection jars. The total specimen may be homogenized in its collection jar.

12. Centrifuge at moderate speed for 5 minutes or allow the bottles to stand until the layers separate. (The concentration of the ethanol is such that, after the mixture has been shaken, the petroleum ether and the acid ethanol layers separate rapidly.)
13. Transfer (use volumetric pipette) 25 ml of the petroleum ether layer to a 50 ml beaker or Erlenmeyer flask.
14. Add 3 or 4 glass beads (or small piece of filter paper) and evaporate the petroleum ether. Do not evaporate to dryness or char the residue. When the ether layer is gone, the odor will no longer be evident.
15. Add 10 ml of neutral 95% ethyl alcohol.
16. Titrate the fatty acids with 1.00 N NaOH from a microburette, using thymol blue as an indicator (4 drops).

Calculation

The calculations are based upon the assumption of an average molecular weight of 265 for fatty acids in stool:

A = Milliliters of 1.00 N NaOH used in titration
B = Total weight (in grams) of blended stool
S = Grams of blended feces taken for analysis

$$\frac{A \times 265 \times 1.04 \times 2 \times B}{1000 \times S} =$$

$$\frac{A \times 0.551 \times B}{S} = \text{Grams of fat in 72-hour stool}$$

The correction factor 1.04 must be used, since the petroleum ether layer increases 1% in volume when shaken with alcoholic hydrochloric acid and 3% of the amount of fatty acids remains in solution in the acid alcohol layer.

Precautions

1. Use either glass beads or a small piece of filter paper to prevent bumping (irregular boiling). Do not use pumice, which absorbs fatty acids.
2. It is not necessary to evaporate the petroleum ether quantitatively, since small amounts do not affect the titration.
3. The extraction is complete after 1 minute. Excessive shaking, beyond a 1-minute period, does not extract more fat.
4. The molecular weight of the fatty acids depends on the kind of fat consumed. For general purposes (both in cases of normal fat excretion and in steatorrhea) 265 was chosen, with palmitic acid being the main component.
5. Weigh and label the jars before sending any to the patient.
6. Weighings must be accurate. If accurate technique is used, the method has an accuracy of ± 2%.

Normal values

Normal values are expressed as percent of fat retained or "% retention." For normal, healthy adults, a fat retention of 95% to 98% is found. For example, if a patient has ingested 150 g of fat during the 3-day stool collection period, then a total stool fat of 3 to 7.5 g would be considered normal.

REFERENCES

1. Frazer, A. C.: The malabsorption syndrome, with special reference to the effects of wheat gluten. In Sobotka, H., and Stewart, C. P., editors: Advances in clinical chemistry, vol. 5, New York, 1962, Academic Press, Inc.
2. Kleiner, I. S., and Orten, J. M.: Biochemistry, ed. 6, St. Louis, 1962, The C. V. Mosby Co.
3. van de Kamer, J. H.: Total fatty acids in stool. In Seligson, D., editor: Standard methods in clinical chemistry, vol. 2, New York, 1958, Academic Press, Inc.
4. van de Kamer, J. H., ten Bokkel Huinink, H., and Weijers, H. A.: Rapid method for the determination of fat in feces, J. Biol. Chem. **177**: 347, 1949.

• 13

• Cerebrospinal fluid

Cerebrospinal fluid is a clear colorless fluid, having a specific gravity of 1.004 to 1.008. The fluid, being chiefly a dialysate from the blood, maintains a pH of 7.35 to 7.40. The protein content is extremely low, with no fibrinogen; the glucose is approximately two thirds that of the blood sugar; and the chloride is about 25% higher than the plasma chloride. The total volume of cerebrospinal fluid is between 120 and 150 ml.

The brain, which is the largest and most complex mass of nervous tissue in the body, may be regarded as the elaborate end of the cerebrospinal axis (Fig. 13-1), and it occupies the entire cranial cavity. A narrow central canal that extends the length of the spinal cord is continued anteriorly into the brain where it widens out to form the ventricles. There are five ventricles—one of which is not a true ventricle, since it has no connection to the ventricular system. The two lateral ventricles are located, one in each of the cerebral hemispheres, under the corpus callosum that connects the two hemispheres. The cavity of the lateral ventricles is large and may become overdistended with cerebrospinal fluid in certain pathologic conditions. The third ventricle, located behind the lateral ventricles, is connected with them by small openings and with the fourth ventricle by a slender canal, the cerebral aqueduct. The fourth ventricle is located in front of the cerebellum, behind the pons varolii and medulla oblongata.

The spinal cord and brain are covered by three layers of connective tissue, known as meninges. One of these, the dura mater, is the outermost tough membrane of fibrous connective tissue. The cranial dura mater is arranged in two layers, the outer layer adhering to the bones of the skull and the inner one covering the brain. Except where they separate to form sinuses for the passage of venous blood these layers are closely connected. The spinal dura mater is a loosely applied membrane fastened to the bony neural arches of the vertebrae. The cranial and spinal dura mater form one complete membrane.

The pia mater, the layer that is innermost and closely applied to the brain and cord, is a vascular covering and nutritive membrane. In the brain it is attached to the cerebral cortex, passing into various fissures and protruding into the ventricles. In the spinal column this delicate membrane (containing many small blood vessels) is attached to the surface of the spinal cord.

Between the dura mater and the pia mater is the arachnoid mater, a serous membrane consisting of a delicate network of fibrous tissue that contains many blood vessels. The network of the arachnoid is filled with spinal fluid. A bloody spinal fluid received in the laboratory may be due to subarachnoid (beneath the arachnoid membrane, within the network of blood vessels and spinal fluid) hemorrhage. Inflammation of the coverings is known as meningitis.

Cerebrospinal fluid is formed in the choroid plexuses, tufts of capillary blood vessels in the cerebral ventricles, and moves through the ventricular system and

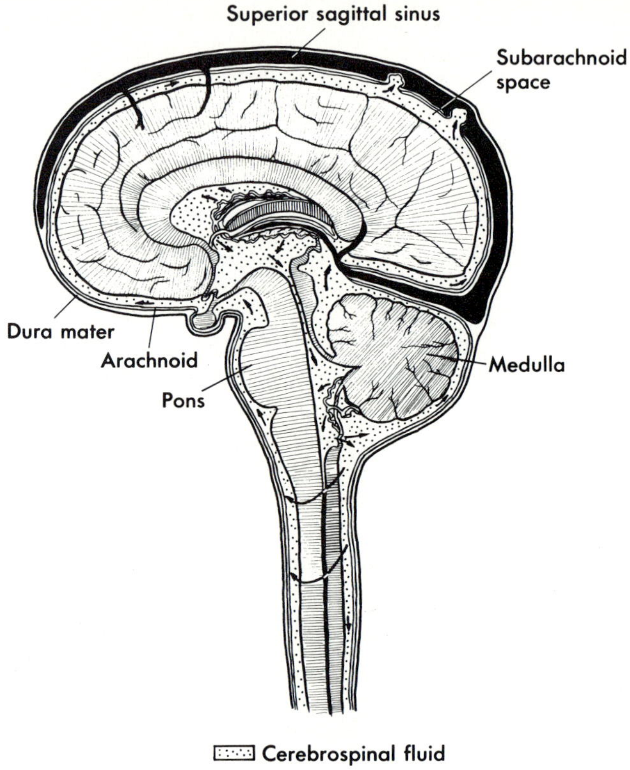

Fig. 13-1. Cross section of brain and spinal cord. (After Netter.)

out of three small openings in the roof of the fourth ventricle into the subarachnoid cavity. It circulates over the cerebral hemispheres and downward over the spinal cord, carrying nutriment to the nerve cells and removing wastes. Spinal fluid fills the central canal, ventricles, and space between the arachnoid and pia mater. The return from the subarachnoid space is through arachnoid villi of the dural sinuses to the venous blood.

The spinal fluid has three functions:

1. To protect the brain from injury by acting as a fluid buffer.
2. To act as a medium of exchange for the transfer of dialyzable material between the bloodstream and the tissues of the brain and the spinal cord.
3. To equalize pressure between the brain and the spinal cord. The pressure usually remains constant, supporting the soft bulk of the central nervous system. However, since this pressure is closely related to the blood pressure, coughing, sneezing, or any stress that causes a rise in venous pressure will also increase the cerebrospinal fluid pressure.

COLLECTION OF CEREBROSPINAL FLUID

Spinal fluid is obtained by making a spinal puncture and this is performed by a physician. The patient should lie on one side with his hips and knees drawn to arch the back. A local anesthetic is usually applied. The tap for fluid is made by the insertion of a needle about 10 cm long, having a bore of 1 to 1.5 mm and provided with a stylet, into the small of the back between the third and fourth lumbar vertebrae.

Blood may be picked up, on or in the spinal needle, during passage through the tissue. Therefore, the first few drops may be discarded or placed in a tube for microbiologic examinations. The usual practice is to collect fluid in three sterile small screw-top tubes, labeled in order of collection. The first tube is delivered to the microbiology laboratory for a smear and culture; the second, to the hematology laboratory for a cell count; and the third, to the chemistry laboratory for chemical examinations. Any fluid remaining in the latter tube is forwarded to the serology laboratory for specified tests.

PHYSICAL APPEARANCE OF SPINAL FLUID WITH VARIOUS DISEASES

Xanthochromia. The spinal fluid is a clear yellow, showing rapid coagulation. There are no blood pigments present, but large amounts of globulin. This disorder results from a tumor pressing on the spinal cord and affecting the flow of spinal fluid. A pocket of fluid collects and stagnates, with these results. Xanthochromia may appear after a cerebral hemorrhage.

Acute meningitic infection. The fluid may appear slightly to very turbid until there is an appearance of almost pure pus.

Tuberculous meningitis. The fluid is clear, although upon standing for 12 to 24 hours, a pellicle will form. With many cases of acute meningitis, tuberculous meningitis, and suppurative meningitis a pellicle of fibrin and blood cells forms in the fluid when the specimen is allowed to stand.

Suppurative meningitis. Pellicle formation is extremely rapid. It is difficult to work with this type of material because of its speedy coagulation.

Spinal fluid contains glucose, protein, and chloride. In disease, the concentration of these materials may vary, and white cells and bacteria may also be present.

The composition of the fluid is usually altered in the following diseases: meningitis, poliomyelitis, encephalitis, and latent syphilis.

MICROSCOPIC EXAMINATION

This includes an examination for microorganisms and a count of the different types of cells.

Smear

In disease, the spinal fluid may contain the following organisms: staphylococci, streptococci, meningococci, tubercle bacilli, and influenza bacilli.

These organisms may be identified by a smear and culture. The procedures follow.

Procedure

1. **Centrifuge the spinal fluid at high speed for 5 minutes.**
2. **Pour off and discard the supernatant fluid.**
3. **Using a bacteriologic loop, smear some of the sediment over a glass slide.**
4. **Warm the bottom of the slide to "fix" the bacteria.**
5. **Stain with Gram's stain, using the usual gentian violet, Gram's iodine, acetone, and safranin. (If examination for tubercle bacilli is requested, make another smear and stain for AFB.)**
6. **Identify the organisms, using a textbook on bacteriology as a guide.**
7. **Report the organisms found.**

Culture

The culture should be made while the spinal fluid is fresh. If this is not possible, store the fluid in the 37° C incubator. This is permissible, since spinal fluid itself is a good culture medium.

In many hospitals, the routine procedure for a spinal culture begins with streaking the fluid onto the following media: (1) blood agar plate or slant and (2) chocolate agar plate or slant. (These media are usually stored in a refrigerator and should be warmed to room temperature before use.)

Procedure

1. **The sediment that was used to prepare the smear may be used to make the culture. If this sediment is not available, centrifuge a sample of the spinal fluid at high speed for 5 minutes. Pour off and discard the supernatant fluid. Use the sediment for the culture.**
2. **Obtain a blood agar plate or slant. Warm gently over a flame.**
3. **Using a bacteriologic loop, streak the plate or slant with the spinal fluid sediment.**
4. **Place in a 37° C incubator.**
5. **Obtain a chocolate agar plate or slant. Warm gently over a flame.**
6. **Using the bacteriologic loop, streak the plate or slant with the spinal fluid sediment.**
7. **Place the plate or slant in the CO_2 jar. Light the candle. Place the lid on the jar so that it is tightly secured. The candle, of course, will go out. This creates an increased CO_2 tension, which aids in the growth of meningococci.**
8. **Place the CO_2 jar in the 37° C incubator.**
9. **Twenty-four hours later, examine the plates or slants for growth, make slides, and identify the organisms as directed in textbooks on bacteriology.**

Cell count

The spinal cell count consists of the total cell count and the differential cell count. Since the cells disintegrate rapidly, the counts should be made while the fluid is fresh—preferably within an hour following the withdrawal of fluid.

Spinal fluid often contains contagious material. Consequently, take particular care in handling the specimens. In making transfers of the fluid, it is a wise policy to use medicine droppers.

Total cell count

The cells found in spinal fluid are usually white cells, called leukocytes. Sometimes, however, red cells produced by the puncture itself are seen. These may cause a slight increase in the count. The normal values for the total cell count are 0 to 10 cells/mm^3. Increased values, such as 20 to 1000 cells, may be found in meningitis, encephalitis, poliomyelitis, and latent syphilis.

Procedure

1. **Mix the spinal fluid thoroughly by tapping the bottom and sides of the test tube.**
2. **Draw the spinal fluid counting solution to the 0.5 mark and draw the spinal fluid to mark 11 in a white counting pipette, thus making a dilution of the spinal fluid of 19:20. Mix thoroughly. Transfer a small portion (2 or 3 drops) to a counting chamber, which is generally used for counting blood cells.**
3. **Using either the 4 mm or the 16 mm objective, count the cells in the *entire* ruled area. This will be the nine large squares as seen with the low power (16 mm objective) of the microscope.**

4. After the count is made, the counting chamber should be soaked in alcohol to prevent contagion.

Calculation

The total area of the counting chamber consists of nine large squares or 9 mm^2. Since the counting chamber is 0.1 mm deep, this will be a volume of 0.9 mm^3 ($9 \times 0.1 = 0.9$). The report, however, is to be given as the number in 1.0 mm^3. Therefore, multiply the number counted by 1.0/0.9 (or 10/9).

Example: The number of cells counted in the entire ruled area was 9. Multiplying this by 10/9:

$$9 \times 10/9 = 10 \text{ cells/mm}^3$$

Spinal fluid counting solution (Tuerk solution)

1. Pipette 1.0 ml of glacial acetic acid into a volumetric flask and dilute to 100 ml with distilled water.
2. Add a few drops of alcoholic gentian violet to color.
3. Stopper and mix thoroughly.
4. Filter several times, until clear.

Differential cell count

If the total cell count is within the normal range of 0 to 10 cells/mm^3, there is no point in making a differential cell count. If, however, the count is elevated, a differential cell count should be made.

The white cells usually found in spinal fluid are lymphocytes and neutrophils. Lymphocytes are usually predominant in tuberculous meningitis (late stage), congenital syphilis, poliomyelitis (late stage), epidemic encephalitis, cerebral abscess (unruptured), and syphilitic meningitis; and neutrophils predominate in influenza meningitis, pyogenic meningitides, cerebral abscess (ruptured), and poliomyelitis (early stage).

Procedure

1. Mix the specimen thoroughly by tapping the bottom and sides of the test tube.
2. With a medicine dropper, place a small drop of the spinal fluid on a glass slide. Using another slide as a spreader, spread the fluid over the slide as for a blood smear and allow to dry.
3. Stain with Wright's stain.
4. Using the oil-immersion objective of the microscope, record the number of lymphocytes and segmented cells seen while counting 100 white cells.
5. Report the percent found.

Example: 40 lymphocytes and 60 segmented cells are recorded while counting 100 white cells. Therefore, there are 40% lymphocytes and 60% segmented cells.

CHEMICAL EXAMINATION

The chemical constituents of the cerebrospinal fluid that have clinical significance are total protein, glucose, and chloride. However, changes in these may be the result of metabolic alterations and not necessarily a reflection of intracranial disturbances.

Protein

The protein fractions of spinal fluid are similar to those of blood serum. Fibrinogen is normally absent. In many pathologic states associated with an increase in permeability of the blood-brain barrier, such as bacterial meningitis, protein in the spinal fluid is increased. Current practice generally requires a quantitative determination of the total protein. However, since the qualitative tests for globulin will indicate the presence of abnormal increases which are always associated with pathologic changes, they are included.

Globulin

The three most common qualitative globulin tests are the Pandy test, Ross-Jones test, and Nonne-Apelt test.

Reagents needed

Pandy's reagent	**(D-253)**
Ammonium sulfate, saturated	**(D-254)**

Pandy test

1. **Pipette 1 ml of Pandy's reagent into a small test tube.**
2. **Using a medicine dropper, add a few drops of the spinal fluid.**
3. **If an abnormal amount of globulin is present, the solution will become *markedly* turbid.**
4. **Report the test as positive or negative, disregarding a faint turbidity, which may be found in normal spinal fluid.**

Ross-Jones test

1. **Pour about 2 ml of a saturated solution of ammonium sulfate into a small test tube.**
2. **Using a medicine dropper, overlay with about 1 ml of the spinal fluid.**
3. **If an abnormal amount of globulin is present, the following will take place: within a few minutes, a grayish white ring will form at the junction of the two liquids.**
4. **Report the test as positive or negative.**

Nonne-Apelt test

1. **Pour about 1 ml of a saturated solution of ammonium sulfate into a small test tube.**
2. **Add about 1 ml of the spinal fluid.**
3. **Mix and set aside for about 3 minutes.**
4. **If an abnormal amount of globulin is present, a markedly cloudy precipitate will form.**
5. **Report the test as positive or negative.**

Total protein

The normal values for total protein in spinal fluid are 15 to 45 mg/100 ml. Increased values are found in meningitis and latent syphilis.

If red cells are present in a sample of cerebrospinal fluid, the fluid must be centrifuged and decanted before an analysis is done. Cells will greatly elevate the values. Elevated proteins are found in most types of meningeal inflammation, latent syphilis, brain tumor, subarachnoid hemorrhage, and sometimes in contamination of blood as mentioned above. The blood contamination results from a trauma during the withdrawal of the fluid.

Meulemans method

Principle

Trichloroacetic acid is added to the spinal fluid and the protein is precipitated as a fine suspension. The turbidity of the suspension is measured in a colorimeter and compared with a standard.

Reagents needed

Trichloroacetic acid, 3% (D-255)

Working standard:

Prepare fresh daily. Transfer 0.5 ml of serum with a known protein concentration to a 100 ml volumetric flask. Dilute to volume with saline, stopper, and mix thoroughly.

To determine the protein concentration of this solution in milligrams per 100 ml, multiply the serum protein concentration in grams per 100 ml by 5. The 5 is obtained by the following computation:

$$\frac{0.5 \times 1000}{100}$$

For example, control values of the serum standard × 5 = mg/100 ml solution value:

6.6 g/100 ml × 5 = 33 mg/100 ml
6.8 g/100 ml × 5 = 34 mg/100 ml
7.0 g/100 ml × 5 = 35 mg/100 ml
7.2 g/100 ml × 5 = 36 mg/100 ml

Procedure

1. Set up the following tubes:
   Blank
   5.0 ml of 3% trichloroacetic acid
   Standard
   1.0 ml of working standard
   4.0 ml of 3% trichloroacetic acid
   Unknown
   1.0 ml of spinal fluid
   4.0 ml of 3% trichloroacetic acid
2. Mix and let stand for 10 minutes.
3. Remix and read absorbance (A) at 450 mμ (or read % T and convert to A)

Calculations

$$\frac{\text{A of unknown}}{\text{A of standard}} \times \text{Concentration of standard} = \text{mg protein/100 ml}$$

Glucose

The normal values for glucose in spinal fluid are approximately two thirds that of the blood glucose, or 40 to 80 mg/100 ml of fluid. Increased values may accompany brain tumors. Decreased values are seen in most types of meningitis.

The determination should be made within an hour after the withdrawal of spinal fluid, because glucose decomposes upon standing. Refer to Chapter 6 for the quantitative analysis for glucose.

Chloride

The chloride is expressed in terms of sodium chloride. The normal values are 695 to 760 mg of sodium chloride/100 ml of spinal fluid, or 119 to 130 mEq/L. Decreased values are found in acute meningitis, particularly in tuberculous meningitis.

Refer to Chapter 7 for the quantitative analysis for chloride.

Tryptophan test

The tryptophan test is used for tuberculous meningitis.

Reagents needed

Formaldehyde, 2% (D-256)

Sodium nitrite, 0.06% (D-257)

Procedure

1. Pour about 2 ml of *clear* spinal fluid into a large test tube.
2. Add 15 ml of concentrated hydrochloric acid.
3. Add 2 or 3 drops of a 2% solution of formaldehyde.
4. Mix and set aside for 5 minutes.
5. Using a medicine dropper, overlay with 2 ml of an 0.06% solution of sodium nitr*ite.*
6. Set aside for 3 minutes.
7. In tuberculous meningitis, a violet ring forms at the junction of the two liquids. Disregard a yellow or brown ring.

14

The kidney

The English physician and clinician Richard Bright first introduced urine examinations as a routine procedure for the physicians in Guy's Hospital (1827). Bright is known universally now for his description of chronic nephritis, Bright's disease. The physicians soon developed the analysis of urine to a fine art, and before the turn of the century many procedures were published on the macroscopic and microscopic examination of urine. Thomas Addis brought the microscopic study of the urinary sediment to a peak of perfection by the introduction of quantitative methods of study, which he felt were necessary for an accurate evaluation of renal disease.

In the 1930s physicians began to realize that too much faith had been placed upon the reading of the urinary sediment during the earlier years of medicine. When Henry Christian, an authority on renal disease, stated that it was impossible to determine during life the anatomic changes in the kidney that would be found after the patient's death, the importance of the urine examination seemed to decline in popularity. With the development of new blood tests and mechanical equipment for registering various body functions, urine analysis became the "stepchild" of the laboratory.

During the last few years, however, new kidney function tests have been developed to aid the physician in determining the anatomic changes in the kidney during life, so the importance of urinary examination is again recognized. Other body malfunctions also are determined with the help of urinary studies.

Discussion of the kidneys and the analysis of urine is divided into the following three sections: (1) renal function tests, (2) renal insufficiency and dialysis, and (3) routine urine examination.

Renal function

The success of dialysis and kidney transplants in saving lives of patients who formerly would have died from uremia has, because of the chemical abnormalities characteristic of renal insufficiency, increased the laboratory testing of blood and biologic fluids as a guide in diagnosis and prognosis. Hemodialysis and peritoneal dialysis will remove from or add to solutes in the blood and tissue fluids as required for survival, either relieving the kidneys until they are able to resume normal functioning or substituting for them by use of frequent dialysis. The awareness of the general public concerning availability of these lifesaving devices and the use of artificial kidneys in the home as well as the hospital have required an education program for patients and their families as well as for hospital personnel associated with renal programs.

The two kidneys serve to regulate the composition of the blood and maintain the internal composition of the body compatible with life. They are located retroperitoneally in the back of the abdominal cavity, one on each side of the vertebral column at approximately the L-3 level. The kidneys are major excretory organs for eliminating from the blood normal metabolic end products, excess or unwanted substances, and chemical poisons. Some of these may cause injury to the tissues and alter the normal elimination procedure.

ANATOMY OF THE KIDNEYS (Fig. 14-1)

Each kidney measures about 12 × 6.5 × 4 cm and weighs between 115 and 170 g. Externally it is shaped like a bean with a depression at its inner edge called the hilus where the renal artery from the aorta enters the organ and the renal vein leaves it to join with the vena cava. Nerve fibers enter along the artery. The hilus is also partially taken up by the pelvis, a funnel-shaped structure that is formed by fusion of the calyces and tapers to form the ureter, which then courses downward to empty into the bladder.

If a kidney is cut longitudinally, two distinct zones may be seen on the cut surface, the outer or cortex and the inner or medulla. The cortex contains glomeruli, which are the filtering units of the kidney. The medulla of each kidney contains about one million microscopic tubules that converge into a number of pyramidal areas, which then empty into the calyceal system and, thence, the renal pelvis.

Each tubule originates from a malpighian corpuscle, the outer wall of which is known as Bowman's capsule while the capillary network within is called the glomerulus. The filtrate within the capsule is plasma water extruded by hydro-

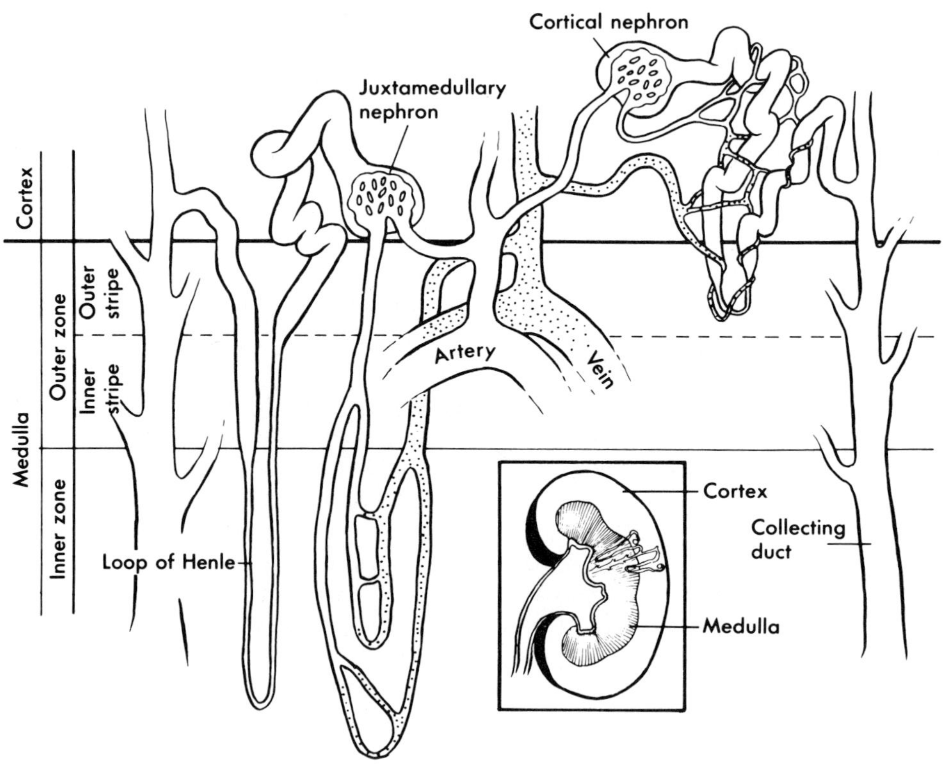

Fig. 14-1. Diagram of cortical and juxtamedullary nephrons with a cross section of the kidney.

static pressure from the thin-walled capillary network of the glomerulus; it empties into the tubule which then evolves as a series of convolutions (proximal or first convoluted tubule) in the cortex. The tubule then straightens and dips downward into the medulla as a thin descending segment, making a U turn known as Henle's loop, to rise again into the cortex where the wall structure changes and the tubule enlarges. It again becomes convoluted (distal or second convoluted tubule) near its own glomerular unit before it straightens out to join with others into the collecting ducts. Many ducts combine, forming channels (papillary ducts of Bellini) to carry the urine through the pyramids to the kidney pelvis. By this arrangement a large surface area of the tubules comes in contact with a meshwork of very small blood vessels. Each glomerulus, with its particular tubule, is an essentially independent functional unit and is referred to as a *nephron.*

The kidney itself synthesizes small amounts of hippuric acid and ammonia. Otherwise, all known urinary constituents are brought to it by the blood and are excreted or reabsorbed unchanged. Each kidney is supplied with blood from the abdominal aorta through the renal artery that enters at the hilus. Normal renal blood flow is about 25% of total circulation. The renal artery divides into several branches and when these reach the border between the cortex and medulla they divide laterally to form the arch or arcuate arteries. From the latter, interlobular arteries enter the cortex, forming at intervals small afferent arteries and then arterioles, which form the capillary network of the individual glomeruli. There are some 50 capillary loops formed that coil and intertwine. The capillaries reunite to pass out of the glomerulus and form the efferent arteriole. This efferent arteriole divides into another network of capillaries, surrounding the tubule in the cortex and the loop of Henle in the medulla. They eventually unite to form the interlobular and medullary veins, from which the blood flows into the arcuate veins between the cortex and medulla. The arcuate veins form interlobar veins that merge into the renal vein, which then leaves the kidney at the hilus and empties into the inferior vena cava.

The blood supply to the kidneys is about 25% of the cardiac output, or approximately 1160 ml/min in men and 940 ml/min for women. It is under pressure that is exerted by contraction of the heart, and over 10% of the fluid is removed as glomerular filtrate. Glomerular filtration rate is about 127 ml/min and 118 ml/min for males and females, respectively.

PHYSIOLOGY OF THE KIDNEYS

In the formation of urine, kidneys perform a series of functions. They excrete end products of protein metabolism (urea, uric acid, creatinine) and about one half of the water eliminated from the body, maintain normal concentration of essential blood constituents by excreting those not needed and reabsorbing the ones required, and help preserve normal acid-base balance of body fluids.

By physical forces such as ultrafiltration and diffusion fluid is removed from the blood in the glomerular capillaries into the filtration chamber of the capsules. The filtrate has essentially the same composition as protein-free plasma although a small amount of protein may pass through the glomeruli and be reabsorbed in part by the tubules. Filtration is dependent upon the ability of the hydrostatic pressure of the blood to overcome the colloid osmotic pressure of the blood proteins and the tissue tension in the capillaries. There is decreased filtration when the blood pressure falls or a lowering of cardiac output results from shock, surgery, dehydration, or cardiac decompensation, and also when renal damage causes a reduction in the total number of functioning glomeruli. When glomeruli are damaged, as in glomerulonephritis, plasma albumin leaks into the filtrate.

In the normal adult between 120 and 130 ml of glomerular filtrate are produced per minute. If this filtrate were excreted unchanged, death would occur rapidly. Approximately 97% of the water, and—selectively—other constituents such as electrolytes, glucose, and amino acids are reabsorbed during the passage of the filtrate down the tubule. Assuming 125 ml of filtrate is produced per minute (180 L each day) and that it is principally protein-free plasma, the major constituents removed per day from the blood may be approximated as shown in Table 14-1. These figures illustrate the enormous reabsorptive action of the kidneys, whereby most of the constituents filtered from the blood are promptly returned to it by absorption at the tubules.

As the filtrate moves from the glomerulus downward through the tubule, selective reabsorption by tubular cells of glucose, amino acids, sodium, and potassium, but not waste materials, occurs by active cellular transport. About 97% of the filtered water is passively reabsorbed, resulting in a urine with a high concentration of waste materials. Reabsorption of organic compounds such as glucose and amino acids is essentially complete in the proximal tubule. It is the function of the kidneys to conserve these but not to regulate their concentration in the blood. Thus, needed constituents are restored to the blood by passage through the cells lining the tubules.

Glucose is transferred by means of a transport system that utilizes the coupling of glucose with phosphate. When there is an excess of glucose in the glomerular filtrate, the capacity of the tubular transport system is exceeded and glycosuria will result. About 85% of the sodium is actively reabsorbed; with it the associated chloride anion is passively reabsorbed, as is water, the latter as a result of the osmotic gradient produced by the sodium movement. The osmotic concentration of the fluid entering the descending limb of Henle's loop is the same as that of the original glomerular filtrate but it progressively increases to

Table 14-1. Comparison of average amounts of biochemical constituents in glomerular filtrate and excreted urine

Constituent	In 180 L of filtrate	In 24-hour urine collection
Fluid volume (L)	180	1.5
Sodium (g)	580	4.5
Potassium (g)	30	2.5
Chloride (g)	650	7.
Calcium, diffusible (g)	9	0.2
Magnesium, diffusible (g)	4	0.08
Urea (g)	60	30
Glucose (g)	170	Below detectable limits

the bend in the depths of the medulla. In the ascending limb of the loop of Henle, sodium is actively reabsorbed (and with it chloride, passively) but the loop is impermeable to water so that because of solute loss the filtrate is dilute by the time it reaches the distal convoluted tubule. In the presence of antidiuretic hormone (ADH), water moves passively from the urine as it passes through the distal tubule and collecting ducts in response to concentration gradients, yielding a urine with concentration similar to that of the interstitial tissue fluid at the tip of the papilla. In the absence of ADH, the distal tubule is impermeable to water, which will therefore be retained although sodium reabsorption may continue, depending upon need of the body to produce a dilute urine. One third to one half of the filtered load of urea is passively reabsorbed along the length of the tubule, mainly by diffusion in response to osmotic gradients.

Certain tubular cells are capable of removing additional waste materials from the surrounding peritubular fluid and capillaries and transferring it to the tubular lumen. By this action, the tubules supplement glomerular filtration in removal of waste materials (urea, creatinine) from the blood.

CHEMICAL TESTS OF RENAL FUNCTION[22]

The basic activities of the kidneys that are measurable by chemical means are glomerular filtration, tubular reabsorption and excretion, and renal blood flow. When glomeruli or tubules are damaged or destroyed, they are not replaced. Any action that results in alteration or destruction of kidney tissue may be critical because of accumulation, in blood and other body fluids, of materials normally eliminated by the kidneys.

Chemical tests of renal function are used to determine the nature and extent of renal dysfunction. However, they give no indication as to whether the damage is anatomic or functional. The latter may result from shock, congestive heart failure, severe dehydration, or other conditions in which the normal blood supply to the kidneys is altered. These tests provide a good index of renal function if the influence of various extrarenal factors is considered—such as the decrease of normal renal function measurements with increasing age, severe muscular exercise, and marked reduction in dietary protein.

Glomerular filtration

The ability of the glomeruli to filter soluble constituents from the blood can be measured by clearance tests. These are based upon the rate at which kidneys remove a substance from the plasma. Clearance is an expression of the volume of blood plasma containing the amount of a substance excreted into the urine in one minute. By measuring the volume of urine excreted per minute (V) and the concentration of a constituent in blood (B) and in urine (U) samples obtained about the same time, the clearance of that substance may be calculated, UV/B.

An ideal test substance for glomerular filtration should be able to move freely across the glomerular membrane, maintaining the same composition in the filtrate as in the plasma, and be neither reabsorbed nor secreted by the tubules. Inulin and mannitol are two compounds that meet these requirements. However, these are not used in routine clinical studies because the procedure is tedious and time-consuming and because catheterization of the patient is required.

Two clearance tests generally employed are those for urea and creatinine.

While the excretion of these endogenous substances is influenced by tubular processes, their clearance rates as determined by standard procedures give an approximation of the glomerular filtration that is sufficient for clinical purposes as an indication of renal function. They do not yield an accurate measure of glomerular filtration rate.

Inulin clearance[1,7,19]

Inulin, a vegetable starch (fructose polysaccharide), is not metabolized in the body but is excreted into the urine by glomerular filtration and neither reabsorbed nor excreted by the tubules. Inulin clearance is a measure of glomerular filtration and is used as the reference test.

Collection of specimens

1. The patient should be fasting, but during the 2-hour period prior to the test he should drink about a liter of water. Insert a urethral catheter to provide continuous drainage of the bladder. Before starting the infusion obtain a blood sample and some urine to use as blanks.
2. Administer intravenously a priming dose of 50 mg of inulin per kilogram of body weight and follow immediately with a sustaining solution that contains 70 ml of 10% inulin diluted to 500 ml with saline, given at about 4 ml/min, or the equivalent of the expected rate of renal clearance of inulin. Where there is impaired renal function, decrease the amount for maintenance accordingly.
3. After an equilibrium period of 30 to 45 minutes see that the bladder is emptied and washed with sterile saline. Collect three accurately timed urine specimens over an interval of approximately 20 minutes. At the end of each collection period inject 20 ml of sterile saline and 20 ml of air into the bladder to wash out all of the urine. Combine the washing with the specimen. Obtain blood samples in heparinized tubes about the midpoint of each urine collection period for plasma inulin determinations.

Reagents needed[11]

(All chemicals are reagent grade.)

Trichloroacetic acid, 10% aqueous (D-164)

Hydrochloric acid, concentrated

Indole-3-acetic acid reagent, 0.5% in 96% ethyl alcohol
Store in the refrigerator.

Inulin standard, 0.05 mg/ml
Prepare a paste of 12.5 mg of inulin in several drops of water and add 50 to 75 ml of hot (90° C) distilled water. Transfer to a 250 ml volumetric flask quantitatively, add sufficient benzoic acid to saturate, and dilute to volume.

Procedure[11]

1. Into a centrifuge tube place 1.0 ml of plasma. Add slowly 4.0 ml of 10% trichloroacetic acid. Allow to stand 10 minutes and filter.
2. Into test tubes in duplicate pipette plasma filtrate, 1.0 ml; urine diluted 1:100 with water, 1.0 ml; standard, 1.0 ml; and water (reagent blank), 1.0 ml.
3. To each tube add 0.2 ml of indole-3-acetic acid reagent and 8.0 ml of concentrated hydrochloric acid.
4. Incubate in a 37° C water bath for 75 minutes.
5. Cool to room temperature and read absorbance at 520 mμ against water. The color intensity increases with standing. Read as rapidly as possible in the order in which the tests were set up.

Calculations

Plasma: $$\frac{\text{A unknown} - \text{A plasma blank}}{\text{A standard} - \text{A reagent blank}} \times 0.05 \times \frac{100}{0.2} = \text{mg inulin/100 ml plasma}$$

Urine: $$\frac{\text{A unknown} - \text{A urine blank}}{\text{A standard} - \text{A reagent blank}} \times 0.05 \times \frac{100}{0.01} = \text{mg inulin/100 ml urine}$$

Clearance corrected to average body surface area:

$$C_{In} = \frac{UV}{P} \times \frac{1.73}{S}$$

where U = mg inulin/ml urine
V = ml urine excreted/min
P = mg inulin/ml plasma
S = body surface area in square meters (Fig. 14-2)
1.73 = average body surface area in square meters

Normal values

The range of normal values for inulin clearance found by different investigators varies. Calculated per 1.73 m^2 of body surface, it is approximately[20]:

Male	125 ± 15 ml/min
Female	110 ± 15 ml/min

Comments

By this method inulin is hydrolyzed to fructose and the products of hydrolysis react with the indole-3-acetic acid to form a purple color. Interference from glucose is less than with other methods, and the color intensity follows Beer's law.

A number of procedures for the determination of inulin have been proposed based upon the reactions of the products formed by hydrolysis. Addition of diphenylamine gives a blue color,[2,10,15] resorcinol a red color,[7,18] vanillin a red color, and indole-3-acetic acid a purple color.[11] Each method has limitations and the largest source of error appears to be the glucose present in the specimens. Many modifications have been suggested to eliminate the glucose or to correct for it.

Inulin clearance is the most accurate measure available of glomerular filtration. However, physicians are reluctant to permit the use of procedures involving catheterization of a patient because of the risk of urinary infection. This is especially true in the presence of some renal diseases.

Urea clearance

Urea, the most abundant end product of protein metabolism, is distributed throughout the intracellular and extracellular body fluids in equal concentrations. Since most of it is excreted from the body by the kidneys, the clearance of this waste product is a measure of kidney function, principally the effectiveness of glomerular filtration.

The concentration of urea in the glomerular filtrate is the same as in the plasma from which it was filtered. About one half is reabsorbed into the blood as the filtrate passes through the tubule. With the exception of production rate, the most important factor influencing the level of urea in blood is glomerular filtration rate (GFR). An increased blood urea level indicates a breakdown in the excretory functions of the kidney.

The clearance of urea measures the volume of plasma required to supply the quantity of urea excreted in a urine specimen in 1 minute. Each minute, the kidney clears or removes urea from the determined volume of plasma. Actually the kidneys never remove all of the urea from the blood and the level in the

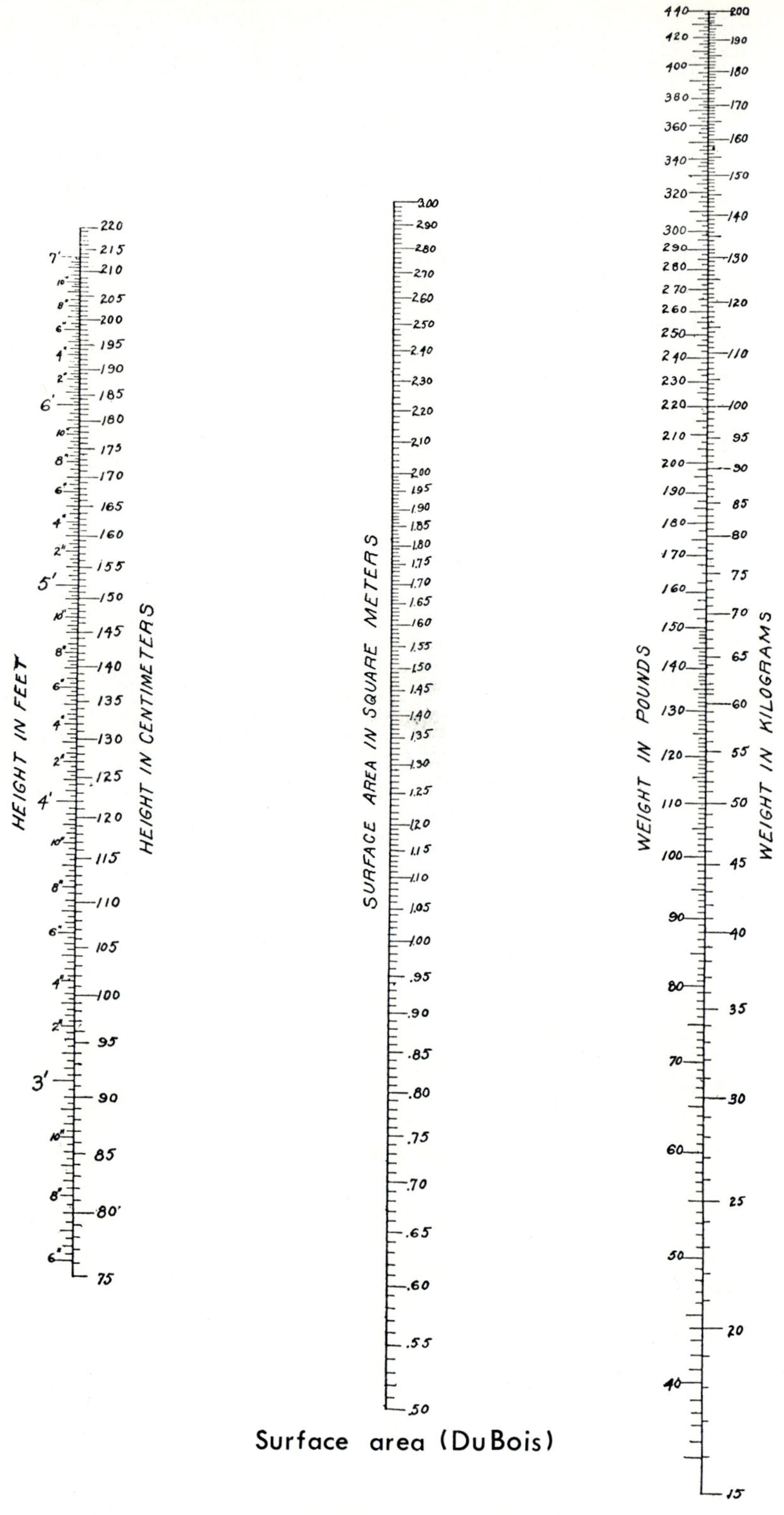

Fig. 14-2. Chart for estimating surface area of body. (From Boothby, W. M., and Sandiford, R. B.: Boston Med. Surg. J. **185**:337, 1921.)

renal vein is lower than in the renal artery, but the concept is a practical means of assessing renal efficiency. Since the rate of excretion is dependent upon urinary flow or volume, two tests have been devised; maximum urea clearance, where the flow rate of the urine is high, and standard urea clearance, where the urine flow is less than 2.0 ml/min.

Moller-McIntosh-Van Slyke method[14]

Because exercise interferes with the test, the patient is instructed not to exercise before or during the test. He is requested to eat a light breakfast, which may include water, but not coffee or tea, and should include little or no protein. The specimens are then collected in the manner given below. Two urine specimens are obtained so that one may serve as a check on the other. In the collection of urine it is essential that the patient completely empty the bladder and that the exact time interval between collections be measured.

Collection of specimens

1. **One hour after breakfast, give the patient a glass of water to drink. Ask him to empty the bladder completely and to discard the urine. Record the exact time at which the urine is voided.**
2. **Exactly 1 hour later, ask the patient to empty the bladder completely and to save the urine. Record the time at which this urine is collected and label the urine: 1-hour specimen. (If the patient cannot void at the prescribed time, the exact period of time that elapses should be recorded on the specimen. For example, if the patient voids 8 minutes after the prescribed time, the specimen should be labeled: 1-hour 8-minute specimen.)**
3. **Make a venipuncture and collect blood for a blood plasma urea nitrogen determination.**
4. **Give the patient another glass of water.**
5. **Exactly 1 hour after the collection of the first urine specimen, tell the patient to empty the bladder completely and to save the urine. Label: 2-hour specimen.**

Procedure

1. **Measure the volume of the 1-hour specimen of urine.**
2. **Calculate the milliliters passed per minute (ml/min) and record the results. Example: urine is collected at the prescribed 1-hour or 60-minute interval, the urine specimen measures 120 ml, and 120 divided by 60 equals 2 ml/min.**
3. **Measure the volume of the 2-hour specimen of urine.**
4. **Calculate the ml/min and record results.**
5. **Run a urea nitrogen determination on the blood and record results.**
6. **Run a urea nitrogen determination on the two urine specimens and record results.**

Calculations

1. **The following data, obtained above, must be on hand: 1-hour urine specimen, ml/min, and urea nitrogen value; 2-hour urine specimen, ml/min and urea nitrogen value; and plasma sample, urea nitrogen value.**
2. **If the patient is a child or an unusually small adult, see Note A, p. 340, before continuing.**
3. **The urea clearance of the 1-hour urine specimen and that of the 2-hour urine specimen are calculated separately. The two should not differ by more than 15%.**
4. **If the volume of urine excreted is more than 2 ml/min, use formula A below and make the calculations. If the volume of urine excreted is 2 ml or less/min, use formula B below and make the calculations.**

Table 14-2. Square roots of small numbers

V	$\sqrt{V}$	V	$\sqrt{V}$	V	$\sqrt{V}$	V	$\sqrt{V}$
0.2	0.45	0.7	0.84	1.2	1.1	1.7	1.3
0.3	0.55	0.8	0.89	1.3	1.14	1.8	1.34
0.4	0.63	0.9	0.95	1.4	1.18	1.9	1.38
0.5	0.71	1.0	1.00	1.5	1.23	2.0	1.42
0.6	0.78	1.1	1.05	1.6	1.27	2.1	1.45

Formula A:

$$\text{Urea clearance (in percent of normal)} = \frac{\text{Urine urea nitrogen (mg/100 ml)}}{\text{Plasma urea nitrogen (mg/100 ml)}} \times V \times \frac{100}{75}$$

where V = ml of urine per minute
and 75 = average normal value

Formula B:

$$\text{Urea clearance (in percent of normal)} = \frac{\text{Urine urea nitrogen (mg/100 ml)}}{\text{Plasma urea nitrogen (mg/100 ml)}} \times \sqrt{V} \times \frac{100}{54}$$

where V = ml of urine per minute
and 54 = average normal value

5. If formula B is used, the $\sqrt{V}$ will be needed. For convenience, the square roots of various small numbers are given in Table 14-2. If V = 0.2, the $\sqrt{V} = 0.45$, etc.
6. Sample calculations here are based on the following data obtained from a urine specimen and the plasma specimen:

   Volume of urine, 138 ml
   Rate of excretion (ml/min), 2.3
   Urine urea nitrogen, 244 mg/100 ml
   Plasma urea nitrogen, 12 mg/100 ml

   Since the volume of urine excreted per minute is more than 2 ml, formula A must be used:

$$\text{Urea clearance (in percent of normal)} = \frac{\text{Urine urea nitrogen (mg/100 ml)}}{\text{Plasma urea nitrogen (mg/100 ml)}} \times V \times \frac{100}{75}$$

$$= \frac{244}{12} \times 2.3 \times \frac{100}{75}$$

$$= 62.3\%$$

Note A:

1. If the patient is a child or an unusually small adult, a correction for size is essential.[13] In such cases, the milliliters excreted per minute must be multiplied by a size correction factor. This is derived in the following manner.
2. Obtain the patient's height and weight. Refer to Fig. 14-2 and place a ruler at the height and weight. Read the surface area from the middle column.
3. Example: Patient is 4 feet tall and weighs 65 pounds. Placing a ruler at these points gives a surface area of 0.97 m^2.
4. To obtain the size correction factor, divide the surface area into 1.73. (The 1.73 is the surface area of the average man.) Example:

$$\text{Patient's surface area} = 0.97\ m^2$$

$$\text{Size correction factor} = \frac{1.73}{\text{Surface area}} = \frac{1.73}{0.97} = 1.78$$

5. Now multiply this size correction factor by the milliliters of urine passed per minute. Use this corrected value for the milliliters passed per minute and continue with the calculations.

Normal values

Maximum average	75 ml/min (64 to 99)
Minimum average	54 ml/min (40 to 68)

Comment

The most common sources of error are incomplete emptying of the bladder and inaccurate timing.

Clinical findings

The clearance rate is lower than the filtration rate because of tubular reabsorption. The amount reabsorbed depends upon the flow rate of the urine in the tubules and the concentration gradients for urea between the urine and blood. When the urine flow is rapid and large, more of the filtered urea is excreted in the urine and the clearance is higher.

When kidney damage occurs, the remaining functioning nephrons increase their capacity to maximum. As GFR decreases, the blood urea concentration rises gradually although it remains within normal limits until the GFR is approximately 40% of normal. Additional loss of filtration capacity causes a rapid increase in blood urea. Thus in chronic renal failure a measurement of blood urea is not a sensitive indicator of renal function, since marked kidney damage can occur without elevating blood urea above the normal range. However, when the urea clearance is decreased, there is renal impairment.

With refinement in other techniques, urea clearance is less often requested because of the inherent variability. Most clinicians depend upon clearance of creatinine, which is described below.

Creatinine clearance[8,21]

The blood level of creatinine, formed from muscle creatine and phosphocreatine, depends upon skeletal muscle mass and is relatively constant in any individual from day to day. In the same manner as urea most of the creatinine is excreted in the kidney by glomerular filtration. Serum creatinine concentration rises when filtration rate falls. Protein variations in the diet have a slight but minimal effect. A 24-hour urine specimen is preferred although shorter periods, 2 or 6 hours, may be used.

Collection of specimens

1. At the beginning of the test ask the patient to empty the bladder completely and to discard the urine. Record the exact time at which the urine was voided.
2. Instruct the patient to collect all urine voided for the next 24 hours in a single container in which 15 ml of toluene has been placed. At the end of the 24-hour period ask the patient to void and add this urine to the total collection.

3. **Near the end of the urine collection draw a fasting blood sample for serum creatinine determination.**
4. **Record the height and weight of the patient for calculating surface area of the body.**

Procedure

1. **Measure the volume of the 24-hour urine specimen.**
2. **Determine the creatinine content of the urine and serum.**

Calculations

$$C_{cr} = \frac{UV}{P} \times \frac{1.73}{S}$$

where C_{cr} = endogenous creatinine clearance corrected to average body surface area
U = mg creatinine/ml urine
V = ml urine excreted/min
P = mg creatinine/ml serum
S = body surface area in square meters (Fig. 14-2)
1.73 = average body surface area in square meters

Normal values

Male	**105 ml/min (72 to 141)**
Female	**95 ml/min (74 to 130)**

Comments

The method generally employed for determining creatinine, the Jaffé reaction, whereby an orange-yellow color is formed with an alkaline picrate solution, is not specific. Normally some 10% of the reaction is due to noncreatinine chromogen. However, the results are considered acceptable for routine clinical purposes.

Clinical findings

Creatinine clearance is used as an estimate of glomerular filtration. It appears to be the most practical test available for clinical evaluation of renal function and in general is comparable to inulin clearance tests. In normal persons the ratio of C_{cr} to C_{in} is about 1.1 to 1.0. It is useful in following the progress of patients with renal insufficiency.

In advanced renal disease endogenous creatinine clearance is significantly higher than the filtration rate. Apparently creatinine is partly excreted by the renal tubules in such a situation. At this point the ratio of C_{cr} to C_{in} may be 1.4 or 1.5 to 1.0. However, this does not occur until there has been extensive renal damage and therefore the error in measurement is not large enough to be misleading.

Tubular functions

Without the reabsorptive activity of the tubules, the high rate of filtration by the glomeruli would rapidly eliminate all water and water soluble constituents from the body. By reabsorption and secretion the tubules maintain the volume, osmolality, and electrolytes of body fluids within normal limits. While they act to conserve water and essential solutes in a manner dependent upon the dietary intake and storage requirements, they excrete waste products and other substances not needed by the body.

Tubular reabsorption is designated as being either active or passive. Active transport refers to the transfer of a constituent such as glucose from a lower to a higher concentration, as from a low level in urine to a much higher level in blood. Energy is expended in the process. Reabsorption is passive when the substance diffuses down a gradient of chemical activity or electric potential, that is, from a high concentration in the urine to a lower one in the plasma, for example, urea.

Another term used in relation to the active absorption of certain substances is threshold. Some constituents of blood are not normally found in urine until their concentration exceeds a certain amount or threshold level. A constituent such as glucose that is completely reabsorbed from the tubular fluid at normal blood levels is described as having a high threshold. The capacity of the tubules to reabsorb glucose is not saturated until the blood level exceeds approximately 180 mg/100 ml. Below that level all glucose in glomerular filtrate is actively reabsorbed. At saturation of the tubular capacity for reabsorption of glucose, the rate of reabsorption becomes constant and is designated as the tubular maximum for glucose (T_{mG}) or maximal reabsorptive capacity for glucose. Other T_m-limited substances are phosphate, sulfate, amino acids, uric acid, proteins, and certain organic acids. If a given mechanism transports more than one substance, competition for the carrier may occur.

The hypertonic urine at the tip of Henle's loop becomes hypotonic in the ascending limb. The volume and concentration of urine is largely controlled by antidiuretic hormone (ADH). ADH is secreted into blood by the neurohypophysis in response to blood osmolality shifts. The concentration or dilution of urine occurs in the distal convolution and the collecting tubules. As the tubular fluid enters the distal tubule, it is hypotonic, becoming isotonic by the time it leaves. Further reabsorption of water continues during passage through the collecting ducts under the influence of ADH. In the absence of antidiuretic hormone, the urine remains hypotonic during its passage from the ascending limb of Henle to the renal papilla.

Concentration tests[3]

Regulation of the concentration of urine is one of the principal functions of the renal tubules and is often diminished in renal disease. Damaged kidneys have fewer nephrons to excrete a larger amount of solute. The formation of hypertonic urine by the distal tubule and collecting duct is fundamental to the concentrating mechanism. Therefore the ability to concentrate may be diminished before retention of nitrogenous waste products occurs in certain diseases such as polycystic kidneys, pyelonephritis, and hydronephrosis; since these conditions tend to damage the medulla earlier than the cortex of the kidney.

Measurements of concentrating ability are carried out by determining either osmolality or specific gravity of urine under standardized or known conditions. Anatomic or environmental conditions can cause urine flow to fluctuate from 0.25 to 25 ml/min. However, the osmolality tends to be inversely proportional to the volume when the tubules are functioning normally. The determination of milliosmolal concentration, calculated from freezing point depression, is a direct measure of the total excretion of solutes and of the concentration. The results can be compared with those of serum. If higher than serum, active concentrating ability is indicated.

If the urine is free from glucose and protein, measurement of specific gravity is sufficiently accurate for most clinical purposes. Since urea and salts are the principal solutes in urine, normal dietary intake must be maintained when the test is performed. An appropriately sealed refractometer may be used to determine specific gravity if the volume of urine is too small for the standard urinometer.

Mosenthal test

The Mosenthal test measures the specific gravity of urine at various intervals under specified conditions.

Collection of specimens

1. On the day of the test, as well as the day before, the patient is to have a full regular diet and is to drink at least a pint of fluids at each meal, but he is not to eat or drink between meals.
2. For the morning of the test, he is instructed to completely empty the bladder before breakfast at 8 AM and to discard the urine. The test is usually started at 8 AM.
3. He is then to completely empty the bladder into separate labeled containers at exactly 10 AM, 12 noon, 2 PM, 4 PM, 6 PM, and 8 PM.
4. Collect in a single container all the urine voided between 8 PM and 8 AM.
5. Bring all specimens to the laboratory.

Procedure

1. Measure the volume of each urine specimen.
2. Determine the specific gravity of each specimen.
3. Report the volume and specific gravity of each specimen. See sample report:

Specimen	10 AM	12 noon	2 PM	4 PM	6 PM	8 PM	8 PM to 8 AM
Volume (ml)	150	156	170	165	200	110	300
Specific gravity	1.017	1.020	1.014	1.022	1.012	1.010	1.018

Interpretation

Normal values are indicated by a fluctuation in the specific gravity of at least 7 points; that is, there is at least a 7-point difference between the lowest and the highest specific gravity. In kidney disease the specific gravity fluctuates only a few points and the volume of the night urine 8 PM to 8 AM) is usually over 600 ml. The test is useful for detecting slight degrees of renal dysfunction. Ability to concentrate is frequently abnormal in chronic glomerulonephritis, but it may be normal in acute glomerulonephritis if filtration rate is not impaired. Polyuria or nocturia interferes with the accuracy of the test.

Fishberg test[9]

The Fishberg test measures the specific gravity of urine at various intervals under specified conditions.

Collection of specimens

1. The patient is told to eat an evening meal consisting of a high-protein diet and not more than 200 ml of fluid.
2. After the evening meal, no food or fluid is to be taken until the completion of the test on the following morning.
3. The patient is further instructed to void at will during the night. In the morning, he should empty the bladder into separate containers at 8 AM, 9 AM, and 10 AM. The three specimens are then brought to the laboratory.

Procedure

1. Measure the volume of each urine specimen.
2. Determine the specific gravity of each specimen.
3. Report the volume and specific gravity of each specimen. See sample report:

Specimen	8 AM	9 AM	10 AM
Volume (ml)	210	190	150
Specific gravity	1.020	1.026	1.025

Interpretation

If kidneys are healthy, at least one of the specimens should have a specific gravity over 1.024. With diseased kidneys all the specific gravities are below 1.020. Achievement of a high specific gravity depends upon both the availability of excretable solute and a maximal stimulus to conserve water. The test is a sensitive indicator of early renal dysfunction. Patients with advanced renal disease may show a fixed urinary specific gravity value between 1.010 and 1.012, the same as an ultrafiltrate of plasma. In edema and congestive heart failure low urinary specific gravity levels may be found, simulating renal impairment.

Dilution tests

In renal disease the ability of the kidneys to dilute urine is impaired. With normal renal function, following a water load, most of the water ingested will be excreted within 4 hours, and the specific gravity of at least one hourly urine specimen will be below 1.003. In renal impairment a low specific gravity is not reached and the urine volume remains low. However, since the ability to concentrate is lost before the ability to dilute, this type of test is of little value in assessing early change in renal function.

Tubular excretory function

Substances such a *p*-aminohippuric acid, Hippuran, and Diodrast are actively excreted by the tubule cells and may be employed to measure tubular secretory function. The limit or maximum capacity of the tubules to reabsorb or secrete these compounds is termed the tubular excretory maximum (T_m). Para-aminohippuric acid (PAH) is most frequently used in renal function studies. At very high blood levels of PAH the transport system will operate at maximum capacity or saturation. The plasma concentration must be above saturation to force the tubule cells to excrete at their maximum capacity over a sufficient period of time to conduct clearance studies. The high constant blood level can be achieved only by a sustained infusion of a solution of PAH. By determining the maximum amount of PAH that can be excreted per minute by the tubule cells and the percentage of PAH removed from the blood during passage through the kidneys the tubular excretory maximum can be calculated. It is an index of functional tubular tissue, since the excretion of PAH diminishes as the mass of tubular cells is reduced. In normal renal function, the maximal tubular excretory capacity for PAH (T_{mPAH}) corrected to 1.73 m^2, the average surface body area, is[20]:

Men	79.8 ± 16.7 mg/min
Women	77.2 ± 10.8 mg/min

The test is not used as a clinical method, since it is time-consuming, difficulty is encountered in maintaining saturation of the plasma with the test compound, and application is limited to one transport mechanism.

Renal blood flow

The kidneys are dependent upon the blood supply for transport to and from them of biologic constituents that are selectively either eliminated from the body into the urine or reabsorbed into the blood. Specific ions or molecules are actively transported in the exchange between the tubular fluid and the blood, while other solutes and water passively diffuse between the lumen and blood according to osmotic or electric potential gradients.

The total renal blood flow is that volume which enters the renal arteries each minute (approximately 1200 ml). The *effective* renal blood flow, or renal plasma flow, is that amount which perfuses the *functional* renal tissue each minute. The latter may be measured by determining PAH clearance and the effective renal blood flow be calculated by applying the hematocrit value. At low plasma levels about 90% to 93% of PAH or of Diodrast is removed in one passage through the kidneys. Agreement is lacking among investigators concerning the disposition of the remaining 7% to 10%.[4] Suggested explanations have been (1) that it is in the blood flowing through inactive or nonfunctioning tissue or (2) that there is a complex formed between PAH and glucose which has a lower clearance rate than PAH. Variation in effective renal plasma flow is great even when corrected to standard body surface area. One standard deviation for any population is nearly 20% of the total mean value.

Phenolsulfonphthalein (PSP), a dye that is readily excreted by the tubules, may be used also to determine renal blood flow. The test lacks sensitivity in that there can be 40% renal impairment before detection by use of the phenol red, but the ease with which the test can be performed has made it popular.

PAH clearance[7]

Since PAH is not normally present in blood, sufficient amounts must be administered to maintain a constant blood level during the clearance test. Below 5 mg/100 ml the PAH clearance of plasma is not dependent upon the concentration. Some investigators[12] prefer a screening procedure that employs a single intramuscular injection of 8 ml of 20% PAH-Na solution and 4 ml of an anesthetic containing epinephrine. The patient is asked to drink several glasses of water and the clearance determination is started after 30 to 50 minutes. Two timed clearance periods of 45 to 60 minutes are used where urine and blood specimens are obtained in the same manner as for other clearance tests.

A more exact PAH clearance may be carried out by using the procedure as described for inulin clearance. Frequently the two clearances are performed simultaneously by including the proper amounts of both chemicals in the priming and infusion solutions.

Collection of specimens

The preparation of the patient, administration of PAH, and collection of urine and blood samples are the same as for inulin clearance. A priming solution of 8 mg of the sodium salt of PAH per kilogram of body weight is administered and followed by a sustaining infusion containing 8 ml of a 20% PAH solution diluted to 500 ml with saline administered at a rate of 4 ml/min.

Procedure

The procedure is the same as that employed for the determination of sulfonamides (p. 505) except the standard solution is prepared from *p*-

aminohippuric acid. For the analyses plasma and urine diluted 1:100 are used. The calculations are the same as for inulin clearance.

Normal range

The normal PAH clearance at low plasma levels corrected to 1.73 m² of body surface as a measure of renal plasma flow is as follows:

Male:	700 ± 135 ml plasma cleared/min
Female:	600 ± 100 ml plasma cleared/min

The effective renal blood flow may be calculated from the plasma clearance by applying the patient's hematocrit.

Male:	1275 ± 245 ml blood cleared/min
Female:	1090 ± 180 ml blood cleared/min

Clinical findings

A decreased effective renal blood flow will be noted when (1) the cardiac output is reduced or arterial blood pressure is lowered as in congestive heart failure or shock; (2) the renal vascular system is damaged as in glomerulonephritis, intercapillary glomerulosclerosis, or renal arteriosclerosis; (3) the mass of functioning kidney tissue is decreased as in polycystic disease, pyelonephritis, renal hypoplasia, or malignancy; or (4) there is constriction of the afferent glomerular arterioles in conditions such as essential hypertension.

An increased renal blood flow may be found following the administration of pyrogenic substances.

Phenolsulfonphthalein (PSP) excretion test[5,6,17]

Phenolsulfonphthalein is a nontoxic dye that binds reversibly to the albumin fraction of plasma. When administered most of it is excreted by the renal tubules and the rate of excretion depends principally upon the renal plasma flow. Therefore as used in low concentration in a test procedure it provides an indication of renal blood flow.

Collection of specimens

The nurse and physician usually have the task of instructing the patient, injecting the dye, and collecting the urine specimens. This consists of the following:

1. The patient is told not to take any stimulants (coffee, tea, or tobacco) for 2 hours prior to the test and also during the test.
2. The patient voids and then is given several glasses of water to drink to promote urine flow. The test is started about 30 minutes later.
3. Exactly 1 ml (6 mg) of sterile phenolsulfonphthalein is injected intravenously and the time noted and recorded.
4. Exactly 15, 30, 60, and 120 minutes after injection urine is collected. The time of collecting each sample is recorded.
5. The specimens are then brought to the laboratory.

Reagents needed

Sodium hydroxide, 10% (2.5 N) (D-146)

For method B

PSP stock standard, 120 mg/500 ml (D-259)

Working standard, 1.2 mg/500 ml

Pipette 5.0 ml of stock standard into a 500 ml volumetric flask, add 1.0 ml of 10% NaOH, and dilute to the mark with distilled water. Mix thoroughly.

Procedure

1. Measure and record the volume of each specimen.
2. Add 10% sodium hydroxide dropwise to each specimen until the maxi-

mum red color is brought out. About 5 ml is usually required. The dye is colorless in acid urine and red in alkaline urine. Check the pH. It should be higher than 9.0.

3. Dilute the urines according to volumes, as follows:
   a. If the volume is less than 50 ml, dilute to 250 ml with water.
   b. If the volume is between 50 and 450 ml, dilute to 500 ml with water.
   c. If the volume is above 450 ml, dilute to 1000 ml with water.
4. Mix each specimen.
5. Determine the percent of dye in each specimen by one of the two following methods:

Method A: Compare with Dunning standards representing known percentages of the dye.

Method B: Compare in a photoelectric colorimeter or spectrophotometer. Filter about 10 ml of the urine. Pipette 6.0 ml of filtered urine into a cuvette and 6.0 ml of diluted working standard into another cuvette. Read absorbance (A) against a water blank at 520 mμ.

Calculation

$$\frac{\text{A of unknown}}{\text{A of standard}} \times 1.2 = \text{mg PSP in specimen}$$

NOTE: If diluted urine volume is 250 ml, divide by 2. If diluted urine volume is 1000 ml, multiply by 2. The sum of the mg PSP in the specimens is the total PSP excreted. Since 6 mg is usually injected, calculate the percent excreted as follows:

$$\frac{\text{Total mg PSP}}{6} \times 100 = \text{percent excreted}$$

Normal values

Time (minutes)	*Percent PSP excreted*	
	Average	*Range*
15	35	28-51
30	17	13-24
60	12	9-17
120	6	3-10

The 15-minute value is the most important for evaluating renal function as some damaged kidneys are able to excrete normal amounts from the blood in 2 hours. It is important that the patient be loaded with water before starting the test to minimize the errors in emptying the bladder.

Clinical findings

Excretion is decreased in congestive heart failure, chronic nephritis, cirrhosis with normal kidneys, cystitis, pyelonephritis, prostatic obstruction, and by edema. In marked renal damage the excretion of PSP tends to vary directly with urine volume. Mild nephritis, essential hypertension, and nephrosclerosis cause some delay in excretion. In chronic renal insufficiency the excretion decreases as the disease progresses. Excretion of PSP may be normal in acute glomerulonephritis even when glomerular filtration is reduced and plasma urea is elevated.

Bile or blood in the urine will interfere and must be removed before analysis for the PSP content is made. Bromsulphalein will also interfere and a period of 24 hours should be allowed between the two tests.

Blood chemistry in renal disease[16,23]

Measurements of the concentration of various chemical constituents in blood are used to assess renal function. However, because of their large reserve capacity the kidneys can maintain normal blood levels when only one half of one kidney is functioning normally.

In the initial stages of renal insufficiency the involvement may be located primarily in either the glomeruli or the tubules, or in both. As a disease progresses, destruction of nephron units occurs until finally the quantity of functioning kidney mass is inadequate to maintain normal body chemistry. In chronic renal failure the nephrons are destroyed without the possibility of reversal, but blood tests do not reflect the condition until extensive kidney damage has occurred with the accompanying retention of nitrogenous metabolic products.

In acute renal failure, depending upon the cause, the nephrons may be only temporarily injured and the process can be reversed if proper therapy is initiated before permanent damage occurs. In these cases the changes in blood constituents do provide an indication of the degree of renal dysfunction.

Urea

Factors other than renal function affect the urea concentration in blood. A patient on a restricted-protein diet or one with severe liver disease may have markedly reduced kidney function, yet a normal blood urea level. Since urea is synthesized in the liver, a malfunctioning liver will produce less urea and the level in the blood may be normal even in the presence of renal damage. A high-protein diet or rapid cell destruction, as in intestinal hemorrhage, may increase the blood urea above the normal concentration range.

Blood urea concentration does not start to increase until glomerular filtration has fallen below 50%. However, the test is useful in serial determinations as a guide during treatment of renal failure, and a sudden rise or elevated value should be considered as indicating probable impairment of renal function.

Creatinine

The concentration of creatinine in serum is a more sensitive index of renal function than is urea. It depends essentially upon muscle mass and is largely independent of protein intake. In a situation where there is severe muscle wasting, the blood level may be exceptionally low. For each individual the muscle mass is relatively constant and therefore the serum creatinine concentration varies inversely with glomerular filtration. A markedly elevated value is indicative of severe renal insufficiency.

Potassium

In chronic renal insufficiency, elevated serum potassium levels are not present until the final stage of the disease. Except when oliguria or anuria occurs, the kidneys are able to maintain a normal level of urinary potassium by tubular secretion. Hyperkalemia may be found in patients with decreased urine output when there is an increased supply of potassium such as occurs in tissue breakdown or exogenous potassium administration. In acute renal failure, an elevated serum potassium is a most important chemical finding because of the ion's toxic effect on the heart.

Potassium depletion may be the result of, or a cause of, renal disease. It may occur postoperatively when losses arising from surgical trauma are not replaced, from administration of diuretics or adrenal steroids with inadequate intake, and from renal losses. Hypokalemia can produce tubular damage with impairment of the ability of the kidneys to concentrate.

Sodium

As a result of decreasing ability of the kidneys to concentrate, the serum sodium concentration is often reduced in progressive renal failure. This appears to result from the inability of the functioning tubules, because of the increased osmotic load, to bring about the necessary reabsorption of sodium from the tubular fluid. In the final stage of chronic renal failure, or when the urine flow is very low, serum sodium is increased and the patient is then maintained on a restricted low-sodium diet.

Bicarbonate

A characteristic finding in renal disease is a lowered plasma bicarbonate, or metabolic acidosis, resulting from the accumulation of acid waste products. Normally the kidneys eliminate acid by combining it either with ammonia or with buffers, principally sodium dihydrogen phosphate, to form titratable acid. The phosphate buffer is obtained from inorganic phosphate in glomerular filtrate, derived from serum. In chronic renal insufficiency, the ability of the kidneys to provide the required ammonia and phosphate buffer decreases as the disease progresses.

Calcium and phosphate

In chronic renal failure the plasma calcium concentration is frequently below normal. However, factors other than serum phosphate concentration appear to influence the level, which may remain normal even with increased serum phosphate or may be decreased with little change in phosphate concentration.

In the later stages of chronic renal disease, there is often marked elevation of serum phosphate. However, this does not occur until there is marked decrease of renal function. The metabolic bone disease, renal osteodystrophy, which is associated with azotemia may cause malacia, sclerosis, or cystic changes in bone; but there is no consistent calcium or phosphate pattern that is characteristic of the condition.

Magnesium

Hypermagnesemia is frequently observed in advanced chronic renal failure. However, normal or low values have been found for some patients. It has been suggested that the calcium level may influence the magnesium concentration. The fact that the ratio of bound to ionized serum magnesium is increased in uremia may be important.

Proteins

Hypoproteinemia is characteristic of patients with chronic renal insufficency. It may result from loss of large quantities of protein in the urine, dietary restriction of protein to control azotemia, and defects in protein metabolism.

Renal insufficiency and dialysis

CHARLEY F. GUTCH, M.D.

The clinical syndrome of renal insufficiency is the result of a decrease in renal function, due to injury or disease, so that metabolic wastes are not adequately removed from the body by urinary excretion. Complicated disturbances of water and electrolyte balance accompany the retention of catabolic products.

Renal insufficiency may be either acute or chronic. Acute renal failure develops rapidly—within hours or a few days—and is usually manifested by oliguria (24-hour urine output less than 400 ml). Often it is the result of acute tubular necrosis, which is a potentially reversible lesion, if the patient can be kept alive for the 2 to 3 weeks required for recovery to begin. The mechanism of acute tubular necrosis is unclear, but decreased perfusion—due to renal vasoconstriction, or shunting of blood away from the kidneys—seems to play a major role. Shock or severe hypoxia, pancreatitis, head injury, severe dehydration, carbon monoxide intoxication, trauma, and septicemia are frequent precipitating factors. Other causes include crush syndrome (with production of hemoglobinemia or myoglobulinemia), sickle cell crisis, hemolysis after transurethral resection, mismatched transfusion, severe burns, and hemorrhagic fever.

Acute renal failure (not always limited to tubular lesions only) may also result from specific nephrotoxic substances—mercury, bismuth, phosphorus; some antibiotics; organic compounds such as carbon tetrachloride; and from mushroom poisoning.

Rapid onset of oliguria, or even anuria (urine output less than 100 ml/24 hr), is often a feature of acute, fulminating glomerulonephritis; necrotizing arteriolar disease; or the consumption coagulopathies. Such lesions, unlike acute tubular necrosis, are often not reversible, although sufficient return of function to maintain life may be regained. Renal function patterns, urine sediment, and renal biopsy serve to permit differentiation.

Renal failure from acute tubular necrosis has an overall mortality rate of 50% with conservative treatment. Early and adequate dialysis has improved the outlook for patients whose oliguria follows hypotension or hemolysis; it has also greatly reduced their morbidity. The mortality of persons with renal failure as a result of extensive trauma remains high even with dialysis, although such patients die of sepsis or other complications rather than uremia.

Chronic renal insufficiency may follow acute or subacute glomerulonephritis, which progresses relentlessly to loss of functional renal mass. However, not infrequently the patient may be unaware of the underlying condition for many years. Symptoms may not become apparent until there is marked elevation of serum urea and creatinine, by which time the functional kidney mass is down to 25% to 50% of normal. Insidious chronic renal insufficiency of this sort may follow an unrecognized glomerulonephritis of many years' duration.

Pyelonephritis, the other major cause of chronic renal failure, is the result of direct bacterial invasion of kidney parenchyma. It is often indolent and unrecognized until renal failure supervenes.

Polycystic disease, a congenital anomaly in which functional renal tissue is gradually compressed by fluid-filled cysts, often remains undetected until the fourth or fifth decade of life, when kidney failure begins to manifest itself.

Diabetic glomerulosclerosis is not infrequently an end-stage complication of long-standing diabetes, particularly the diabetes that has its onset in childhood or adolescence. Persons with gout may develop a nephropathy as a result of urate deposits in renal parenchyma. People with primary hypertension may develop arteriolar nephrosclerosis, in which damage to renal arterioles causes loss of function due to ischemia.

Obstruction of the lower urinary tract (such as prostatic enlargement) or other anatomic defect may lead to damming back of urine, resulting in striking distention of the renal pelvis and calyces (hydronephrosis) and decreased renal function due to the pressure. There are many other, less common disorders that can cause insidious loss of renal function to the point of chronic renal failure.

The body can adjust to rather severe biochemical disturbances, particularly if these have been gradual in development. Persons can maintain an active life with less than 10% of normal renal function.

DIALYSIS

Dialysis is defined as the separation of solutes and/or crystalloids by differential diffusion across a porous (semipermeable) membrane between two solutions. When there is a concentration gradient between the two solutions, involving one or more substances, differential diffusion occurs. Natural or artificial membranes may be used as the semipermeable screen in clinical application of dialysis.

In the reversible renal failure of acute tubular necrosis, dialysis therapy can remove metabolic wastes and minimize biochemical abnormalities until function of the patient's own kidneys returns. Occasionally the nature of the basic renal lesion is not apparent in a patient with acute renal failure, and dialysis may be undertaken while appropriate diagnostic procedures are initiated.

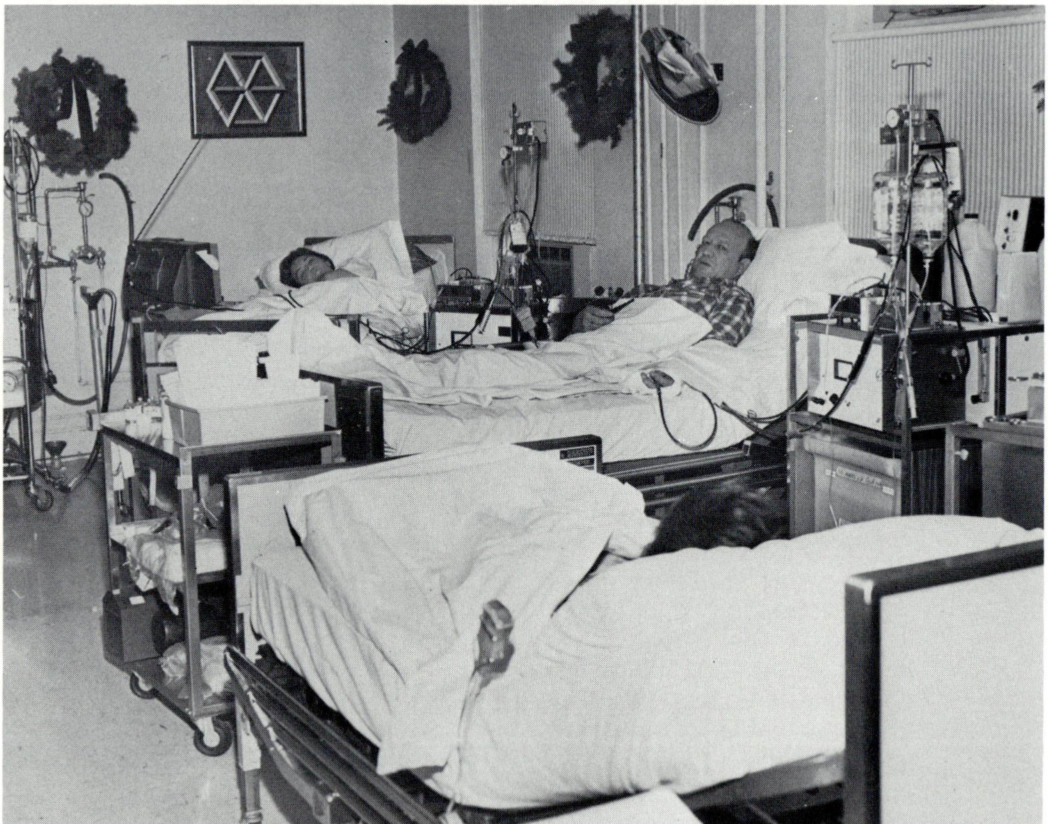

Fig. 14.3. Chronic Dialysis Center, University of Colorado Medical Center, Denver.

Additionally, some patients with permanent renal insufficiency, of such severity as to be incompatible with life, can be maintained by regular dialysis treatment for prolonged periods. Several hundred such people are now alive after three, four, or more years of regular maintenance hemodialysis, even though they have little or no renal function of their own. Such patients have numerous biochemical abnormalities, many of which are even now poorly understood. Their lives are not normal but can be useful and satisfying (Figs. 14-3 and 14-4).

Hemodialysis

The first hemodialysis apparatus, or artificial kidney, was designed by Abel, Rowntree, and Turner in 1913.[24] Cylindrical tubes of collodion were used as the semipermeable membrane. These inventors did not have adequate anticoagulants to keep blood from clotting and they had difficulty controlling the electrolyte concentration of the bath, so that their apparatus could not be used clinically. Thirty years later (1943) Kolff developed and used the first practical artificial kidney, employing a length of cellophane tubing wound in a spiral manner on a rotating drum. The rotating type of kidney was described in 1943 by Kolff and Berk,[34] and detailed plans were published by Kolff in 1947.[33] Several other types of hemodialyzers were developed about the same time. In 1947, Alwall[25] (Sweden) developed an efficient apparatus, and in the United States Skeggs and Leonards[41] devised a layer type dialyzer utilizing cellophane sheets placed between grooved rubber plates. In 1955 Kolff[35] perfected the twin-coil dialyzer, which was then mass produced by Travenol Laboratories and was the prototype of present coil dialyzers. The Kiil dialyzer, a simplified layer design,

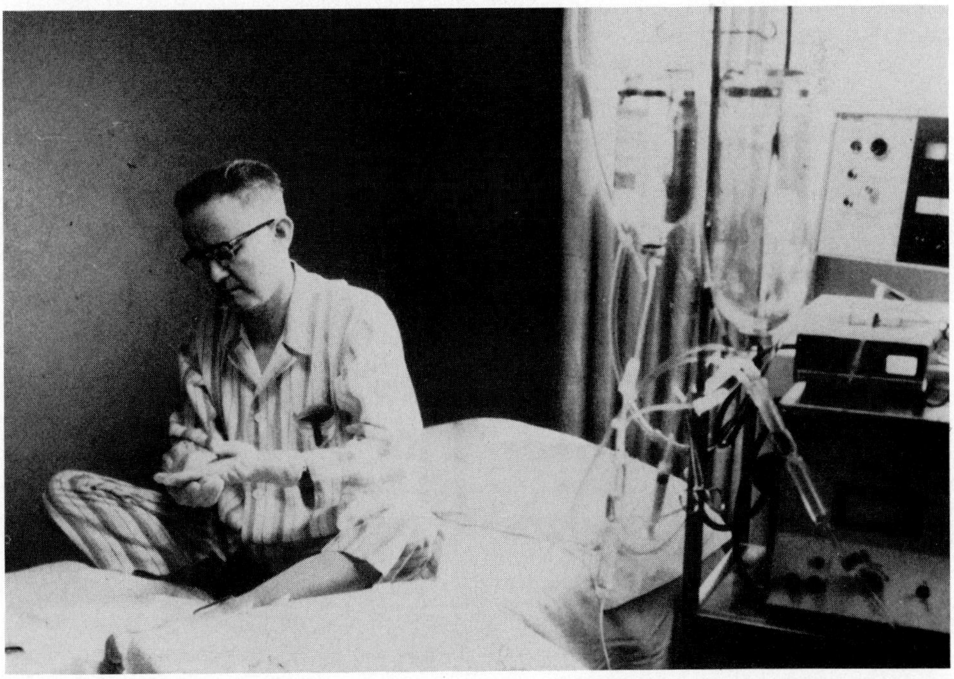

Fig. 14-4. Home dialysis patient preparing to attach himself to the dialysis unit, University of Colorado Medical Center, Denver.

was modified in 1960 by Scribner's group in Seattle.[39] The Kiil and the Kolff coil are the two most widely used dialyzers currently.

Initially, hemodialysis was considered suitable only for patients with reversible renal failure. Since only one or two dialyses were usually performed, access to blood vessels was obtained by cannulating an artery and a vein in an arm or leg with metal, glass, or plastic cannulae. Scribner and co-workers[40] realized that if repeated dialysis was to be done for chronic patients, permanent and easier means of access to the blood vessels was needed. They devised indwelling cannulae of Teflon and Silastic, with an external connecting loop outside the skin. Creation of this external arteriovenous fistula prevented clotting when not in use, yet the arterial and venous tubes could be attached to the artificial kidney whenever needed.

In recent years intensive research has gone on to develop smaller and more economical hemodialysis devices. Cuprophane, a copper-treated cellulose membrane, is thinner than cellophane and more permeable. It is in common use in both Kiil and coil dialyzers. Other types of membranes are being investigated, including some formed from collagen from animal sources. A hollow cellulose-fiber kidney—somewhat reminiscent of the original apparatus of Abel, Rowntree, and Turner but using very tiny capillary tubes and having a small blood volume—has been developed and used clinically.

Dialysate solutions

Hemodialysis involves the passage of blood along one side of a semipermeable membrane while a special bath solution flows along the other side. The membrane has pores of 25 to 30 Angstrom units' diameter, so that red cells, leukocytes, and serum proteins do not pass through; but solutes—including the metabolic wastes such as urea, creatinine, and uric acid, and other undesired products—diffuse through. On the bath side of the membrane is a solution of electrolytes prepared to approximate the concentration of plasma water. Thus, desired electrolytes are maintained in equilibrium on both sides of the membrane, while wastes and unwanted ions such as sulfate and phosphate pass from the blood to the area of lesser concentration, the bath.

The several varieties of dialysis fluid delivery systems may be classified into three general categories:

1. Recirculating system. In recirculating systems a quantity of dialyzing fluid (usually 100 L) is continually recirculated through the dialysis apparatus. This bath is changed every 2 or 3 hours, to maintain a gradient of waste product from blood to bath and to control bacterial growth in the bath.
2. Single-pass system. For the single-pass systems, fresh dialysis fluid flows through the dialyzer only once, passing on to the waste collector.
3. Recirculating single-pass system. The recirculating single-pass system combines the other two. A small recirculating volume (8 to 12 L) is continually replenished with fresh dialysis fluid at a slower rate. The volume used varies with each system and with the duration of dialysis.

Fluid for dialysis may be prepared by the single-batch method, whereby a given volume is made up and checked for proper electrolyte concentration immediately before use. The tanks for such systems are usually of 100, 200, 300, or 385 L capacity. For acute patients, fluid prepared with tap water is usually satisfactory. However, when chronic patients are dialyzed repeatedly, the chemicals in the raw water (usually sodium, calcium, magnesium, carbonate, sulfate, and, sometimes, fluoride) may assume significance. Filtering, softening, or deionization may be necessary to ensure uniformity.

The dialysis bath may be prepared using reagent grade chemicals. A common formula for the Kolff 100 L unit is as follows:

Component	(g/100 L)	Na^+	K^+	Ca^{++} (mEq/L)	Mg^{++}	Cl^-	HCO_3^-
NaCl	570	97				97	
$NaHCO_3$	300	36					36
KCl	30		4			4	
$CaCl_2$	28			5		5	
$MgCl_2$	7.5				1.5	1.5	
Total		133	4	5	1.5	107.5	36

Dextrose or invert sugar is added to a concentration of 200 to 400 mg/100 ml to provide appropriate osmolality. Lactic acid is used to adjust the pH to 7.4, and 10% CO_2 with 90% oxygen is bubbled through the fluid continually to maintain the carbonate concentration.

When the quantity prepared is other than 100 L, appropriate adjustments must be made in the amount of each chemical used. The calcium and magnesium chloride should be dissolved in a separate container of warm water and added to the solution of sodium chloride and bicarbonate after these are dissolved, the lactic acid added, and the tank nearly full, in order to avert precipitation of the calcium and magnesium.

Most dialysis centers now employ commercially prepared concentrates that, appropriately diluted, yield the desired concentration of electrolytes. Sodium acetate is used in these rather than bicarbonate to achieve stability. This may slow correction of a patient's acidosis slightly since the acetate must be metabolized by the body to bicarbonate. Also in prolonged dialysis, there may be a tendency to metabolic alkalosis at the end.

Commercial concentrates may be supplied in bulk (50 gallons) or in premeasured containers for specific machines. There are a number of formulations available with various concentrations of sodium, calcium, potassium, and magnesium to meet varying needs of chronic dialysis patients. It is extremely important to double check all concentrate containers and to verify that the correct dilution is obtained.

Verification of the dialysis bath is of critical importance to the well-being of the patient. Several deaths have resulted from bath errors. Bath checks should include (1) pH, (2) test for solute concentration, either osmolality by freezing point depression or refractivity on a high-grade refractometer, and (3) accurate titration for total chloride content since sodium and chloride are the major ions present. The ionic content may alternatively be registered by an electric conductivity probe. Conductivity (reciprocal of resistance) is an accurate index of ion content of the solution if the instrument is temperature compensated. All tests should be recorded as a permanent record and instruments and methodology verified against known standards at frequent intervals.

Many large dialysis centers, and some home dialysis equipment, use an on-line dialysis fluid–mixing system. In these units, concentrate is metered and continually mixed with water, usually by a set of interconnected proportioning pumps. Thus a fresh supply of appropriately mixed fluid is constantly provided. These systems most commonly are used with layer type dialyzers where the duration of the run and quantity of dialysate used are greater than in coil systems. Such central proportioning equipment has built-in filters, heaters, one or more conductivity controls, and temperature controls. These are designed to operate within very narrow limits and to be "fail-safe" in that flow of dialysis fluid is stopped and alarm indicators activated if there is any deviation from preset limits.

Dialysis equipment and procedures

The setup of two types of dialyzers will be described as examples of procedures for extracorporeal hemodialysis therapy.

Twin-coil dialyzer (Fig. 14-5)

The coil and the arterial and venous tubing sets are separately packaged and are factory sterilized. Arterial blood enters the coil at the inner core and leaves at the peripheral. Because of the internal resistance in the coil, a pump is necessary to propel the blood through the unit.

Assembly (Fig. 14-6)

1. **Soak the coil for 5 minutes in warm (37° C) water or dialysate.**
2. **Fasten the coil (with inflatable cuff, if required) in the canister.**
3. **Using sterile technique, connect the arterial set to the inlet side of the coil.**

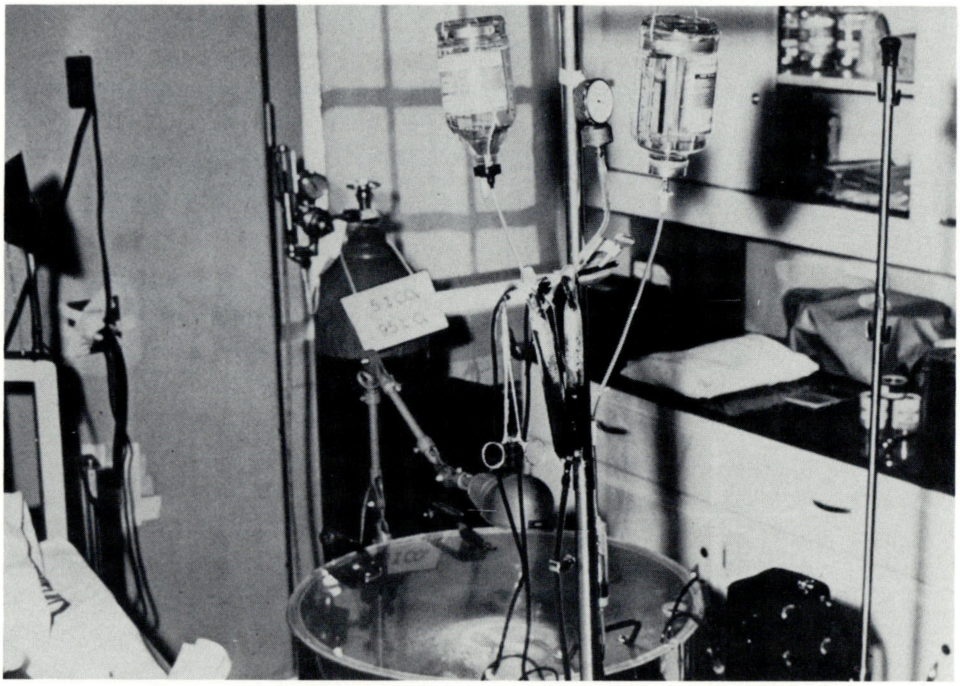

Fig. 14-5. Hemodialysis with twin-coil dialyzer using 100 L tank.

4. Position the pumping segments in the blood pump and clamp.
5. Suspend from a standard one or more liter bottles of heparinized saline and attach the infusion tubing to the priming tube inlet.
6. Attach the venous set to the outlets of the coil.
7. Suspend a bubble trap chamber from an IV standard and attach one side-arm tubing to the positive pressure monitor.
8. Prepare the dialyzing fluid in the tank at the required temperature (39° C) with a continuous bubbling of the CO_2-O_2 gas mixture (10% to 90%) if using a bicarbonate bath. Verify the dialysate bath as described previously and start the recirculating pump.

Testing

1. Slowly pump sterile, heparinized saline from the infusion set through the coil at a rate not to exceed 75 ml/min until the saline reaches the bubble trap on the outlet side. Rinse the coil with 500 to 1000 ml of saline.

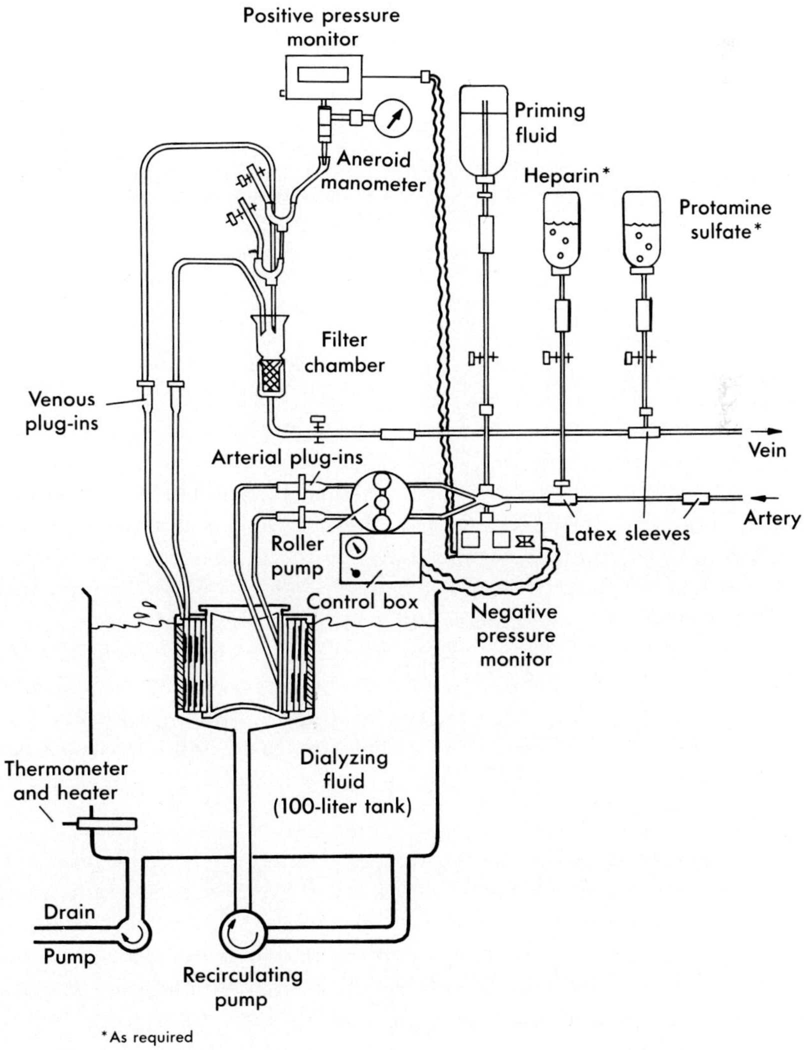

Fig. 14-6. Schematic of twin-coil system in operation.

2. **Cross-clamp the outlet tubing below the bubble trap.**
3. **When the manometer pressure reaches 200, stop the blood pump.**
4. **Pressure will drop slowly, 5 to 10 mm/min. A precipitous drop indicates a leak or inadequate pump occlusion.**

Connecting patient to dialyzer

1. **Adjust the bubble trap pressure to 70 to 100 mm Hg.**
2. **Fill the proximal arterial line with saline from the infusion set.**
3. **Attach the arterial line to the patient's arterial cannula.**
   a. **If transfusion is to be used for priming, attach heparinized blood to side arm of arterial set and start pumping. When blood fills the venous line, clamp, stop the pump, and attach the venous line to the venous cannula. Remove the clamps from both lines and start the blood pump slowly.**
   b. **If the patient is to be bled for priming, attach the arterial line to the arterial cannula and start the blood pump very slowly. When blood reaches the venous line, attach to the venous cannula.**
   c. **For saline priming, attach both arterial and venous lines to appropriate cannulae and start the blood pump slowly.**
4. **As soon as blood enters the coil, watch the canister for evidence of blood leakage.**
5. **Attach the heparin infusion line to the arterial line or give heparin intermittently as determined by clotting times.**
6. **During the first 15 minutes carefully monitor the patient's blood pressure. The blood pumping rate may be gradually increased to 200 to 300 ml/min. Adjust the level of the blood in the bubble trap and the pressure on the manometer by a screw-clamp below the bubble trap. With increased pressure more ultrafiltration results.**

Coil dialyzers require a high rate of bath flow over the membrane and this is accomplished by a pump delivering approximately 30 L/min. In the simplest type of coil dialyzer the bath is made up in a 100 L tank and recirculated through the coil. The bath is changed at intervals of 2 to 3 hours to maintain the diffusion gradient from blood to bath. To replace the dialyzing fluid during the dialysis of a patient, the blood pump is kept running while the bath fluid is removed from the tank with the exhaust pump. The recirculating pump is stopped before all of the fluid is out of the tank and is not restarted until the new batch of dialyzing fluid is prepared and in the tank. Each new batch of dialyzing fluid must be verified. If it is not isotonic before being pumped through the coil, hemolysis will occur in the cellulose tubes of the coil.

In the recirculating single-pass (RSP) coil dialyzer the coil is immersed in a small (8 L) compartment with a recirculating pump. Fresh dialysate from a separate tank is continually added at a slower rate, usually 500 ml/min, while an equal quantity of used dialysate runs off through an overflow.

Kiil dialyzer (Fig. 14-7)

The modified Kiil dialyzer consists of three rigid plastic boards that fit together in sandwich fashion with two sheets of membrane (cellophane or cuprophane) between each layer of the boards. Layer type dialyzers have less resistance than the coil dialyzers and the patient's arterial pressure is usually sufficient to provide adequate blood flow. Limited thickness of the blood film, uniform distribution, and avoidance of laminar flow are critical factors in dialyzer efficiency. The Kiil and other layer type dialyzers employ the single pass of dialysate at flow rates of 300 to 1500 ml/min.

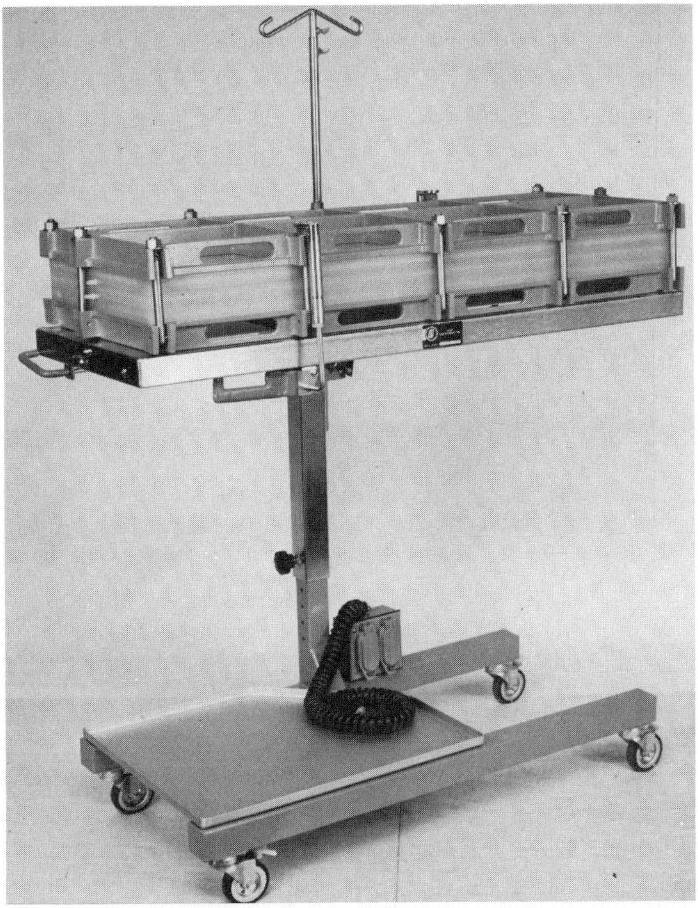

Fig. 14-7. Kiil dialyzer. (Courtesy Cobe Laboratories, Inc., Denver.)

Dialyzer assembly

Membrane preparation

1. Fill the special membrane pan with hot water and carefully lay the membranes on the water.
2. Soak the membranes for at least 15 minutes.

Stacking and clamping

1. Center the bottom dialyzer board on the bottom clamp with grooves upward.
2. Remove one wet membrane from the pan, stretch gently, and place carefully onto the board.
3. Place a blood port in the port housing at each end of the board.
4. Layer a second membrane over the first.
5. Position the middle board.
6. Repeat the procedure with a second set of membranes and blood ports.
7. Position the top board with grooves downward.
8. Place the top dialyzer clamp on the assembled boards and clamp the dialyzer, torquing the bolts to a designated pressure.

Pressure testing

1. Clamp off all tubes on one end of dialyzer.
2. Test each blood compartment separately.
3. With manometer attached, pump each blood compartment to 260 mm

Hg air pressure. After allowing 2 minutes for the membranes to stretch, the rate of fall of pressure should not be greater than 6 mm/min.
4. Test the dialysate compartments in a similar fashion.

Sterilizing

1. With dialyzer in a vertical position, fill both blood compartments from below, to displace air, with 4% formalin solution.
2. Fill dialysate compartments in a similar fashion and clamp all tubes.
3. Minimum sterilization time is 2 hours; maximum, 10 days.

Preparation for dialysis

1. Attach the dialysate hoses to a source of warm water and flush the compartments for 30 minutes at a flow rate of 500 ml/min.
2. Attach blood tubing sets and, after step 1 is complete, allow 800 to 1000 ml of sterile heparinized saline to flow by gravity through the blood compartments.
3. Near the end of the saline rinse, check the fluid with a Clinitest tablet to determine if all formalin has been removed.

Connecting patient to dialyzer

1. Check the dialysate supply and set the flow to *on*.
2. Connect the arterial line to the patient's arterial cannula.
3. When the venous line is filled with blood, connect to the venous cannula.
4. Check the dialysate outflow for evidence of blood (leak).
5. Set the heparin infusion pump to deliver to the arterial line.
6. Attach arterial pressure monitor tubing to arterial line drip chamber.
7. Adjust negative pressure at dialysate outflow to govern ultrafiltration.

Clinical application

Acute patients are dialyzed according to the severity of their biochemical disturbances, with frequent monitoring of serum values. Chronic patients are dialyzed more routinely, usually three times a week for 5 to 10 hours at a time, depending upon the equipment.

Peritoneal dialysis

Peritoneal dialysis accomplishes essentially the same things as does hemodialysis.[31] The lining of the peritoneal cavity, although a living membrane, acts in many ways similarly to the inert semipermeable membrane of the artificial kidney. The clearance of some metabolic products of large molecular size such as creatinine, and uric acid, is somewhat greater than by hemodialysis.

The first human use of peritoneal dialysis for removal of substances ordinarily excreted by the kidney was reported by Ganter[29] in Germany in 1923. Subsequent clinical use was not encouraging, although animal work showed that transperitoneal to blood transfer of solutes and vice versa, did occur and that uremic animals showed improvement when treated by peritoneal dialysis. As with hemodialysis, increasing knowledge of body fluid and electrolyte balance and of methods for accurate measurement has made the procedure more physiologic and safe. The advent of effective antibiotics served to reduce the risk of peritonitis. By 1946 a group of investigators[28] in Boston described a technique of continuous peritoneal lavage with a balanced electrolyte solution, which was effective in treating uremia. Technique problems consisted of "streaming" of the fluid, resulting in contact with only a fraction of the surface area of the peritoneum, and walling off by omentum. Grollman and co-workers[32] in 1951 described a technique of intermittent peritoneal dialysis that was simple, efficient, and economical in the use of fluid. This was essentially the same technique that is used today. However, clinical acceptance was slow until 1959 when, with their co-workers, Doolan[27] and Maxwell[37] separately reported effective results from intermittent peritoneal dialysis.

The procedure has been made generally available by the commercial preparation of prepackaged sterile fluids of appropriate electrolyte concentration, pH, and osmolality. Such fluids contain sodium, calcium, magnesium, and chloride in concentrations approximating that of normal plasma. Lactate or acetate is included, which, when metabolized, provides a suitable source of bicarbonate. Dextrose is furnished, usually in concentration of 1.5 or 7 g/100 ml, to provide the desired osmolality. The standard solution with 1.5 g dextrose/100 ml has a measured osmolality of approximately 350 mOsm/L, while the 7% dextrose solution exceeds 600 mOsm/L (measured by freezing point, the osmolality is lower than the theoretical value). Glucose molecules diffuse more slowly across the peritoneal membrane than do those of water and electrolytes. Therefore the hypertonic dextrose solution serves to remove fluid from the blood and prevent or relieve overhydration. Potassium may be added to dialysis fluid; if hyperkalemia is a problem, it is omitted. Heparin may be added to reduce fibrin deposition in the catheter, and antibiotics may be included at the discretion of the physician.

Patient preparation (Fig. 14-8)

In the usual case, the patient's abdomen is prepared as for a paracentesis. A small (1 cm) skin incision is made in the midline below the umbilicus, and a trocar inserted until the tip is felt to pass into the peritoneal cavity. A plastic or silicone rubber catheter is placed through the trocar, which is then withdrawn. Bleeding is usually minimal. The catheter may be secured with a snug purse-string skin suture, or simply secured with tape. The fluid administration set is connected to the catheter and to the bottles of fluid.

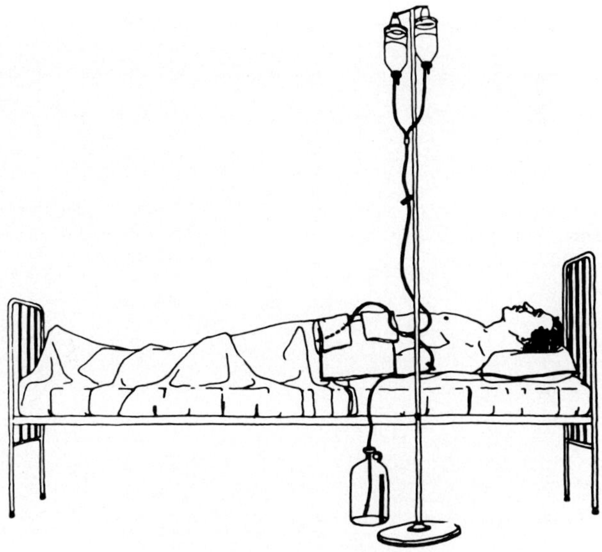

Fig. 14-8. Peritoneal dialysis.

Dialysis fluid

Commercially prepared sterile dialysis solutions in 1 L bottles are used. The fluid should be warmed to approximately body temperature before infusion. The customary volume of fluid for each exchange for an adult patient is 2 L. Smaller volumes are used for children or small infants (initial volumes of 300 to 700 ml can usually be employed with youngsters under 1 year of age).

Patient dialysis

Utilizing a high IV standard, 2 L of fluid can be instilled within 7 to 10 minutes. The fluid is left in the peritoneal cavity for 45 to 60 minutes and then drained by gravity through the closed tubing system into a container on the floor.

Urea concentration in the dialysis fluid will reach 65% to 80% of that in serum in 45 to 60 minutes, while approximately 30% to 50% of the dextrose will have been absorbed. The usual period for drainage of the fluid is 10 minutes, or until the *brisk flow* tapers off. If the patient initially has no ascitic fluid present, it is common to get incomplete return of the first two or three exchanges of dialysis fluid. Barry and co-workers[26] pointed out that it is often necessary to establish a pool of fluid in the abdomen, following which drainage is usually quite free, with return of 2 L or more in 10 minutes. Continued slow or erratic drainage may be due to obstruction of the catheter by omentum, or to plugging with fibrin clot, and may require manipulation or repositioning of the catheter. Such difficulties are less frequent with Silastic catheters[30] (Fig. 14-9) than the more rigid, small plastic ones.

The exchanges are repeated continually for such period of time as the condition of the patient indicates. Five to six exchanges of 2 L each will remove the normal 24-hour accumulation of metabolic wastes of an average size adult who is not catabolic. If the patient is severely azotemic initially, many more exchanges obviously will be required. A careful running record of the volumes

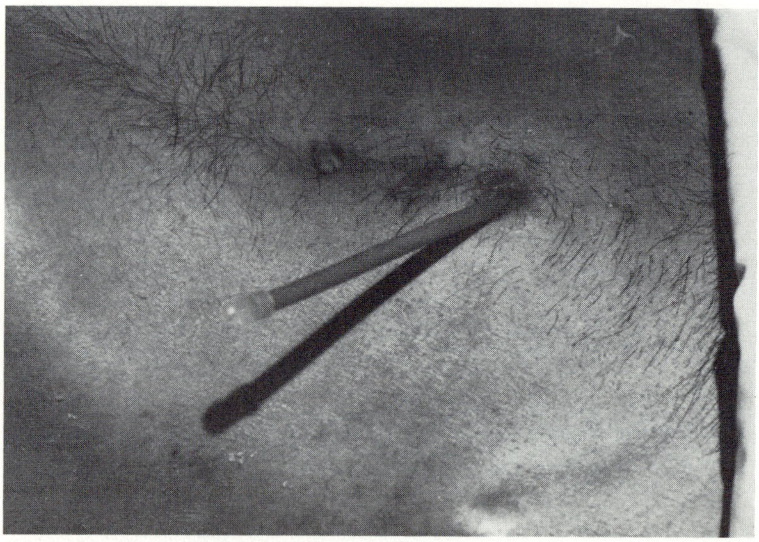

Fig. 14-9. Silastic catheter.

infused and removed, as well as medications and vital signs, is essential for monitoring the patient's course.

The use of 7% dextrose solutions for peritoneal dialysis is only rarely indicated, the prime indication being found in the severely waterlogged patient with incipient pulmonary edema. Such hyperosmotic solutions attract vast quantities of fluid rapidly and can result in hypovolemia and shock. When used, the 7% solution should not be left in the abdomen longer than half an hour, and rarely should more than two or three such exchanges be used. On the other hand, severely azotemic persons (those with a BUN of 150 mg/100 ml) may have a serum osmolality sufficiently high that there is an inadequate gradient for removal of water from the body if only 1.5% solutions are used. Use of a bottle each of 1.5% and 7% dextrose solutions (yielding an effective dextrose concentration of 4.25%) will almost always result in the desired rate of fluid removal.

Clinical application

Peritoneal dialysis removes metabolic wastes more slowly than does hemodialysis—a 12- to 15-hour peritoneal dialysis will be approximately equal to a 4-hour hemodialysis, using the commonly employed coil equipment, or 6 to 8 hours of Kiil dialysis without a blood pump. This is often actually advantageous to the stable patient, since the water and solute shifts between the various bodily compartments is more gradual and less likely to be accompanied by adverse symptoms attributable to the dialysis procedure.

Peritoneal dialysis in diabetic individuals is difficult and generally unsatisfactory because of inability to metabolize the large load of glucose. The acidosis of azotemia contributes to insulin resistance. Dialysis solutions in which sorbitol is the osmotic agent are available and have been suggested for use with diabetics as well as other patients. However, this sugar is slowly metabolized and can, over the course of a prolonged dialysis, diffuse throughout extracellular fluid, resulting in a hyperosmolar syndrome.

Several types of apparatus have been designed to automate peritoneal dialysis. Employing a system of pumps, solenoid valves, and automatic timers, such equipment greatly reduces the amount of nursing time needed for conducting the exchanges. On-site preparation of dialysis fluid, using flash sterilization or Millipore filtration, is a feature of such apparatus as those designed by Tenchoff and associates[42] and by McDonald.[38]

TRANSPLANTATION

For the person with end-stage renal failure, the prospect of receiving a transplanted kidney, which might function normally, is a goal that dialysis therapy cannot approach. However, the immunologic defenses of the body are such that homograft tissues (grafts between different individuals of the same species) are rejected by these defense mechanisms. Rejection of a kidney homograft may occur within the first few days or weeks or months later. It is manifested by edema, cellular infiltration of the parenchyma, proliferation of vascular endothelium, and fibrinoid necrosis of arterial and arteriolar walls. The result is loss of the transplanted kidney. Only when the transplant takes place between identical twins is rejection avoided.

Homograft reaction may be suppressed by irradiation and by some drugs, but in man it is never completely overcome. The amount of irradiation necessary for immunosuppression is dangerously close to the lethal dose. Numerous drugs have been investigated as immunosuppressive agents. Of these, azathioprine (Imuran; Burroughs Welcome) and prednisone have proved most useful. More recently the use of antilymphocyte globulin (ALG), prepared from serum of horses that have been injected with human lymphoid tissue, has made possible the use of smaller and less toxic amounts of the other suppressive agents.

Tissue-matching techniques have been devised and are used to determine the best "match" between recipient and possible donor. Donor and recipient sera and leukocytes are crossmatched and are matched against a panel of cells and sera in a manner somewhat akin to red blood cell typing and matching.

Upward of 3,000 renal homografts have been done throughout the world. Using current immunosuppressive chemotherapy and antilymphocyte globulin, and with careful tissue typing technique, persons who receive a transplanted kidney from a related living donor now have an approximately 90% chance of having that kidney survive and function at least 2 years. It is projected that the 5-year survival may approach the 90% figure also. Data for cadaveric transplantation (in which the donor kidney is taken from a person after death) is less complete but, with the best methods of matching and suppressive therapy, may approach a 2-year survival rate of 70%.

DIALYSIS FOR ACUTE POISONING OR DRUG INTOXICATION

Dialysis is an effective means of removing many poisons from the body. However, dialysis is never a substitute for appropriate and intensive supportive therapy. Once emergency care has been established (adequate airway, circulatory support, gastric lavage, etc.), the question of indications for dialysis should be raised. Substances that are completely or largely excreted by the kidney are theoretically dialyzable.

If possible, the intoxicating substance or substances should be identified and quantitative levels in body fluids determined.

The decision to dialyze must be based on the clinical picture, and not solely upon serum levels. If the substance cannot be verified, or if quantitation in the laboratory may require considerable time but there is evidence from the history that the drug ingested is one that meets the criteria for dialysis, the procedure should be instituted. When one or more of the substances ingested is dialyzable, the duration of coma can be reduced, and morbidity and mortality decreased.

A patient may be considered a candidate for dialysis when he is poisoned with an agent (agents) that meet the criteria for dialysis and one or more of the following clinical indications are present:

1. Hypotension, not responsive to adjusting circulatory volume
2. Apnea, or marked respiratory depression
3. Significant hyperthermia
4. Severe shift in acid-base balance
5. Significant electrolyte shift
6. Hyperosmolality

According to Maher and Schreiner,[36] standards for deciding on the applicability of dialysis for poisoning are as follows:

1. The substance should diffuse through a dialyzing membrane at a reasonable rate.

2. A significant quantity of the poison should be in serum water, or in body compartments which rapidly equilibrate with serum water.

3. The severity of clinical intoxication should be directly related to the blood concentration and to duration of exposure to the poison.

4. The amount of the poison that will be removed by dialysis must add significantly to the quantity that will be removed by normal mechanisms (metabolism, conjugation, eliminating by bowel and kidney) under the physiologic circumstances at the time.

It should be appreciated that renal excretion of substances which meet these criteria can often be greatly accelerated by forced diuresis (high fluid intake, mannitol, and/or pharmacologic diuretic). When dialysis is deemed appropriate, the use of a high-efficiency apparatus (in terms of quantity of solute removed per unit time) is indicated. A coil dialyzer is the preferred hemodialysis unit. Indeed two coils may be connected in series, or a coil and a Kiil dialyzer operated in series, etc. Peritoneal dialysis is appreciably slower in removing almost all poisons than is hemodialysis. However, a peritoneal dialysis can be started and under way in less than 30 minutes after the decision to dialyze is made, whereas in most institutions 2 or 3 or more hours usually elapse before a hemodialysis can be started. Maher reported use of all three modalities—forced diuresis, peritoneal dialysis, and hemodialysis—simultaneously in one case of poisoning!

The *Transactions of the American Society of Internal Organs* each year includes an updating of available information on dialyzable poisons. This information should be on hand at all major poison control centers.

In general, ingestion of such agents as methyl alcohol, ethylene glycol and propylene glycol, sulfonamides, sodium chlorate, acetone, carbon tetrachloride, heavy metals in soluble compounds, and mushrooms proved toxic (*Amanita phalloides*) should call for immediate dialysis. Early removal may prevent the severe complications of these agents, such as blindness, liver necrosis, or renal failure.

In most other intoxications, the indications are relative and depend upon the clinical situation. Agents falling into this category include barbiturates, salicylates, chloral hydrate, amphetamines, antibiotics, boric acid, thiocyanate, meprobamate, isoniazid, and others. Glutethimide (Doriden) is a special problem because of its partitioning in body fat; dialysis is generally reserved for the most seriously ill patient who appears to be worsening in spite of all other supportive measures.

Dialysis is not of benefit for excess intake of the phenothiazines, chloridazepoxide, diazepam, Imipramine, and other thymoleptics, dextro propoxyphene (Darvon), or the anticholinergic medications.

Routine urine examination

A routine examination of the urine consists of the tests listed in Fig. 14-10. These tests are discussed on the following pages.

URINE COLLECTION

For routine qualitative clinical studies a random voided urine specimen is suitable. The first urine voided in the morning is usually preferred because of the concentration of the constituents and, therefore, the tendency to reveal abnormalities. However, glycosuria and albuminuria are less likely to be revealed by this sample. The specimen should be delivered to the laboratory and ex-

Name: Mr. John Doe	6-4-70
Color	straw
Transparency	cloudy
Reaction	acid
Specific gravity	1.020
Albumin	3+
Sugar	negative
Acetone	negative
Microscopic examination.	Centrifuged specimen shows 4 to 6 granular casts per high dry field

Fig. 14-10. Sample urine analysis report.

amined as soon as possible. Bacterial growth causes rapid decomposition and alteration of urinary constituents. When a delay is unavoidable, the specimen should be refrigerated or a few drops of toluene added as a preservative. For preservation of the structure of cells and casts in urine, 40% formalin may be added. The microscopic examination to be valid should be performed when the urine is fresh. Hydrochloric acid is a suitable preservative when inorganic constituents other than chlorides are being investigated.

When quantitative measurements such as electrolytes, proteins, electrophoresis or hormones are requested, a 24-hour collection is necessary. The patient should be instructed to empty his bladder at a specific time (for example, 7 AM) and discard this urine. He should save all urine voided up to and including the 7 AM specimen on the following day. Each specimen as voided should be placed in a container having the preservative required for the test procedure being conducted.

Catheterization is rarely used for collection of urine specimens as it is difficult to prevent contamination of the bladder and the upper urinary tract during passage of a catheter. This may result in a genitourinary tract infection and pyelonephritis.

The quantities of preservatives that may be used for routine qualitative examinations per 100 ml of urine are 6 to 8 drops of formalin, 0.2 g of boric acid, and 0.1 g of thymol.

For quantitative chemical analysis sufficient toluene to form a layer on top of the specimen is satisfactory for most examinations. Preservation by acidification with sulfuric or hydrochloric acids (1 ml of concentrated acid per 100 ml of urine) may be used for a majority of the inorganic constituents and nitrogen. When special tests are requested, such as hormone assays, refer to the recommended procedures described with each method.

The bottles for the collections should be clean and dry before the preservatives are added.

COLOR

The color of the urine often indicates urinary abnormalities. Some of these colors and their usual causes are given in Table 14-3.

Procedure

1. **Mix urine by inverting or swirling the container.**
2. **Note and record the color.**
3. **If the color is dark amber, the urine should be shaken to produce foam. If the foam is yellow, a test for bile should be run.**

Table 14-3. Usual causes of urinary colors

Color	Usual cause
Straw to amber	Urochrome, a pigment found in normal urine
Colorless	Reduced concentration
Silvery sheen or milky appearance	Pus, bacteria, or epithelial cells
Smoky brown	Blood
Black	Melanin
Port-wine	Porphyrins
Yellow foam	Bile or medications
Orange, green, blue, or red	Medications

Table 14-4. Rapid urine tests using tape, dipsticks, and tablets

Test	Method
Multitest	
Protein, glucose	Uristix
Protein, glucose, pH	Combistix
Protein, glucose, pH, occult blood	Hema-Combistix
Protein, glucose, pH, occult blood, ketones	Labstix
Acetoacetic acid, acetone	Acetest, Ketostix
Single tests	
Glucose	Clinistix, Tes-Tape
Sugar	Clinitest, Galatest
Protein	Albustix, Albutest, Bumintest
Bilirubin	Ictotest
Occult blood	Hemastix, Occultest
Salicylate	Phenistix
Phenylpyruvic acid	Phenistix

RAPID TESTS OF THE URINE

Due to time-consuming and demanding techniques required with many of the chemical tests for urine analyses, the modern tablet, dipstick, and tape test for urinalysis are coming to the aid of the physician for a bedside check of the patient or for use in the office (Table 14-4). These new tests are sensitive and must be used with recognition of their limitations. Each tape, dipstick, or tablet is designed for a special purpose, and if each is used according to the manufacturer's specifications and instructions, accurate results may be obtained.

The strips are made with selected cellulose of standard porosity. The tips are impregnated with chemicals that will react with various substances in the urine, producing colors that can be measured against standard color charts. Some color reactions are timed. This requirement is given in the instructions and *must* be followed exactly!

Precautions

1. **When not in use, the tablets or strips must be kept in their bottles, tightly stoppered to keep out the moisture.**
2. **Bottles that are so marked should be placed in refrigeration.**
3. **If the strips (dipsticks) are kept too long in the urine or in the urine stream, the chemicals impregnated in the cellulose layer will dissolve and inaccurate results will be obtained.**
4. **Directions must be carefully followed when there is more than one reaction on a strip. For example, on Combistix the sections indicating pH, protein, and glucose are separated by water-impermeable barriers of plastic.**

TRANSPARENCY

The transparency of urine may be classified as clear or cloudy. The degree of transparency depends upon the amount of suspended materials: the more suspended materials, the greater the cloudiness. In normal urine, the main causes of cloudiness are crystals and epithelial cells. In pathologic urine, the principal causes of cloudiness are pus, blood, and bacteria.

If the urine is clear, it may inform the physician that the patient is not concentrating the urine, that is, removing metabolic waste materials. On the whole, however, this test has little significance, for both normal and pathologic urine may be either clear or cloudy.

Procedure

1. **Mix urine.**
2. **Note degree of transparency.**
3. **Report as clear or cloudy. Also report any abnormal contents such as clumps or mucus or pieces of tissue.**

REACTION

Freshly voided urine may be either acid or alkaline, the normal pH range being 4.0 to 8.0. If allowed to stand, acid urine becomes alkaline because urea decomposes and liberates ammonia. Increased acidity of urine is found in fevers and diabetes. Increased alkalinity is seen in alkali therapy and in retention of urine.

Labstix and Combistix, in the pH portion of the strip, are impregnated with a combination of the acid-base indicators, bromthymol blue and methyl red. Distinctive colors are produced at different levels within the urinary pH range of 5 to 9.

Procedures

The reaction of urine may be found by any of the tests given below.

Neutral litmus paper test

1. **Dip a piece of neutral litmus paper into the urine.**
2. **After the paper absorbs urine, note color of paper.**
3. **If the paper is pink, report Acid. If there is no color change, report Neutral. If paper is blue, report Alkaline.**

Nitrazine paper test

1. **Dip a piece of nitrazine paper into the urine.**
2. **After paper absorbs urine, note color of paper.**
3. **Compare with color chart that comes with the nitrazine paper.**
4. **Report pH value.**

pH range paper test

1. **Dip a piece of pH range paper into the urine.**
2. **After paper absorbs urine, note color of paper.**
3. **Compare with the color chart that comes with the pH range paper.**
4. **Report pH value.**

DISORDERS AND THE URINE pH[43]

Tests for pH of urine should be done a short time after the specimen has been voided. As the urine stands, it will become alkaline, due to the decomposition of urea into ammonia. However, when fresh urine is persistently acid or alkaline, this may be due to one of the following disorders:

A. Consistently acid urine
   1. Tuberculosis of the kidney
   2. Pyrexia
   3. Methyl alcohol poisoning due to the formation of formic acid
   4. Metabolic disorders
      a. Phenylketonuria
      b. Alkaptonuria
   5. "Metabolic" acidosis due to severe diarrhea, starvation, diabetic ketosis, uremia, and certain types of poisoning
   6. "Respiratory" acidosis due to conditions in which there is CO_2 retention

B. Consistently alkaline urine
   1. Certain genitourinary infections
   2. "Metabolic" alkalosis due to overdosage of sodium bicarbonate, potassium citrate, or other alkalies
   3. "Respiratory" alkalosis due to hyperventilation or cardiac failure
   4. Bronchial carcinoma (in some cases)
   5. Cushing's syndrome or hyperaldosteronism
   6. The use of certain diuretics, such as Diamox and chlorothiazide, which are carbonic anhydrase inhibitors

SPECIFIC GRAVITY

Specific gravity is a term used to compare the weight of a liquid with the weight of water. For example, sulfuric acid weighs 1.840 g/ml, whereas water weighs 1.000 g/ml. Since sulfuric acid weighs 1.840 times as much as water, the specific gravity is 1.840.

Because of its dissolved substances, urine weighs more than water, the normal specific gravity being 1.008 to 1.030. Increased values are found in acute nephritis, fevers, and diabetes mellitus (because of the dissolved sugar). Decreased values are found in chronic nephritis and diabetes insipidus (because of the copious volume).

Measuring the specific gravity of the urine is the most convenient way of measuring the concentrating and diluting function of the kidney, meaning the ability of the kidney to produce urine of a specific gravity greater or less than 1.010.

Procedure

1. **Mix the urine by inverting or swirling the container.**
2. **Pour into a urinometer cylinder (Fig. 14-11).**

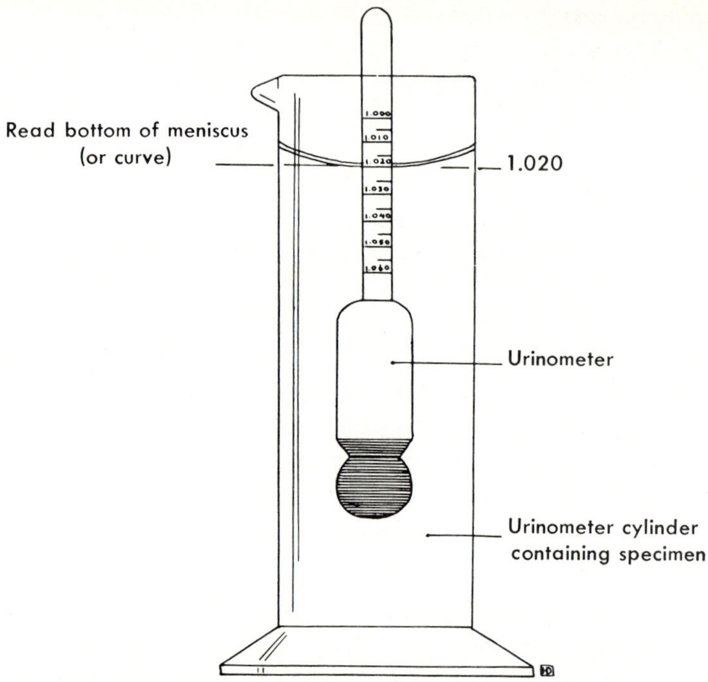

Fig. 14-11. Urinometer used to measure the specific gravity.

3. Place the urinometer in the urine and spin it as you would a top.
4. When it comes to rest, take the reading as indicated in Fig. 14-11. When the reading is taken, the urinometer must *not* be touching the sides of the cylinder.
5. Record reading.
6. If there is not sufficient urine to fill the cylinder, some laboratories report "qns" (quantity nonsufficient). The temperature-compensated hand refractometer will measure the specific gravity of a minute volume (1 to 2 drops) of urine.

    In addition to a refractometer, an osmometer may be employed to measure the concentration of dissolved solutes in urine. Both methods are especially useful when very small amounts are available. The refractive index of a solution is dependent upon the concentration of the dissolved substances, a relationship that approaches linearity for a pure solution. The osmolality measures the number of molecules and ions in a solution and is usually performed by a freezing point depression method. The normal range for an adult with adequate fluid intake is 500 to 850 mOsm per kilogram of water. Using Long's coefficient, determine the total solids per liter by multiplying the last two figures of the specific gravity of urine by 2.66.

PROTEIN IN URINE

Albumin is a protein. The presence of albumin in the urine is referred to as albuminuria. There are two types of albuminuria: false and true. In false albuminuria, the albumin may be caused by excessive exercise, blood from the bladder, or vaginal discharge. In true albuminuria, the albumin leaves the bloodstream, seeps through the kidneys, and enters the urine. True albuminuria points to a flaw in the filtering system of the kidneys and consequently aids the physician in diagnosing kidney disorders.

Several of the rapid screening tests for urinary protein (Labstix, Combistix, Albustix, Albutest) utilize the principle of protein error of a pH indicator. Bromphenol blue is yellow at pH 3 but in the presence of protein the color is a greenish blue. For the paper strips, the proper pH in the protein portion is maintained with citrate buffer. The protein reagent area is sensitive to albumin and globulin but less sensitive to hemoglobin and Bence Jones protein. The insensitivity to myeloma proteins is a disadvantage. A highly buffered alkaline urine will give a positive reaction; therefore the pH must be checked for each specimen.

For the tablet form, salicylic acid–buffered cellulose is used with the bromphenol blue indicator. When a drop of urine is placed on the tablet and followed by 2 drops of water, a green color on the surface of the tablet is the result of adsorbed protein.

The majority of the other tests for urinary protein depend upon protein precipitation with agents such as heat and acetic acid, nitric acid, sulfosalicylic acid, and trichloroacetic acid. In acidified solutions, albumin and globulin are coagulated by heat in the presence of inorganic salt. Proteoses are also precipitated. Phosphate salts, if present, precipitate out and form a white cloud when urine is heated. The addition of an acid, however, dissolves the salts.

Sulfosalicylic and trichloroacetic acids precipitate protein in urine with a turbidity that may be measured by a spectrophotometer, colorimeter, nephelometer, or by eye and compared with known standards. Bumintest tablets may be used to prepare a 5% sulfosalicylic acid solution and the results may be graded from trace to 4+. Substances that are also precipitated by sulfosalicylic acid are urinary proteose, polypeptides, tolbutamide metabolite, and x-ray contrast media.

Procedures for albumin in urine

If it is necessary to prepare any of the reagents used in the following tests, take the number that follows the reagent and refer to Appendix D.

Heat and acetic acid test

Reagent needed

Acetic acid, 5% (D-260)

Procedure

1. **Pour about 10 ml of urine into a large Pyrex test tube.**
2. **Using a test tube holder, boil the *upper* portion over a flame as illustrated in Fig. 14-12. If a cloudy white precipitate forms in the heated portion, it is either phosphate salts of albumin. To decide which, proceed as follows.**
3. **Add a few drops of the 5% acetic acid. If the precipitate is phosphate salts, it will be dissolved by the acid. If the precipitate is albumin, however, it will not dissolve.**
4. **Report the test as negative, 1+, 2+, 3+, or 4+, depending upon the amount of precipitate (Fig. 14-13).**

Purdy test

Reagents needed

Sodium chloride, saturated (D-261)
Acetic acid, 50% (D-262)

Procedure

1. **Pour about 5 ml of urine into a large Pyrex test tube.**
2. **Add 1 ml of the saturated solution of sodium chloride. (This prevents the precipitation of mucin, if it should be present.)**

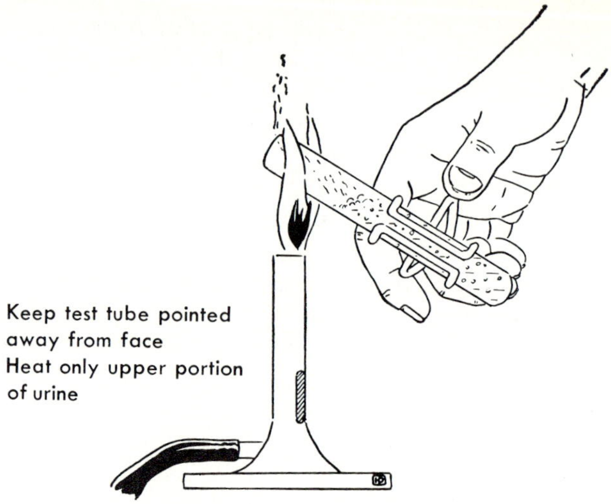

Fig. 14-12. Heating urine with a flame.

3. Add 5 to 10 drops of the 50% acetic acid. Mix.
4. Using a test tube holder, heat the *upper* portion over a flame as illustrated in Fig. 14-12. If a white precipitate forms, albumin is present.
5. Report the test as negative, 1+, 2+, 3+, or 4+, depending upon the amount of precipitate (Fig. 14-13).

Osgood-Haskins test

Reagents needed

Acetic acid, 50% (D-262)
Sodium chloride, saturated (D-261)

Procedure

1. Pour about 5 ml of urine into a large Pyrex test tube.
2. Add 1 ml of the 50% acetic acid.
3. Add 3 ml of the saturated solution of sodium chloride. (This prevents the precipitation of mucin, if it should be present.)
4. Using a test tube holder, heat the *upper* portion over a flame as illustrated in Fig. 14-12. If albumin is present, a white precipitate will form.
5. Report the test as negative, 1+, 2+, 3+, or 4+, depending upon the amount of precipitate (Fig. 14-13).

Sulfosalicylic acid test

Reagent needed

Sulfosalicylic acid, 20% (D-263)

Procedure

1. Pour about 5 ml of urine into a test tube.
2. Add 1 drop, per milliliter of urine, of the 20% sulfosalicylic acid. Mix. If albumin is present, a white cloud will form.
3. Report the test as negative, 1+, 2+, 3+, or 4+, depending upon the amount of precipitate (Fig. 14-13).

Exton test

Reagent needed

Exton's reagent (D-264)

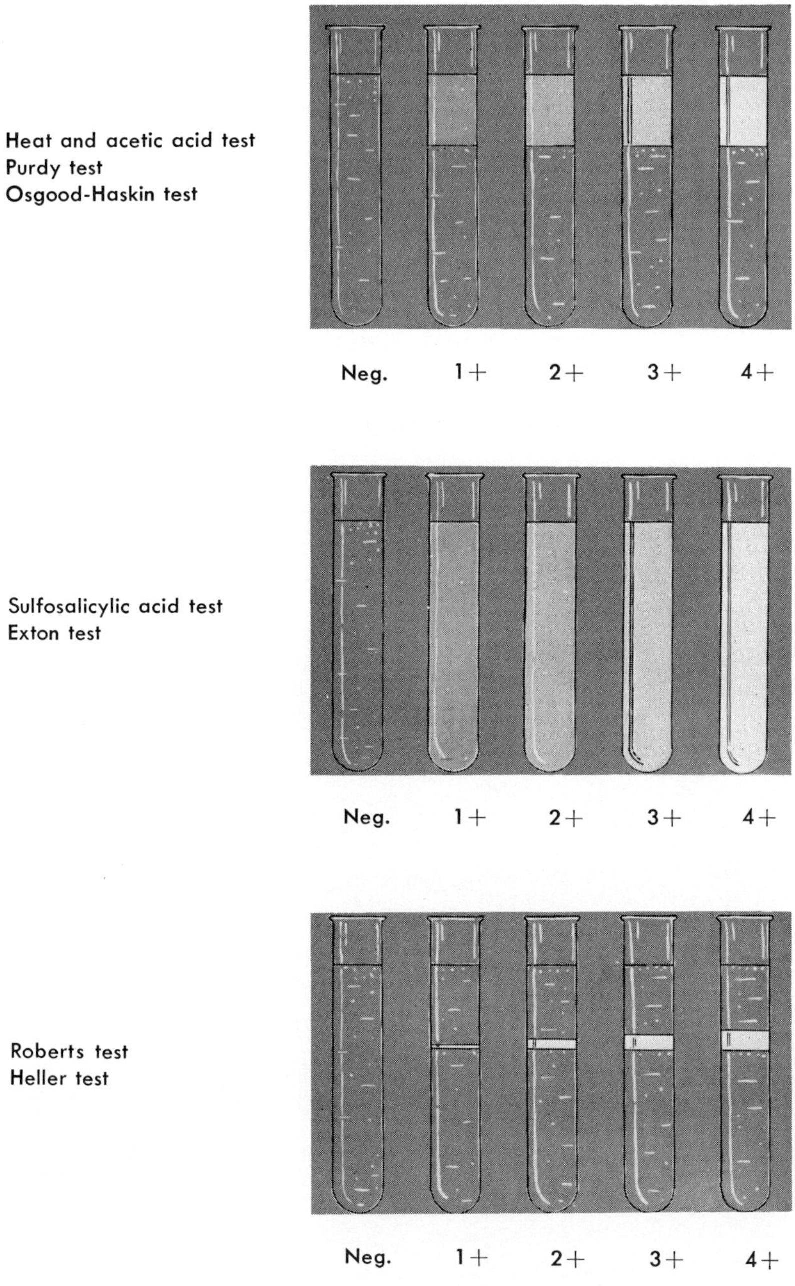

Fig. 14-13. Positive and negative results in testing for albumin.

Procedure

1. Mix equal parts of urine and Exton's reagent. If no cloudiness forms, albumin is absent.
2. If cloudiness appears, heat gently, but do not boil.
3. If cloudiness remains or increases, albumin is present.
4. Report the test as negative, 1+, 2+, 3+, or 4+, depending upon the amount of precipitate (Fig. 14-13). The reading should be made while the solution is warm because proteoses produce a cloud upon standing.

Roberts test

Reagent needed

Roberts' reagent (D-265)

Procedure

1. Follow either method A or method B:
   Method A: Pipette a few milliliters of Roberts' reagent into a test tube. Tilt the tube to about a 30-degree angle. Using a medicine dropper, overlay with clear urine by allowing the urine to flow *slowly* down the sides of the test tube (Fig. 14-14).
   Method B: Pour about 4 ml of urine into a test tube. Using a 1 ml serologic pipette, pipette out 1 ml of the Roberts' reagent. Place the tip of the pipette in the bottom of the test tube containing the urine. Allow the pipette to drain slowly (Fig. 14-14).
2. If albumin is present, within a few minutes a white ring will form at the junction between the two liquids.

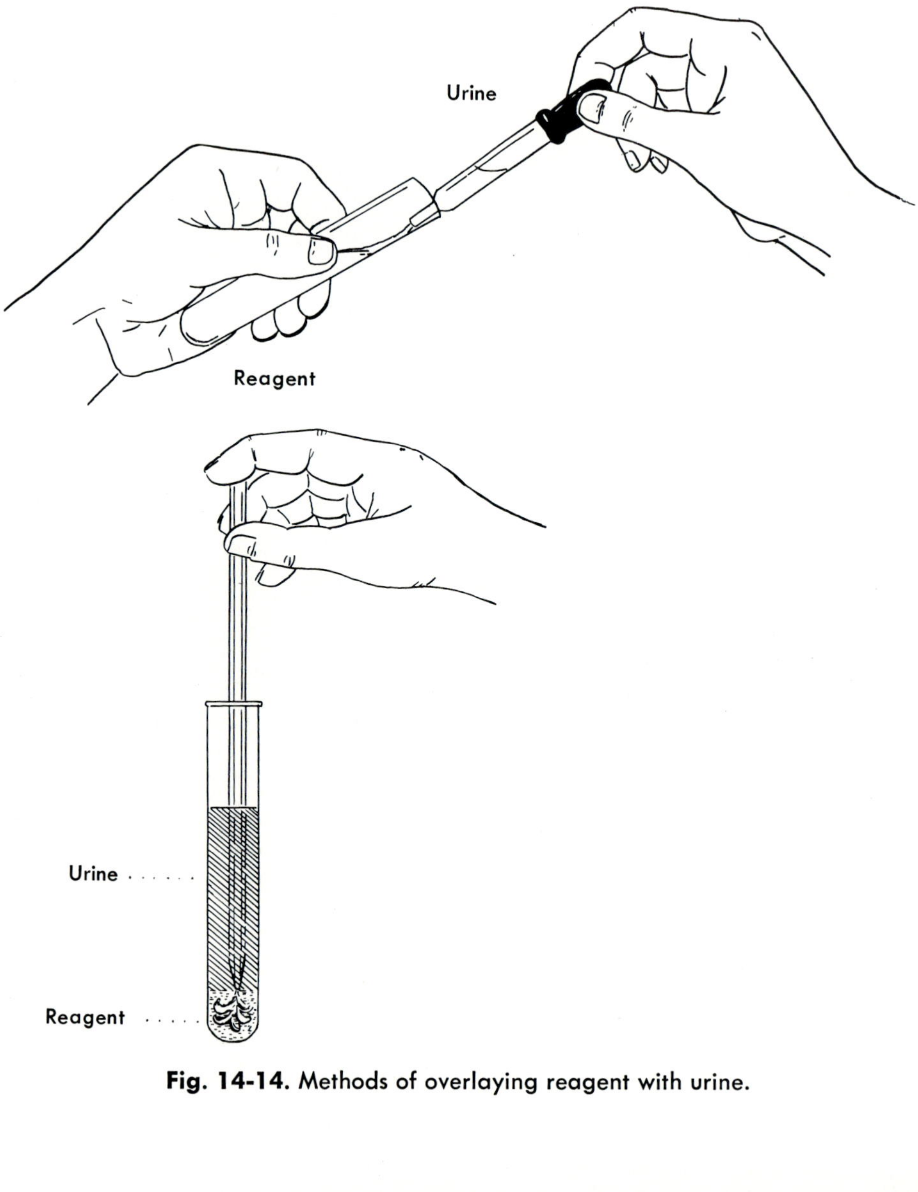

Fig. 14-14. Methods of overlaying reagent with urine.

3. Report the test as negative, 1+, 2+, 3+, or 4+, depending upon the amount of precipitate (Fig. 14-13).

Heller test

Reagent needed

Concentrated nitric acid

Procedure

1. Place about 3 ml of concentrated nitric acid in a small test tube.
2. Tilt the tube to about a 30-degree angle (Fig. 14-14). Using a medicine dropper, overlay with urine by allowing the urine to flow *slowly* down the sides of the tube.
3. If albumin is present, within a few minutes a white ring will form at the junction between the two liquids.
4. Report the test as negative, 1+, 2+, 3+, or 4+, depending upon the amount of precipitate (Fig. 14-13).

Quantitative protein

A quantitative protein determination may be of some help to the physician in following the course of a kidney disease.

For the quantitative determination of protein in urine, the following methods are commonly used: (1) Esbach's method, (2) Tsuchiya's method, (3) Shevky-Stafford rapid approximate sedimentation method, (4) Exton's method, and (5) Kingsbury's method. The first three methods call for the precipitation of the protein by an acid. The amount of precipitation is then measured in a special tube. The latter two methods also call for the precipitation of the protein by an acid, but the amount of precipitation is then compared with a set of standards.

Procedures for quantitative protein

Esbach method

Reagents needed

Hydrochloric acid, 5% (D-266)
Esbach's reagent (D-267)

Procedure

1. Acidify the urine with the 5% hydrochloric acid to pH 5.
2. Filter.
3. Fill the Esbach tube to "U" mark with the filtered urine.
4. Add the Esbach reagent to the "R" mark.
5. Close the tube with a rubber stopper and invert several times to mix.
6. Place in an upright position for 24 hours.
7. Read off the height of the coagulation. This will be the number of grams per 1000 ml of urine. To obtain percentage, divide reading by 10.

Tsuchiya method

Reagents needed

Hydrochloric acid, 5% (v/v) (D-266)
Tsuchiya's reagent (D-268)

Procedure

1. Acidify the urine with 5% hydrochloric acid, to pH 5.
2. Filter.
3. Fill an Esbach tube to the "U" mark with the filtered urine.
4. Add the Tsuchiya reagent to the "R" mark.
5. Close the tube with a rubber stopper and invert several times to mix.

6. Place in an upright postion for 24 hours.
7. Read off the height of the coagulation. This will be the number of grams per 1000 ml of urine. To obtain percentage, divide reading by 10.

Shevky-Stafford[44] rapid approximate sedimentation method

The following procedure deviates in a few details from the original description. The first variation is in the dilution of the urine. Nephritic urines are usually diluted tenfold. However, urines yielding very small amounts of protein should be diluted much less or not at all for accurate results. Occasionally a urine will be found with more than 2.8% protein. If a 1:10 dilution is not adequate, use a 1:20 dilution.

Reagent needed

Tsuchiya's reagent (D-268)

Procedure

1. Pipette 4.0 ml of diluted urine into a special graduated centrifuge tube.* Since the 4 ml mark on the tube may serve as a measurement, the serologic pipette is used for convenience, to avoid spilling the urine and to allow better control for the proper measurement to the 4 ml mark.
2. Add Tsuchiya's reagent until the tube is filled to the 6.5 ml mark.
3. Stopper and invert slowly three times to mix.
4. Let stand for 10 minutes.
5. Centrifuge for 10 minutes (use timer) at 1800 rpm.
6. Read the volume of precipitate on the scale.

Calculation

Milliliters of precipitate read × 7.2 × Urine dilution = Grams of protein/liter of urine

Exton method

Reagents needed

Acetic acid, 5% (D-260)
Exton's reagent (D-264)
Exton's standard tubes†

Procedure

1. Acidify the urine with 5% acetic acid to a pH 5.
2. Place 3 ml of the acidified urine in a test tube that is the same size as the Exton standard tubes.
3. Add 3 ml of Exton's reagent. Mix.
4. Let stand 5 minutes.
5. Warm gently—but do not boil.
6. Compare with a set of Exton's standard tubes, giving report in mg, of protein/100 ml of urine.

Kingsbury method

Reagent needed

Sulfosalicylic acid, 3% (D-269)
Kingsbury's standard tubes†

*Special equipment needed: The graduated centrifuge tube is a special calibrated tube that may be obtained from the following distributors: Phipps & Bird, Inc., Coffman & Ladimer, Palo Laboratories, Howe & French, George T. Walker.

†May be purchased from a local medical supply house.

Procedure

1. Filter the urine and place 2.5 ml in a test tube that is the same size as the Kingsbury standard tubes.
2. Add 7.5 ml of 3% sulfosalicylic acid.
3. Mix and let stand for 5 minutes.
4. Compare with Kingsbury's standard tubes, giving the report in mg of protein/100 ml of urine.

Proteoses

Proteoses are intermediate products in the digestion of proteins. Under normal conditions, proteoses are not found in the urine. They may be present, however, in cancer, osteomalacia, pneumonia, diphtheria, and atrophy of the kidneys.

Reagents needed

Acetic acid, 5%	(D-260)
Trichloroacetic acid, 10%	(D-164)

Procedure

1. Acidify the urine with 5% acetic acid, using nitrazine paper as an indicator. Mix.
2. Pour 10 ml of the acidified urine into a small beaker and bring to a boil.
3. Filter the urine while it is still hot in order to remove any albumin or globulin that may be present.
4. Allow the filtrate to cool.
5. Add several drops of 10% TCA to the filtrate. If proteoses are present, a white cloud will form.

Bence Jones protein

This protein, discovered by Bence Jones, has the following peculiar characteristics: it precipitates out at a temperature of about 50° C, partially or completely disappears at 100° C, and then reappears upon cooling to room temperature. Its presence in the urine is considered a significant diagnostic aid in multiple myeloma, a malignant tumor of the bone marrow.

Reagent needed

Acetic acid, 5%	(D-260)

Procedure

1. Pour about 10 ml of the urine into a test tube.
2. Place the thermometer and the test tube containing the urine in a beaker of water.
3. Heat gently and watch both the thermometer and the urine.
4. Bence Jones protein is present if *all* the three following conditions are met: (a) At about 50° to 60° C, a white cloud will appear, which turns into a white precipitate and usually clings to the sides of the tube. (b) Upon the addition of a few drops of 5% acetic acid and further heating to the boiling point, the precipitate will partially or completely disappear. (c) If the urine is filtered while hot, using a heated funnel and test tube, the white cloud will reappear as the solution cools to room temperature. (NOTE: If a heated funnel and tube are not used, a false positive may be read.)

GLUCOSE OR SUGAR (REDUCING SUBSTANCES)

Glucose and other carbohydrates are normally present in urine in quantities too small to be measured by the usual test procedures. The presence of measurable quantities of urinary glucose (glycosuria) may occur as a temporary condition following general anesthesia and administration of certain drugs, in pregnancy, after shock or due to emotional disturbances. However, persistent glycosuria may be diagnostic of diabetes mellitus. Other sugars that may appear in the urine under certain conditions are lactose, fructose, pentose, and galactose.

The following discussion covers some methods that may be used to test for glucose.

In Benedict's method and the Clinitest method, glucose reduces copper. The reduced copper then forms a colored compound, cuprous oxide. As the particle size of cuprous oxide increases, the color of the precipitate varies from green to yellow to orange.

The Clinitest tablet includes all of the reagents and the heating mechanism for reduction of the copper by sugar or other reducing substances. Sodium carbonate and a small amount of citric acid act as an effervescent couple to facilitate rapid solution of the tablet. The sodium carbonate yields CO_2, which forms a protective shield over the heated solution. This prevents oxidation of the reducing sugar in the hot alkaline solution by atmospheric oxygen. If oxidation does occur (the solution being shaken and the shield broken) during the reaction, the readings will be low or negative even though the specimen may contain 0.25% glucose. Sodium hydroxide produces the alkaline medium for the reaction, and the necessary heat is obtained from the heat of solution of the sodium hydroxide and from reaction between the sodium hydroxide and citric acid. The heating process takes place as the tablet dissolves, keeping a suitable temperature, not underheating, overheating, or caramelizing the specimen.

Clinistix, Combistix, and Labstix are strips, impregnated with the enzyme glucose oxidase, peroxidase, and *o*-tolidine, that produce a color reaction when moistened with a solution containing glucose. In the presence of oxygen and glucose oxidase, glucose forms gluconic acid and hydrogen peroxide. The hydrogen peroxide in the presence of peroxidase oxidizes the colorless dye, *o*-tolidine, to its blue oxidized form. The test is specific for glucose and sensitive. Therapeutic doses of ascorbic acid or certain antibiotics may inhibit the color development of enzymatic tests.

In the Galatest method, glucose reduces bismuth oxide to a gray or black precipitate of metallic bismuth.

For the Tes-Tape method, the color change depends upon an enzyme system reaction that is specific for glucose. The color varies from light green to deep blue.

Benedict method

Qualitative test

Reagent needed

Benedict's qualitative reagent (D-270)

Procedure

1. Place about 8 drops of urine in a test tube. (A few milliliters of urine may be poured into a test tube and the urine discarded by inverting the tube. Then quickly place the tube in a an upright position. The amount of urine prevented from draining is about 8 drops.)
2. Add about 5 ml of Benedict's qualitative reagent.
3. Place tube in a boiling water bath for 5 minutes. If glucose is present,

Fig. 1	Fig. 2	Fig. 3	Fig. 4	Fig. 5
Negative	Trace of Sugar (+)	1 Percent Sugar (++)	More Than 1 Percent (+++)	More Than 2 Percent (++++)

Plate I. Benedict qualitative test for sugar in urine. (From Essentials in the management of diabetes mellitus, Eli Lilly & Co., Indianapolis, Ind.)

it will reduce the copper in the Benedict's solution. The color will vary from green to yellow to orange, depending upon the amount of glucose present.

4. Report the test as negative, 1+, 2+, 3+, or 4+ (Plate I).

Quantitative test

A quantitative urinary sugar test may aid the physician in determining the severity of a diabetic case and also in following its course during treatment.

For the quantitative determination of sugar in urine, the method of Benedict is commonly used. In this test, the following takes place. Urine, containing glucose, is slowly added to 5 ml of a copper solution of known strength. The amount of urine required to cause complete reduction of the copper is recorded. This figure is then used in a formula to calculate the percent of glucose present.

NOTE: The Somogyi-Nelson (quantitative) method is described on pp. 144 and 145.

Reagents needed

Sodium carbonate
Benedict's quantitative reagent (D-270)

Procedure (Benedict)

1. Use a 250 ml beaker and half fill with water. Heat to boiling.
2. Pipette *exactly* 5 ml of Benedict's *quantitative* reagent into a large Pyrex test tube. Add 1 to 2 g of sodium carbonate. Mix.
3. Place the tube in the boiling water and allow the solution to come to a boil.
4. Get a 1 ml serologic pipette having 100 graduations or a burette. Pipette out exactly 1 ml of urine and add it drop by drop to the boiling Benedict's solution until the blue color completely disappears. Take the pipette or burette reading and make the calculation as follows:

Calculations

The Benedict's solution is made so that 1.0 ml of a 1% glucose solution will exactly neutralize the 5 ml of Benedict's solution. To check the strength of the Benedict's solution, proceed as follows. Using the analytical balance, weigh out 1.000 g of glucose. Transfer to a 100 ml volumetric flask and add distilled water to the 100 ml mark. Mix. Use this 1% glucose solution in place of the urine and continue as in the procedure. It should require 1.0 ml. If it does not, take the number of milliliters required and substitute that figure in the formula for the 1.0 shown here (milliliters required for neutralization). To find the percentage of glucose in the urine, use this formula:

$$\frac{1.0 \text{ (ml required for neutralization)}}{\text{ml used in titration}} \times 1\% = \% \text{ glucose in g/100 ml}$$

Example: If 0.5 ml of urine is used in the titration:

$$\frac{\text{ml required for neutralization}}{\text{ml used in titration}} \times 1\% = \% \text{ glucose}$$

$$\frac{1.0}{0.5} \times 1\% = 2\% \text{ glucose}$$

Clinitest method

Reagent needed

Clinitest tablets*

*May be purchased from your local supply house or from Ames Co.

Procedure

1. Place 5 drops of urine in a test tube.
2. Add 10 drops of distilled water.
3. Add 1 Clinitest tablet. If there is no glucose present, the solution will be blue. If glucose is present, however, the color will vary from green to yellow to orange, depending upon the amount of glucose present.
4. Report the test as negative, 1+, 2+, 3+, or 4+.

Galatest method

Reagent needed

Galatest powder*

Procedure

1. Place a small mound (about one half the size of a dime) of the Galatest powder on a piece of white paper. NOTE: Keep the vial tightly closed when not in use.
2. Add a small drop of urine. If glucose is present, the powder will change color. The color will vary from gray to black. The blacker the color, the greater the concentration of glucose.
3. Report the test as negative, 1+, 2+, 3+, or 4+.

Tes-Tape method

Reagent needed

Tes-Tape†

Procedure

1. Tear off a piece of Tes-Tape.
2. Dip one end into the urine.
3. When moist, remove from urine.
4. Continue holding tape and allow 1 minute to elapse.
5. Compare with color chart that comes with Tes-Tape.
6. Report the test as negative, 1+, 2+, 3+, or 4+.

CARBOHYDRATES

This section considers tests for lactose, levulose, galactose, and pentose, as well as a fermentation test for glucose, and a phenylhydrazine test.

Lactose

Shortly before and shortly after the birth of a baby, a woman's urine may give false positive tests for glucose. This is often caused by the presence of lactose, which reacts similarly to glucose. The two sugars may be distinguished by a test introduced by Max Rubner.

Rubner test

Reagents needed

Lead acetate
Ammonium hydroxide, concentrated

Procedure

1. Mix the urine and measure 10 ml into a large test tube.
2. Add 3 g of lead acetate, shake, and filter.
3. Bring the filtrate to a boil, add 2 ml of concentrated ammonium hydroxide, and boil again. If lactose is present, the solution turns brick-red and

*May be purchased from your local supply house or from Denver Chemical Manufacturing Co., Inc.

†May be purchased from your local supply house or from Eli Lilly & Co.

a red precipitate forms. If glucose is present, the solution also turns red but the precipitate is yellow.

Levulose

Levulose is often referred to as fructose. Levulose reacts similarly to glucose but may be distinguished from it by the following test of Seliwanoff.

Seliwanoff test

Reagents needed

Ethyl alcohol

Seliwanoff's reagent: Place 0.05 g of resorcinol in a 125 ml Erlenmeyer flask. Add 60 ml of distilled water and 30 ml of concentrated hydrochloric acid. Mix.

Procedure

1. **Mix the urine and place 5 drops in a small test tube.**
2. **Add 5 ml of Seliwanoff's reagent and boil for not more than 30 seconds.**
3. **Centrifuge, pour off the fluid, and add 3 ml of ethyl alcohol. Mix. If levulose is present, a bright red color will be produced.**

Galactose

Galactose may be found in the urine of nursing infants having disturbances of the digestive tract. It may be distinguished from other sugars (with the exception of lactose) by the following mucic test.

Mucic test

Reagent needed

Nitric acid, concentrated

Procedure

1. **Mix the urine and pour 50 ml into a small beaker.**
2. **Add 10 ml of concentrated nitric acid and boil until the volume is reduced to about 10 ml.**
3. **Add 10 ml of water and let stand overnight. If galactose (or lactose) is present, a fine white precipitate will form.**

Pentose

Pentose may be found in the urine when the diet is rich in fruit juices, grapes, prunes, and plums. It has no pathologic significance.

Tauber test

Reagent needed

Tauber's reagent: Dissolve 1 g of benzidine in 25 ml of glacial acetic acid. Mix.

Procedure

1. **Mix the urine and pipette 0.5 ml into a test tube.**
2. **Add 2.5 ml of Tauber's reagent and boil vigorously for 2 minutes.**
3. **Cool by immersing tube in cold water.**
4. **Add 5 ml of distilled water and mix. If pentose is present, a pink to red color is produced. Disregard any yellowish brown color.**

Glucose

Fermentation test

The fermentation test may be used to distinguish glucose from lactose and pentose. When glucose and yeast are mixed, fermentation occurs and carbon dioxide is given off. This, however, does not occur with lactose and pentose.

Reagents needed

Fresh yeast
Glucose, 5%

Procedure

1. Place a small test tube in an inverted position in each of three large test tubes (see Fig. 14-15, which shows either positive or negative result for tube 3).
2. Label these tubes as follows: Positive control, Negative control, and Test.
3. *Test:* Boil the urine to be tested. Cool. Pour 25 ml of the urine into a small beaker and add one third of a cake of fresh yeast. Make an emulsion by mixing thoroughly. Pour into the tube labeled Test. Stopper and slowly invert so that the small tube becomes *completely* filled.
4. *Positive control:* Pour 25 ml of 5% glucose into a small beaker and add one third of a cake of fresh yeast. Make an emulsion and pour into the tube labeled Positive control. Stopper and slowly invert so that the small tube becomes *completely* filled.
5. *Negative control:* Make an emulsion with 25 ml of water and the remaining third of the yeast cake. Pour this into the tube labeled Negative control. Stopper and slowly invert so that the small tube becomes *completely* filled.
6. Plug the three tubes with cotton and place in a 37° C incubator for 2 hours.
7. Remove the tubes from the incubator. The tube labeled Positive control should contain gas in the small tube (Fig. 14-15), whereas the tube labeled Negative control should not contain gas in the small tube. If this is not so, repeat the test with fresh yeast. If this is so, look at the tube marked Test. If this tube contains gas in the small tube, the test is positive for glucose. However, if this tube does not contain gas, the test is negative for glucose.

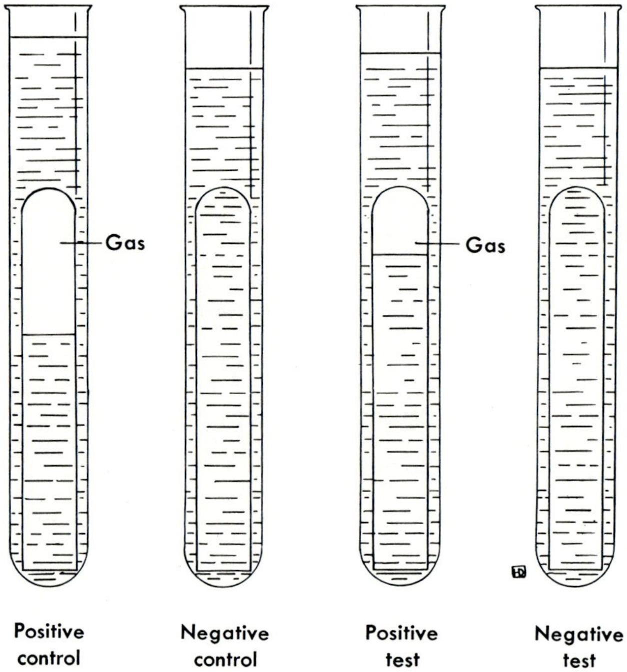

Fig. 14-15. Reading the results in a fermentation test.

Phenylhydrazine test

Kowarsky test as modified by Blumel

The phenylhydrazine test helps to distinguish glucose from other sugars occurring in urine. When sugars are treated with osazone, they form characteristic crystalline structures.

Reagents needed

Phenylhydrazine
Acetic acid, glacial
Sodium chloride, saturated

Procedure

1. **Place 5 drops of pure phenylhydrazine in a large test tube.**
2. **Add 10 drops of glacial acetic acid and 1 ml of a saturated solution of sodium chloride. Mix.**
3. **Add 4 ml of urine and 4 ml of distilled water. Mix.**
4. **Add about 4 glass beads to prevent bumping and boil until the volume is lowered to about 2 or 3 ml. Allow to cool for 30 minutes.**
5. **Using a medicine dropper, place a sample of the sediment on a glass slide and examine with the low-power (16 mm) objective of the microscope. If glucose is present, yellow needlelike crystals (usually arranged in clusters) may be seen. These crystals are phenylglucosazone.**

ACETONE

When glucose is not oxidized, as in diabetes, it is believed that the body endeavors to compensate by oxidizing fats. This is referred to as abnormal fat catabolism.

During abnormal fat catabolism, diacetic acid is produced. When diacetic acid decomposes, it liberates acetone and β-oxybutyric acid:

Abnormal fat catabolism ⟶ Diacetic acid
↓
(Decomposition)
↙ ↘
Acetone β-Oxybutyric acid

Acetone in the urine is significant, for it often indicates acidosis—a decrease in the alkali reserve. This could mean approaching death.

Acetone may be found in the following conditions: prior to diabetic coma, after anesthesia, in toxemias of pregnancy, in fevers, and in gastrointestinal disorders.

In discussions of acetone, textbooks use various terminology, and it is well to remember the following terms: (1) the presence of acetone in the urine is frequently referred to as ketosis; (2) diacetic acid is often called acetoacetic acid; (3) the three substances—diacetic acid, acetone, and β-oxybutyric acid—are often referred to as acetone or ketone bodies.

In the presence of acetone, sodium nitroprusside forms a complex compound that is purple to red in color. This is the basis for the following tests: Acetest, Ketostix, acetone test (Denco), Lange test, Rothera test, and Rantzman test.

Acetest is a tablet containing nitroprusside, phosphate buffers, glycine, and lactose. A purple color develops when a drop of urine containing acetone is placed on the surface. Ketostix and the ketone portion of Labstix contain a stable alkaline phosphate-buffered nitroprusside reagent that includes glycine. When the strips are dipped into urine containing ketone bodies, the characteristic purple color appears. BSP dye and large amounts of phenylpyruvate in urine will give a false positive reaction.

In the presence of acetone, salicylaldehyde also forms a complex compound that is purple to red in color. This is the basis of the Frommer test.

If the acetone test is positive, many laboratories run a test for diacetic acid. In diabetes, it is believed that the presence of diacetic acid is an additional indication of approaching coma.

Procedures for acetone in urine

Acetest

Reagent needed

Acetest tablets*

Procedure

1. Place an Acetest tablet on a piece of white paper.
2. Place a drop of urine on the tablet. If acetone is present, a lavender to purple color develops within 30 seconds. Compare with color chart that comes with the reagent.
3. Report the test as positive or negative.

Acetone test (Denco)

Reagent needed

Acetone test (Denco)†

Procedure

1. Shake the vial containing the powdered reagent.
2. Place a small mound (about the size of a small pea) on a piece of white paper.
3. Place a drop of urine on the powder. If acetone is present, a lavender to purple color develops.
4. Report the test as positive or negative.

Lange test

Reagents needed

Glacial acetic acid
Ammonium hydroxide, 28% (full-strength)
Sodium nitroprusside, saturated (D-271)

Procedure

1. Pour about 5 ml of urine into a test tube.
2. Add 5 drops of glacial acetic acid and a few drops of a saturated solution of sodium nitroprusside. Mix.
3. Tilt the tube to about a 30-degree angle. Using a medicine dropper, *slowly* overlay with 28% (full-strength) ammonium hydroxide. A purple or purplish red ring at the junction of the two liquids indicates the presence of acetone.
4. Report the test as positive or negative.

Rothera test

Reagents needed

Ammonium sulfate
Ammonium hydroxide, 28% (full strength)
Sodium nitroprusside reagent (D-272)

---

*May be purchased from your local supply house or from Ames Co.

†This is a powder contained in a vial. It may be purchased from your local supply house or from Denver Chemical Manufacturing Co., Inc.

Procedure

1. Pour about 5 ml of fresh urine into a test tube.
2. Add about 2 g of ammonium sulfate. Mix to saturate the urine.
3. Add 2 drops of sodium nitroprusside reagent. Mix.
4. Tilt the tube to about a 30-degree angle and, using a medicine dropper, *slowly* overlay with 28% (full-strength) ammonium hydroxide. The test is positive for acetone (or diacetic acid) if a purple to red color appears at the junction between the two liquids. The color may take up to 15 minutes to appear but must not fade. Disregard any brown or orange colors.
5. Report the test as positive or negative.

Rantzman test

Reagents needed

Ammonium hydroxide, 28% (full-strength)
Rantzman's reagent (D-273)

Procedure

1. Pour about 3 ml of urine into a test tube.
2. Add 1 ml of Rantzman's reagent. Mix.
3. Tilt the tube to about a 30-degree angle and *slowly* overlay with 28% ammonium hydroxide. If acetone is present, a purple ring appears within a few minutes at the junction of the two liquids.
4. Report the test as positive or negative.

Frommer test

Reagents needed

Sodium hydroxide, 40% (D-274)
Alcoholic solution of salicylaldehyde, 10% (D-275)

Procedure

1. Pour about 10 ml of urine into a test tube.
2. Add 2 or 3 ml of 40% sodium hydroxide.
3. Add 10 drops of a 10% alcoholic solution of salicylaldehyde.
4. Place in a 70° to 75° C water bath for 10 minutes. If acetone is present, an intense purplish red color is formed.
5. Report the test as positive or negative.

DIACETIC ACID

As was previously mentioned, diacetic acid results from abnormal fat catabolism, which may occur in severe diabetes. Its presence in the urine may forewarn the physician of approaching coma.

Either Gerhardt's or Lindemann's test is used to test for diacetic acid. In Gerhardt's test, a reaction between diacetic acid and ferric chloride produces a red color. In Lindemann's test, a reaction between diacetic acid and iodine decolorizes the iodine.

Procedures for diacetic acid in urine

Gerhardt test

Reagents needed

Concentrated nitric acid
Ferric chloride, 10% (D-241)

Procedure

1. Pour about 5 ml of *fresh* urine into a test tube.
2. Add 10% ferric chloride drop by drop until no further precipitation is produced.

3. Centrifuge.
4. Transfer the supernatant fluid to a test tube. Add a few more drops of 10% ferric chloride to this supernatant fluid. If no color is produced, the test is negative. If a red color is produced, however, it is due either to diacetic acid or to a drug. To decide which, proceed as follows.
5. Pour about 5 ml of *fresh* urine into a beaker. Add 5 ml of distilled water and 2 drops of concentrated nitric acid.
6. Boil down to 5 ml. Cool.
7. Add a few drops of the 10% ferric chloride. If *no* red color is produced, the test is *positive* for diacetic acid. The boiling drives off or oxidizes the diacetic acid but not the drug.

Lindemann test

Reagents needed

Chloroform
Acetic acid, 30% (D-276)
Lugol's solution (D-277)

Procedure

1. Pour about 10 ml of *fresh* urine into a large test tube.
2. Add 5 drops of 30% acetic acid.
3. Add 5 drops of Lugol's solution.
4. Add 3 drops of chloroform. Shake gently.
5. The chloroform layer *does not* change color in the *presence* of diacetic acid. Consequently, if the chloroform layer does not change color, the test is positive for diacetic acid.

BETA-OXYBUTYRIC ACID

Beta-oxybutyric acid is a reduction product of diacetic acid. When found in the urine, it is usually accompanied by both diacetic acid and acetone and, like these, its presence indicates the abnormal catabolism of fats.

Hart method

Reagents needed

Acetic acid, glacial
Hydrogen peroxide, 3%
Ammonium hydroxide, concentrated
Sodium nitroprusside, saturated (freshly prepared) (D-271)

Procedure

1. Mix urine. Pour 20 ml into a small beaker. With a marking pencil, mark the level to which the urine comes (the 20 ml level). Add 20 ml of distilled water and a few drops of glacial acetic acid. Drive off any acetone or diacetc acid by boiling down to the 20 ml level.
2. Add 10 ml of distilled water. Mix. Get two large test tubes. Label as 1 and 2. Pour 10 ml of the mixture into each tube. Add 1 ml of 3% hydrogen peroxide to tube 1. (This oxidizes any β-oxybutyric acid present to diacetic acid and acetone.) Add none to tube 2. Warm gently and allow to cool.
3. To each tube add 10 drops of glacial acetic acid and 10 drops of a freshly prepared concentrated solution of sodium nitroprusside. Mix both tubes by inversion.
4. Using a medicine dropper, overlay both tubes with concentrated ammonium hydroxide. If tube 1 shows a reddish purple ring at the junction of the two liquids, the test is positive for β-oxybutyric acid. Tube 2 should show no ring.

MICROSCOPIC EXAMINATION

The microscopic examination consists of obtaining, examining, and reporting the urinary sediments. The sediments, which are examined with a microscope, are magnified either 100 or 450 times.

Knowing the type and quantity of sediments aids the physician in diagnosing diseases of the urethra, bladder, and kidneys. Some significant sediments and their usual causes are given in Table 14-5. The pus cells mentioned in the table are often referred to as white blood cells or WBC. The red cells are often reported as red blood cells or RBC. The casts are solidified sediments that have been forced out of urinary tubules. The *Trichomonas* is a parasite that is some-

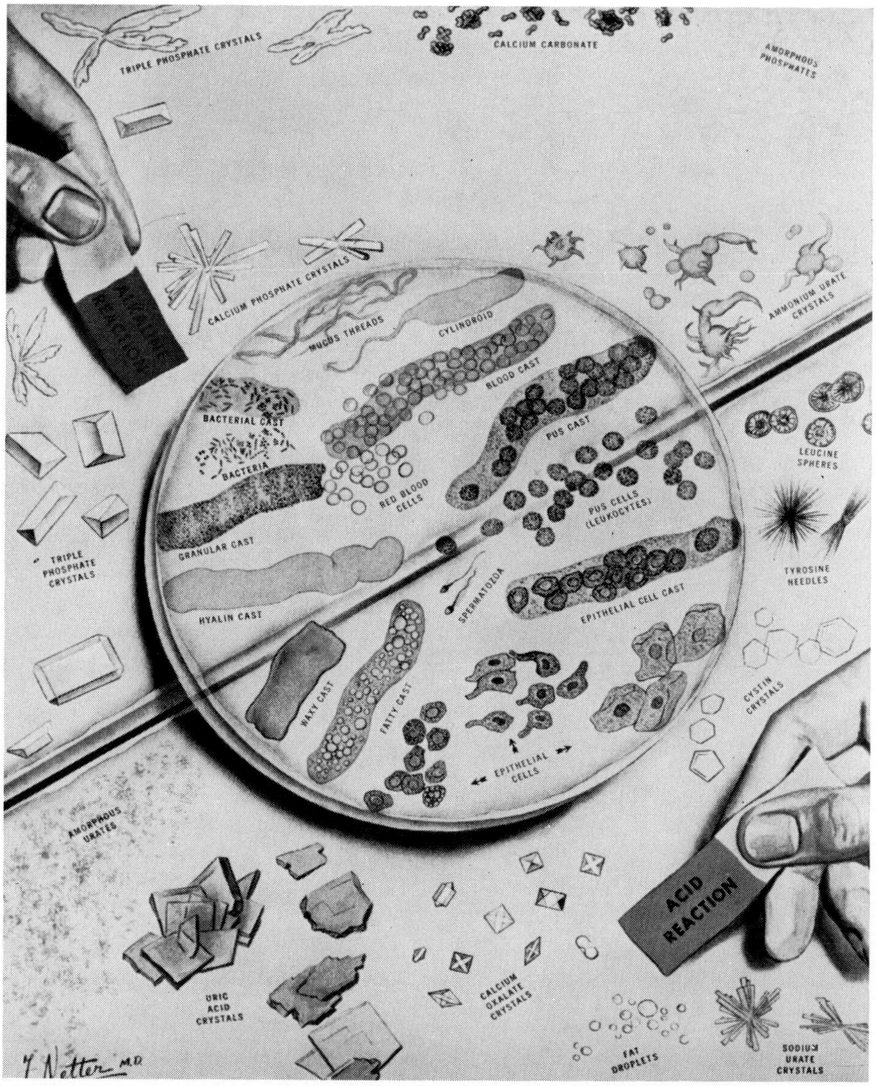

Fig. 14-16. Urinary sediments. (From Sharp & Dohme Seminar, November, 1940, Sharp & Dohme, West Point, Pa.)

Table 14-5. Some significant sediments and their usual causes

Sediment	Usual causes
Pus	Inflammation of urethra, bladder, vagina, or kidneys
Red cells	Bleeding from bladder, uterus, or kidneys
Casts	Nephritis
Bacteria	Bacterial infection of urethra, bladder, vagina, or kidneys
Trichomonas	*Trichomonas* infestation of vagina

times found in the urine of women. With the exception of *Trichomonas,* the sediments are illustrated in Fig. 14-16.

The manner of performing and reporting the microscopic examination varies from laboratory to laboratory. For example, some laboratories obtain the sediment by centrifuging the urine; other laboratories do not centrifuge the urine. Some laboratories report the number of sediments per low-power field (magnification 100); other laboratories report the number of sediments per high dry field (magnification 450). Some laboratories report all sediments seen; other laboratories report only the more significant sediments.

In performing the microscopic examination, guard against the following sources of error:

1. **Failure to see casts. If the light is too bright, casts are very difficult to see. Therefore, always lower the source of light.**
2. **Failure to distinguish red cells from pus cells. If in doubt, place a drop of 5% acetic acid on the sediment. If the cells are red cells, they will hemolyze. If the cells are pus cells, it is possible to see their nuclei.**
3. **Failure to distinguish red cells from fat globules. The red cells are uniform in size, whereas fat globules usually come in various sizes.**

A suggested procedure for the microscopic examination is given below.

Procedure for microscopic examination

1. **Mix the urine specimen by inverting or swirling the container.**
2. **Place about 5 ml of urine into a test tube. Centrifuge at high speed for 5 minutes and decant the supernatant fluid.**
3. **The grayish white precipitate (urinary sediment) may be loosened by gently shaking or tapping the bottom of the tube.**
4. **Place the sediment on a glass slide. This step is illustrated in Fig. 14-17, where three different specimens are being placed on one slide.**
5. **When the sediment is on the slide, smear it over the surface of the slide with the lip of the test tube. This "spreading out" will prevent the high dry objective of the microscope from touching the sediment and eliminates the use of coverslips. However, coverslips are preferable.**
6. **Place the slide on the stage of a microscope as illustrated in Fig. 14-18.**
7. **Lower the source of light by lowering the condenser (Fig. 14-18).**
8. **Switch the low-power (16 mm or 10×) objective into position so that it is directly over the sediment.**
9. **Bring the field into focus by using the coarse and fine adjustments.**
10. **Using the movable stage, scan four or five fields to get a general idea of the contents.**

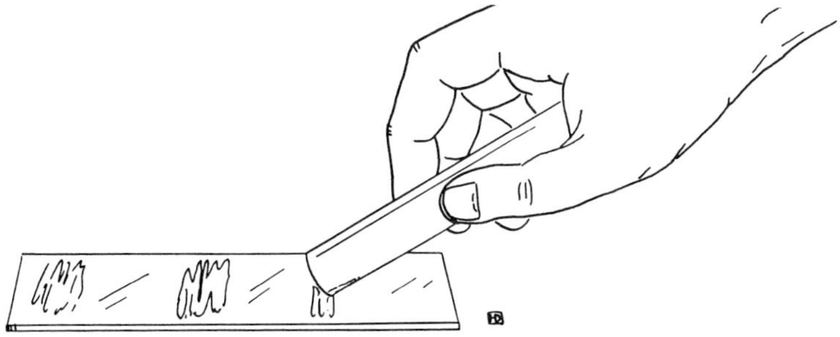

Fig. 14-17. Spreading urinary sediments on slide.

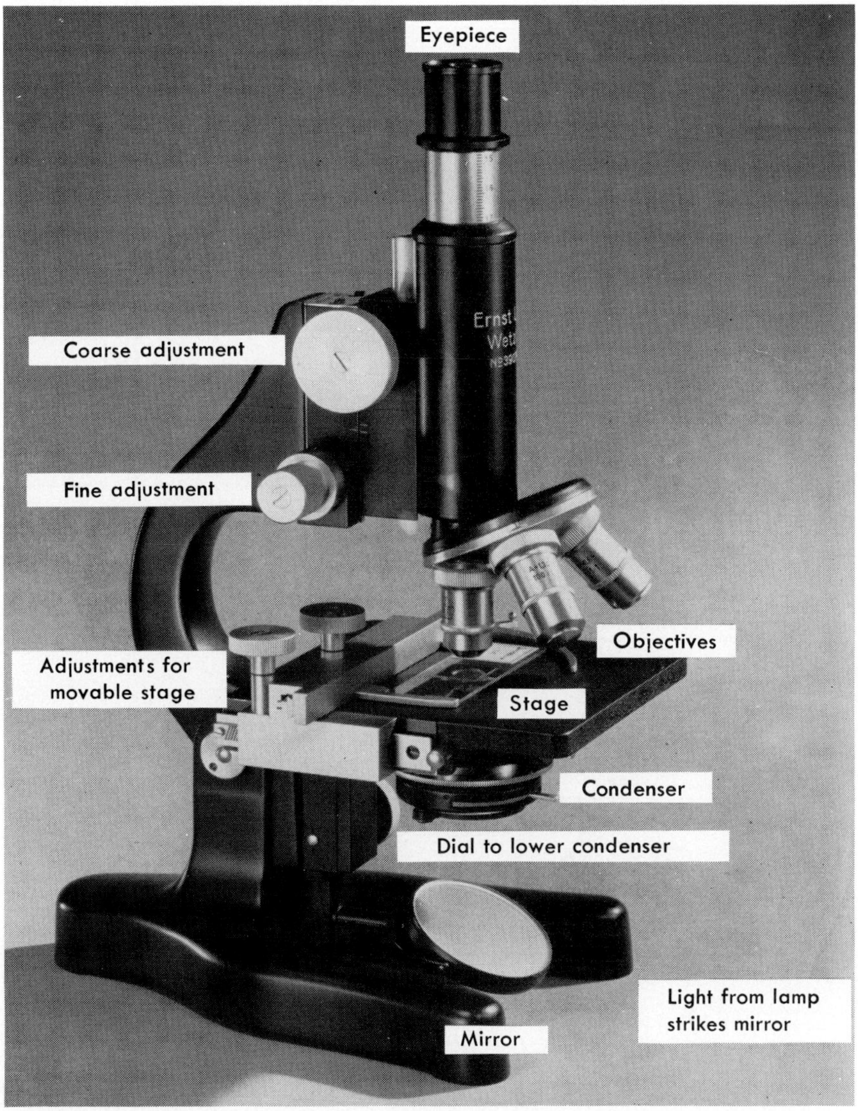

Fig. 14-18. Microscopic examination of urinary sediments. (Courtesy E. Leitz, Inc., New York, N. Y.)

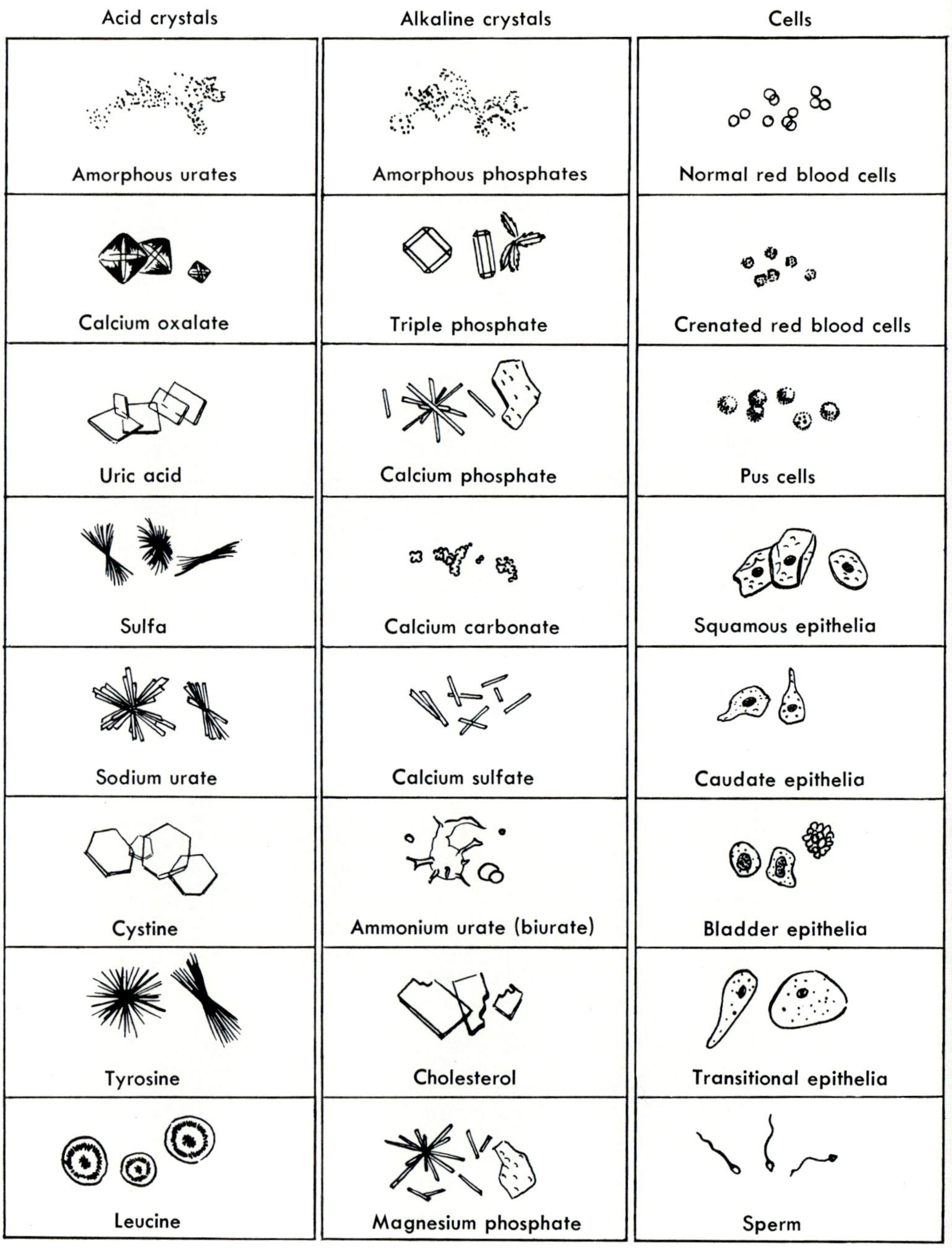

Fig. 14-19. Sediments found in urine. (Drawings by Robert Pribbenow and Samuel Taylor.)

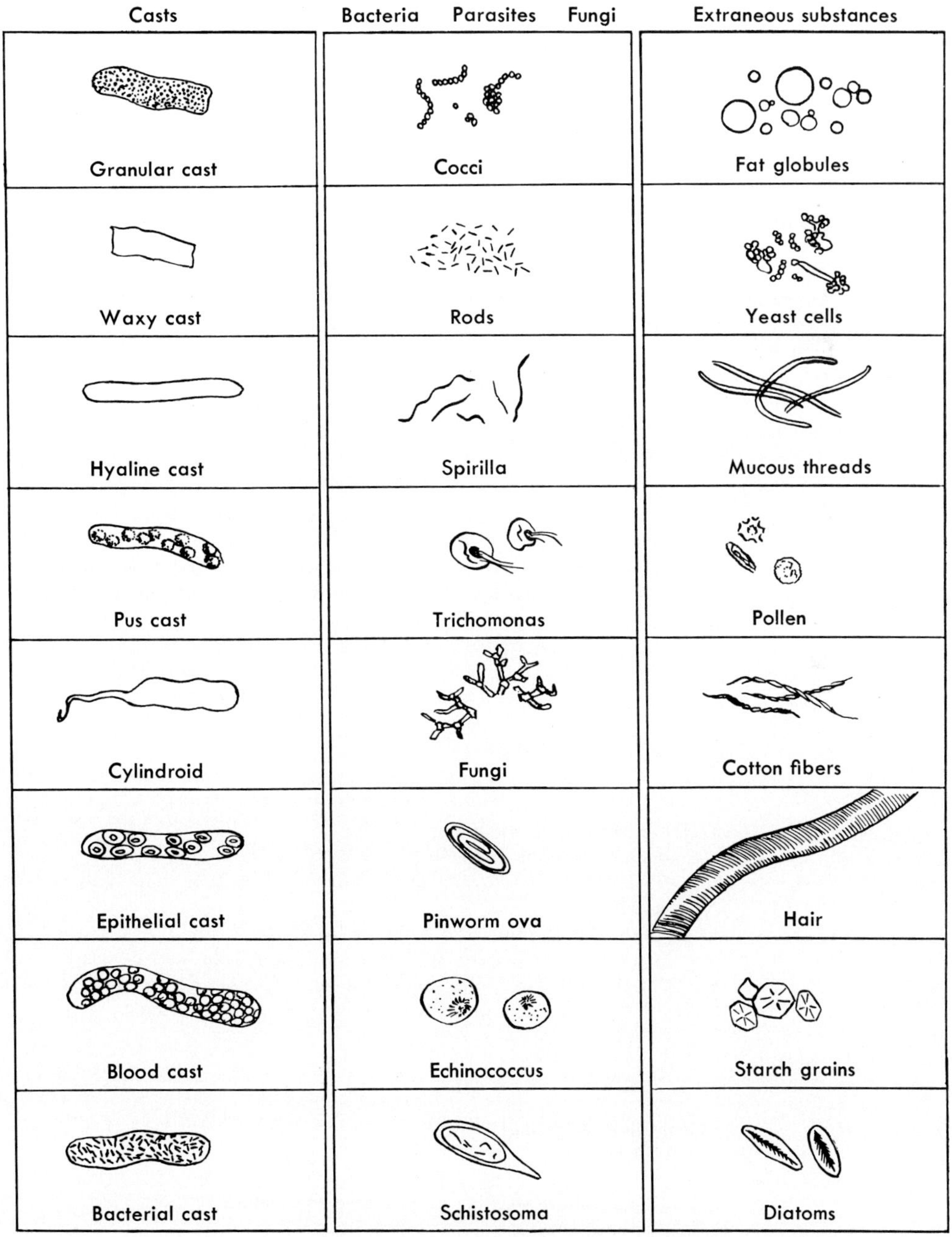

Fig. 14-19, cont'd. For legend see opposite page.

11. Switch to the high dry (4 mm or 45×) objective.
12. Identify the contents, using Figs. 14-16 and 14-19 as guides.
13. Make the report according to the practice of your particular laboratory.
14. Sample microscopic report: Centrifuged specimen shows 10 to 12 RBC and 0 to 2 granular casts per high dry field. Bacteria present.

BILIRUBIN (BILE)

Bilirubin is a red bile pigment. It is found in the urine in infectious hepatitis, obstructive jaundice, and many other liver diseases. In liver disease, bilirubin may be detected in the urine even before the blood level is elevated.

The majority of tests for bilirubin depend upon the oxidation of bilirubin to various colored compounds. Among these compounds are biliverdin, which is green, bilicyanin which is blue, and choletelin, which is yellow.

Ictotest is a test system consisting of an asbestos-cellulose mat and a reagent tablet containing sulfosalicylic acid with a stabilized diazonium salt, *p*-nitrobenzene diazonium *p*-toluene sulfonate (bilazo). Bilirubin in urine reacts with the reagent to form a blue or purple color.

Procedures for bilirubin in urine

Ictotest

Reagent needed

Ictotest reagent tablets*

Procedure

1. Place 5 drops of urine on a special mat that comes with the reagent.
2. Put 1 Ictotest tablet in the center of the moistened area.
3. Place 2 drops of water on the tablet so that the water just flows over the sides. If a bluish purple color appears on the mat around the tablet, the test is positive for bilirubin.
4. Report the test as positive or negative.

Foam test

Reagent needed

None

Procedure

1. Obtain two small test tubes.
2. Pour about 5 ml of the urine that is to be tested into one tube.
3. Pour about 5 ml of normal urine into the other tube. This will be used for a color comparison.
4. Shake both test tubes. Note the color of the foam. If the foam has a yellow color that persists for several minutes, bile may be present and one of the following tests should be run. (Certain drugs, such as Pyridium and Serenium, also impart a yellow color to the foam.)

Nitric acid test

Reagent needed

Concentrated nitric acid

Procedure

1. Mix the urine by swirling or inverting the container.
2. Filter about 20 ml.
3. Allow filter paper to partially dry.

*May be purchased from your local supply house or from Ames Co.

4. Place a few drops of concentrated nitric acid on the filter paper. A display of colors—yellow, green, blue—is positive for bile.

Gmelin test

Reagent needed

Concentrated nitric acid

Procedure

1. Carefully pour a few milliliters of concentrated nitric acid into a test tube.
2. Tilt the tube to about a 30-degree angle.
3. Using a medicine dropper, overlay with urine. A band of colored rings at the junction of the two liquids is positive for bile.

Harrison test

Reagents needed

Barium chloride, 10% (D-246)
Fouchet's reagent (D-278)

Procedure

1. Pour about 20 ml of urine into a small flask.
2. Add 10 ml of 10% barium chloride.
3. Mix and let stand for a few minutes.
4. Filter.
5. Place a few drops of Fouchet's reagent on the filter paper. A blue to green color is positive for bile.

Smith test

Reagent needed

Iodine, ethyl alcohol solution (D-279)

Procedure

1. Pour about 5 ml of clear urine into a test tube.
2. Tilt the tube to about a 30-degree angle.
3. Using a medicine dropper, overlay with the ethyl alcohol solution of iodine. An emerald green color at the junction of the two liquids is positive for bile.

UROBILINOGEN

Urobilinogen is a colorless compound derived from bilirubin. Small traces are normally present in urine. Large amounts are found in liver disease, lead poisoning, hemolytic anemia, and other diseases of a hemolytic nature.

Urobilinogen is formed in the intestine by the bacterial action on bilirubin. This is reabsorbed in the small intestine and excreted partly by the kidney and partly by the liver. With hemolytic anemia, where there is an increased destruction of the red blood cells, and in parenchymal liver disease, the urinary urobilinogen is increased. With complete obstructive jaundice, there may be no urobilinogen in the urine, since bilirubin cannot reach the intestine to be changed to urobilinogen.

For the determination of urobilinogen in urine, the following methods are commonly used: (1) Ehrlich qualitative test, (2) Ehrlich semiquantitative test, and (3) Schwartz-Watson test. In all these methods, a reaction takes place between urobilinogen and Ehrlich's reagent (*p*-dimethylaminobenzaldehyde) to produce a cherry red color. The extent of the reaction may be compared with standards.

Procedures for urobilinogen in urine

Ehrlich qualitative test

Reagent needed

Ehrlich's reagent (D-280)

Procedure

1. Pour about 5 ml of *freshly* voided urine into a test tube.
2. Add 1 ml of Ehrlich's reagent. Mix.
3. Let stand for 5 minutes. A cherry red color is positive for abnormal amounts of urobilinogen. Disregard a pink color.

Ehrlich semiquantitative test

Reagent needed

Ehrlich's reagent (D-280)

Procedure

1. Get five large test tubes.
2. Number them 1 to 5 and place in a test tube rack.
3. Pipette the following amounts of *freshly* voided urine and distilled water into the tubes.

Tube	Urine (ml)	Water (ml)	Dilution
1	1.0	9.0	1:10
2	0.5	9.5	1:20
3	0.2	9.8	1:50
4	0.1	9.9	1:100
5	0.05	9.95	1:200

4. Add 1 ml of Ehrlich's reagent to each tube.
5. Mix and let stand for 5 minutes.
6. Note color. A pink color in the first two tubes (that is, through dilutions 1:20) is normal. A pink or red color in tube 3, 4, or 5 indicates abnormal amounts of urobilinogen.
7. Report the highest dilution giving a pink or red color, stating that it is normal for the test to be positive through dilutions 1:20.
8. Example: If tubes 1, 2, 3, and 4 are pink, the report is made as follows: Test for urobilinogen is positive through dilutions 1:100 (NOTE: It is normal for the test to be positive through dilutions 1:20).

Schwartz-Watson test

Reagents needed

Ehrlich's reagent (D-280)
Sodium acetate, saturated (D-281)
Sodium acetate, crystals
Chloroform

Procedure

1. Pipette 1.0 ml of fresh urine into a test tube.
2. Add 1.0 ml of Ehrlich's reagent. Mix.
3. Add 4.0 ml of supersaturated aqueous solution of sodium acetate. Mix.
4. Test with Congo red paper. Paper reaction:
   a. If red, continue to step 5.
   b. If not red, add sodium acetate crystals (shake to dissolve) until a positive result of red is obtained.
5. Add 3.0 ml of chloroform.
6. Mix carefully by inverting twice and releasing the stopper; then *shake*

vigorously. (NOTE: If the tube is not shaken vigorously, the test will produce a false positive reaction.)

7. Results
   a. *Negative*—the aqueous layer is not colored.
   b. *Positive*—the aqueous layer forms a purple-red aldehyde reaction.
   c. NOTE: *Porphobilinogen* will be indicated by a red-brown aldehyde layer.
8. Results with ultraviolet light:
   a. *Uroporphyrin* and *coproporphyrin* are fluorescent in ultraviolet light.

Diazo substances

Diazo substances are unknown substances that give a color reaction with Ehrlich's diazo reagent. They may appear in the urine in febrile conditions, measles, typhoid fever, and tuberculosis.

Reagents needed

Ehrlich's diazo reagent	(D-282)
Ammonium hydroxide	

Procedure

1. Pipette 2.0 ml of urine into a small test tube and add 2.0 ml of Ehrlich's diazo reagent. Shake well.
2. Using a medicine dropper, overlay with 2.0 ml of full-strength (28%) ammonium hydroxide. If a red ring appears at the junction of the two liquids, the test is positive for diazo substances. Disregard a yellow or orange color.

UROBILIN

When urine stands, urobilinogen is oxidized to a brown pigment known as urobilin. Small amounts of urobilin are normally found in urine. Large amounts may be seen in diseases of the liver and in diseases of a hemolytic nature.

To determine urobilin in urine, the method of Schlesinger is commonly used. In this method, calcium chloride is added to the urine to precipitate bile. The urine is then filtered. Lugol's solution is added to the filtrate to convert urobilinogen to urobilin. A solution of zinc acetate is then added. This reacts with the urobilin to produce a green fluorescence.

Procedures for urobilin in urine

Reagents needed

Calcium chloride, 10%	(D-283)
Lugol's solution	(D-277)
Zinc acetate, saturated alcoholic solution	(D-284)

Procedure (Schlesinger)

1. Mix the urine and pour 12 ml into a small flask. (Measure with a graduated cylinder.)
2. Add 3 ml of 10% calcium chloride.
3. Mix and filter to remove any bile present.
4. Take 10 ml of the filtrate and add a few drops of Lugol's solution.
5. Mix and add 10 ml of a saturated alcoholic solution of zinc acetate.
6. Mix and filter. Let filtrate stand about an hour.
7. Examine in sunlight, concentrating the sunlight on the solution with a hand lens. The presence of a green fluorescence is positive for urobilin.

PORPHYRINS

The porphyrins are complex organic compounds. Small amounts may be found in normal urine. Increased amounts are found in porphyria, a metabolic disease in which porphyrins are retained in pathologic quantities in the blood and other tissues, and in feces and urine. Urine containing increased amounts of porphyrins usually turns dark red upon standing.

The two porphyrins of interest are coproporphyrin and uroporphyrin.

Reagents needed

Acetic acid, glacial
Ethyl ether
Hydrochloric acid, 5% (v/v) (D-266)

Procedure

1. **Mix the urine, pour 25 ml into a separatory funnel, and add 10 ml of glacial acetic acid.**
2. **Extract the mixture twice with 50 ml portions of ether, saving the *waste fluids* from the extractions, since they will be used later in the procedure.**
3. **Combine the two ether extractions and wash with 10 ml of 5% hydrochloric acid, saving the *washings.***
4. **Examine the *washings* under ultraviolet light. If there is a strong fluorescence, the test is positive for coproporphyrin.**
5. **Examine the urinary *waste fluids* (which resulted from the above ether extractions) under ultraviolet light. If there is a strong red fluorescence, the test is positive for uroporphyrin.**

PORPHOBILINOGEN

Porphobilinogen is a precursor of uroporphyrin. It is found in the urine in acute intermittent porphyria.

Reagents needed

Ehrlich's reagent, modified
Dissolve 0.7 g of *p*-dimethylaminobenzaldehyde in a mixture of 100 ml of distilled water and 150 ml of concentrated hydrochloric acid. Mix well and store in a brown bottle.
Hydrochloric acid, concentrated
Sodium acetate, saturated (D-281)
Chloroform

Procedure

1. **Obtain a fresh sample of urine and acidify to pH 4 or 5 with concentrated hydrochloric acid, using nitrazine paper as an indicator.**
2. **Pipette 3 ml of the urine into a large test tube and add 3 ml of the modified Ehrlich's reagent. Mix by inversion.**
3. **Add 6 ml of saturated sodium acetate and mix.**
4. **Add 3 ml of chloroform and mix thoroughly.**
5. **Let settle for a few minutes. If the water fraction (upper layer) is red, the test is positive for porphobilinogen. If the chloroform fraction (lower layer) is red, it indicates the presence of urobilinogen or indole.**

HIPPURIC ACID SYNTHESIS

A liver function (hippuric acid synthesis) test is based on the principle that a healthy liver will convert benzoic acid to hippuric acid. The latter is then excreted in the urine. When the liver is not functioning properly, however, this process of conversion and excretion is decreased.

Quick's method

Reagents needed

Sodium benzoate	
Oil of peppermint	
Ammonium sulfate	
Glacial acetic acid	
Hydrochloric acid, concentrated	
Congo red indicator	(Table C-3)
Phenolphthalein indicator	(D-208)
Sodium hydroxide, 0.2 N	(D-201)

Procedure

1. The patient is instructed to eat a light breakfast. One hour after breakfast, he is told to empty the bladder and discard the urine. He is then given a dose of 6 g of sodium benzoate dissolved in 30 ml of water and flavored with a little oil of peppermint. This is followed with a half glass of water.
2. Collect all urine voided within the next 4 hours. Ask the patient to empty the bladder at the end of the 4-hour period and add this to the collection.
3. Measure the volume of urine. If the volume is 150 ml or less, it does not have to be boiled down. However, if the volume is more than 150 ml, acidify with a few drops of glacial acetic acid and boil down to 150 ml.
4. Add 5 g of ammonium sulfate for each 10 ml of urine. Stir to dissolve. Filter.
5. Add concentrated hydrochloric acid to the filtrate until the indicator Congo red turns blue (pH 3.0). Usually about 2 ml is needed. Stir for several minutes to aid crystallization. Place in the refrigerator for 1 hour.
6. Using a Buchner filter and suction, filter off the crystalline hippuric acid. Wash the container with ice-cold distilled water and pour the washings into the Buchner filter. Wash the precipitate several times with ice-cold distilled water.
7. Carefully transfer the precipitate and filter paper to a beaker. Add enough distilled water to cover and heat gently to dissolve the crystals. Add a few drops of phenolphthalein indicator and titrate to a permanent pink color with 0.2 N sodium hydroxide. Record the milliliters of sodium hydroxide used.

Calculations

One milliliter of the 0.2 N sodium hydroxide is equivalent to 0.0358 g of hippuric acid. Therefore:

$$\text{ml of NaOH used} \times 0.0358 = \text{g of hippuric acid}$$

Some hippuric acid, however, still remains in solution. To correct for this, take your answer obtained above and add 0.1 g for each 100 ml of urine used. If you wish to express the hippuric acid in terms of benzoic acid, multiply your corrected answer by 0.68.

Normal values

The normal values are 3 g or more of benzoic acid in 4 hours or, if sodium benzoic acid is given intravenously, the normal value is at least 1 g of benzoic acid in 1 hour.

INDICAN

Indican is a product of putrefaction. Small amounts are normally present in the urine. Large amounts are found in heavy-protein diets and in diseases involving stagnation of the intestinal contents, such as cancer of the stomach, peritonitis, and prolonged constipation.

Obermayer method

Reagents needed

Chloroform
Ferric chloride
Hydrochloric acid, concentrated
Preparation of the indican reagent before using: weigh 0.5 g of ferric chloride on the analytical balance and dissolve in 250 ml of concentrated hydrochloric acid. Mix.

Procedure

1. Mix urine specimen and pipette 5.0 ml into a large test tube.
2. Pipette 5.0 ml of the indican reagent into the tube and mix by inverting the tube several times.
3. Add about 3 ml of chloroform and mix by inverting several times. If abnormal amounts of indican are present, the chloroform layer will become a deep blue color.

MUCIN

Mucin is the chief constituent of mucus, the watery secretion that covers mucous membranes. Small amounts of mucin are found in normal urine. Increased amounts are present in inflammations of the urinary and vaginal tracts.

Reagent needed

Acetic acid, glacial

Procedure

1. Mix urine and transfer 3 ml to a small test tube.
2. Add 3 ml of water and mix.
3. Add 0.5 ml of glacial acetic acid and mix gently. If mucin is present, a white cloud will form.

MELANIN

The word "melanin" is Greek for black. When present in the urine, melanin usually occurs as a colorless compound. Upon standing, however, it oxidizes to a black pigment that is readily discernible. Melanin is not present in normal urine but may be found in the urine of patients with melanotic tumors.

For the determination of melanin, the following methods are commonly used: (1) ferric chloride test and (2) bromine test. Both methods call for the oxidation of the colorless compound (melanogen) to the colored compound (melanin).

Procedures for melanin in urine

Ferric chloride test

Reagent needed

Ferric chloride, 10% (D-241)

Procedure

1. Mix the urine and pour about 8 ml into a small flask.
2. Add about 6 drops of 10% ferric chloride. Mix.
3. Allow to stand 30 minutes. If the solution turns black, melanin is present.

Bromine test

Reagent needed

Bromine water (D-285)

Procedure

1. Mix the urine and pour about 25 ml into a small flask.
2. Add about 25 ml of the bromine reagent. Mix.
3. Allow to stand 30 minutes. If the solution turns black, melanin is present.

OCCULT BLOOD

Occult blood is hidden blood; that is, blood which has dissolved or hemolyzed. When it occurs in the urine, the condition is referred to as hemoglobinuria. It may be found in the following disorders, all of which are accompanied by excessive red cell destruction: transfusion reactions, severe burns, and various chemical poisonings.

For the detection of occult blood, the following tests are commonly used: (1) benzidine test, (2) *o*-tolidine test, (3) guaiac test, and (4) Occultest. The principle of these tests is as follows. The hemoglobin of the blood reacts with peroxide to liberate oxygen. The liberated oxygen then reacts with an organic reagent to produce a colored compound.

Occultest is a tablet containing *o*-tolidine, strontium peroxide, calcium acetate, tartaric acid, and sodium bicarbonate. The interaction of hydrogen peroxide with *o*-tolidine catalyzed by the peroxidative activity of blood produces a characteristic blue color. Hemastix and the occult blood portion of Labstix contain *o*-tolidine and cumene hydroperoxide, which when dipped into urine containing occult blood develop the blue color of oxidized *o*-tolidine. Large quantities of ascorbic acid may inhibit the sensitivity of this test.

Procedures for occult blood in urine

Benzidine test

Reagents needed

Hydrogen peroxide, 3%
Benzidine, saturated, in glacial acetic acid (D-239)

Procedure

1. Mix the urine and pour about 2 ml into a test tube.
2. Add 3 ml of a saturated solution of benzidine in glacial acetic acid. Mix.
3. Add 1 ml of 3% hydrogen peroxide. Mix. A green to blue color indicates a positive reaction.

Orthotolidine test

Reagents needed

Hydrogen peroxide, 3%
o-Tolidine reagent (D-286)

Procedure

1. Place about 1 ml of urine in a test tube.
2. Add 1 ml of the *o*-tolidine reagent.
3. Add about 10 drops of 3% hydrogen peroxide. Mix. A green to blue color is a positive reaction.

Guaiac test

Reagents needed

Glacial acetic acid
Guaiac reagent (D-287)

Procedure

1. Pour about 5 ml of urine into a test tube.
2. Add a few drops of glacial acetic acid.

3. Mix. Tilt the tube to about a 30-degree angle and slowly overlay with the guaiac reagent. A green to blue color at the junction of the two liquids is a positive reaction.

Occultest

Reagent needed

Occultest tablets*

Procedure

1. Place a piece of filter paper (which comes with the reagent bottle) on a clean surface.
2. Place an Occultest tablet on the filter paper.
3. Place a large drop of urine on the tablet so that the urine just flows over the sides. If a blue color appears on the filter paper around the tablet, the test is positive for occult blood.

HEMOSIDERIN

Hemosiderin is a pigment derived from hemoglobin. It may be found in the urine in hemolytic anemia and other conditions associated with the presence of hemoglobin.

Ham method

Reagent needed

Aqueous solution of ammonium sulfide, 30%

Procedure

1. Mix the urine, pour about 6 ml into a test tube, and centrifuge.
2. Place a drop of the sediment on a glass slide.
3. Add a drop of 30% aqueous solution of ammonium sulfide and mix with the sediment, using an applicator stick.
4. Examine with the low-power (16 mm) objective of the microscope. If hemosiderin is present, it will appear as jet black granules of various sizes. The granules may be separate or may be included in epithelial cells or casts.

CALCIUM

Calcium is normally present in the urine. Increased amounts are found in hyperparathyroidism. Decreased quantities are observed in hypoparathyroidism.

The qualitative test of Sulkowitch is commonly used. In this test, a solution of ammonium oxalate is added to the urine. If calcium is present in excessive amounts, it drops out of solution as a heavy white precipitate of calcium oxalate.

Qualitative test

Reagent needed

Sulkowitch reagent (D-288)

Procedure (Sulkowitch)

1. Pour 5 ml of urine into a test tube.
2. Add 5 ml of the Sulkowitch reagent.
3. Mix by inverting the tube several times.
4. Allow to stand 3 minutes.
5. If a fine white cloud appears, the calcium content is normal.
6. If no white cloud appears, the calcium content is decreased.

---

*May be purchased through your local medical supply house or ordered directly from Ames Co.

7. If a heavy white milky precipitate forms, the calcium content is increased.
8. Report the calcium content as normal, decreased, or increased.

Quantitative test

The urine of the average person contains 50 to 400 mg of calcium/day. Increased values may be found in hyperparathyroidism, whereas decreased values are seen in rickets.

Wang method

Reagents needed

Trichloroacetic acid (TCA), 20%	(D-289)
Sodium acetate, 20%	(D-290)
Ammonium hydroxide, 10% (w/v)	(D-291)
Acetic acid, 10% (v/v)	(D-205)
Sulfuric acid, 2 N	(D-292)
Ammonium oxalate, 0.1 M	(D-293)

Potassium permanganate, 0.01 N (D-181 diluted 1 to 10)

Wash solution: Place 80 ml of distilled water in a large flask; add 80 ml of ether and 80 ml of alcohol; then add 5 ml of ammonium hydroxide. Mix.

Procedure

1. Collect a 24-hour specimen of urine from the patient. Measure and record the volume. Mix. Pour about 100 ml into a flask. Add 25 ml of the 20% trichloroacetic acid. Add 2 g of acid-washed charcoal. Shake well and set aside for 15 minutes.
2. Filter. Pipette 5 ml of the filtrate into a 15 ml centrifuge tube. Add 1 ml of the 20% sodium acetate solution. Mix by tapping the bottom side of the tube. Add 1 ml of the 0.1 M ammonium oxalate solution. Mix again.
3. Adjust the acidity to pH 5 as follows. Find the approximate pH with nitrazine paper. If above pH 5, add the 10% acetic acid dropwise until the solution is green to bromcresol green or pink to methyl red. If below pH 5, add the 10% ammonium hydroxide dropwise until the solution is green to bromcresol green or pink to methyl red. Mix again and place in the refrigerator overnight.
4. The next morning, centrifuge for 15 minutes at high speed. With a steady deliberate motion, invert the tube, thus discarding the supernatant fluid and leaving the precipitate in the bottom of the tube. Keep the tube in an inverted position, place on a piece of filter paper, and let drain for a few minutes.
5. Wash the precipitate *twice* as follows. Add 2 ml of the wash solution (preparation given above). Mix well by tapping the bottom sides of the tube. Wash down the inside of the tube with another 2 ml of the wash solution. Centrifuge for 15 minutes.
6. With a steady deliberate motion, invert the tube, thus discarding the supernatant fluid and leaving the precipitate in the bottom of the tube. Keep the tube in an inverted position, place on a piece of filter paper, and let drain for a few minutes.
7. Place the tube (right side up) in an oven at 85° to 100° C for 1 hour. This removes organic solvents.
8. Add 2 ml of the 2 N (approximate) sulfuric acid. Mix to dissolve the precipitate. Place in a boiling water bath for 2 minutes. Remove from water bath. Titrate while hot (over 70° C) with the 0.01 N. potassium permanganate. The end point is a faint pink color that persists for at least 1 minute. Record the milliliters of permanganate used.

Calculations

One milliliter of the 0.01 N potassium permanganate is equivalent to 0.2 mg of calcium. Therefore:

$$\frac{\text{mg of calcium}}{(\text{24-hour specimen})} = \frac{\text{ml of urine (24-hour specimen)}}{4} \times 0.2 \times \text{ml of } KMnO_4 \text{ used} \times \text{Factor}$$

a. The 4 is the ml of urine used in the test (5 ml of a 4 to 1 mixture of urine and 20% trichloroacetic acid).

b. The factor is a number telling how strong or how weak the potassium permanganate solution is in comparison to an exact 0.01 N $KMnO_4$ solution. The factor is determined in preparing the solution.

Sample calculation: If the 24-hour specimen measures 1000 ml, the number of milliliters of 0.01 N potassium permanganate used is 2, and the factor is 1.0, the calculation is performed as follows:

$$\text{mg of calcium (24-hour specimen)} = \frac{1000}{4} \times 0.2 \times 2 \times 1.0 = 100 \text{ mg}$$

SULFATES

The normal values for total urinary sulfates in a 24-hour period are 0.6 to 1.2 g. Increased values are found in acute fevers. Decreased values are seen in conditions associated with decreased metabolic activity.

Folin method

Reagents needed

Hydrochloric acid, concentrated
Barium chloride, 5% (D-294)

Procedure

1. Collect a 24-hour specimen of urine. Mix. Pour 50 ml into a 250 ml flask. Add 4 ml of concentrated hydrochloric acid. Mix. Place a watchglass or small beaker over the mouth of the flask and boil gently for ½ hour.
2. Cool the flask in cold water. Add sufficient distilled water to make the total volume equal 150 ml. Mix. Without stirring the solution, add dropwise 10 ml of a 5% solution of barium chloride. This precipitates the sulfur as barium sulfate. Set aside for 1 hour.
3. Handling the crucible with tongs, weigh a Gooch crucible on the analytical balance. Record the weight. Shake the solution and then filter it through the weighed Gooch crucible. Wash the precipitate with 200 ml of cold distilled water. Dry in an oven. Ignite, cool, and weigh.

Calculations

a. To obtain the weight of the barium sulfate, subtract the weight of the crucible before filtering from the weight of the crucible after ignition.

b. To find the weight of sulfur in this amount of barium sulfate, use the following formula:

$$\text{Weight of sulfur} = \text{Weight of } BaSO_4 \times \frac{32.064}{233.402}$$

The 32.064 is the atomic weight of sulfur and 233.402 is the atomic weight of barium sulfate.

c. To find the amount of sulfur in the 24-hour specimen, use this equation:

$$\text{Weight of sulfur in 24-hour specimen} = \text{Weight of sulfur in sample} \times \frac{\text{ml of 24-hour specimen}}{\text{ml used in test}}$$

Sample calculations: A 24-hour specimen measured 1000 ml and 50 ml was used in the test. The weight of the Gooch crucible before filtering was 5.000 g. The weight of the Gooch crucible after ignition was 5.364 g.

a. The weight of the barium sulfate is found by subtracting the weight of the Gooch crucible before filtering from the weight after ignition. The weight of the barium sulfate is therefore 5.364 g minus 5.000 g which is 0.364 g.

b. The weight of sulfur in 0.364 g of barium sulfate is found by using the formula:

$$\text{Weight of sulfur} = \text{Weight of } BaSO_4 \times \frac{32.064}{233.402}$$

$$= 0.364 \times \frac{32.064}{233.402}$$

$$= 0.05 \text{ g}$$

This is the weight of sulfur in the 50 ml of urine that was used in the test.

c. To find the weight in the 1000 ml of the 24-hour specimen:

$$\text{Weight of sulfur in 24-hour specimen} = \text{Weight of sulfur in sample} \times \frac{\text{ml of 24-hour specimen}}{\text{ml used in test}}$$

$$= 0.05 \times \frac{1000}{50}$$

$$= 1.0 \text{ g}$$

TITRATABLE ACIDITY

The titratable acidity of urine is the total acid present in a 24-hour specimen. The normal values are 200 to 500 ml of 0.1 N sodium hydroxide. Increased values are found in acidosis.

Folin method

Reagents needed

Phenolphthalein indicator	(D-208)
Potassium oxalate	
Sodium hydroxide, 0.1 N	(D-233)

Procedure

1. Collect a 24-hour specimen of urine, preserving the collections by storing them in the refrigerator.
2. Measure and record the volume in milliliters.
3. Mix. Pipette 25 ml into a small flask. Add a few drops of a 1% alcoholic solution of phenolphthalein. Add 10 g of finely pulverized potassium oxalate. Shake well.
4. Using 0.1 N sodium hydroxide, titrate to a permanent faint pink color. Record the milliliters of 0.1 N sodium hydroxide used for the titration.

Calculation

Calculate the total acidity as follows:

$$\text{Total acidity} = \text{ml of 0.1 N NaOH used in titration} \times \frac{\text{ml of urine in 24-hour specimen}}{\text{ml of urine used in titration}}$$

Sample calculation: The 24-hour specimen measured 1000 ml; 25 ml of urine was used in the titration; and 5.0 ml of the 0.1 N NaOH was used in the titration.

$$\text{Total acidity} = 5.0 \times \frac{1000}{25} = 200 \text{ ml of 0.1 N NaOH}$$

TOTAL SOLIDS

The average person excretes about 60 g of solids/day, the output varying mainly with diet, exercise, and the functioning ability of the kidneys. If the diet and exercise are controlled, the amount of solids excreted gives a rough index of the functioning ability of the kidneys.

Procedure

1. To estimate the grams of solids in a specimen of urine, multiply the last two figures of the specific gravity by the volume in milliliters and then multiply by the factor 0.0026.
2. Example: The specific gravity is 1.018 and the volume is 1500 ml.

$$\text{Grams of solids} = \frac{\text{Last 2 figures}}{\text{of sp. gr.}} \times \frac{\text{Volume}}{\text{in ml}} \times 0.0026$$

$$= 18 \times 1500 \times 0.0026$$

$$= 70.2$$

NOTE: In most methods for the estimation of total solids, Long's coefficient (2.6) is used in the numerator and 1000 is used in the denominator. The above factor of 0.0026 is simply Long's coefficient divided by 1000.

VITAMIN C

The urine of the average person contains about 10 to 30 mg of vitamin C (ascorbic acid) per day. Decreased amounts are found in scurvy.

ADDIS COUNT

The Addis count is an estimate of the number of casts, red cells, and white cells in a 12-hour specimen of urine. The normal values are given below. In nephritis, the normal values are greatly exceeded.

Casts	0 to 5000
Red cells	0 to 1,000,000
White cells	0 to 1,000,000

Procedure

Instructions to the patient

The patient is given a clean jar containing 0.5 ml of formalin, with the following instructions:

1. On the day of the test, eat a normal breakfast.
2. Throughout the day and night, continue a normal diet, with the following exception: *Do not drink anything* or eat extra fruit.
3. At 8 PM completely empty the bladder and discard the urine.
4. Collect all urine from 8 PM to 8 AM.
5. Void at 8 AM and add this to the collection.
6. Bring specimen to the laboratory.

Laboratory procedure

1. Mix the urine specimen *thoroughly.*
2. Pour into a large graduate and measure the amount. Record the volume in milliliters.
3. Mix urine again. Pipette *exactly* 10 ml of the mixed urine to a centrifuge tube and centrifuge for 6 minutes at 1800 rpm (medium speed).
4. Pipette out *exactly* 9.0 ml and discard, thus leaving 1.0 ml of urine containing the sediment.
5. Mix the sediment thoroughly by tapping the bottom sides of the tube.
6. Use a blood-counting chamber. With a white cell pipette transfer a sample of the sediment to the counting chamber. Do this in the same manner as for a white cell count.
7. Place the counting chamber on the stage of a microscope, bring into

focus with the low power (16 mm objective), and switch to high dry power (4 mm objective).

8. Now count the casts, red cells, and white cells in the *entire* ruled area (9 mm^2). This represents a volume of 0.9 mm^3 or 0.0009 ml.

Calculation

Calculate the number excreted in the 12-hour specimen:

$$\text{Number} = n \times \frac{V_1}{V_2} \times \frac{V_3}{V_4}$$

n = Number counted in 0.0009 ml
V_1 = Volume of 12-hour specimen
V_2 = Volume of urine centrifuged
V_3 = Volume of mixed sediment used
V_4 = Volume of counting chamber

Example: The 12-hour specimen measured 1000 ml, and 10 casts were counted in the entire ruled area (in 0.0009 ml).

$$\text{Number} = n \times \frac{V_1}{V_2} \times \frac{V_3}{V_4}$$

$$= 10 \times \frac{1000}{10} \times \frac{1.0}{0.0009}$$

$$= \frac{1000}{0.0009}$$

$= 1{,}111{,}111$ casts in 12-hour specimen

AMYLASE

The normal values for urinary amylase (or diastase) depend upon the method of determination. Greatly increased values are found in acute pancreatitis. See page 282 for procedure.

CHLORIDE

The average person excretes 10 to 15 g of chloride (expressed as sodium chloride) per day. Decreased values may be found in acute infections, severe diarrhea, nephritis with edema, and dehydration conditions. See page 181 for procedure.

CALCULI

Calculi, the Latin word for stones, are formed by the crystallization of salts. Their formation is often started by an infection in the urinary system and is enhanced by sedentary habits and the presence of urinary salts that are difficult to dissolve, two of the chief offenders being calcium oxalate and calcium phosphate. In mild cases, the stones may be passed in the urine; in severe cases, surgery is essential.

Procedure

To perform the various tests, follow the procedures given in Table 14-6.

Reagents needed

Ammonium molybdate solution
Dissolve 3.5 g of ammonium molybdate in 75 ml of water. Add 25 ml of concentrated nitric acid and mix.

Hydrochloric acid, 10% (v/v)
Add 10 ml of concentrated hydrochloric acid to 90 ml of water. Mix.

Table 14-6. Analysis of urinary calculi

Stone	Physical characteristics	Procedure	Positive results
Urates or uric acid	1. Multiple, smooth, round, without luster 2. Single stones may have bumpy surface, miniature craters 3. Crushed stones are yellow	1. Pulverize stone 2. Place in small evaporating dish 3. Add several drops of concentrated nitric acid 4. Heat gently and evaporate to dryness 5. Add 2–3 drops of concentrated NH_4OH	Deep yellow, orange-red, or crimson, turning purple with NH_4OH, or bluish violet
Phosphates	1. Rather large compact balls 2. Also appear as large friable masses 3. Color: clay or chalky 4. Note: white color, porous, corallike formations are usually calcium oxalate, *not* phosphate	1. Pulverize stone 2. Use microscope slide instead of evaporating dish 3. Add 4–5 drops of ammonium molybdate solution 4. Warm slide over flame	Distinct yellow precipitate forms; this is $(NH_4)_3PO_4 \cdot 12MnO_3$
Carbonate	Same appearance as phosphate	1. Pulverize stone 2. Use a dark background for easier visibility 3. Place a larger quantity of stone on a slide 4. Add 8–10 drops of 10% HCl	Foaming effervescence
Oxalates	1. Smooth luster or 2. Irregular, buff-colored stone with elaborate crystalline structure 3. Occasionally the structure may appear loose and porous	1. Pulverize stone 2. Mix with an equal portion of resorcinol 3. Add 3 drops of concentracted H_2SO_4	Dark blue-green color
Cystine	1. Pale yellow or white granules 2. Appearance similar to calcium oxalate	1. Pulverize stone 2. Add 1 drop of NH_4OH and 1 drop of NaCN 3. Wait 5 minutes 4. Add 2–3 drops of sodium nitroprusside	Upon standing, a beet-red color may change to orange-red

Sodium cyanide, 5%

Weigh 25.0 g of sodium cyanide and transfer to a 500 ml volumetric flask. Add 1 ml of concentrated ammonium hydroxide. Dissolve and dilute to volume with water.

Sodium nitroprusside, 5%

Weigh 5.0 g of sodium nitroprusside (nitroferricyanide) and transfer to a 100 ml volumetric flask. Dissolve and dilute to volume with water. Mix. Store in a brown bottle.

5-HYDROXYINDOLEACETIC ACID (SEROTONIN)

Carcinoid syndrome with its various symptoms can be confused with other disorders. In 1954, it was proposed that carcinoid tumors secreted a large volume of serotonin, which might account for the variety of symptoms such as chronic diarrhea, cutaneous flushes, valvular disease of the heart, and respiratory distress. Since 5-hydroxyindoleacetic acid (5-HIAA) is a urinary metabolite of serotonin (5-hydroxytryptamine), the measurement of the 5-HIAA is now a diagnostic procedure.

Nitrosonaphthol reagent and nitrous acid are added to urine, and if 5-HIAA is present a purple color is produced. Some medications, however, interfere with the reactions. If a rose color or a color not quite purple is observed, a check for medication should be made. A positive test is a definite purple—no other color.

The patient should be off *all* medication for 72 hours before the urine is collected. Excessive eating of bananas may result in a false positive. Following are drugs that affect the test.

False positives	*False negatives*
Aspirin	Chlorpromazine groups
Phenacetin	Phenergan
Thorazine Spansules	Promethazine HCl
	Mephentermine sulfate
	Thorazine
	Compazine
	Pacatal
	Trilafon
	Sparine
	Azulfidine
	Pamine

Qualitative test

Reagents needed

Sulfuric acid, 2 N	**(D-292)**
Nitrosonaphthol reagent, 0.1%	**(D-295)**
Sodium nitrite, 2.5%	**(D-296)**

Ethylene chloride (Fisher, purified)

Hydrochloric acid, concentrated

Nitrous acid reagent

Mix just before using! Pipette 0.2 ml of 2.5% sodium nitrite into a test tube. Add 4.8 ml of 2 N sulfuric acid. Mix.

Procedure

1. **Acidify either a 24-hour specimen or a random urine specimen to pH 3 with concentrated HCl.**
2. **Pipette 1.0 ml of urine into test tube.**
3. **Add 0.5 ml of nitrosonaphthol solution. Mix.**
4. **Add 0.5 ml of nitrous acid reagent. Mix.**
5. **Let stand for 10 minutes.**

6. Add 5.0 ml of ethylene chloride. Shake vigorously!
7. Centrifuge for 5 minutes at 3000 rpm.
8. A purple color in the upper urine layer indicates a positive reaction which should be further investigated.

Quantitative test

Refer to Chapter 15.

PHENYLKETONURIA[43]

Phenylketonuria is an inherited disorder of phenylalanine metabolism, which will cause severe mental retardation. This disorder usually appears between the second and sixth week of the infant's life. The urines of all infants are now usually tested for phenylpyruvic acid weekly up to the age of 6 months, if they have convulsions, severe diaper rashes, or eczema. Normal children are checked either biweekly or monthly during this period of 6 months. Phenylketonuria is diagnosed by the excess phenylpyruvic acid found in urine or serum. Urine will also contain phenylacetic and phenyllactic acid.

By early treatment with diets low in phenylalanine, mental defects may be avoided or corrected. If undetected, abnormal metabolism of the amino acid (phenylalanine) will impair the normal development of the brain.

Qualitative methods

Ferric chloride method

Reagents needed

Sulfuric acid, dilute (1 N)	(D-184)
Ferric chloride, 5%	(D-299)

Procedure

1. Place 5.0 ml of fresh urine into a test tube and acidify with 2 drops of diluted sulfuric acid.
2. Add 5 drops of 5% ferric chloride solution. Mix.
3. Positive result: solution turns to a bluish green.

NOTE: The ferric chloride method has been used also for determination of acetoacetic acid. Positive result: solution turns to red. Other substances may interfere with the test.

Phenistix test

The reagent strip reacts rapidly with the phenylketones in urine, the color of the strip changing from gray to a distinctive blue-gray. The strip is either dipped into a fresh specimen of urine or pressed against the wet diaper. Within 30 seconds the color is developed on the strip and it is then compared with the color scale whose reference points are zero and 15, 40, and 100 mg/100 ml of phenylketone.

Reagents on the strip

Ferric ammonium sulfate
  Supplies the ferric ions, which react with phenylpyruvic acid in a properly acidified medium

Cyclohexylsulfamic acid
  Provides the optimum acidity for the above reaction

Magnesium sulfate
  The magnesium ions are incorporated to prevent the urine phosphates from interfering with the color reaction.

Procedure

1. Dip the test strip in fresh urine and remove immediately, or press against a wet diaper to moisten.
2. At the end of 30 seconds compare the color against the reference points on the color chart.

NOTE: The only other substance giving a typical blue-gray color of phenylpyruvic acid in a 30-second interval is β-imidazolepyruvic acid. This is excreted in the urine due to the deficiency of the enzyme histidine α-deaminase, a rare disorder of children. Substances not affecting Phenistix are as follows:

Phenylacetate	Thymol
Phenyllactate	Toluene
Phenylalanine	Baby powders
Pyruvate	Baby oils
Lactate	Antiseptics
Tyrosine	Homogentisic acid
Tryptophan	Acetoacetic acid

URIC ACID

The urine of the average person contains about 250 to 750 mg of uric acid/day. Increased amounts are found after an attack of gout. (See pp. 86 and 157.)

UREA NITROGEN

The normal urine urea nitrogen is 6 to 17 g/24 hr. Decreased values may be found in acute nephritis, acidosis, and cirrhosis of the liver. Increased values may be seen in febrile conditions. (See pp. 82 and 160.)

CREATININE

The normal range for 24-hour urine creatinine is 20 to 26 mg/kg for males and 14 to 22 mg/kg for females. Creatinine excretion is decreased in disorders associated with muscular atrophy and muscular weakness. It is increased with increased tissue catabolism, as in fever. The concentration is not dependent upon diet or diuresis. (See pp. 85 and 152.)

CREATINE

The 24-hour urine specimen of the average person contains 0 to 200 mg of creatine. Increased values are found in fever, malnutrition, and pregnancy. (See p. 153.)

REFERENCES

Renal function

1. Alving, A. S., and Miller, B. F.: A practical method for the measurement of glomerular filtration (inulin clearance), Arch. Intern. Med. **66**:306, 1940.
2. Alving, A. S., Rubin, J., and Miller, B. F.: A direct colorimetric method for the determination of inulin in blood and urine, J. Biol. Chem. **127**:609, 1939.
3. Alving, A. S., and Van Slyke, D. D.: The significance of concentration and dilution tests in Bright's disease, J. Clin. Invest. **13**:969, 1934.
4. Brodwall, E. K.: Renal extraction of PAH in normal individuals, Scand. J. Clin. Lab. Invest. **16**:1, 1964.
5. Chapman, E. M.: Further experience with the fractional 'phthalein test, New Eng. J. Med **214**:16, 1936.
6. Chapman, E. M., and Halstead, J. A.: The fractional phenolsulphonephthalein test in Bright's disease, Amer. J. Med. Sci. **186**:223, 1933.
7. Dick, A., and Davies, C. E.: Measurement of the glomerular filtration rate and the effective renal plasma flow using sodium thiosulphate and p-aminohippuric acid, J. Clin. Path. **2**: 67, 1949.

8. Enger, E., and Blegen, E. M.: The relationship between endogenous creatinine clearance and serum creatinine in renal failure, Scand. J. Clin. Lab. Invest. **16**:273, 1964.
9. Fishberg, A. M.: Hypertension and nephritis, ed. 5, Philadelphia, 1954, Lea & Febiger.
10. Harrison, H. E.: A modification of the diphenylamine method for determination of inulin, Proc. Soc. Exp. Biol. Med. **49**:111, 1942.
11. Heyrovsky, A.: A new method for the determination of inulin in plasma and urine, Clin. Chim. Acta **1**:470, 1956.
12. Josephson, B., and Ek, J.: The assessment of the tubular function of the kidneys. In Sobotka, H., and Stewart, C. P., editors: Advances in clinical chemistry, vol. 1, New York, 1958, Academic Press, Inc.
13. McIntosh, J. F., Moller, E., and Van Slyke, D. D.: Studies of urea excretion. III. Influence of body size on urea output, J. Clin. Invest. **6**:467, 1929.
14. Moller, E., McIntosh, J. G., and Van Slyke, D. D.: Studies of urea excretion. II. Relationship between urine volume and the rate of urea excretion by normal adults, J. Clin. Invest. **6**:427, 1928.
15. Rolf, D., Switshen, A., and White, H. L.: A modified diphenylamine procedure for fructose or inulin determination, Proc. Soc. Exp. Biol. Med. **72**:351, 1949.
16. Rosenheim, M. L., and Ross, E. J.: Chronic renal failure. In Black, D. A. K., editor: Renal disease, Philadelphia, 1963, F. A. Davis Co.
17. Rowntree, L. G., and Geraghty, J. T.: An experimental and clinical study of phenolsulphonphthalein in relation to renal function in health and disease, Arch. Intern. Med. **9**:284, 1912.
18. Schreiner, G. E.: Determination of inulin by means of resorcinol, Proc. Soc. Exp. Biol. Med. **74**:117, 1950.
19. Smith, H. W.: The kidney, New York, 1951, Oxford University Press.
20. Smith, H. W.: Principles of renal physiology, New York, 1956, Oxford University Press.
21. Tobias, G. J., McLaughlin, R. F., Jr., and Hopper, J., Jr.: Endogenous creatinine clearance, a valuable clinical test of glomerular filtration and prognostic guide in chronic renal disease, New Eng. J. Med. **266**:317, 1962.
22. Van Slyke, D. D.: Renal function tests, New York J. Med. **41**:825, 1941.
23. Wills, M. R.: Biochemical consequences of chronic renal failure: a review, J. Clin. Path. **21**:541, 1968.

Renal insufficiency and dialysis

24. Abel, J. J., Rowntree, L. G., and Turner, B. B.: On the removal of different substances from the circulating blood of living animals by dialysis, J. Pharmacol. Exp. Ther. **5**:275, 1914.
25. Alwall, N.: On the artificial kidney. I. Apparatus for dialysis of blood in vivo, Acta Med. Scand. **128**:317, 1947.
26. Barry, K. G., Shambaugh, G. E., Goler, D., and Matthews, F. E.: A new flexible cannula and seal to provide prolonged access to the peritoneal cavity for dialysis, Trans. Amer. Soc. Artif. Intern. Organs **9**:105, 1963.
27. Doolan, P. D., Murphy, W. P., Jr., Wiggins, R. A., Carter, N. W., Cooper, W. C., Watten, R. H., and Alpen, E. L.: An evaluation of intermittent peritoneal lavage, Amer. J. Med. **26**:831, 1959.
28. Fine, J., Frank, H. A., and Seligman, A. M.: The treatment of acute renal failure by peritoneal irrigation, Ann. Surg. **124**:857, 1946.
29. Ganter, G.: Ueber die Beseitigung giftiger Stoffe aus dem Blute durch Dialyse, München Med. Wschr. **70**:1478, 1923.
30. Gutch, C. F.: Peritoneal dialysis, Trans. Amer. Soc. Artif. Intern. Organs **10**:406, 1964.
31. Gutch, C. F., Stevens, S. C., and Watkins, F. L.: Periodic peritoneal dialysis in chronic renal insufficiency, Ann. Intern. Med. **60**:289, 1964.
32. Grollman, A., Turner, L. B., and McLean, J. A.: Intermittent peritoneal lavage in nephrectomized dogs and its application to the human being, Arch. Intern. Med. **87**:379, 1951.
33. Kolff, W. J.: New ways of treating uremia; the artificial kidney, peritoneal lavage, intestinal lavage, London, 1947, J. & A. Churchill, Ltd.
34. Kolff, W. J., and Berk, H. T. J.: Artificial kidney, dialyzer with great area, Geneesk Geds **5**:21, 1943; Acta Med. Scand. **117**:121, 1944.
35. Kolff, W. J., and Watschinger, B.: Further development of a coil kidney; disposable artificial kidney, J. Lab. Clin. Med. **47**:969, 1956.
36. Maher, J. F., and Schreiner, G. E.: Current status of dialysis of poisons and drugs, Trans. Amer. Soc. Artif. Intern. Organs **15**:461, 1965.
37. Maxwell, M. H., Rockney, R. E., and Kleeman, C. R.: Peritoneal dialysis. I. Technique and applications, J.A.M.A. **94**:917, 1959.
38. McDonald, H. P.: An automatic peritoneal dialysis machine for hospital or home peritoneal dialysis: preliminary report, Trans. Amer. Soc. Artif. Intern. Organs **15**:108, 1969.
39. Scribner, B. H., Caner, J. E. Z., Buri, R., and Quinton, W.: Technique of continuous hemodialysis, Trans. Amer. Soc. Artif. Intern. Organs **6**:88, 1960.
40. Quinton, W., Dillard, D., Cole, J. J., and

Scribner, B. H.: Eight months' experience with Silastic-Teflon bypass cannulas, Trans. Amer. Soc. Artif. Intern. Organs **8**:236, 1962.
41. Skeggs, L. T., and Leonards, T. R.: Studies of an artificial kidney. I. Preliminary results with a new type of continuous dialyzer, Science **108**:212, 1948.
42. Tenchoff, H., Shilipeter, G., Van Paasschen, W. H., and Swanson, E.: A home peritoneal dialysate delivery system, Trans. Amer. Soc. Artif. Intern. Organs **15**:103, 1969.

Routine urine examination

43. Kark, R., Lawrence, J., Pollak, V., Pirani, C., Muehrcke, R., and Silva, H.: A primer of urinalysis, ed. 2, New York, 1963, Paul B Hoeber, Inc., Medical Book Department, Harper & Row, Publishers.
44. Peters, J. P., and Van Slyke, D. D.: Quantitative clinical chemistry, vol. 2, Baltimore, 1956, The Williams & Wilkins Co.

• 15

• Hormones

Hormones are chemical compounds that are secreted into the blood or tissue fluid by the ductless glands of the endocrine system and are transported to other parts of the body where they act upon target tissue or organs. The glands of the endocrine system are located in different parts of the body and include the pituitary (anterior and posterior), thyroid, parathyroid, adrenal (medulla and cortex), islets of Langerhans of the pancreas, gonads (testes, ovaries, placenta), thymus, and pineal body. (See Fig. 15-1.) Their activity is not that of discrete units but of an integrated system where dysfunction of one of the glands may cause changes in the function or response of other glands.

Since hormones are carried in the blood, they are available to all parts of the body. Their function is to maintain homeostasis within the body. Various hypotheses concerning the mode of action of hormones have been reviewed by Karlson and Sekeris.[11] Permeability of cell membranes and enzyme induction are regulatory mechanisms of cell metabolism acted upon by hormones. For example, some of the effects of insulin on carbohydrate distribution and utilization can be explained by its action on permeability of cells. The antidiuretic hormone acts primarily on permeability of cells in regulating passage of materials across the walls. Cortisol appears to be an inducer of enzymes associated with gluconeogenesis from amino acids. It is suggested that this may explain the effect of cortisol on carbohydrate metabolism.

When the hormones have completed their function, they are usually inactivated by conjugation, degradation, or other changes before excretion. Laboratory measurements are performed to determine whether hormones or their metabolites in body fluids are increased or decreased relative to a normal level since certain pathologic conditions are associated with defined changes.

PITUITARY GLAND (HYPOPHYSIS)

The pituitary gland is a small mass, weighing about 0.5 g and located in a cavity of the sphenoid bone, below the base of the brain. It is divided into anterior and posterior lobes which are functionally unrelated. The blood supply for the anterior pituitary lobe is derived from two sources: (1) branches of the superior hypophyseal artery and (2) a system of portal veins. The portal system arises from a network of blood vessels formed in the median eminence from branches of the superior hypophyseal artery. These capillaries are in contact with the nerve fibers from the hypothalamus and act as the means of transport for the neurohumors. From this network blood is collected into hypophyseal portal veins, carried down the anterior surface of the pituitary stalk into the sinusoidal capillaries, and brought into contact with the hormone-secreting cells of the anterior lobe. The posterior lobe receives its blood from branches of the inferior hypophyseal arteries.

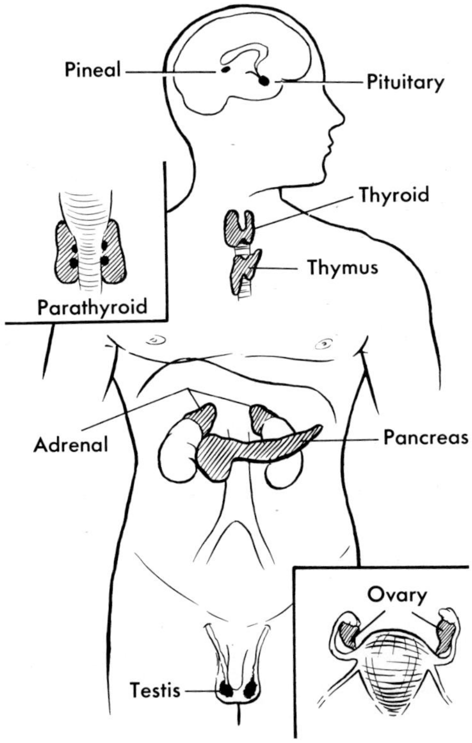

Fig. 15-1. Endocrine glands.

Anterior pituitary (adenohypophysis)

The anterior pituitary gland has been referred to as the master gland because of its influence on some of the other endocrine organs by the secretion of several trophic (to nourish) hormones. The activity of the gland is influenced by the concentration of circulating peripheral hormone through a feedback mechanism. For example, the anterior pituitary elaborates the thyroid-stimulating hormone (TSH) that causes the thyroid gland to produce and secrete thyroxine. An excess of thyroxine in the blood inhibits the production of TSH and establishes a balance by feedback control. In a similar manner the pituitary is stimulated to release adrenocorticotropic hormone (ACTH), which acts on the adrenal cortex to produce and secrete steroids into the bloodstream. When cortisol in the circulating blood reaches a certain level, it inhibits the release of ACTH from the pituitary. If exogeneous ACTH is administered, the adrenal will increase the output of steroids and the pituitary will be inhibited. However, the control mechanism is not effective when there are disturbances in the feedback regulation such as are found with ACTH accumulation for certain types of adrenal overfunction associated with adrenal hyperplasia (Cushing's syndrome).

Seven major hormones secreted by the anterior pituitary are somatotropin, thyrotropin, corticotropin, melanotropin, follicle-stimulating hormone, luteinizing hormone, and prolactin.

Human growth hormone (HGH)

Somatotropin or HGH is a polypeptide that stimulates bone and cartilage growth, metabolism of carbohydrate and fat, and protein synthesis.[6,7] It is influenced by other hormones such as thyroxine and insulin. The plasma level may be measured by radioimmunoassay[10,14] and for an adult in a fasting basal state it is less than 3 mμg/ml.[5] The activity and metabolic condition of the subject markedly affect the blood concentration.

The physiologic effect of decreased or increased secretion depends upon the age of the individual. An increased elaboration may be manifested as gigantism in a child and as acromegaly or enlarged hands, feet, frontal bone, and jaw in the adult. Since other organs of the body are influenced by the pituitary, excess concentrations usually produce changes such as enlargement of the thyroid, altered sexual function, and diabetes.

Dwarfism results from the absence of growth hormone in a child. In the adult decreased secretion in growth hormone produces less obvious changes in metabolism.

Thyroid-stimulating hormone (TSH)

Thyrotropin or TSH, a glycoprotein, stimulates the thyroid gland to elaborate thyroxine. Secretion of TSH by the anterior pituitary is partially influenced by a thyrotropin-releasing factor from the hypothalamus. Under normal conditions an equilibrium exists between the circulating hormones of the pituitary and thyroid for supporting physiologic requirements of the body.

Measurement of the concentration of thyrotropin in plasma may be performed by radioimmunoassay and the normal level has been found to be less than 1 to 15 mμg TSH/ml.[15] A marked increase occurs in primary hypothyroidism between 18 and 180 mμg/ml, but in hypothyroidism secondary to hypopituitarism TSH is undetectable. In hyperthyroidism TSH is not elevated. There is present a long-acting thyroid-stimulating substance (LATS), an immunoglobulin, so called because upon being administered intravenously it is cleared from the blood more slowly than TSH. It is measured by bioassay and usually none is detected in normal serum.

Adrenocorticotropic hormone (ACTH)

Corticotropin or ACTH, a peptide, acts on the adrenal cortex, stimulating it to secrete hormones; in experimental situations it influences carbohydrate and fat metabolism. The plasma concentration of ACTH may be determined by radioimmunoassay but application has been limited because of the small amount present, 0.3 to 0.7 mU/100 ml.[8] There is a diurnal variation, the level being highest about 6 AM and lowest about 10 PM. Increased blood concentrations are produced by stress and low cortisol output. ACTH is increased in Cushing's syndrome due to pituitary disease and in nonendocrine tumors releasing ACTH.

Melanocyte-stimulating hormone (MSH)

MSH or melanotropin was so named because it darkens the skin of certain animals. Increased amounts have been found in the plasma and urine of patients with Addison's disease and with ACTH-secreting tumors.[5] Its presence in pituitary and tumor tissue has been demonstrated.[1] MSH can cause darkening of the skin of humans when corticoids are absent such as in Addison's disease.

Gonadotropic hormones

The gonadotropic hormones secreted by the anterior pituitary are glycoproteins and are not sex-specific. There are two—follicle-stimulating hormone (FSH) and luteinizing (LH) or interstitial cell–stimulating hormone (ICSH). The FSH and LH act primarily on the gonads and are the most sensitive trophic hormones in response to damage of the pituitary gland. For this reason, the individual determinations provide a means of evaluating anterior pituitary function or differentiating primary from secondary hypogonadism.

Follicle-stimulating hormone

FSH causes development of the normal growth of the follicles in the ovary and production of estrogens. It stimulates spermatogenesis in the testes. The majority of the assays at this time are performed by the use of bioassay procedures although radioimmunoassay methods[3,9,13] are being developed for research investigations. The gonadotropin of human urine (HPG) used in bioassays is usually a mixture of FSH and LH. The assay consists of extracting the hormone from urine, concentrating it, injecting the extract into immature female rats or mice, and weighing the ovaries or uteri.[2,4] To decrease the synergistic effect of LH, human chorionic gonadotropin may be added to the test extract.

Specimen collection

A 24-hour urine specimen is collected in a 1-gallon bottle with toluene as a preservative or is maintained under refrigeration without a preservative. If a low value is expected, a 48-hour urine specimen should be requested. The assay should be carried out as soon as possible after the collection.

Reagents needed

Kaolin mixture, Cook's, for absorbing hormones from urine
Acetic acid, glacial, reagent grade
Ammonium hydroxide, 2 M
Dilute 135 ml of concentrated (14.8 M) ammonium hydroxide, reagent grade, to 1 L.
Acetone, reagent grade
Sodium chloride solution, 0.9%
Dissolve 9 g of NaCl, reagent grade, in distilled water and dilute to 1 L.

Procedure

1. **Filter the entire urine collection and, if less than 1 L, dilute to that volume with distilled water.**
2. **Adjust the pH to 4.5 with glacial acetic acid (pH paper), add 20 g of dry kaolin mixture, stir for 15 seconds, and allow to stand for 30 minutes.**
3. **Filter the mixture on a Buchner funnel, using suction, and discard the filtrate.**
4. **Wash the kaolin cake within the filter with 2 L of distilled water acidified to pH 4.5 with glacial acetic acid. Discard the washing.**
5. **Transfer the kaolin cake to a 600 ml beaker, add 100 ml of 2 M ammonium hydroxide, stir for 2 minutes, and allow to stand for 10 minutes to elute the hormone.**
6. **Filter on a Buchner funnel and wash the cake with 50 ml of distilled water. Combine the ammonium hydroxide eluate with the water wash. Discard the kaolin cake.**
7. **Adjust the pH of the eluate to 5.5 with glacial acetic acid (pH paper) and add two volumes of acetone (300 ml).**
8. **After stirring, place the beaker in the refrigerator for 30 minutes.**
9. **Centrifuge in 50 ml tubes, for 15 minutes at 2000 rpm and discard supernatant.**
10. **Allow the acetone to evaporate from the precipitate and, using a mortar and pestle, suspend the precipitate in 16 ml of 0.9% sodium chloride solution.**

11. Using two weanling rats, 21 to 23 days old, inject 2 ml subcutaneously once daily for 4 days.
12. Perform necropsy on the fifth day and remove the ovaries.
13. Weigh the ovaries (Roller-Smith micro torsion balance) and average the weight of the ovaries from the two animals.

Calculations

A dose-response curve, variation of ovarian weight as a function of the logarithm of the dose expressed in rat units (RU), was set up by Albert[4] from experimental data. From the curve the units of HPG can be read directly.

Ovarian weight (mg)	*Equivalent (RU of HPG)*
22	5
59	10
81	15
95	20
131	40
167	80
202	160

An equation of the curve was prepared[2] and the same values are obtained by substituting the mean weight of two pairs of ovaries as follows:

$$X = \text{antilog}\,\frac{W + 60.7}{119.1} = \text{Rat units}$$

where X is the units of gonadotropin and W is the mean weight of the two pairs of ovaries. The units are corrected to the 24-hour basis. If a 24-hour urine specimen is used, each rat receives the equivalent of a 12-hour specimen and the value must be multiplied by 2.

Normal values

Children, before puberty	Below detectable limits
Normal men	3 to 35 RU
Normal young women	0 to 35 RU
Postmenopausal women	14 to 205 RU

Comments

A pair of ovaries from an uninjected rat will weigh less than 15 mg. If the average weight of the ovaries from animals receiving the urinary extract is less than 15 mg, the titer is zero.

A preparation from postmenopausal urine is available as a standard for general gonadotropin assays. It is obtainable from the Endocrinology Study Section, National Institutes of Health, Bethesda, Md.

The administration of estrogens, androgens, progestogens, and corticosteroids to a patient will affect the urinary output of gonadotropins. Such therapy should be withheld about a month before starting the collection of urine for assay.

Clinical findings

The urinary output of gonadotropins will be elevated in men and women with hypogonadism. The concentration will be decreased with increased levels of estrogens and hypofunction of the hypothalamus and pituitary.

Luteinizing hormone

The luteinizing hormone stimulates maturation of the follicle tissue, ovulation, and development of the corpus luteum with the production of progesterone.

The feedback mechanism or balance is between LH and progesterone although some estrogen appears necessary to initiate LH release. In the male, LH stimulates the interstitial cells of the testes to produce testosterone.

Bioassay of the hormone in urine may be carried out by injecting extracts to be tested, intravenously, into immature pseudopregnant rats and determining the depletion of ovarian ascorbic acid. Radioimmunoassay methods[12] that are specific and sensitive are available for measuring LH in serum or urine; serum or plasma is preferred because of sensitivity and because of the presence in urine of inhibitory material. Normal LH levels in plasma vary between 0.6 and 3.2 mμg/ml (1 mμg is equivalent to 8 mU of the Second International Reference Preparation of Human Menopausal Gonadotropin). Plasma levels for children before puberty are either low or below detectable limits. In men the level remains low until the age of 70, when an increase is apparent. The normal low level in women does not increase until menopause. Plasma from castrated men or women or postmenopausal women contains 4.5 to 10.5 mμg/ml. Elevated LH levels are found in amenorrhea resulting from ovarian failure.

Prolactin

Prolactin stimulates secretion of milk in the mammary gland. However, it is dependent upon the presence of other hormones to achieve this function. Information concerning prolactin secretion in man is limited because the pigeon crop sac method of assay lacks sensitivity and specificity.

Posterior pituitary (neurohypophysis)

The posterior pituitary is part of a neurosecretory system. The hormones vasopressin and oxytocin are produced in the hypothalamic nuclei and are transported to the posterior pituitary where they are released into the blood. Assay requires complicated physiologic methods that are not suitable for general diagnostic laboratories.[16]

Vasopressin

Vasopressin or antidiuretic hormone (ADH) has as its major physiologic function antidiuretic activity although it does cause elevation of blood pressure and does excite the smooth musculature of the intestine. It acts in the kidney tubules, promoting reabsorption of water. When production is impaired, diabetes insipidus results in which large quantities of diluted urine are excreted. The condition may be treated by injections of the hormone.

Oxytocin

Oxytocin stimulates contraction of the smooth muscle of the uterus and the lactating mammary gland.

THYROID GLAND

The thyroid gland consists of two lobes and a connecting isthmus, is located at the sides of the trachea below the larynx, and weighs about 20 g. Normal thyroid function depends largely on the secretion of the thyroid-stimulating hormone by the anterior pituitary. The following iodinated hormones are formed in the thyroid gland: 3-monoiodotyrosine, 3,5-diiodotyrosine, 3,3′-diiodothyronine, 3,3′,-5′-triiodothyronine, 3,5,3′-triiodothyronine (T_3), and 3,5,3′,5′-tetraiodothyronine

(T_4) or thyroxine. The function of the gland is the control of the normal growth and development of the body and its metabolic rate.

Ingested iodine is absorbed from the gastrointestinal tract as iodide and the inorganic iodide is removed from the circulation by the thyroid gland at a clearance rate of about 35 to 40 ml of plasma per minute, or approximately 50% of the iodide from the blood moving through the gland. The thyroid cells concentrate and oxidize the iodide to free iodine that combines with the tyrosine of thyroglobulin, a complex glycoprotein. Tyrosyl residues are iodinated first to the mono form and then to diiodotyrosine, and two diiodotyrosine residues combine to form thyroxine. Storage of the hormone occurs in the colloid of the gland located in the follicles of the lobules. The colloid consists mainly of thyroglobulin, other proteins, and water.

Hormones secreted by the thyroid gland have a characteristic iodine content. They exist in the gland as units of the peptide chain of thyroglobulin but are circulated in the blood, noncohesively bound to serum proteins. Degradation to the amino acids and secretion by the gland are controlled by the thyrotropic hormone (TSH) from the anterior pituitary. The iodinated thyroglobulin is hydrolyzed by proteolytic enzymes, releasing iodothyronines into the cells. The two active iodothyronines are the amino acids thyroxine (T_4) and 3,5,3′-triiodothyronine (T_3).

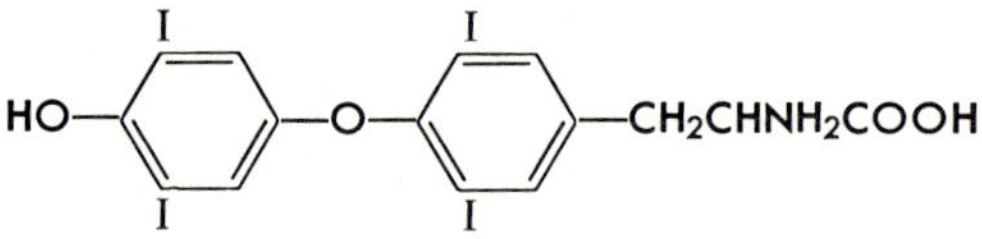

Thyroxine (T_4)

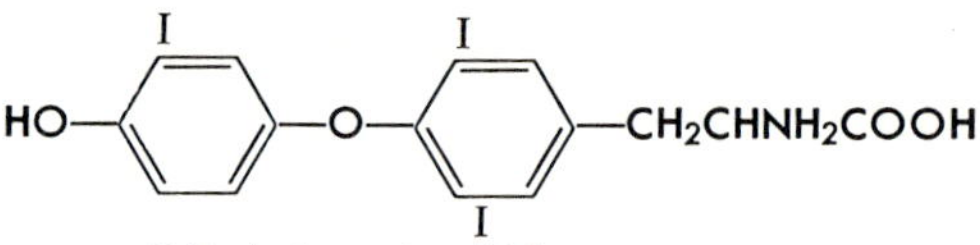

Triiodothyronine (T_3)

Thyroxine accounts for more than 90% of the hormonal iodine circulating in the blood and is transported in combination with a glycoprotein, thyroxine-binding globulin (TBG), that has an electrophoretic mobility between the α_1 and α_2 globulins. A small amount is bound with prealbumin (TBPA) and albumin, and about 0.05% remains free in circulation. Normally about one third of the available binding sites on TBG are occupied by T_3 and T_4; the remainder are unoccupied. The triiodothyronine is less firmly bound to plasma protein and is much more active physiologically than thyroxine. However, the concentration of T_3 in blood, less than 5% of the hormonal iodine, is so low that it does not affect significantly the iodine level. Biologically its effect is four to five times that of thyroxine. The free hormone causes the characteristic action in the target cells. The specific mechanism and location of action for the free hormones are not known. If the four iodine molecules of thyroxine are replaced by hydrogen, all of the stimulating action on cell metabolism is lost.

Of the many laboratory methods that aid in assessing the activity of the thyroid gland no one test may be designated as giving the complete information needed in clinical evaluations.[20] Since the level of circulating hormone is related directly to the activity of the gland, procedures for determining the hormone levels are used in studying the functioning and physiology of the organ. The characteristics of the hormone that provide a basis for methodology are the concentration of iodine in the molecule, the equilibrium that exists in vivo and in vitro between the free and the protein-bound hormone, and the specific protein binding of the hormone.

Protein-bound iodine and butanol-extractable iodine

The protein-bound iodine (PBI) or serum-precipitable iodine is used extensively as a measure of serum thyroxine. Red blood cells contain about 25% of the amount of protein-bound iodine found in serum. When proteins of serum are precipitated with the usual precipitating reagents, the hormone is quantitatively coprecipitated since approximately 99% of the iodothyronine is bound to serum proteins. Precipitation is performed to eliminate contaminants, but nonhormonal iodine such as contrast materials used in x-ray examinations and drugs containing iodine do interfere. The serum may also be treated with an ion exchange resin to remove nonbound iodide. The precipitate or treated serum is digested and oxidized with acid (wet-ash method) or is dried, oxidized, and ashed in a muffle (dry-ash method) to convert the organic iodide to iodate for the iodine analysis. In an effort to reduce the error introduced into the PBI procedure by nonhormonal iodine, a butanol-extractable iodine (BEI) method was developed. The test is based upon the solubility of T_3 and T_4 in butanol and removal from the organic extract of inorganic iodide and the monoiodotyrosine and diiodotyrosine by use of alkali. Since the method is time-consuming and organic iodine compounds interfere, it has little advantage over the PBI procedure.

Thyroxine by column[19,25]

Interference by exogenous organic iodine compounds may be reduced but not eliminated by employing a thyroxine-by-column method. The procedure utilizes an anion exchange resin that removes T_4, T_3, other iodoamino acids when present, and iodides from an acidified (pH 3.8) solution of the specimen. By acidified wash solutions the column is freed of residual protein, some organic iodine compounds used in diagnosis and therapy, and mono and di iodotyrosines if present. Elution of T_3 and T_4 is accomplished by using a 50% acetic acid solution (pH 1.4). Inorganic iodine remains on the column. Three fractions are collected and the iodine content of each fraction is determined. Several procedures, with[25] and without incineration,[24,26] have been proposed for determining the iodine content of the eluate. The organic iodine compounds that have been found to interfere are Floraquin, Itrumil, Orabilex, Priodax, Telepaque, Teridax, and Viscidol.

Thyroxine by competitive protein binding (Murphy-Pattee)

Murphy and Pattee[22] developed a more specific method for thyroxine assay using the specific binding properties of thyroxine-binding globulin. It is based upon the equilibrium between free and bound thyroxine in plasma. Since the measurement is that of total thyroxine, neither inorganic nor organic iodine contaminants interfere.

A measured volume of pooled serum, diluted with barbital buffer and saturated with ^{125}I-thyroxine, is added to a dried ethanol extract of the patient's serum. In plasma diluted with a barbital buffer solution, T_4 is preferentially bound by TBG. After incubation and equilibration dry resin is added to absorb the unbound thyroxine. Displacement of the radioactive T_4 from the protein-binding sites on TBG provides a measure of the amount of nonradioactive thyroxine in the unknown. The radioactive T_4 not bound to the TBG is removed with a resin, and the radioactivity associated with the supernatant or resin is measured. Increased thyroxine values are obtained if patients are pregnant or are receiving estrogen or antiovulatory drugs. Low values are found in hypoproteinemia and when patients are receiving Dilantin and triiodothyronine. The Dilantin competes with thyroxine for TBG-binding sites at the thyroxine levels.

Radioactive iodide (RAI) uptake

The radioactive iodide uptake method is a measure of the rate at which the thyroid gland removes iodine from the circulation, relative to the efficiency of the kidney. It is not necessarily a measure of hormone release and it depends upon blood flow, iodine availability, and normal biosynthesis. A tracer dose of ^{131}I is administered orally and the radioactivity measured over the thyroid after 3, 6, and 24 hours. The result is related to the dose administered. Uptake is usually increased in hyperthyroidism for the shorter time periods but may be normal at 24 hours. In hypothyroidism the 24-hour uptake is decreased but the 3- and the 6-hour uptake may be normal. The plasma protein-bound radioiodine 24 hours after administration of the dose may also be determined as an indication of hyperthyroidism and hypothyroidism. The method is limited, as antithyroid and iodine-containing drugs and organic iodine x-ray contrast media interfere.

T_3 resin uptake

A continuous equilibrium exists in vivo and in vitro between free and protein-bound thyroid hormones. Free T_4 will easily replace bound T_3 but the reverse reaction is insignificant. The T_3 uptake test, utilizing radioiodine-labeled triiodothyronine in vitro, is based upon the competition for TBG-binding sites in the patient's serum between the labeled T_3 added to the serum and the patient's endogenous thyroid hormone. The labeled compound is added to a plasma sample and the unbound T_3 is adsorbed on media such as resin, charcoal, Sephadex, or red blood cells. The solid material is removed by centrifugation and washed. Distribution of radioactivity between the plasma and resin is determined. T_3 is bound to the TBG molecule at places not occupied by T_4. When there is a high level of endogenous thyroid hormone, the binding sites in the plasma are relatively full and the radioiodine tends to be attached to the resin. When the endogenous hormone level is decreased, the radioiodine tends to be bound in the plasma. The T_3 uptake test provides an indirect measurement of the unbound thyroxine-binding globulin and an index of thyroid function. The procedure is rapid and not affected by inorganic iodine. In conditions where TBG is being produced in increased amounts, such as pregnancy and estrogen therapy, values will be in the hypothyroid range. The T_3 uptake is increased when TBG production is decreased in conditions such as nephrosis and liver disease, for patients receiving ACTH or corticosteroid therapy where fewer unbound sites are available, and by the action of drugs that release thyroxine from its binding sites such as salicylates, diphenylhydantoin and phenylbutazone. More radioactive T_3 is bound by the serum proteins of hypothyroid patients and therefore, a low T_3 uptake by the resin results.

Free thyroxine

The physiologic action of thyroxine depends upon the concentration of the unbound hormone, and a measurement of this fraction would be useful for assessment of the activity of the thyroid gland. Levels of free and bound thyroxine in the circulation depend upon the quantity of thyroxine and the availability of thyroxine-binding globulin. Dialyzing a serum sample after addition of radioactive T_4 and analyzing the dialysate for thyroxine will give a measure of the free thyroxine in the serum.[28] The test is important in assessing thyroid status in hypoproteinemia, in pregnancy, and in patients who are using antiovulatory birth control compounds or receiving drugs such as Dilantin where the protein-bound iodine and total thyroxine levels may provide misleading information.[27] In pregnancy and when estrogens are administered, free thyroxine remains normal in the euthyroid state. Clark and Horn[18] proposed a "free thyroxine index" obtained from the product of the total serum thyroxine and the T_3 uptake. In euthyroid states with pregnancy or with estrogen therapy the index is within the normal range.

Protein-bound iodine (PBI)

Ideally, the PBI analysis should be done in an isolated room away from laboratory traffic and contaminants. Precautions should be taken against iodine and mercury. Analyses with reagents containing these compounds *must* be kept away from the PBI area.

The sources of inorganic iodine available to the patient are numerous. Although the precipitate is washed, if the inorganic iodine level is very high not all may be washed clear, since some may be attached or adsorbed to the protein. Therefore, falsely elevated results will appear with no apparent clinical reason. For a list of some of the contaminants, refer to Table 15-1. However, bromides and meprobamate do not affect PBI results.

The automated method is discussed thoroughly in *Practical automation.**

Two common manual methods for the determination of protein-bound iodine are given: the Barker-Soley-Humphrey modification of the "dry-ash" method and a "wet-digestion" method.

A modification may be used for these two procedures. The inorganic iodine may be extracted from the serum with an ion exchange resin, Amberlite IRA-410, c.p. This may be done by using the following procedure. It is recommended that serum standards be included when this modification is employed.

1. **Mix about 0.5 g of resin with about 4 ml of serum and allow the tube to stand for 5 minutes; then centrifuge and decant the serum.**
2. **Proceed with the method of choice, eliminating only the washing of the precipitated protein.**

Normal values
4.0 to 8.0 μg/100 ml

Some disorders that affect the protein-bound iodine levels in serum or plasma are listed in Table 15-2.

Barker-Soley-Humphrey modification[17]

Iodine exists in the blood in two forms: organic (protein-bound) and inorganic. The procedures routinely used at the present time require preliminary precipitation of the protein and washing of the proteins to remove the inorganic iodine. The organic (hormonal) iodine is precipitated with the serum or plasma

*White, Wilma L., Erickson, Marilyn M., and Stevens, Sue C.: Practical automation, St. Louis, 1968, The C. V. Mosby Co.

Table 15-1. PBI contaminants[7a,19]

Contaminants causing elevated levels	Duration
Radiographic contrast media	
Short duration	3 days to 2 weeks
Diodrast	2 weeks
Hippuran	
Hypaque	1 to 4 days
Neo-Iopax	
Pyelombrine	2 weeks
Salpix (for hysterosalpingography)	
Skiodan	
Urokon	4 weeks
Medium duration	May be evident as long as 6 months
Biligrafin	3 weeks
Cholografin	3 to 4 months
Dionosil	1 to 5 months
Orabilex	4 to 9 weeks
Priodax	3 to 6 months
Telepaque	6 to 12 weeks
Long duration	
Lipiodol	1 to 2 years
Pantopaque	3 to 12 months
Teridax	Many years (estimated 33) and will cross placental barrier, causing high values in offspring for years
Tincture of iodine on skin at site of puncture where blood is drawn	Specimen contaminated
Antidandruff preparations, cosmetics, suntan lotions, and toothpastes containing iodine	
Amebicides, analgesics, antiseptics, cough preparations, cold medicines, dietary foods, food supplements, lozenges and vitamin preparations containing iodine	Except as indicated, 3 to 30 days after source is removed; dosages of about 0.2 to 0.6 g of KI daily will increase PBI by 1.3 to 2.2 μg/100 ml
Betadine	
Bromo quinine	
Calcidin	
Calcidrine	
Chiniofon	
Dainite KI	
Dietene	
Diodoquin	
Enfamil	
Erythrosine (tetraiodofluorescein)	3 months
Felsol	
Floraquin	30 days
Hyquin	6 to 9 months
Isuprel compound elixir	
Metrecal	30 days
Norisodrine	
Ornade	
Pediacof	
Quadrinal	
Sobee	

Table 15-1. PBI contaminants—cont'd

Contaminants causing elevated levels	Duration
Tuss-Ornade	
Vioform	6 to 9 months
Estrogens	2 to 4 weeks
Premarin	
Tace	
Oral contraceptives	
C-Quens	
Enovid	
Norlestrin	
Oracon	
Ortho-Novum	
Ovulen	
Other compounds	
BSP (Bromsulphalein)	4 days
Darbid (isopropamide iodide)	
Delvex (dithiazanine iodide)	
5–Iodothiouracil	
Itrumil	
Lugol's solution	10 to 30 days
Pathilon (tridihexethyl chloride)	
Perphenazine	
Pyrazinamide	
Synthroid (*l*–thyroxine)	
Trilafon (perphenazine)	

Contaminants causing decreased levels	Duration
ACTH	2 weeks
Amphenone	
Androgens	
Anions (SCN^- and ClO_4^-)	
Corticosteroids	2 weeks
Dilantin (diphenylhydantoin)	7 to 10 days
Dinitrophenol	
Exchange transfusion (newborn)	
Mercurial diuretics (*not* if alkaline ash technique used for PBI analysis)	1 to 2 weeks
Para-aminosalicylic acid (PAS)	
Phenylbutazone	
Radioactive iodine (therapeutic dose)	
Resorcinol	
Salicylates (6 g/day) affect results as follows	
Euthyroids—PBI decrease of 1.4 μg/100 ml	
Hyperthyroids—PBI decrease of 4.2 μg/100 ml	
Hypothyroids—PBI decrease of 0.5 μg/100 ml	
Sulfonylurea compounds	
Carbutamide	
Chlorpropamide	
Tolbutamide	
Thiouracils (noniodinated)	
Triiodothyronine	2 to 4 weeks
Triparanol	

Continued.

Table 15-1. PBI contaminants—cont'd

Substances that do not affect PBI levels
Antibiotics (with exception of PAS)
Anticoagulants
Dicumarol
Heparin
Antihistamines
Parabromdylamine (Dimetane)
Antihypertensive agents (with possible exception of reserpine)
Chemotherapeutic agents
Floxuridine
5–Fluorouracil
Methotrexate
Nitrogen mustard
Vincristine
Dexedrine
Digitalis
Diuretic agents
Acetazolamide (Diamox)
Alkaline ash method of PBI determination not affected by mercurial diuretics
Chlorothiazides (Diuril, e.g.)
Gonadotropins
Tranquilizers (with exception of perphenazine and possibly reserpine)
Barbiturates
Chlorpromazine
Hydroxyzine
Imipramine
Isocarboxazid
Meprobamate
Promazine
Promethazine
Trifluoperazine

Table 15-2. Disorders affecting the PBI

When increased	When decreased
Hyperthyroidism	Hypothyroidism
Acute thyroiditis	Cretinism
Subacute thyroiditis	Myxedema
Chronic thyroiditis	Panhypopituitarism (Simmonds' disease)
Hepatitis (normal 4 weeks)	Malnutrition
Pregnancy (third week)	Hypoproteinemia
	Nephrosis
	Cirrhosis

proteins. The organic material is destroyed by incineration in the presence of sodium carbonate, converting organic iodine to inorganic iodine. The inorganic iodine is quantitatively measured by means of the catalytic action of iodine on the reduction of ceric ammonium sulfate by arsenious acid, the yellow ceric ions being decolorized.

Equipment needed

Pyrex test tubes (15 × 125 mm)
Marktex Tech-Pen with heat-resistant ink
Drying oven and baskets
Muffle furnace
Spectrophotometer
Constant-temperature water bath
Centrifuge

Reagents needed

Zinc sulfate, 10% (D-300)
Zinc sulfate working solution (D-301)
Sodium hydroxide, 0.5 N (D-34)
Sodium carbonate, 4.0 N (D-302)
Hydrochloric acid, 2.0 N (D-303)
Sulfuric acid, 7.0 N (D-304)
Sulfuric acid, dilute (D-305)
Iodine standards (D-306)

a. Concentrated stock standard, 100 μg/ml
b. Dilute stock standard, 0.2 μg/ml
c. Dilute working standard, 0.04 μg/ml
Pipette 10.0 ml of dilute stock standard into a 50 ml volumetric flask. Dilute to mark with double-distilled water. Stopper and mix thoroughly.

Ceric ammonium sulfate, 0.02 N (D-307)
Sodium arsenite, 0.1 N (D-308)
Antifoam solution (D-309)

Procedure

Number rimless, acid-washed Pyrex test tubes (15 × 125 mm) in duplicate, using a Marktex Tech-Pen. This type of ink withstands the extreme furnace temperature. For the first rack of twenty-four test tubes, include one set (two tubes) of reagent blanks and two sets (four tubes) of commercial control. For successive racks, include one set of reagent blanks and one set of either a commercial control or a pooled serum that has been assayed for quality control.

Precipitation

1. Pipette 1.0 ml of double-distilled water into the blanks and 1.0 ml of controls and plasma into the proper tubes. Use Ostwald-Folin pipettes.
2. Add 8.0 ml of the diluted zinc sulfate to all tubes. Use a serologic pipette or, for large volumes of work, a Brewer automatic pipette is helpful.
3. The serum and zinc sulfate (diluted) are mixed either with a Vortex mixer or with glass stirring rods of 3 mm diameter.
4. Leave the rods in the tubes and add 1.0 ml of 0.5 N NaOH to each tube.
5. Mix the contents until they are homogeneous.
6. Remove each rod with a rotary movement, scraping off any adhering protein.
7. Place the rods on either a numbered, corrugated edging strip or a strip of paper toweling that can be discarded after use. The idea is to keep the stirring rods off the table and separated.

Washing the precipitate

1. Centrifuge the tubes for 10 minutes at 3000 rpm.
2. With the precipitate well packed by centrifugation, invert the tubes, pouring off the supernatant. Shake the tubes to dispose of the small amount of water remaining.
3. Pipette 10.0 ml of double-distilled water into the tubes. A 5 ml Cornwall semiautomatic pipette is very helpful.
4. Place the stirring rods into the correct tubes and homogenize the protein and water thoroughly. Leave no "lumps."
5. Centrifuge the tubes for 10 minutes at 3000 rpm.
6. The washing of the precipitate is done three times. Record when each washing is completed, so there will be no mistake in the number of times the precipitates have been washed.

Drying the precipitate

1. To each tube, add 1.0 ml of 4 N sodium carbonate plus 1 drop of antifoam solution.
2. Mix well with the stirring rods, keeping the precipitate in the lower half of the tube.
3. Rotate each rod as it is removed, to scrape off any adhering protein.
4. Place the tubes in a slanted position in a small basket.
5. Place in a 100° C drying oven until dry or leave overnight. An electric drying block may be used if one is available. (NOTE: If the antifoam solution is used, the temperature may be elevated to 110° C for an approximate drying time of 2 hours. If the antifoam is not used, the temperature must be kept at 100° C or the precipitate will bubble out of the tube.)

Incineration

The tubes are prepared in the following way to prevent cross-contamination.[29]

1. Tear off or cut a piece of heavy aluminum foil about 10 inches long and 4½ inches wide. Allow one sheet for each pair of tubes.
2. Place the pair of tubes at least 2½ inches from the top of the foil. The sides are then folded together and the foil is wrapped up from the bottom. The rising vapors are held within the aluminum foil cylinder, preventing contamination of other specimens.
3. Stand the foil-wrapped tubes upright in a steel bucket and incinerate for 3 hours at 600° ± 10° C.
4. Using a steel shovel, remove the tubes from the furnace and cool to room temperature.
5. Unwrap the tubes and discard the foil.

Extracting the iodide from the ash

1. Add 2.0 ml of 2 N HCl carefully—to avoid excessive effervescence—and rotate each tube slowly while adding the acid to wash down the sides of the tubes. Allow to stand for 5 minutes until the effervescence ceases.
2. Add 5.0 ml of dilute H_2SO_4 to each tube.
3. Mix contents by flipping the tube.
4. Centrifuge for 10 minutes at 1700 rpm to pack the insoluble material.

NOTE: For adding the HCl and H_2SO_4, either a 2 ml and 5 ml Machlett Auto-Pipet or all-glass automatic zero burette may be used.

Determination of the iodide

1. For each unknown, number a spectrophotomer tube.
2. Treat the first ten spectrophotometer tubes as follows:
   a. Tubes 1 and 2. Pipette 3.0 ml aliquots of the reagent blanks.
   b. Tubes 3 to 6. Keep empty for standards.
   c. Tubes 7 to 10. Pipette 3.0 ml aliquots of the commercial control.
3. Continue to pipette a 3.0 ml aliquot of the unknown into each tube.

4. To the tubes containing the aliquots, add 5.0 ml of double-distilled water. A 5 ml Cornwall pipette is accurate and time-saving for this step.
5. Treat the standards as follows:
   a. Prepare a pH-adjusting solution:
      (1) Pipette 4.0 ml of 2 N HCl into an Erlenmeyer flask.
      (2) Add 4.0 ml of 7 N H_2SO_4 to the flask.
      (3) Add 4.0 ml of double-distilled (d-distilled) water and swirl flask to mix.
      (4) Add 2.0 ml of 4 N Na_2CO_3, swirl to mix, and set aside for effervescence to cease.

   This solution will adjust the pH of the standards to equal that of the unknowns.

   b. To the "standard" spectrophotometer tubes add the following:

Tube	Iodine concentration of standard (μg/ml)	Volume of working standard (ml)	Volume of water (ml)	Volume of pH solution (ml)
3	0.02	0.5	4.5	3.0
4	0.04	1.0	4.0	3.0
5	0.06	1.5	3.5	3.0
6	0.08	2.0	3.0	3.0

6. Add 0.5 ml of 0.1 N sodium arsenite to each tube. Mix well.
7. Place tubes in a 40° C water bath for 10 minutes.
8. Tape an Erlenmeyer flask containing ceric ammonium sulfate solution to a corner of the water bath, with the reagent below the water level. Allow 10 minutes for the reagent to come to bath temperature.
9. Set the timer for 30 minutes. Add 1.0 ml of ceric ammonium sulfate solution to each tube at *exactly* 30-second intervals and mix. Use a fast-flowing 1 ml volumetric pipette.
10. *Exactly* 15 minutes later, read the tubes at exactly 30-second intervals against a water blank at 420 mμ wavelength.

Calculations

Prepare a standard curve by plotting the absorbance readings of the standards against their iodide concentrations on linear graph paper. The 3 ml aliquot of iodide extract used in the reading portion of the test represents 3/7 ml of serum or plasma. By multiplying the iodide concentration of each of the standards, 0.02, 0.04, and 0.06, and 0.08 μg/ml, by 7/3, one obtains values that represent 4.67, 9.33, 14.00 and 18.7 μg/100 ml, respectively. By plotting the absorbance readings against these values, the unknowns may be determined directly in μg/100 ml of serum or plasma. If the reagent blank reading falls above the 0.0 μg/100 ml point, determine the amount of iodide in the blank and subtract it from the unknown values to obtain the corrected values in μg/100 ml of serum or plasma.

Wet-digestion method (chloric acid)[23,30,31]

There are three wet-digestion methods available for protein-bound iodine determinations. These are distinguished by the oxidizing agents utilized: permanganate acid, chromic acid, and chloric acid. One of the latest methods for PBI determinations is the automated method of Technicon's AutoAnalyzer, which utilizes perchloric acid as the oxidizing agent.

The method described is based on the chloric acid digestion of the precipitated protein of serum or heparinized plasma in which a small volume of chromic acid is used as a catalyst. During this period of digestion the organic iodine is oxidized to iodate. Upon the addition of arsenious acid the iodate is then reduced to iodide. Also, any excess chromate giving color is reduced to a clear solution. With the timed addition of ceric ammonium sulfate, the arsenious acid and the ceric ammonium sulfate react, reducing the ceric ions to cerous ions. This results in a loss of color as the reduction takes place. The iodide acts as a catalyst. The higher the concentration of iodide, the faster is the reaction or loss of color. This is easily observed with grossly elevated specimens.

Extreme care must be taken with reagents and prevention of contaminating materials entering the PBI area. Commercial controls or pooled serum of various concentrations must be analyzed with each group of tests. *Never* use yesterday's curve. *Always* run a set of standards and controls with each day's run.

Precaution. The digestion process must take place in a strong-drawing fume hood. If possible, the hood should have a system for washing the unit down after the digestion is completed.

Apparatus needed

Round-bottom centrifuge tubes, 40 ml
Electrolytic beakers, 400 ml
Cuvettes (19 × 105 mm)

*Reagents needed**

Perchloric acid, 10%	(D-310)
Chloric acid (store in refrigerator)	(D-311)
Chromic acid	(D-312)
Arsenious acid, 0.2 N	(D-313)
Ceric ammonium sulfate, 0.04 N	(D-314)
Sulfuric acid, 10 N with 100 mg NaCl/ml	(D-315)
Iodate solution, stock standard, 100 μg/ml	(D-316)

Working iodate standard, 0.10 μg/ml

Using a volumetric pipette, pipette 1.0 ml of stock standard into a 1 L volumetric flask. Dilute to mark with double-distilled water. Mix. Prepare fresh each week!

Procedure

Precipitation of protein

1. Using a volumetric pipette, measure 3.0 ml of serum or plasma into a 40 ml round-bottom centrifuge tube.
2. Using a burette, slowly add 25 ml of 10% perchloric acid solution while you agitate the tube.
3. Centrifuge the tubes for 30 minutes at 2000 rpm.
4. If only the PBI is to be determined, *decant and discard* the supernatant. However, for the determination of serum-soluble iodide, add an aliquot of the supernatant directly to the arsenious acid.
5. After discarding the supernatant, again add 25 ml of 10% perchloric acid solution to the precipitated protein.
6. Using a glass stirring rod, resuspend the precipitate and, when the protein is homogeneous, centrifuge for 30 minutes.
7. Discard the supernatant. At this point you may either continue with the analysis or freeze the precipitated protein.

Digestion-oxidation

1. For consistent results (step 8) it is important that *exactly* 15 glass beads (5 mm) be added to each 400 ml electrolytic beaker. The beads are to prevent excessive bumping during the evaporation. Knowing the exact number of beads in each beaker will be invaluable when you

*All chemicals must be of analytic reagent purity. Double-distilled water must be used in making all aqueous solutions.

make the visual estimation of the fluid volume remaining in each beaker.

2. Standards including the blank may be prepared at this time. Prepare four 400 ml electrolytic beakers by adding to each.
   a. 15 beads (5 mm)
   b. 25 ml of chloric acid
   c. 1 ml of chromic acid
3. The four standards are made up for each run consisting of the blank, containing no added iodate, and the following beakers, containing the given amount of working standard:
   a. 1.0 ml, which is equivalent to 0.1 μg of iodate
   b. 1.5 ml, which is equivalent to 0.15 μg of iodate
   c. 2.0 ml, which is equivalent to 0.2 μg of iodate
4. Transfer the precipitated protein to the 400 ml beakers containing the beads. To be assured of a complete transfer of the precipitated protein from the centrifuge tube to the beaker, the following procedure is used.
   a. Divide the 25 ml of *chloric acid* into three washings of 10, 10, and 5 ml volumes.
   b. With a pipette, carefully add 10 ml of chloric acid, washing down the sides of the tube. If this does not loosen the precipitate "coin," carefully rim the outside of the coin to loosen it from the centrifuge tube. The coin usually slips out easily.
   c. With the second 10 ml, wash down the sides of the tube and with a glass stirring rod (for each tube) disengage any protein clinging to the sides of the tube.
   d. The last 5 ml of chloric acid is used to completely rinse the tube and stirring rod of any protein material.
5. Using a 5 ml serologic pipette, transfer 1.0 ml of *chromic acid* to the beakers containing the protein precipitate but not to the blank and standards. You have already added chromic acid to these beakers.
6. Evaporate *all* of the beakers on a hot plate with a surface temperature of about 150° C. When the crackling (due to decomposition of chloric acid) ceases and the white fumes appear, pipette 0.5 ml of chloric acid into each beaker.
7. The color of the solution will be orange, so during the process of evaporation 1 or 2 drops of chloric acid must be added from time to time in order to maintain the chromium in its *orange* quadrivalent state. This is very important, because the reduction of chromium to the green trivalent state for more than a few seconds is associated with the loss of iodine.
8. Evaporate to near-dryness, not to complete dryness or iodine may be lost. The final volume should be less than 0.5 ml. This estimation of volume is done visually, and with a standard number of beads in each beaker there is negligible volume error. Take a beaker with 15 beads, add less than 0.5 ml of water. Use 0.4 ml of water, tip the beaker at an angle, and you will see a small amount of liquid on either side of the beads. Until you are experienced with this type of measurement, fix a beaker with the water for guidance.

Spectrophotometry

1. Allow the beakers to cool slightly; then immediately add 10.0 ml of distilled water to each, washing down the sides of the beaker. Swirl the beaker to effect solution (to mix thoroughly the 10.0 ml of water and the ± 0.4 ml of dissolved digestate).
2. Cool to room temperature.
3. Using volumetric pipettes, pipette a 4.0 ml aliquot from each beaker into its respectively numbered cuvette.
4. Pipette 2.0 ml of 0.2 N arsenious acid and 0.5 ml of the sulfuric acid–sodium chloride reagent. Mix well.
5. Place the tubes in a 30° C controlled-temperature water bath for 30 minutes. NOTE: Never attempt to read more than ten tubes at a time.

6. Set a timer, and at 30-second intervals add 0.5 ml of 0.04 N ceric ammonium sulfate solution to each of the tubes. Mix well.
7. Using a Coleman Universal Spectrophotometer at a 420 mμ wavelength, set the water reference blank at 100% transmittance. The tubes are then read at 30-second intervals. Each tube is read in 4- or 5-minute intervals.

Calculation

1. Calculate the rate of reaction by subtracting the spectrophotometer reading closest to 35% transmittance from the reading closest to 60% transmittance (T) and dividing by the time (t) interval in minutes (Table 15-3).
2. Construct a graph by plotting $\Delta T/t$ against the known amounts of iodine in the aliquots of the standards (Fig. 15-2).
3. Read the iodine concentration in the aliquots of the unknown samples directly from the graph (see example).
4. Express the PBI in μg/100 ml.

$$\frac{I_2 \text{ concentration in aliquot}}{\text{Volume sample} \times \text{aliquot\%}} \times 100 = \mu\text{g of PBI/100 ml}$$

$$\frac{0.047}{3 \times 0.40} \times 100 = 3.9\ \mu\text{g of PBI/100 ml}$$

Table 15-3. Protein-bound iodine

Date: 4-23-70								
Time	Bl.	0.1	0.15	2.0	I	II	III	IV
4	19½	21¼	21½	23¼	20	22	24	21¾
8	20¾	24¾	27	30	24¾	28½	33	26½
12	22½	29	31½	38½	29¾	37	45	32½
16	24	34¼	38¼	48	34¾	45½	57½	40¼
20	25¾	39½	46	57	40½	55		48
24	27¾	45	52¾		46¼			56
28	29¼	50¼			52¾			
32	31¾							
$\Delta T/t$	0.46	1.33	1.77	2.25	1.50	2.25	3.06	1.95
$\mu g I_2$					.047	.080	.117	.067
μg/100ml					3.9	6.6	9.7	5.5
%T	31.75	50.25	52.75	57.0	52.75	55.00	57.50	56.00
	20.75	34.25	31.50	30.0	34.75	37.00	33.00	32.50
Δ%T	11.00	16.00	21.25	27.0	18.00	18.00	24.50	23.50
t	24	12	12	12	12	8	8	12

5. Provided 3.0 ml of serum is precipitated and a 4.0 ml aliquot of dissolved digestate is used in the analysis, the factor 83 may be used; therefore:
   I_2 concentration in aliquot $\times$ 83 = μg of PBI/100 ml

Thyroxine by competitive protein-binding analysis[21,22]

The method is based upon the specific binding properties of thyroxine-binding globulin. Since measurement of iodine is not involved, exogenous organic iodine contaminants do not interfere. Thyroxine is extracted from the patient's serum with ethyl alcohol and then allowed to react with a measured quantity of serum TBG saturated with radioactive T_4. The displacement of radioactive T_4 from the TBG is proportional to the concentration of T_4 in the patient serum.

Reagents needed

Radioactive thyroxine

Thyroxine-^{131}I in 50% propylene glycol, with an initial specific activity of 26 to 60 mc/mg, or thyroxine-^{125}I in 50% propylene glycol, with an initial specific activity of 50 to 250 mc/mM, may be purchased. The solutions are stable for 3 months in a freezer.

Stable thyroxine standard

Use Eltroxin (sodium *l*-thyroxine pentahydrate) as 100 μg tablets. Crush 10 tablets and add 2 ml of 0.1 N NaOH. Dilute to 200 ml with 70% ethyl alcohol and mix thoroughly. Dilute 2.28 ml with 95% ethyl alcohol to 200 ml for the working standard (0.05 μg thyroxine/ml). The solution is stable for 3 months in the freezer.

Anion exchange resin, Rexyn 202 (Cl-SO_4), Fisher, analytical grade, medium porosity

Soak in barbital buffer (pH 8.6, ionic strength 0.075) overnight and rinse with buffer several times until the pH is approximately 8.6. The supernatant is discarded and the resin dried in an oven. Approximately 0.5 ml quantities of resin are dispensed with a plastic spoon.

Ethyl alcohol, 95%

Store in a brown bottle.

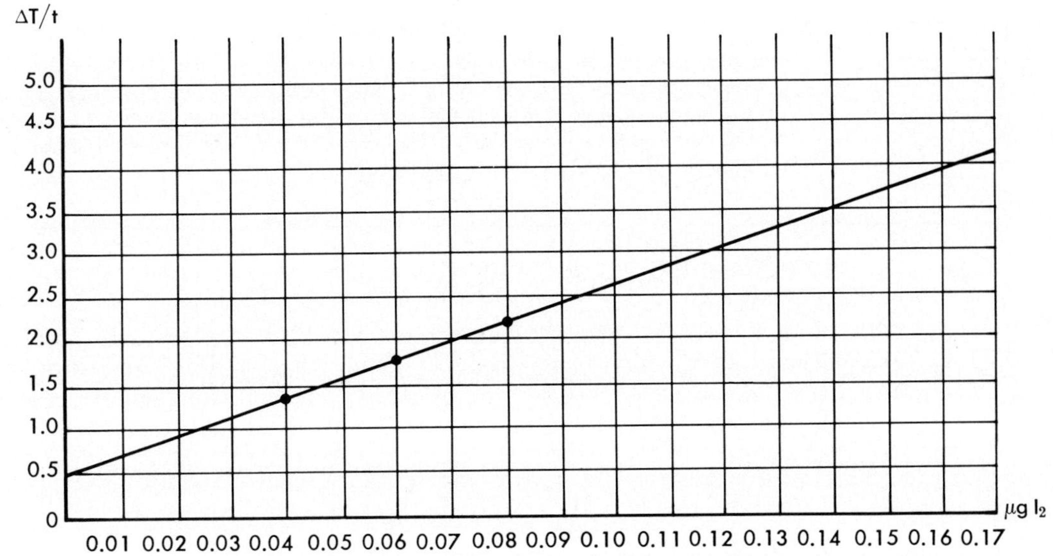

Fig. 15-2. Example of a PBI standard curve.

Sodium hydroxide, 0.1 N
Propylene glycol
Phenol, 1%
Ethyl alcohol, 70%
Radioactive thyroxine solution

Add 5 to 25 mμc (according to convenience for counting) to a solution containing 15 ml pooled serum, 5 ml 1% phenol, 5 ml propylene glycol, and barbital buffer (0.1 M) to make 500 ml. The solution may be stored for several weeks at 5° C.

Barbital buffer, 0.1 M.

Procedure

1. Into small test tubes pipette 1.0 ml of serum and 2.0 ml of 95% ethyl alcohol. Mix vigorously, using a Vortex mixer.
2. Cover the tubes and centrifuge for 5 minutes at 2500 rpm.
3. Into clear tubes suitable for use with the counter transfer in duplicate 0.3 ml of the supernatant and evaporate to dryness in a 45° C water bath with a gentle stream of air.
4. In the same manner prepare thyroxine standards by pipetting into tubes 0, 0.1, 0.2, 0.3, and 0.4 ml of the standard solution containing 0.05 μg thyroxine/ml and evaporating to dryness. These are set up in duplicate or triplicate.
5. To the dried residue in each tube add 1.0 ml of the radioactive thyroxine solution. Shake gently and incubate in a 45° C water bath for 8 minutes.
6. Shake to dissolve the residue and place in an ice water bath for 10 minutes.
7. Immediately after removing from the ice bath add 1 spoonful of resin and shake the rack vigorously with a horizontal motion, using an automatic shaker, for 1 minute.
8. Buffer, 3.0 ml at refrigerator temperature is added quickly to each tube and mixed by shaking for a few seconds. The resin settles rapidly.
9. Remove approximately 80% of the buffer immediately by aspirating and add another 3.0 ml to the resin. Mix by shaking for a few seconds, allow the resin to settle, and remove most of the supernatant.
10. Determine the activity of the resin, c.p.s.

Calculations

Prepare a standard curve by plotting the time in seconds per 10,000 counts versus thyroxine concentration in micromilligrams. The thyroxine in the serum sample (equivalent to 0.1 ml of serum) is obtained from the curve and converted to μg/100 ml of serum. The thyroxine iodine may be calculated as 65.3% of this value.

Comments

Incomplete extraction of the thyroxine from the serum into alcohol can be a major source of error. Timing of the shaking of the resin with the thyroxine solution is important since the reaction is irreversible. Prolonged shaking will release some of the thyroxine originally bound to the TBG for adsorption by the resin and too little shaking will result in some unbound material not being taken up by the resin. Removal of the first 3 ml buffer wash should be accomplished as soon as possible, since standing will cause falsely elevated values.

Normal values

Normal values for various measurements are summarized in Table 15-4.

Table 15.4. Thyroid function tests, summary of normal values

Method	Normal value
PBI	4.0 to 8.0 μg/100 ml serum
BEI	3.2 to 6.2 μg/100 ml serum
T_4 by column, as iodine	2.9 to 6.4 μg/100 ml serum
Thyroxine	4.0 to 11.0 μg/100 ml serum
Thyroxine as iodine	2.6 to 7.1 μg/100 ml serum
Thyroxine, free	1.0 to 2.1 mμg
T_3 uptake	10% to 14.6%

Clinical findings

Thyroxine deficiency is found in cretinism, myxedema, and some forms of goiter. Cretinism, a congenital defect caused by hypothyroidism in infancy, is characterized by retardation in mental, physical, and sexual development. In myxedema, a condition resulting from atrophy or removal of the thyroid gland during adulthood, low metabolic rate, loss of hair, apathy, and thickening of the skin resembling edema are observed. Some forms of goiter are associated with hypothyroidism. When a defect of thyroid physiology occurs in which there is an inadequate output of thyroid hormone, the thyroid gland enlarges. Thyroid cells increase in size and number to compensate in part for the deficient activity of the cells. If the defect is relatively slight, hyperfunction may produce normal hormone levels and the enlarged thyroid gland is classified as a simple goiter. Iodine is required for elaboration of the thyroid hormone; when foods and drinking water are deficient in iodine, the cells cannot synthesize the hormone. With decreased thyroxine secretion the subject is lethargic, passive, and sensitive to cold.

Hyperthyroidism or thyrotoxicosis occurs when there is overproduction of the hormone by the thyroid gland. Some physical characteristics of the disease are increased metabolic rate and heart rate, excessive sweating, heat intolerance, gastrointestinal upsets, protrusion of eyeballs, insomnia, and increased appetite accompanied by weight loss and nervousness. Diffuse enlargement of the thyroid gland occurs in exophthalmic goiter (Graves' disease). In toxic adenoma, the enlargement is uneven as a result of the tumors. Recent investigations suggest that excessive activity of the thyroid gland may not be the cause of clinical hyperthyroidism but, rather, the result of action of LATS (long-acting thyroid stimulator).

PARATHYROID GLANDS

The parathyroid glands consist of two pairs of discrete small masses embedded in thyroid tissue, two on each side of the neck. They secrete the parathyroid hormone, a polypeptide, which influences calcium and phosphate metabolism. This hormone increases calcium absorption from the intestines, regulates excretion in the urine, and stimulates release from the bone. When there is a drain on the blood calcium, the parathyroid glands are stimulated; if sufficient dietary calcium is not supplied, bones will be demineralized. The primary function of the glands is to maintain proper concentration of ionized calcium in the blood.

Secretion of parathyroid hormone is controlled by a feedback mechanism related to levels of circulating ionized calcium.[32] Another hormone that influences calcium levels is thyrocalcitonin, a polypeptide from the thyroid gland, which is secreted in response to hypercalcemia.

Blood levels[35] and urinary excretion[33] of parathyroid hormone may be determined through bioassay by injecting an extract that contains the hormone into parathyroidectomized rats. The most sensitive procedure is a radioimmunoassay method which is technically difficult.

Hyperparathyroidism or an increase in parathyroid hormone production may occur (1) when there is defect in the normal feedback mechanism by plasma calcium concentration or (2) when there is association with compensatory enlargement or tumors of the gland. Blood calcium is abnormally high and there may be withdrawal of calcium from the bones, causing cyst formation and weakening of the bones. Renal tubular cells are sensitive to alterations in circulating parathyroid hormone. The associated renal disease is characterized by renal colic and hematuria resulting from renal calculi.

Hypoparathyroidism or deficiency of the hormone is usually the result of parathyroidectomy. There is a decrease of calcium ion levels in blood, and tetanic convulsions eventually appear because of changes in electrolyte balance in muscular and nervous tissue. Thus by means of its control of blood calcium the hormone regulates nerve and muscle irritability.

Upon injection of the hormone into the body or in hyperparathyroidism there is increased excretion of phosphorus in the urine. The parathyroid hormone decreases tubular reabsorption of phosphorus (TRP). Measurement of TRP is used as an aid in early diagnosis of primary hyperparathyroidism.

Tubular reabsorption of phosphate (TRP)[34]

Patient preparation

It is recommended that patients be prepared for this test by the ingestion of at least a quart of either whole or skim milk each day for 3 days. This eliminates the influence of phosphate depletion in the results. The test should be performed in the morning with the patient in a fasting state. The bladder is emptied at the start of the test and this urine is discarded. The ingestion of water should be encouraged during the test period.

Collection of specimens

Urine: **Urine is collected for a timed 4-hour period.**

Serum: **Venous blood is obtained at the midpoint of the urine collection or at the end of the 2-hour period.**

Procedure

1. **Measure the volume of urine specimen.**
2. **Determine the concentration of inorganic phosphorus and creatinine in the blood serum and urine specimens.**

Calculations

$$100\left(1-\frac{\text{Urine phosphorus}\times\text{Serum creatinine}}{\text{Serum phosphorus}\times\text{Urine creatinine}}\right)=\text{TRP (as percent of filtered phosphate that is reabsorbed)}$$

where the concentrations are expressed as milligrams per milliliters

Clinical findings

In primary hyperparathyroidism the TRP is usually decreased and may vary between 41% and 82%. The test does not differentiate between primary hyperparathyroidism and malignant tumors associated with hypercalcemia. Renal insufficiency, also, will decrease TRP.

ISLETS OF LANGERHANS OF THE PANCREAS

The endocrine tissue of the pancreas constitutes approximately 1% of the gland and consists of about two million islets of Langerhans. Glucagon-secreting alpha cells and insulin-secreting beta cells are located in the islets.

Insulin

Insulin is a polypeptide hormone stored in granules within the beta cells. When it is released, the granules migrate to the cell surface and are discharged into the extracellular space.

The major function of insulin is the control of carbohydrate metabolism whereby blood glucose concentrations are maintained at a normal level. When the supply of insulin is deficient or lacking, the peripheral tissues decrease glucose use, there is increased hepatic gluconeogenesis, blood glucose levels increase above the threshold, glycosuria occurs, and an osmotic diuresis results. Because of this increased urinary flow, or output, dehydration and hemoconcentration may become evident, followed by shock, hypotension, diminished renal flow, anuria, and death. Insulin deficiency also causes a mobilization of depot fat into the blood as free fatty acids and a decrease in protein synthesis. Ketone bodies, acetoacetic acid and β-hydroxybutyric acid, appear in the blood since synthesis of fatty acids stops at this stage in the liver cells and in the adipose tissue.

When the concentration of glucose in the blood moving through the pancreas attains a certain level, insulin production is increased. Other agents that will stimulate insulin secretion are amino acids, secretin, fructose, glucagon, sulfonylureas, pancreazymin, and gut glucase, a material released by the intestinal tract, which cross-reacts immunologically with glucagon. Inhibitors of insulin release are epinephrine, norepinephrine, and the diazoxide and thiazide drugs. Insulin lowers blood glucose by increasing the oxidation of carbohydrates and the deposition of glycogen in muscles and liver.

Plasma and urine insulin levels may be determined by radioimmunoassay procedures.[36,38,40,42] The result is usually expressed in IU, an international unit that by definition refers to the amount of insulin activity in 0.04082 mg of the International Standard Preparation. The measurement is of value in some disorders of carbohydrate metabolism.[37,39,41]

After oral ingestion of glucose, blood insulin in normal adults increases rapidly to a maximum in less than an hour and then decreases to the fasting level in 2 to 3 hours. Diabetes mellitus, a pathologic condition characterized by elevated blood glucose and urinary excretion of glucose, is the result of insulin deficiency. Close relatives of diabetics have been found to have blood insulin levels lower than normal.

Hyperinsulinism has been shown to be characteristic of obesity. It may or may not be increased in cases of insulinoma as the tumors frequently produce insulin intermittently. Elevated plasma insulin levels have been found in acromegaly. When hypoglycemia is associated with nonpancreatic tumors, the plasma insulin concentration is usually within the normal range.

The symptoms of diabetes mellitus can be relieved by injection of insulin. The hormone cannot be given orally because it is inactivated by digestion with proteolytic enzymes. A number of types of insulin are available for clinical use—including regular insulin, crystalline insulin, protamine zinc insulin, NPH insulin, globin insulin, and lente insulin. The chief difference is the time of action.

Glucagon

Glucagon is a polypeptide that acts in the liver to promote glycogenolysis and when administered will cause the pancreatic islets to release insulin. Glucagon-secreting tumors of the pancreas have been described.

ADRENAL GLAND

The superior pole of each kidney is covered by an adrenal gland in the manner of a cap that weighs between 5 and 7 g. The glands are composed of two separate parts: about 80% consists of the outer shell or cortex and the remainder is an inner core or medulla. They are attached to and separated from the kidneys by a layer of fat. The two parts arise from different types of tissue and have unrelated functions.

Adrenal cortex

The cortex is indispensable for life and has as its function the maintenance of homeostasis. It produces steroid hormones that control water and electrolyte balance as well as influence the metabolism of carbohydrate, protein, and fat. Structurally the cortex consists of three separate layers or zones: zona glomerulosa, the outer and thinnest layer, produces mineralocorticoids and is influenced very little by adrenocorticotropin; zona fasciculata, the central and thickest

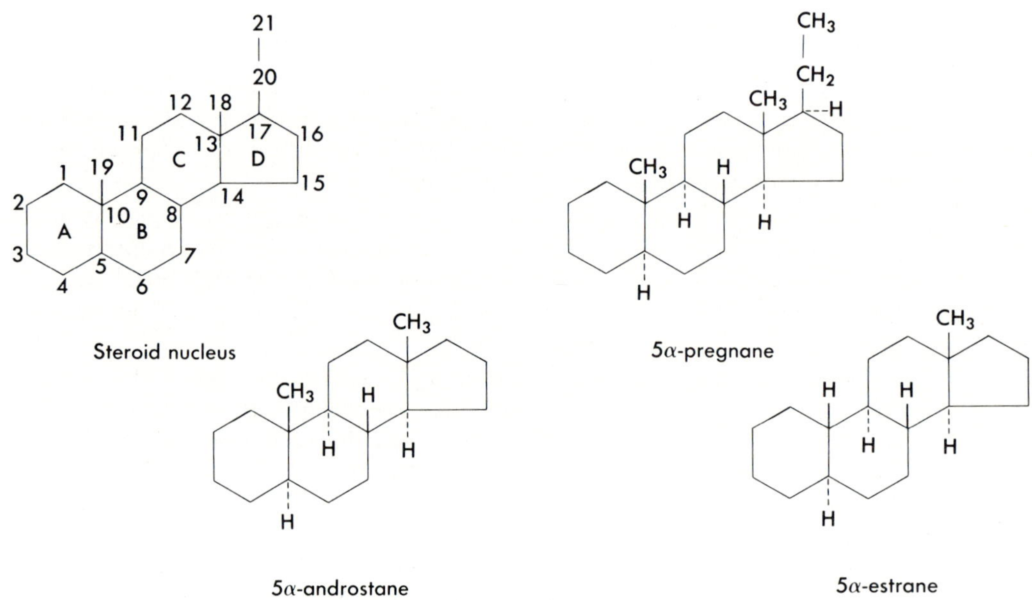

Fig. 15-3. Basic nucleus of steroid groups.

layer, and zona reticularis, the inner layer, are responsible for the production of glucocorticoids and androgens. For the normal male, about two thirds of the urinary 17-ketosteroids are produced by the adrenal cortex, the remainder being synthesized in the testes. In the female, most of the androgens excreted in the urine as 17-ketosteroids originate in the zona reticularis. Activity, function, and growth of the two inner zones are controlled by corticotropin from the anterior pituitary gland.

Steroid hormones: structure, nomenclature, and biosynthesis

All steroid hormones have in common a cyclopentanoperhydrophenanthrene ring structure. The four rings are identified by the letters A, B, C, and D, and the carbon atoms by numbers 1 to 21. The natural steroids, with the exception of

Table 15-5. Major steroid hormones and their metabolites

Trivial name	Systematic name
Carbon-18 compounds	
Estrone	1,3,5(10)estratrien-3-ol-17-one
Estradiol-17β	1,3,5(10)estratrien-3,17β-diol
Estriol	1,3,5(10)estratrien-3,16α,17β-triol
Carbon-19 compounds	
Androsterone	5α-androstan-3α-ol-17-one
Dehydroepiandrosterone	5-androsten-3β-ol-17-one
Etiocholanolone	5β-androstan-3α-ol-17-one
11-hydroxyetiocholanolone	5β-androstan-3α,11β-diol-17-one
11-hydroxyandrosterone	5α-androstan-3α,11β-diol-17-one
11-ketoetiocholanolone	5β-androstan-3α-ol-11,17-dione
11-ketoandrosterone	5α-androstan-3α-ol-11,17-dione
Androstenedione	5α-androstan-3,17-dione
Testosterone	4-androsten-17β-ol-3-one
Epitestosterone	4-androsten-17α-ol-3-one
Carbon-21 compounds	
Cortisol	4-pregnen-11β,17α,21-triol-3,20-dione
Cortisone	4-pregnen-17α,21-diol-3,11,20-trione
Corticosterone	4-pregnen-11β,21-diol-3,20-dione
11-deoxycortisol	4-pregnen-17α,21-dione-3,20-dione
21-deoxycortisol	4-pregnen-11β,17α-diol-3,20-dione
Progesterone	4-pregnen-3,20-dione
Aldosterone	4-pregnen-18al-11β,21-diol-3,20-dione
Tetrahydrocortisol	5β-pregnan-3α,11β,17α,21-tetrol-20:one
Tetrahydrocortisone	5β-pregnan-3α,17α,21-triol-11,20-dione
Tetrahydrocorticosterone	5β-pregnan-3α,11β,21-triol-20-one
Tetrahydro S	5β-pregnan-3α,17α,21-triol-20-one
Cortol	5β-pregnan-3α,11β,17α,20α,21-pentol
Cortolone	5β-pregnan-3α,17α,20α,21-tetrol-11-one
Pregnanetriol	5β-pregnan-3α,17α,20α-triol
Pregnanediol	5β-pregnan-3α,20α-diol
Pregnanolone	5β-pregnan-3β-ol-20-one
Pregnenolone	5-pregnen-3β-ol-2-one

the estrogens, have two angular methyl groups that are attached to carbon atoms 10 and 13. Except where a carbon atom joins two rings or is attached to another substituent, each carbon is saturated by two hydrogen atoms. Substituted groups may be found on carbon atoms 3, 10, 11, 13, 16, and 17. The angular methyl groups occupy positions 18 and 19.

Steroid hormones may be classified as adrenocortical steroids, progesterones, androgens, and estrogens. They may also be grouped (Table 15-5) according to the number of carbon atoms in the compounds: corticosteroids, including progesterones, have 21 carbon atoms; androgens, 19 carbon atoms; and estrogens, 18 carbon atoms. The parent hydrocarbon of each—respectively, pregnane, androstane, estrane—is used as the basis for systematic naming (Fig. 15-3).

Differences between the various compounds may result from replacement of a hydrogen atom by a hydroxyl or a ketone group or the reverse, or by different spatial arrangements of substituents. A compound may be referred to as a *dehydro* compound if two hydrogen atoms are lost. For example, when two hydrogen atoms are lost from C-11 of corticosterone, the new compound is dehydrocorticosterone. If an oxygen atom is lost, the new compound may be termed a *deoxy* compound. Deoxycorticosterone is produced when an oxygen atom is lost from the hydroxyl group on C-11 of corticosterone. (See Fig. 15-4.) The presence of various groups in the compounds are frequently indicated by the follow-

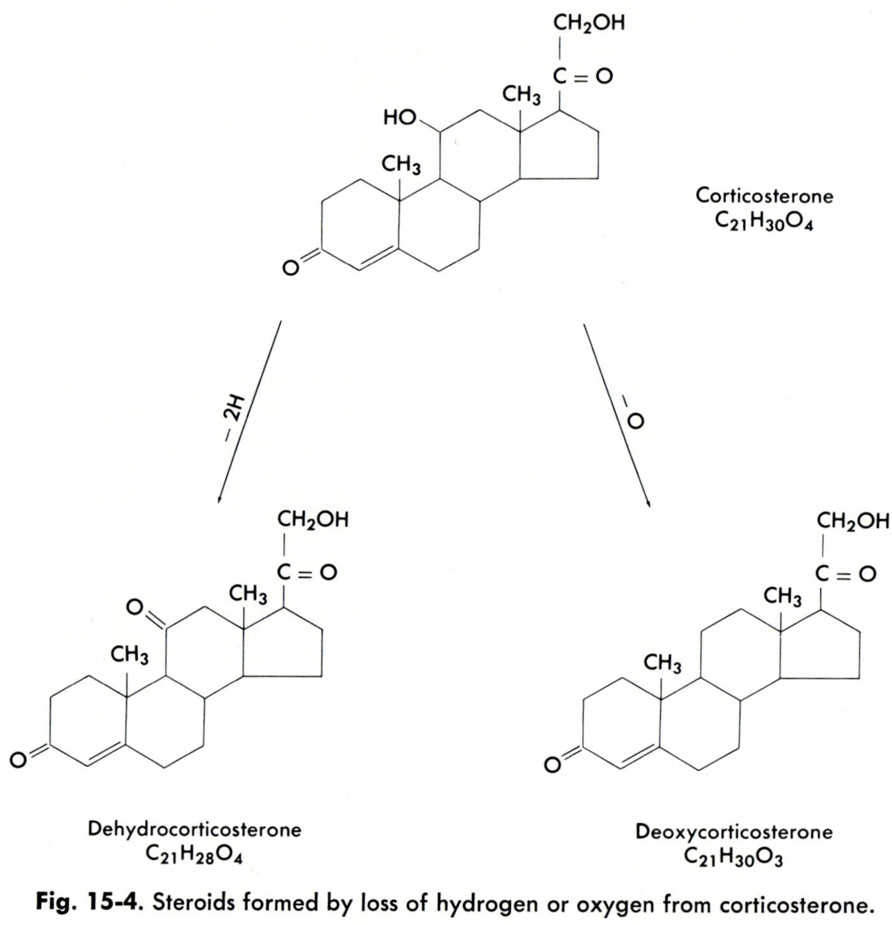

Fig. 15-4. Steroids formed by loss of hydrogen or oxygen from corticosterone.

ing terms: for a hydroxyl group, *hydroxy* and *oxy* (prefix) and *ol* (suffix), and for a ketone group *oxo* and *keto* (prefix) and *one* (suffix).

Unsaturation is denoted by the symbol Δ and is followed by a number indicating location. A saturated nucleus is designated by the suffix *-ane,* one double bond by *-ene,* and two double bonds by *-diene.*

Stereoisomerism provides for a variety of steroids, and a slight structural change may produce a compound with very different physiologic effect. Isomerism describes molecules having the same number and kinds of atoms but arranged in different configurations. Some of the terms employed in steroid nomenclature to indicate different spatial configurations are the following: *cis, trans,* α, β, *allo, epi,* and *iso.* Geometric or cis-trans isomerism describes the position of a substituent relative to the plane of the molecule. Two substituents of a cyclohexane ring can be either on the same side of the plane of the ring (cis) or on opposite sides (trans). Two saturated rings may be fused in two ways, cis or trans. For all naturally occurring steroids four rings are fused and both isomers are found in the attachment of rings A and B, whereas rings BC and CD are attached trans. The steroids exist in what is known as the chair conformation.

When a substituent atom is projected to the rear or underside of a steroid molecule, it is considered to be α-oriented to the methyl group at C-10 and is indicated by a dotted bond. In contrast, a β-oriented substituent will be to the front or above the molecule and is indicated by a solid bond. For example, the only difference between androsterone and etiocholanolone is the orientation of the hydrogen atom on C-5 (Fig. 15-5). It is α-oriented in the androsterone molecule and β-oriented in the etiocholanolone molecule. In the term 3-trans-hydroxy, the relationship is expressed that the OH group attached to C-3 is trans with respect to the methyl group attached to C-10. The prefix allo (denotes one of two possible isomers) is applied to 5α-compounds as opposed to 5β-compounds that are considered belonging to a normal series. For example, allopregnanediol is 5α-pregnan-3α,20α-diol and pregnanediol is 5β-pregnan-3α,20α-diol. Epi or iso indicates an isomer that differs only in spatial arrangement of a group on one carbon atom or, as stated by Forsham,[54] compounds with unnatural OH groups. For example:

Androsterone	5α-androstan-3α-ol-17-one
Epi(iso)androsterone	5α-androstan-3β-ol-17-one
Dehydroepi(iso)androsterone	5-androsten-3β-ol-17-one
Cortisol	4-pregnen-11β,17α,21-triol-3,20-dione
Epicortisol	4-pregnen-11α,17α,21-triol-3,20-dione

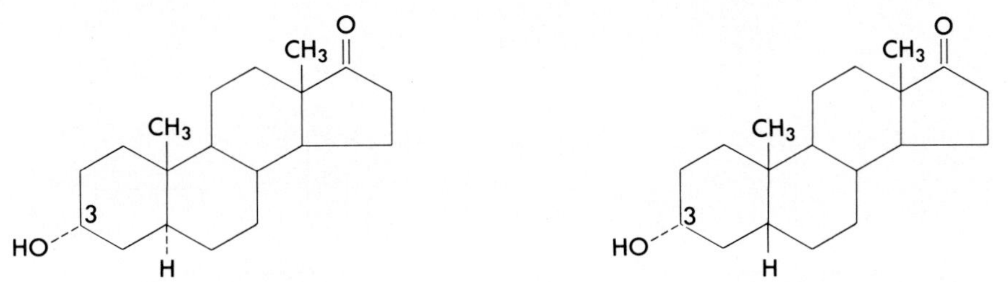

Androsterone
3α-hydroxy-5α-androstan-17-one

Etiocholanolone
3α-hydroxy-5β-androstan-17-one

Fig. 15-5. Steroids differing only by orientation of hydrogen atom on C-5.

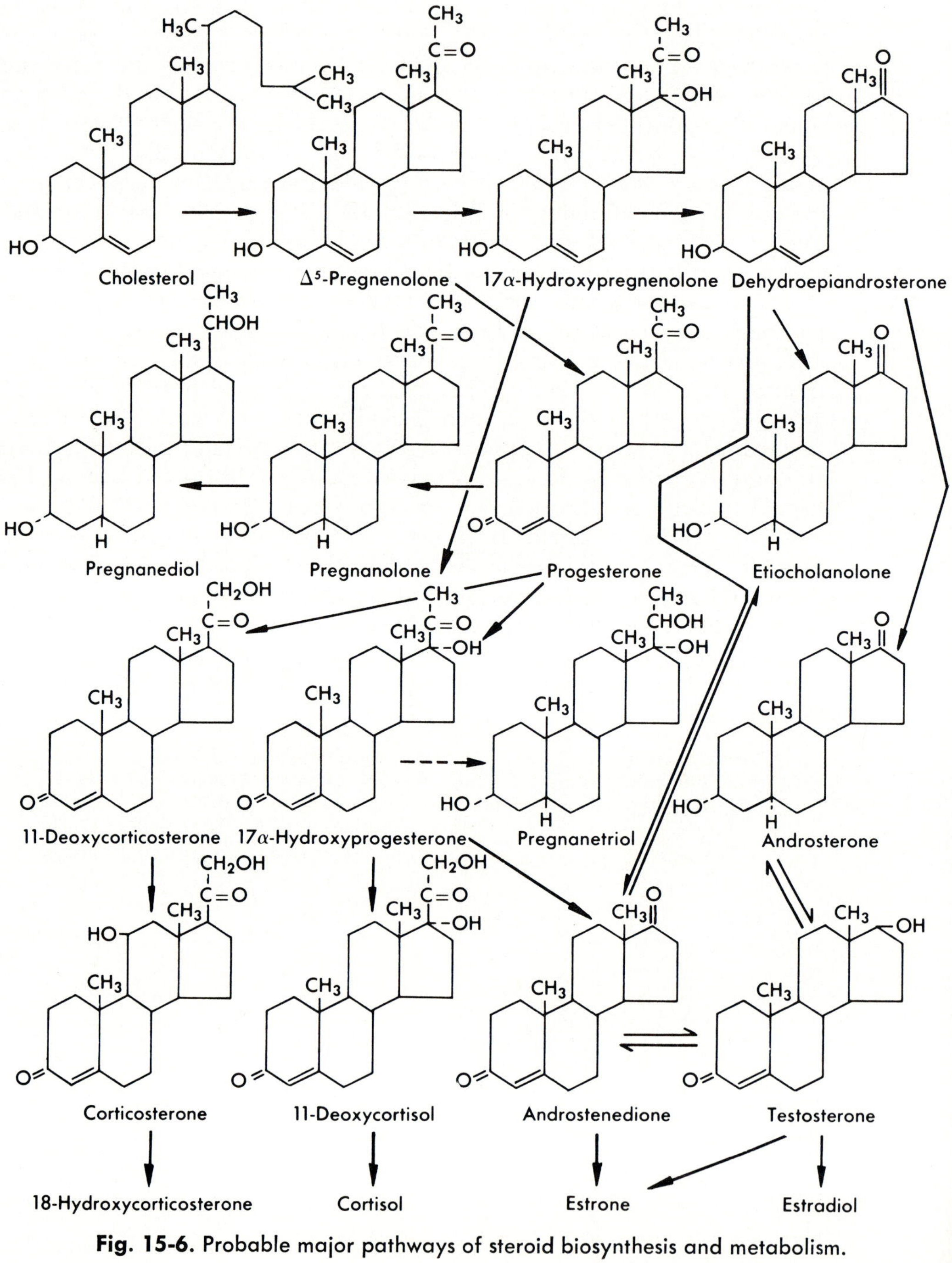

Fig. 15-6. Probable major pathways of steroid biosynthesis and metabolism.

Synthesis of steroids proceeds by enzymatic action from acetate by way of cholesterol (Fig. 15-6). The pathways are similar regardless of whether the hormone is produced in the adrenal cortex or in ovaries or testes.

Nearly 50 steroids have been isolated from the adrenal cortex but cortisol, corticosterone, and aldosterone and their metabolites produce the major activity attributed to the gland. The formation of pregnenolone is partly controlled by ACTH and the subsequent conversions by specific enzyme systems. The enzymes 3β-hydroxysteroid dehydrogenase and an isomerase convert pregnenolone to progesterone. The hydroxylation of progesterone can occur in two ways and usually proceeds in the order of C-17, C-21, and C-11. If hydroxylation occurs

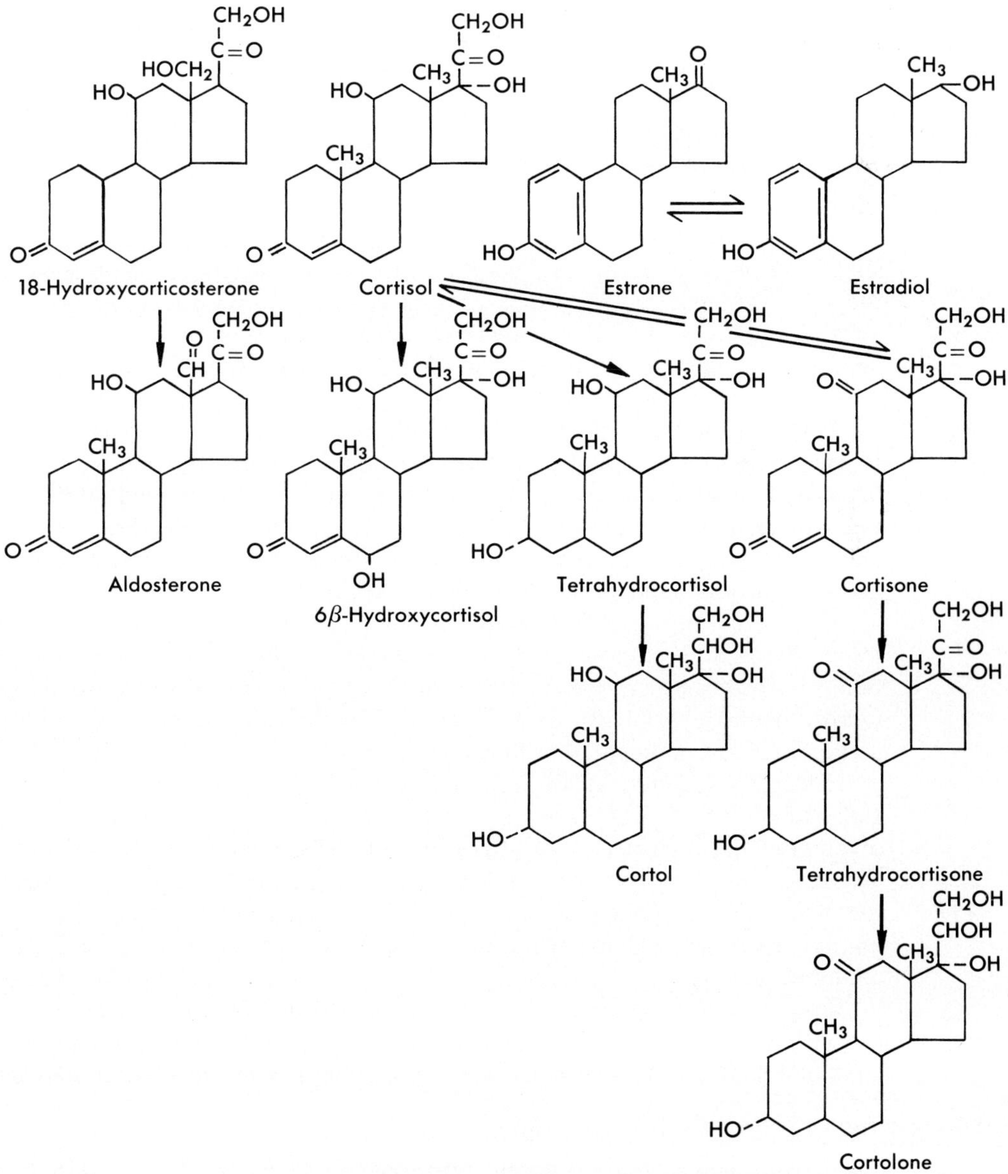

Fig. 15-6, cont'd. For legend see opposite page.

at C-17, then hydroxylation will follow on C-21 and C-11β to yield cortisol. If it does not take place at C-17, hydroxylation at C-21 and C-11β will form corticosterone. For aldosterone the accepted pathway is corticosterone, 18-hydroxycorticosterone, and aldosterone.

The secretory activity of the adrenal cortex is controlled by the adrenocorticotropin, and cortisol has an important role in the basal rate of corticotropin secretion. When free cortisol is decreased, corticotropin increases and an increase in free cortisol will decrease corticotropin secretion. In the absence of stress, about 10% of the plasma cortisol is free and the remainder bound to transcortin, an alpha-1 glycoprotein, also known as corticosteroid-binding globulin (CBG). A small amount is also bound to albumin. The concentration of CBG is increased during estrogen treatment and late pregnancy but free cortisol is elevated only slightly. Corticosterone and progesterone are also bound to CBG in a similar manner.

Metabolism of the adrenocortical steroids occurs mainly in the liver where reduction of the A ring between carbon atoms 4 and 5 produces the dihydro derivatives, which are rapidly reduced to the tetrahydro compounds. Additional reduction occurs that involves the keto group on C-20, yielding cortol and cortolone.

The metabolites of cortisol are conjugated with glucuronic acid, and to a slight extent with phosphoric and sulfuric acids, water-soluble forms in which they are excreted in the urine. The major metabolites found in the urine are tetrahydrocortisol, tetrahydrocortisone, allotetrahydrocortisol, cortol, and cortolone. Only a very small amount of the 17-hydroxycorticosteroids appears in the urine in the active, unconjugated form.

Angiotensin II stimulates the secretion of aldosterone. In the 11-18 hemiacetal form it is reduced in the liver to tetrahydroaldosterone and this metabolite conjugated with glucuronic acid. In the liver and kidneys an acid-labile conjugate of aldosterone is also formed, an 18-glucuronide. Excreted in the urine is a small amount of free aldosterone along with the conjugates, tetrahydroaldosterone 3-glucuronide and aldosterone 18-glucuronide.

The major androgens secreted by the adrenal cortex are dehydroepiandrosterone (free and ester, some being conjugated with sulfuric acid before secretion), testosterone, androstenedione, and 11β-hydroxyandrostenedione. They are usually bound by globulin in the plasma. Metabolism occurs in the liver and the metabolites are conjugated to form water-soluble glucuronides and sulfates.

Estrogens secreted by the adrenal cortex include estradiol-17β, estrone, and 18-hydroxyestrone. They are apparently synthesized from testosterone and androstenedione. In plasma more than half of the estrogens are bound to albumin. There are also present estrogen-binding beta globulins. High concentrations of the free estrogens appear in bile, indicating this as one of the pathways for excretion. Estrogens can be metabolized to 2-methoxy derivatives and 16-oxygenated derivatives.

Progesterone is synthesized as indicated in Fig. 15-6 and metabolized chiefly in the liver. The principal metabolites are pregnanediol, allopregnanediol, pregnanolone, and allopregnanolone, the latter being present in highest concentration.

Laboratory tests of adrenocortical function

Measurements of the different hormones secreted by the adrenal cortex and of their metabolites found in the urine are used as a means of evaluating adrenocortical function. In routine analysis the chemical estimations are for groups of structurally or chemically related compounds rather than determination of individual steroids. Assaying for a single steroid requires that the compound be isolated and purified prior to the final measurement.

The specimen for analysis is usually a 24-hour urine collection even though this is more difficult and inconvenient to obtain than blood and the completeness of the collection often questionable. Analysis of a blood specimen, while more reliable, is a measure of the steroid level only at the time of drawing and does not reflect diurnal variations nor any temporary changes that may occur in the subject.

Androgens and 17-ketosteroids (17-oxosteroids)

Androgens are substances that are capable of stimulating male genitalia, axillary and pubic hair, and certain other secondary sex characteristics. Although there has been a tendency to consider all 17-ketosteroids in a group as androgens, the terms are not synonymous. The most active androgen, testosterone (4-androsten-17β-ol-3-one), is not a 17-ketosteroid. Testosterone precursors dehydroepiandrosterone and androstenedione are 17-ketosteroids and are found in the adrenal cortex.

The neutral 17-ketosteroids are a group of steroid compounds having 19 carbon atoms and a ketone group at position 17 of the steroid nucleus. They are derived from 5α-androstane and etiocholanolone (5β-androstane). Androsterone, epiandrosterone, and dehydroepiandrosterone are mildly androgenic but etiocholanolone is biologically inactive. The metabolites present in highest concentration in the urine are androsterone, etiocholanolone, dehydroepiandrosterone, 11β-hydroxyandrosterone, 11-ketoandrosterone, 11β-hydroxyetiocholanolone, and 11-ketoetiocholanolone. These may be classified into two fractions, using as examples the three major steroids: an alpha fraction (androsterone, etiocholanolone) and a beta fraction (dehydroepindrosterone). Normally 85% to 95% of the 17-ketosteroids are in the alpha fraction.

The neutral 17-ketosteroids[52] are excreted into the urine as water-soluble conjugates, primarily as glucosiduronate and sulfate esters. Hydrolysis is required to obtain the extractable free steroids and this is accomplished routinely by treating a sample with a mineral acid at an elevated temperature. Hydrochloric and sulfuric acids have been used. No entirely satisfactory method for hydrolysis is available. The use of acid causes some destruction and alteration of steroids as well as formation of pigments and nonspecific chromogens that interfere in the final colorimetric reading. Methods have been developed for differential hydrolysis of the sulfate and glucuronides but these are not practical as routine procedures.

Benzene, carbon tetrachloride, and diethyl ether have been the most widely used solvents for extraction of the free steroids from the hydrolyzate. Each has advantages and disadvantages. For example, ether may be preferred because of steroid solubility, ease of removal at low temperatures, and less tendency than other solvents to form emulsions. However, it must be free of peroxides and it is highly flammable.

For the final measurement, the Zimmermann reaction is generally employed. It is based upon the formation of a reddish purple color when a 17-ketosteroid reacts with *m*-dinitrobenzene in an alkaline solution. This action occurs only if the ketone group is adjacent to a CH_2 group. Maximum absorption is evident at 520 mμ when the ketone is in position 17. For compounds with ketones in other positions, the color is less intense and the maximum does not occur at 520 mμ.

17-Ketosteroids, urinary[68,70]

Specimen collection

The urinary 17-ketosteroids appear to be relatively stable. The collection may be made without a preservative if it is kept in a refrigerator. Otherwise, a 24-hour urine specimen is collected in a glass bottle containing 15 ml of toluene to prevent bacterial contamination. Significant deterioration will not occur for several weeks if a specimen is stored at 0° to 4° C. For longer storage periods, preserve at −10° C.

Reagents needed

All chemicals are reagent grade.
Hydrochloric acid, concentrated
Ethyl ether (Squibb), for anesthesia, or redistilled and peroxide-free
  Each batch should be tested for peroxides by dissolving a few crystals of potassium iodide in 5 ml of water and shaking for 1 minute with 5 ml of the ether. A yellow color indicates the presence of peroxides.
Sodium hydroxide, 10% (w/v)
Dichloromethane, redistilled
Methyl alcohol, absolute, spectro quality
m-Dinitrobenzene (Sigma)
  Specially prepared for the Zimmermann reaction in steroid determinations, 1% in absolute ethanol. Prepare fresh daily.
Benzyl trimethyl ammonium methoxide, 40% in methanol
Ethyl alcohol, absolute, ketone-free and aldehyde-free
  To purify add 4 g of *m*-phenylenediamine hydrochloride to 1 L of absolute alcohol and allow to stand in the dark about a week with occasional shaking. Distill, using an all-glass apparatus, and discard the first and the last 10%. Store the purified alcohol in the dark.
Stock standard
  Dissolve 100 mg of dehydroepiandrosterone in absolute methanol and dilute to 100 ml (1000 μg/ml).
Working standard (200 μg/ml)
  Dilute 10 ml of stock standard to 50 ml with absolute methanol.

Procedure

1. Measure total volume of 24-hour collection of urine and filter an aliquot.

Hydrolysis

2. Into 50 ml round-bottom centrifuge tubes pipette *in duplicate:*
   Unknown: 5 ml of filtered urine
   Control: 5 ml of control urine
3. Add to each tube 1.5 ml of concentrated hydrochloric acid.
4. Heat the tubes in a water bath at 80° to 85° C for 12 minutes. Cool in running tap water.

Extraction and washing

5. To each tube add 20 ml of ethyl ether and shake by hand for 30 seconds. Discard the aqueous layer by aspirating.
6. Add 10 ml of 10% NaOH and shake for 10 seconds by hand. Discard the NaOH layer by aspirating.
7. Add 10 ml of distilled water and shake for 10 seconds. Cover tubes with foil and centrifuge for 5 minutes at 1500 rpm.

Color development

8. Into 40 ml round-bottom tubes place:
   Unknown: 10 ml of ether extract
   Control: 10 ml of ether extract
   Standard (duplicate): 0.1 ml of working standard solution (20 μg)
   Blank (duplicate): Dry tubes
9. Evaporate to dryness in a water bath at 40° to 45° C under a stream of air. Wash down the sides of the tubes with 0.3 ml of absolute ethyl alcohol and evaporate to dryness.
10. To the dried residue in each tube add 0.1 ml of 1% *m*-dinitrobenzene in absolute ethyl alcohol and mix to dissolve the residue; add 0.2 ml of benzyl trimethyl ammonium methoxide and mix.
11. Place tubes in a water bath at 25° C for 90 minutes. Keep in the dark by covering with a black cloth.
12. Add to each tube 3.0 ml of 50% ethanol in water and mix; add 3.0 ml of dichloromethane and mix vigorously for 10 seconds.
13. Allow the tubes to stand in the dark until the two layers separate and the bottom layer becomes crystal clear.
14. Measure absorbance (A) of the lower layer against the reagent blank at 480, 520, and 560 mμ.
15. Apply the Allen correction.

Calculations

$$\frac{\text{Corrected A unknown}}{\text{Corrected A standard}} \times \text{Standard concentration (mg)} \times$$

$$\frac{\text{Total urine in volume (ml)}}{\text{Aliquot urine in extract}} = \text{mg 17-ketosteroids in 24-hour specimen}$$

17-Ketosteroids, urinary—alternate method[70]

Reagents needed

Refer to previous method.

Potassium hydroxide, 5 N

This solution must be standardized.

Ethyl alcohol, 80% aqueous solution (v/v)

Prepared from purified reagent.

m-Dinitrobenzene, 2% in absolute ethyl alcohol

Prepare fresh daily.

Procedure

Steps 1 to 9 and 14 and 15 are the same as for the previous method.

10. To the dried residue in each tube add 0.5 ml of alcoholic-KOH reagent (prepared immediately before use by mixing 2 parts absolute ethyl alcohol with 3 parts 5 N KOH) and 0.2 ml of 2% *m*-dinitrobenzene reagent.
11. Place tubes in a water bath at 25° C for 60 minutes. *Keep in dark;* cover with a black cloth.
12. Add to each tube 4.0 ml of 80% ethyl alcohol.
13. Measurement of the absorbance and calculations are the same as described for the first method.

Comments

The free steroids are relatively insoluble in an aqueous solution but soluble in organic solvents. A solvent consisting of a mixture of equal parts of benzene and petroleum ether[68] may be used in place of ether to extract the ketosteroids. (CAUTION: Benzene is highly toxic and, if used, adequate ventilation must be provided.)

The presence of nonspecific substances that contribute color to that produced by the 17-ketosteroids in the Zimmermann reaction is a major source of error.

The alkaline wash removes acidic and phenolic substances including estrogenic pigments and chromogens from the crude extract. Thus the material remaining in the organic solvent has been referred to as the neutral fraction. Upon extraction of the Zimmerman color complex with dichloromethane, nonspecific brown pigments are retained in the water-alcohol layer.

The addition of formalin to the urine before hydrolysis has been used to decrease the chromogens formed that interefere with the Zimmermann reaction. However, a recent study by Goldzieher and de la Pena[55] showed that in the presence of formalin the destruction of free dehydroepiandrosterone is increased and the recovery of ketosteroid sulfates is lower.

For the Allen correction, spectrophotometric readings are made at three wavelengths, one on each side of the maximum for the specific constituent being assayed. Maximum absorbance is at 520 mμ for the Zimmermann chromogens when the ketone group is at C-17. The corticosteroids testosterone and progesterone develop a slight color with the reagent but the maximum absorbance is near 420 mμ. Corticosterone, pregnanedione, and pregnanolone yield complexes that absorb maximally at 490 mμ. The Allen correction is valid as long as the absorbance of the interfering substances is linear. This has been shown to be not completely true for urine extracts free of 17-ketosteroids.

The use of benzyl trimethyl ammonium methoxide in place of KOH in the Zimmermann reaction makes possible the extraction of the color complex by dichloromethane without the turbidity that is difficult to remove. In the presence of this base, the color complex formed by the principal 17-ketosteroids has approximately the same intensity. Fading is less than with KOH.

Administration of the following drugs has been reported[45,78] to interfere with the 17-ketosteroid determination: cortisol, dexamethasone, Equanil (meprobamate), Librium (chlordiazepoxide hydrochloric acid), Monase (etryptamine acetate), Noludar (methyprylon), paraldehyde, Pyridium (phenazopyridine hydrochloride), quinine, seconal, Serpasil (reserpine), Thorazine (chlorpromazine), and Valmid (ethinamate).

Normal values

The normal values are as follows:

Men	6-18 mg/24 hr
Women	5-15 mg/24 hr

Using a modified Zimmermann procedure, Kaiser obtained, for 519 female and 948 male healthy subjects, the values in Table 15-6.

Clinical findings

The determination of 17-ketosteroids is one of the tests used in assessing some adrenocortical, pituitary, and gonadal activity. In the normal male approximately two thirds of the urinary 17-ketosteroids are derived from the adrenal cortex and the remainder from the testis. In the normal female almost all of this group of hormones arises from the adrenal cortex with little or none from the ovaries. Hence the level of 17-ketosteroids in the urine of the female is lower than that of the male. The level of excretion also varies with age.

Extensive studies on the excretion of 17-ketosteroids in various pathologic states have been carried out because for many years this was the only method available to assess adrenal cortex function. However, it is possible to make only

Table 15-6. Data on 17-ketosteroids, urine, values from Kaiser

Age (years)	Men (mg/24 hr)	Women (mg/24 hr)
10—15	3.8— 7.8	4.3— 8.7
15—20	6.3—12.7	5.8—11.6
20—25	7.4—14.8	6.1—12.2
25—30	7.2—14.6	5.8—11.7
30—35	6.8—13.7	5.4—10.8
35—40	6.3—12.7	4.8— 9.7
40—45	5.8—11.6	4.3— 8.7
45—50	5.2—10.6	3.8— 7.7
50—60	4.2— 9.4	3.2— 6.9
60—70	3.4— 7.6	2.9— 6.2

generalizations because of the wide variations in the values reported. Investigation has shown that the method used for routine testing whereby a group of compounds is assayed may yield a total 17-ketosteroid value consisting of up to 90% nonspecific chromogens. The test is useful in screening patients for marked changes in adrenocortical activity.

A decrease in excretion of total 17-ketosteroids is usually associated with panhypopituitarism, pituitary dwarfism, Addison's disease, myxedema, Klinefelter's syndrome, anorexia nervosa, and advanced liver disease. When Cushing's syndrome is caused by a benign tumor, the urinary 17-ketosteroids tend to be low; when the tumor is malignant, the level is usually high. An increased excretion of total 17-ketosteroids is generally found in adrenocortical hyperplasia, congenital adrenal hyperplasia, and hirsutism. The determination has been shown to be of little diagnostic value in thyrotoxicosis, hypogonadism, Stein-Leventhal syndrome, Turner's syndrome, and mammary carcinoma.

In general, there is agreement between the urinary 17-ketosteroid and 17-hydroxycorticosteroid levels. The exceptions arise in certain pathologic conditions such as benign tumor of the adrenal gland where the urinary 17-ketosteroids are usually within normal limits and the 17-hydroxycorticosteroids are elevated.

Allen color test for DHA[43]

Increased amounts of dehydroepiandrosterone are found in the urine in certain pathologic conditions. The Allen color test may be used as a screening procedure for adrenocortical tumors. Hydrolysis and extraction are carried out in the same manner as for total 17-ketosteroids. Color is developed by adding an ethanol–sulfuric acid reagent that reacts with ketosteroids having a 3-hydroxyΔ^5 structure.

Reagents needed

All chemicals are reagent grade.
Hydrochloric acid, concentrated
Ethyl ether (Squibb) for anesthesia, or redistilled and peroxide-free
Sodium hydroxide, 10% (w/v)
Ethyl alcohol, absolute, ketone-free and aldehyde-free
Purify by the procedure described in the method for 17-ketosteroids.

Ethyl alcohol, 95% aqueous (v/v) and 90% (v/v)
Alcohol–sulfuric acid reagent
Add 12 volumes of concentrated sulfuric acid to 3 volumes of 90% ethyl alcohol.
Sulfuric acid, concentrated
Methyl alcohol, absolute, spectro quality
Stock standard
Dissolve 100 mg of dehydroepinandrosterone in absolute methanol and dilute to 100 ml (1000 μg/ml).
Working standard (200 μg/ml)
Dilute 10 ml of stock standard to 50 ml with absolute methanol.

Procedure

1. Measure total volume of 24-hour collection of urine and filter an aliquot.

Hydrolysis

2. Into 50 ml round-bottom centrifuge tubes pipette in duplicate 10 ml of filtered urine and 3 ml of concentrated HCl.
3. Heat the tubes in a water bath at 80° to 85° C for 12 minutes. Cool in running tap water.

Extraction and washing

4. To each tube add 20 ml of ethyl ether and shake by hand for 30 seconds. Discard the aqueous layer by aspirating.
5. Add 10 ml of 10% NaOH, shake for 10 seconds by hand. Discard the NaOH layer by aspirating.
6. Add 10 ml of distilled water and shake for 10 seconds. Cover tubes with foil and centrifuge for 5 minutes at 1500 rpm.

Color development

7. Into 40 ml round-bottom tubes place:
   Unknown: 10 ml of ether extract
   Standard (duplicate): 0.1 ml of working standard (20 μg)
   Blank: Dry tubes
8. Evaporate to dryness in a water bath at 40° to 45° C under a stream of air. Wash down the sides of the tubes with 0.3 ml of absolute ethyl alcohol and evaporate to dryness.
9. To each tube add 2 ml of alcohol–sulfuric acid reagent and heat for 12 minutes at 55° C.
10. Cool in running tap water for 1 minute, add 3 ml of 95% ethyl alcohol, and mix.
11. Measure the absorbance (A) against the reagent blank at 560, 600, and 640 mμ.
12. Apply the Allen correction.

Calculations

$$\frac{\text{Corrected A unknown}}{\text{Corrected A standard}} \times \frac{20}{1000} \times \frac{\text{Total urine volume (ml)}}{\text{Aliquot urine in extract (5 ml)}} =$$

mg 3β-hydroxy-17-ketosteroids (calculated as DHA) in 24-hour specimen

Comments

The blue color is not specific for dehydroepiandrosterone but the other compounds that react similarly are present in small amount and do not interfere significantly with the test.

Clinical findings

The test has limited application in differentiating patients with adrenal tumor from those with adrenal hyperplasia. In some cases of adrenal carcinoma the level of urinary 17-ketosteroids is elevated and dehydroepiandrosterone is primarily responsible for the increase. Allen cited 4 out of 5 cases of adrenal tumor where an extract of the urine failed only once to produce a grossly positive blue color and 7 cases of female pseudohermaphroditism that gave a negative reaction even though the amount of total ketosteroid was markedly increased. In cases of adult adrenal carcinoma the ratio of beta to alpha isomers (β:α) may range from 0.28 to 4.0. The beta fraction consists mostly of dehydroepiandrosterone and the alpha fraction of androsterone and etiocholanolone. Normally the β:α ratio is less than 0.2. The test is considered significant when the ratio exceeds 0.4.

Corticosteroids

The term corticosteroids includes the active adrenal hormones related to cortisol and corticosterone. Although there is some overlapping of activity, this group of steroids may be divided functionally into mineralocorticoids and glucocorticoids.

Mineralocorticoids—aldosterone

Mineralocorticoids were so named because of their influence on electrolyte metabolism and water balance in the body. The most active steroid of the group is aldosterone. It causes sodium retention and potassium loss, which are enhanced by the glucocorticoids.

Procedures for the determination of aldosterone and its metabolites are not suitable for routine clinical laboratories. The one most frequently employed is that of Kliman and Peterson,[58] which is a double isotope derivative method. The residue from an organic extract of hydrolyzed urine is acetylated, using tritiated acetic anhydride, and a known amount of ^{14}C-labeled aldosterone diacetate is added for correction of losses. The procedure requires two paper chromatographic systems, oxidation of the aldosterone diacetate, and further purification using paper chromatography before measuring the final radioactivity. Normal excretion for patients on a daily intake of 120 mEq of sodium is between 4 and 17 μg/24 hr.

Primary hyperaldosteronism, a curable cause of hypertension, is usually caused by an adrenal tumor. Because prognosis is favorable when the tumor is removed, distinguishing these patients from the large number of patients with essential hypertension is necessary. However, diagnosis or treatment cannot be made from the determination of the aldosterone alone. Secondary aldosteronism must be ruled out, since there are many factors that stimulate aldosterone secretion. A normal person when placed on a low-salt diet will show a decrease in sodium excretion and an increase in aldosterone excretion. Increased excretion also occurs in pregnancy, nephrotic edema, dehydration, cardiac failure, decompensated hepatic cirrhosis, and diuretic therapy. When obtaining a specimen for analysis the patient should be on a high-salt diet and not receiving thiazide diuretics.

Laboratory values are used to support clinical findings in an effort to exclude secondary aldosteronism. An erroneous diagnosis could result in the removal of normal adrenal glands. Forsham[54] has stated that "the definitive diagnosis of primary aldosteronism is made on the basis of a low or absent renin titer in the plasma in an upright individual in whom aldosterone levels on a normal salt intake are elevated." Renin, an enzyme formed in the kidneys, acts on a plasma substrate to produce angiotensin, a vasopressor. The measurement of renin is tedious and performed by very few laboratories but the level of plasma renin is high in secondary aldosteronism and low in primary aldosteronism.

Glucocorticoids

Glucocorticoids, cortisol and corticosterone, promote gluconeogenesis from protein, maintaining liver glycogen and blood glucose. They stimulate fat synthesis, depress response to injury, increase renal blood flow, and maintain blood pressure. Cortisol, the most abundant corticosteroid, is a weak mineralocorticoid but because of its high concentration it accounts for approximately one fourth of the total mineralocorticoid effect.

Measurement of cortisol or its metabolites in plasma and urine is used in assessing adrenocortical function. These have a hydroxyl group and a two-carbon side chain on C-17 and are usually designated as 17-hydroxycorticosteroids.

Corticosteroids, plasma

Cortisol, a 17-hydroxycorticosteroid, is the major secretory product of the adrenal cortex. The steroid in second highest concentration, corticosterone, is approximately one tenth that of cortisol. Measurements of blood corticosteroids may include both of these or show mainly cortisol. In blood, the adrenocorticoids are found conjugated as glucuronides or in the free unconjugated form. Both forms may be loosely associated with plasma proteins. For routine studies, the unconjugated cortisol is usually determined and the total plasma corticosteroids rarely requested.

Plasma or serum analysis is preferred to whole blood because of the difficulties encountered when extracting whole blood with organic solvents. Since the cortisol concentration is lower in the red cells, plasma must be removed as soon as possible after drawing the blood.

In clinical laboratories there are three general types of procedures in use for assaying plasma corticosteroids. One method[67] employs the sulfuric acid–phenylhydrazine reagent described by Porter and Silber that determines the 17,21-dihydroxy-20-ketosteroids, which in plasma is principally cortisol. Other compounds than cortisol may be determined, but they are present usually in such low concentration that the procedure is considered specific for plasma cortisol. A more sensitive technique is one that measures fluorescence[49,51,76] in a sulfuric acid medium of corticosteroids with the 11-hydroxy configuration. Both cortisol and corticosterone have this pattern. A more recent and highly specific method[48,63] for determination of cortisol and corticosterone is based upon the steroid-binding properties of corticosteroid-binding globulin (CBG, transcortin). The technique is also referred to as competitive protein-binding radioassay.

Specimen collection

Two blood specimens are usually obtained from a patient, morning and afternoon. By venipuncture 10 ml is drawn into heparinized tubes. The morning specimen is drawn between 7 and 9 AM from a fasting patient and the afternoon specimen between 4:30 and 5 PM before the evening meal. Less interference from lipids will be encountered if the blood is obtained when the patient is in a fasting state or 3 or more hours after a meal. The time of drawing should be recorded because of the normal diurnal variation of cortisol secretion. The specimen is delivered to the laboratory without delay for separation of the plasma from the cells. If the plasma is to be stored before analysis, it should be kept frozen. However, with the fluorometric procedure no signifi-

cant changes in results have been noted upon storing for 5 days at 5° C and for 3 days at 25° C. This is not true for other methods of analysis.

Reagents needed

All chemicals are reagent grade.

Sulfuric acid, concentrated

Methylene chloride, spectro quality, redistilled

Ethyl alcohol, absolute, purified by procedure described for 17-ketosteroids

Petroleum ether

Sodium hydroxide, 0.1 N

Dissolve 4 g of sodium hydroxide in distilled water and dilute to 1 L.

Sulfuric acid–alcohol reagent

Place a flask in an ice bath with 1 volume of ethyl alcohol and add 3 volumes of concentrated sulfuric acid slowly with mixing. The reagent is stable for about a week. The blank reading should not exceed 20% of that obtained for the average normal specimen.

Stock standard

Dissolve 50 mg of hydrocortisone in 5 ml of absolute alcohol and dilute to 250 ml with distilled water (200 μg/ml).

Diluted stock standards

Dilute 10 ml of stock standard to 200 ml with distilled water (10 μg/ml).

Working standard (0.25 μg/ml)

Place 5 ml of the diluted stock standard in a 200 ml volumetric flask and dilute to volume with distilled water.

Procedure[49,51,64]

1. Into 40 ml glass-stoppered centrifuge tubes pipette 3 ml of plasma (unknown, control), 3 ml of distilled water, and 20 ml of petroleum ether. Shake vigorously for 30 seconds and centrifuge for 5 minutes. Discard petroleum ether by aspirating.

Extraction

2. Into 40 ml glass-stoppered centrifuge tubes pipette in duplicate
   Unknown: 2 ml of diluted-washed plasma
   Control: 2 ml of diluted-washed plasma
   Standard: 2 ml of working standard
   Blank: 2 ml of distilled water
3. Add to each tube 15 ml of methylene chloride. Shake vigorously for 30 seconds and centrifuge for 5 minutes. Discard the aqueous plasma by aspirating.

Washing

4. To each tube containing the methylene chloride extract add 1 ml of 0.1 N NaOH. Shake for 15 seconds and centrifuge for 5 minutes. Discard the NaOH layer immediately by aspirating.

Development of fluorescence

NOTE: Since development of fluorescence requires careful timing, the fluorometer, timer, mixer, centrifuge, glassware, and reagents must be set to prevent any delay in operations.

5. Into 40 ml glass-stoppered centrifuge tubes (marking duplicate tubes 1 and 2) pipette 8 ml of the methylene chloride extract. Divide into sets of four tubes.
6. Develop fluorescence in blank and standard tubes first. Set the clock at 5 minutes but do not start. Add 2.5 ml of sulfuric acid–ethanol reagent to blank tube 1 and start the clock. Continue with the addition of the reagent to blank tube 2, standard tube 1, and standard tube 2.
7. Stopper tubes and mix on a Vortex mixer for 10 seconds. Centrifuge

for 1 minute and stop the centrifuge, using the brake. Discard the methylene chloride (top) layer by aspirating.

8. Transfer the acid-alcohol layer to a fluorometer cuvette and determine fluorescence at exactly 5 minutes. Read blank tube 1, blank tube 2, standard tube 1, and standard tube 2 in that order. With a Turner fluorometer use as the primary filters 2A + 47B and as the secondary filter 2A−15. The instrument is set using distilled water.
9. Follow the same procedure as in steps 6 through 8 in developing fluorescence in the control and unknown tubes.

Calculations

$$\frac{F_u - F_b}{F_s - F_b} \times \text{Standard concentration} \times \frac{100}{\text{Volume plasma used}} =$$

μg 17-hydroxycorticosteroids/100 ml plasma (calculated as cortisol)

Comments

All glassware used in this procedure must be washed in 50% nitric acid.

de Moor and co-workers[49] reported that under the conditions described for this method 95% of the fluorescence resulted from cortisol and corticosterone. When read at 5 minutes, both corticoids showed the same fluorescence and interference from nonspecific substances was minimal. Washing with petroleum ether and sodium hydroxide removes some of the lipids, pigments, and other interfering material.

Fluorometric methods lack specificity and investigators have suggested various modifications to correct for or partially eliminate interfering substances extracted from plasma by methylene chloride. A reagent containing water, sulfuric acid, and ethanol has been used and fluorescence determined in 13 minutes.[60] The presence of two groups of fluorogens in the plasma extracts has been reported,[75] the one being nonspecific substances where fluorescence increased linearly with time and the other cortisol and corticosterone where fluorescence remained constant between 8 and 20 minutes after extraction into the acid reagent. By reading fluorescence after 8 and 16 minutes a corrected value for plasma corticosteroids was obtained. A reagent consisting of 30 N sulfuric acid has been used and fluorescence allowed to develop for 30 to 60 minutes.[73] To increase specificity, other workers have applied methods whereby cortisol and corticosterone are separated in water and carbon tetrachloride[77] or in benzene and water.[46] However, elimination of the problem of interference by nonspecific fluorescence has not been achieved.

In procedures measuring fluorescence elevated values will occur if patients are receiving drugs that are naturally fluorescent such as quinine, quinidine, Aureomycin or other tetracycline antibiotics, or vitamin B complex. Patients receiving contraceptive drugs may also show elevated values as the estrogens cause an increase in the corticosteroid-binding protein.

Normal values

In normal individuals the range for plasma corticosteroids drawn between 8 and 9 AM and assayed by the fluorometric method is 17 to 26 μg/100 ml. The level in the 5 PM specimen is approximately one half that at 8 AM.

de Moor and co-workers[49] reported the following plasma corticoid levels for normal hospitalized patients, 15 to 60 years of age, the blood drawn at the times indicated, as the mean ±S.E.M.:

8 AM	22.3 ± 0.27 μg/100 ml plasma	(673 subjects)
9 AM	19.5 ± 0.80	(30 subjects)
10 AM	16.2 ± 0.61	(30 subjects)
Noon	12.3 ± 0.25	(36 subjects)
8 PM	11.9 ± 1.2	(36 subjects)
Midnight	9.8 ± 0.4	(81 subjects)

These values were for persons who followed the evening sleep and day activity routine. A reversal of the day-night or activity-sleep schedule resulted in a reversal of the plasma cortisol pattern.[66] The erythrocyte corticoid level of blood drawn between 8 and 9 AM from 139 subjects was 6.90 ± 0.22 μg (S.E.M.) per 100 ml.[49]

Clinical findings

Plasma corticosteroid concentration increases as pregnancy progresses (largely because of the increased amount of corticosteroid-binding globulin), in conditions of stress such as shock and surgery, and with administration of drugs such as ACTH and vasopressin. The levels are high in Cushing's syndrome and there is no diurnal variation.

Plasma corticosteroid levels may be lowered by dexamethasone. They are usually low in patients with Addison's disease and panhypopituitarism.

17-Hydroxycorticosteroids, urinary

The major C-21 metabolites found in the urine are tetrahydrocortisol, tetrahydrocortisone, tetrahydro S, pregnanetriol, 11-ketopregnanetriol, cortols, cortolones, 6β-hydroxycortisol, and Δ^5pregnanetriol. These are conjugated principally with glucuronic acid although there are some sulfates. A small amount of free cortisol is also excreted.

There are two basic procedures in general use for the determination of 17-hydroxycorticosteroids. One method utilizes enzyme hydrolysis and the reaction of the 17-hydroxycorticosteroids with the phenylhydrazine–sulfuric acid reagent to measure Porter-Silber chromogens. The second method, a determination of 17-ketogenic steroids, employs sodium metaperiodate to convert corticosteroids to 17-ketosteroids for final assay by the Zimmermann reaction.

Specimen collection

A 24-hour urine collection is obtained preferably without a preservative and kept at about 2° to 4° C. If refrigeration is not possible, 15 ml of toluene may be added as preservative.

Porter-Silber chromogens[69,72,74]

The Porter-Silber chromogens are cortisol metabolites possessing a 17,21-dihydroxy acetone side chain at C-17 and they represent about one third of the urinary metabolites of cortisol. Steroids such as compound S (11-deoxycortisol) and its tetrahydro metabolite that are not metabolites of cortisol but have the dihydroxyacetone side chain are also measured.

Reagents needed

All chemicals are reagent grade.
Chloroform, redistilled
Sodium hydroxide, 0.1 N
Dissolve 4 g of sodium hydroxide pellets in distilled water and dilute to 1 L.

Phenylhydrazine hydrochloride, recrystallized

A product especially prepared for steroid analysis is available from Sigma Chemical Co. Otherwise, purify in the following manner. Add 100 g to 200 ml of warm (60° to 70° C) distilled water and stir frequently during the 1 to 3 hours required to dissolve the phenylhydrazine. Heat 1 L of ethyl alcohol to boiling and add to the dissolved phenylhydrazine in water. While hot, filter quickly through No. 2 filter paper. Cool in the refrigerator and collect the crystals on a sintered-glass filter (medium). Repeat the procedure twice, dissolving the crystals in less water each time. The last collection of crystals should be washed with cold ethyl alcohol and dried over $CaCl_2$ in vacuo. Determine the purity by a melting point measurement 240° to 243° C.

Sulfuric acid, 62% (22 N)

Add 310 ml of concentrated acid to 190 ml of distilled water. The flask should be kept cool under running cold water from a tap during addition of the acid.

Blank reagent

Prepare the quantity required for the daily batch by mixing 7.5 parts sulfuric acid reagent (62%) with 5 parts of methyl alcohol.

Methyl alcohol, absolute, spectro quality

Phenylhydrazine–sulfuric acid–methyl alcohol reagent

Dissolve 12 mg of phenylhydrazine hydrochloride in 12.5 ml of the blank reagent. Prepare fresh daily.

β-Glucuronidase (Sigma, type II)

Prepare the required amount daily by using 250 Sigma units/ml of 0.067 M phosphate buffer, pH 6.5.

Phosphate buffer, pH 6.5

Solution I

Dissolve 9.078 g of KH_2PO_4 in distilled water and dilute to 1 L, 0.067 M primary phosphate solution.

Solution II

Dissolve 9.465 g of Na_2HPO_4 in distilled water and dilute to 1 L, 0.067 M secondary sodium phosphate solution.

Prepare 1 L of 0.067 M phosphate buffer, pH 6.5, by mixing 650 ml of solution I with 350 ml of solution II.

Stock standard

Dissolve 50 mg of hydrocortisone in 5 ml of absolute ethyl alcohol and dilute to 250 ml with distilled water (200 μg/ml).

Working standard (10 μg/ml)

Dilute 10 ml of stock standard to 200 ml with distilled water.

Procedure

1. Measure total volume of 24-hour collection of urine and filter an aliquot.

Hydrolysis

2. Into 50 ml round-bottom centrifuge tubes pipette *in duplicate:*
   Unknown: 2.0 ml of filtered urine
   Control: 2.0 ml of filtered urine
3. Add to each tube 1.0 ml of β-glucuronidase solution and mix. Check the pH; it should be 6.5. If not, adjust using 0.1 N NaOH. Add 0.1 ml of chloroform.
4. Incubate for 18 to 24 hours at 37° C.

Extraction and washing

5. Into 50 ml round-bottom centrifuge tubes pipette in duplicate 2.0 ml of working standard, 10 μg/ml, and 1.0 ml of distilled water.
6. To all tubes from steps 4 and 5 add 25 ml of chloroform and shake for 15 seconds by hand. Discard the aqueous layer by aspirating.
7. Add 2.0 ml of 0.1 N NaOH, shake for 10 seconds, and centrifuge for 5 minutes. Discard the alkali wash by aspirating.

Color development

8. Into one set of 12 ml centrifuge tubes transfer from tubes of step 7 5.0 ml of chloroform extract. Add 1.0 ml of phenylhydrazine–sulfuric acid–methyl alcohol reagent and mix for 10 seconds on a Vortex mixer. Centrifuge for 5 minutes and discard the chloroform layer by aspirating.
9. Into a second set of 12 ml centrifuge tubes transfer from tubes of step 7, for blanks, 5.0 ml of chloroform extract. Add 1.0 ml of blank reagent and mix for 10 seconds on a Vortex mixer. Centrifuge for 5 minutes and discard the chloroform layer by aspirating.
10. Incubate tubes in a water bath at 60° C for 1 hour.
11. Allow to cool to room temperature.
12. Using capillary pipettes transfer the samples to micro-cuvettes.
13. Determine absorbance (A) at 370, 410, and 450 mμ against the blank reagent.
14. Subtract the blank readings from the color readings and determine corrected absorbance (A), using Allen corrections:

Calculations

$$\frac{\text{Corrected A unknown}}{\text{Corrected A standard}} \times \text{Standard concentration (mg)} \times \frac{\text{Total urine volume (ml)}}{\text{Aliquot urine used}} = \text{mg 17-hydroxycorticosteroids in 24 hours}$$

Comments. Hydrolysis is necessary to split the water-soluble glucosiduronidate complexes to free the steroids in a solvent-soluble form. Mild hydrolytic procedures such as enzyme hydrolysis with β-glucuronidase are used, since corticosteroids are very labile in hot mineral acids.

Chloroform or methylene chloride may be used as the solvent in the procedure. The wash with dilute sodium hydroxide removes some of the nonspecific substances that interfere with the final color development. However, it should be carried out rapidly to prevent loss of corticosteroids.

In the phenylhydrazine–sulfuric acid reagent the 17,21-dihydroxy-20-ketosteroids rearrange to form 21-aldehydes, which then react with the phenylhydrazine to yield a yellow phenylhydrazone having a maximum absorption at 410 mμ. Phenylhydrazine also reacts at the 3 position of cortisol and with other steroids without the dihydroxyacetone side chain but with the 3-keto-Δ^4 A ring. These products have absorption maxima about 340 mμ.

Some of the drugs that have been reported[45,47,59,78] to interfere with the determination are quinine, colchicine, paraldehyde, Dexedrine, Equanil (meprobamate), Librium (chlordiazepoxide HCl), Monase (etryptamine acetate), seconal, Serpasil (reserpine), Thorazine (chlorpromazine), Valmid (ethinamate), dexamethasone, ACTH, Atarax, and Mandelamine.

Normal values. de Moor and associates[50] reported that differences in body weight accounted for most of the sex differences in corticoid excretion. It was suggested that the values should be expressed in units of mg/100 kg/24 hr. Results may also be calculated per gram of creatinine with the concentration of the latter given as mg/kg/24 hr.

Values for 17-hydroxycorticosteroids in urine have been reported as Porter-Silber chromogens—for men 3 to 15 mg/24 hr and for women 2 to 13 mg/24 hr; or as 3.1 ± 1.1 mg/m^3/24 hr. The nomogram (Fig. 14-2) of DuBois gives the surface area from the subject's height and weight.

Clinical findings. Cushing's syndrome or hyperadrenocorticism is the result of hyperfunction of the adrenal cortex. This may be caused by adrenal hyperplasia, adenoma, or carcinoma. The clinical picture of the patient may reveal typical physical characteristics such as obesity with thin extremities, moon facies, thin skin with bruising and striae, hirsutism, menstrual disturbances, testicular atrophy, hypertension, skin pigmentation, mental depression, osteoporosis, and often diabetes mellitus. Usually the adrenals are enlarged and a tumor is found in about 30% of the patients.

In adrenal hyperplasia the primary source of the disorder may be either the pituitary or the adrenal cortex. Congenital cases, precocious genital development (male) and pseudohermaphroditism (female), are due to genetic defects that cause complete or partial absence of enzymes required for steroid synthesis. Several of the defects that have been recognized are a C-20 block with lipoid hyperplasia of the adrenals, 3β-hydroxysteroid dehydrogenase deficiency, C-11 block that leads to an increase in 11-deoxycortisol and 11-deoxycorticosterone, C-18 hydroxylation defect with decrease in aldosterone secretion, and C-21 block, the most common type, with increased excretion of 17-ketosteroids and pregnanetriol. The C-21 hydroxylase defect accounts for most of the patients with the syndrome of virilization alone. It is not complete, and in mild cases the levels of cortisol are normal or nearly normal. When the deficiency is severe, little or no cortisol is formed. Adult adrenal virilizing syndrome is usually unrecognized in males. In females, a woman previously normal shows development of masculine characteristics. These patients may have adrenal tumors or a delayed congenital adrenal hyperplasia with an associated increased excretion of 17-ketosteroids. Simple or idiopathic hirsutism is of unknown cause and usually does not appear to have the above described defects in cortisol synthesis.

Adrenocortical insufficiency or Addison's disease is characterized physically by weakness, hypotension, increased pigmentation, and sensitivity to stress. Some cases are the result of infections such as tuberculosis or neoplastic processes but mostly the cause is unknown.

In the liver most adrenal steroids are inactivated and conjugated with glucuronic or sulfuric acid. Therefore when there is liver damage, an increase in free and a decrease in conjugated adrenal steroids occur in the blood and urine. Kidney damage causes an accumulation of the conjugates in the blood and thereby influences the rate of conjugation taking place in the liver.

17-Ketogenic steroids

The most useful chemical measurement of hydroxycorticosteroid excretion is by the procedure for Porter-Silber chromogens or compounds with the 17,21-dihydroxy-20-keto configuration (Fig. 15-7). However this group does not include all 17-hydroxycorticosteroids and there are pathologic conditions where the total level is important. By chemical action all 17-hydroxycorticosteroids may be converted to and measured as 17-ketosteroids. These are referred to as 17-ketogenic steroids.[53,61,62]

Addition of borohydride reduces endogenous 17-ketosteroids to hydroxyl compounds, and sodium metaperiodate oxidizes the 17-hydroxycorticosteroids to ketosteroids. Hydrolysis of glucuronides occurs during periodate oxidation.

Reagents needed

All chemicals are reagent grade.
Ethyl ether (Squibb) for anesthesia or redistilled, peroxide-free
Ethyl alcohol, absolute, purified
Refer to reagents for 17-ketosteroids.

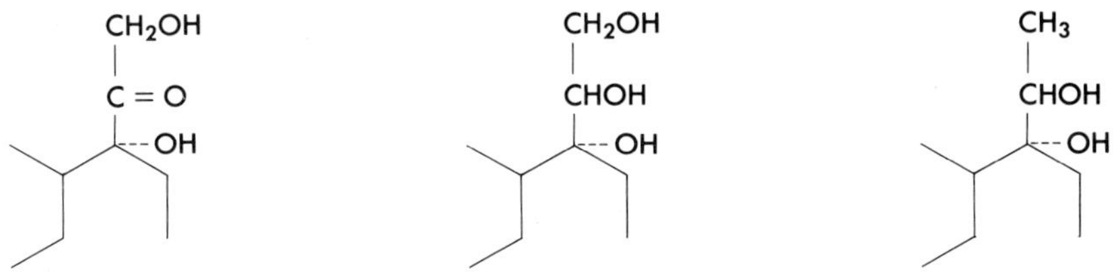

Dihydroxyacetone
Cortisol
Cortisone
Compound S

Glycerol
Cortol
Cortolone

17,20-glycol
Pregnanetriol

Fig. 15-7. Chemical groups at C-17 for 17-hydroxycorticosteroids and examples.

Sodium hydroxide, 4% (1 N), 40% (10 N)
Acetic acid, 5.7%, 25%
Sodium borohydride, 5% in 0.1 N NaOH, freshly prepared
Sodium metaperiodate, 10%, freshly prepared
Sodium dithionite (hydrosulfite, $Na_2S_2O_4 \cdot 2H_2O$)
m-Dinitrobenzene (Sigma)
Specially prepared for the Zimmermann reaction. Prepare fresh daily 2% solution in absolute ethyl alcohol.
Potassium hydroxide, 5.0 N (aqueous), standardized
Standard solution
Dissolve 20 mg of dehydroisoandrosterone in 100 ml of absolute ethyl alcohol (200 μg/ml)

Procedure

1. Measure total volume of 24-hour collection of urine and filter an aliquot.
2. Test for the presence of glucose with Uristix. If present, double the amount of borohydride for each 2% of glucose and increase accordingly the 25% acetic acid.

Reduction

3. Into 50 ml round-bottom centrifuge tubes pipette in duplicate:
   Unknown: 5 ml of filtered urine (2 ml diluted to 5 ml with distilled water when the 17-OHCS is above 15 mg/24 hr)
   Control: 5 ml of control urine
4. Adjust pH of samples to 7.0 and add 0.5 ml of 5% sodium borohydride reagent (freshly prepared) for glucose-free urine.
5. Allow to stand overnight at room temperature.
6. Add 0.25 ml of 25% acetic acid to decompose excess sodium borohydride for glucose-free urine. Allow to stand 15 minutes.

*Recovery**

Oxidation

7. Add 2 ml of 10% sodium metaperiodate (freshly prepared) for glucose-free urine and 8.0 ml for glucose-positive urine (6 ml of 10% sodium metaperiodate solution is approximately equivalent to 5 ml of a 2% glucose solution).
8. Neutralize by adding 0.8 ml of 1 N NaOH for glucose-free urine or 1.0 ml for glucose-positive urine (pH must be 6.7 to 7.0); check with indicator paper.

---

*Recoveries: Etiocholanolone and 11-β-hydroxyetiocholanolone (alcoholic solution) may be added to reduced urine after the destruction of the excess borohydride to test their stability and separation in the presence of other urinary constituents. C-21 steroids (alcoholic solution) may be added to the urine prior to borohydride reduction.

9. Incubate at 37° C for 1 hour and then add 0.12 ml of 10 N NaOH.
10. Incubate at 37° C for 15 minutes and cool to room temperature.

Extraction and wash

11. To each tube add 20 ml of ethyl ether and shake by hand for 30 seconds. Discard the aqueous layer by aspirating.
12. Add 10 ml of 10% NaOH solution containing approximately 5% sodium dithionite (hydrosulfite) and shake by hand for 10 to 15 seconds. Discard the NaOH wash layer by aspirating.
13. Wash twice with 10 ml of distilled water for 10 to 15 seconds and discard the aqueous layer by aspirating; after the final wash cover the tubes with foil and centrifuge for 5 minutes at 1500 rpm.

Color development

14. Into 40 ml round-bottom centrifuge tubes pipette:
    Unknown: 10 ml of ether extract
    Control: 10 ml of ether extract
    Standard (duplicate): 0.1 ml of standard solution
    Blank (duplicate): Dry tubes
15. Evaporate to dryness in a water bath at 40° to 45° C under a stream of air.
16. To the dried residue in each tube add 0.5 ml of alcoholic-KOH reagent (prepared immediately before use by mixing 2 parts absolute ethyl alcohol with 3 parts 5 N KOH) and 0.2 ml of 2% *m*-dinitrobenzene reagent.
17. Place tubes in a water bath at 25° C for 60 minutes. *Keep in dark,* covering with a black cloth.
18. Add to each tube 4.0 ml of 80% ethyl alcohol.
19. Measure absorbance (A) against the reagent blank at 480, 520 and 560 mμ.
20. Apply the Allen correction.

Calculations

$$\frac{\text{Corrected A unknown}}{\text{Corrected A standard}} \times \text{Standard concentration (mg)} \times \frac{\text{Total urine volume (ml)}}{\text{Aliquot urine in extract}} = \text{mg 17-ketogenic steroids/24 hr}$$

Comments. The two reagents that have been used to oxidize 17-hydroxycorticosteroids to 17-ketogenic steroids are sodium bismuthate and sodium metaperiodate. The poor reproducibility of results using bismuthate has been attributed to nonuniformity in oxidizing capabilities of the bismuthate from batch to batch and the hot acid hydrolysis required to complete the hydrolysis of all conjugates.

Sodium metaperiodate is a mild oxidant that eliminates the need for acid hydrolysis. It is stable and specific for oxidation of adjacent hydroxyl groups. The ketosteroids formed are mostly etiocholanolone, 11-hydroxyetiocholanolone, and a small amount of 11-hydroxyandrosterone. By the borohydride periodate method the color equivalent depends on the alkali used. With potassium hydroxide the 11-hydroxyetiocholanolone gives a color approximately the same as dehydroepiandrosterone.

When the sodium borohydride is added to the samples, the pH should not be less than 7 as the compound breaks down rapidly in an acid solution and slowly in a neutral solution. The exact pH is not critical and may vary between 7 and 10.

Normal values. For men, normal values range from 5 to 23 mg/24 hr and for women, from 3 to 15 mg/24 hr. Between the ages of 20 and 50 the maximum rate of excretion occurs and there is some decline with increasing age.

Clinical findings. The normal range appears to be broader than for the Porter-Silber chromogens, limiting the usefulness in diagnosis of Cushing's disease. Since this is a measure of the total 17-hydroxycorticosteroids, the determination is useful in those conditions where abnormal metabolites not measured by the Porter-Silber technique are found. For example, in some cases of carcinoma of the adrenal gland and adrenal hyperplasia elevated 17-ketogenic steroids are observed while the Porter-Silber chromogens are normal. Increased output has been noted in some cases of obesity, virilism, and precocious puberty of adrenocortical origin. A 17-ketogenic steroid level more than 5 mg/24 hr over the Porter-Silber chromogens is considered significant. Decreased levels usually accompany Addison's disease and hypopituitarism.

Pregnanetriol

Measurement of pregnanetriol is used for differential diagnosis of abnormal intermediary metabolism of adrenocortical steroids. It is derived from 17α-hydroxyprogesterone and is present in the urine as a glucuronide. Hydrolysis is effected with β-glucuronidase. The free steroid is extracted with methylene chloride, separated by chromatographing on a column of alumina, and eluted by using 10% ethyl alcohol in benzene. It is converted to a sulfuric acid chromogen and assayed by reading the peak absorption at 435 mμ.[44,57]

Reagents needed

All chemicals are reagent grade.

Sodium acetate buffer, pH 4.5, 1.0 M

Dissolve 28 g of sodium acetate trihydrate or 16.4 g of anhydrous sodium acetate in water and dilute to 200 ml. Adjust to pH 4.5 with glacial acetic acid.

Ketodase, beef liver β-glucuronidase solution, 5000 Fishman units/ml (Warner-Chilcott)

Methylene chloride, spectro quality

Benzene, spectro quality

Ethyl alcohol, absolute, purified, as described in the procedure for 17-ketosteroids

Ethyl alcohol, 2% in benzene

Ethyl alcohol, 10% in benzene

Sodium sulfite, anhydrous, granular

Aluminum oxide, alumina, activated, chromatographic, powdered, catalyst grade (Harshaw Chemical Co.)

Activate by heating at 350° to 400° C for 4 hours, cool to 100° C, and store in a desiccator. Prepare for chromatography by shaking 100 g of anhydrous alumina with 4.0 ml of distilled water (= 3.85%). Allow to equilibrate and store in a well-closed bottle to keep water content constant.

Sulfuric acid, concentrated

Sodium hydroxide, 0.1 N

Acetic acid, 50% aqueous solution (v/v)

5β-pregnane-3α,17α,20α-triol

Stock standard

Dissolve 20 mg in absolute ethyl alcohol and dilute to 100 ml (200 μg/ml). Store in an amber bottle at room temperature.

Working standard

Dilute 5 ml of stock standard with 15 ml of absolute ethyl alcohol (50 μg/ml). Prepare monthly and store in an amber bottle at room temperature.

Procedure

1. Measure total volume of 24-hour collection of urine and filter an aliquot.

Hydrolysis

2. Into 50 ml round-bottom centrifuge tubes pipette in duplicate:
   Unknown: 10 ml of filtered urine
   Control: 10 ml of control urine
3. Add to each tube 1 ml of acetate buffer and 1 ml of Ketodase solution. Mix on a Vortex mixer for 10 seconds and check pH. If pH is not 4.5, adjust with 50% acetic acid or 0.1 N NaOH.
4. Incubate at 37° C for 36 to 48 hours, or at 47° C for 18 to 24 hours.

Extraction and washing

5. Prepare two standard tubes containing 25 μg of pregnanetriol (0.5 ml of working standard) and 10 ml of distilled water.
6. To all tubes from steps 4 and 5 add 30 ml of methylene chloride and shake vigorously by hand for 3 minutes to effect the extraction. Discard the aqueous layer by aspirating. If a heavy emulsion occurs, centrifuge for 5 minutes.
7. Wash each extract twice with 10 ml of 0.1 N NaOH (three times if the urine is highly pigmented). Discard the NaOH wash by aspirating.
8. Rinse two times with 10 ml of distilled water. Centrifuge and discard the aqueous layer by aspirating.
9. Transfer 20 ml of the methylene chloride extract to a 40 ml round-bottom centrifuge tube and evaporate to dryness under a stream of air in a water bath at 35° to 45° C.

Chromatography

10. Use columns 300 × 10 mm fitted with Teflon stopcocks and having 50 ml reservoirs. With the stopcock closed, pour approximately 15 ml of benzene into each column and add alumina, small amounts at a time, allowing it to slowly drift to the bottom, to a height of 4 cm. Wash the sides of the column with 5 ml of benzene in order to wash down the alumina adhering to the glass. Place a plug of glass wool on top of the alumina. Fill the column with benzene, open the stopcock, and allow the benzene to flow through the column until the meniscus reaches the top of the alumina. *Never* allow the column to drain dry. Place a 125 ml Erlenmeyer flask under the column and cover with foil.
11. Dissolve the residue from step 9 in 5 ml of benzene, mix 15 seconds, using a Vortex mixer, and transfer quantitatively to a column. Rinse the tube with 5 ml of benzene and transfer to the column.
12. Allow the extract to pass through the column and rinse the column with 10 ml of benzene. Discard all benzene that has passed through the column.
13. Rinse column with 40 ml of 2% ethyl alcohol in benzene and discard this fraction.

Elution

14. Add 40 ml of 10% ethyl alcohol in benzene to the column and collect the eluate in a 50 ml round-bottom centrifuge tube.
15. Evaporate the eluate to dryness under a stream of air in a water bath at 35° to 45° C. Wash down the sides of each tube with 2 ml of ethyl alcohol and evaporate to dryness to remove all traces of benzene.

Color development

16. Prepare two standard tubes for recovery calculations by pipetting into 50 ml round-bottom tubes 0.5 ml of working standard (25 μg) and evaporating to dryness.
17. To all tubes from steps 15 and 16 add 15 mg of sodium sulfite and 3.5

ml of concentrated sulfuric acid. *Without mixing* allow the tubes to stand at room temperature for 30 minutes.

18. After the 30-minute period mix the contents in the tubes. Allow to stand an additional 60 to 90 minutes and determine absorbance (A) at 400, 435, and 470 mμ against concentrated sulfuric acid as the blank. If the concentration of pregnanetriol in the unknown is too high, dilutions may be made at this step, using concentrated sulfuric acid.
19. Apply the Allen correction.

Calculations

$$\frac{\text{Corrected A unknown}}{\text{Corrected A extracted standard}} \times \text{Standard concentration} \times \frac{\text{Total volume urine}}{\text{Aliquot urine used}} = \text{mg pregnanetriol/24 hr}$$

Comments. The method is specific. Enzyme hydrolysis is necessary, since acid hydrolysis destroys pregnanetriol. The color remains unchanged for 1½ to 2½ hours.

Clinical findings. Large quantities of pregnanetriol are present in urines of subjects with congenital adrenocortical hyperplasia secondary to 21-hydroxylase deficiency. The decreased elaboration of cortisol and increased ACTH secretion result in accumulation of pregnanetriol because of enzyme deficiency. Usually the Porter-Silber chromogens are within the normal range and the 17-ketosteroid excretion is increased. For normal values, by age, see Table 15-7.

Compound S (11-deoxycorticosteroid)

Compound S (11-deoxycortisol, cortexolone) is present only in very small quantities in normal blood and urine. Upon administration of Metopirone the 11β-hydroxylation of 11-deoxycortisol is inhibited, causing increased concentration of plasma 11-deoxycortisol and deoxycorticosterone and decreased plasma cortisol concentration. The lowered plasma cortisol level stimulates production of ACTH which is not controlled by 11-deoxycortisol. This provides a means of evaluating the feedback mechanism of the hypothalamic-pituitary-adrenal axis. Subjects with hypopituitary function will not show increased levels of ACTH and 11-deoxycortisol.

Although 11-deoxycortisol is determined with the Porter-Silber chromogens and the 17-ketogenic steroids, there is no way to designate an increase in the 17-hydroxycorticosteroid concentration as being caused by compound S. The tetrahydro derivatives of the 11-deoxycortisol and 11-deoxycorticosterone may be extracted with carbon tetrachloride and assayed by using the phenylhydrazine–sulfuric acid reagent.[56]

Table 15-7. Normal pregnanetriol values

Age (years)	Range (mg/24 hr)
Less than 6	0.0 to 0.2
6 to 16	0.3 to 1.1
Over 16	0.2 to 3.5

Reagents needed

Except for the following, the reagents are the same as those for Porter-Silber chromogens.

Carbon tetrachloride, reagent grade

Stock standard

Dissolve 10 mg of tetrahydro S in 5 ml of absolute ethyl alcohol and dilute to 100 ml with distilled water (0.1 mg/ml).

Working standard

Dilute stock standard 1:10 with distilled water (10 μg/ml).

Procedure

1. Measure total volume of 24-hour collection of urine and filter an aliquot.

Hydrolysis

2. Into 50 ml round-bottom centrifuge tubes pipette in duplicate:
   Unknown: 2.0 ml of filtered urine
   Add 1.0 ml of β-glucuronidase solution and mix. Check the pH; it should be 6.5 (indicator paper). If not, adjust by using 0.1 N NaOH. Add 0.1 ml of chloroform.
3. Incubate for 18 hours at 37° C.

Extraction and washing

4. Into 50 ml round-bottom centrifuge tubes pipette in duplicate 2.0 ml of working standard, 10 μg/ml, and 1.0 ml of distilled water.
5. To all tubes from steps 3 and 4 add 25 ml of carbon tetrachloride and shake for 15 seconds by hand. Centrifuge for 5 minutes and discard the aqueous layer by aspirating.
6. Add 2.0 ml 0.1 N NaOH, shake for 10 seconds, and centrifuge for 5 minutes. Discard the alkali wash by aspirating.

Color development

7. Into 12 ml centrifuge tubes transfer 5.0 ml of carbon tetrachloride extract and add 1.0 ml of phenylhydrazine–sulfuric acid–methyl alcohol reagent; mix for 10 seconds on a Vortex mixer.
8. Centrifuge for 5 minutes and discard the carbon tetrachloride layer by aspirating.
9. Incubate tubes in a water bath at 60° C for 1 hour.
10. Allow to cool to room temperature and transfer samples to microcuvettes, using capillary pipettes.
11. Determine absorbance (A) at 370, 410, and 450 mμ against a blank of phenylhydrazine–sulfuric acid reagent.
12. Determine corrected absorbance (A), using Allen correction.

Calculations

$$\frac{\text{Corrected A unknown}}{\text{Corrected A standard}} \times \text{Standard concentration (mg)} \times \frac{\text{Total urine volume (ml)}}{\text{Aliquot urine used}} = \text{mg compound S in 24 hr}$$

Clinical findings. Following an infusion of Metopirone, normal subjects excreted in 24 hours more than 8.0 mg of compound S metabolites whereas patients with untreated panhypopituitarism and adrenal insufficiency causing 11β-hydroxylation deficiency excreted less than 2.0 mg.

Adrenocortical function tests

Many factors influence the production and secretion of steroids. When functioning normally the adrenal cortex is largely controlled by the ACTH secretion of the pituitary gland and a balance is maintained between the level of ACTH and cortisol in the blood.

In establishing or confirming a diagnosis of hypofunction or hyperfunction of the adrenal cortex, compounds that stimulate or suppress the activity are used in combination with assays of blood and urine for steroids. The three chemicals employed routinely for this purpose are ACTH, Metopirone, and dexamethasone.

ACTH stimulation test

The ACTH stimulation test provides a specific measure of the response of the adrenal gland to stimulation by adrenocorticotropin or of the functional cortisol reserve. There are several different procedures but the 8-hour intravenous method appears to be the one generally employed.

Two 24-hour urinary collections are obtained prior to and one collection on the day following the test for baseline 17-ketosteroids and 17-hydroxycorticosteroids or 17-ketogenic steroids. The test consists of adding 25 IU of ACTH to 500 ml of normal saline and administering the solution to a patient continuously by infusion for *exactly* 8 hours. The timing for the infusion is critical. Urine is collected during the administration and is continued for a total 24-hour period. It is assayed for 17-hydroxycorticosteroids or 17-ketogenic steroids.

For a normal subject the urinary steroid level is raised threefold to fivefold above the baseline. There is no increase of excretion in Addison's disease or primary adrenocortical failure, very little increase over the already high concentrations for patients with Cushing's syndrome or adrenocortical tumors, and little if any response in congenital adrenal hyperplasia. There is a definite increase in adrenocortical hyperplasia caused by pituitary tumors.

Rapid ACTH screening procedures have been proposed whereby plasma is assayed for cortisol before and following intravenous infusion of 25 IU of the drug. There is a marked increase in blood cortisol levels for normal subjects but in adrenal insufficiency or Addison's disease the response is very slight.

Metopirone test

Metopirone (metyrapone, SU 4885), an inhibitor, tests the response of the pituitary gland to a decrease in circulating cortisol. It selectively inhibits 11β-hydroxylase, the enzyme for 11-hydroxylation of steroids, and thus partially blocks the pathway to cortisol. The cortisol level is reduced and because of the feedback mechanism ACTH secretion is increased. The inhibition of 11β-hydroxylase and stimulation of ACTH secretion result in an increased production of 11-deoxycortisol (compound S) which is measured in the urine.

The dose of Metopirone usually administered is 750 mg given orally every 4 hours for 6 doses. A 24-hour urinary collection is obtained before the test for the baseline level of 17-hydroxycorticosteroids or 17-ketogenic steroids, as well as a collection during drug administration that is continued for a total 24-hour period. For the specimen started at the same time as the test, a normal subject will show a twofold or fourfold increase in corticosteroid excretion over the baseline value. It is preferable to determine the concentration of the metabolite of 11-deoxycortisol, tetrahydro S. No response will be observed when there is diminished adrenocortical reserve or adrenocortical tumor. In adrenal hyperplasia the output is increased.

Dexamethasone suppression test

An adrenal suppression test may be employed to differentiate Cushing's syndrome caused by an adrenal tumor from adrenal hyperplasia. ACTH secretion is suppressed with a synthetic glucocorticoid, dexamethasone. The drug is administered orally in doses of 0.5 mg at 6-hour intervals on two successive days and 2.0 mg at 6-hour intervals on the two following days. Prior to the test two 24-hour baseline control urine specimens are collected and a specimen on each second day of suppressive therapy (0.5 and 2.0 mg, respectively) for measurement of 17-hydroxycorticosteroid or 17-ketogenic steroid levels.

Normal subjects show a decrease of about one half in urinary corticosteroid levels after administration of 0.5 mg of dexamethasone. In bilateral adrenal hyperplasia suppression will be observed only after repetition of the procedure with the larger dose. Normal subjects show suppression to less than 3 mg/day on 2 mg doses but patients with Cushing's syndrome do not suppress. No suppression even with larger doses is obtained in cases of adrenocortical tumors.

An overnight suppression has been reported[65] as a rapid screening test for Cushing's syndrome whereby 1 mg of dexamethasone is administered orally between 11 and 12 PM and the cortisol is measured in an 8 AM blood specimen of the following morning. There is a question as to the cortisol level at which Cushing's may be excluded. Sawin and associates[71] have suggested the values of 5 to 6 μg/100 ml, or less. Levels obtained for acutely ill patients and those taking estrogens may be misleading.

Adrenal medulla

The hormones secreted by the adrenal medulla are important to the clinical laboratory because chemical measurement of these hormones is vital in the diagnosis of adrenal medullary tumors. Pheochromocytoma, a tumor that develops in chromaffin tissue, is detected through the increased amounts of vasoactive pressor substances that are released and excreted. Although this is not a common pathologic condition, it is of clinical importance because death may result if the tumor is not removed.

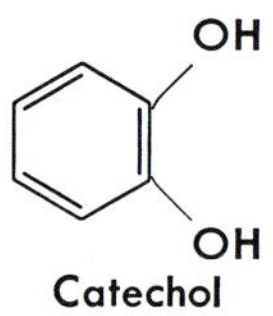

Catechol

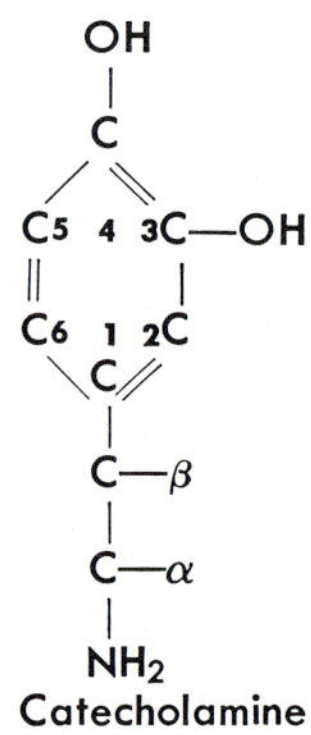

Catecholamine

The adrenal medulla secretes a group of hormones referred to as catecholamines—substances that contain a catechol nucleus and an amine. These are mostly dopamine and its metabolic products epinephrine (adrenaline) and norepinephrine (noradrenaline). They are synthesized from tyrosine (Fig. 15-8), which is hydroxylated to form dopa (dihydroxyphenylalanine) and then decarboxylated to yield dopamine (dihydroxyphenylethylamine, hydroxytyramine). The side chain of dopamine is hydroxylated to give norepinephrine and a methyl group is introduced into the amino group to produce epinephrine. The conversion of norepinephrine to epinephrine is limited to those cells which have available phenylethanolamine-N-methyl transferase and this occurs almost exclusively

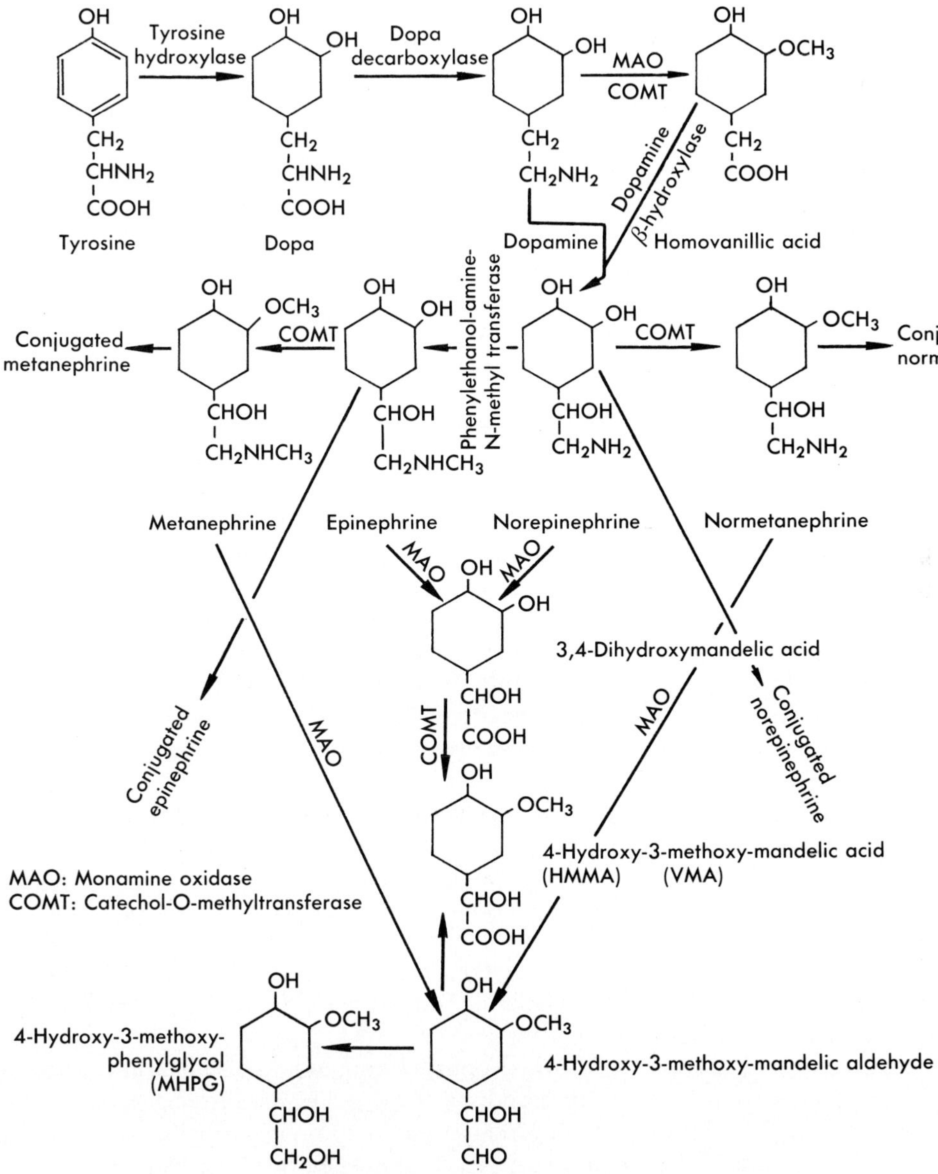

Fig. 15-8. Synthesis and metabolism of catecholamines.

in the adrenal glands. Epinephrine has been called the emergency hormone, since production and secretion increase rapidly during conditions of stress.

Catecholamines are also synthesized in the brain, sympathetic ganglia, and in peripheral tissue sympathetic nerve endings, and chromaffin cells. In the adrenal medulla and sympathetic nerve endings are granules or granulated vesicles that act as storage units for catecholamines. There is continuous secretion of small amounts of catecholamines but the largest quantities are released in spurts in response to various nervous and physiologic stimuli. They are rapidly metabolized, most catecholamines remaining in the blood for only a few circulations, and less than 5% are excreted in the urine in free or conjugated forms. In urine, the metabolite of epinephrine and norepinephrine in highest concentration is 4-hydroxy-3-methoxymandelic acid. Metanephrine and normetanephrine in the conjugated forms are also present in easily measurable quantities. Other metabolites that have been found in the urine are 3,4-dihydroxymandelic acid and 4-hydroxy-3-methoxy-phenylglycol. The metabolite of dopamine, 4-hydroxy-3-methoxy-phenylacetic acid (homovanillic acid, HVA), is useful in detecting dopamine-secreting tumors such as neuroblastomas.

Adrenal medullary function may be assessed by determining the levels of catecholamines and their metabolites in urine specimens. Since synthesis is controlled by the rate of secretion, measurement of the metabolites provides an index of catecholamine secretion or synthesis. The determinations are employed routinely as an aid in the diagnosis of pheochromocytoma.

Catecholamines

Urinary catecholamines usually refers to a measurement of the free epinephrine and norepinephrine in a timed (24-hour) specimen of urine. Although these adrenal hormones are excreted in both the free and conjugated forms, determinations that include the conjugated forms have greater interference from nonspecific substances. A random urine sample taken after an episode of hypertension may have an elevated value that would be undetected because of dilution in a 24-hour specimen. Blood levels may also be measured but the information can be misleading because of intermittent secretion of catecholamines by tumors. These analyses are of value when location of a tumor is not possible by any other means, and for this purpose blood is obtained from veins draining a suspected area.

Fluorometric methods that are specific for catecholamines are generally employed for their measurements. Two different chemical procedures have been developed for formation of fluorescent compounds of adrenaline and noradrenaline: (1) conversion to trihydroxyindole (THI) derivatives and (2) condensation with ethylenediamine (EDA) to form quinoxaline derivatives. The THI method is more specific and is used for routine analysis. The necessary preliminary purification of the catecholamines from urine is carried out using alumina or ion exchange chromatography.

The catecholamines are adsorbed onto alumina, eluted with acetic acid, oxidized by potassium ferricyanide to adrenochrome and noradrenochrome, and rearranged to adrenolutine and noradrenolutine in a strongly alkaline solution. The latter compounds give a yellowish green fluorescence in ultraviolet light.[79,81,84,87]

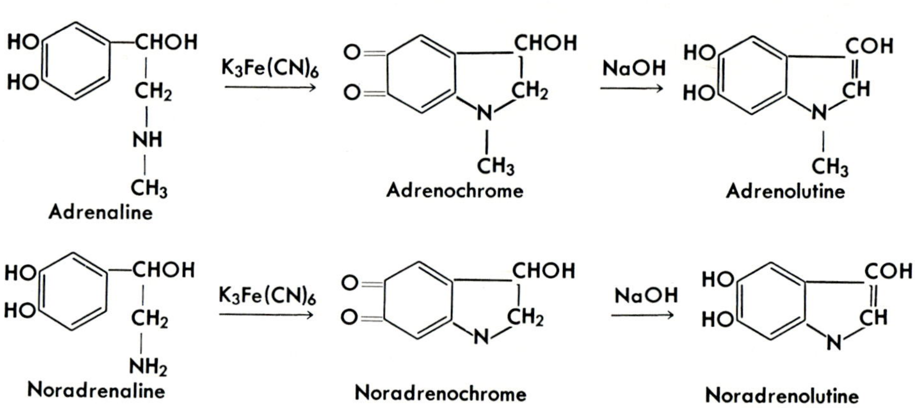

Specimen collection

A 24-hour urine specimen is collected in a glass bottle containing 15 ml of 6 N hydrochloric acid to provide a final pH below 3.0. At this pH urinary catecholamines are stable for several days at room temperature and for months in the refrigerator. The urine must be kept acid during the entire collection as catecholamines deteriorate rapidly in neutral or alkaline solutions. If a single voided specimen is collected during or immediately after an episode of hypertension it should be preserved at a pH between 2 and 3 with a small amount of hydrochloric acid.

Reagents needed

All chemicals are reagent grade.

Alumina

(Brockman, activity I, 80-200 mesh for chromatography) Wash with distilled water to remove the fine dust, dry at 100° C, and heat for 2 hours at 200° C.

Disodium ethylenediamine tetraacetate (EDTA)

Sodium hydroxide, 10 N (40%)

Sodium hydroxide, 5 N (20%)

Sodium hydroxide, 0.5 N (2%)

Acetic acid, CH_3COOH, 0.2 M

Dilute 11.5 ml of glacial acetic acid to 1 L with distilled water.

Potassium ferricyanide, $K_3Fe(CN)_6$, 0.25%

Dissolve 250 mg in water and dilute to 100 ml. Prepare fresh each week and store in the refrigerator.

Disodium hydrogen phosphate buffer, 0.5 M, Na_2HPO_4

Dissolve 35.5 g of the anhydrous salt in water and dilute to 500 ml. Store in the refrigerator.

Zinc sulfate, $ZnSO_4$, 0.25%

Dissolve 250 mg in water and dilute to 100 ml.

Ascorbic acid–sodium hydroxide reagent

Prepare immediately before using: 20 mg of ascorbic acid + 0.2 ml of distilled water + 10 ml of 10 N NaOH.

Hydrochloric acid, 0.1 N

Dilute 8.6 ml of concentrated (11.6 N) hydrochloric acid to 1 L.

Hydrochloric acid, 6 N

Dilute concentrated hydrochloric acid 1:1.

Stock standard (100 μg norepinephrine/ml)

Dissolve 20.1 mg of *l*-arterenol bitartrate in 0.1 N HCl and dilute to 100 ml. Stable for at least 6 months in the refrigerator.

Working standard (0.25 μg norepinephrine/ml)

Into a 200 ml volumetric flask pipette 0.5 ml of stock standard solution and 20 ml of 0.1 N HCl. Dilute to volume with water. Prepare fresh each week.

Procedure

1. Measure the total volume of urine and filter an aliquot.
2. Into 50 ml beakers (in duplicate) pipette
   Unknown: 25 ml of filtered urine
   Control: 25 ml of filtered urine control
3. Add to each beaker 200 mg of EDTA.
4. Adjust urine to pH 8.5, using 5 N NaOH to pH 7 to 7.5 and 0.5 N NaOH for the final adjustment, while stirring with a magnetic stirrer. A pH meter should be used although narrow-range pH paper may be employed.

Adsorption and elution

5. Prepare chromatographic columns, using 8 × 300 mm columns having a 50 ml reservoir, by adding dry alumina to a height of 4 cm. Cover with a small plug of glass wool.

6. Pour the urine onto the columns. It should flow through at a rate of 1 to 2 ml/min.
7. Wash each column with 20 ml of distilled water.
8. Elute the catecholamines with 20 ml of 0.2 M acetic acid.

Oxidation and fluorescence

9. Prepare 15 × 180 mm test tubes as follows:
   Reagent blank (duplicate): 2.0 ml of 0.2 M acetic acid + 1.0 ml of distilled water.
   Standard (duplicate): 2.0 ml of 0.2 M acetic acid + 1.0 ml of working standard.
   Tests and control (three tubes each):
      Tube 1: 2.0 ml of eluate + 1.0 ml of distilled water
      Tube 2 (blank): 2.0 ml of eluate + 1.0 ml of distilled water
      Tube 3 (internal standard): 2.0 ml of eluate + 1.0 ml of working standard
   Group tubes into sets not to exceed twelve tubes (two determinations in duplicate).
10. To each tube add 0.8 ml of phosphate buffer, pH 8.9, to give a pH of 6.5 in the tubes (check the pH, using narrow-range paper) and 0.1 ml of zinc sulfate reagent, 0.25%.
11. To tubes containing reagent blank, control blank, and test blank add 1.0 ml of ascorbic acid–NaOH reagent.
12. To tubes containing standard, control, and test (including internal standard tubes) add 0.1 ml of potassium ferricyanide reagent and allow to stand *exactly 2 minutes* (use an interval timer); then add 1.0 ml of ascorbic acid–NaOH reagent, mix, and read the fluorescence within 10 minutes.
13. To blank tubes containing ascorbic acid–NaOH reagent add 0.1 ml of potassium ferricyanide reagent and mix.

Measurement and calculation

14. Read fluorescence in a fluorometer or spectrophotofluorometer against a water blank. For filter fluorometers use an activation wavelength of 405 mμ and a fluorescent wavelength of 510 mμ.

$$\frac{F_u}{F_{std}} \times \text{Standard concentration} \times \frac{\text{Volume eluate}}{\text{Aliquot eluate}} \times \frac{\text{Total volume urine}}{\text{Aliquot urine}} =$$

$$\frac{F_u}{F_{std}} \times 0.25 \times \frac{20}{2} \times \frac{TV}{25} = \mu g \text{ catecholamines in 24-hour specimen or volume received (calculated as norepinephrine)}$$

where F_u = fluorescence of unknown (sample—blank)

F_{std} = fluorescence of standard (standard—sample)

Comments

Adsorption occurs rapidly at pH 8.0 to 8.5 and, since catecholamines are unstable at this pH, the time for passage through the columns must be as brief as possible. Usually less than 15 minutes is required.

If urine acidified below pH 1.5 is maintained at room temperature, hydrolysis of the conjugated catecholamines may occur resulting in increased values for free catecholamine determinations.

Some of the catecholamine derivatives, if present, will also be adsorbed on the alumina. However, acetic acid will not elute dopa and with the use of potassium ferricyanide as the oxidizing agent no interference is obtained from dopamine, normetanephrine, and metanephrine.

The addition of EDTA prevents the formation of a gelatinous calcium-magnesium-phosphate precipitate that will interfere with the reading of fluorescence.

Interference will occur from fluorescent drugs in the urine if a patient is receiving Aureomycin or other tetracycline antibiotics, quinine, quinidine, or large amounts of the vitamin B complex.

Exogenous catecholamines prescribed as vasopressors, bronchodilators, and nasal decongestants may elevate the levels in the urine. Drugs that have been reported to interfere with the determination are acetylcholine, amphetamine, chlorpromazine, cocaine, ephedrine, epinephrine, guanethidine, isoproterenol (Isuprel), mandelamine, metanephrine, methyldopa (Aldomet), reserpine, tyramine, and Uritone (Urotropin, methenamine). Biliary pigment in urine causes interference. For patients with pheochromocytoma catecholamines may be depressed during the early administration of antihypertensive drugs. Increased catecholamine excretion may be observed following vigorous exercise and in cases of progressive muscular dystrophy and myasthenia gravis.

Fractionated catecholamines: epinephrine and norepinephrine

After determining the presence of a pheochromocytoma, a quantitative differential determination of urinary epinephrine and norepinephrine may provide an indication of the site of the tumor.[82,83] The location of the lesion will be usually in the adrenal gland if both of the amines are increased. If the increase is mostly norepinephrine, the tumor may be expected to be in or near one of the adrenal glands in about two thirds of the cases and extra-adrenal for the remainder.

Procedures that are available for differential determination of norepinephrine and epinephrine are paper chromatography, column separation, oxidation at different pH, and application of different activation and fluorescence spectra of the lutines. The use of two different filter combinations eliminates two oxidation steps required when two pH levels are employed.[80] The concentration of each amine is calculated by applying simultaneous equations.

Reagents needed

Same as for catecholamines
Standard solution No. 2:
Stock standard (100 μg epinephrine/ml)
  Dissolve 18.2 mg of *l*-epinephrine *d*-bitartrate in 0.1 N HCl and dilute at 100 ml. Stable for at least 6 months in the refrigerator.
Working standard (0.25 μg epinephrine/ml)
  Into a 200 ml volumetric flask pipette 0.5 ml of stock standard solution and 20 ml of 0.1 N HCl. Dilute to volume with distilled water. Prepare fresh each week.

Procedure

1 to 8. Same as for catecholamines.
9. Prepare 15 × 180 mm test tubes as follows:
  Reagent blank (duplicate): 2.0 ml of 0.2 M acetic acid + 1.0 ml of distilled water
  Standard norepinephrine (duplicate): 2.0 ml of 0.2 M acetic acid + 1.0 ml of NE working standard
  Standard epinephrine (duplicate): 2.0 ml of 0.2 M acetic acid + 1.0 ml of E working standard
  Tests and controls (four tubes each)
    Tube 1: 2.0 ml of eluate + 1.0 ml of distilled water
    Tube 2 (blank): 2.0 ml of eluate + 1.0 ml of distilled water
    Tube 3 (NE internal standard): 2.0 ml of eluate + 1.0 ml of NE working standard

Tube 4 (E internal standard): 2.0 ml of eluate + 1.0 ml of E working standard

10 to 13. Same as for catecholamines.

Measurement and calculation

14. Using a Turner fluorometer and two different sets of filters, read the fluorescence of the solution in all tubes against a water blank.

	Filter set I	Filter set II
Primary	405	2A and 47B
Secondary	2A-12	2A-15

The following symbols may be used for setting up the equations:

B = Test or control blank
U = Test or control
N = Test or control + NE internal standard
A = Test or control + E internal standard

Filter set I: $n_I = (N_I - U_I) \times 4$ = Fluorescence/μg NE added
Filter set II: $n_{II} = (N_{II} - U_{II}) \times 4$ = Fluorescence /μg NE added
Filter set I: $a_I = (A_I - U_I) \times 4$ = Fluorescence/μg E added
Filter set II: $a_{II} = (A_{II} - U_{II}) \times 4$ = Fluorescence/μg E added
Filter set I: $u_I = U_I - B_I$
Filter set II: $u_{II} = U_{II} - B_{II}$

$$Y = \frac{u_I a_{II}/a_I - u_{II}}{n_I a_{II}/a_I - n_{II}} = \text{Norepinephrine in 2.5 ml urine}^*$$

$$X = \frac{u_{II} - Yn_{II}}{a_{II}} = \text{Epinephrine in 2.5 ml urine}^*$$

$Y/2.5 \times$ total volume = μg norepinephrine/24 hr

$X/2.5 \times$ total volume = μg epinephrine/24 hr

4-Hydroxy-3-methoxymandelic acid (HMMA) (VMA)

The quantitative determination of the catecholamine metabolite 4-hydroxy-3-methoxymandelic acid[86] provides a more reliable measure of catecholamine secretion than epinephrine and norepinephrine, since it is present in larger quantities in urine. Some of the urinary pigments are removed by adsorption on activated magnesium silicate; the HMMA is extracted with ethyl acetate and oxidized to vanillin, using periodate. By extracting the oxidized mixture with toluene many of the interfering substances are left in the aqueous layer. The vanillin is extracted from the toluene by a potassium carbonate solution.

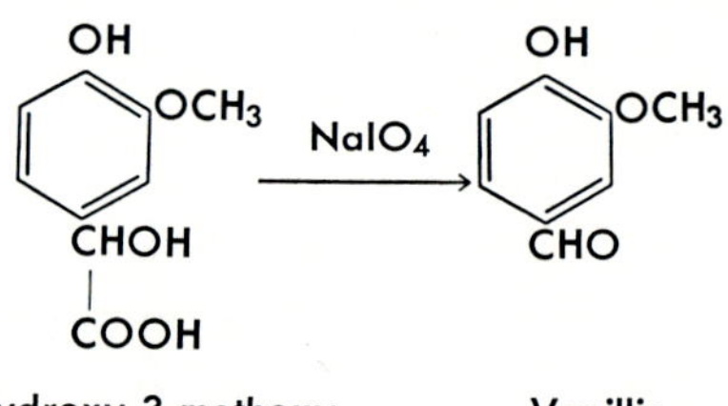

4-Hydroxy-3-methoxy-mandelic acid   Vanillin

Specimen collection

A 24-hour urine specimen is collected in a glass bottle containing 15 ml of 6 N hydrochloric acid. It is stored in the refrigerator.

---

*Assuming that a standard containing 0.25 μg/ml and an equivalent of 2.5 ml of urine are used in the procedure. If other quantities are employed, substitute those values in the equations.

Reagents needed

All chemicals are reagent grade.
Sodium metaperiodate, $NaIO_4$, 2% in distilled water
Stable for 1 week at 25° C.
Sodium metabisulfite, $Na_2S_2O_5$, 10% in distilled water
Prepare fresh each day.
Hydrochloric acid, 6 N
Dilute concentrated hydrochloric acid 1:1 with distilled water
Potassium carbonate, 1 M
Dissolve 138.2 g of anhydrous K_2CO_3 in distilled water and dilute to 1 L.
Hydrochloric acid, 0.01 N
Dilute 0.86 ml of concentrated hydrochloric acid to 1 L.
Acetic acid, 5 N
Dilute one part of glacial acetic acid with two parts of distilled water.
Potassium dibasic phosphate, K_2HPO_4, 3 M
Dissolve 52.2 g in distilled water and dilute to 100 ml, pH 7.5.
Ethyl acetate
Toluene
Sodium chloride
Stock standard (1 mg/ml)
Dissolve 50 mg of 4-hydroxy-3-methoxymandelic acid in 0.01 N HCl and dilute to 50 ml. The solution is stable for at least 3 months at 3° C.
Working standard (10 μg/ml)
Dilute the stock standard 1:100 with distilled water. Store in refrigerator and prepare fresh each week.

Procedure

Extraction of 4-hydroxy-3-methoxymandelic acid

1. Treat approximately 20 ml of unknown and control urine specimens with a small amount of Florisil in a beaker and filter.
2. Into 50 ml polypropylene centrifuge tubes with caps (in duplicate) place:
   Unknown and control: 4 ml of urine and 2 ml of distilled water
   Reagent blank: 6 ml of distilled water
   Standard: 2 ml of working standard and 4 ml of distilled water
3. To each tube add 0.5 ml of 6 N HCl; NaCl (solid) to saturate, approximately 3 g; and 30 ml of ethyl acetate.
4. Mix on a Vortex mixer for 1 minute.
5. Centrifuge for 5 minutes at 1500 rpm.
6. Transfer 20 ml of the ethyl acetate layer (top) to another 50 ml plastic centrifuge tube and add 3 ml of 1 M K_2CO_3.
7. Mix on a Vortex mixer for 30 seconds.
8. Centrifuge for 5 minutes at 1500 rpm.
9. Discard ethyl acetate layer (top) by aspirating.
10. Into 50 ml glass centrifuge tubes with stoppers transfer 1.0 ml of K_2CO_3 layer. Set up unknown blanks and control blanks for each specimen, using 1.0 ml of the K_2CO_3 layer.

Oxidation

11. To tubes with extract from unknowns, controls, reagent blanks, and standards add 0.1 ml of sodium metaperiodate solution. To tubes with extract for unknown and control blanks add 0.1 ml of sodium metabisulfite solution.
12. Incubate at 50° C for 30 minutes.
13. To the tubes containing periodate add 0.1 ml of sodium metabisulfite solution to reduce residual periodate. To the tubes incubated with the bisulfite add 0.1 ml of sodium metaperiodate.

Extraction of vanillin

14. To each tube add 0.3 ml of 5 N acetic acid, 0.6 ml of 3 M potassium phosphate buffer (check the pH with narrow-range paper; it should be between 7.5 and 8.0), and 15 ml of toluene.
15. Mix with a Vortex mixer for 30 seconds.
16. Centrifuge for 5 minutes at 1500 rpm.
17. Into a 40 ml glass centrifuge tube place 10 ml of toluene layer and 1.5 ml of 1 M K_2CO_3 solution.
18. Mix on a Vortex mixer for 30 seconds.
19. Centrifuge for 5 minutes and discard the toluene layer (top) by aspirating.

Measurement and calculation

20. Transfer the potassium carbonate layer to a microcuvette and determine the absorbance (A) at 360 mμ against water.

$$\frac{\text{A unknown} - \text{A unknown blank}}{\text{A standard} - \text{A reagent blank}} \times \text{Standard concentration} \times$$

$$\frac{\text{Total volume}}{\text{Aliquot used}} = \frac{\text{A unknown} - \text{A unknown blank}}{\text{A standard} - \text{A reagent blank}} \times 0.02 \times$$

$$\frac{\text{Total volume}}{4} = \text{mg 4-hydroxy-3-methoxymandelic acid (VMA)/volume}$$

Comments

The absorbance is determined at 360 mμ instead of 347 to 350 mμ, the vanillin peak, because many urines contain a substance that is oxidized by periodate to give an absorption peak about 347 mμ.

Procedures based upon conversion of 4-hydroxy-3-methoxymandelic acid to vanillin appear to have better specificity and reproducibility than some of the other methods employed for routine screening. They are relatively free from interference by many of the drugs and dietary constituents that affect other techniques. A thin-layer chromatographic procedure for VMA is described in Chapter 19.

Some of the widely used screening tests, including those employing the diazo color reaction, lack specificity and may provide misleading elevated values. These are subject to interference by diet (coffee, tea, vanilla, bananas, citrus fruits) and by drugs such as aspirin, sulfa compounds, morphine, Aldactone, chlorpromazine, pentobarbital, and reserpine.

Metanephrines

Included in the group of major metabolites of catecholamines are metanephrine and normetanephrine. They are excreted mainly as conjugated derivatives and their assay provides another measurement of catecholamine secretion. The procedure[85] involves hydrolysis, adsorption of the metabolites on Amberlite CG-50, elution with ammonium hydroxide, oxidation to vanillin by periodate, and final determination as total metanephrine.

Specimen collection

A 24-hour urine specimen is collected in a glass bottle containing either 15 ml of 6 N hydrochloric acid or 15 ml of toluene.

Reagents needed

All chemicals are reagent grade.

Sodium metaperiodate, $NaIO_4$, 2%
Stable for 1 week at 25° C or for 1 month at 4° C.

Sodium metabisulfite, $Na_2S_2O_5$, 10%
Prepare fresh each day.

Hydrochloric acid, 6 N

Dilute concentrated hydrochloric acid 1:1 with distilled water.

Ammonium hydroxide, NH_4OH, 4 N

Dilute 28 ml of concentrated (14.3 N) ammonium hydroxide to 100 ml with distilled water.

Amberlite CG-50, 100-200 mesh, pH 6.0 to 6.5

Suspend 100 g of the resin in 300 ml of distilled water, stir for 10 minutes, allow to settle for 30 minutes, and discard the supernatant solution. The washing and settling procedure is repeated (approximately five times) until the supernatant is clear after a 15-minute settling period. The resin is then cycled through the sodium form. It is suspended in 300 ml of distilled water and over a period of 15 minutes 200 ml of 10 N NaOH is added with constant stirring and the stirring continued for 2 hours. The resin is allowed to settle for 15 minutes, the supernatant is decanted, and the resin is washed with distilled water several times by decantation. It is converted back to the acidic form by adding 500 ml of 6 N HCl and stirring for 30 minutes. The excess acid is removed by washing several times with distilled water and decanting the supernatant. The resin is reconverted to the sodium form in the manner described above. The washed sodium form of the resin is suspended in 100 ml of distilled water, and with continuous stirring glacial acetic acid is added until the solution pH is 6.0 to 6.5. The pH must remain constant during 30 minutes of continuous stirring. The resin is added to the columns in this form. After use, the resin may be regenerated by converting to the acidic form and then the sodium form.

Stock standard (100 μg/ml)

Dissolve 10 mg of metanephrine or normetanephrine in distilled water and dilute to 100 ml. The solution is stable at 4° C for several weeks.

Working standard (10 μg/ml)

Dilute the stock standard 1:10 with distilled water. Store in refrigerator and prepare fresh each week.

Procedure

Hydrolysis of conjugated metanephrines

1. Into 50 ml beakers place in duplicate
   Test: 10 ml of filtered urine
   Control: 10 ml of control urine
2. Adjust samples to pH 0.9 with 6 N HCl, using a pH meter.
3. Transfer to 40 ml glass-stoppered tubes.
4. Place in a boiling water bath for 20 minutes.
5. Cool solutions, adjust to pH 6.0 to 6.5 with 1 N NaOH using a pH meter, and dilute to approximately 20 ml with distilled water.

Adsorption and elution

6. Prepare chromatographic columns: Amberlite CG-50 is added to columns to make beds 1 × 5 cm. Wash with 7 ml of water to remove the buffer.
7. Pass through columns, at a rate of 0.5 ml/min.
   a. Hydrolyzed samples
   b. 1.0 ml of working standard in approximately 20 ml of water
8. Wash columns with 15 to 20 ml of water.
9. Elute metanephrines with 15 ml of 4 N ammonium hydroxide. Discard the first 2 ml and collect the next 10 ml.

Oxidation

10. Into test tubes place
    Test (unknown, control, standard): 4.0 ml of eluate and 0.1 ml of periodate solution
    Blank (unknown, control, standard): 4.0 ml of eluate and 0.1 ml of distilled water
    Reagent blank: 4.0 ml of 4 N NH_4OH and 0.1 ml of periodate solution

Measurement and calculations

11. **Determine the absorbance (A) of each solution at 360 mμ against water.**

$$\frac{\text{A test} - \text{A blank} - \text{A reagent blank}}{\text{A standard} - \text{A blank} - \text{A reagent blank}} \times \mu\text{g standard} \times \frac{\text{24-hr urine volume}}{\text{Sample volume}} = \mu\text{g metanephrines/24 hr}$$

Comments

The oxidation of the metanephrines by periodate in ammonium hydroxide is completed in about 30 seconds and the vanillin formed is stable for several hours.

This is a rapid screening test for pheochromocytoma but there is interference from such drugs as chlorpromazine, imipramine, methyldopa, and pargylone.

Homovanillic acid (HVA)

Neuroblastoma and ganglioneuroma are malignant sympathetic tumors usually found in children. They occur most frequently in the adrenal gland but may arise from other neural crest tissue. They are dopamine-secreting tumors and the determination of homovanillic acid, a metabolite of dopamine, is of value for their detection. However, the HVA methods available are not suitable for routine screening for the neuroblastoma group of tumors.

Normal values

A summary of the normal range of catecholamine and metabolite concentrations is given in Table 15-8.

Clinical findings

Adrenal medullary insufficiency does not produce any recognized clinical syndromes. Pheochromocytomas, tumors involving the chromaffin cells of the adrenal medulla and of other areas in the body, frequently synthesize and secrete large quantities of catecholamines. The physical symptoms often observed are severe or intermittent hypertension associated with hypermetabolism, headache, excessive sweating, weight loss, and elevated fasting blood glucose level. Approximately 80% of all pheochromocytomas are located in the adrenal gland.

Table 15-8. Normal catecholamine and metabolite concentrations

Urine	
Catecholamines	Up to 100 μg/24 hr
	Up to 18 μg/100 ml, random sample
Norepinephrine	10 to 80 μg/24 hr
Epinephrine	0 to 20 μg/24 hr
Metanephrines, total	0.3 to 0.9 mg/24 hr
3-Methoxy-4-hydroxy-mandelic acid (VMA)	0.5 to 7.0 mg/24 hr
Plasma	
Catecholamines	Less than 1.5 μg/L

Measurement of urinary catecholamines and/or VMA should be done on all young patients with essential hypertension and on older patients where there is some suspicion that a pheochromocytoma may exist, since a diagnosis of pheochromocytoma may not be indicated by clinical examination. It is found in less than 1% of hypertensive patients. However, hypertension that results from pheochromocytoma is reversed by removal of the tumor. The ratio of urinary catecholamines to 3-methoxy-4-hydroxymandelic acid has been shown to indicate the size of the tumor mass.

The determination of 3-methoxy-4-hydroxymandelic acid is unaffected by many of the drugs that interfere with catecholamine assays. For screening purposes it is not necessary to withdraw reserpine, thiazides, ganglionic blocking agents, or guanethidine from patients being treated with these. Misleading values may be obtained when monamine oxidase inhibitors are administered; these will cause a decrease in urinary 3-methoxy-4-hydroxymandelic acid and an increase in metanephrines.

In pheochromocytoma, catecholamine levels may vary between 300 μg and 4.0 mg/24 hr or up to 150 μg/100 ml for a random specimen; 3-methoxy-4-hydroxymandelic acid values may range from 11 to 250 mg/24 hr.

Neuroblastomas, highly malignant tumors of infants and young children, are characterized by a high urinary excretion of 3-methoxy-4-hydroxymandelic acid and homovanillic acid. Diarrhea is the most usual symptom but hypertension may occur. Ganglioneuromas are benign sympathetic tumors which also occur most commonly in children. Affected individuals may present with chronic diarrhea, wasting, and abdominal distention that disappear with the removal of the tumor.

Summarizing, in pheochromocytoma there are increased epinephrine, norepinephrine, metanephrine, normetanephrine, and 3-methoxy-4-hydroxymandelic acid; in neuroblastoma and ganglioneuroma, increased homovanillic acid, dopa, dopamine, and 3-methoxy-4-hydroxymandelic acid; in melanoma, increased homovanillic acid; and in familial dysautonomia, increased homovanillic acid and decreased 3-methoxy-4-hydroxymandelic acid.

GONADAL HORMONES

The testes and ovaries elaborate gonadal hormones that, along with small amounts from the adrenal cortex, produce and maintain secondary sex characteristics, control the reproductive cycle, and promote the growth and development of secondary sex glands. They also have a generalized anabolic effect which includes the stimulating of protein synthesis.

Testosterone

Testosterone, the principal androgenic testicular hormone, is produced mainly by the interstitial cells of the testes. Small amounts of other androgenic substances, androstenedione and dehydroepiandrosterone, are also secreted by the testes. The activity of the testes is under the control of the pituitary gonadotropins.

Testosterone is converted in the body to androsterone, etiocholanolone, and other metabolites. It is transported bound to serum albumin and a beta globulin. In the liver conjugation with glucuronic and sulfuric acids occurs. The ovary secretes the androgen androstenedione and appears capable of producing testosterone.

For measurements of plasma and urine testosterone, procedures based upon the double isotope principle described by Kliman and Peterson[95] are in general use. This consists of adding ^{14}C-labeled testosterone to each sample as an internal standard, acetylating with tritium-labeled acetic anhydride, purifying by chromatography, and assaying for ^{14}C and tritium by liquid scintillation spectrometry. The following are normal ranges for adult males: plasma, 0.3 to 1.0 μg/100 ml, and urine, 50 to 200 μg/24 hr. Those for adult females are plasma, less than 0.1 μg/100 ml, and urine, less than 30 μg/24 hr.

Since metabolites of testosterone are also metabolites of 11-deoxygenated adrenocortical steroids, the usefulness of the 17-ketosteroid determinations as an indication of Leydig cell function is limited. In applying a measurement of urinary testosterone as an index of androgenic function consideration must be given to the fact that the hormone is derived from sources other than the testes.

In testicular failure or deficiency of testosterone production before puberty the normal pubertal changes do not develop. In the adult male continuous production of the hormone is necessary for the maturation of the sperm and the activity of the secondary sex glands. Testosterone excretion may be increased in women with adrenogenital syndrome, some cases of hirsutism, and Stein-Leventhal syndrome associated with virilization, and in children with congenital adrenal hyperplasia. Testosterone excretion may be decreased in hypogonadism and Klinefelter's syndrome in males.

Estrogens and progestins

Female sex hormones may be classified physiologically into two groups, estrogens or follicular hormones and progestins or corpus luteum hormones.

Estrogens

Estrogens are steroid hormones having 18 carbon atoms and an aromatic A ring. They are produced primarily in the ovaries and during pregnancy in the placenta. Smaller amounts are formed in the adrenal cortex and testes.

A number of estrogens have been isolated but the three naturally occurring ones usually measured for investigations of ovarian function are estrone, estradiol-17β, and estriol. Estradiol-17β appears to be the important estrogen released by the ovary. Conjugation with glucuronic and sulfuric acids takes place in the liver.

Estrogens are responsible for the proliferation of uterine mucosa. In the normal menstruating female, estrogen excretion varies with the menstrual cycle, with maximum occurring during the preovulatory phase. There is a gradual increase during the first trimester of pregnancy and a rapid rise during the third trimester, which is approximately 80% to 90% estriol.

The major pathway of biosynthesis of estrone and estradiol-17β is shown in Fig. 15-6. Suggested pathways for estriol are as follows:

16α-hydroxydehydroepiandrosterone → 16α-hydroxyandrostenedione → 16α-hydroxyestrone → Estriol

5-androsten-3β,16α,17β-triol → 16α-hydroxytestosterone → Estriol

Estradiol-17β ⇌ Estrone → 2-hydroxyestrone → 2-methoxyestrone

Estrone → 16α-hydroxyestrone → Estriol

Estrone → 16β-hydroxyestrone → 16-epiestriol

Estriol ⇌ 16-oxoestradiol-17β ⇌ 16-epiestriol

A number of methods are available for the determination of total and fractionated estrogens. In assaying blood or urine from normal males and nonpregnant females the procedures are tedious and time-consuming because of the low concentrations of estrogens. The classic method of Brown[88] with various modifications has been most commonly used. However the procedure is not applicable for estrogen levels below 2 to 3 μg/24-hour specimen. This sensitivity is not adequate for determination of estrogens in certain pathologic states and in postmenopausal women. The final measurement using gas chromatography[98] or fluorescence[91,99] provides more sensitivity but is subject to greater interference from nonspecific contaminants and thus requires highly purified extracts.

With the exception of pregnancy, a determination of the total estrogen is sufficient for most clinical investigations. The measurement may provide an indication of whether amenorrhea is the result of pituitary, ovarian, or end-organ failure. Estrogen deficiency may be the result of either ovarian or pituitary failure. If it happens after puberty there will be cessation of menstruation, and sterility along with regression of some secondary sexual characteristics. Hyperestrogenism or elevated estrogen levels that may be the result of estrogen-producing ovarian tumors will cause early maturation in prepubertal girls. Testicular tumors may produce feminization and cause increase in urinary estrogens. Normal values are given in Table 15-9.

Estrogen determinations are used widely to assess fetal status during pregnancy. An indication of fetal distress may be based upon decreased estriol production as demonstrated when estriol levels do not rise rapidly or begin to decrease after the second trimester of pregnancy.

Estriol in pregnancy urine

The high concentration of estriol in normal pregnancy urine permits the use of a small volume of specimen for analysis. The less urine needed, the lower the concentration in the hydrolysate of nonspecific urinary substances that interfere with the assay. In the method described[93,97] the estrogen conjugates are hydrolyzed with acid and extracted with ether. After purification of the extract, solvent partition is used to separate estriol from estrone and estradiol-17β, and the Kober reagent is added for the colorimetry.

Table 15-9. Urinary estrogen levels of normal subjects[92,100]

	Estrone (μg/24 hr)	Estradiol-17β (μg/24 hr)	Estriol (μg/24 hr)	Total (μg/24 hr)
Male	3.5(0.8 to 7.1)	0.9(0.4 to 1.7)	6.1(1.5 to 10.4)	4 to 25
Female				
Nonpregnant	2 to 25	0 to 10	2 to 30	4 to 60
Onset of menstruation	5	2	6	
Ovulation peak	20	9	27	
Luteal maximum	14	7	22	
Postmenopausal	1.2(0.2 to 3.1)	0.4(0.0 to 0.9)	4.1(0.6 to 10.4)	1 to 16

Specimen collection

A 24-hour urine specimen is collected in a glass or polyethylene bottle without a preservative. The analysis should be performed as soon as possible after the specimen is received. It may be stored for several days at 4° C without deterioration occurring but when it is to be held for longer periods, the specimen must be frozen.

Reagents needed

All chemicals are reagent grade.
Ethyl ether (Squibb) for anesthesia or redistilled, peroxide-free
n-Heptane, redistilled, saturated with distilled water
Toluene, redistilled, saturated with distilled water
Sodium bicarbonate, saturated solution
  Prepare fresh each day.
Sodium bicarbonate reagent, pH 10.5
  Add 15 ml of 20% NaOH solution to 100 ml of saturated sodium bicarbonate solution. Prepare fresh daily.
Sodium hydroxide, 20%
Sodium hydroxide, 8%
Ethyl alcohol, 95%
Hydrochloric acid, concentrated
Alundum, 60 mesh
Sodium nitrate, crystal
p-Quinone, purified
Hydroquinone, purified
Sulfuric acid, concentrated
Kober color reagent
  Cool 200 ml of distilled water in a 2 L flask in an ice bath and add 760 ml of concentrated sulfuric acid slowly while stirring. Dilute to 1 L after cooling to room temperature. Add and dissolve 10 mg of sodium nitrate and 20 mg of *p*-quinone in the liter of diluted acid. Warm in a water bath at 100° C until a light green color just appears. Immediately add 20 g of hydroquinone and heat in a boiling water bath for 45 minutes, shaking occasionally until solution is complete. The reagent is placed in the dark at room temperature for about a week and then filtered through a sintered-glass funnel. It is stored in an amber glass bottle at room temperature.
Standard solution
  Dissolve 10 mg of estriol in 95% ethyl alcohol and dilute to 100 ml (100 μg/ml).

Procedure

1. Measure total volume of 24-hour collection of urine and check for glucose, using Uristix. Filter an aliquot of the urine. If glucose is present, hydrolyze, using an enzyme (β-glucuronidase); otherwise hydrolyze with acid.

Hydrolysis (acid)

Unknown and control specimens: For the last trimester of pregnancy use 2% of the total volume of urine and dilute to 80 ml with distilled water in a 250 ml round-bottom distilling flask. Add approximately 2 g of alundum and heat to boiling under a reflux condenser. Add 16 ml of concentrated HCl and allow to boil for 30 minutes.

Cool under running tap water and transfer to a 100 ml volumetric flask. Dilute to volume.

Hydrolysis (enzymatic)

Into 50 ml round-bottom centrifuge tubes pipette in duplicate:
  Unknown: 20 ml of filtered urine
  Control: 20 ml of filtered urine

Add to each tube 4.0 ml of phosphate buffer (pH 6.5) containing β-glucuronidase (1200 units) and mix. Check the pH, which should be 6.5. If it is not, adjust by using 0.1 N NaOH. Refer to reagents for Porter-Silber chromogens. Incubate for 18 to 24 hours at 37° C.

Dilute hydrolysate to 25 ml in a volumetric flask. Calculate the amount of hydrolysate equivalent to one fifth of 2% of the original urine diluted to 100 ml. Dilute this aliquot to 20 ml and proceed as directed for acid-hydrolyzed urine.

Extraction and washing

2. Into 125 ml separatory funnels pipette in duplicate:
   Unknown: 20 ml of hydrolysate
   Control: 20 ml of hydrolysate
   Standard: 19 ml of distilled water + 0.4 ml of estriol standard + 1.0 ml of concentrated HCl
3. Add to each funnel 1 g of NaCl and 40 ml of ethyl ether. Shake for 1 minute by hand and discard the aqueous layer.
4. Wash the ether extract with 8 ml of sodium bicarbonate reagent (pH 10.5) and discard the washing.
5. Shake the extract with 2 ml of 8% NaOH and add 8 ml of saturated sodium bicarbonate. After mixing for 30 seconds discard the aqueous layer.
6. Remove the remaining alkali from the ether layer by washing with and then discarding 2 ml of saturated sodium bicarbonate solution. Wash two times with 1 ml of distilled water or until neutral.
7. Into 40 ml round-bottom centrifuge tubes place 30 ml aliquot of the washed ether extract.
8. Evaporate to dryness in a water bath at 40° to 45° C under a stream of air. Wash down the sides of the tubes with 1 ml of 95% ethyl alcohol, allow to stand in a 50° C water bath for 15 minutes, and then evaporate to dryness.

Partition of estriol fraction

9. Dissolve the dried ether extract residue in 0.8 ml of hot 95% ethyl alcohol and add 5 ml of *n*-heptane saturated with water and mix; 5 ml of toluene saturated with water and mix; 20 ml of distilled water and shake for 1 minute.
10. Centrifuge for 5 minutes and discard the organic solvent layer. The estriol is in the aqueous layer.
11. Into 50 ml round-bottom centrifuge tubes place 10 ml aliquot of the aqueous layer, 1 g of NaCl, and 20 ml of ethyl ether. Shake for 1 minute.
12. Into 40 ml round-bottom centrifuge tubes transfer 15 ml aliquot of ether extract. Evaporate to dryness in a water bath at 40° to 45° C under a stream of air.
13. Prepare direct standards for recovery calculations by adding 0.2 ml of standard to 40 ml tubes and evaporating to dryness.

Color development

14. Add 3 ml of Kober reagent to each tube from steps 12 and 13 and to two clean tubes that serve as blanks.
15. Heat tubes in boiling water for 20 minutes. Stir frequently with a glass rod during the first 5 minutes and occasionally thereafter.
16. Cool in an ice bath and add 1.0 ml of distilled water. Mix.
17. Reheat tubes in a boiling water bath for 10 minutes and then cool rapidly in an ice bath.
18. Determine absorbance (A) at 460, 500, and 540 mμ, reading against the reagent blank.
19. Apply the Allen correction.

Calculations

$$\frac{\text{Corrected A unknown}}{\text{Corrected A extracted standard}} \times \text{Standard concentration} \times \frac{\text{Total volume urine}}{\text{Aliquot urine used}} = \text{mg estriol/24 hr}$$

Comments

Acid hydrolysis causes some destruction of estrogens but this effect is reduced by diluting the urine prior to analysis. An enzyme preparation, Glusulase, which contains both glucuronidase and sulfatase, has been used successfully for the hydrolysis. While the time required for the action is longer than with acid, the advantages are that there is no significant destruction of the estrogens and less production of interfering substances. However, urine contains variable amounts of inhibitors for both enzymes and therefore acid hydrolysis is used in the routine procedure.

Glucose has a destructive effect on estrogens during acid hydrolysis.[89,90] A tenfold dilution prior to hydrolysis may be used but enzyme hydrolysis is preferable.

Drugs that have been reported as interfering with some of the estrogen procedures are phenolphthalein, cascara, senna, stilbesterol, meprobamates, large doses of cortisone, and mandelamine.

Normal values

For normal values during pregnancy see Table 15-10.

Clinical findings

Serial estriol excretion levels are necessary for evaluating fetal-placental function in abnormal pregnancies as daily values fluctuate markedly. Steroids are conjugated and metabolized by the fetus. Estriol in pregnancy is derived from 16a-hydroxydehydroepiandrosterone of fetal origin. Estriol is conjugated by the fetus.

Decreased excretion of estriol may be found in fetal distress resulting from various diseases such as diabetes mellitus, toxemia, pre-eclampsia, or placental insufficiency. Persistently low or decreasing levels after the thirty-second week suggest the need for premature delivery as a means of fetal survival. When dia-

Table 15-10. Estrogen excretion in urine of normal pregnant women, range and mean[96,97]

Pregnancy (weeks)	Estrone (mg/24 hr)	Estradiol-17β (mg/24 hr)	Estriol (mg/24 hr)
20–25	0.33–1.1 (0.73)	0.16–0.30 (0.23)	3.7– 9.9 (6.5)
26–30	0.40–2.0 (0.98)	0.25–0.47 (0.36)	7.5–13.3 (10.0)
31–35	0.55–1.4 (1.0)	0.29–0.61 (0.38)	6.7–17.0 (12.5)
36–40	0.58–1.8 (1.2)	0.32–0.60 (0.45)	14.9–31.2 (19.8)

betics have lower than normal concentrations of estriol but show a constant tendency for increase, the pregnancy can be allowed to proceed to term without anticipating complications.[89] Estriol measurements provide a method for the management of complicated pregnancies.

Progestins

Progesterone, the major steroid possessing progestational activity, is secreted by the corpus luteum during the normal menstrual cycle and by the placenta, as its primary source, during pregnancy. Smaller amounts are derived from the adrenal cortex and testis. It is responsible for the endometrium changes that occur during the second half of the menstrual cycle, and secretion is highest at ovulation. If fertilization occurs, the major function is to maintain pregnancy. Production of progesterone will cease about the twenty-third or twenty-fourth day of the menstrual period if the ovum is not fertilized. It is secreted throughout pregnancy, during which time it inhibits muscle contraction and subsequent ovulation. Progesterone is essential for development of the maternal portion of the placenta and of the mammary glands. The highest levels appear about the thirtieth week of pregnancy. Plasma progesterone may be measured by competitive protein-binding radioassay.[63] Normal values per 100 ml are nonpregnant females, less than 2.0 μg; pregnant females, 2 to 28 μg; and males, less than 0.8 μg.

The major metabolic products of progesterone are pregnanediol, allopregnanediol, pregnanedione, and pregnanolone. Pregnanediol excretion levels provide an indication of endogenous production of progesterone, since it is present in highest concentration. It is excreted in the urine conjugated with glucuronic acid.

Pregnanediol

Pregnanediol is measured by the yellow color produced with sulfuric acid. The method[94] involves hydrolysis, solvent extraction, and alkaline wash of the extract to remove phenolic steroids. The reaction is not specific; various purification methods include thin-layer, paper, or column chromatography.

Specimen collection

A 24-hour urine specimen is collected in a glass or polyethylene bottle preferably without a preservative. The assay should be performed as soon as possible although during storage for several days at 4° C no appreciable deterioration appears to take place. For longer periods of storage, the specimen must be frozen.

Reagents needed

Reagents are the same as those used for pregnanetriol determinations with the addition of the following:

Ethyl alcohol, 0.05% in benzene

5β-pregnane-3α,20α-diol standard

Stock standard

Dissolve 40 mg in absolute ethyl alcohol and dilute to 100 ml (400 μg/ml). Store in an amber bottle at room temperature.

Working standard

Dilute 5 ml of stock standard with 15 ml of absolute ethyl alcohol (100 μg/ml). Prepare monthly and store in an amber bottle at room temperature.

Procedure

Hydrolysis, extraction, and washing

Proceed as in steps 1 through 11 for pregnanetriol (p. 460) except

that a standard containing 50 μg/ml (0.5 ml of working standard) pregnanediol is carried through the procedure. If the patient is pregnant, use only 5 ml of urine for the analysis.

Chromatography

Prepare the columns in the same manner as for pregnanetriol.

12. Allow the extract to pass through the column and rinse the column with 10 ml benzene. Discard all benzene that passed through the column.
13. Rinse the column with 40 ml of 0.05% ethanol in benzene and discard the washing.

Elution

14. Add 40 ml of 2% ethyl alcohol in benzene to the column. Collect the eluate in a 50 ml round-bottom centrifuge tube.
15. Evaporate the eluate to dryness under a stream of air in a water bath at 35° to 45° C. Wash down the sides of each tube with 2 ml of ethyl alcohol and evaporate to dryness to remove all traces of benzene.

Color development

16. Prepare two standard tubes for recovery calculations by pipetting into 50 ml round-bottom tubes 0.5 ml of working standard (50 μg) and evaporating to dryness.
17. To all tubes from steps 15 and 16 add 15 mg of sodium sulfite and 3.5 ml of concentrated sulfuric acid. *Without mixing* allow the tubes to stand at room temperature for 30 minutes.
18. After the 30-minute period mix the contents in the tubes. Allow to stand an additional 60 to 90 minutes and determine absorbance (A) at 390, 425, and 460 mμ against concentrated sulfuric acid as the blank. If the concentration of pregnanediol in the unknown is too high, dilutions may be made at this step, using concentrated sulfuric acid.
19. Apply the Allen correction.

Calculations

$$\frac{\text{Corrected A unknown}}{\text{Corrected A extracted standard}} \times \text{Standard concentration} \times \frac{\text{Total volume urine}}{\text{Aliquot urine used}} = \text{mg pregnanediol/24 hr}$$

Comment

Pregnanediol and pregnanetriol may be performed on the same hydrolysate and extract. Pregnanediol is eluted from the column first by using a different concentration of alcohol in benzene.

Normal values

Normal values of urinary pregnanediol concentration are given in Table 15-11.

Clinical findings

Pregnanediol excretion is increased during the secretory (luteal) phase of the menstrual cycle and reaches a maximum during the third trimester of pregnancy. Abnormalities in menstruation and pregnancy may be associated with decreased secretion of pregnanediol.

Progesterone synthesis that occurs in the placenta does not appear to involve the fetus. Thus fetal disorders may be present when the pregnanediol excretion levels are normal but can be indicated by altered estriol excretion. In pregnancy low values are often followed by abortions and in fetal death pregnanediol levels decrease rapidly.

Table 15-11. Normal values for urinary pregnanediol

Male	0.38–1.42 mg/24 hr
Female	
Nonpregnant	0.5 –7.0
Proliferative phase	0.5 –1.5
Luteal phase	2.0 –7.0
Postmenopausal	0.2 –1.0
Pregnant: weeks of amenorrhea	
10–12	5–15
12–18	5–25
18–24	13–33
24–28	20–42
28–32	27–47

Pregnanediol excretion is usually decreased in amenorrhea, anovular menstruation, and pre-eclamptic toxemia of pregnancy. Elevated values may be caused by some adrenal tumors, retained placental tissue following parturition, or corpus luteum cysts.

SEROTONIN AND 5-HYDROXYINDOLEACETIC ACID

Serotonin (5-hydroxytryptamine) is a hormone normally produced by argentaffin cells. These cells are found in the mucous membrane of the gastrointestinal tract, appendix, bronchus, gallbladder, and ducts of the pancreas. Carcinoid tumors arising from argentaffin cells produce excessive amounts of serotonin. The metabolic alteration is the increased conversion of tryptophan to serotonin. Tryptophan is hydroxylated to 5-hydroxytryptophan which is decarboxylated to 5-hydroxytryptamine (serotonin). The latter is converted by oxidative deamination by the enzyme monamine oxidase to 5-hydroxyindoleacetic acid (5-HIAA).

In the circulating blood, serotonin is carried by the platelets. As an indication of increased production of the hormone, determination of the concentration of the metabolic product, 5-HIAA, in the urine is used. The method[101] is based on the development of a purple color specific for 5-hydroxyindoles with nitrous acid and nitrosonaphthol. Interfering chromogens are removed with ethyl acetate.

5-Hydroxyindoleacetic acid

Reagents needed

Chloroform, redistilled

Hydrochloric acid, 6 N

Dilute concentrated HCl 1:1 with distilled water.

Hydrochloric acid, 1 N

Dilute 8.6 ml of concentrated (11.6 N) HCl to 100 ml with distilled water.

Sodium hydroxide, 1 N

Dissolve 4 g sodium hydroxide pellets in distilled water and dilute to 100 ml.

Sodium chloride, crystals, AR

Ethyl ether (Squibb or Mallinckrodt) for anesthesia or redistilled and peroxide-free

Phosphate buffer, pH 8.0

Solution I: Dissolve 9.078 g of KH_2PO_4 in distilled water and dilute to 1 L, 0.067 M primary phosphate solution.

Solution II: Dissolve 9.465 g of Na_2HPO_4 in distilled water and dilute to 1 L, 0.067 M secondary sodium phosphate solution.

Prepare 0.067 M phosphate buffer, pH 8.0, by mixing 26.5 ml of solution I with 473.5 ml of solution II.

Nitrosonaphthol reagent, 0.1%

Dissolve 100 mg in 95% ethyl alcohol and dilute to volume.

Sodium nitrite, 2.5%

Dissolve 250 mg in distilled water and dilute to 10 ml.

Nitrous acid reagent

Mix 0.5 ml of 2.5% sodium nitrite solution with 12.5 ml of 3 N sulfuric acid.

Sulfuric acid, 3 N

Dilute 8.4 ml of concentrated (17.8 M) acid to 100 ml with distilled water.

Ethyl acetate, AR

Sulfuric acid, 0.1 M

Dilute 1.4 ml of concentrated acid to 250 ml with distilled water.

Standard solution

Dissolve 20 mg of 5-hydroxyindoleacetic acid (free acid) in 0.1 M sulfuric acid and dilute to 100 ml (50 μg/250 μl). Stable if frozen. Use only the first day after thawing.

Procedure

A 24-hour urine specimen is collected in a glass bottle containing 15 ml of 6 N hydrochloric acid.

1. Measure the urine and filter an aliquot.
2. Into 40 ml plastic centrifuge tubes pipette, in triplicate:
   Unknown: 10 ml of filtered urine
   Control: 10 ml of filtered urine control
   Standard: 250 μl of standard + 10 ml of normal saline
3. Using a pH meter, adjust the pH to 2.0 using as little 1 N HCl or 1 N NaOH as possible.

Washing and extraction

4. To each tube add 3.5 g of sodium chloride and 25 ml of chloroform. Shake on a Vortex mixer for 1 minute and centrifuge for 5 minutes at 1500 rpm.
5. Discard the chloroform layer by aspirating, being careful not to lose any of the aqueous layer.
6. Repeat the chloroform wash by adding to each tube 25 ml of chloroform. Shake on a Vortex mixer for 45 seconds and centrifuge for 5 minutes at 1500 rpm. Discard the chloroform layer by aspirating.
7. To each tube add 3.5 g of sodium chloride and 20 ml of ether. Shake for 1 minute on a Vortex mixer and centrifuge for 5 minutes at 1500 rpm.
8. Transfer 15 ml of ether layer to 40 ml glass-stoppered centrifuge tubes and add 1.5 ml of phosphate buffer.
9. Stopper the tubes and shake on a Vortex mixer for 1 minute. Centrifuge for 5 minutes at 1500 rpm. Discard the ether layer by aspirating.

Color development

10. Set up three direct standard tubes by pipetting into 40 ml glass-stoppered centrifuge tubes 1.5 ml of phosphate buffer and 250 μl of standard solution.

11. Divide tubes from steps 9 and 10 into two groups:
    a. Group I:
       Blanks: one tube from the set of three for each standard, extracted standard, control, and unknown.
       Reagent blank: set up by adding 1.5 ml of phosphate buffer to a 40 ml glass-stoppered centrifuge tube.
    b. Group II: Other two tubes from the set of three for each standard, extracted standard, control, and unknown.
12. To tubes of Group I (blanks) add 0.5 ml of nitrosonaphthol reagent and mix; 0.5 ml 3 N sulfuric acid and mix.
13. To tubes of Group II (5-HIAA measurement) add 0.5 ml of nitrosonaphthol reagent and mix; 0.5 ml of nitrous acid reagent and mix.
14. Incubate all tubes at 37° C for 10 minutes and allow to cool to room temperature.
15. Add to each tube 3.3 ml of ethyl acetate. Shake, centrifuge, and discard ethyl acetate by aspirating.
16. Repeat the ethyl acetate washing as described in step 15.
17. Transfer the aqueous phase to a microcuvette and determine absorbance (A) at 540 mμ, reading against a reagent blank.

Calculations

$$\frac{\text{A unknown} - \text{A unknown blank}}{\text{A extracted standard} - \text{A extracted standard blank}} \times \text{Standard concentration (0.05 mg)} \times \frac{\text{Total volume urine}}{\text{Aliquot urine used (10 ml)}} = \text{mg 5-HIAA in specimen}$$

Normal values

1 to 7 mg/24 hr

Comments

In this procedure it is considered important to have the urine sample adjusted to pH 2 and salt saturation during the first $CHCl_3$ wash.

In determining urinary 5-HIAA, the following have been reported as interfering: acetanilide, bananas, chlorpromazine, mephenesin, methocarbamol, and phenothiazine.

Clinical findings

Excessive production of serotonin is characteristic of carcinoid tumors and determination of the concentration of 5-HIAA in the urine is the most useful method of detection. However, some patients with the carcinoid syndrome do not have increased levels of 5-HIAA in the urine and some patients without the syndrome may exhibit increased concentrations.

GENERAL COMMENTS ON HORMONE ANALYSIS

The rapid turnover and low concentrations of hormones in biologic fluids cause difficulties in their measurement. One source of error in analysis of urine is incomplete collection of the 24-hour urine specimen. A creatinine determination may be used as an indication of the completeness of the collection. The rate of elimination is related to muscle mass and remains relatively constant in any one individual but varies considerably between subjects. The normal range for men is 20 to 26 mg/kg/24 hr; for women it is 14 to 22 mg/kg/24 hr.

If possible, all medication should be withheld from the patient for at least 48 hours before a specimen for analysis is collected. Information is available on the effect of some drugs, but for many that physicians now prescribe this is unknown. Liver and kidney function tests should not be scheduled during periods of urine collection for hormone assays.

There is considerable diurnal variation. Therefore, when blood samples are collected, the time must be recorded; and when urine collections are made, careful attention must be given to obtaining the entire volume for the 24 hours.

All determinations should be performed in duplicate. Standards, control specimens, and reagent blanks should be carried out with every group of determinations. Glassware must be cleaned carefully in chromic acid cleansing mixture or, where fluorometry is employed, 50% nitric acid, and then be thoroughly rinsed with tap water and distilled water.

Substances that react with a reagent but have a different absorption maximum are partially eliminated by using the Allen correction. This correction assumes that interfering substances have a maximum absorption other than that of the compound being determined and that their absorption curve is a straight line in the wavelength range applied.

Each extraction and purification step causes small losses of hormonal material. The amount of losses should be determined by measuring the simultaneous recovery of known amounts of the hormone. The use of radioactive tracers to correct for losses increases the accuracy but not the specificity of the methods. Solvent and reagent contamination must be corrected for, by the performance of blank determinations.

REFERENCES

General

Eisenstein, A. B., editor: The adrenal cortex, Boston, 1967, Little, Brown & Co.

Gray, C. H., and Bacharach, A. L., editors: Hormones in blood, vols. I and II, New York, 1967, Academic Press, Inc.

O'Neal L. W.: Surgery of the adrenal glands, St. Louis, 1968, The C. V. Mosby Co.

Williams, R. H., editor: Textbook of endocrinology, ed. 4, Philadelphia, 1968, W. B. Saunders Co.

Pituitary gland (hypophysis)

1. Abe, K., Island, D. P., Liddle, G. W., Fleischer, N., and Nicholson, W. E.: Radioimmunologic evidence for the α-MSH (melanocyte stimulating hormone) in human pituitary and tumor tissue, J. Clin. Endocr. **27:** 46, 1967.
2. Albert A.: Gonadotropins, Fertil. Steril. **10:** 60, 1959.
3. Albert A.: Bioassay and radioimmunoassay of human gonadotropins, J. Clin. Endocr. **28:**1683, 1968.
4. Albert A., Mattox, V. R., and Mason, H. L.: Laboratory tests in clinical endocrinology. In Williams, R. H., editor: Textbook of endocrinology, ed. 4, Philadelphia, 1968, W. B. Saunders Co.
5. Daughaday, W. H.: The adenohypophysis. In Williams, R. H.: editor: Textbook of endocrinology, ed. 4, Philadelphia, 1968, W. B. Saunders Co.
6. Daughaday, W. H., and Parker, M. L.: The pituitary in disorders of growth, D.M., p. 1, Aug., 1962.
7. Daughaday, W. H., and Parker, M. L.: Human pituitary growth hormone, Ann. Rev. Med. **16:**47, 1965.

7a. Davis, P. J.: Factors affecting the determination of the serum protein-bound iodine, Amer. J. Med. **40:**918, 1966.

8. Demura, H., West, C. D., Nugent, C. A., Nakagawa, K., and Tyler, F.: A sensitive radioimmunoassay for plasma ACTH levels, J. Clin. Endocr. **26:**1297, 1966.
9. Faiman, C., and Ryan, R. J.: Radioimmunoassay for human follicle stimulating hormone, J. Clin. Endocr. **27:**444, 1967.
10. Glick, S. M., Roth, J., Yalow, R. S., and Berson, S. A.: Immunoassay of human growth hormone in plasma, Nature **199:**784, 1963.
11. Karlson, P., and Sekeris, C. E.: Biochemical mechanisms of hormone action, Acta Endocr. **53:**505, 1966.
12. Odell, W. D., Ross, G. T., and Rayford, P. I.: Radioimmunoassay for luteinizing hormone in human plasma or serum: physiological studies, J. Clin. Invest. **46:**248, 1967.
13. Raiti, S., and Blizzard, R. M.: Measurement of immunologically reactive follicle stimulating hormone in human urine by radioimmunoassay, J. Clin. Endocr. **28:**1719, 1968.
14. Utiger, R. D.: Extraction and radioimmunoassay of growth hormone in human serum, J. Clin. Endocr. **24:**60, 1964.
15. Utiger, R. D.: Radioimmunoassay of human plasma thyrotropin, J. Clin. Invest. **44:**1277, 1965.
16. Walker, J. M.: Oxytocin and vasopressin. In Gray, C. H., and Bacharach, A. L., editors: Hormones in blood, vol. I, New York, 1967, Academic Press, Inc.

Thyroid gland

17. Barker, S. B., and Humphrey, M. J.: Clinical determination of protein-bound iodine in plasma, J. Clin. Endocr. **10**:1136, 1950.
18. Clark, F., and Horn, D. B.: Assessment of thyroid function by the combined use of the serum protein-bound iodine and resin uptake of ^{131}I-triiodothyronine, J. Clin. Endocr. **25**: 39, 1965.
19. Henry, R. J.: Clinical chemistry: principles and technics, New York, 1964, Paul B. Hoeber, Inc., Medical Book Department, Harper & Row, Publishers.
20. Lamberg, B. A.: Thyroid function tests, Acta Endocr. (supp.) **124**:153, 1967.
21. Murphy, B. P.: The determination of thyroxine by competitive protein-binding analysis employing an anion-exchange resin and radiothyroxine, J. Lab. Clin. Med. **66**:161, 1965.
22. Murphy, B. E., and Pattee, C. J.: Determination of thyroxine utilizing the property of protein-binding, J. Clin. Endocr. **24**:187, 1964.
23. O'Neal, L. W., and Simms, E. S.: Determination of protein-bound iodine in plasma or serum, Amer. J. Clin. Path. **23**:493, 1953.
24. Passen, S., and von Saleski, R.: A semiautomated method which does not require digestion for the determination of serum thyroxine iodine (T_4), Amer. J. Clin. Path. **51**:166, 1969.
25. Pileggi, V. J., Lee, N. D., Golub, O. J., and Henry, R. J.: Determination of iodine compounds in the serum. I. Serum thyroxine in the presence of some iodine contaminants. J. Clin. Endocr. **21**:1272, 1961.
26. Pileggi, V. J., and Kessler, G.: Determination of organic iodine compounds in serum. IV. A new nonincineration technic for serum thyroxine, Clin. Chem. **14**:339, 1968.
27. Rosenbaum, J. M., Kreig, A. F., Henry, J. B., Mozley, J. M., and McAfee, J. G.: Thyroid function evaluation in patients with increased or decreased thyroxine-binding protein, Amer. J. Clin. Path. **50**:336, 1968.
28. Sterling, K., and Brenner, M. A.: Free thyroxine in human serum: simplified measurement with the aid of magnesium precipitation, J. Clin. Invest. **45**:153, 1966.
29. White, W. L.: The prevention of cross-contamination in PBI determinations, Clin. Chem. **9**:365, 1963.
30. Zak, B., Koen, A. M., and Boyle, A. J.: Normal and abnormal values of protein-bound iodine, Amer. J. Clin. Path. **23**:603, 1953.
31. Zak, B., Willard, H. H., Myers, G. B., and Boyle, A. J.: Chloric acid method for determination of protein-bound iodine, Anal. Chem. **24**:1345, 1952.

Parathyroid glands

32. Aurbach, G. D., and Potts, J. T., Jr.: Parathyroid hormone, Amer. J. Med. **42**:1, 1967.
33. Eliel, L. P., Chanes, R., and Hawrylko, J.: Urinary excretion of parathyroid hormone in man: effects of calcium loads, protein-free diets, adrenal cortical steroids and neoplastic disease, J. Clin. Endocr. **45**:445, 1965.
34. Reiss, E., and Alexander, F.: The tubular reabsorption of phosphate in the differential diagnosis of metabolic bone disease, J. Clin. Endocr. **19**:1212, 1959.
35. Stoerk, H. C., Aceto, R. M., and Budzilovich, T.: Parathormone levels in serum of patients with hyperparathyroidism, J. Clin. Endocr. **26**:668, 1966.

Islets of Langerhans of the pancreas

36. Berson, S. A., and Yalow, R. S.: Immunoassay of protein hormones. In Pincus, G., Thimann, K. V., and Astwood, E. B., editors: The hormones, vol. 4, New York, 1964, Academic Press, Inc.
37. Berson, S. A., and Yalow, R. S.: Some current controversies in diabetes research, Diabetes **14**:549, 1965.
38. Jorgensen, K. R.: Evaluation of the double antibody radioimmunoassay of insulin and the determination of insulin in plasma and urine in normal subjects, Acta Endocr. **60**: 327, 1969.
39. Jorgensen, K. R.: Radioimmunoassay of insulin in plasma and urine in obese subjects and in diabetic patients, Acta Endocr. **60**: 719, 1969.
40. Morgan, C. R., and Lazarow, A.: Immunoassay of insulin: two antibody system. Plasma insulin levels of normal, subdiabetic and diabetic rats, Diabetes **12**:115, 1963.
41. Seltzer, H. S., Allen, E. W., Herron, A. L., Jr., and Brennan, M. T.: Insulin secretion in response to glycemic stimulus: relation of delayed initial release to carbohydrate intolerance in mild diabetes mellitus, J. Clin. Invest. **46**:323, 1967.
42. Yalow, R. S., and Berson, S. A.: Immunoassay of endogenous plasma insulin in man, J. Clin. Invest. **39**:1157, 1960.

Adrenal gland

43. Allen, W. M., Hayward, S. J., and Pinto, A. A.: A color test for dehydroisoandrosterone and closely related steroids of use in the diagnosis of adrenocortical tumors, J. Clin. Endocr. **10**:54, 1950.
44. Bongiovanni, A. M., and Eberlein, W. R.: Critical analysis of methods for measurement of pregnane-3-alpha,17-alpha,20-alpha-triol in human urine, Anal. Chem. **30**:388, 1958.

45. Borushek, A., and Gold, J. J.: Commonly used medications that interfere with routine endocrine laboratory procedures, Clin. Chem. **10**:41, 1964.
46. Braunsberg, H., and James, V. H. T.: The determination of adrenocortical steroids in blood: Observations on the reliability of a simple fluorimetric method for cortisol, J. Endocr. **25**:309, 1962.
47. Braverman, L. E., Reinstein, H., Loayze, H., and Pomfret, D.: Formaldehyde-forming drugs and urinary 17-hydroxycorticosteroids, Clin. Chem. **14**:374, 1968.
48. Daughaday, W. H.: The binding of corticosteroids by plasma proteins. In Eisenstein, A. B., editor: The adrenal cortex, Boston, 1967, Little, Brown & Co.
49. de Moor, P., Osinski, P., Deckx, R., and Steeno, O.: The specificity of fluorometric corticoid determinations, Clin. Chim. Acta **7**:475, 1962.
50. de Moor, P., Steeno, O., Meulepas, E., Hendrikx, A., Delaere, K., and Ostyn, M.: Influence of body size and of sex on urinary corticoid excretion in a group of normal young males and females, J. Clin. Endocr. **23**:677, 1963.
51. de Moor, P., Steeno, O., Raskin, M., and Hendrikx, A.: Fluorimetric determination of free plasma 11-hydroxycorticosteroids in man, Acta Endocr. **33**:297, 1960.
52. Engel, L. L.: The assay of urinary neutral 17-ketosteroids. In Glick, D., editor: Methods of biochemical analysis, vol. 1, New York, 1954, Interscience Publishers, Inc.
53. Few, J. D.: A method for the analysis of urinary 17-hydroxycorticosteroids, J. Endocr. **22**:31, 1961.
54. Forsham, P. H.: The adrenal cortex. In Williams, R. H., editor: Textbook of endocrinology, ed. 4, Philadelphia, 1968, W. B. Saunders Co.
55. Goldzieher, J. W., and de la Pena, A.: The effect of formalin and the duration of acid hydrolysis on free and conjugated urinary 17-ketosteroids, Clin. Chem. **14**:1125, 1968.
56. Henke, W. J., Doe, R. P., and Jacobson, M. E.: A test of pituitary reserve utilizing intravenous SU-4885 with a new method for extraction of 11-desoxycorticosteroid, J. Clin. Endocr. **20**:1527, 1960.
57. Klein, D., and Chernaik, J. M.: Urinary pregnanediol and pregnanetriol: Spectrophotometric determination following aluminum oxide column chromatography, Clin. Chem. **10**:170, 1964.
58. Kliman, B., and Peterson, R. E.: Double isotope derivative assay of aldosterone in biological extracts, J. Biol. Chem. **235**:1639, 1960.
59. Lupovitch, A.: Phenothiazine interference and the Porter-Silber reaction, Clin. Chem. **14**:179, 1968.
60. Mattingly, D.: A simple fluorimetric method for the estimation of free 11-hydroxycorticoids in human plasma, J. Clin. Path. **15**:374, 1962.
61. Metcalf, M. G.: A rapid method for measuring 17-hydroxycorticosteroids in urine, J. Endocr. **26**:415, 1963.
62. Metcalf, M. G.: Quantitative analysis of urinary 17-oxygenic steroids using sodium metaperiodate oxidation, Clin. Chem. **12**:258, 1966.
63. Murphy, B. E.: Some studies of the protein-binding of steroids and their application to the routine micro and ultramicro measurement of various steroids in body fluids by competitive protein-binding radioassay, J. Clin. Endocr. **27**:973, 1967.
64. Nielsen, E., and Asfeldt, V. H.: Studies on the specificity of fluorimetric determination of plasma corticosteroids ad modum de Moor and Steeno, Scand. J. Clin. Lab. Invest. **20**:185, 1967.
65. Nugent, C. A., Nichols, T., and Tyler, F. H.: Diagnosis of Cushing's syndrome; single dose dexamethasone suppression test, Arch. Intern. Med. **116**:172, 1965.
66. Perkoff, G. T., Eik-Nes, K., Nugent, C. A., Fred, H. L., Nimer, R. A., Rush, L., Samuels, L. T., and Tyler, F. H.: Studies of diurnal variation of plasma 17-hydroxycorticosteroids in man, J. Clin. Endocr. **19**:432, 1959.
67. Peterson, R. E.: Measurement of plasma or serum cortisol. In Sunderman, F. W., and Sunderman, F. W., Jr., editors: Lipids and the steroid hormones in clinical medicine, Philadelphia, 1960, J. B. Lippincott Co.
68. Peterson, R. E.: Determination of urinary neutral 17-ketosteroids. In Seligson, D., editor: Standard methods of clinical chemistry, vol. 4, New York, 1963, Academic Press, Inc.
69. Porter, C. C., and Silber, R. H.: A quantitative color reaction for cortisone and related 17,21-dihydroxy-20-ketosteroids, J. Biol. Chem. **185**:201, 1950.
70. Rappaport, F., Fischl, J., and Pinto, N.: A rapid method for the estimation of urinary 17-ketosteroids, Clin. Chem. **6**:16, 1960.
71. Sawin, C. T., Bray, G. A., and Idelson, B. A.: Overnight suppression test with dexamethasone in Cushing's syndrome, J. Clin. Endocr. **28**:422, 1968.
72. Silber, R. H.: Free and conjugated 17-hydroxycorticosteroids in urine. In Seligson, D., editor: Standard methods of clinical chemstry, vol. 4, New York, 1963, Academic Press, Inc.

73. Silber, R. H.: Fluorimetric analysis of corticoids. In Glick, D., editor: Methods of biochemical analysis, vol. 14, New York, 1966, Interscience Publishers, Inc.
74. Silber, R. H., and Porter, C. C.: Determination of 17,21-dihydroxy-20-ketosteroids in urine and plasma. In Glick, D., editor: Methods of biochemical analysis, vol. 4, New York, 1957, Interscience Publishers, Inc.
75. Spencer-Peet, J., Daly, J. R., and Smith, V.: A simple method for improving the specificity of the fluorometric determination of adrenal corticosteroids in human plasma, J. Endocr. **31**:235, 1965.
76. Sweat, M. L.: Sulfuric acid induced fluorescence of corticosteroids, Anal. Chem. **26**: 773, 1954.
77. van der Vies, J.: Individual determination of cortisol and corticosterone in a single small sample of peripheral blood, Acta Endocr. **38**:399, 1961.
78. Wirth, W. A., and Thompson, R. L.: The effect of various conditions and substances on the results of laboratory procedures, Amer. J. Clin. Path. **43**:579, 1965.

Medulla

79. Anton, A. H., and Sayre, D. F.: A study of the factors affecting the aluminum oxide-trihydroxyindole procedure for the analysis of catecholamines, J. Pharmacol. Exp. Ther. **138**:360, 1962.
80. Catecholamines in urine. In Manual of fluorometric clinical procedures, Palo Alto, Calif., 1961, G. K. Turner Associates.
81. Crout, J. R.: Catecholamines in urine. In Seligson, D., editor: Standard methods of clinical chemistry, vol. 3, New York, 1961, Academic Press, Inc.
82. Crout, J. R., and Sjoerdsma, A.: Catecholamines in the localization of pheochromocytoma, Circulation **22**:516, 1960.
83. Crout, J. R., and Sjoerdsma, A.: Turnover and metabolism of catecholamines in patients with pheochromocytoma, J. Clin. Invest. **43**:94, 1964.
84. Jacobs, S. L., Sobel, C., and Henry, R. J.: Specificity of the trihydroxyindole method for determination of urinary catecholamines, J. Clin. Endocr. **21**:305, 1961.
85. Pisano, J. J.: A simple analysis for normetanephrine and metanephrine in urine, Clin. Chim. Acta **5**:406, 1960.
86. Pisano, J. J., Crout, J. R., and Abraham, D.: Determination of 3-methoxy-4-hydroxymandelic acid in urine. Clin. Chim. Acta **7**:285, 1962.
87. Sobel, C., and Henry, R. J.: Determination of catecholamines (adrenalin and noradrenalin) in urine and tissue, Amer. J. Clin. Path. **27**:240, 1957.

Gonads

88. Brown, J. B.: The determination and influence of the natural estrogens. In Sobotka, H., and Stewart, C. P., editors: Advances in clinical chemistry, vol. 3, New York, 1960, Academic Press, Inc.
89. Greene, J. W., Jr., Smith, K., Kyle, G. C., Touchstone, J. C., and Duhring, J. L.: The use of urinary estriol excretion in the management of pregnancies complicated by diabetes mellitus, Amer. J. Obstet. Gynec. **91**: 684, 1965.
90. Hobkirk, R., Alfhein, A., and Bugge, S.: Hydrolysis of estrogen conjugates in diabetic urine, J. Clin. Endocr. **19**:1352, 1959.
91. Hobkirk, R., and Metcalfe-Gibson, A.: Urinary estrogens. In Seligson, D., editor: Standard methods of clinical chemistry, vol. 4, New York, 1963, Academic Press, Inc.
92. Hobkirk, R., and Nilsen, M.: Urinary estrogens in non-pregnant human subjects measured by modifications of Bauld's method, Steroids **3**:453, 1964.
93. Huang, W. Y.: A rapid and reliable method of urinary estriol determination in pregnancy, Steroids **11**:453, 1968.
94. Klein, D., and Chernaik, J. M.: Urinary pregnanediol and pregnanetriol: Spectrophotometric determination following aluminum oxide column chromatography, Clin. Chem. **10**:170, 1964.
95. Kliman, B., and Peterson, R. E.: Double isotope derivative assay of aldosterone in biological extracts, J. Biol. Chem. **235**:1639, 1960.
96. Paulsen, C. A., editor: Estrogen assays in clinical medicine, Seattle, 1965, University of Washington Press.
97. Stevens, S. C.: Estriol in pregnancy urine, publication pending.
98. Thijssen, J. H. H., and Veeman, W.: A gas chromatographic method for the measurement of small amounts of estrogens in urine, Steroids **11**:369, 1968.
99. van Baelen, H., Heyns, W., and de Moor, P.: Measurement of urinary estrogens using adsorption on sephadex, J. Clin. Endocr. **27**: 1056, 1967.
100. Zarrow, M. X., Yochem, J. M., and McCarthy, J. L.: Experimental endocrinology, a sourcebook of basic techniques, New York, 1964, Academic Press, Inc.

Serotonin

101. Anton, A. H., and Sayre, D. F.: Improving the specificity of the nitrosonaphthal procedure for urinary 5-hydroxyindoleacetic acid, Clin. Chem. **13**:1017, 1967.

• 16

• Radioimmunoassay of hormones

JAMES N. HEINS, M.D., and MARY E. COMFORT, B.S.

The measurement of peptide hormones and glycoproteins in biologic fluids has presented almost insurmountable problems to the clinical chemist. Most are present in 10^{-12} to 10^{-8} molar amounts, a few to several hundred millimicrograms per milliliter, and they exist in solution with hundreds of similar peptides of much higher concentration. The advent of purified hormone preparations and sophisticated isotopic techniques led to the development of the radioimmunoassay of insulin in 1960.[1] Subsequent investigators have applied these basic techniques to insulin and to many other hormones: human growth hormone, thyroxine, ACTH, luteinizing hormone, etc. The method utilizes the ability of isotopically labeled hormone to form reversible complexes with specific antibody that has been generated in animals against the same or heterologous hormone. Competition for binding sites on the antibody by natural unlabeled hormone in the test sample leads to the displacement of label from the antigen-antibody complexes. Measurement of the liberated hormone or bound labeled product after suitable separation steps permits quantitation of the unknowns. Standard curves are generated by the addition of known amounts of pure unlabeled hormone to similar reaction tubes. Isotopes of iodine, ^{125}I, ^{127}I, and ^{131}I, are most often used for labeling because of short half-lives, safety in handling, and availability of salts of high specific activity.

Refined immunoassays capable of measuring small amounts of hormones in blood of normal patients require labeled hormones of high specific activity, antibody with high specificity for the hormone to be tested, and separation techniques for efficient removal of the free labeled hormone from that which is antibody-bound.

IODINATION OF HORMONES

Preparation of the labeled hormone must avoid the hazards of iodine-induced distortion of the immunologically active site and radiation damage to the hormone molecule. This has been accomplished by the use of carrier-free isotopes, limiting the substitutions on tyrosyl side chains of the molecules and shortening the iodination reaction by rapid separation of the iodinated product from unbound isotope. Despite low-temperature storage of the labeled hormones, deterioration occurs through release of radioiodine and "endogenous" radiation damage and limits shelf life to 2 to 4 weeks. The iodination technique which is most used is a modification of that of Hunter and Greenwood.[2]

Materials needed

Sodium iodide-125 (Isoserve) carrier-free
Specific activity should be about 150 to 250 mc/ml.
Phosphate buffer, 0.5 M, pH 7.6
Chloramine-T, 35 mg in 10 ml of 0.05 M phosphate buffer, pH 7.6
Sodium *m* bisulfite, 25 mg in 10 ml of 0.05 M phosphate buffer
Human serum albumin, 3%
Bovine serum albumin, 1%, in 0.01 M phosphate buffer, pH 7.6, and 0.15 M NaCl
Sephadex G-50 (coarse beads)
Freshly prepared column, 0.9 cm × 21 cm, suspended in 0.05 M barbital buffer, pH 8.6, coated with 200 μl of 30% bovine serum albumin, and washed with barbital buffer.
Barbital buffer, 0.1 M
Dissolve in water 2.31 g of barbituric acid and 20.6 g of sodium barbital; dilute to 1 L, pH 8.6.
Pure hormone
Prepared in a solution of 1 mg/ml with a few drops of 0.01 N NaOH and then water to volume. Seal about 40 μl in ampules and freeze.

Procedure

1. In a small culture tube, 10 × 75 mm, add 20 μl of hormone, 1 mg/ml solution, 100 μl of 0.5 M phosphate buffer, and 10 or 20 μl (1.5 to 2 mc) of Na ^{125}I; mix well by shaking. Add 25 μl of chloramine T and shake gently for 45 seconds. Add 100 μl of sodium *m* bisulfite.
2. Transfer to column and dilute with 0.05 M barbital buffer. Collect 8 or 9 drops per tube containing a drop of 3% human serum albumin. Count on a Geiger counter and place the three tubes containing the hormone-^{125}I (comes off the column in 4 to 5 ml eluate) in an ice bucket.

The first peak contains iodinated hormone. The two or three tubes are pooled and diluted about four times with 1% BSA-PO_4-NaCl buffer. About 1 ml aliquots are placed in small tubes and immediately frozen. On day of assay, the 1 ml aliquot is thawed and placed on a column, 1.5 cm × 55 cm, of Sephadex G-100 beads, particle size 40 to 120 μ, packed in 0.05 M barbital buffer and coated with 100 μl of 30% bovine serum albumin. Elute with 0.05 M barbital buffer. Collect about 2 ml eluates. The first

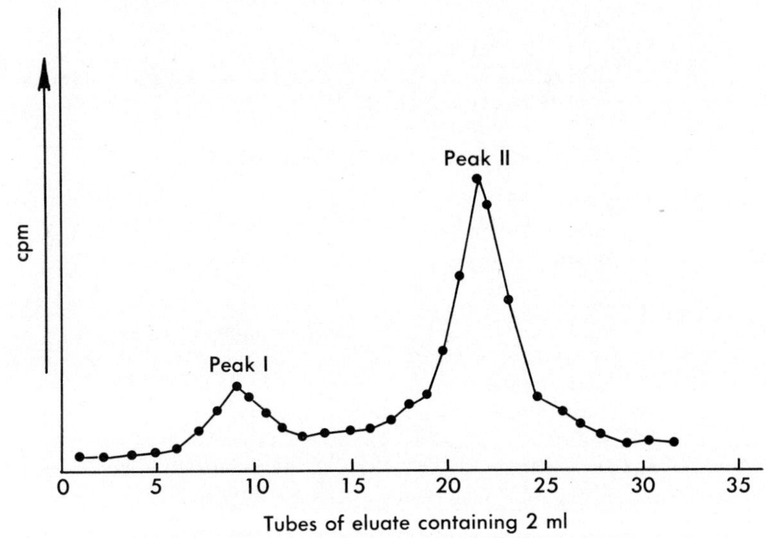

Fig. 16-1. Purification of iodinated hormone.

peak contains the aggregated molecules, while the second peak has the best iodinated hormone for the assay (Fig. 16-1). The three or four tubes from peak II are pooled and diluted to about 10,000 c.p.m./100 μl and kept cold until added to assay.

METHOD FOR HUMAN GROWTH HORMONE[3,4]*

Materials needed

Test tubes, 10 × 75 mm
Disposable glass tubes are used for incubation and in turn are inserted into larger plastic carrier tubes to be counted in a well-type gamma counter.

Diluting solution
3% bovine serum albumin in 0.01 M phosphate buffer and 0.15 N NaCl, pH 7.6.

Anti-HGH
Guinea pig anti–human growth hormone titrated to give the best displacement for the range of standards used (see below).

Carrier
Normal guinea pig gamma globulin titrated to give maximum precipitability.

Second antibody
Goat or rabbit anti–guinea pig gamma globulin also titrated for maximum precipitability.

HGH-^{125}I
Diluted to 10,000 counts/100 μl with diluting solution.

Standard HGH solution
Pure HGH (standard grade) is prepared in a solution of 1 mg/ml with a few drops of 0.01 N NaOH and then water to volume. This stock A is then diluted 1:100 with 1% BSA-PO_4-NaCl buffer, making a 500 mμg/50 μl stock B. Stock B is then diluted 1:10 with 1% BSA-PO_4-NaCl, making a 50 mμg/50 μl working stock. All stocks are sealed and frozen in ampules. From the working stock, dilutions ranging from 0.1 mμg/50 μl to 5 mμg/50 μl are made and run in triplicate.

Procedure

First day

1. Number disposable assay tubes and place in plastic racks.
2. To each tube add 50 μl of standard or unknown serum (unknowns are done in duplicate).
3. Add 100 μl of HGH-^{125}I.
4. Add 0.3 ml of 3% BSA-PO_4-NaCl buffer with an automatic syringe.
5. Add 100 μl of anti-HGH.
6. Add 100 μl of carrier.
7. Shake and incubate for 72 hours at 4° C.

Third day

8. Add 100 μl of second antibody, shake very well, and incubate overnight.

Fourth day

9. Add 3 ml of cold phosphosaline buffer to each tube, centrifuge at 5° C for 30 minutes at 1000 × g or more. Remove the supernatant with gentle suction.
10. Count precipitates in gamma counter.
11. Construct a standard curve, c.p.m. versus mμg per tube (50 μl); from the curve read the values of the unknowns directly by number (Fig. 16-2). Convert to micrograms per milliliter.

---

*Modified and used by Metabolism Division, Washington University School of Medicine, St. Louis, Mo.

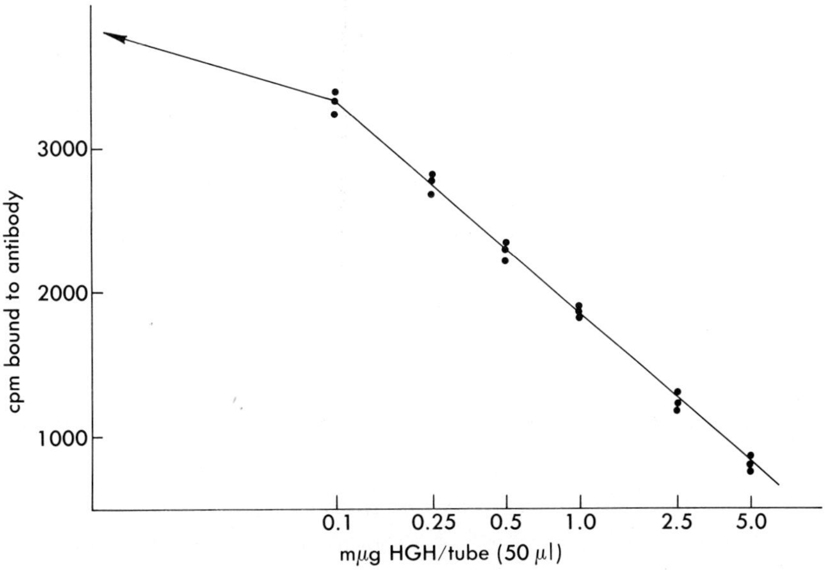

Fig. 16-2. Standard curve.

Titration of antibody

Reagents needed

BSA-PO_4-NaCl buffer, 3%
HGH-^{125}I (as described above)
Carrier
Second antibody
Anti-HGH, diluted with 3% BSA-PO_4-NaCl, 1:1000 to 1:100,000

Procedure

1. Tubes with identical amounts of HGH-^{125}I are run with each anti-HGH dilution, and the anti-HGH dilution that precipitates approximately 50% of the HGH-^{125}I counts is selected.
2. The procedure is repeated for dilutions of carrier and second antibody.

ALTERNATE SEPARATION TECHNIQUES

Coated charcoal immunoassay[5]

Charcoal coated with a large molecule has the property of binding small molecules almost instantly while not adsorbing the same molecules if they are bound to larger carriers, such as plasma proteins. This behavior can be utilized to effect quantitative separation of antibody-bound and free hormones during immunoassay.

Materials needed

Buffer
Dissolve 14.714 g of sodium barbital and 9.714 g of sodium acetate in distilled water and dilute to 500 ml. Add 1800 ml of 0.85% sodium chloride solution and 100 ml of 0.1 N hydrochloric acid to adjust the final pH to 7.4.

Charcoal suspension
Norit A, pharmaceutical grade decolorizing carbon, 5 g/100 ml buffer.

Dextran solution
Dextran 80, of 80,000 average molecular weight, 0.5 g/100 ml solution with buffer.

Dextran-coated charcoal

Mix equal volumes of charcoal suspension and dextran solution. Shake for 10 seconds and store at 4° C. Shelf life is approximately 1 month.

Albumin buffer

Human serum albumin is added to buffer to final concentration of 350 mg/100 ml.

Procedure

1. The assay is run in duplicate with standards as above.
2. After incubation to reach equilibrium conditions, 2 ml of the dextran–charcoal suspension is added to each tube.
3. Tubes capped with parafilm are inverted repeatedly for 10 seconds. Centrifuge for 15 minutes at 3000 rpm to obtain a solid charcoal button.
4. Supernatant fluid is decanted into counting tubes and counted in a well-type scintillation counter.
5. A standard curve is generated and calculation of samples is carried out as before.

Chromatoelectrophoresis[1]

Free hormone will bind quantitatively to cellulose while antibody-bound molecules do so less well. Appropriate types of filter paper strips can be used to effect such a separation.

Materials needed

Paper electrophoresis tanks with constant-voltage power supply in cold room

Veronal buffer, 0.1 ionic strength, pH 8.6

Whatman 3MM filter paper strips, 37 × 550 mm

Paper strip scanning gamma counter

Normal guinea pig serum

Planimeter

Procedure

1. At termination of incubation 200 μl of normal guinea pig serum is added to all tubes to prevent trailing.
2. Paper strips are dipped in buffer and blotted. Electrophoresis tanks are placed 10 cm apart. Immerse about 10 mm of each end of the paper strips into the buffer tanks.
3. A sufficient amount of sample is spotted at the origin to provide accurate counting statistics and electrophoresed at 150 to 250 volts from the constant-voltage source (current approximately 3 milliamperes per strip).
4. Free hormone adsorbs to cellulose while antibody-bound hormone migrates in the electric field.
5. Most runs can be terminated at 1 to 1½ hours. The albumin peak has moved approximately 3 inches in this period of time and antigen-antibody complexes about 2¼ inches.
6. The strips are dried at 115° C for 45 minutes without being touched except at the 10 mm ends.
7. Damaged labeled hormone migrates with the antibody-bound hormone nonspecifically attached to protein. This technique does not separate bound hormone from free hormone. Internal controls for damaged labeled hormone use nonimmune guinea pig serum as a set of controls and standards. Any iodine found to migrate as a complex is assumed to be radiation-damaged molecules, and those counts are subtracted from total counts before calculation of bound (B) to free (F) ratios.
8. The area under the free hormone peak is measured by planimetry and subtracted from the total counts per strip. The calculation of standard hormone concentrations are expressed in graphic form as the ratio of B/F hormone versus micro units of hormone in the controls.
9. B/F ratios of unknown samples are read off the abscissa of the standard curve in microunits.

REFERENCES

1. Yalow, R. S., and Berson, S. A.: Immunoassay of endogenous plasma insulin in man, J. Clin. Invest. **39**:1157, 1960.
2. Hunter, W. M., and Greenwood, F. C.: Preparation of iodine-131 labeled growth hormone of high specific activity, Nature **194**:495, 1962.
3. Schlach, D. S., and Parker, M. L.: A sensitive double antibody immunoassay for human growth hormone in plasma, Nature **203**:1141, 1964.
4. Utiger, R. D., Parker, M. L., and Daughaday, W. H.: Studies on human growth hormone. I. A radio-immunoassay for human growth hormone, J. Clin. Invest. **41**:254, 1962.
5. Herbert, V., Lau, K. S., Gottlieb, C. W., and Bleicher, S. J.: Coated charcoal immunoassay of insulin, J. Clin. Endocr. **25**:1375, 1965.

17

Toxicology

A number of deaths result each year from acute poisoning, exclusive of alcohol and carbon monoxide. When the illness of a patient is the result of a toxic substance, identification may be sufficient but a quantitative estimation often is very important. Among the more commonly encountered causes of toxicity are barbiturates, bromides, salicylates, glutethimide (Doriden), mercuric chloride, and thiocyanates.

During the treatment of disease, the physician prescribes various drugs for the patient. Many of these drugs must be maintained at certain blood levels in order to be effective. If this blood level is greatly exceeded, however, the drug may have a toxic effect. In these cases, serial quantitative measurement of the concentration in blood is usually requested.

There are other situations when the physician may need to know whether the prescribed drugs are being utilized or simply excreted in the urine, or whether the patient is taking drugs without his knowledge. In certain of these instances, a procedure that indicates the presence or absence of the specific compound will satisfy the physician.

Ethyl alcohol

Ethyl alcohol determinations are required to establish the degree of alcohol intoxication of a patient and to differentiate pathologic states that mimic but are totally unrelated to alcoholism. Ethyl alcohol is rapidly absorbed by the body and distributed in proportion to the water content of each tissue or fluid. The liver metabolizes approximately 90% and the remainder is eliminated through the kidneys and breath.

Several methods are available for the determination of blood alcohol levels. One type commonly used is that of distilling, aerating or allowing the alcohol to diffuse into potassium dichromate and sulfuric acid. An enzymatic method is also available, which uses alcohol dehydrogenase and requires NAD as coenzyme. These methods are not as specific as the gas chromatographic procedure that is in general use where a large number of these analyses are performed.

Distillation methods[9] will provide reliable results if the necessary precautions are followed. Ethanol in the distillate may be determined by oxidation with acid dichromate or acid permanganate.

Specimen collection

For the blood sample, obtain venous blood in a tube containing 150 mg of sodium fluoride and 100 mg of sodium citrate for 10 ml of blood. Ethyl alcohol must not be used to cleanse the skin before the venipuncture. Aqueous germicidal solutions are satisfactory. For the urine sample, obtain two specimens. Have the patient empty his bladder for the first collection and then 30 minutes later have him empty his bladder

again. The second specimen should correlate with the blood level. If not analyzed immediately, store samples in a refrigerator. For storage periods exceeding 2 days, freeze the specimens.

Reagents needed

Picric acid solution, 1% (D-317)

Standard dichromate reagent

Dissolve 3.333 g of potassium dichromate in 500 ml of water, place in a running tap water bath to keep cool, and add slowly 521 ml of 36 N (concentrated) sulfuric acid. Dilute to approximately 1 L and when the solution is at room temperature make the final volume adjustment. Store in a glass-stoppered bottle (not polyethylene) in the refrigerator.

Antifoam A emulsion (Dow Corning)

Ethyl alcohol standards

Calculate the concentration by weight of the pure ethanol (200 proof) from standard density tables and the temperature of the alcohol. For example, at 23° C, 2.0 ml of 200 proof ethyl alcohol diluted to 1 L yields a solution that contains 0.16 g/100 ml. A standard curve is prepared by adding, respectively, 10.0, 7.5, 5.0, 2.5, and 0 ml of the standard solution to 10.0 ml of alcohol-free blood and carrying out the procedure as described for a sample. The absorbance (or % T on semilog paper) of each standard is plotted against the corresponding ethanol concentration. The alcohol concentration in the blood or urine is reported as grams per 100 ml.

Procedure

An all-glass distilling apparatus must be used.

1. Into a 300 ml Erlenmeyer distillation flask pipettte 10.0 ml of whole blood or urine. Rinse the pipette with 10 to 20 ml of water and include with the sample.
2. Add to the flask 10 to 15 ml of 1% picric acid and 5 to 10 drops of Antifoam A.
3. Connect the flask to the distillation apparatus and heat gently so that 10.0 ml of distillate is obtained in 10 to 12 minutes. Mix the distillate.
4. To a test tube add 3.0 ml of standard dichromate reagent and 1.0 ml of distillate. Mix.
5. Place tube in a boiling water bath for 15 minutes.
6. Allow to cool to room temperature and read the absorbance at 600 mμ against a reference solution containing 3.0 ml of the dichromate reagent and 1.0 ml of water similarly heated.

Calculations

7. Ethanol content is determined from a previously prepared standard graph.

Comments

If the concentration of ethanol in the sample is greater than 0.20%, the distillate must be diluted before adding it to the dichromate reagent. The final calculations are corrected for the dilution factor.

To confirm a positive finding or that the dichromate reduction is the result of ethanol, additional tests must be made for the presence of methanol, formaldehyde, isopropanol, and paraldehyde. There are other substances that would produce the reaction but their presence is rare.

Clinical findings

Blood alcohol concentrations below 100 mg/100 ml are generally considered subclinical. Above this level some degree of intoxication is present, with confusion apparent at more than 200 mg/100 ml, stupor above 300 mg, and coma above

400 mg. However, these responses will vary somewhat with different individuals.

Qualitative tests for methanol, formaldehyde, and paraldehyde[9]

In order to rule out volatile reducing agents other than ethyl alcohol in the distillate previously described the following qualitative tests must be conducted when a positive test is found.

Reagents needed

Chromotropic acid, 0.5%	(D-318)
Sulfuric acid, concentrated	
Sulfuric acid, 10%	(D-319)
Potassium permanganate, 5%	(D-320)
Sodium bisulfite, 0.15 M	(D-321)
p-Hydroxybiphenyl, 1% in 0.5 N NaOH	(D-322)
Copper sulfate, 4%	(D-323)

Procedure

Detection of methanol and/or formaldehyde

1. Into a test tube pipette 1.0 ml of distillate, 1.0 ml of 10% sulfuric acid, and 10 drops of potassium permanganate solution. Allow to stand for 15 minutes.
2. Add sodium bisulfite solution dropwise until the purple color disappears.
3. Place tube in an ice bath and add 0.2 ml of chromotropic acid and 4.0 ml of concentrated sulfuric acid.
4. Shake to mix and place in a boiling water bath for 15 minutes.
5. Appearance of a purple color indicates the presence of methanol and/or formaldehyde.

Detection of formaldehyde

6. Into a test tube pipette 1.0 ml of distillate and place in an ice bath.
7. Add 0.2 ml of chromotropic acid and 4.0 ml of concentrated sulfuric acid.
8. Shake to mix and place in a boiling water bath for 15 minutes.
9. Appearance of a purple color indicates the presence of formaldehyde.

Detection of paraldehyde

10. Into a test tube pipette 1.0 ml of distillate, add 1 drop of 4% copper sulfate and 6 ml of concentrated sulfuric acid, and mix.
11. With constant shaking add 0.2 ml of *p*-hydroxydiphenyl reagent.
12. Place tube in a boiling water bath for 3 minutes.
13. Cool to room temperature.
14. In the absence of formaldehyde a pink color indicates paraldehyde, which can also be detected by its characteristic odor.

Comments

The distillate should be checked for acetone, using one of the tests described in Chapter 14. Occasionally a patient in a diabetic coma may be misdiagnosed as intoxicated from alcohol.

Barbiturates

The extensive use of barbiturates as sedatives has resulted in overingestion as the cause of a number of near-fatal or fatal toxic reactions. The patient is frequently unconscious and laboratory analyses must be relied upon to indicate the type of drug and body concentration.

The ultraviolet spectrophotometric methods available are based upon the study of absorption curves at pH 10 and pH 13.4 over the range of 230 to 270 mμ. The method of Broughton[4] utilizes the characteristics of 5:5′-disubstituted barbiturates in the following three forms in solution: (1) the un-ionized form in acid solution has negligible selective light absorption between 230 and 270 mμ; (2) an ionized form at pH 9.8 to 10.5 has an absorption maximum at 240 mμ but no minimum in the range of 230 to 270 mμ; and (3) a second ionized form at pH 13 to 14 has an ultraviolet absorption maximum between 252 and 255 mμ and a minimum between 234 and 237 mμ.

Specimen collection

Whole blood is preferred for the analysis although serum or plasma may be used. When urine is analyzed, it must be recognized that certain of the barbiturates are metabolized in the body, especially by the liver, and are excreted as nondepressant metabolites. Analysis of stomach contents may be of value if aspiration is done shortly after ingestion of the drug.

Reagents needed

Sodium hydroxide, 0.45 N (D-324)

Chloroform, reagent grade, redistilled

Phosphate buffer, pH 7.4, 0.5 M

Dissolve in water 57.4 g of anhydrous disodium hydrogen phosphate and 18.06 g of anhydrous potassium dihydrogen phosphate, mix, and dilute to 1 L. Preserve with a few drops of chloroform.

Boric acid solution

Dissolve 37.2 g of boric acid and 45 g of potassium chloride in water and dilute to 1 L.

Phenobarbital stock standard

Dissolve 50 mg of phenobarbital in 0.45 N sodium hydroxide and dilute to 50 ml (1 mg/ml).

Phenobarbital working standard

Dilute 2 ml of the stock solution to 100 ml with 0.45 N sodium hydroxide (2 mg/100 ml). These solutions must be prepared just before using as barbiturates are unstable in alkaline solution.

Procedure

1. **Into a 60 ml separatory funnel place 5.0 ml of oxalated blood.**
2. **Extract three times, 3 to 5 minutes each time, with 30 ml of chloroform.**
3. **Filter the combined chloroform extract through Whatman 41 filter paper into a 125 ml separatory funnel. Wash the paper with a small amount of chloroform.**
4. **Wash the chloroform extract twice with 5 ml portions of phosphate buffer in two different separatory funnels. Discard the washings.**
5. **Add 10 ml of 0.45 N sodium hydroxide and shake vigorously for 1 minute.**
6. **After the phases have separated, the aqueous (sodium hydroxide) layer is centrifuged.**
7. **For detection of barbiturates, use solutions A and B.**
   Solution A, pH 13.4, place in test tubes
   ***Unknown:* 2 ml of NaOH layer + 2 ml of 0.45 N NaOH**
   ***Standard:* 2 ml of working standard + 2 ml of 0.45 N NaOH**
   ***Blank:* 4 ml of 0.45 N NaOH**
   Solution B, pH 10, place in test tubes
   ***Unknown:* 2 ml of NaOH layer + 2 ml of boric acid solution**
   ***Standard:* 2 ml of working standard + 2 ml of boric acid solution**
   ***Blank:* 2 ml of 0.45 N NaOH + 2 ml of boric acid solution**
   Measure absorbance (A) of the unknown and of the standard samples against the respective blanks in 1 cm silica cuvettes at 260 mμ. If the absorbance of the unknown is too high, it may be diluted with 0.45 N NaOH.

Calculations

8. **The difference between the absorbance of solution A and solution B is taken as a measure of the barbiturate concentration.**

$$\frac{\text{A unknown}}{\text{A standard}} \times \text{Standard concentration} \times \text{Dilution aqueous layer} =$$

Barbiturate in the blood (expressed as mg phenobarbital/100 ml blood)

For the 5 ml blood specimen in the method as described:

$$\frac{\text{A unknown}}{\text{A standard}} \times 4 = \text{mg phenobarbital/100 ml}$$

Comments

Interfering substances are extracted by the procedure but they do not have the characteristic change in absorbance when the pH is shifted.

By the use of phenobarbital as a standard, an error may be introduced if other barbiturates are present. If the other barbiturate is known, it may be used as the standard.

If identification of the barbiturates is required, use the technique of thin-layer chromatography as described in Chapter 19.

Clinical findings

Barbiturates vary in duration of action and in rate of absorption from the stomach and small intestine. Long-acting barbiturates are absorbed slowly and the short-acting ones rapidly.

- Long-acting
  - Barbital (Veronal, Medinal)
  - Phenobarbital (Luminal)
- Intermediate-acting
  - Amobarbital (Amytal)
  - Butabarbital (Butisol, Soneryl)
- Short-acting
  - Pentobarbital (Nembutal)
  - Secobarbital (Seconal)

Variations in response, from patient to patient, may sometimes be caused by

Table 17-1. Correlation of barbiturate blood levels (mg/100 ml) with patient's clinical condition

Patient's condition	Phenobarbital	Pentobarbital	Secobarbital	Amobarbital
Awake, mildly sedated, competent	1.0–4.0	0.5–1.0	0.5–1.0	0.5–1.5
Sedated, reflexes present, prefers sleep	4.0–5.0	1.0–2.0	1.0–2.0	1.5–3.0
Comatose, reflexes present	5.0–8.0	2.0–3.0	2.0–3.0	3.0–5.0
Comatose, reflexes absent	8.0–11.0	3.0–4.0	3.0–4.0	5.0–7.0
Comatose, reflexes absent, circulatory and respiratory difficulty	Above 11	Above 4	Above 4	Above 7

interfering substances and synergistic effects of other compounds. Alcohol will enhance the pharmacologic action of the barbiturates. In general, the patient's clinical condition will correlate with barbiturate blood levels as shown in Table 17-1.

Barbiturates, rapid method[2,11]

This procedure utilizes the fact that derivatives of barbituric acid react with mercury in an alkaline medium to form a complex that is soluble in chloroform.

Reagents needed

Chloroform, reagent grade
Mercuric nitrate solution, stock
Dissolve 1.0 g of mercuric nitrate in 100 ml of distilled water and add a few drops of nitric acid to acidify the solution.
Mercuric-bicarbonate reagent
Dissolve 7 g of sodium bicarbonate in 50 ml of distilled water and add 2.5 ml of the mercuric nitrate stock solution. Filter any undissolved particles or precipitate. Prepare fresh before using.
Dithizone solution
Dissolve 3.2 mg of dithizone (diphenylthiocarbazone) in chloroform and dilute to 100 ml.
Hydrochloric acid, 1 N
Phenobarbital standard
Dissolve 4.0 mg of phenobarbital in distilled water and dilute to 100 ml.

Procedure

1. Into separate glass-stoppered test tubes place:
   Unknown: 1 ml of serum
   Blank: 1 ml of water
   Standard: 1 ml of standard
2. Add to each tube 0.03 ml of 1 N HCl and 10 ml of chloroform. Shake for 1 minute.
3. Into each tube place a rolled piece of Whatman 42 filter paper, to absorb the water, and invert the tube.
4. Decant the chloroform into another tube and add 0.5 ml of mercury-bicarbonate reagent. Shake for 1 minute.
5. Insert a rolled piece of filter paper as before and invert the tube.
6. Into a test tube pipette a 5 ml aliquot of the chloroform extract and add 0.5 ml of dithizone solution.
7. Determine the absorbance (A) of unknown and of standard at 605 mμ against a chloroform blank.
8. Subtract each sample from the blank reading and use these figures for the calculations.

Calculations

$$\frac{\text{A unknown}}{\text{A standard}} \times 4 = \text{mg barbiturate in 100 ml serum (calculated as phenobarbital)}$$

Comments

If the concentration of the mercury in the solution is so high that all dithizone is used up, add 5 ml of chloroform and 0.5 ml of dithizone. This will be apparent from the orange color of the sample. The result should then be multiplied by 2.

Drugs that interfere with the method are Malysol, Doriden, and some hydantoin derivatives, such as diphenylhydantoin.

Bromides

Bromide is used in the treatment of epilepsy and also as a sedative. It is normally present in the blood, being found to the extent of 0 to 1.5 mg/100 ml of serum. If the blood level of bromide exceeds 136 mg/100 ml, toxic symptoms may occur.

Wuth method[12]

A trichloroacetic acid protein-free filtrate is made, using serum. Gold chloride is added to the filtrate, producing a brown color due to the production of gold bromide. The color is obtained by the replacement of the chloride ion in the gold chloride by the bromide ion. The absorbance of the brown color is directly proportional to the bromide concentration.

Reagents needed

Gold chloride, 0.5% in water	(D-325)
Trichloroacetic acid (TCA), 10% in water	(D-164)
Bromide stock standard, 20 mg/ml	(D-326)
Bromide working standard	

Pipette 1.0 ml of stock standard into a 100 ml volumetric flask, using a volumetric pipette. Dilute to mark with 10% TCA.

Procedure

1. Pipette 2.0 ml of serum into a tube containing 8.0 ml of 10% TCA. Stopper and mix by inverting.
2. Allow to stand for 10 minutes; then centrifuge for 10 minutes at 3000 rpm. Filter the supernatant.
3. Prepare the following tubes:
   Reagent blank
   Pipette 5.0 ml of 10% TCA.
   Standards
   Pipette 1.0 ml of bromide working standard.
   Add 4.0 ml of 10% TCA.
   Unknown
   Pipette 5.0 ml of filtrate.
4. Add 1.0 ml of gold chloride solution to all tubes and mix.
5. Allow to stand for 5 minutes and read absorbance (A) against the reagent blank at 440 mμ.

Calculation

$$\frac{\text{A unknown}}{\text{A standard}} \times 20 = \text{mg bromide/100 ml}$$

Normal range

0 to 1.5 mg/100 ml

Carbon monoxide

If carbon monoxide is inhaled, it unites with hemoglobin and thereby prevents oxygen from reaching the cells. If sufficient quantities are inhaled, death occurs due to asphyxiation (suffocation). This test may tell the physician whether or not the patient has inhaled large amounts of carbon monoxide.

Procedure (Sayers-Yant)

1. Make a venipuncture and withdraw a few milliliters of blood. Expel into a test tube containing an anticoagulant of sodium or potassium citrate. Gently invert each tube several times to dissolve the citrate. A control must be run; so draw a sample of blood from a normal person, using the same anticoagulant.

2. Using a spatula, place a small bit of pyrogallic acid in a beaker. Add an equal amount of tannic acid. Mix.
3. Label three small test tubes: patient, positive control, and negative control.
4. Pipette 2 ml of distilled water into each tube.
5. Add 3 drops of the patient's blood to the patient's tube. Add 3 drops of normal blood to each of the other tubes. Gently mix. Bubble some illuminating gas through the positive control tube in order to saturate it with carbon monoxide.
6. Using a spatula, add a small bit of the pyrogallic and tannic acid mixture to each tube. Gently mix. Let stand 15 minutes.
7. Readings: Blood containing normal hemoglobin turns grayish brown. Blood containing carbon monoxide retains a pink color.

Doriden

Doriden (glutethimide) is a sedative and hypnotic drug that is frequently prescribed instead of barbiturates. However, intoxication with it is not uncommon, and fatalities have increased. Diagnosis of Doriden poisoning should be confirmed by measurement of the drug levels in blood and urine. By the use of thin-layer chromatography glutethimide in biologic samples may be estimated semiquantitatively in less than an hour.[7]

Reagents needed

Acetone, analytic reagent
Benzene, analytic reagent
Chloroform, analytic reagent
Developing solvent
    Mix 95 parts benzene with 5 parts acetone.
Detection reagent
    Prepare a saturated aqueous solution of mercurous nitrate by shaking 1 g of $HgNO_3 \cdot H_2O$ in 100 ml of distilled water for at least 5 to 10 minutes. The reagent must be freshly prepared before using.
Eastman Chromagrams, type K301R, 20 × 20 cm
Glutethimide standard
    Dissolve 10 mg of the compound in 25 ml of chloroform. Prepare fresh weekly.

Procedure

1. Into a 60 ml separatory funnel pipette 2 ml of blood and 10 ml of chloroform. Shake vigorously for 5 minutes.
2. Into a 15 ml centrifuge tube filter the chloroform layer and wash the filter and funnel each with 0.5 ml of chloroform.
3. Evaporate the chloroform in the centrifuge tube with the aid of a stream of nitrogen or air. The tube may be placed in a water bath at 30° to 35° C to increase the rate of evaporation.
4. Dissolve the residue in 0.25 ml of chloroform.
5. *Chromatography*
    a. Use the Eastman Chromagrams without activating and cut to a 10 × 20 cm size.
    b. Mark points for sample application along the 20 cm length, 1 cm from the edge and 2 cm apart.
    c. Mark the length of solvent travel by drawing a line across the top, 8.5 cm from the point of sample application.
    d. Into a 600 ml beaker place 30 ml of the developing solvent and cover.
    e. Apply 25 μl samples of the blood extract to the marked Chromagram.
    f. Apply to five other points glutethimide standard solution in volumes of 5, 10, 20, 30, and 40 μl. The size of the spots should be uniform and approximately 0.5 cm in diameter.

g. After the spotting, adsorbent is scraped from the 10 cm edge so that approximately 1 mm of the film baking is exposed.
h. The Chromagram is then formed in the developing beaker with the coated side facing into the beaker and the bottom edge dipping into the solvent.
i. The beaker is covered and the solvent allowed to ascend to the top line.
j. The Chromagram is then removed and air dried.

6. Into a large Petri dish pour the detecting reagent and gently draw the Chromagram through the solution, facing the adsorbent downward. If the Chromagram remains in the solvent, the binder will dissolve and the adsorbent will fall off.
7. Place the Chromagram face up on a sheet of white paper. Glutethimide produces a gray-black spot with the reagent.
8. Compare the blood extract with the glutethimide standards. If glutethimide is present, it will appear at the same distance from the point of application as the standard. If detected, it is compared with the standards for size and color intensity of the spots.

Clinical findings

Doriden action is similar to that of barbiturates. In fatal cases the ingestion has been 10 g or more; this is equivalent to a blood concentration of 3 mg/100 ml, or more.[8] The absorption is variable because of its poor solubility in water. However, after absorption glutethimide is rapidly taken up by the body adipose tissue and then gradually released into the blood. Because of its solubility in fat globules, females are more susceptible to toxic reactions than males.

Salicylates

The salt, sodium salicylate, is used in the treatment of rheumatic fever. The acid, salicylic acid, occurs in the blood during treatment, the therapeutic level being at least 25 mg/100 ml of serum. Neither the salt nor the acid is normally present in the blood.

Trinder method[10]

The Trinder method utilizes a color reagent containing ferric nitrate, mercuric chloride, and hydrochloric acid, which precipitates the proteins and simultaneously reacts with the salicylate, giving a purple color. Plasma, serum, whole blood, urine, or cerebrospinal fluid may be analyzed by this method.

Reagents needed

Color reagent (ferric-mercuric reagent)	(D-327)
Salicylate solution, stock, 1 mg/ml	(D-328)
Salicylate solution, standard, 0.05 mg/ml	(D-329)
Salicylate solution, standard, 0.20 mg/ml	(D-330)

Procedure

1. Prepare the following tubes:

Reagent blank

Pipette 1.0 ml of distilled water into tube.
Add 5.0 ml of ferric-mercuric reagent.
Stopper and mix by inverting.

Standards

Pipette 1.0 ml of standard, 0.05 mg/ml, into one test tube.
Pipette 1.0 ml of standard, 0.20 mg/ml, into another test tube.

To both tubes, add 5.0 ml of ferric-mercuric reagent.
Stopper and mix by inverting.

Unknown

Pipette 1.0 ml of serum, cerebrospinal fluid, or urine into a test tube.
Add 5.0 ml of ferric-mercuric reagent.
Stopper and mix by inverting; then filter.

2. Read absorbance (A) against the reagent blank at a wavelength of 540 mμ.

Calculations

Calculate with the standard reading closer to the unknown.
Using the standard 0.05 mg/ml:

$$\frac{\text{A test}}{\text{A standard}} \times 0.05 \times \frac{100}{1} =$$

$$\frac{\text{A test}}{\text{A standard}} \times 5 = \text{mg salicylates/100 ml}$$

Using the standard 0.20 mg/ml:

$$\frac{\text{A test}}{\text{A standard}} \times 0.2 \times \frac{100}{1} =$$

$$\frac{\text{A test}}{\text{A standard}} \times 20 = \text{mg salicylates/100 ml}$$

The results are expressed in mg of sodium salicylate/100 ml, since the standards contain sodium salicylate. It may be requested that the results be reported as salicylic acid. This may be done by use of the molecular weight of sodium salicylate, which is 160, and the molecular weight of salicylic acid, which is 138. Divide the molecular weight 138 by the molecular weight 160 and multiply by the sodium salicylate value to obtain the salicylic acid value. Set up the formula as follows:

$$\text{Sodium salicylate} \times \frac{138}{160} =$$

$$\text{Sodium salicylate} \times 0.86 = \text{Salicylic acid}$$

or

$$\text{Salicylic acid} \times \frac{160}{138} =$$

$$\text{Salicylic acid} \times 1.16 = \text{Sodium salicylate}$$

Normal range

1 mg or less of sodium salicylate/100 ml of plasma

Toxic level

Above 30 mg of sodium salicylate/100 ml of plasma

Sulfonamides

The sulfonamides are used to combat many types of infection. They found their greatest use during World War II, just prior to the advent of penicillin. There are about ten different sulfa drugs. Their optimum blood levels during therapy depend upon the particular drug, the disease, and the patient. When present in excessive amounts, they tend to crystallize out in the kidneys and may cause severe damage to these organs.

When the sulfonamides are present in the blood, the distribution between the cells and plasma is not equal. The plasma concentration is higher; oxalated whole blood should be used, therefore, since the optimum drug levels are expressed in concentration of the drug in whole blood.

Bratton-Marshall method[3]

Whole blood is deproteinized with trichloroacetic acid, and the sulfanilamide is measured in the filtrate, by means of the diazo reaction.

Reagents needed

Trichloroacetic acid (TCA), 5%	(D-219)
Sodium nitrite working solution B, 0.5%	(D-212)
Ammonium sulfamate, 0.5%	(D-331)
Coupling reagent (naphthyl ethylenediamine solution)	(D-332)
Sulfanilamide stock standard, 0.1 mg/ml	(D-333)

Sulfanilamide working standard, 0.001 mg/ml

Into a 100 ml volumetric flask, pipette 1.0 ml of stock standard solution and dilute to mark with 5% TCA. Stopper and mix thoroughly.

Procedure

1. Pipette 19.5 ml of 5% TCA into a 125 ml Erlenmeyer flask.
2. Add 0.5 ml of oxalated whole blood to the flask, stopper, and shake to mix.
3. Filter into another Erlenmeyer flask.
4. Prepare the following tubes:

Blank

Pipette 3.0 ml of 5% TCA.
Add 0.1 ml of sodium nitrite.
Add 1.0 ml of ammonium sulfamate solution.
Add 2.0 ml of coupling reagent.
Stopper and mix by inverting.

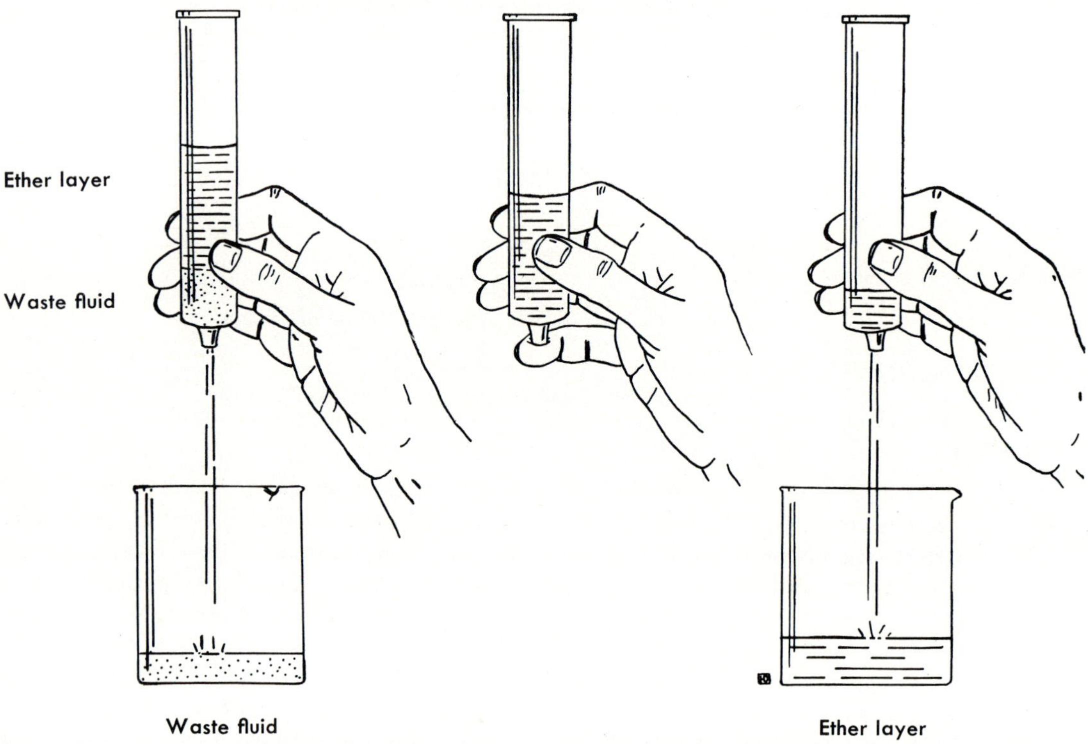

Fig. 17-1. Using the barrel of a syringe to separate ether from waste fluid.

Standard and unknown

Pipette 3.0 ml of working standard (0.003 mg of sulfanilamide) into a test tube.

Pipette 3.0 ml of blood filtrate (0.075 ml of blood) into the other test tube.

5. Add 0.1 ml of sodium nitrite to each tube; mix.
6. Let stand for 3 minutes.
7. Add 1.0 ml of ammonium sulfamate to each tube; mix.
8. Let stand for 2 minutes.
9. Add 2.0 ml of coupling reagent and mix well.

Read absorbance (A) against the blank with a wavelength of 540 mμ.

Dilution

If the concentration of the unknown is extremely elevated, use 1.5 ml of filtrate and 1.5 ml of 5% TCA. Multiply the result by 2.

Calculation

$$\frac{\text{Reading test}}{\text{Reading standard}} \times 0.003 \times \frac{100}{0.075} =$$

$$\frac{\text{Reading test}}{\text{Reading standard}} \times 4 = \text{mg sulfanilamide/100 ml blood}$$

NOTE: Although the use of separate calibrations for each of the sulfa compounds is preferable, it is possible to calculate the values of the various compounds. With the sulfanilamide used as a standard, the value obtained is multipled by a factor that gives the concentration of the drug being analyzed. The factors are calculated from the ratios of the molecular weights of the various drugs to that of the sulfanilamide:

Sulfanilamide	1.00
Sulfapyridine	1.45
Sulfathiazole	1.48
Sulfamethylthiazole	1.56
Sulfadiazine	1.45
Sulfadimethylpyrimidine	1.62
Sulfaguanidine	1.35
Sulfanilylsulfanilamide	1.90
Sulfasuccinylsulfathiazole	2.06
Sulfamethazine	1.30

DETECTION OF DRUGS IN URINE

Many of the following tests call for the use of a separatory funnel to separate waste fluids from ether or chloroform. If in an emergency there is not a separatory funnel available, the two liquids may be separated by using the barrel of a 10 or 20 ml syringe. The technique is illustrated in Fig. 17-1.

In ether extractions of blood or aqueous solutions, the ether layer will be on top. In chloroform extractions, the chloroform layer will be on the bottom.

Acetanilid

Acetanilid relieves pain and reduces fever. It is used in the treatment of neuralgia and rheumatism.

Yvon method

Reagents needed

Chloroform
Mercurous nitrate

Procedure

1. Mix urine and pour about 15 ml into a separatory funnel.
2. Add about 15 ml of chloroform, stopper, and shake.
3. Allow to stand until the two layers are separated.
4. Carefully allow the bottom chloroform layer to run out into a 100 ml beaker.
5. Gently boil down this chloroform layer to about 4 ml.
6. Add about 0.5 g of mercurous nitrate. If acetanilid is present, a green color will form.

Antipyrine

Antipyrine relieves pain and reduces fever. It is used in the treatment of neuralgia and rheumatism.

Fieux method

Reagents needed

Sulfuric acid, concentrated
Sodium metaphosphate
Sodium nitrate, saturated (D-334)

Procedure

1. Mix urine and pour about 20 ml into a small flask.
2. Add 1 ml of concentrated sulfuric acid and 2 g of sodium metaphosphate.
3. Mix and filter.
4. Add 5 drops of a saturated solution of sodium nitrate to the filtrate. If antipyrine is present, a green color will be produced.

Chloral hydrate

Chloral hydrate is used as a hypnotic in tetanus, insomnia, labor, mania, and many other conditions.

Todd-Sanford method

Reagents needed

Hydrochloric acid, concentrated
Ether
Benedict's qualitative reagent (D-270)

Procedure

1. Mix urine and transfer 100 ml to a 250 ml beaker.
2. Boil off 75 ml, leaving 25 ml. Cool.
3. Add 5 ml of concentrated hydrochloric acid and mix.
4. Transfer to a separatory funnel.
5. Add 15 ml of ether and shake well.
6. Separate the two layers, saving the ether layer (top) in a small beaker.
7. Using a hot plate, evaporate the ether just to dryness. Allow to cool.
8. Add 5 ml of distilled water and mix.
9. Run a qualitative Benedict test for glucose on this solution. If the glucose test is positive, chloral hydrate is present in the urine.

Iodine

Iodine may be found in the urine after treatment with iodides.

Todd-Sanford method

Reagents needed

Sulfuric acid, concentrated
Nitric acid, concentrated
Chloroform

Procedure

1. Mix urine and measure 5 ml into a small test tube.
2. Add 4 drops of concentrated sulfuric acid and mix.
3. Add 4 drops of concentrated nitric acid and mix.
4. Add 2 ml of chloroform and shake well. If iodine is present, the chloroform layer will become a pink to reddish violet color.

Morphine

Morphine is a form of opium used to relieve pain.

Todd-Sanford method

Reagents needed

Ammonium hydroxide, concentrated
Ether
Sulfuric acid, concentrated
Formalin

Procedure

1. Mix urine and measure 15 ml into a small beaker.
2. Add ammonium hydroxide until the urine is alkaline to litmus or nitrazine paper.
3. Transfer to a separatory funnel, add 15 ml of ether, and shake well.
4. Separate the two layers, collecting the ether layer (top) in a small beaker.
5. Using a hot plate, evaporate the ether just to dryness.
6. Cautiously pipette 1 ml of concentrated sulfuric acid into a test tube; add 1 drop of formalin to this test tube and mix.
7. Add a few drops of this mixture to the ether residue in the beaker. If morphine is present, a color will be produced, which changes from red to violet to blue.

Phenolphthalein

Phenolphthalein is sometimes used as a purgative.

Reagent needed

Ammonium hydroxide, concentrated

Procedure

1. Add about 5 ml of concentrated ammonium hydroxide to about 25 ml of urine. If phenolphthalein is present, a pink to red color will be produced.

Quinine

Quinine has many uses, but it is probably best known for its effectiveness in the prevention and treatment of malaria. It is also used to reduce fever and to stimulate uterine contraction in labor.

Todd-Sanford method

Reagents needed

Ammonium hydroxide, concentrated
Ether
Ethyl alcohol, 10%
Sulfuric acid, 10% (D-319)
Iodine, ethyl alcohol solution (D-279)

Procedure

1. Mix urine and add ammonium hydroxide dropwise until the urine is alkaline.

2. Transfer 100 ml to a separatory funnel, add 15 ml of ether, and shake well.
3. Separate the two layers, collecting the ether layer (top) in a small beaker.
4. Using a hot plate, evaporate the ether just to dryness.
5. Add 2 ml of 10% alcohol, 1 drop of 10% sulfuric acid, and 1 drop of an ethyl alcohol solution of iodine. Mix.
6. Heat gently and allow to cool. If quinine is present, green crystals will form.

Salicylic acid and acetylsalicylic acid

Salicylic acid and acetylsalicylic acid (aspirin) are used to relieve pain and reduce fever.

Siebold-Bradbury method

Reagents needed

Potassium carbonate	
Lead nitrate, saturated	(D-335)
Ferric chloride, 10%	(D-241)

Procedure

1. Mix urine and add potassium carbonate until the urine is alkaline to litmus or nitrazine paper.
2. Measure 25 ml into a small flask.
3. Add 25 ml of a saturated solution of lead nitrate, mix, and filter.
4. To the filtrate, add about 5 drops of a 10% solution of ferric chloride. If salicylic acid or acetylsalicylic acid is present, a violet color will be produced.

POISONS

This section considers tests for the toxic substances arsenic and mercury. Although salts of these elements are used in certain medications, arsenic and mercury are usually considered industrial poisons. More detailed discussions may be found in texts on toxicology.

Arsenic

Reinsch test

Reagents needed

Hydrochloric acid, concentrated
Copper foil
Ethyl alcohol
Ether

Procedure

1. Mix urine and transfer 40 ml to a small beaker.
2. Add 10 ml of concentrated hydrochloric acid and mix.
3. Add a 1-inch square of clean copper foil and boil for a few minutes. If arsenic is present, a gray or bluish film is deposited on the copper. To confirm the presence of arsenic, proceed with steps 4, 5, and 6.
4. Allow to cool. Remove the copper foil and dip successively into water, alcohol, and ether. Allow to dry.
5. Roll up the foil like a cigarette and insert in a glass tube.
6. Gently heat over a Bunsen burner. The arsenic will sublime to the cooler parts of the tube. The crystals may be seen with a magnifying glass and will appear as eight-sided figures.

Mercury

Reinsch test

Reagents needed

Hydrochloric acid, concentrated
Copper foil
Ethyl alcohol
Ether

Procedure

1. Mix urine and pour 40 ml into a small beaker.
2. Add 10 ml of concentrated hydrochloric acid and mix.
3. Place a 1-inch square of copper foil in the acidified urine and warm to 50° C.
4. Set aside overnight. If mercury is present, it will be deposited on the copper foil as a bright lustrous mirror. To confirm the presence of mercury, proceed with steps 5, 6, and 7.
5. Remove the copper foil and dip successively into water, alcohol, and ether. Allow to dry.
6. Roll up the foil like a cigarette and insert in a glass tube.
7. Heat gently over a Bunsen burner. If mercury is present, it will be deposited on the cooler portion of the glass tubes. It can be seen with a magnifying glass and will appear as fine metallic globules.

Lead

For lead by atomic absorption refer to Chapter 10.

Plasma lithium by flame photometry[5a]

In recent years, lithium salts have been used in the treatment of mental patients of the manic group. Since levels above 2.0 mEq Li/L of blood may cause dangerous side effects, it is essential to follow such treatment with periodic plasma lithium determinations.

Methods have been described for the Kipp and EEL flame photometers, which require deproteinization of plasma.[1,6] This is a time-consuming process and there is some doubt as to whether some of the lithium ions are absorbed by the precipitated proteins. Automated methods are available that employ dialysis but in laboratories where requests for lithium determinations are few and infrequent, these are impractical.

The method described here employs a greater dilution of the plasma (1:25 instead of 1:5), which obviates the need for deproteinization. The use of a serum blank and standard minimizes background interference. The emission is linear in the range of 0 to 2.0 mEq Li/L of plasma; hence only one standard is necessary.

Since many laboratories already have a Coleman flame photometer, the only additional accessory needed is a Coleman lithium filter (21-211), which isolates the 671 mμ emission line of lithium. The lower limit of detection with this filter is 0.0014 mEq Li/L.[5]

Reagents needed

Lithium stock standard, 10 mEq Li/L	(D-336)
Lithium working standard, 1 mEq Li/L	(D-337)
Sterox SE, 1% (Coleman reagent 1)	(D-338)
Sterox SE, 0.02% (Coleman reagent 5)	(D-339)

Procedure

Either serum or heparinized plasma may be used.

1. Prepare the following in 25 ml volumetric flasks.

Blank: 0.5 ml of reagent 1 and 1.0 ml of lithium-free serum
Standard: 0.5 ml of reagent 1, 1.0 ml of lithium-free serum, and 1.0 ml of lithium working standard (1 mEq/L)
Unknown: 0.5 ml of reagent 1 and 1.0 ml of unknown plasma or serum

2. Dilute each flask to volume with distilled water and mix well.
3. Place the lithium filter in position, atomize a portion of the blank solution, and set the galvanometer to zero.
4. Atomize a portion of the standard and set the galvanometer to the midpoint on the *black* scale (5 mEq on the K-Ca scale).
5. Atomize a portion of the diluted unknown plasma and record reading.

Calculation

$$\frac{\text{Reading unknown}}{\text{Reading standard}} \times 1 = \frac{\text{Reading unknown}}{5} = \text{mEq Li/L plasma}$$

Comments

The blank and the standard solutions must be prepared fresh daily. Serum used in the blank and standard must be lithium-free and must have normal electrolyte and protein values. If obtaining this type of serum presents a problem, a commercial control serum such as Versatol, with values in the normal range, may be used.

When eleven replicate determinations were run on each of three specimens, 0.57, 1.05, and 1.60 mEq Li/L, coefficients of variation of 1.4%, 1.3%, and 1.3%, respectively, were obtained.

REFERENCES

1. Barrow, G. R.: The estimation of lithium in blood, J. Med. Lab. Techn. **17**:236, 1960.
2. Bjorling, C. O., Berggren, A., and Willman-Johnson, B.: Determination of barbituric acid derivatives as mercury complexes, J. Pharm. Pharmacol. **11**:297, 1959.
3. Bratton, A. C., and Marshall, E. K., Jr.: New coupling component for sulfanilamide determination, J. Biol. Chem. **128**:537, 1939.
4. Broughton, P. M. G.: A rapid ultraviolet spectrophotometric method for the detection, estimation and identification of barbiturates in biological material, Biochem. J. **63**:207, 1956.
5. Coleman Technical Report, T-174B, Determination of lithium with the Coleman flame photometer, 1965.

5a. Erickson, M. E.: The estimation of plasma lithium using a Coleman model 21 flame photometer, Publication pending, 1969.

6. Helm, H. J. v.d., and Andriesse, D.: The determination of lithium in serum during therapy with lithium salts, Clin. Chim. Acta **6**:747, 1961.
7. Korzun, B. P., Brody, S. M., Keegan, P. G., Luders, R. C., and Rehm, C. R.: Rapid chromatographic method for the identification and estimation of glutethimide (Doriden) in blood, J. Lab. Clin. Med. **68**:333, 1966.
8. Maher, J. F., Schreiner, G. E., and Westervelt, F. B., Jr.: Acute glutethimide intoxication. I. Clinical experience (twenty-two patients) compared to acute barbiturate intoxication (sixty-three patients), Amer. J. Med. **33**:70, 1962.
9. Sunshine, I.: Alcohol in biological materials. In Seligson, D., editor: Standard methods of clinical chemistry, vol. 3, New York, 1961, Academic Press, Inc.
10. Trinder, P.: Rapid determination of salicylate in biological fluids, Biochem. J. **57**:301, 1954.
11. Wallenius, G., Zaar, B., and Lausing, E.: A rapid method for determination of barbiturate in serum or urine, Scand. J. Clin. Lab. Invest. **15** (supp. 69): 252, 1963.
12. Wuth, O.: Rational bromide treatment; new method for its control, J.A.M.A. **88**:2013, 1927.

• 18

• Special tests

THE PREGNANCY TEST

The blood and urine of pregnant women contain a chemical substance that is known as human chorionic gonadotropic hormone. When this hormone is injected into female mice, rats, or rabbits, the ovaries of the animals become enlarged and bloody. This is the principle of the pregnancy test. However, recently developed immunologic tests have replaced the biologic assays.

In addition to normal pregnancy, the hormone may be present in the following conditions: cancer of the ovary, beginning of menopause, abnormal pregnancies, abortions, and cystlike moles. The hormone may also be present in men with malignant tumors of the testes. Consequently, the pregnancy test may be used not only to reveal an early pregnancy, but also to help diagnose a pathologic condition.

Immuno-plate pregnancy test (HCG)[2]

This gel-diffusion procedure is a simple, rapid, and minimal-cost test done on urine specimens. It is based on an immunologic reaction that gives results similar to those obtained by the usual animal test.

Urine contains high levels of chorionic gonadotropin from 6 weeks after the last normal menstrual cycle through the first trimester of pregnancy. This is the hormone previously described as injected into the animal. A rapid concentration based on the Yale technique[2] makes it possible to use a specimen obtained at any time of the day, although the first morning specimen would have the highest concentration. A measured amount of urine is placed in a well in an agar plate containing an antiserum to human chorionic gonadotropin. As this concentrate (urine) diffuses into the gel, a zone of precipitation is produced, the diameter of which is related to the concentration of the hormone in the specimen.

*Reagents and supplies needed**

1. **Disposable plastic plates, each containing anti–human chorionic gonadotropic serum in agar, with sodium azide 0.1% as a preservative. Each plate should be individually sealed in a plastic envelope and consist of three wells: one for a positive control, and two for the urine test.**
2. **A plastic vial containing 24 capillary pipettes and 2 rubber bulbs.**
3. **A vial containing 0.5 ml of positive control (200 IU of human chorionic gonadotropin/ml in a buffered saline with sodium azide 0.1% for a preservative).**
4. **Dropper bottle containing 1 ml of eluting fluid (dilute ammonium hydroxide containing bromthymol blue).**
5. **Dropper bottle containing 2.5 ml of adsorbent suspension (kaolin in acetic acid–acetate buffer). An excess is provided to facilitate mixing before use.**

---

*These materials may be purchased in a kit form from the Hyland Laboratories or obtained from your local dealer.

Procedure

1. Pipette 2.0 ml of clear urine into a small tube (12 × 75 mm).
2. Mix adsorbent suspension by shaking vigorously. Add 2 drops to the urine and mix well.
3. Mix frequently for 2 minutes; then centrifuge sufficiently (about 5 minutes) to pack kaolin particles.
4. Discard supernatant fluid by inverting the tube. Either shake off the last drop or remove with absorbent tissue.
5. Add 1 drop of eluting fluid and bring kaolin into suspension, either by flicking the tube with the finger or by using a vibrator.
6. Centrifuge for 5 minutes. Do not discard supernatant.
7. Using a capillary pipette, fill one well of the *immuno-plate* with the supernatant until level with the agar surface.
8. Fill one well of each plate used, to the level of the agar surface, with the positive control.
9. In the third well, use a urine specimen from a nonpregnant person.
10. Place the plastic cover on the plate and incubate at 37° to 40° C for 4 to 5 hours.
11. Examine plate for zones of precipitation around wells, using indirect (reflected) or oblique light against a dark background as when observing slide or tube agglutination tests. A zone, halo, or ring of precipitation appears as a translucent circular area around a well, faint but sharply demarked at its outer edge. A magnifying device such as a hand lens will be helpful in recognizing the zone.
12. A halo or ring of precipitation of any diameter, either smaller or larger than the one produced by the positive control, is a positive reaction.

Precautions

1. Refrigerate urine until it is tested or freeze the specimen if it is to be stored for a prolonged time.
2. Refrigerate the immuno-plates but do *not* store in freezer. Do not use the plates if the agar gel is dry or shrunken excessively.
3. The eluting fluid is an alkaline pH (blue or greenish blue). If it appears acid (yellow), do not use the fluid, since the proper elution of chorionic gonadotropin from the kaolin adsorbent will occur *only* if the elute is alkaline.
4. The wells should be completely filled but not overfilled, since there will be a deposit formation on the surface of the agar that could be mistaken for a positive reaction.

Reactions

1. With a rapid-reacting positive specimen, distinct zones of precipitation may show before 4 hours of incubation.
2. After the first trimester, pregnant women frequently excrete too little of the hormone for a positive result to be detected.
3. False positive results may be found in nonpregnant patients with chorioepithelioma or hydatidiform mole. With these two disorders chorionic gonadotropin is usually excreted.

FETAL CELL DIFFERENTIAL STAIN

EUGENE G. HAMILTON, M.D., and JOSEPH R. SIMMLER, A.S., B.S., M.A.*

The determination and identification of fetal cells is important for several reasons. The staining of fetal cells finds its greatest application in determining fetomaternal bleeds, both qualitatively and quantitatively. Other applications include the following:

1. To determine whether cells obtained in a bloody amniocentesis are fetal or maternal.

*Method devised by Mr. Simmler and application discussed by Dr. Hamilton.

2. To determine the percentage of adult-type cells present in a baby that has had intrauterine transfusions.
3. In cases of antepartum bleeding to determine whether any of the blood comes from the fetus.
4. In adult hemoglobinopathies to determine the presence of hemoglobin F.

The following staining procedure is based on the principle that the acid buffer system differentially elutes the hemoglobin from the adult cells, leaving the fetal cells intact. Subsequent staining reveals only the fetal cells while the adult cells are unaffected.

*Reagents needed**

Ethanol, 80% (v/v)
Citrate buffer system

Solution A
Dissolve and dilute 28.4 g of disodium phosphate to 1 L.

Solution B
Dissolve and dilute 21.01 g of citric acid to 1 L.
Add 24.7 ml of solution A to 75.3 ml of solution B.

Fetal cell stain
Dissolve 0.1 g of erythrosine in 100 ml of distilled water.

*Reagents in a kit form can be obtained from Simmler & Son, Inc.

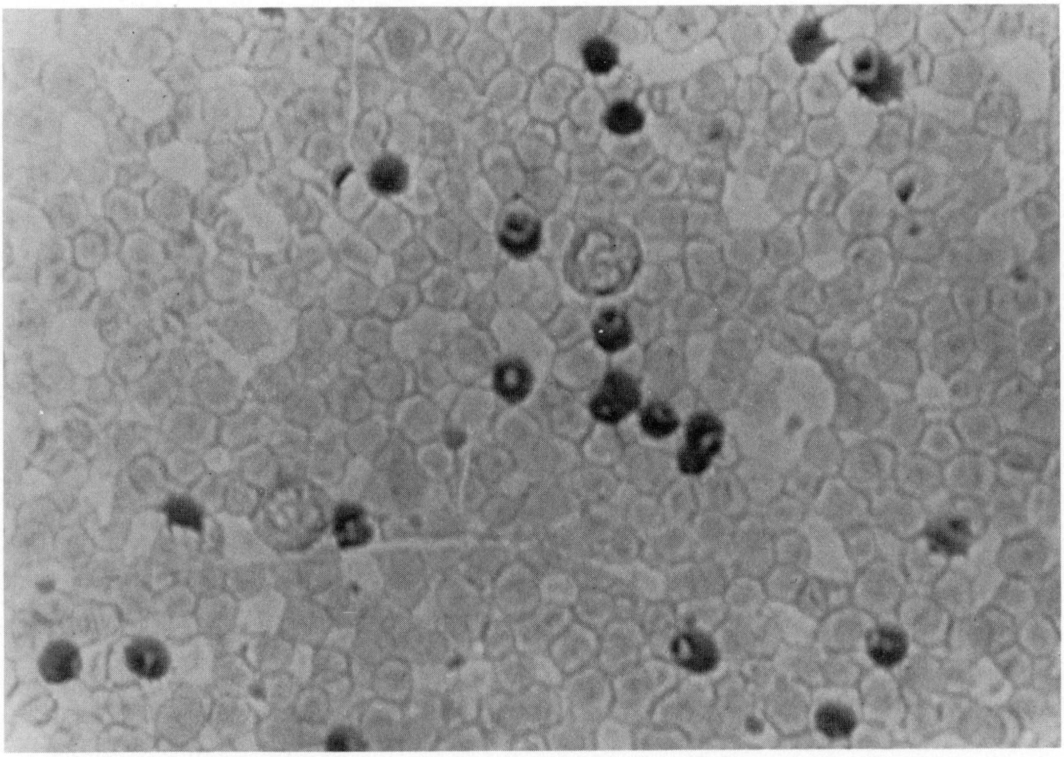

Fig. 18-1. Photomicrograph of a stained blood smear containing fetal and adult cells. Fetal cells stain red against a background of adult ghost cells. High power, dry objective.

Staining procedure

1. Fix the air-dried thin smear, made from fresh blood, in 80% ethanol for 5 minutes (the alcohol can be used several times).
2. Wash the slide well and allow it to dry.
3. Stand the slide vertically in a Coplin jar containing the citrate buffer which is at a temperature between 33° and 38° C. Let the slide set in the buffer for 3 minutes. After the first minute pull the slide out and slowly replace for 2 minutes more.
4. Wash the slide well. Place the slide on a staining rack and cover the slide with the stain. Stain for 3 minutes.
5. Wash and dry the slide. Examine the slide under high dry objective.

Findings

The fetal red cells containing the hemoglobin F are stained red while the adult cells are barely evident as unstained ghost cells. (See Fig. 18-1.)

Comments

Only fresh blood or fluids should be used. It is recommended that from time to time smears be made from a mixture of fetal cells and adult cells to check the activity of the reagents. The reagents can be used several times.

SERUM IRON[1]

*Bothwell-Mallett method**

Iron circulates in blood plasma, bound to the β_1 globulin. The absorption of the iron occurs in the upper part of the small intestine and perhaps in the stomach. The role of iron in the body is closely associated with that of hemoglobin. Disorders related to abnormal values are listed in Table 18-1.

Table 18-1. Disorders affecting the serum iron

When increased	When decreased
Pernicious anemia	Iron deficiency anemia
Aplastic and hemolytic anemia	Malignancy
Hemochromatosis	Uremia
Acute hepatitis	Chronic infections
Polycythemia	Nephrotic syndrome
	Extensive burns

Most methods for the analysis of serum iron are done in three basic stages.

Principle of method of determination

1. The iron is usually split from its protein by an acid (HCl). The acidified serum is allowed to stand or incubate 15 minutes to several hours (according to the method used), for the iron to separate from its protein.
2. The free iron is isolated by the precipitation of proteins with trichloroacetic acid. The tube is allowed to stand for 15 minutes for complete precipitation.

---

*Modified by the Hematology Department, Washington University School of Medicine, St. Louis, Mo.

3. The color complex is developed with a 2,2′-dipyridyl at a pH range of 3 to 9. Ascorbic acid is used as the reducing agent, and the solution is buffered with saturated sodium acetate.

Reagents needed

Hydrochloric acid, 2.0 N	(D-303)
Trichloroacetic acid (TCA), 20%	(D-289)
Ascorbic acid, crystalline	
2,2′-Dipyridyl, 0.4%	(D-340)
Sodium acetate, saturated	(D-281)
Iron stock standard, 100 mg/100 ml (stable indefinitely)	(D-341)
Iron working standard, 100 μg/100 ml (stable)	

Pipette 0.5 ml of stock standard (100 mg/100 ml) into a 500 ml volumetric flask. Dilute to mark with iron-free water. Stopper and mix by inverting.

Procedure

1. To the following iron-free tubes:
   Blank
      Pipette 3.0 ml of 2 N HCl.
      Add 4.0 ml of double-distilled water.
   Standard
      Pipette 3.0 ml of 2 N HCl.
      Add 4.0 ml of working standard (100 μg/100 ml).
   Unknown
      Pipette 3.0 ml of 2 N HCl.
      Add 4.0 ml of serum.
2. Using glass stirring rods, mix all tubes thoroughly.
3. Let stand for 15 minutes.
4. Add 2.0 ml of 20% TCA to the tubes.
5. Stir vigorously for 45 seconds.
6. Let stand for 15 minutes.
7. Centrifuge the tubes for 20 minutes at 3000 rpm (since the blank and standard have no precipitate, they will not have to be centrifuged).
8. To *each* spectrophotometer tube (blank, standard, and unknown):
   a. Pipette 6.0 ml of supernatant.
   b. Add 30 mg of crystalline ascorbic acid.
   c. Add 0.5 ml of 2,2′-dipyridyl, 0.4%, to each tube.
   d. Add 2.5 ml of saturated sodium acetate.
   e. Shake the tubes to mix thoroughly.
   f. Let stand for 5 minutes to develop color fully.
9. Read the reagent blank, standard, and unknowns against a distilled water blank at a wavelength of 520 mμ. Convert the readings to absorbance (A).

Calculation

$$\frac{(\text{A unknown} - \text{A blank})}{(\text{A standard} - \text{A blank})} \times 100 = \mu\text{g Fe}/100\text{ ml}$$

Normal range

Women	50 to 170 μg/100 ml
Men	61 to 193 μg/100 ml

MACROGLOBULINEMIA

Macroglobulins consist of proteins with a molecular weight exceeding 1,000,000, and a sedimentation constant of 20 or more. Under normal conditions these account for 3% to 5% of the total serum proteins. Under abnormal conditions, as in Waldenström's macroglobulinemia, these proteins comprise over 20% of the total proteins.[4] Ultracentrifugation is required for the definite diagnosis of this disease. A simple qualitative test may be done, due to the solubility of macroglobulins in electrolyte solutions and their precipitation by dilution. High serum viscosity and a rapid erythrocyte sedimentation rate may also suggest macroglobulinemia.

Sia test (water-dilution test)

Reagent needed

Saline, 0.9% (D-22)

Procedure

1. **Fill a small test tube with distilled water.**
2. ***Drop* 1 drop of serum into the distilled water. Use either pipette or dropper.**
3. **Negative results: the serum either disperses or forms a slight haziness.**
4. **Positive results: a heavy precipitate forms, dropping to the bottom of the tube when macroglobulins are present.**
5. **If positive, centrifuge and pour off supernatant.**
6. **Redissolve the precipitate in a small amount of 0.9% sodium chloride (saline) solution.**
7. **Reprecipitate by adding distilled water to the small volume of saline. This suggests the presence of macroglobulins.**

False positive reactions may occur with the Sia test. The test usually shows negative results when small or even moderate numbers of macroglobulins are present. Consistent positive results are not always found with large quantities of macroglobulin.

CRYOGLOBULINEMIA[3]

Cryoglobulins are a group of proteins that precipitate from cooled serum and redissolve upon warming or returning to room temperature. Cryoglobulinemia is usually found in patients having multiple myeloma, although it may be found associated with other conditions or by itself. The cryoglobulins may be detected by a variety of methods based upon the effects of temperature and their solubility. Two simple methods are given below. The specimen *must not* be refrigerated before the test.

Cold-precipitable globulins

Procedure

1. **Place fresh serum in the refrigerator for a minimum of 4 hours or overnight.**
2. **If there is a gelatinous precipitate at the end of the chilling period, proceed with step 3.**
3. **Bring to room temperature by putting tube into a beaker of water at room temperature.**
4. **Cryoglobulins will redissolve upon warming.**

Erythrocyte sedimentation rate—at 37.5° and at 10° C

1. **In the cold at 10° C, the red cells do not settle out, due to the precipitation and gel formation of cryoglobulins.**
2. **For red cells warmed to 37.5° C, the sedimentation rate is usually abnormally rapid.**

REFERENCES

1. Bothwell, T. H., and Mallett, B.: The determination of iron in plasma or serum, Biochem. J. 59:599, 1955.
2. Hyland HCG immuno-plate pregnancy test, Los Angeles, 1963, Hyland Division, Travenol Laboratories, Inc.
3. Page, L. B., and Culver, P. J.: A syllabus of laboratory examinations in clinical diagnosis, rev. ed., Cambridge, Mass., 1961, Harvard University Press.
4. Waldenström, J.: Abnormal proteins in myeloma, Advances Intern. Med. 5:298, 1952.

• 19

• Chromatography

JOSEPH R. SIMMLER, A.S., B.S., M.A.

DEVELOPMENT OF CHROMATOGRAPHY

Chromatography is a separation process by which the components of a mixture are separated by differentially interacting between two physical phases that contact each other in a countercurrent fashion. This general definition applies to paper, thin-layer, column, and gas chromatography. The greatest application has been in the field of organic chemistry, and to an increasing extent for routine analyses in clinical and medical laboratories. The principles and application were first described by a Russian scientist, Michael Tswett, in 1906 and published in a detailed monograph in 1910. Apparently the separation technique lay dormant for many years, since there was very little chromatographic work published between 1910 and 1931. Tswett separated the complex mixtures of many botanic extracts by adsorption chromatography. Since the time of Tswett's original work several different techniques have been introduced: ion exchange, partition, emulsion, gel permeation, and gas chromatography.

Paper chromatography, which is essentially partition chromatography, was first introduced in 1944 by Consden, Gorden, and Martin. The first description of a kind of thin-layer chromatography was by Ismailov and Shraiber in 1938 and preceded the introduction of partition chromatography on paper strips by several years. The initial application of paper chromatography was much greater than that of thin-layer chromatography. While many lipophilic (nonpolar) substances were separable on adsorption columns, the many hydrophilic (polar) substances such as the nucleic acids, polysaccharides, and proteins were difficult to separate until Martin and Synge introduced partition chromatography. The development from partition chromatography on a column to the introduction of partition chromatography on paper strips took only three years. However, the "laying-out" of a Tswett adsorption column to an open-face column or thin-layer took many years. The development and application of thin-layer chromatography quite obviously trailed that of paper chromatography. The latter method was rapidly applied and in a short time the number of publications reporting use of this separation technique was in the hundreds. However, the application of thin-layer chromatography by Williams and Kirchner, and particularly by Stahl, soon popularized the use of thin-layers. Today the earlier situation is reversed. This is not surprising, since thin-layer chromatography has several advantages over paper chromatography. The thin-layer method is faster and the separated components are more compact, thus giving a higher sensitivity or degree of detection. The resolution or degree of separation is better, and corrosive detecting reagents such as sulfuric acid may be used. The choice of adsorbents for thin-layer use is extensive. Also, prepared plates of uniform adsorbent thickness and of different adsorbents can now be obtained commercially at reasonable prices. It is for these reasons that thin-layer chromatography will be stressed over paper chromatography.

Paper strips that act as physical supports for silicic acid, alumina, ion exchange resins, or cellulose, having been treated chemically, have resulting properties different from those of the original paper and cannot be classified under paper chromatography. These paper-supported adsorbents have the same advantages as thin-layers on glass plates.

Column chromatography has application where there is a need for more of the separated components than can be obtained by preparative thin-layer chromatography. About 1 g of a mixture can be chromatographed on a 20 × 20 cm thin-layer preparative plate. The automation of column chromatography and the use of fraction collectors, solvent feed, and sample zone detectors have made column chromatography very convenient but it is still a time-consuming procedure.

If the components of a mixture can be volatilized at a reasonable temperature or can be converted to volatile compounds, it may be best to separate the mixture by gas chromatography. The desirable features of gas chromatography are the component detectors and the readout. The recordings of the separated components can be quantitated by measuring the area under the curve or tracing. The resolved components on a thin-layer can in many cases be automatically read-out by use of densitometers or reflectometers, but the results are not as accurate or as quantitative as in gas chromatography. Good recoveries, 95% or more, of the separated components on a thin-layer can be obtained by removing the adsorbent and component from the plate, separating the component from the adsorbent, and measuring the component by some analytic method —usually, spectrophotometry. The recovery of separated components on a column would be done in essentially the same fashion or the components be moved out of the column by continued elution and the collected fractions analyzed, usually by evaporating the solvent and weighing the residue. The purity of the residues can be ascertained by thin-layer chromatography.

THEORY OF CHROMATOGRAPHY

In paper, thin-layer, column, and gas chromatography, three different principles are applied: adsorption, partition, and ion exchange. The development of the chromatographic system, or the process of separation of the components of a mixture, is primarily by elution and displacement (p. 524) and very rarely by frontal displacement. In almost every instance the components of a mixture are in contact with a solid adsorbent or stationary liquid phase on some kind of support medium, and the components are caused to migrate along the support medium by being confronted by massive (relative to the amount of solutes) gas or liquid mobile phase. The components reside in each phase for some part of the overall time required to develop the chromatogram. The amount of time spent in the respective phases is related to the partition coefficient or adsorption coefficient of the solute or component.

In adsorption chromatography the solutes interact between the solid stationary phase, such as silicic acid or alumina, and a mobile liquid phase. In partition chromatography the solutes interact between a stationary liquid phase, on an inert solid support, and a mobile liquid phase. The separation of gaseous components of a volatile mixture by interaction with a solid adsorbent in a column and a mobile gas phase flowing through the column is called gas-solid chromatography (GSC) and when the stationary phase is a liquid, it is called gas-liquid chromatography (GLC).

The solutes of a mixture will reside for a portion of the overall time of development in one phase and the remaining amount of time in the other phase. During the time the solutes are in the stationary phase, it is assumed that they are stationary and when they are in the mobile phase, they move at the same speed as the mobile phase. The distance the solutes move is equal to the time spent in stationary phase multiplied by the rate of movement of the mobile phase: $D_s = t_s \times V_m$.

The distance the solute moves divided by the distance the mobile phase moves on a thin-layer or paper chromatogram is defined as the R_f value; therefore $R_f = \frac{\text{D solute}}{\text{D solvent}}$. The R_f value is a characteristic property of a solute and is related to its molecular structure. Thus solutes having different molecular composition and structure will have different R_f values. In gas chromatography the retention time, R_t, is a characteristic value of the volatile solute and is measured from the time the solute is injected into the system until the time it leaves the column and is detected. Separation of those solutes of a mixture having different R_f or R_t values should then be feasible. The R_f value, while a characteristic and intrinsic value of a solute under a given set of conditions, is difficult to reproduce and the cause of the variation is unpredictable.

Another term that is related to the R_f value is the R value. The R value is equal to the fraction of the solute in the mobile phase at any one time. The R value is also related to the velocity of migration of the solute zone and to the partition coefficient $\alpha = \frac{C_s}{C_m}$ in which C_s and C_m are the concentrations of the solute in the stationary phase and the mobile phase, respectively. In any volume element, the amount of solute in the mobile phase divided by the amount of the solute in the stationary phase is $\frac{A_m C_m}{A_s C_s}$ or $\frac{A_m}{\alpha A_s}$ where $\frac{A_m}{A_s}$ is the relative volume of the mobile phase compared to the stationary phase (partition chromatography). The ratio $\frac{A_m}{\alpha A_s}$ is equal to the fraction R of solute in the mobile phase divided by 1 − R in the stationary phase or $\frac{R}{1-R}$. Equating the ratios $\frac{R}{1-R} = \frac{A_m}{\alpha A_s}$ or $\alpha = \frac{A_m(1-R)}{A_s R}$ and solving for R gives $R = \frac{A_m}{A_m + \alpha A_s} = R_f$.

This equation is commonly employed for R_f. One error in relating the R_f value to the R via the equation is that the velocity of the solvent front may be different from the velocity of the solvent at the solute zone, and this is usually the case. Another error is assuming that A_m (the amount of solvent) is the same through the chromatogram or paper strip. This is actually not the case.

Fortunately the solvent movement and solvent concentration along the chromatogram have compensating effects and the overall error contribution to the validity of the equation $R_f = \frac{A_m}{A_m + \alpha A_s}$ is very small. From this equation it can be seen that the R_f value of the solute is inversely related to the partition coefficient. The R_f value of a solute in a particular two-phase liquid system can be predicted by determining the partition coefficient of the solute. For example, the amount of solute distributed between equal volumes of water and benzene gives a ratio of 10; that is, ten times as much solute is found in the water phase as compared to the amount of solute found in the benzene phase. The partition coefficient then is equal to 10. If we assume $A_m = A_s$, we find the volume of the stationary phase equal to the volume of the mobile phase in the volume element; the R_f is equal to 0.10 by calculation. A low R_f value of the solute is what would be predicted if the solute were more soluble in the water phase and the water phase were the stationary phase.

The calculation of the R_f value in adsorption systems is more complicated than in partition systems. This is due to the difficulty of measuring the distribution coefficients as well as the fact that distribution coefficients between solids and liquids or gases are not constant. As the amount of solute is increased, the partition coefficient remains constant; but this is usually not true of the distribution coefficient. The variation of the distribution coefficient on a thin-layer plate is evident by the shape of the solute spot as it moves up the plate. In some cases there is a tailing of the spot or a trailing of the solute behind the major concentration of the solute. The reverse condition, in which the major concentration of the solute spot lags behind, is called bearding. These effects are primarily due to the association and dissociation of the solute in the two different media with a corresponding difference of solubility or adsorptive properties. More comprehensive discussions of the theory of chromatography are available.[1-3]

A plot of the concentration of the solute in the stationary phase versus the concentration in the mobile phase seldom results in a linear relationship because of molecular association, change in degree of ionization, and diffusion in the liquid phase or liquid phases. A picture of an ideal and an actual situation would appear as shown in Fig. 19-1. A profile of the resulting zone or spot assumes the shapes as indicated in Fig. 19-2.

The effectiveness of separation is related to the zone shape and size. The

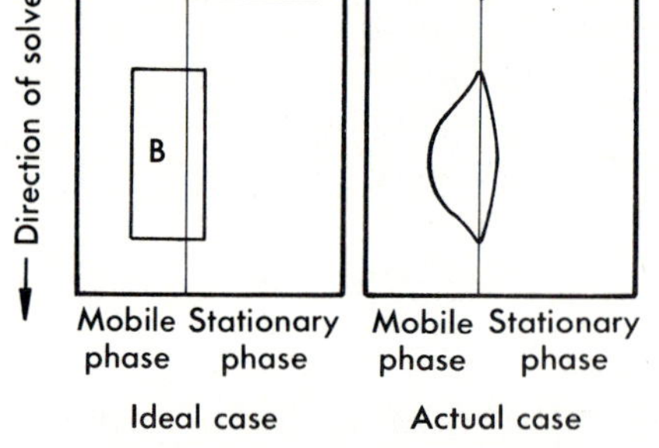

Fig. 19-1. Distribution of the solutes of an ideal and an actual condition.

zone spreading is due primarily to ordinary diffusion, eddy diffusion, and local nonequilibration.

1. Ordinary diffusion is related directly to the length of time the chromatogram is allowed to develop. Ideally the zone is elliptic in shape, with the larger axis in the direction of the solvent flow (Fig. 19-3).
2. Eddy diffusion occurs as the liquid moves past the particles of the adsorbent. At one moment the liquid passes readily and in another moment it is greatly restricted. This phenomenon is related to the square root of the particle size, $\sqrt{dp}$.
3. Local nonequilibration is due to (a) rate of flow of the mobile phase and (b) rate of attainment of equilibration between the two phases.

Other parameters, which affect these causes of zone or spot shape, are temperature, pressure, thickness of the adsorbent layer, amount of solute, and degree of saturation of the atmosphere in the chamber with the mobile solvent. In addition to determining the zone shape, these effects also influence the R_f value. It should be pointed out that as the plates become less active with storage time, the R_f value will also change (adsorption chromatography).

Since the zone shape is the sum of the parameters listed, this compromise should be attained for maximum results: the solvent velocity should not be too

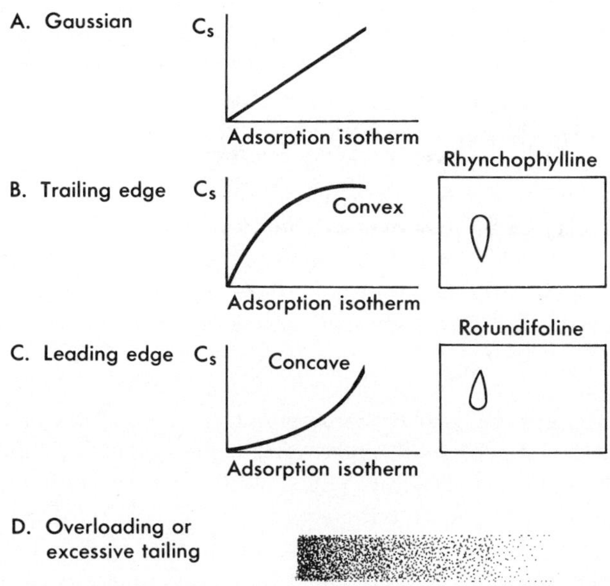

Fig. 19-2. Concentration of the solute versus its adsorption isotherm.

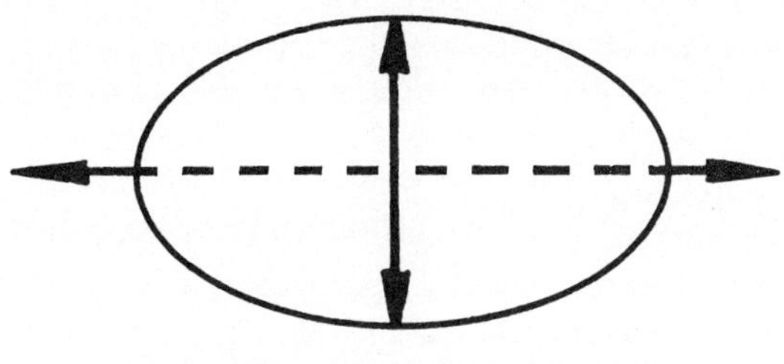

Fig. 19-3. Elliptic shape of zone.

fast, and the adsorbent particle size should not be small enough to excessively inhibit the solvent flow.

In partition chromatography it appears that the solute zones are even more vulnerable to diffusion effects than in adsorption chromatography because two liquid phases are involved.

Another cause for distortion of zone shape and change in R_f is related to the displacement effect of a large amount of one component on a lesser amount of another component when the two have very similar R_f values.

It is actually these same parameters, which operate to a lesser degree on thin-layers, that account for better results than on paper strips. The superiority is not due to some intrinsic property of the cellulose. When paper is powdered and the cellulose spread as a thin-layer, it acts chromatographically very much like other thin-layer adsorbents in that it has a much shorter development time and the detection threshold is higher.

GENERAL ASPECTS OF CHROMATOGRAPHY

In the separation technique called chromatography there are two physical phases, a stationary phase and a mobile phase. These will be considered separately although no chromatographic system can be described without considering both phases, as well as the components of the mixture to be separated.

Stationary phase

The stationary phase in adsorption chromatography is a solid, such as silicic acid or alumina, and the mobile phase is a liquid. In partition chromatography the stationary phase is a liquid, usually affixed to a supporting solid phase, such as silica gel or cellulose, and the mobile phase is a liquid. The two liquid phases will have limited solubility in each other.

While different solids—calcium sulfate, magnesium oxide, calcium carbonate, and activated charcoal, to name a few—have been used as adsorption media, silicic acid, alumina, and magnesium silicate are the adsorption solid phases most widely used. Experience has shown that the particle size of the adsorbents for thin-layer chromatography should be between 5 and 70 μ in diameter. For column chromatography the particle size should be between 50 and 200 μ in diameter. If the particles are too large, the zones on the column or the spots on the plate are diffused and poorly separated. When the particles of the adsorbent are too small, development is very slow. Silicic acid, silica gel, and silica are synonymous terms for powdered solids having the general formula $SiO_2 \cdot XH_2O$. Silica powders consist of three-dimensional siloxane structures (Si-O-Si) with surface silanol groups (Si-OH). On the silanol groups there is a mono- or multi-molecular layer of adsorbed water. The water may be reversibly removed by heating at 100° C. At temperatures above 170° C an irreversible thermal decomposition of the silanol groups occurs, liberating fixed water. The adsorbent property or activity of the silicic acid depends on the content of silanol groups, free water, pore size, shape, and particle size. Diatomaceous earth is a completely hydrated silicic acid having practically no adsorption characteristics and it is used exclusively for partition chromatography.

The activity of an adsorbent is roughly related to its hardness; the harder the material, the greater the activity. Talc is a very weak adsorbent but silicic acid, which is much harder than talc, is an active adsorbent. The adsorptive properties of silicic acid are due to the electrostatic forces that are responsible for the crystal lattice structure of crystalline silicic acid. The electrostatic forces are, for adsorption purposes, on the surface of the crystal and are due to residual ionic forces of more or less uncombined ions. The activity sites on the crystal are along the edges, corners, cracks, and defects of the crystal. The electrostatic forces of the adsorbent induce dipole moments* and increase the dipole moment of polar compounds. The binding of compounds onto the surfaces of adsorbents is thus due to the ion-dipole or to dipole-dipole forces.

Alumina used as an adsorbent has several forms. The adsorption activity sites of alumina may be due to residual ionic alumina, normal basic Al-OH, and aluminate $Al\text{-}O^-$ or $Al\text{-}O\text{-}H^+$, depending on how the alumina is prepared. The latter two forms act as ion exchangers. The acidic form of the alumina is used to separate organic acids and dyestuffs while basic alumina is used to separate amino acids, alkaloids, and amines. The basic grade is also used as a dehydrating agent, such as for the removal of water from ether. The alumina also removes peroxides in the ether. Neutral alumina is used for the separation of ketosteroids, vitamins, carotenoids, ketols, and lactones. Basic alumina very likely has the greatest application. Acetone as an eluent should not be used with basic alumina, since it causes the acetone to condense to form diacetone alcohol. However, acetone can be used with neutral alumina. The latter is used to dehydrate acetone.

Although the various forms of alumina and silicic acid can be prepared in the laboratory for chromatographic application, it is much more convenient and less expensive to purchase the prepared adsorbents. Sources and the names of some of the distributors or manufacturers of adsorbents are indicated in footnotes and addresses listed in Appendix E. The adsorbents from a given supplier can usually be used for both thin-layer and column chromatography. This is convenient, since the chromatographic system developed for thin-layer chromatography will generally apply to column chromatography.

The activity of the silicic acid and alumina and other adsorbents is a function of their water content. Techniques for establishing the activity of alumina have been described which involve the use of several organic dyes, such as azobenzene, *p*-methoxyazobenzene, sudan yellow, sudan red, and *p*-aminoazobenzene.

The activity grade, running from grade I which is highly active to grade V which is least active, is established by the R_f values of these dyes when a mixture of the dyes is developed on alumina plates, without a binder and with carbon tetrachloride as the eluent. The R_f values of the dyes have been previously established on alumina of known designated activity. This is called the Brockman-Schodder scale of activity. Activated silicic acid has an activity comparable to alumina of grade II or III. Grade I alumina is prepared by heating the alumina to redness, and silicic acid is usually activated by heating to 110° C for 30 to 60 minutes. When a water slurry is made of the adsorbent, it is completely deactivated and must be activated by heating the prepared plates. Various ratios of adsorbents having different activities can be combined to give desired activities. To partially deactivate silicic acid, for instance, diatomaceous earth is added to the silicic acid.

*Dipole moment. A chemical dipole arises when an unequal charge distribution exists between the atoms within a molecule. This results in the molecule's having a positive and a negative charged end.

Glass fiber paper, obtainable from Whatman and other sources, can be prepared of adsorption chromatogaphy. The glass fiber paper is impregnated with silicic acid by a process of dipping the paper into a dilute solution of potassium silicate, removing the excess liquid, drying, and then dipping the paper into hydrochloric acid. After it is washed until neutral, the paper strip is dried at 100° C. By the same method, but with an aluminum chloride solution and ammonium hydroxide, a glass fiber paper impregnated with alumina can be prepared. ChromARsorb* is essentially the same as this product, in that the silica and alumina adsorbents are supported by a network of glass fibers intimately incorporated with the adsorbents to give the quality of a sheet or paper.

Various other materials can be added to the adsorbents for special applications. For the separation of very low polar compounds or unsaturated hydrocarbons, a silver nitrate solution is used in making the adsorbent slurry to prepare the thin-layer plates. The silver ions interact with the unsaturated bonding of the compounds.

A binder is usually added to the adsorbents. The binders aid in affixing the adsorbents to the plates and in giving some degree of mechanical strength to the thin-layer. Some binders are glue, starch, and plaster of paris. Plaster of paris, a calcium sulfate hemihydrate, is by far the most commonly used binder and can readily be made by heating a reagent grade of calcium sulfate dihydrate for several hours between 140° and 180° C.

Sometimes other chemicals are added to the adsorbent to impart desired acidic or basic properties, or buffers are added to maintain a given pH value during the development of the plate. Boric acid, sodium borate, and sodium arsenite, for instance, have been incorporated into the adsorbent to aid in the resolution of glycols, lipids, and triglycerides. Ortho-hydroxyquinoline has been used to resolve some metal ions which form a complex with the *o*-hydroxyquinoline. The addition of silver nitrate to the adsorbent permits the separation of compounds containing double bonds.

A desirable additive to practically all the adsorbents is a fluorescing agent. Many organic compounds (all aromatic compounds) will adsorb shortwave ultraviolet light and their presence on the plate can be readily ascertained as a dark area against the fluorescing background. In some cases, as little as 0.1 μg can be detected. There is also the added feature that the compounds can be detected without being chemically altered.

The stationary phase in partition chromatography is a liquid supported on a solid medium. For the most part, the stationary liquid phase is water with the supporting medium being cellulose, kieselguhr, silica, or alumina. The activity sites of the alumina or silica when used as the support for partition chromatography are completely saturated with the highly polar stationary water phase and their adsorptive characteristics are thus neutralized. The stationary liquid phase for the usual or normal partition chromatography is a very polar liquid, such as water, while the mobile phase is an essentially nonpolar liquid, such as benzene. A few systems are given in Table 19-1. When the stationary phase is an essentially nonpolar liquid and the mobile phase a polar liquid, the system is termed reverse phase partition chromatography. Examples of this type of partition are presented in the second half of Table 19-1.

---

*Mallinckrodt Chemical Works.

Table 19-1. Solvent systems for partition chromatography

Stationary phase	Mobile phase
Normal partition	
Water	Benzene–ethanol
Water	*n*–Butanol
Water and sodium hydroxide	Chloroform–ethanol
Water buffered at pH 5	Water–alcohol–butanol
Formamide	Benzene–heptane–chloroform
Water buffered at pH 6.5	Ethanol–water
Water buffered with sodium acetate	Ethyl acetate–propanol–water
Polyethylene glycol	Isopropyl ether–formic acid–water
Reverse phase partition	
Undecane	Acetic acid
Tetradecane	Acetic acid–water
Paraffin	Acetic acid–water
Silicone oil	Methyl cyanide–acetic acid–water
Petroleum	Acetone–methyl cyanide

Table 19-2. Elutropic solvent series

Solvent	Dielectric constant
Water	80
Methanol	33
Ethanol	24
Acetone	21
n–Propanol	20
n–Butanol	17
Ethyl acetate	6.0
Chloroform	4.8
Diethyl ether	4.3
Trichloroethylene	3.4
Toluene	2.4
Benzene	2.3
Carbon tetrachloride	2.2
Cyclohexane	2.0
n–Hexane	1.9

(Arrow from Carbon tetrachloride up to Acetone: Increasing elution power ↑)

Mobile phase

In any chromatographic system there are two phases, which relate to each other in a countercurrent fashion, with the solute residing in one phase or the other for a given period of time. The distribution of the solutes of a mixture between the two phases is a function of their polarity or dipole moment and in adsorption chromatography, once the solid adsorption phase has been established, it is usually a simple matter to find the proper solvent, the mobile liquid phase, that will resolve the solutes. As the polarity of the solvent is increased, the greater the elution and/or displacement effect it has. A list that shows increasing polarity of solvents can be arranged and is called an elutropic series (Table 19-2). Depending on the dipole moments of the solutes, a solvent can be found that will separate the solutes. However, if there is a significant difference of the dipole moments of the solutes, no one system of solvent or solvents will satisfactorily separate the solutes by a single development. Other development techniques, which can separate solutes having very different dipole moments can be resorted to, and will be discussed in the practical aspects of chromatography.

The tendency of a solute to dissolve in and move with the mobile phase is called elution. For elution to take place the polarity of the solvent should be such that its solvating effect on the solute can satisfactorily overcome the adsorptive power of the adsorbent. Movement of the solutes along the adsorbent can also take place by displacement. In this case, the polarity of the solvent is such that it competes with the solute for the adsorption sites of the adsorbent, a case of pushing the solute along. Elution is usually thought of as occurring when the solvents are nonprotonic such as the hydrocarbons, ethers, and carbonyl compounds. Displacement usually occurs when the more protonic solvents such as the alcohols and amines are used as the developing solvent. No doubt both phenomena occur during development, and from a practical point of view the elution concept suffices for interpretation of the results.

The dielectric constant of the mobile liquid phase can be adjusted to almost any value by mixing one or more of the solvents, provided they are miscible. The dielectric values are not arithmetically additive; that is, the dielectric constant of equal volumes of acetone and benzene is not one half the sum of the dielectric constants of these two solvents. A small amount of acetone added to the benzene increases the dielectric constant to a greater extent than might be expected. The dielectric constant of a combination of solvents is more an exponential relationship than an arithmetic one. As might be expected, there are several combinations of solvents that will have the same dielectric constant; practically, however, it is best to get the desired dielectric constant by combining the proper proportions of the solvents having the nearest dielectric constant values.

The choice of the mobile liquid phase for partition chromatography is somewhat less involved than it is for adsorption chromatography. The first consideration is that the two liquid phases must be only slightly miscible in one another. For two liquids to be immiscible, their dielectric constants should be quite different—one solvent being much more polar than the other solvent.

The movement of the solutes of a mixture by partition chromatography is a function of the solubility of the solute in the liquid phases, and the ratio of the concentrations of the solute in the two liquid phases is termed the partition coefficient. If the solute is more soluble in the stationary polar liquid phase, it will spend a greater proportion of the overall development time in that phase and consequently have a low R_f value. Conversely, if the solute is more soluble in the mobile less polar liquid phase, it will have a high R_f value.

MOLECULAR STRUCTURE AND CHROMATOGRAPHIC RELATIONSHIP

The chromatographic behavior of a compound is related to its polarity. The polarity is due to the elemental composition and the structural composition of the compound. Thus the polarity is a function of the number and kind of functional groups that make up the compound. The very polar compounds move through a chromatographic system slowly or have low R_f values; conversely, compounds having little or no polar properties move rapidly through the chromatographic systems and thus have high R_f values. The hydrocarbons are essentially nonpolar (lipophilic) and are hardly adsorbed at all. Unsaturated hydrocarbons are more strongly adsorbed, the greater the number of double bonds they contain. Hydrocarbons having conjugated double bonds are more strongly adsorbed than hydrocarbons having isolated double bonds. The hydrocarbons can be separated on highly active adsorbents with nonpolar developing solvents or by reverse phase chromatography. By incorporating silver ions in the silicic acid adsorbent it is possible to separate the hydrocarbons and, in some cases, the members of a homologous series. The adsorption affinity of the hydrocarbons increases as functional groups are added. The adsorption affinity increases with the functional group in the following sequence: $—CH_3$, —O— alkyl, C=O, $—NH_2$, —OH, —COOH. As an example, if a mixture of organic compounds were chromatographed on silicic acid with benzene used as the developing solvent, esters and ethers should be in the upper part of the plate, the aldehydes and ketones in the middle of the plate, and the alcohols and acids in the bottom of the plate.

Hydrogen bonding also plays a part in the chromatographic behavior of molecules. Molecules that have intrahydrogen bonding are retained less strongly than isomeric compounds without intrahydrogen bonding.

CHOICE OF KIND OF CHROMATOGRAPHY

Initially, when confronted with the problem of what kind of chromatography should be used to separate the components of a mixture, reliance on experience, literature reference, or possibly some shrewd and careful preliminary experimentation may be necessary. If the mixture is volatile and thermally stable, gas chromatography should be the method of choice.

Since thin-layer and column methods are closely related and involve the same kind of chromatography, the choice of one or the other would be determined by the end purpose. The preliminary investigation to find a suitable chromatographic system would be by thin-layer (TLC). The chromatographic system found best by thin-layer could then be directly applied to column chromatography in the event that larger amounts of the separated components were needed for further study.

The kind of chromatography to be used depends on the type of compounds being separated. If the compounds are closely related structurally but differ only in molecular weight, as in a homologous series, it would be best to try to separate them by partition chromatography, relying on the differences of solubility that usually prevail in a homologous series. On the other hand, the adsorption characteristics of compounds having similar composition, such as isomers, but not having the same structural arrangement can be quite different and this could be the basis of their separation. However, if a mixture is very complex, preliminary separation of major groups may have to be made and the separated groups further analyzed. Experimentally, adsorption chromatography is quicker and has a higher capacity than does partition chromatography.

Generally very polar compounds such as carbohydrates, amines, and nucleotides are best separated by partition chromatography while the less polar compounds are best separated by adsorption chromatography. To generalize, as the

Text continued on p. 537.

Table 19-3. Choice of chromatographic system

Kind of compound	Adsorption	Partition
Long-chain aliphatic ketones	+	
Aliphatic and aromatic aldehydes and ketones	+	
C_2 to C_{12} aliphatic acids		+
Aromatic aldehydes and ketones	+	
C_4 to C_6 dicarboxylic acids		+
Monohydric alcohols		+
Vanillin and other substituted benzaldehydes	+	
Alkaloids	+	
Phenols and cresols		+
Dihydric and trihydric alcohols		+
Alkaloids and barbiturates	+	
Cardiac glycosides	+	
Sulfonamides	+	
Amino acids		+
Penicillins		+
Sugars and derivatives		+
Miscellaneous lipids	+	
Glycerides	+	

Table 19-4. Sensitivity of methods of separation to structural differences in the compounds to be separated*

Structural differences	Adsorption	Partition	Ion exchange
Molecular size	+ to (+)	+ to ++	+ to ++
Structural isomers			
Chains, rings	+ to (+)	+	–
Chains, branching	+ to (+)	(+)	(+)
Position of double bonds	+ to (+)	–	–
Steric isomers			
Cis-trans at a double bond	+ to (+)	+	+
Cis-trans at a ring	++ to (+)	+	+
Optical isomers	(+)	(+)	+ to (+)
Number of double bonds	++	+	–
Conjugation of double bonds	++ to +	+	+
Number of nonpolar substituents	+ to (+)	+ to (+)	+ to (+)
Number of polar substituents	++	++	++
Polarity of substituents	++	++	++

Ratings: ++, very good; +, good; (+), fair; –, not investigated.

*From Turba, F.: Chromatographische Methoden in der Proteinchemie, Berlin, 1954, Springer-Verlag.

Table 19-5. Some applications of thin-layer chromatography by classes of compounds

Compounds	Adsorbent	Developing solvent system	Detection	Reference
		Alkaloids		
Fifty-four alkaloids	Silica gel G	$CHCl_3$-acetone-diethylamine (5:4:1)	Potassium platinate	4
		$CHCl_3$-diethylamine (9:1)		
		Cyclohexane-$CHCl_3$-diethylamine (5:4:1)		
		Cyclohexane-diethylamine (9:1)		
		Benzene-ethylacetate-diethylamine (7:2:1)		
	Alumina G	$CHCl_3$		
		Cyclohexane-$CHCl_3$ (3:7) plus 0.05% diethylamine		
	Silica gel G–NaOH	Methyl alcohol		
Indoles	Silica gel G	Methylacetate-2-propanol-NH_4OH (45:35:20)	Van Urk reagent	5
			Prochaska reagent	
		$CHCl_3$–ethyl ether (95:5)	2,4-dinitrophenylhydrazine	
Alkaloids and barbiturates	Silica gel G	Methanol	Dragendorff's reagent	6
		$CHCl_3$–ethyl ether (85:15)	I_2 and KI in acetic acid	
			Fluorescein	
			$FeCl_3$ in acetic acid	
			$KMnO_4$ in acetic acid	
Opium alkaloids	Alumina buffered at pH 5 with acetate	Water-ethanol-butanol (1:1:9)	UV light	7
Opium alkaloids	Silica gel G	Benzene-methanol (8:2)	Dragendorff's reagent	8
		Aldehydes and ketones and their derivatives		
Long-chain aliphatic ketones	Silica gel G	Benzene-ether mixtures	Phosphomolybdic acid	9
		Toluene-ether mixtures		
		Petroleum ether–ether mixtures		
Vanillin and other substituted benzaldehydes	Silica gel G	Petroleum ether–ethyl acetate (2:1)	Hydrazine sulfate in HCl	10
		Hexane–ethyl acetate (5:2)	Methanolic KOH	
		$CHCl_3$–ethyl acetate (98:2)	Methanolic KOH followed by diazotized sulfonic acid	
		Decalin–CH_2Cl_2–methyl alcohol (5:4:1)		

Continued.

Table 19-5. Some applications of thin-layer chromatography by classes of compounds—cont'd

Compounds	Adsorbent	Developing solvent system	Detection	Reference
		Aldehydes and ketones and their derivatives—cont'd		
2,4-dinitrophenyl hydrazones of aldehydes	Silica gel G	Benzene–petroleum ether (3:1) Benzene–ethyl acetate for aromatic aldehydes	Colored compounds	11
2,4-dinitrophenyl hydrazones of hydroxyaldehydes and ketones	Aluminum oxide G Silica gel G	Toluene–ethyl acetate (3:1), (1:1)	Colored compounds	12
		Amines and amine derivatives		
Aliphatic amines	Silica gel G	Ethanol (95%)-NH_4OH (25% NH_3) (4:1) Phenol-water (8:3) Butanol–acetic acid–water (4:1:5)	Ninhydrin	13
	Silica gel G buffered at pH 6.8	Ethanol-water (70:30)		
Diphenylamine derivatives	Silica gel G	$CHCl_3$ Toluene Benzene	$NaNO_2$ and H_2SO_4	14
Pteridines	Silica gel G	Dimethylformamide-water (190:10), (20:180) 5% citric acid NH_4OH (25% NH_3) Many others	UV light	15
		Carbohydrates and glycosides and their derivatives		
Simple sugars	Silica gel G buffered with boric acid	Benzene–acetic acid–methanol (1:1:3) Methyl ethyl ketone–acetic acid–methanol (3:1:1)	Naphthoresorcinol	16
Simple sugars	Silica gel G	Propanol-concentrated NH_4OH–water (6:2:1)	Ninhydrin Aniline hydrogen Phthalate	17
Sugar acetates and inositol acetates	Silica gel G	Benzene with 2% to 10% methanol	Water Ferric hydroxamate	18
Cardiac glycosides	Silica gel G	$CHCl_3$-methanol (9:1) $CHCl_3$-acetone (65:35)	H_2SO_4 and acetic anhydride	19

Table 19-5. Some applications of thin-layer chromatography by classes of compounds—cont'd

Compounds	Adsorbent	Developing solvent system	Detection	Reference
		Carboxylic acids and their derivatives		
Dicarboxylic acids	Silica gel G	Benzene–methanol–acetic acid (45:8:4) Benzene–dioxane–acetic acid (90:25:4)	Bromphenol blue acidified with citric acid	20
p-Hydroxybenzoic acid esters	Silica gel GP	Pentane–acetic acid (88:12)	UV light	21
		Dyes		
Food dyes	Cellulose powder MN 300 G and 300 MN	Aqueous sodium citrate (2.5%)–NH_4OH(25% NH_3) (4:1) Propanol–ethyl acetate–acetic acid (6:1:3) Tertiary butanol–propanoic acid–water (50:12:38)	Colored compounds	22
		Essential oils		
Essential oils	Silica gel G	Benzene-$CHCl_3$ (1:1)	$SbCl_2$ in $CHCl_3$	23
Oils of sage	Silica gel G	Benzene	Vanillin in H_2SO_4	24
		Lipids—class separation		
Miscellaneous lipids	Silica gel G	Petroleum ether–ethyl ether–acetic acid (90:10:1)	2′, 7′-dichlorofluorescein	25
Serum lipids and phospholipids	Silica gel G	$CHCl_3$-methanol-water (80:25:3) Petroleum ether–ethyl ether–acetic acid (90:10:1), (60:40:1)	2′, 7′-dichlorofluorescein	26
Plasma lipids	Silica gel G	$CHCl_3$-methanol-water (65:24:4)	Bromthymol blue–ammonia Ninhydrin Fuchsin-H_2SO_4 Ferric hydroxamate	27
Brain tissue lipids	Silica gel G	$CHCl_3$-methanol-water (24:7:1) $CHCl_3$ CCl_4	Ammonium molybdate –$HClO_4$ Phosphomolybdic acid Ninhydrin Dragendorff's reagent Diphenylamine	28 29

Continued.

Table 19-5. Some applications of thin-layer chromatography by classes of compounds—cont'd

Compounds	Adsorbent	Developing solvent system	Detection	Reference
		Essential oils—cont'd		
Lipids of feces and fecoliths	Silica gel G	Petroleum ether–ethyl ether–acetic acid (80:20:1)	50% H_2SO_4	30
		Lipids—fatty acids and their derivatives		
Fatty acids	Silica gel G impregnated with undecane	Acetic acid–methyl cyanide (1:1) Solution 1 with 0.5% Br_2	Copper acetate–dithiooxamide	31
Fatty acid methyl esters	Silica gel G	Hexane–ethyl ether–acetic acid (85:15:1)	2′, 7′-dichloro-fluorescein	32
		Lipids—glycerides		
Glycerides	Silica gel G	$CHCl_3$-benzene (70:30)	Concentrated H_2SO_4	33
Mono- di- tri-glycerides	Silica gel G	Petroleum hydrocarbon (40-60 C)–diethyl ether (70:30)	50% H_2SO_4	34
		Lipids—phospholipids, sulfolipids, and glycolipids		
Phospholipids Sulfolipids Glycolipids	Silica gel Silica gel G Silica gel plus 10% ammonium sulfate	Various proportions of $CHCl_3$-methanol-water Propanol-ammonia Butanol-pyridine-water	I_2 vapor 50% H_2SO_4 Ninhydrin Dragendorff's reagent Ammonium molyb-date–$HClO_4$	
		Phenols and alcohols and their derivatives		
Cyclohexanol derivatives	Silica gel G	Benzene Benzene–ethyl acetate (80:20)	Water	35
Phenols	Silica gel G plus oxalic acid	Hexane–ethyl acetate (4:1), (3:2) Benzene–ethyl ether (4:1) Benzene		36
Long-chain aliphatic alcohols	Silica gel G	Petroleum ether–ethyl ether–acetic acid (90:10:1)	50% H_2SO_4	37
3,5-dinitro-benzoate of alcohols and phenols	Silica gel G	Benzene–petroleum ether (1:1) Hexane–ethyl acetate (85:15), (75:25)	Colored compounds	38

Table 19-5. Some applications of thin-layer chromatography by classes of compounds—cont'd

Compounds	Adsorbent	Developing solvent system	Detection	Reference
		Amino acids and proteins and their derivatives		
Amino acids	Cellulose powder MN 300	Upper phase of butanol–water–acetic acid (4:5:1) Pyridine–methyl ethyl ketone–water (15:70:15) Methanol-water-pyridine (80:20:4)	Ninhydrin	39
Polypeptides and derivatives	Silica gel G	Butanol saturated with 0.1% NH_4OH	Ninhydrin	40
		Steroids		
Miscellaneous steroids	Silica gel G	Benzene Benzene–ethyl acetate (9:1), (2:1) Cyclohexane–ethyl acetate (9:1), (19:1) 1,2-dichloroethane Others	$SbCl_3$ in $CHCl_3$	41
Thirteen miscellaneous steroids	Silica gel GP	$CHCl_3$–ethyl acetate (3:2), (3:1) Benzene–ethyl acetate (1:1), (9:1)	I_2 vapors	42
	Aluminum oxide GP	Hexane–ethyl acetate (1:1)		
Estrogens	Silica gel G	Benzene-ethanol (9:1)	$SbCl_5$ in $CHCl_3$	43
Pregnanetriols in urine	Unbound aluminum oxide	Benzene-ethanol (9:1)	70% H_3PO_4	44
Cholesterol	Silica gel G	Hexane–ether–acetic acid (50:50:2)	$HClO_4$	45
Cholesterol ethers	Silica gel G	Benzene–petroleum ether (40:60)	50% H_2SO_4	46
Bile acids, free and conjugated	Silica gel G	Acetic acid–isopropyl ether–amyl acetate–CCl_4–benzene (5:30:40:20:10) for free acids Propionic acid–isoamyl acetate–water–propanol (15:20:5:10) for conjugated acids	Phosphomolybdic acid	47

Continued.

Table 19-5. Some applications of thin-layer chromatography by classes of compounds—cont'd

Compounds	Adsorbent	Developing solvent system	Detection	Reference
		Amino acids and proteins and their derivatives—cont'd		
Corticosteroids	Unactivated silica gel G	$CHCl_3$-methanol-water (485:15:1), (188:12:1), (90:10:1)	50% H_2SO_4	48
Aldosterones from urine	Silica gel G	Cyclohexane-2-propanol (7:3) $CHCl_3$–glacial acetic acid (8:2)	Tetrazolium blue	49
		Terpenoids		
Forty-nine terpenoids	Silicic acid, starch-bound	Hexane	Fluorescein H_2SO_4-HNO_3	50
Carotenoids	Silica gel G + calcium hydroxide (1:4) Silica gel G + calcium hydroxide (1:4) impregnated with paraffin	Benzene–petroleum ether (1:1) Above solution with 1% methanol Methanol saturated with paraffin	Colored compounds	51
		Vitamins		
Carotenes and fat–soluble vitamins A, D, E, and K	Aluminum oxide, unbound	Methanol Carbon tetrachloride Xylene Eleven others	70% $HClO_4$	52
Tocopherol mixtures	Silica gel G Aluminum oxide D5	$CHCl_3$ Benzene	Ceric sulfate Phosphomolybdic acid	53
		Miscellaneous		
Flavonoids and coumarins of lemon	Silica gel, starch-bound, with phosphor	$CHCl_3$–acetone mixtures	UV light	54
Inorganic ions, potassium and magnesium	Silica gel G	Ethanol-methanol (1:1) plus 1% acetic acid	Acid violet 6BU	55
Insecticide, thiophosphate esters	Aluminum oxide, starch-bound	Heptane-acetone (10:1)	H_5IO_6-$HClO_4$-V_2O_5	56

number of functional groups increases, the polarity of the molecule also increases but as the molecular weight and carbon content increase, the polarity of the molecule decreases.

As a first approximation to the kind of chromatography to be used some generalities apply and can be used as a guide to preliminary experimentation. These are given in Table 19-3.

If they are known, the structural differences of the compounds of a mixture should also be considered in selecting a chromatographic method. The structural differences and a rating of the chromatographic methods are given in Table 19-4. Application of thin-layer chromatography to the classes of compounds is given in Table 19-5. This table includes the kind of chromatography, compound class, adsorbent, developing solvent system, and visualization technique of the separated components. An idea of the kind of thin-layer chromatography that may be advantageous to try may be obtained from the table.

In some respects it may be just as easy to try a few systems instead of doing extensive literature research. This can be readily done by spotting the sample on the thin-layer plate and introducing the developing solvent from a capillary pipette into the center of the spot. The developing solvent should be one taken from the top, middle, and bottom (decreasing polarity) of an elutropic solvent series. If the sample is colorless, the separated constituents can be visualized by some of the general visualizing agents such as sulfuric acid, sulfuric acid–chromate mixture, and iodine vapor. If the compounds are aromatic, the incorporation of an ultraviolet fluorescing agent in the adsorbent will reveal these compounds when the plate is exposed to ultraviolet radiation. The major constituent(s) should move about half the radius of the circle that the developing solvent makes around the spot. The same technique can of course be followed for partition chromatography. From this preliminary investigation a combination of two or more solvents, which straddles the desired distance, may be suggested. Another rapid method to check out a chromatographic system is to use small thin-layer plates made from microscopic slides and develop the plate in a Coplin staining jar containing a few milliliters of the solvent.

There are other kinds of chromatography such as ion exchange chromatography and gel permeation chromatography, but this chapter is intended to be primarily an introduction to the most commonly used chromatographic methods and to some of their practical applications. Briefly, however, gel filtration should be considered for trying to separate very large molecules where the differences in size of the molecules can be the basis of their separation. Amino acids, peptides, and proteins can be separated according to their molecular weights on layers of Sephadex, a gel filtration medium.[57]

Components of a mixture which have different ionic characteristics can be separated by ion exchange chromatography. The nucleic acid derivatives and proteins are such a class of compounds.[58] The stationary phase in ion exchange chromatography is for the most part organic compounds that are practically insoluble in highly polar solvents such as alcohol or water. There are two main types of ion exchange media—(1) the ion exchange resins, made from polymers so synthesized as to contain an anion exchanger group such as a quaternary ammonium or a cation exchanger group such as carboxylic or sulfonic acid, and (2) the ion exchange celluloses, made by reacting cellulose with various reagents. The latter results in functional groups being attached to the cellulose by ether or ester linkages; each functional group acts as a cation or anion exchanger site, depending on the functional group. The anion exchangers called DEAE (diethylaminoethyl cellulose) and ECTEOLA (epichlorohydrin triethanolamine cellulose) are used for the separation of a number of proteins, peptides, enzymes, hormones, and nucleic acids. The cellulose cation exchangers contain carboxymethyl and phosphate as the functional groups.

The application of a high voltage, low current, through the above media is called ionophoresis and electrophoresis and has been used to separate amino acids, peptides, and other classes of compounds. Thin-layer electrophoresis combines the separation features of thin-layer chromatography and electrophoresis.

Column chromatography

In column chromatography the sample, as a solution, is placed on a column of adsorbent in a narrow band. The adsorbent is supported for the most part in a glass tube. (For clinical application refer to Chapter 15, methods for catecholamine, pregnanediol, and pregnanetriol.) The developing solvent is then placed in a reservoir over the column of adsorbent, passing through the column by force of gravity. Sometimes pressure is applied to increase the flow rate of the solvent through the column. The components of the sample will ideally separate into narrow bands as the components preferentially move down the column under the influence of the moving solvent. Small portions or fractions of the issuing solvent, called the elutrient, are collected. If the mixture consisted of wholly colored components the degree of separation can be readily ascertained as they move down the column and it is an easy matter to collect these colored components as they come out of the column. In most cases the components are colorless and the elutrient is collected in small fractions which are analyzed. There are devices for detecting the passing of the separated components, which are based on a change in some physical property of the elutrient such as conductivity, absorbance, or refractive index—to name a few.

From a preliminary TLC investigation of the sample some rules can be presented for adapting the thin-layer chromatography to column chromatography. The solvent system for adsorption TLC should be of such a composition that the main component of the sample mixture has an R_f of not more than 0.3. Such a solvent system is used to pack the column and to resolve the sample mixture on the column. When TLC by partition chromatography is used to separate the components, a direct application of the TLC solvent system can be made to column chromatography but again the major components should have R_f of 0.5 or less. Generally the chromatographic columns are supported in glass tubes that are tapered at the bottom. The length of the tube should be no less than ten times the diameter of the tube and may be as long as 100 times the diameter. The size of the glass tube is determined by the amount of sample to be applied. A wad of cotton or glass wool is inserted into the tube down to the taper. The operation of the column can be improved by adding a bulb at the top of the glass tube, which serves as a reservoir for the developing solvent, and instead of a cotton plug, a sintered-glass disk or a Teflon filter disk can be used. Also a stopcock at the exit end of the glass tube aids in the control of the solvent flow. For optimum flexibility a glass tube with a ball socket at one end, which can be attached to a ball socket receptacle that accommodates a sintered-glass disk and valve control, can be constructed. The ball socket joint is secured with a pinch or screw clamp. The ball socket joint must be free of particles of adsorbent; otherwise the joint will leak, especially if the column is put under pressure. Chromatographic columns of various designs can be obtained from the laboratory supply houses.

Adsorbent

Different materials have been used as column packing but the most widely used are alumina and silicic acid, and, to a lesser extent, cellulose. Generally the particle size of the alumina or silicic acid is larger for column chromatography than for thin-layer chromatography. The particle size is somewhere between 80 and 200 mesh. The activity and kinds of alumina and silicic acid have already been discussed (refer to pp. 524 and 525).

The column for adsorption chromatography can be prepared in either of two ways: a wet method or a dry method. By the dry method a small amount of dry adsorbent is added at a time to the column and tapped or pressed firmly into the tube. The pressing is done with a long-handle plunger with a metal or glass disk at the end. Small amounts of adsorbent are added at a time until the desired column height is obtained. In the wet packing technique the adsorbent is mixed with a portion of the developing solvent or with some member of an elutropic series which is less polar than the solvent system that will be used to develop the column. Small increments of the slurry are added and the column is tapped with a rubber stopper or the end of a pencil to facilitate the settling of the adsorbent and to release entrapped air. For partition chromatography the stationary phase—silicic acid, alumina, or cellulose—is intimately mixed with the polar solvent in a mortar or in a powder blender. A slurry is made by using the developing solvent which has been saturated with the stationary liquid phase. The amount of stationary phase to be incorporated in the supporting medium depends on several factors. Small-size particles of silicic acid will hold more liquid than large-size particles. Kieselguhr, a diatomaceous earth, consists of very small particles of water-saturated silicic acid and is therefore particularly good for partition chromatography and reverse phase chromatography, but it will hold less of the stationary phase than small particle size silicic acid. Cellulose columns are difficult to prepare. The cellulose slurry should be added in small increments in order to obtain uniform packing. For most purposes the weight ratio of the stationary liquid phase to the supporting medium is 0.5 to 1 part stationary phase to support medium.

Application of the sample

The sample is introduced onto the column by dissolving it in the developing solvent. The developing solvent for partition chromatography is saturated with the same liquid that is in the stationary phase. The sample solution is carefully added to the column so as not to disturb the supporting medium. A piece of filter paper or a sintered-glass disk may be placed on the surface of the support medium. If the sample does not dissolve in a small amount of the developing solvent, a larger volume of solvent can be used and evaporated to dryness in the presence of a small amount of the supporting medium. The dry powder containing the sample is then quantitatively transferred to the column. This technique of sample introduction gives tight bands.

Solvent system

The solvent system to be used for column development is determined as given in the section on the mobile phase (p. 527). From the preliminary thin-layer investigation for suitable chromatographic system, the proper solvent system is determined and can be applied directly for development on the column. The developing solvent is usually added from some reservoir which is a part of the glass column. Usually better and sharper separations take place with a slow flow rate. However, if the flow rate is too slow, diffusion becomes excessive and the bands may overlap. On the other hand, if the flow rate is too fast, there is a tendency for the solvent to channel. Also, the optimum operational conditions are so distorted by an excessive flow rate that the ideal distribution of the component between the two phases cannot be attained. The proper flow rate is best determined by experimentation. The solvent flow rate can be changed during development but it is advisable not to stop the flow rate for any length of time. If pressure is applied to increase the flow rate, a constant pressure must be maintained; otherwise the column tends to separate on decreasing of the pressure. The polarity of the solvent can be changed during development (gradient elution) but the change must be done gradually or stepwise, going down the solvents of an elutropic series. Again in partition chromatography the developing solvent must be saturated with the solvent used in the stationary phase. A convenient way to gradually change the composition of the developing solvent is to place one solvent reservoir containing the more polar solvent over the less polar developing solvent reservoir attached to the column (Fig. 19-4). The solvents are mixed in reservoir B with a magnetic stirrer. A third solvent can be introduced in a like fashion. For complete recovery of the sample the column can be given a final elution with water, methanol, or methanol containing 5% glacial acetic acid. The solvents in all cases must be of the highest purity and, in some cases, the solvent can be first purified by percolating it through a column of the same support medium used in the developing or working column.

Detection

There are several ways of detecting the components as they come out of the column. In most cases the components are colorless and therefore some chemi-

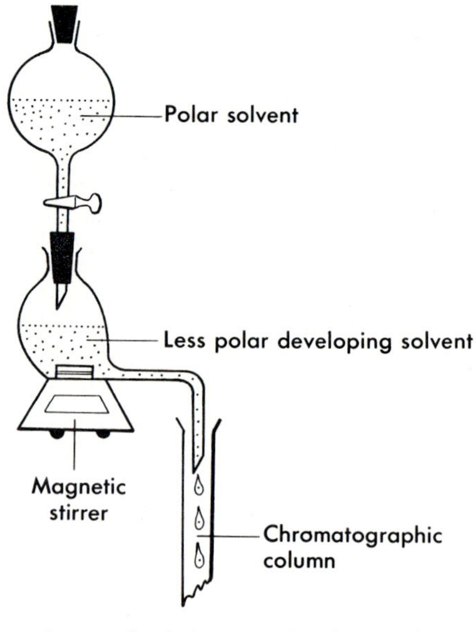

Fig. 19-4. Setup for gradual change of polarity of developing solvent.

cal or physical property of the solute must be exploited to reveal its exit from the column. Portions of the effluent are collected by either volume or time factors. There are available fraction collectors that count the number of drops and, after a preset number of drops has been collected in a receiving tube, bring up another tube to receive the next preset number of drops. The receiving tubes are in a turntable that rotates, aligning a new tube under the column outlet after each fraction. Another device for triggering the turntable is a fixed-volume siphon that, after a certain volume of effluent has been collected, will activate the servomotor driving the turntable. A different mechanism* for detecting components in the effluent is based on measurement of the amount of light reflected from the interface of the flowing stream and a glass surface. The amount of reflected light is related to the refractive index of the liquid and the refractive index of the glass. If the liquid and glass have the same refractive index, no light is reflected and, of course, the instrument cannot detect the component in the effluent stream. If gradient elution is used to develop the columns, the baseline of the recording device must be continually adjusted to compensate for the adsorption of the changing composition of the developing solvent. Many constituents such as nucleotides, phenylalanine, steroids, and hormones, can be detected as they come off the column, by ultraviolet flow analyzers.† The elutrient carrying the ultraviolet-adsorbing constituent passes through a flow cell of a spectrophotometer at the end of the column. The spectrophotometer feeds into a recorder. The recorder makes a continuous strip chart reading and indicates the test tube in the turntable collector of the constituents as they come off the column. If gradient elution is performed, the equipment will automatically compensate for the change in the ultraviolet adsorption characteristics of the solvent system.

A simple way to monitor effluent fractions is by thin-layer chromatography. Initially every fraction should be analyzed so a qualitative profile can be obtained to indicate where the major constituents' peaks are located in relation to the fraction collected.

In some cases the components of the mixture are separated in the column and the development is stopped short of passing the first component from the column. The column is then pushed from the supporting glass tube and the position of the components identified by streaking the column with an appropriate reagent, viewing under ultraviolet radiation, or cutting the column into small portions and eluting with a very polar solvent. The subsequent extract is evaporated to dryness and the residue weighed or subjected to further analysis.

Thin-layer chromatography

Practical aspects

Plate preparation

The following discussion on the preparation of thin-layer plates may be considered academic, since thin-layer plates can be purchased that are equally as good as handmade plates and in some respects are better. It is also less expensive to purchase prepared plates. Besides being readily available, prepared plates are more likely to have a uniform deposit of the adsorbent layer, which allows for a more accurate quantitative evaluation of the spot by either area measurement or densitometric or reflectrometric measurements. However, there are several rea-

*Waters Associates, Inc.

†Canal Industrial Corp.

Table 19-6. Ratio of adsorbent to water

Adsorbent	Supplier	Adsorbent/water (g/ml)
Silica gel G series	Brinkmann	30/60–65
Silica gel N series	Brinkmann	30/60–70
Silica gel H series	Brinkmann	30/80–90
Silica gel D series	Gelman	50/100
Silica gel TLC series	Mallinckrodt	30/60–70
Aluminum oxide G series	Brinkmann	30/40
Aluminum oxide, Woelm	Aluphram	35/40
Aluminum oxide D series	Gelman	20/60
Cellulose 300 and 300G series	Brinkmann	15/90
Cellulose CC41, Whatman	Reeve Angel	30/60
Kieselguhr	Brinkmann	30/60–65

sons why it is desirable to know how to prepare thin-layer plates, such as usefulness for evaluating the feasibility of other than the usual adsorbents or a special buffered system and for special investigational work. The description of preparing thin-layer plates with silicic acid will serve as a general approach to preparing plates with practically any kind of adsorbent. For the most part, the particle size that is generally acceptable for thin-layer chromatography lies between 5 and 70 microns. Regardless of what adsorbent-solvent system is used, the weight of the solid adsorbent to the liquid should be such that a slurry results, which can be poured or pulled over a glass plate, leaving a uniform deposit. If the mixture is too fluid, lines and fissures will form; and if too thick, the mixture will become firm before settling uniformly. The latter is especially true if a binding agent, such as calcium sulfate, has been added. It is a case of trial and error until the proper weight of adsorbent to liquid volume is obtained. Some suggested weight ratios of adsorbent to liquid volume are given in Table 19-6. These adsorbent to water ratios are suggested for large-plate (5 × 20 cm to 20 × 20 cm) preparation using commercially available apparatus to coat the plates, such as the Camag* or the Stahl-De Saga.† To coat microscope slides that are convenient for qualitative purposes, such as monitoring an organic reaction or checking a solvent system, the liquid used to prepare the slurry is usually other than water. Rather extensive use of microscope slides as adsorbent layer supports was made by Peifer.[59] As an example, the silica gel adsorbents are made into a slurry by using 35 g of solid with 100 ml of chloroform-methanol (2:1, v/v). Two microscope slides, held back to back, are dipped into the slurry, separated, and allowed to air-dry. The plates may be further conditioned by exposing them to a gentle stream of steam. The steamed plates can be used as such for partition

*Gelman Instrument Co. or Arthur H. Thomas Co.
†Brinkmann Instruments, Inc.

chromatography or be activated for adsorption chromatography by drying at 110° C for 30 minutes.

For most thin-layer chromatography three plate sizes are used: 5 × 20 cm, 10 × 20 cm, and 20 × 20 cm. The larger the plate, the more samples can be analyzed. The plates must be absolutely clean. They should be washed with a detergent or cleaning solution, rinsed well with tap water and then distilled water, and dried. It is suggested that before use the plates be wiped with hexane to remove any trace of grease.

The adsorbent slurry is applied by any one of three techniques. The simplest technique for a single-plate preparation is merely to pour the slurry onto the middle of the plate and tilt the plate to evenly cover the surface. A somewhat uniform layer results. This technique requires a great deal of practice before even fair reproduction of adsorbent layer thickness can be achieved.

Another method is to place narrow strips of tape along the edges of the plate, adding more strips of tape to obtain the layer thickness desired, and to pour the slurry primarily at one end of the plate. With a glass rod that extends beyond the edges of the plate, gently and evenly distribute the slurry over the plate. The glass rod rides on the two strips of tape. A fairly good adsorbent layer thickness can be achieved by this technique.

Generally several thin-layer plates are made at one time. The plates, up to six that are 20 × 20 cm, and combinations of narrower-size plates, but 20 cm long, are placed in an aligning tray. The aligning tray is usually made of a plastic material and is used to guide the slurry applicator over the positioned plates. The technique of pulling the slurry over the plates is called the Stahl-De Saga technique. When the aligning tray is used to guide the plates as they are pushed under the stationary slurry reservoir, the method is called the Camag technique of plate preparation. The Camag technique gives particularly uniform layer thickness. By the Stahl-De Saga technique the adsorbent layer is subject to variation due to warping of the aligning tray or nonuniform plate thickness. There are apparatuses, however, that accommodate for plate thickness, and a thickness variation of ± 0.01 mm can be obtained. These plate-leveling devices are obtainable commercially.* The stepwise application of the slurry by the Stahl-De Saga technique using a fixed-depth (250 μ) applicator and a regular aligning tray is shown in Fig. 19-5. For the preparation of plates having adsorbent layers of other than the usual 250 μ thickness, applicators are available that have adjustable gates permitting layer thickness up to 2 mm. Layers of this thickness would be used for preparative thin-layer chromatography.

The freshly applied slurry should be allowed to set up for about 30 minutes and, if it is to be used for adsorption chromatography, the plates should be activated for at least 1 hour at 110° C. After the plates have cooled, the excess adsorbent on plate edges is removed and the plates are stored in a desiccator. A convenient desiccator can be made from a large plastic bag containing a couple of bottles of a drying agent. The carrying tray that contains the plates is placed in the plastic bag and the bag is closed with a large paper clamp.

Sample application

The sample can be applied to the plate in one of several ways, depending on the intent or purpose. Initially, in lieu of the concentrations of the unknown

---

*Quickfit, Inc., and Shandon Scientific Co., Inc.

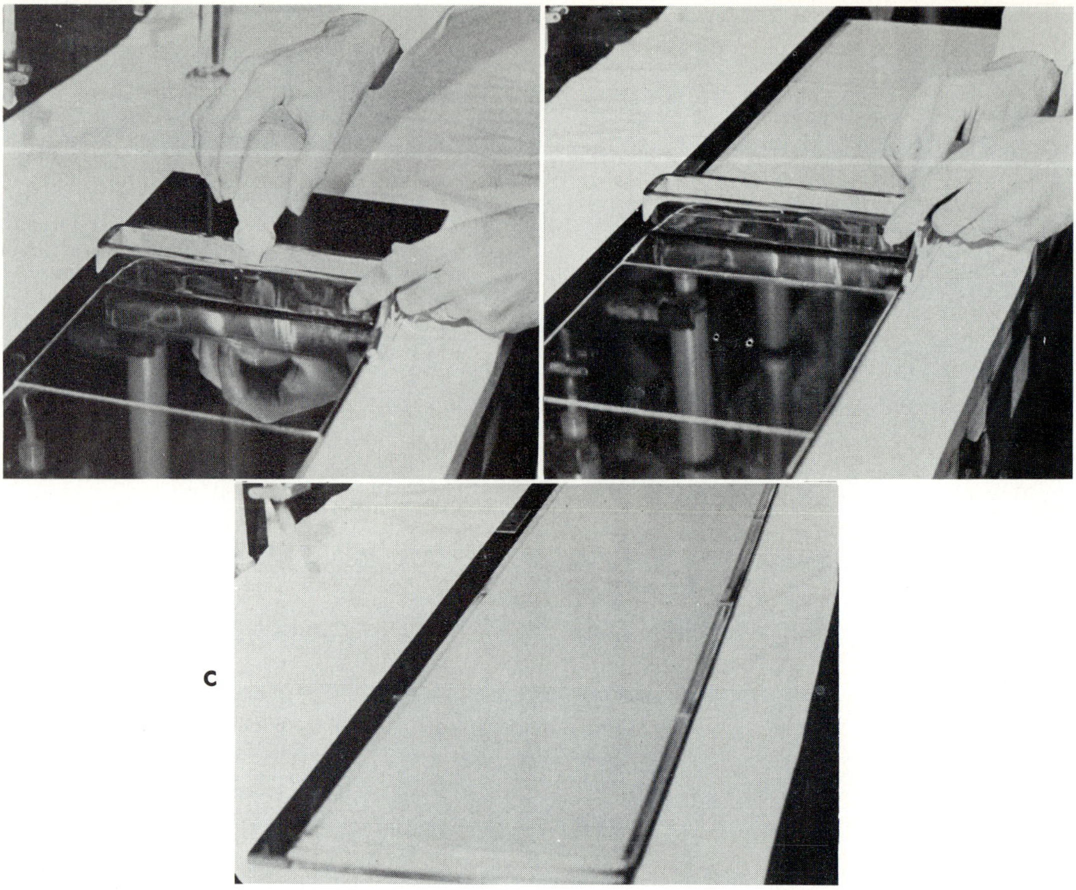

Fig. 19-5. A, Introducing silica gel slurry into applicator trough. **B,** Drawing slurry over plates, Stahl-De Saga technique. **C,** Coated plates in aligning tray.

in the sample, it is advisable to spot 5, 20, and 100 μl. The small amount of sample will very likely give sharper separations but the large-size sample will reveal minor constituents. The sample can be conveniently spotted with disposable micropipettes or capillaries.* The sample is drawn into the tube by capillary action and when touched to the adsorbent layer delivers essentially the rated volume of the capillary. Microsyringes† can be used to spot samples of practically any size. Application of large volumes of solutions from the syringe, greater than 10 μl, which have a low surface tension, such as hexane or benzene, usually have to be made in increments of 5 to 10 μl droplets. Sample spotting or streaking for preparative thin-layer chromatography is difficult to do by hand. However, there are several devices commercially available for streaking large volumes of samples in a quantitative fashion. Bobbitt and associates[60] have described a device used in their laboratory for hand streaking thin-layer preparative plates that is of interest. It consists of a 2 ml pipette. The wick tip is made by looping a fine Nichrome wire and inserting it into the end of the pipette. A piece of string is

*Available from Gelman, Helena Laboratories, and Kensington Scientific Corp.
†Hamilton Co.

Fig. 19-6. Assembled S chamber with plate in position and set in solvent trough.

then tied to the end of the wire loop and the loop is pushed into the pipette. The string is cut to a length of 7 to 8 mm. The rate of flow can be controlled by the degree of tilt of the pipette.

Another method of adding a sample for preparative purposes is to apply a series of spots, one just touching the other, across the length of the plate. The plate is then developed a short distance with a very polar solvent, such as methanol, the spots thus being compressed into a thin straight line. The plate is dried and developed normally.

Development of the plate

For the most part the plates are developed by ascending chromatography, in which the solvent moves up the plate. There are other methods of development, such as circular, horizontal, and descending, but these methods do not have any particular advantage over the ascending development technique. Developing chambers can be any convenient-size jar or container; but for rapid saturation of the atmosphere in the jar, with the developing solvent vapor, the developing chamber should be just large enough to accommodate the plate. For rapid saturation of the atmosphere, the chamber is lined with filter paper. In some cases, however, a sharper resolution will result if no attempt is made to saturate the atmosphere. The S chamber or sandwich chamber atmosphere (Fig. 19-6) is very rapidly saturated because of the very small space that exists between the cover plate and the thin-layer plate. In the S chamber the two plates are separated a distance of a few millimeters by cardboard or Teflon spacers on three sides and are held together by a couple of spring clamps. The clamped plates are then placed in a slotted trough containing the developing solvent. The order of separation of the components of a mixture is the same when development is in a tank chamber or an S chamber, but the R_f values are usually not the same.

Several different developing techniques can be used and it is by experimenting with these particular techniques that the best method can be determined.

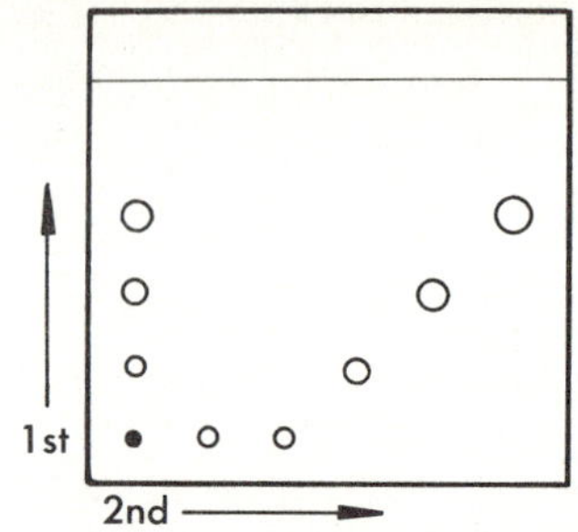

(Black dot represents initial spotting of sample)

Fig. 19-7. Idealized two-dimensional developments.

The developing method used must by necessity relate to the kind of sample. If the sample consists of components having a wide difference in polar properties, it may be best to resort to *multiple development*. In this case a solvent is found that will move the least polar component about half the way (R_f value of 0.5) and just about move the most polar component from the origin of spotting. Subsequent drying and developing the plate again in the same solvent system will further resolve the low R_f value components without running the initial R_f value of 0.5 component into the solvent front. The multiple-development technique is equivalent to allowing the solvent to move approximately twice the distance on a larger plate. The advantage of multiple development lies in the fact that the greatest resolution of the components of a mixture takes place in the first one third of the distance the solvent moves.

Continuous development, in a regular developing chamber or an S chamber, can be readily performed by merely extending the thin-layer plate outside the chamber and thus allowing the solvent to evaporate off the end of the thin-layer plate. Again a solvent system should be so chosen that the least polar component does not move up the plate too rapidly. If the components of a mixture have properties that are quite different, it may be best to resort to *two-dimensional* development. In this case a solvent system is used that will resolve the less polar components, and after drying the plate, redevelop it in a more polar solvent system to resolve the polar constituents. The plate is turned 90 degrees prior to development with the more polar solvent; thus the solvent moves at a right angle to the direction of the first solvent. The sample is spotted at one end of the plate, and only one sample can be developed per plate. Ideally the results should appear as given in Fig. 19-7.

Detection of the separated components

There are two classes of detectors: the universal and the specific. The universal detectors in common usage are concentrated sulfuric acid, sulfuric acid–potassium dichromate (cleaning solution), sulfuric acid–nitric acid, perchloric acid, and iodine. The acids are sprayed onto the plate and the plate is warmed for a few minutes between 100° and 150° C. Most organic compounds will appear as black dots because of the charring by the acids. Volatile organic compounds will not be detected. Iodine is a nondestructive detector of a great many organic compounds. A developing chamber is ideal for use as an iodine chamber. A few grams of iodine in a 50 ml beaker can be placed in the developing chamber and, after the chamber is saturated with the iodine, it is ready for use. The iodine-adsorbing organic compounds appear as dark brown spots against a tan background. On removal of the plate from the iodine chamber the spots should be outlined because the iodine evaporates rapidly. Many organic compounds adsorb ultraviolet light and are readily and conveniently detected if a fluorescent agent is incorporated into the adsorbent layer. The UV-adsorbing components appear as dark spots against a green or white fluorescing background. In some instances an amount as small as 0.1 μg can be detected. With a few exceptions, the organic compounds can be detected without destruction—which is an advantage if quantitative recovery of the separated component is desired.

A few specific agents will be cited here. For a more extensive list refer to the detection column in Table 19-5. Carboxylic acids are detected by spraying with a slightly ammoniac solution of bromcresol green. The acids appear as yellow spots against a green background. Aldehydes and ketones appear as yellow to red spots when they are sprayed with an acid solution of 2,4-dinitrophenylhydrazine. A good reagent for the detection of steroids, lipids, and vitamins is a solution of antimony trichloride in chloroform. Dragendorff's reagent, which is a mixture of bismuth subnitrate and potassium iodide, is used to detect alkaloids and organic bases in general. Phenols are detected when sprayed with a solution of ferric chloride. Many inorganic cations can be detected by spraying with solutions of hydrogen sulfide, ammonium sulfide, or ammoniac 8-hydroxyquinoline.

Quantitative thin-layer chromatography

There are two approaches to quantitatively measuring the separated components. In one case the measurement is made on the plate and in the other case, the sample spot is removed from the plate and subsequently assayed.

Quantitation on the plate is achieved by measuring the spot size or by using optical densitometry. Many attempts have been made to correlate a measured parameter of the spot size and the weight of the substance. However, it was Purdy and Truter[61] who found a linear relationship that exists between the square root of the area of the spot and the logarithm of the weight (sample weight between 1 and 80 μg) of the substance

$$\sqrt{A} = K \log_{10} M$$

where K is a proportionality constant. Only two standard solutions, one made from the dilution of the other, are required to determine the equation of a straight line. Equal volumes of the two standards and the unknown are chromatographed simultaneously, and the weight of the compound in the sample is calculated from the equation

$$\log W = \log W_s + \left\{\frac{\sqrt{A} - \sqrt{A_s}}{\sqrt{A_s} - \sqrt{A_d}}\right\} \log D$$

where the subscripts s and d refer to the standard and the diluted standard, respectively, and D is the dilution factor. Great care must be exercised when performing thin-layer chromatography for quantitative evaluation. It is particularly important that the same volume of standards and sample be applied to the plate, preferably with a microsyringe of $\pm$ 1% of the rated delivery amount. The spot area must also be carefully visualized. The area of the spot is measured by placing a sheet of tracing paper over the chromatogram and tracing the area of the spots. The chromatogram can be photographed and the area of the spots measured from an enlargement of the photograph. The area can also be determined with the use of a planimeter or by taking the mean of three measurements of the diameter of the circular spot. The mean coefficient of variation by this method of Purdy and Truter has been found to be between 2% and 7%.

A very simple but not as precise a technique is to spot the sample and a series of standards. A visual comparison of the unknown spot size to the nearest standard spot size provides a quantitative estimation of the unknown to about 90% of the actual amount present.

There is a linear relationship between the extinction area of a spot on the densitometric tracing of a chromatogram and the weight of the material in the spot. This relationship allows for the direct quantitative determination of the compound on the plate by densitometry. There are instruments available* for scanning and measuring the spots. The readings are conveniently traced on recording charts. The areas under the curves are automatically integrated, which gives a rapid quantitative evaluation of the material in the spot. Careful chromatographic techniques must be adhered to if valid quantitative results are desired. Some of the scanning devices make correction for variation of the thickness and color of the plates by use of a double beam—one beam scans the sample path and the other beam scans the area adjacent to the sample path. The errors involved in densitometry vary widely and depend on several conditions, but recoveries of 95 ± 15% are feasible.

The second approach for quantitative analysis of the resolved spots involves removal of the spot material and adsorbent from the plate. The spot material is then eluted from the adsorbent and subsequently analyzed. This technique is more accurate but is time-consuming and subject to several restrictions. The plates are spotted with both the unknown and a standard sample and developed in the usual fashion. The spot positions are located, preferably by a nonchromogenic method such as fluorescence or ultraviolet adsorption, with use of a phosphor in the adsorbent. If a chromogenic method must be used, the unknown solution is spotted at one end of the plate and, after development, the plate is covered except for the end sample spot. This area is sprayed to reveal the position of the unknowns. Scoring the plate vertically every 2 cm to form channels helps to locate the unknown and to confine the path of the samples.

The unknown spot area, standard sample spot area, and an area equivalent to and adjacent to the sample spot area are quantitatively removed. One technique for the quantitative removal is to completely remove the adsorbent layer up to and around the outlined spot area. The remaining spot area, which is rectangular in shape, is carefully removed by scraping with a small flexible plastic ruler or a razor blade and a camel-hair brush. The plate is slightly inclined away from the operator and the scrapings are directed down onto a piece of glassine paper under the plate. A small vacuum cleaner device (shown in Fig. 19-8) can also be used to remove the adsorbent and spot from the glass plate.

The sample is then eluted from the adsorbent with a very polar solvent such as methanol or ethanol. The elutrient is taken to volume in a small volumetric flask and read by ultraviolet spectrophotometry or some other appropriate analytic method. A blank is prepared by taking from the plate an area of adsorbent

*American Instrument Co., Inc., National Instrument Laboratories, Inc., Photovolt Corp., and Schoeffel Instrument Corp.

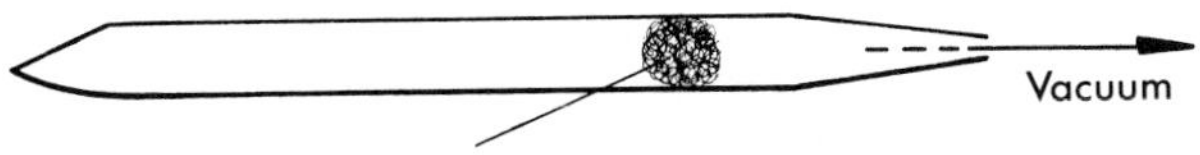

Fig. 19-8. Vacuum cleaner device.

comparable to the sample area and adjacent to the sample spot. Adsorption due to the blank is corrected for and the unknown is compared to the standard for quantitative evaluation, assuming that Beer's law prevails. Spencer and Beggs[62] further elaborate on the conditions required for quantitative spectrophotometry in working from thin-layer plates.

Clinical application

Thin-layer chromatography is used to analyze many different kinds of biologic samples such as blood, urine, other body fluids, and tissue for various classes of compounds such as carbohydrates, amino acids, lipids, hormones, steroids, drugs, and drug metabolites. Most of the methods do not lend themselves to routine clinical laboratory application because the procedures are rather involved and time-consuming. In some cases it is not known if the findings by the thin-layer method have any pathologic significance. Some thin-layer procedures that have found use on a routine basis will be described.

Rapid assay for urinary porphyrins

Porphyria is a disorder of pigment metabolism characterized by an overproduction of porphyrins. The porphyrins are excreted in the urine (Chapter 14). The disease may occur as congenital porphyria or as acute porphyria. Acute porphyria may be precipitated by some toxic agent, such as barbiturate or lead poisoning. Small amounts of uroporphyrin and coproporphyrin are normally found in urine. The finding of abnormal porphyrin excretion in the urine is usually indicative of one or the other kind of porphyria and is of aid in clinical diagnosis of this disease.

The procedure is the thin-layer method described by Scott and co-workers.[63] The test has three levels of interpretation:

1. Rapid screening assay for elevated porphyrins
2. Qualitative assay for separated uroporphyrin and coproporphyrin
3. Quantitative assay for uroporphyrin and coproporphyrin

Reagents and apparatus needed

All the reagents used are analytic grade and further purification is not necessary.

The standard for routine fluorometric measurements is made by dissolving 5.0 ± 0.1 mg of chromatographically pure coproporphyrin in 1.5 N HCl and diluting to 1000 ml. A 5 ml quantity of this solution is diluted to 100 ml with 1.5 N HCl. A solution of 1.5 N HCl is made by diluting 125 ml of concentrated HCl to 1000 ml. The final concentration of the routine standard is 0.05 μg/ml.

The developing solution is made by combining chloroform, methanol, ammonium hydroxide (≈25% NH_3), and water in the ratio 12:12:3:2, respectively.

Fluorometric measurements are made of the silica gel taken from the thin-layer plate and suspended in a mixture of two parts glycerol and one part water. The plates are made from commercially available silica gel

containing calcium sulfate binder, or prepared plates can be used. Standard 20 × 20 cm plates are coated with silica gel of about 250 μ thickness, by the Stahl-De Saga method. The plates are allowed to dry overnight and can be used directly without activation. One of several brands of prepared plates can be used but the plates should be tried first to note which gives the best results.

The porphyrin spots on the thin-layer chromatogram are visualized with a high-intensity ultraviolet lamp containing a 100-watt mercury vapor spot bulb.* The fluorescence of the porphyrins is measured by using either a Farrand model A or the Turner model 111 fluorometer with a high-sensitivity door. The primary filter is a Corning No. 5113 and the secondary filter is an Eastman Kodak No. 29A. The same filter combination can be used for both instruments.

Procedure

Preparation of the urine sample

To 8 ml of urine in a 12 ml centrifuge tube, add 0.2 ml of 8 N HCl. Add (0.2 to 0.3 ml) a saturated solution of sodium acetate to the mixture, to a pH of 3 as indicated by Hydrion paper. Add 2.0 ml of *n*-butanol and shake vigorously. Centrifuge the mixture for 10 to 15 minutes to clearly separate the phases.

At this stage, as a quick screening test, the butanol layer is examined under ultraviolet light. Normal urine gives a barely perceptible fluorescence. For further quantitative analysis, smaller amounts of the sample of the elevated porphyrins should be taken and the volume made up to 8.0 ml with water.

Thin-layer chromatography

Spot 60 μl of the butanol layer onto the thin-layer plate, starting about 2 to 2.5 cm from the left-hand edge of the plate and about 2 cm from the bottom edge. With the aid of a stream of warm air from a hair drier to dry the spot, apply the 60 μl of butanol in increments of about 10 μl to maintain a spot size of about 7 mm or less. The spot should be completely dry before development and the spotting should be done in subdued light. As many as eight samples can be spotted on a 20 × 20 cm plate.

Place the plate into the developing chamber containing the solvent mixture. Allow the solvent to move up the plate 10 cm. This takes about 30 to 45 minutes. Remove the plate, or plates, since two plates can be developed at a time, and dry them in a forced-air hood for about 30 minutes. Do not dry them much longer than 30 minutes, for prolonged standing results in loss of fluorescence.

The separated porphyrins are visualized by illumination with the ultraviolet lamp and the spots outlined with a needle. The uroporphyrin remains at the origin of spotting and the coproporphyrin is a sharp band at an R_f of 60 to 65. At this point a more refined screening test result can be given, since the uroporphyrin and coproporphyrin can be related to a normal and to each other. The relative amount of each porphyrin is based on its intensity of fluorescence.

To quantitate the porphyrins, carefully scrape the spot onto a piece of glassine paper and transfer the scrapings into a 10 × 75 mm test tube used for fluorometric measurement. A similar-size area of silica gel containing no fluorescence is removed below the origin and used as a blank. Glycerol solution (2 ml) is added to the test tubes, and the tubes are shaken vigorously. Allow the tubes to stand for 20 minutes and shake gently just before reading.

*Ultra-Violet Products, Inc.

The 0.05 μg/ml of coproporphyrin in 1.5 N HCl is used as a full-scale reference to adjust the fluorometer. Since the porphyrins adsorbed on the silica gel are not stable enough to be used as standards, the fluorometric readings must be related to the coproporphyrin standard. By putting known amounts of pure uroporphyrin and coproporphyrin through the chromatographic procedure, a conversion factor for each porphyrin is determined. As an example, 0.1 μg each of pure coproporphyrin and uroporphyrin are put through the chromatographic procedure and read in the 2 ml glycerol solution with the fluorometer, which has been set at a galvanometer reading of 10.0, using 2 ml of the 0.05 μg/ml coproporphyrin HCl standard. If the coproporphyrin sample gave a reading of 10.0 on the glavanometer, the conversion factor would be 1.0 and if the uroporphyrin sample gave a reading of 11.7, the conversion factor for uroporphyrins would be 1.17. To calculate the urinary porphyrin excretion use the following formula:

$$\frac{\text{Galvanometer reading}}{\text{Conversion factor}} \times 0.1 \times \frac{\text{24-hour volume (in ml)}}{0.06} \times \frac{1.5}{8.0} = \mu\text{g porphyrin/24 hours}$$

Results

Porphyrin fluorescence is linear to a concentration of 10 μg/ml when measured in HCl. The lower limit of detection is 0.002 μg of either uroporphyrin or coproporphyrin. A reproducibility of ± 5% of the mean value can be obtained. Low recoveries, 80% to 85%, of uroporphyrin added to urine samples and high recoveries, 105% to 111%, of coproporphyrin are typical of the recovery of porphyrins that can be expected.

The amount of silica gel suspended in the glycerol solution is not important. It has little effect on the porphyrin analysis, since it contributes very little fluorescence and has no quenching effect on the porphyrin fluorescence.

The normal range for uroporphyrin is 10 to 50 μg/24 hr and for coproporphyrin it is 50 to 200 μg/24 hr. In the case of coproporphyria, the coproporphyrin content in the urine is very high and many times more than the slightly elevated uroporphyrin content. In the other cases of porphyria—such as acute intermittent, congenital erythropoietic, and acquired—the uroporphyrin content in the urine is very high and many times the elevated coproporphyrin content. In some cases significant amounts of hepta-, hexa-, and penta-carboxylic porphyrins lie between the uroporphyrin and coproporphyrin spots but they can be ignored. Some drugs and their metabolites that fluoresce and could be an interference are usually eliminated by the butanol extraction step in the procedure.

Identification of barbiturates in urine by thin-layer chromatography

On occasion it is important to rapidly identify the barbiturate in a case of suspected barbiturate poisoning. The following thin-layer chromatographic procedure* has been used to identify some of the more common barbiturates in urine. The procedure takes about 1½ hours.

Reagents and equipment needed

Sodium pentobarbital, sodium phenobarbital, sodium Seconal, barbital, and amobarbital are the *barbiturates* used in this procedure. Other barbiturates that may be of interest can be used but their R_f values should be determined, to find any relationships in regard to overlapping. Differ-

*J. R. Simmler, Kirkwood, Mo.

ent staining characteristics that may help in differentiation may also be revealed during the initial chromatographic examination of a barbiturate different from the ones used in this procedure.

For *individual standards, 1 μg/μl,* weigh 0.100 ± 0.005 g of the individual barbiturate, transfer it to a 100 ml glass-stoppered volumetric flask, and add 5 ml of water to dissolve the sodium salts of the barbiturate. Dilute to volume with ethanol or reagent grade alcohol. Reagent alcohol is 5% by volume isopropyl alcohol and 95% ethanol by volume. The amobarbital standard is made by vigorously shaking one capsule, weight 0.172 g, with 10 ml of 1:1 water-alcohol mixture in a 100 ml flask. After complete disintegration of the capsule, dilute the mixture to volume and filter the mixture into another 100 ml flask.

For *mixed standards, 1 μg each/μl,* weigh 0.100 ± 0.005 g of the individual barbiturates and transfer to a 100 ml volumetric flask. There is no point in adding the amobarbital, since it has approximately the same R_f value as the Seconal. Add 5 ml of water and dilute to volume with reagent alcohol. Store the standards in 4 oz polyethylene bottles.

Prepare the *developing solution* by adding 6.0 ml of acetone to 100 ml of benzene. Solvents are of analytic grade. A large volume of developing solution can be made by adding 60 ml of acetone to 1 L of benzene and storing the mixture in an amber glass-stoppered bottle.

The barbiturate spots are visualized by spraying the developed chromatograms with the following *spray reagents:*

1. Saturated solution of mercurous nitrate—place 1.0 g of analytic grade mercurous nitrate into a 4 oz plastic bottle, add 100 ml of water, and shake vigorously. Allow the excess mercurous nitrate to settle out and decant the supernatant liquor into another plastic bottle.
2. S-diphenylcarbazone, 0.10% (w/v)—weigh 0.100 g of S-diphenylcarbazone* and transfer to a 4 oz plastic bottle. Dissolve the reagent in 100 ml of reagent alcohol. Keep the solution in the refrigerator when not in use.
3. Potassium permanganate, 0.1%—dissolve 0.1 g of analytic grade potassium permanganate in 100 ml of water.

Prepared silicic acid plates designated 4GF or glass fiber impregnated sheets, ChromAR sheet 500,† can be used as the *adsorbent.*

As a *developing chamber,* the S or sandwich chamber equipment for developing plates or sheets can be obtained from Distillation Products Industries.

The detecting reagents are conveniently sprayed with an aerosol *spray apparatus* obtained from Warner-Chilcott Laboratories; however, any TLC spraying equipment can be used.

Microcapillary tubes or pipettes of 5 and 10 μl volumes are obtainable from Kensington Scientific Corp. or from Hamilton Co. *Microsyringes* (10 μl) are used for spotting the standards and sample onto the adsorbent layer.

Procedure

Preparation of the sample

Add 1.0 ml of 0.2 N sulfuric acid (3 ml of concentrated sulfuric acid/L) to 25 ml of urine in a 125 ml separatory funnel. Add 25 ml of chloroform to the separatory funnel. Rock the funnel gently back and forth about 70 to 80 times, causing the chloroform and aqueous layers to intimately interact without forming an emulsion. Allow the layers to separate. Drain the chloroform layer into a filter funnel lined with Whatman No. 40, 11 cm paper and filter the chloroform layer into another 125 ml separatory funnel. Repeat the extraction two more times.

---

*Matheson Chemical Co.

†Mallinckrodt Chemical Works.

Add 5.0 ml of 0.45 N sodium hydroxide (18 g of sodium hydroxide/L) to the combined, filtered chloroform layers in the second separatory funnel. Shake the funnel 100 times in the manner indicated above. Allow the layers to separate. Draw off the chloroform layer and discard it. To the caustic aqueous layer add 0.5 ml of 6 N hydrochloric acid (50 ml of concentrated hydrochloric acid and 50 ml of water). Extract the barbiturates with 5.0 ml of chloroform, shaking as indicted above. Transfer the chloroform layer to a 10 ml conic centrifuge tube. Repeat the extraction with another 5.0 ml of chloroform. Combine the chloroform extract in the centrifuge tube.

Evaporate the chloroform layer to dryness. Placing the tube in a hot-water bath will accelerate the evaporation. The temperature of the bath should not exceed 80° C. A stream of air directed into the tube will accelerate the evaporation of the chloroform. Under proper conditions it should not take more than 10 to 15 minutes to evaporate the chloroform. Wash down the sides of the tube with 0.5 ml of reagent alcohol and again evaporate until the volume has been reduced to about 50 μl.

Thin-layer chromatographic procedure

Spot the sample and standards in the following way:

Cut a piece of ChromAR 500 sheet 20 cm in length—the sheet adsorbent is 20 cm wide. Using either the prepared glass thin-layer plate or the sheet adsorbent, spot 5 μl of the mixed barbiturate standard with the micro-capillary pipette onto the adsorbents at a point 2 cm from the left margin of the adsorbent and 2 cm from the bottom edge. See Fig. 19-9. A spotting template, obtainable from many of the scientific apparatus distributing companies, is of great aid in aligning the spots. At a point 2 cm from the right of the first spot and again 2 cm from the bottom edge, spot 10 μl of the sample. Spot the sample in two 5 μl portions, allowing the first spot to evaporate before adding the second 5 μl portion. Follow the sample spot with the individual standards 2 cm apart. If more than one sample is to be analyzed, the mixed standard can be left off the plate. As many as eight spots can be placed on the plate or as many as three samples and five single standards can be placed on one

Fig. 19-9. Technique of spotting sample to thin-layer plate, using capillary pipette.

plate. As a way of screening, six samples and two mixed standard spots can be placed on the plate. While spotting the solutions, direct a stream of warm air from a hair drier onto the plate. This will help keep the spots compact. The spots should be dry before development.

Place the ChromAR sheet on one of the **S** chamber halves, aligning the lower edge of the sheet with the lower edge of the **S** chamber. Place the second half of the **S** chamber on the first half supporting the adsorbent sheet and clamp into position with the two large spring clamps provided. See Fig. 19-6 for view of the assembled chamber set in the solvent trough.

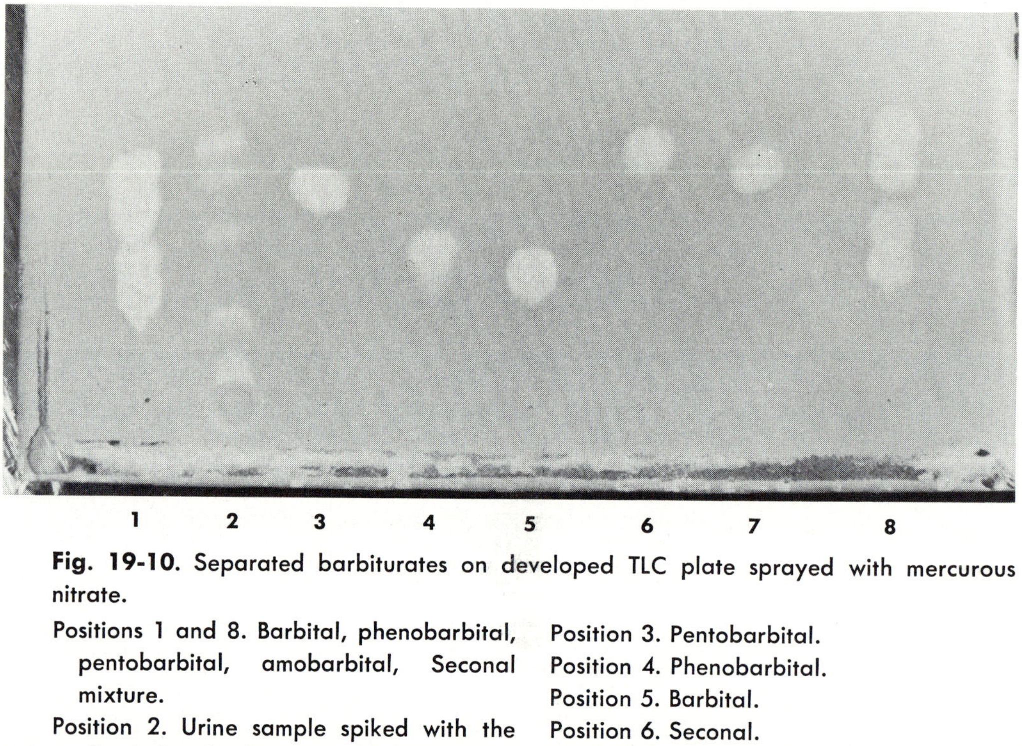

1 2 3 4 5 6 7 8

Fig. 19-10. Separated barbiturates on developed TLC plate sprayed with mercurous nitrate.

Positions 1 and 8. Barbital, phenobarbital, pentobarbital, amobarbital, Seconal mixture.

Position 2. Urine sample spiked with the above five barbiturates. First two spots nearest point of spotting are unknown.

Position 3. Pentobarbital.

Position 4. Phenobarbital.

Position 5. Barbital.

Position 6. Seconal.

Position 7. Amobarbital.

Table 19-7. R_f values and identifying color of the barbiturates

Compound	R_f value		Spray reagent		
	Sheet	*Plate*	*Mercurous nitrate*	*S-diphenyl-carbazone*	*Potassium permanganate*
Barbital	0.23	0.19	White	Violet	——
Phenobarbital	0.29	0.22	White	Violet	——
Pentobarbital	0.37	0.30	White	Violet	——
Amobarbital	0.37	0.35	White	Violet	——
Seconal	0.41	0.37	White	Blue	Brown-yellow

If the prepared plate is used, the space between the two **S** chambers must be increased. This is done by cutting strips of Teflon or cardboard to give a thickness of 2 to 4 mm. The thin-layer plate, spotted with the sample and standards as just indicated, is set on the inside row of glass nipples on the lower edge of one of the halves of the **S** chamber. The strip spacers are put in place along the outer edges of the lower half of the **S** chamber, and the other half of the chamber is placed on the spacers. The clamps are then put in place to hold the **S** chamber together.

The amount of solvent required to develop the ChromAR sheet is 50 ml, and the precoated prepared plate requires 100 ml. The solvent is placed in the trough, covered with the plastic plate, and the **S** chamber is slowly lowered into the solvent, which is allowed to move up the plate 15 cm from the point of spotting the samples. It takes about 15 to 20 minutes for the solvent to move this distance. The chamber is removed from the solvent trough and is opened. The plate or sheet is removed and is dried with the aid of a stream of hot air from the hair drier.

Placing the plate or sheet in a suitable chamber and spraying it with the mercurous nitrate solution will cause the barbiturate spots to become evident. They will appear as white spots against a wet gray background. The partially dry plate may be oversprayed with the S-diphenylcarbazone solution. The purple-colored background will slowly fade away, leaving blue-violet spots.

Comparison of the R_f value and position of the mercurous nitrate barbiturate spot from the urine sample and the corresponding standard barbiturate will, as a first approximation, identify the barbiturate in the sample. The second spraying with the S-diphenylcarbazone, if the spot colors are the same, will further verify the relationship of the unknown barbiturate and the standard. The dry plate, after a spraying with the mercurous nitrate, will, when sprayed with the potassium permanganate solution, reveal the Seconal barbiturate as a brown-yellow spot whereas the other barbiturates are unaffected. This will help differentiate between the closely related Seconal and amobarbital.

A comparison of the R_f values of barbiturates resolved on prepared plates and on sheet adsorbent and the color reactions seen after spraying are given in Table 19-7. See also Fig. 19-10.

Results

The limit of detection, using mercurous nitrate as the detecting reagent, is 0.2 μg. Therefore, starting with a 25 ml of urine sample it is possible to detect 1 μg/25 ml, or 0.04 ppm. The extraction of the barbiturates is evidently quantitative, at least to the limit of detection of the method, since urine samples to which known amounts of barbiturates are added do not reveal any barbiturates on a second extraction.

It should be possible to quantitate the barbiturates by first determining what barbiturate is present in the sample and then rerunning the analysis; compare the sample spot size against the spot size from a series of different concentrations of the barbiturate (semiquantitative) or use a recording, integrating densitometer. Whole blood and other body fluids can also be analyzed for barbiturates by this procedure. A 2 to 10 ml quantity of whole blood or serum should be used. The extraction is the same as it is for urine samples.

Determination of 3-methoxy-4-hydroxymandelic acid in urine

3-methoxy-4-hydroxymandelic acid, generally referred to as vanillmandelic acid or VMA, is a product of catecholamine metabolism (Chapter 15). Patients with pheochromocytoma demonstrate large amounts of VMA in their urine. The thin-layer chromatographic procedure for VMA to be presented here is based on the procedure of Annino and co-workers.[64] It can be helpful in the diagnosis of pheochromocytoma.

Reagents and apparatus needed

Solvent system—*n*-butanol–glacial acetic acid–water (90:20:40)
Florisil*
Silica gel G (Brinkmann)
Dye solution—Place 0.05 g of fast red GG† in 100 ml of water. Filter through glass wool. Prepare fresh each day and protect from light.
Potassium carbonate, saturated solution.
Buffer, borate, 0.1 M—Dissolve 38.1 g of $Na_2B_4O_7 \cdot 10\,H_2O$ in 950 ml of water and adjust pH to 10.2 ± 0.1, using concentrated (18 N) sodium hydroxide. Dilute to 1 L and store in a plastic bottle.
Eluting solution—Mix 15 ml of borate buffer and 10 ml of methanol. Prepare just before use.
Stock standard—Dissolve 25.0 mg of 3-methoxy-4-hydroxymandelic acid in 100 ml of absolute ethanol. Store in a refrigerator.
Working standard—Dilute 4 ml of the stock standard to 10 ml with absolute ethanol. This standard is equivalent to a urinary VMA concentration of 11 mg/L. This value takes into account the extra dilution effected when the urine is treated with HCl.
Apparatus to prepare plates by Stahl-De Saga method (Brinkmann Instruments)

Procedure

Preparation of the plates

Using a 125 ml flask, shake vigorously, for about 1 minute, 30 g of silica gel G and 60 to 65 ml of water. Pour the slurry into the trough and pull it over six 20 × 20 cm plates. Set the gate of the trough applicator at 0.25 mm. Dry the plates overnight and use without activation.

Preparation of the urine sample

Into a 20 × 150 mm tube with a ground glass top, measure 10 ml of urine and 1 ml of concentrated HCl. Swirl to mix; then heat in a boiling water bath for 10 minutes. Cool, add 1 g of Florisil, stopper, and shake for 10 minutes. Centrifuge for 5 minutes.
Place a 4.0 ml portion of the urine so treated in another 20 × 150 mm tube and add 10 ml of ethyl acetate. Stopper, shake for 2 minutes, centrifuge briefly, and with a capillary pipette transfer the acetate (upper) layer to a 50 ml centrifuge tube (wide-mouth).
To the remaining aqueous layer add another 10 ml of ethyl acetate. Repeat the extraction as above and add to the first extract.
Evaporate the combined extracts to dryness in a 60° water bath, using a stream of air to facilitate evaporation. Remove from the bath, add 0.4 ml of absolute ethanol, and mix well to redissolve the extracted VMA.

Chromatography

Draw a line on the plate, 11.5 cm from the bottom. Apply 50 μl of extract and of standard with a syringe, pipette, or other suitable applicator, 1.5 cm from the bottom of the plate and 2 cm from the sides of the plate. Apply as streaks approximately 2 cm long, in small increments, drying with a hot-air hair drier between applications. Six applications may be made on a 20 cm plate.
Place the plate in the tank with the coated side facing the wall of the chamber which has been previously lined with filter paper, cover, and allow the solvent to run until it reaches the limiting line.

---

*Floridin Co.
†General Dyestuff.

Remove the plate from the tank and with a hot-air hair drier dry it in a fume hood until the odor of the solvent is gone.

Spray with the dye solution and then dry with the hot-air hair drier. Spray with saturated potassium carbonate solution and dry again.

Quantitation

Using a pointed instrument (e.g., a dissecting needle), outline the purple VMA spots. Outline an extra area below the standard, approximately equivalent in size to the VMA spots. This area serves as a "blank" and is carried through the subsequent stages. Using a flat-edged spatula, remove the silica gel around the spots; then scrape the silica gel from the outlined area into 16 × 100 mm test tubes.

Add 2.5 ml of eluting solution to each tube and mix well so as to break up the silicic acid gel. Add 0.5 ml of dye solution and mix again. Centrifuge for 5 minutes. Decant the supernatant into 1 cm cuvettes. Read absorbance against the blank at 510 mμ within 20 minutes after addition of the dye.

Calculations

$$\frac{11.0}{\text{Absorbance of standard}} \times \text{Absorbance of test} = \text{Concentration of VMA in mg/L}$$

Clinical findings

Refer to discussions in Chapter 15.

Gas chromatography

Gas chromatography is the latest chromatographic technique to be applied to the separation and analysis of biomedical and industrial specimens. It is no doubt the most sophisticated of the methods of chromatography, at least in regard to instrumentation. A great deal of mechanization has gone into what is in fact no more than the classic application of chromatography except that the mobile phase is a gas. Aside from the separation aspects of gas chromatography, the detection of the resolved components is particularly convenient. There are several different kinds of detectors, each with its degree of sensitivity and application. The detectors not only respond to a change in the composition on the mobile gas phase but also respond to the amount of change. This allows for a rapid quantitative evaluation of the resolved constituents as they come off the column.

In spite of the extensive instrumentation usually associated with gas chromatography, a quite satisfactory device can be made from rather simple equipment. Practically all gas chromatographic instruments consist of the following components:

1. Source of an inert carrier gas and a gas flow regulator
2. Injector port for introducing the sample into the system
3. Long or spiral metal or glass column for supporting the combination solid and liquid phase
4. Furnaces for controlled heating of the injector port and column
5. Detector device and associated electrometer and recorder

Of these components the most important, and the "heart" of the system, is the column. There are two kinds of columns, the packed and the capillary. The packed column is the one most in use at present. The column, made of some inert material such as stainless steel or glass, is packed with a granular solid (usually diatomaceous earth) called the solid support, which is coated with a liquid of low volatility. The supported liquid is called the stationary phase. The carrier gas is usually hydrogen, helium, argon, or nitrogen flowing through the system at a controlled rate. Between the source of the carrier gas and the column is the sample inlet. Usually this area is heated to a higher temperature than the column; thus, rapid vaporization of the injected sample is assured. It is desirable that the sample be introduced as a "plug" of vapor prior to moving into the column. The column also is heated, and the temperature of the column can be programmed to give a fixed rate of temperature increase with time. Precise temperature control of the column is essential for repetitive results, since the vapor pressure of the stationary liquid phase varies with the temperature. The vapor pressure of the liquid phase contributes to baseline effect noted on the recorder.

As the individual separated components come off the column, they pass through a detector, which then gives a signal that is related to the amount of the substance present. There are several kinds of detectors—such as the following types: (1) thermal conductivity, (2) gas density balance, (3) radioactive ionization, (4) flame ionization and thermionic, and (5) coulometric. Differential detectors are almost exclusively used. The detector measures a certain property related to the amount of the individual components in the gas stream. The resulting chromatogram is a gaussian type of curve with the peak of the curve corresponding to the greatest concentration of the substance, and the baseline is due to the pure carrier gas. The distance on the time axis of the chromatogram, from the time the sample is injected into the column to the maximum of the peak, corresponds to the retention time of the particular substance. Some detectors, the thermal conductivity and the gas density balance types, are universal detectors while the other detectors are selective. One of the characteristic properties of the thermionic detectors is its increased response to halogen- and phosphorus-bearing compounds. For further and more detailed discussion of

Table 19-8. Some biochemical applications of gas chromatography

Class of compounds	Sample treatment
Estrogens	Hydrolysis, extraction, acetylation
Corticosteroids	Extraction, paper or thin-layer chromatography, acetylation, and oxidation
Urinary aromatic acids	Extraction, ion exchange, methylation
Amines and amino acids,	Extraction, esterification
aralkylamines and alkylamines	Extraction, injected directly
Bile acids	Extraction, esterification (usually methyl ester)
Carbohydrates	Extraction, trimethylsilyl ethers
Barbiturates	Extraction, injected directly
Organic anesthetics	Extraction, injected directly

the theory, equipment, and application refer to the various texts on gas chromatography included at the end of this chapter.

Gas chromatography of biologic materials is, almost without exception, preceded by some preliminary treatment of the sample.

The application of gas chromatography to biomedical studies has given great impetus to this field. Lipid metabolism has been studied by a combination of thin-layer and gas chromatography. After appropriate treatment of the sample—urine, plasma, or tissue—the lipid classes and subclasses were purified by thin-layer chromatography. The bands were removed from the thin-layer plate and the lipids methylated. The methylated fatty acids were separated and quantitated by means of gas-liquid chromatography. Other biomedical applications of gas chromatography are given in Table 19-8.

The detailed gas chromatographic conditions for the analysis of the various classes given in the table can be obtained from textbooks on gas chromatography given at the end of the chapter.

REFERENCES

1. Heftmann, E., editor: Chromatography, ed. 2, New York, 1967, Reinhold Publishing Corp.
2. Cassidy, H. G.: Fundamentals of chromatography, New York, 1957, Interscience Publishers, Inc.
3. Stahl, E., editor: Thin-layer chromatography; a laboratory handbook, ed. 2, (Transl. by Cambridge consultants), New York, 1965, Academic Press, Inc.
4. Waldi, D., Schnackerz, K., and Munter, F.: Eine systematische Analyse von Alkaloiden aus Dünnschichtplatten, J. Chromatogr. **6**:61, 1961.
5. Stahl, E., and Kaldewey, H.: Spurenanalyse physiologisch aktiver, einfacher Indolderivate, Hoppe Seyler Z. Physiol. Chem. **323**:182, 1961.
6. Machata, G.: Dünnschichtchromatographie in der Toxikologie, Mikrochim. Acta, p. 79, 1960.
7. Mariani, A., and Mariani-Marelli, O.: Separation of the opium alkaloids by means of aluminum plate chromatography, Rendic. Ist. Sup. Sanit. **22**:759, 1959.
8. Newbauer, D., and Mothes, K.: Thin-layer chromatography of poppy alkaloids, Chim. Planta Med. **9**:466, 1961.
9. Marcuse, R.: Separation of alkanals and alkanones with carbonyl groups in different positions by thin-layer chromatography, J. Chromatogr. **7**:407, 1962.
10. Sundt, E., and Saccardi, A.: Thin-layer chromatography and paper chromatography of vanilla flavoring compounds, Food Technol. **16**:89, 1962.
11. Dhont, J. H., and DeRooy, C.: Chromatographic behavior of 2,4-dinitrophenylhydrazones on chromatoplates, Analyst **86**:74, 1961.
12. Anet, E. F.: Thin-layer chromatography of 2,-4-dinitrophenylhydrazine derivatives of hydroxycarbonyl compounds, J. Chromatogr. **9**: 291, 1962.
13. Teichert, K., Mutschler, E., and Rochelmeyer, H.: Plate chromatographic studies of mixtures of natural products, Z. Anal. Chem. **181**:325, 1961.
14. Hansson, J., and Alm, A.: Chromatographie sur couches minces des dérivés de la diphénylamine application à l'analyse des poudres, J. Chromatogr. **9**:385, 1962.
15. Nicolaus, B. J. R.: Anwendung der Dünnschichtchromatographie auf Pteridine, J. Chromatogr. **4**:384, 1960.
16. Pastuska, G.: Untersuchungen über die qualitative und quantitative Bestimmung der Zucker mit Hilfe der Kieselgelschicht-Chromatographie, Z. Anal. Chem. **179**:427, 1961.
17. Weicker, H., and Brossmer, R.: The structural identification of some naturally occurring branched-chain fatty aldehydes, Klin. Wschr. **39**:1265, 1961.
18. Tate, M. E., and Bishop, C. T.: Thin-layer chromatography of carbohydrate acetates, Canad. J. Chem. **40**:1043, 1962.
19. Stahl, E., and Kaltenbach, U.: Dünnschicht-Chromatographie. IX. Schnelltrennung von Digitalis- und Podophyllum-Glycosidgemischen, J. Chromatogr. **5**:458, 1961.
20. Petrowitz, H. J., and Patuska, G.: Über die Kieselgelschicht-Chromatographie gesältigter aliphatischer Dicarbonsäuren, J. Chromatogr. **7**:128, 1962.
21. Gänshirt, H., and Morianz, H.: Untersuchungen zur quantitativen Auswertenung der Dünnschichtchromatographie. I. Arch. Pharm. 293/65, 1065, 1960.
22. Wollenweber, P. J.: Dünnschicht-chromatographische Trennungen von Farbstoffen an Cel-

lulose-Schichten, J. Chromatogr. **7**:557, 1962.
23. Stahl, E., and Trennheuser, L.: Gas-phase chromatography of terpene and hydroxyphenyl propane substances, Arch. Pharm. 293/65:21, 1960.
24. Brieskorn, C. H., and Wenger, E.: Constituents of *Salvia officinalis*. XI. Analysis of ethereal sage oil by gas and thin-bedded chromatography, Arch. Pharm. 293/65:21, 1960.
25. Malins, D. C., and Mangold, H. K.: Fractionation of fats, oils, and waxes on thin layers of silicic acid, J. Amer. Oil Chem. Soc. **37**: 576, 1960.
26. Vogel, W. C., Doizaki, W. M., and Zieve, L.: Rapid thin-layer chromatographic separation of phospholipids, J. Lipid Res. **3**:138, 1962.
27. Habermann, E., Brandtlow, G., and Krusche, B.: Determination of plasma phospholipids using thin-film chromatography, Klin. Wschr. **39**:816, 1961.
28. Honegger, C. G.: Über die Dünnschichtchromatographie von Lipidenuntersuchungen von Gehirngewebe Multiple-Sklerose-Kranker und Normaler, Helv. Chim. Acta **45**:281, 1962.
29. Honegger, C. G.: Über die Dünnschichtchromatographie von Lipidenuntersuchungen von Gehirngewebe aus der weissen Substanz Multiple-Sklerose-Kranker, Helv. Chim. Acta **45**: 2020, 1962.
30. Williams, J. A., Shorma, A., Morris, L. J., and Holman, R. T.: Fatty acid composition of feces and fecaliths, Proc. Soc. Exp. Biol. Med. **105**:192, 1960.
31. Kaufman, H. P., and Khoe, T. H.: Thin-layer chromatography of fats. VII. Separation of fatty acids and triglycerides on plaster of paris plates, Fette Seifen **64**:81, 1962.
32. Ruggieri, S.: Separation of the methyl esters of fatty acids by thin-layer chromatography, Nature **193**:1282, 1962.
33. Crump, G. B.: Rapid characterization of small amounts of oil by thin-layer silica-gel chromatography, Nature **193**:674, 1962.
34. Blank, M. L., and Laudberg, W. D.: Determination of mono, di and triglycerides by molecular distillation and thin-layer chromatography, J. Amer. Oil Chem. Soc. **38**:312, 1961.
35. Gritter, R. J., and Albers, R. J.: Non-destructive zone developing in thin-layer chromatography, J. Chromatogr. **9**:392, 1962.
36. Furukawa, T.: Strip chromatography. X. Separation of phenols, aromatic aldehydes, ketones, and carboxylic acids by means of chromatostrips, Chem. Abstr. **54**:13938, 1960.
37. Subbarao, R., Roomi, M. W., Subbaram, M. R., and Achaya, K. T.: Separation of oxygenated fatty compounds by thin-layer chromatography, J. Chromatogr. **9**:295, 1962.
38. Dhont, J. H., and De Rooy, C.: Thin-layer chromatography of 3,5-dinitrobenzoates, Analyst **86**:74, 1961.
39. Wollenweber, P.: Dünnschicht-chromatographische Trennungen von Aminosauren an Cellulose-Schichten, J. Chromatogr. **9**:369, 1962.
40. Vogler, K., Studer, R. O., Lergier, W., and Lanz, P.: Synthesen in der Polymyxin-Reihe Synthesis des cyclischen Decapeptids 88, Helv. Chem. Acta **43**:1751, 1960.
41. VanDam, M. J. O., Dekleuver, G. J., and Detteus, J. G.: Thin-layer chromatography of weakly polar steroids, J. Chromatogr. **4**:26, 1960.
42. Matthews, J. S., Pereds, V. A. L., and Aguilera, P. A.: Steroids. CCXV. The quantitative analysis of steroids by thin-layer chromatography, J. Chromatogr. **9**:331, 1962.
43. Struck, H.: Trennung und Bestimmung von Oeston, 17β-Oestradiol und 16α, 17β-Oestriol mit Hilfe der Dünnschichtchromatographie, Mikrochim. Acta, p. 634, 1961.
44. Starka, L., and Malikova, J.: A simple method for the separation of pregnanetriols by spread chromatography, J. Endocr. **22**:215, 1961.
45. Peifer, J. J.: A rapid and simplified method of analysis by thin-layer chromatography using microchromatoplates, Mikrochim. Acta, p. 529, 1962.
46. Mahadevan, V., and Lundberg, W. O.: Preparation of cholesterol esters of long-chain fatty acids and characterization of cholesteryl arachidonate, J. Lipid Res. **3**:106, 1962.
47. Hofmann, A. F.: Thin-layer adsorption chromatography of free and conjugated bile acids on silicic acid, J. Lipid Res. **3**:127, 1962.
48. Bennet, R. D., and Heftmann, E.: Thin-layer chromatography of corticosteroids, J. Chromatogr. **9**:348, 1962.
49. Nishikaze, O., and Staudinger, H. J.: Dünnschichtchromatographische Isolierung von Aldosteron aus Harn, Klin. Wschr. **40**:1014, 1962.
50. Kirchner, J. G., and Miller, J. M.: Preparation of terpeneless essential oils, Ind. Eng. Chem. **44**:318, 1952.
51. Winterstein, A., Studer, A., and Rüegg, R.: New results of carotenoid research, Chem. Ber. **93**:2951, 1960.
52. Davidek, J.: Separation of gallic acid and its esters on thin layers of polyamide powder, J. Chromatogr. **9**:363, 1962.
53. Scher, A.: Die Analyse von Tocopherol gemischen mit Hilfe der Dünnschichtchromatographie, Mikrochim. Acta, p. 308, 1961.
54. Horowitz, R. M., and Gentili, B.: Flavonoids of citrus. IV. Isolation of some aglycones from the lemon (*Citrus limon*), J. Organ. Chem. **25**:2183, 1960.
55. Purdy, S. J., and Truter, E. V.: Analysis of

thin-film chromatography, Analyst 87:802, 1962.
56. Petrowitz, H. J., and Pastuska, G.: Über die Kieselgelschicht-Chromatographie gesättigter aliphatischer Dicarbonsäuren, J. Chromatogr. 7:128, 1962.
57. Honegger, C. G.: Dünnschicht-Ionophirese und Dunnschicht-Ionophorese-Chromatographie, Helv. Chim. Acta 44:173, 1961.
58. Helfferich, F.: Ion exchange, New York, 1962, McGraw-Hill Book Co.
59. Peifer, J. J.: A rapid and simplified method of analysis by thin-layer chromatography using microchromatoplates, Mikrochim. Acta, p. 529, 1962.
60. Bobbitt, J. M., Schwarting, A. E., and Gritter, R. J.: Introduction to chromatography, New York, 1968, Reinhold Publishing Corp.
61. Purdy, S. J., and Truter, E. V.: Quantitative analysis by thin-film chromatography, Analyst 87:802, 1962.
62. Spencer, D. R., and Beggs, B. H.: Thin-layer chromatography on silica gel, quantitative analysis by direct spectrophotometry, J. Chromatogr. 21:52, 1966.
63. Scott, C. R., Labbe, R. F., and Nutter, J.: A rapid assay for urinary porphyrins by thin-layer chromatography, Clin. Chem. 13:6, 493, 1967.
64. Annino, J. S., Lipson, M., and Williams, L. A.: Determination of 3-methoxy-4-hydroxymandelic acid (VMA) in urine by thin-layer chromatography, Clin. Chem. 11:905, 1965.

ADDITIONAL REFERENCES

Bayer, E.: Gas chromatography, New York, 1961, American Elsevier Publishing Co., Inc.

Burchfield, H. P., and Storrs, E. E.: Biochemical applications of gas chromatography, New York, 1962, Academic Press, Inc.

Dal Nogare, S., and Juvet, R. S., Jr.: Gas-liquid chromatography, New York, 1962, Interscience Publishers, Inc.

Giddings, J. C.: Dynamics of chromatography, part I, New York, 1965, Marcel Dekker, Inc.

Hashimoto, J.: Thin-layer chromatography, Tokyo, 1962, Hirokawa Publishing Co.

James, A. T., and Morris, L. J., editors: New biochemical separations, New York, 1964, D. Van Nostrand Co., Inc.

Kromen, H. S., and Bender, S. R., editors: Theory and application of gas chromatography in industry and medicine, New York, 1968, Grune & Stratton, Inc.

Lederer, E., and Lederer, M.: Chromatography, a review of principles and applications, ed. 2, New York, 1957, American Elsevier Publishing Co., Inc.

Littlewood, A. B.: Gas chromatography, New York, 1962, Academic Press, Inc.

Mangold, H. K., Schmid, H. H. O., and Stahl, E.: Thin-layer chromatography, methods of biochemical analysis, vol. 12, New York, 1964, Interscience Publishers, Inc.

Marini-Bettolo, G. B., editor: Thin-layer chromatography, New York, 1964, American Elsevier Publishing Co., Inc.

Purnell, H.: Gas chromatography, New York, 1962, John Wiley & Sons, Inc.

Randerath, K.: Thin-layer chromatography, New York, 1966, Academic Press, Inc.

Scott, R. P. W., editor: Gas chromatography, London, 1960, Butterworth & Co., Inc.

Truter, E. V.: Thin-film chromatography, New York, 1963, Interscience Publishers, Inc.

Reference lists are available from the following manufacturers of chromatography equipment:

Camag Bibliography Service. References on TLC obtainable from Gelman Instrument Co., P. O. Box 1448, Ann Arbor, Mich. 48106; or from Arthur H. Thomas Co., P. O. Box 779, Philadelphia, Pa. 19105

Eastman Kodak Distillation Products Industries, Rochester, N. Y. 14603 (dyes for detection and adsorbents on plastic film)

Mallinckrodt Chemical Works, 2nd and Mallinckrodt Sts., St. Louis, Mo. 63160

G. K. Turner Associates, 2524 Pulgas Ave., Palo Alto, Calif. 94303

20

Electrophoresis and immunoelectrophoresis

CARL R. JOLLIFF, B.S.

In the last decade considerable emphasis has been placed on the results obtained by the separation of body fluid proteins by electrophoresis. The clinical laboratory and its personnel should be able to perform at least the basic electrophoretic separations. Considerable apprehension on the part of many laboratory workers has existed in the past toward electrophoresis. This apprehension may be attributed, in part, to the tremendous strides of the last few years that have been taken in this field,[1,2] the myriad separation techniques, and the various media that have become available for the procedure. It is the purpose in this chapter to acquaint the reader with the basic methodology available for the separation of body fluid proteins by electrophoresis, some of the technical difficulties involved, and the more sophisticated procedures and methods that may be asked for by the clinician to better ascertain protein abnormalities exhibited in the various pathologic entities.

ELECTROPHORESIS

Basic concepts

No attempt will be made here to discuss the physical-chemical principles involved in electrophoresis. The reader is referred to a number of excellent references dealing with the subject.[3-15]

In simplest terms, electrophoresis is the movement of colloidal particles in an electric field. The movement of these particles depends on a number of factors: the shape of the particle, its size, and its electric charge; the pH of the buffer with which it comes in contact; the medium on which the separation is taking place; the temperature; and the force of the electric field.

Historically, electrophoresis has progressed from the original work of Tiselius[16] who separated proteins by "moving boundary" or "free" electrophoresis in which no supporting medium was used but, rather, the sample separating layers of buffer to which an electric potential had been applied. By the use of an elaborate optical and measurement system the difference in refractive indices between the separated protein constituents could be photographed and recorded. The cost of equipment, the large amount of sample, and the incomplete separation necessitated establishment of a more practical method of separation.

Electrophoresis on a stabilized medium, or so-called "zone electrophoresis," was a long-awaited answer to the need for a procedure that could be adapted to the clinical laboratory and it is this form of electrophoresis that is in most common use today. Zone electrophoresis utilizes some form of medium that will support the sample of protein-containing material and allow for separation and final staining of the separated components. The media most commonly used are paper, cellulose acetate, starch gel, agar gel, agarose, and polyacrylamide gels. It is with these media that the methods herein included concern themselves. Starch gel and polyacrylamide gels utilize the phenomena of molecular sieving as well as separation by electrophoretic mobility. Some twenty-five or more serum protein components are separated by these systems whereas only five or six are separated by the other media mentioned. For exacting and investigational work the molecular sieving techniques may be required; however, such techniques used in the general clinical laboratory only tend to confuse the situation where known established patterns in disease states are based on the separation of the usual five components in serum protein electrophoresis.

The separation of serum proteins—albumin, $alpha_1$ globulin, $alpha_2$ globulin, beta globulin, and gamma globulin—is for the most part only a screening procedure. In the last few years a combination of electrophoresis and immunodiffusion, the immunoelectrophoretic method, has become most important in the visualization of the specific protein abnormalities. Further, the immunoquantitation of specific components and the establishment of their normal concentrations have become paramount in the diagnosis of immunoglobulinopathies and other protein dyscrasias.

The following methodology is that which is used in our laboratory. It has been found to be quite adequate.

Serum proteins

Paper electrophoresis

Equipment needed

Model R paper electrophoresis system (Beckman), consisting of the following:
Cell with cover
Duostat (Beckman) or other suitable power supply*—2.5 ma/cell, 70 volts
Analytrol (Beckman) or Photovolt densitometer model 542 with integrator (Fig. 20-1)
Drying rack (Beckman 320037)
Folding rack (Beckman 319313)
Dyeing rack (Beckman 320027)
Dyeing and washing trays, five (Beckman 324463)
Sample applicator (Beckman 320005)
Filter paper wicks (Beckman 319329)
Filter paper strips (Beckman 320046)
Blotting paper (Beckman 320039)
Microsyringe (Hamilton 701)

Reagents needed

Buffer—B2 (Beckman 320024) pH 8.6, ionic strength 0.075

Veronal	**2.76 g**
Sodium Veronal	**15.40 g**
Water	**1000 ml**

Dye—B4 (Beckman 320047) bromphenol blue, 1 g/1000 ml absolute methanol
Dye—light green SF (National Aniline), 1 g/1000 ml absolute methanol
Ammonium hydroxide, A.R. (28% to 30%)

Procedure for electrophoresis

1. **Place filter paper wicks in the two wick supports. Number filter paper strips with pencil. Place strips in a small amount of buffer so that they**

*Instrumentation Specialties Co.

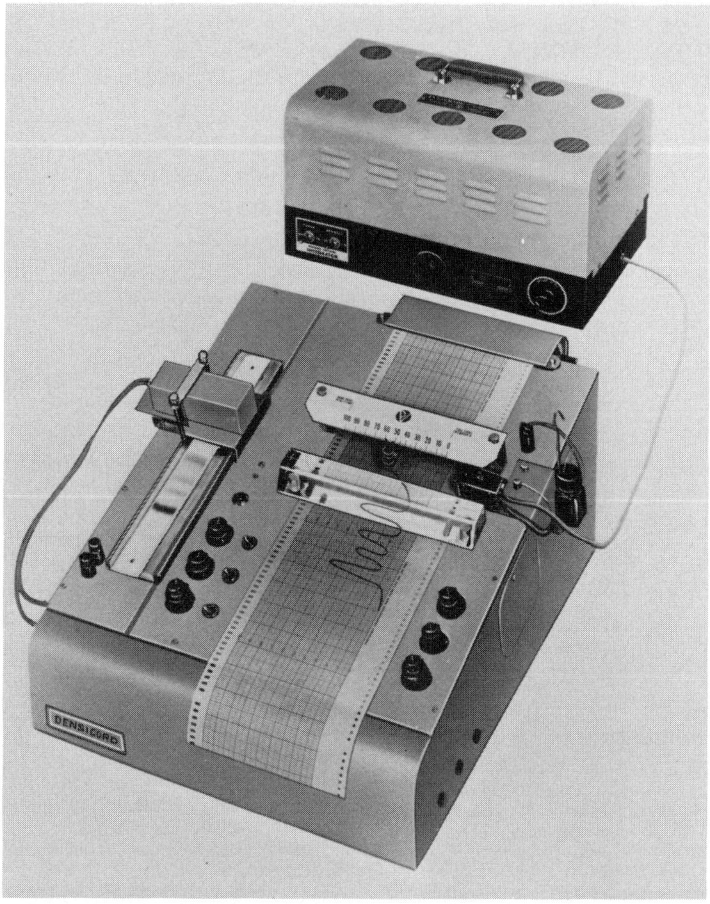

Fig. 20-1. Photovolt densitometer with integrator attachment. (Courtesy Photovolt Corp., New York, N.Y.)

are completely soaked. Eight strips should always be used. Blot with blotting paper until just moist.

2. Place the buffer (same as room temperature or at temperature to be run) into the lower portion of the cell. Tip the cell so that the buffer level is the same on each side. The buffer level should be at the maximum level mark.
3. Place the folding rack in cell over center support. Tighten knurled knobs so that rack is secure. Now carefully place prewetted strips on rack. Make certain strips are taut but do not tear. Ends of strips must lie flat on the wet filter paper wicks.
4. Close cell and allow to equilibrate for several hours. Remove tape from cover of cell. Pipette 0.008 ml of serum onto serum applicator. Be certain serum is distributed equally along the wire. With a gentle motion place serum on strip at midline. Repeat same procedure for all eight strips.
5. Close cell with cover tape. Cover cell with plastic cover and cardboard box to avoid air currents. Turn on power supply to deliver 2.5 ma/cell. Check current for first several minutes and adjust. Sixteen hours is usually adequate time to separate proteins.
6. Turn off power supply. Remove covers, unplug cell, and remove strip

rack. Place rack in a preheated oven (120° to 130° C) until strips are dry.

7. Remove strips from folding rack and place in dyeing rack. Immerse strips in purified methanol for 5 minutes. Remove and place immediately in 0.1% bromphenol blue or 0.1% light green SF in methanol for 30 minutes.
8. Remove from stain and place in three rinses of 5% glacial acetic acid (USP or A.R.) in water (v/v) for 5 minutes each. Blot strips and place on drying rack in oven at 120° to 130° C until dry.

Procedures for quantitation

Scanning

If bromphenol blue has been used to stain the strips, remove the strips from the oven and place them in the dyeing rack. Place rack into a tray containing on the bottom a piece of wick filter paper that has been impregnated with 1 ml of ammonium hydroxide (28% to 30%). Remove after 15 minutes. Protein bands will now appear blue. If light green SF stain is used, the above procedure is not necessary.

Scan the strips on the Beckman Analytrol according to the manual supplied with the instrument. The Photovolt densitometer (542) with integrator (542) may also be used. Carefully follow the directions supplied with the instrument.

For the Analytrol two 500 mμ interference filters are used if bromphenol blue stain is used. Use the B-5 cam. If light green SF is used, change to a 600 mμ filter with the B-5 cam. With the Photovolt densitometer a 500 mμ filter is used for bromphenol blue, and a 595 mμ for light green SF.

Calculate percentages of albumin, etc. from the integrator markings, after following directions for the bisecting of the protein bands.

Determine the total protein of the sample by the method used in your laboratory and calculate the percent of each fraction. See example that follows.

Total protein, 8.0 g/100 ml

Component	Total count	Percent	g/100 ml
Albumin	100	57.1	4.57
$Alpha_1$ globulin	5	3.0	0.24
$Alpha_2$ globulin	20	11.4	0.91
Beta globulin	20	11.4	0.91
Gamma globulin	30	17.2	1.38
Totals	175	100.1	8.02

Elution

For those who do not have a densitometer the bromphenol blue strips may be quantitated by elution. This procedure is satisfactory and in some instances may be more exact than the densitometer method.

Cut strip at lighter region between each band. Cut strip just ahead of the albumin band and just behind the gamma globulin band. Obtain a blank (background) by cutting a strip intermediate between the bands already cut.

Place each segment plus the blank segment in a test tube containing a volume of 0.5% sodium carbonate (anhydrous) in water (w/v); this allows for accurate measurement of absorbance (A) on the spectrophotometer. It may be necessary to use twice the volume of eluent for the albumin band as the other bands. Usually 20 ml for the albumin band and 10 ml for the others will do. The exact volume of sodium carbonate must be recorded.

After elution has taken place, pour off the sodium carbonate into a spectrophotometer cuvette (1 cm light path) and allow filter paper fibers to settle.

Calibrate spectrophotometer with solutions of a known concentration of bromphenol blue in 0.5% sodium carbonate. Read standards at 590 mμ against 0.5% sodium carbonate as a blank. Read eluates against background eluate as the blank.

Convert A readings to concentration by using calibration curve made with standard readings. Determine the amount of bromphenol blue eluted from each paper strip by multiplying the concentration by the volume of eluent. Divide the amount of bromphenol blue bound to each component by the total bound by all components to obtain the percentage of each component. Having found the percentages for each band, perform calculations as in the scanning section for determining the g/100 ml values.

Cellulose acetate electrophoresis

Equipment needed

Cell (Davis, model 117 A) or other types by Gelman and Colab.
Frame (Davis, model 120 F) (Fig. 20-2)
Power supply, 175-volt (ISCO* 490) (Fig. 20-3)
Trays, five plastic (Gelman 7211)
Trays, three stainless, 2 × 8 inch with lids
Glass slab, 6 × 12 inch
Photovolt densitometer with cellulose acetate attachment, 0.1 mm slit, and 525 mμ filter (Photovolt 5073 and 5069)
Blotting paper (Gelman 51290)
Sepraphore III, 1 × 6 inch (Gelman 51003)
Beveled microscope slide (see procedure)
Microsyringe, 10 mμ capacity in 1 mμ divisions (Hamilton 701)
Teflon tubing tips (Turner 310-001)
Marking pen (Scientific Products P-1226)

Reagents needed

High-resolution buffer (Gelman 51104)
Dissolve 1 packet/1200 ml water.
Ponceau S stain (0.1% in 5% acetic acid, aqueous) (Gelman 51281)
Destaining solution: 5% acetic acid in water (v/v)
Dehydrating solution: absolute methanol, A.R.
Clearing solution: 15% acetic acid in absolute methanol, A.R.

Procedure

1. Place about 100 ml of high-resolution buffer into a plastic tray. With the aid of forceps remove the cellulose acetate strips from the box and gently allow strips to fall into the buffer so that they soak from bottom. Do not push them into the buffer. Entrapped air pockets would cause an irregular pattern. As soon as strips are soaked, immerse in buffer for at least 5 minutes.
2. Remove strips and place on blotting paper. Carefully blot. Number strips with marking pen at the end of the strips. (These marks will remain visible throughout the entire procedure and will not wash off.)
3. Place strips on rack so that ends are equidistant from the frame on each side (Fig. 20-2). Tighten the knurled nuts on frame bars to hold strips firmly in place and loosen knurled nuts on side so as to make strips taut.
4. Place high-resolution buffer into cell to maximum level line. Tilt cell so that buffer level is equalized on both sides. Place rack into cell, making certain strips ends are in contact with buffer on both sides.
5. A microscope slide is beveled on both sides at one end so that the width is no more than 15 mm. This works as an excellent applicator.

*Instrumentation Specialties Co.

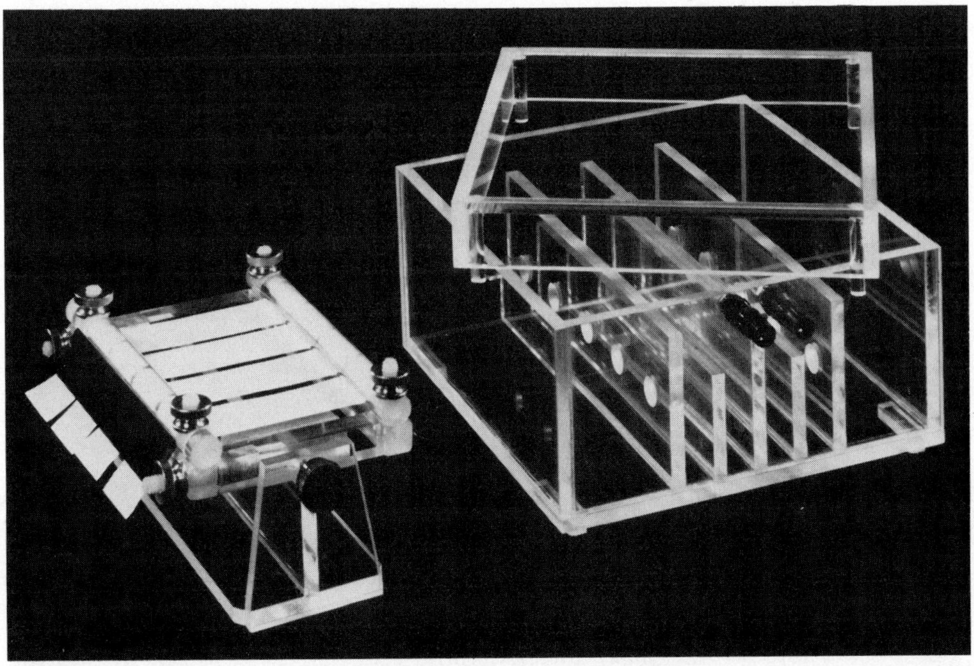

Fig. 20-2. Cell and cellulose acetate frame with stand. (Courtesy C. L. Davis Co., Lincoln, Neb.)

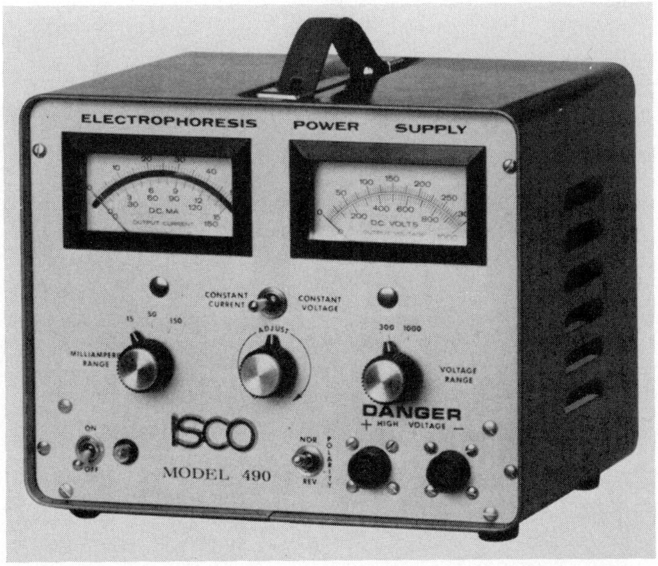

Fig. 20-3. Power supply for constant current as well as constant voltage. Polarity may be switched. (Courtesy Instrumentation Specialties Co., Lincoln, Neb.)

With the aid of a microsyringe fitted with a Teflon tubing tip carefully remove 2 μl of serum from the specimen tube. With care place the sample on the edge of the slide, being certain no serum spreads off the edge onto the side of the slide. Carefully place the slide on the strip, 20 mm in from the cathodic (−) end of the cell frame bar. After all strips are so loaded, place the cover on the cell and run at a constant level of 175 volts for 25 minutes.

6. Remove strips from the cell and place, one at a time, into ponceau S stain as in the buffer, allowing strips to float until uniformly soaked; then allow to stain for 5 minutes.
7. Remove from stain and remove excess stain by allowing strips to swish in aqueous 5% acetic acid for several minutes for each of three successive trays. Place the washed strips on blotting paper and blot until barely moist.
8. The strips may be photographed or used for densitometer tracings at this point. However, we find that clearing is much preferred for densitometer evaluation.
9. The strips may be cleared by placing them (no more than two at a time) in two successive trays of absolute methanol for 5 minutes each and finally in a tray of 15% acetic acid in absolute methanol (v/v) for 5 minutes. After this they are carefully removed and placed on a clean glass slab by allowing the bottom of the strip to first touch the glass and letting the strip gently adhere to the plate. *Do not attempt to blot or wipe the strip until dry.*
10. The strip can be removed for densitometer evaluation by placing a piece of masking tape on each end of the strip and carefully pulling the strip from the glass plate.
11. The strips are then placed on the densitometer carriage and quantitated. (See discussion of paper electrophoresis for quantitation method.) The cleared strips may be stored in a laboratory workbook by simply holding them in place with masking tape at each end on the page.

Micro cellulose acetate electrophoresis

Various manufacturers have available cellulose acetate electrophoresis equipment that allows for separation utilizing micro quantities of sample. Some of the

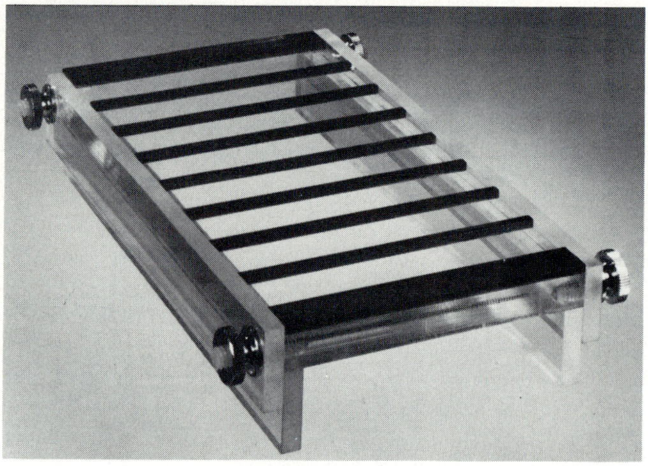

Fig. 20-4. Tray for micro starch gel electrophoresis. (Courtesy C. L. Davis Co., Lincoln, Neb.)

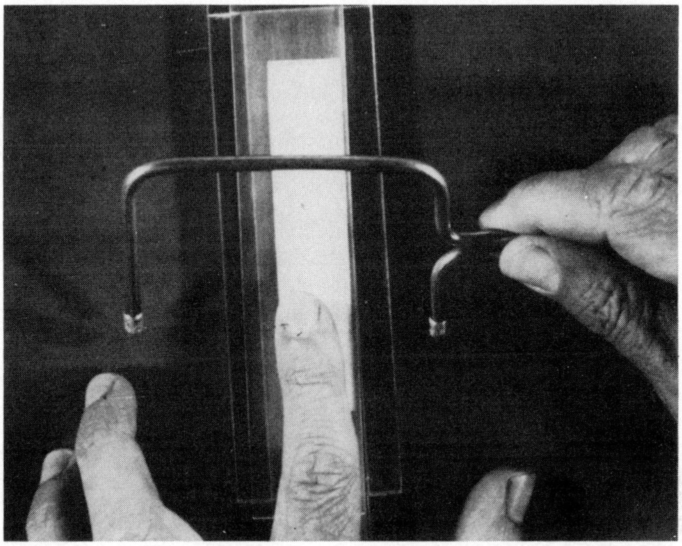

Fig. 20-5. Gel slicing tray showing cutter made from ordinary cheese cutter.

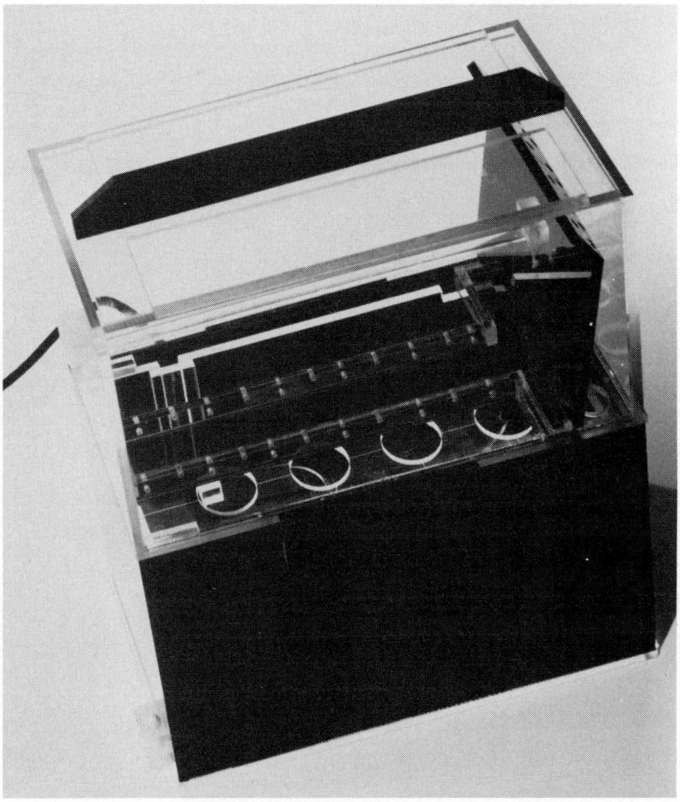

Fig. 20-6. Circulating destaining unit. May be used for starch gel procedure and immunoelectrophoresis. (Courtesy C. L. Davis Co., Lincoln, Neb.)

systems available are Microzone (Beckman), Sepra-Tek (Gelman), Microphore (Brinkmann), and PhoroSlide (Millipore). Complete instructions are supplied with these systems and for best results should be carefully followed.

Micro starch gel electrophoresis

The method[17] presented here is a modification of the Smithes[18] technique. It is included because of its simplicity and for its use in specialized protein studies as well as a method for hemoglobin electrophoresis.

Equipment needed

Electrophoresis cell (Davis 117 A)
(same as for cellulose acetate method)
Micro starch gel tray (Davis 39 S) (Fig. 20-4)
Power supply (ISCO 490)
Slicing tray (Davis 110 ST) (Fig. 20-5)
Destainer (Davis 12 D) (Fig. 20-6)
Washer (Davis 130 SW) (Fig. 20-7)
Slicer (ordinary cheese cutter with fine recorder wire or piano wire)
Filter paper wicks (Beckman 319329)
Razor blade (stainless steel type; both sides ground off so that blade is approximately ½ inch wide)
Filter paper strips, 2 × 20 mm, made from Whatman 3 mm
Small-nosed forceps
Petri dishes

Fig. 20-7. Washing unit. Separate chambers provided for water and saline. May be used for any washing procedure. (Courtesy C. L. Davis Co., Lincoln, Neb.)

Reagents needed

Starch gel, hydrolyzed (Connaught Laboratories, distributed by Fisher Scientific Co.)

Stock buffer, pH 8.8

Tris (hydroxymethyl) aminomethane	30.25 g
Ethylenediamine tetraacetic acid	3.00 g
Boric acid	2.30 g

Distilled water to final volume of 1000 ml

Gel buffer

One part stock buffer is diluted with 2 parts distilled water to prepare the dilute buffer for the gels.

Electrode chamber buffer, pH 8.2

Boric acid	18.5 g
Sodium hydroxide	2.5 g

Distilled water to final volume of 1000 ml

A few crystals of bromphenol blue are added to buffer in order to visualize the borate front during electrophoresis.

Starch gel: 13.3 g starch/100 ml dilute gel buffer

Procedure

Preparation of starch gel trays

Mix 200 ml of gel buffer and 26.6 g of starch, hydrolyzed, in a 500 ml Pyrex suction flask, cook, and degas as described by Smithes.[18] Pour mixture into the starch gel tray, allowing the starch mixture to overflow. With a straight edge (one time only) start at one end of the tray and scrape from one end gate to the other to remove excess starch. As soon as the gel has set (approximately 20 minutes) cover the gels smoothly with Saran wrap. Gels may be used when cool or they may be stored overnight at room temperature, wrapped in Saran.

Application of sample

Remove the Saran wrap from the tray just far enough so that the sample may be inserted 3 cm away from the cathodic (−) end of the gel. After serum has been introduced into the gel, by means of soaking a 2 × 20 mm filter paper strip in the serum sample, place the strip on the edge of a special razor blade—far enough from the edge so that as the edge of the blade is forced into the gel, at right angles to the trough edge and to the gel, the paper remains above the gel. Place the sample approximately 3 cm from the cathodic end of the gel. Gently pull the blade back so that the incised gel is opened and with a pair of fine-nosed forceps push the serum-impregnated filter paper into the slit formed. Carefully pull out the blade, allowing the serum sample strip to remain in the gel. Be certain no bubbles or gaps result between the gel edges and the strip.

Electrophoresis

After all samples have been introduced into the gel, re-cover the gel tray end with Saran wrap. Place the tray into the cell (Fig. 20-4). Remove the end gates carefully and place the wick paper vertically on the wick hollers so they fit tightly against the bare starch gel. The end gates are replaced and the current is adjusted on the power supply to 50 ma (approximately 200 volts). At 10 minutes the power is turned off, the Saran wrap is folded back, and the filter paper strips are removed from the gel. Care is again observed that no air pockets or gaps remain in the gel slit.

The Saran wrap is placed back over the gel, the current is set at 50 ma, and the cell cover is filled with ice cubes and placed on the cell. As the electrophoresis progresses, the bromphenol blue–containing buffer front passes through the gel, staining the albumin component. The albumin band is allowed to migrate 4.5 to 5 cm from the point of application. The current is turned off and the tray removed from the cell.

Slicing the gels

The gels are removed from the tray by first running a sharp edge along the sides of the trough (a small platinum spatula works quite well). Remove the end gates and gently, with a squeeze type wash bottle, force water under the gel, allowing the gel to slide at a slight angle into the cutting tray. The slabs may be marked at this point using serum on the end of a pointed applicator stick. Use a series of dots to designate the gel number. Be certain to mark both ends of the gel as the gels will break at the site of serum application.

Cut the gels once longitudinally with the wire cutter (Fig. 20-5). Plastic wedges of proper thickness placed under the gel in the slicing tray allow various thicknesses of the gels to be cut. The top slice is usually the most smeared and the lower slices are best used for staining. As many as three slices of the slab can be cut without difficulty.

Staining and destaining the gels

The gels are stained in a large Petri dish containing amido black (10 g/1000 ml) dissolved in the following solvent: absolute methanol 5 parts, water 5 parts, and glacial acetic acid 1 part. The gels are allowed to stain for 4 minutes. They are then transferred to a destainer (Fig. 20-6) with the same solvent as above or, if a destainer is not available, to a series of Petri dishes containing the solvent.

Scanning the gels

A scan of the gels is possible with the starch gel tray provided for the Photovolt densitometer (5063). The slow drive used for cellulose acetate scans is also used. A 525 mμ filter is used. No attempt has been made to quantitate serum patterns by using this procedure for protein electrophoresis but rather to establish patterns for specific disease states (Figs. 20-8 to 20-10). We have not found it necessary to clear the gels in glycerol before scanning.

Agar gel electrophoresis (method of Cawley[7])

Equipment needed

Cell—same as for cellulose acetate and starch gel methods
Power supply—same as above
Densitometer—same as above
Drying oven—60° C
Hamilton syringe—same as for cellulose acetate method
Staining trays—same as above
Special razor blade—as in starch gel procedure
Cronar photographic film base, unperforated, 0.004-inch thickness (du Pont)
Pencil, heat-resistant (Blaisdell)

Reagents needed

Barbital buffer, pH 8.6, ionic strength 0.05

Veronal	1.84 g
Sodium Veronal	10.30 g
Distilled water	1000 ml

Fixation solution: 10% (v/v) acetic acid in absolute methanol
Stain: 1 g thiazine red R in 1% aqueous acetic acid (Allied)

Fig. 20-8. Densitometer scan of serum on starch gel. (From Marsh, C. L., and Jolliff, C. R.: Micro starch gel electrophoresis. In Smith, Ivor, editor: Chromatographic and electrophoretic technique, vol. 2, Zone electrophoresis, ed. 2, London, 1968, Wm. Heinemann Medical Books, Ltd.)

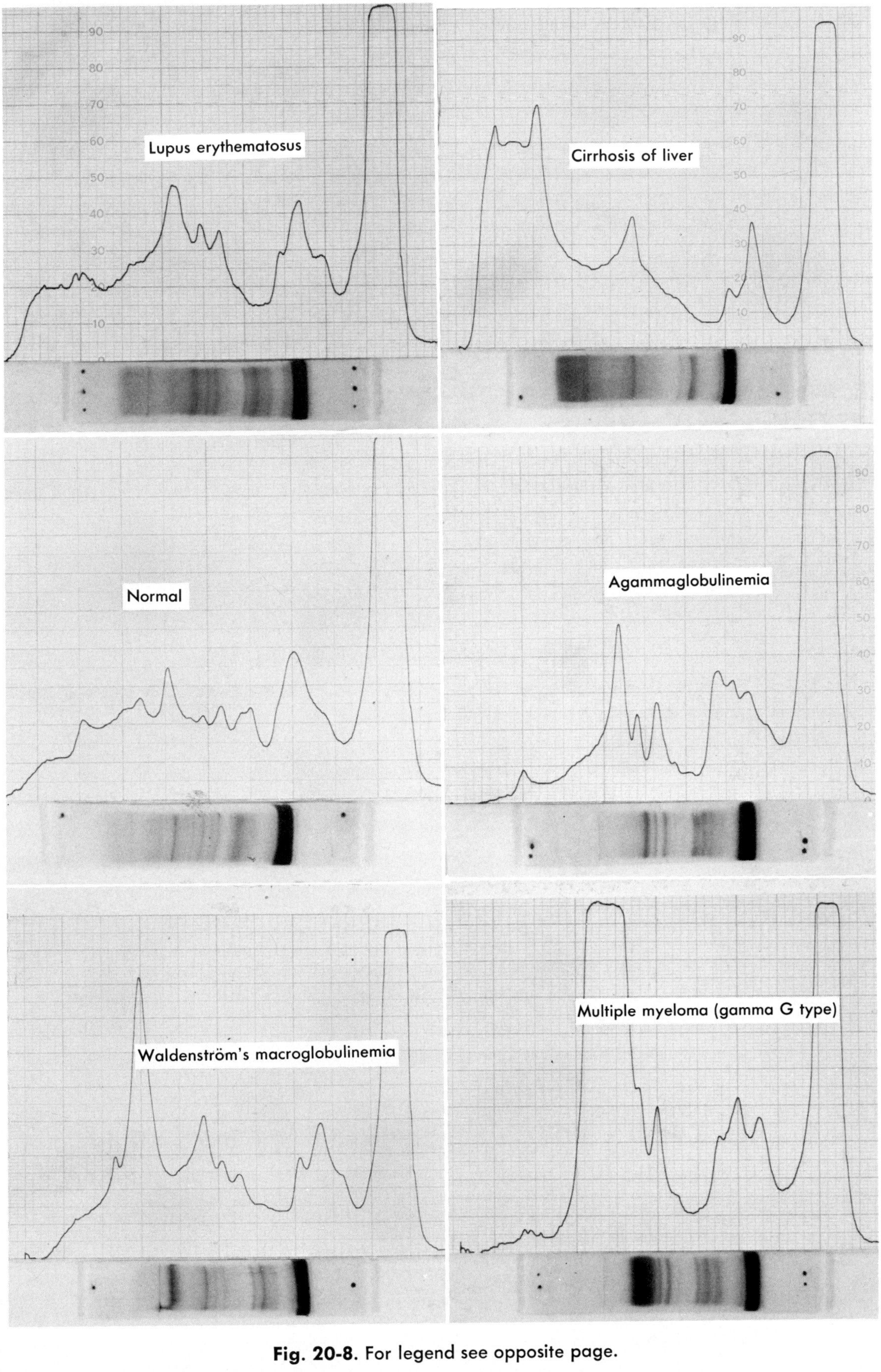

Fig. 20-8. For legend see opposite page.

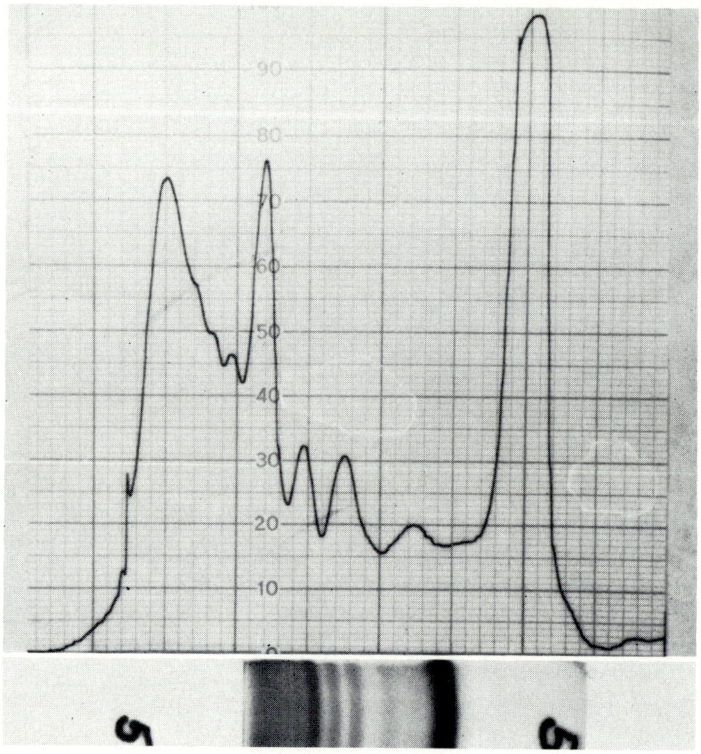

Fig. 20-9. Densitometer scan of serum from patient with gamma A monoclonal gammopathy. Starch gel.

Destaining solution: 2% (v/v) aqueous acetic acid
Agarose (Bausch & Lomb)

Procedure

Preparation of gel strips

Cut Cronar film base into 15 cm strips. Mark the strips on convex side with heat-resistant Blaisdell marking pencil as to number of specimen and point of origin (midpoint).

To 50 ml of barbital buffer add 250 mg of agarose. Dissolve by bringing to a boil, with gentle agitation of the flask. Add 2 drops of a 2% (v/v) aqueous solution of bromphenol blue to the prepared agarose.

With the concave side up, coat the strips by carefully layering 5 ml of the hot agarose over the entire surface of the strip. Allow to harden for at least 10 minutes at room temperature. The strips may be used immediately or may be stored refrigerated in a Tupperware meat keeper that contains moisturized pads.

Introduction of sample and electrophoresis

The serum sample is introduced into the midpoint of the gel strip by a method similar to that described for the micro starch gel procedure except that the sample is not introduced on filter paper; the sample of serum (1.5 μl) is introduced into the slit with a Hamilton microsyringe with Teflon tip. The agarose gel slit is closed, the voltage adjusted to 150 volts, and the run is allowed to proceed until the stained albumin band migrates to the desired distance.

Remove strips from cell and fix in fixative solution for 20 minutes. Place in stain for 30 minutes and destain in three changes of destaining solution.

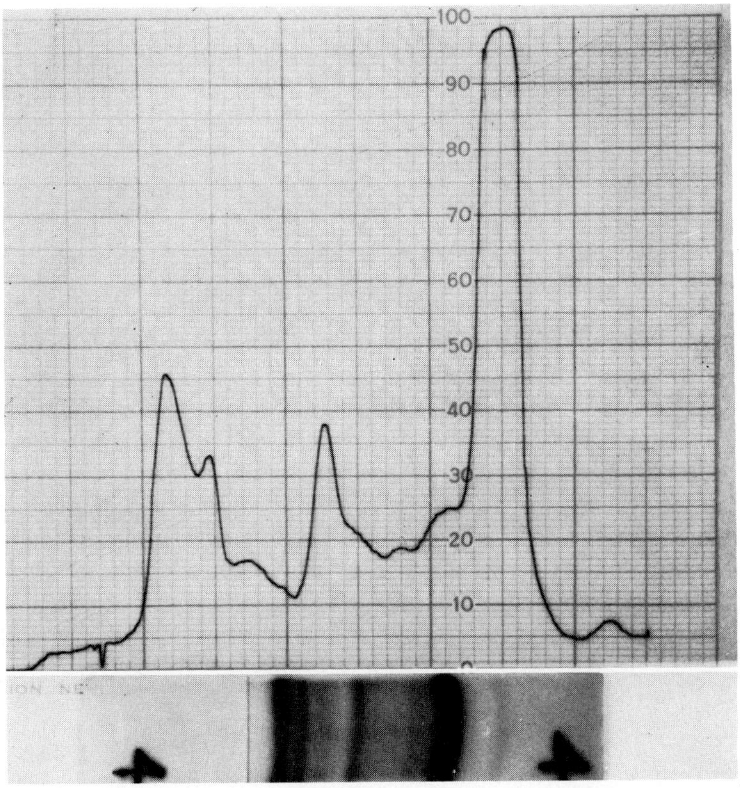

Fig. 20-10. Densitometer scan of serum from patient with gamma G monoclonal gammopathy. Starch gel.

Allow strips to dry at room temperature or in 60° C oven and scan on the Photovolt densitometer, using the cellulose acetate carriage and drive with a 500 mμ filter. Quantitate as for paper or cellulose acetate electrophoresis.

It should be noted that six protein bands are separated by this procedure. The beta globulin fraction is split into a beta$_1$ and a beta$_2$ band (Fig. 20-11).

Polyacrylamide gel electrophoresis

Except for certain methods this medium is not used routinely for serum protein electrophoresis in most clinical laboratories. It has found favor, however, in the chemistry laboratory for the separation of lipoproteins, enzymes, and hemoglobins. Excellent references to this method may be obtained from any of the following sources:

E-C Apparatus Corp., Canalco Inc., Buchler Instruments, Inc.

Plasma lipoproteins

Paper electrophoresis

Equipment needed

The same general equipment is used as in paper electrophoresis of serum proteins.

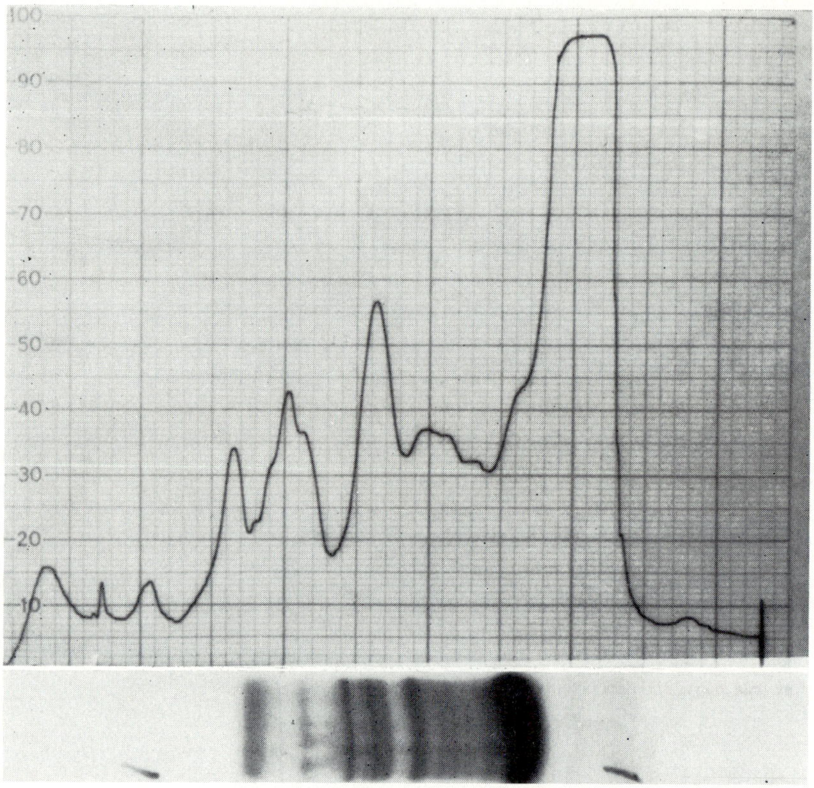

Fig. 20-11. Agar gel electrophoretic pattern of normal human serum (Courtesy Dr. L. P. Cawley, Wichita, Kans.)

Reagents and supplies needed

Filter paper strips (Beckman 320046)
Filter paper wicks (Beckman 319329)
Applicator (Beckman 320005)
Oxford pipette, 20 μl
Buffer: 0.10 M barbital in 0.001 M EDTA

This buffer may be prepared by adding 2 vials of B-1 type electrophoresis buffer, pH 8.6, ionic strength 0.075, plus 0.558 g of EDTA, to 1500 ml of a 1% human or bovine albumin (Hyland or Sigma). Bovine albumin works quite well and is less expensive. A preservative of 5 ml/L of 5% thymol in isopropyl alcohol is added. The buffer may be used for approximately 1 month without changing. It is stored in the refrigerator within the cell between runs. The polarity should be reversed between runs.

Oil red O stain: 750 ml ethanol (reagent grade, isopropyl denatured, satisfactory); 500 ml distilled water; and 0.5 g oil red O (National Aniline) This mixture is brought to a boil in a reflux condensor. After cooling, it is stored in an airtight flask in the incubator at 37° C. The stain should age several days before use. If the stain requires longer than 4 hours to visualize the alpha lipoprotein band on the paper strip, it should be replaced. *(Do not filter stain.)*

Procedure

NOTE: Plasma specimens are stable for 2 weeks at 3° C. Add 1 ml of 0.001 M EDTA in 0.15 M NaCl for blood collection.

1. Follow the procedure as for paper electrophoresis of proteins except for lipoprotein buffer in the cell.
2. Label strips with pencil, marking the application point and identification number. Soak strips in buffer for 15 minutes. Remove excess buffer by blotting.
3. Place strips on rack, being certain they are taut.
4. A 20 μl sample of plasma is applied to the strip with the aid of the Oxford pipette or similar pipette, using the Beckman applicator. Two loadings of the applicator with approximately 10 μl each are applied to each strip. Do not try to load applicator with 20 μl as pooling will occur and strips will not be sharp. Cover the cell.
5. Set constant current to 7.5 ma per cell and run for 16 hours.
6. Remove strips from cell and place in oven at 110° to 120° C for 30 minutes. Place in oil red O stain for 4 hours or until the alpha lipoprotein band becomes faintly visible. Place in tap water to remove excess stain. Dry in oven at 110° to 120° C for 30 minutes.
7. This method will distinguish four lipoprotein components. In order of their position from the point of application to the anode they are: chylomicrons, beta, pre-beta, and alpha. See Fig. 20-15.
8. All plasmas should be run on patients who have observed a 16-hour fast. In this case the chylomicrons will not appear in a normal plasma.

Cellulose acetate electrophoresis

Equipment needed

Same equipment as for serum proteins

Reagents and supplies needed

Buffer
    Add to 1 package of high-resolution buffer (Gelman) 1800 ml of distilled water. To this add 0.685 g of EDTA.

Stain
    Add an excess of oil red O to 500 ml of warm 70% absolute methanol. Age stain for several days at 37° C.

Decolorizing solution
    To 200 ml of 5% aqueous acetic acid add 1 ml of commercial bleach (Clorox).

Sepraphore III, 1 × 6 inch (Gelman)

Procedure

1. Soak strips in lipoprotein buffer, blot, and place on electrophoresis frame. Mark strips with pen as in serum protein method. Place frame in cell.
2. Apply 15 μl of plasma to the strip in about three applications with a Beckman or Gelman applicator.
3. Electrophorese at 175 volts for 25 minutes. Remove strips and place in oil red O for 24 hours. Do not overlay strips in stain.
4. Rinse strips in tap water and decolorize with solution for about 15 minutes.
5. Quantitation of the uncleared strips may be performed if so desired on the Photovolt densitometer.

Polyacrylamide gel electrophoresis

QDL Quick-Disc method

This method has recently been evaluated and was found to be quite time-saving as well as diagnostic (see Fig. 20-16).*

*The procedure may be procured from Canalco Inc.

Hemoglobins

Cellulose acetate electrophoresis

Equipment needed

Same as for serum protein electrophoresis

Reagents and supplies needed

Sepraphore III strips, 6¾ × 6 inch (Gelman 51014)
Hemoglobin buffer (Gelman 51122)
 Refrigerate on preparation and use directly from the refrigerator.
Ponceau S stain—same as for serum proteins
Microscope slide applicator—see protein procedure
Microsyringe—10 μl capacity in 1 μl graduations (Hamilton 701)
Trays
 1 Buffer tray, 8 × 8 inch
 1 Stain tray, 8 × 8 inch
 3 Destaining trays, 8 × 8 inch
 3 Dehydrating trays, 8 × 8 inch
 1 Clearing tray, 8 × 8 inch
Destaining solution: 5% (v/v) acetic acid in water
Clearing solution: 15% (v/v) acetic acid in absolute methanol
Dehydrating solution: absolute methanol, A.R.
Glass plate, 12 × 12 inch
Blotting paper
Controls
 Commercial preparations are available.*

We routinely run the following: A_1A_2 (normal), A_1A_2 (abnormal), C, S, and F.

Procedure

Preparation of sample

1. Obtain red blood cells from anticoagulated blood. (EDTA or double oxalate.) Obtain enough blood so that 1 ml or more of pooled cells after centrifugation will be obtained. Wash the cells three times with 0.9% NaCl.
2. To the packed cells add 1.2 volumes of distilled water. Mix. Add 0.4 volume of toluene. Stopper and shake vigorously for several minutes.
3. Centrifuge at 3000 rpm for 20 minutes. Discard supernatant WBC toluene layer by aspiration.
4. Filter through Whatman No. 3 and determine hemoglobin concentration by the cyanmethemoglobin method. Adjust filtrate to 10 g hemoglobin/100 ml. (Example: if hemoglobin concentration is 14.0 g/100 ml, add 0.40 ml of distilled water to each 1.0 ml of filtrate.) The hemolysate is now ready for electrophoresis and may also be used in the alkali denaturation test (see procedure). The hemolysate is stable for several weeks if refrigerated.

Electrophoresis

1. Soak Sepraphore III strips in buffer and blot. Strips should soak for a period of not less than 30 minutes. Place in cell frame. Lay a straight edge along the strip about 2 cm toward the cathodic end from the middle of the strip.
2. Place 1 μl of hemolysate on the edge of the glass slide with the Hamilton microsyringe fitted with a Teflon tubing tip. Carefully place the hemolysate on the strip and at the same time note each application by marking the strip with the marking pen. A total of 6 samples will fill the strip which will allow for 5 controls and the unknown.
3. Invert the lid and fill with ice cubes. Electrophorese for 75 minutes at 400 volts.

*Hyland Laboratories and Schering Diagnostics.

4. The patterns may be photographed for a permanent record at this time.
5. Stain strips as for serum protein, destain, blot, dehydrate, and clear.
6. Strips may be cut and quantitated as for serum protein method.

Starch gel electrophoresis

Equipment needed

Same as for serum protein electrophoresis

Reagents and supplies needed

Cell buffer

Tris	4.7 g
Glycine	22.6 g
Water to	1000 ml

Add several grains of bromphenol blue.

Buffer for preparation of gel

Tris	36.3 g
HCl	48.0 ml
EDTA	1.0 g

Distilled water to 1000 ml.

Amido black 10 B stain

Amido black 10 B	5 g
Methanol	450 ml
Distilled water	450 ml
Glacial acetic acid	90 ml

First dissolve the stain in the methanol and then add the distilled water. Mix well and filter through glass wool into a dark bottle.

Preparation of the gel

Dilute the buffer with 4 parts distilled water. Weigh 13.3 g of starch, hydrolyzed, and add 100 ml of the diluted buffer.

Procedure

Same as for serum proteins

Hemolysate prepared for cellulose acetate method is satisfactory. It has been found that a concentration of 5 g hemoglobin/100 ml is most satisfactory. Higher concentrations tend to cause blurring of the samples on migration. Hemoglobins are allowed to migrate until the A_1 band is approximately 4.5 to 5.0 cm from the starting slot.

The gels are stained in amido black 10 B for about 5 minutes and are placed in the destainer overnight for adequate removal of the stain.

The hemoglobin bands may be stained with the amido black stain or the benzidine stain described by Smithes.[19]

Procedure for determination of Hb F by alkali denaturation (modified from Singer and associates[20])

Reagents needed

Stock, 0.25 N KOH

Keep in polyethylene bottle in the refrigerator.

KOH, 0.08 N

Dilute 0.25 N KOH just before use.

Ammonium sulfate, 50% saturated

Dilute 500 ml of saturated ammonium sulfate with 500 ml of distilled water. Add 2.5 ml of concentrated HCl.

Procedure

Use the hemolysate as prepared in hemoglobin electrophoresis section. The hemolysate should be at least 10.0 g hemoglobin/100 ml.

Determination of denatured hemoglobin (do in triplicate)

1. Place 1.6 ml of 0.08 N KOH in a 100 × 13 mm test tube.

2. Starting stopwatch simultaneously, add 0.1 ml of the hemolysate. Rinse pipette several times.
3. Shake 10 seconds.
4. Exactly 1 minute after addition of hemolysate, add 3.4 ml of the ammonium sulfate solution.
5. Invert several times to ensure mixture, centrifuge 5 minutes at 1500 rpm, and filter the supernatant through Whatman No. 3, 55 mm.
6. Transfer to a suitable spectrophotometer cuvette and read the A at 540 mμ against reagent blank.

Determination of undenatured hemoglobin (do in triplicate)

1. Add 5 ml of distilled water to spectrophotometer cuvette.
2. Add 20 μl of hemolysate. Let stand 10 minutes.
3. Read against blank of distilled water at 540 mμ.

Calculation of fetal hemoglobin

Let D = average A readings of denatured hemoglobin; let U = average A readings of undenatured hemoglobin:

$$\frac{D}{U \times 5} \times 100 = \text{Percent fetal hemoglobin}$$

Normal values

In the adult = less than 2% total hemoglobin
At birth = 50% to 85% total hemoglobin
At 1 year = less than 15% total hemoglobin
Up to 2 years = less than 5% total hemoglobin

Other serum components

Isoenzymes: Lactic dehydrogenase, leucine aminopeptidase, malic dehydrogenase, creatinine phosphokinase, glutamic oxalacetic transaminase, alkaline and acid phosphatase, glucose-6-phosphate dehydrogenase, amylase, and catalase methodology may be found in a number of references.[7,14,21,22]

Glycoprotein: Consult references 7, 14, and 23.

Haptoglobin: Consult references 7, 9, 14, 17, and 24.

Myoglobin: Consult reference 7.

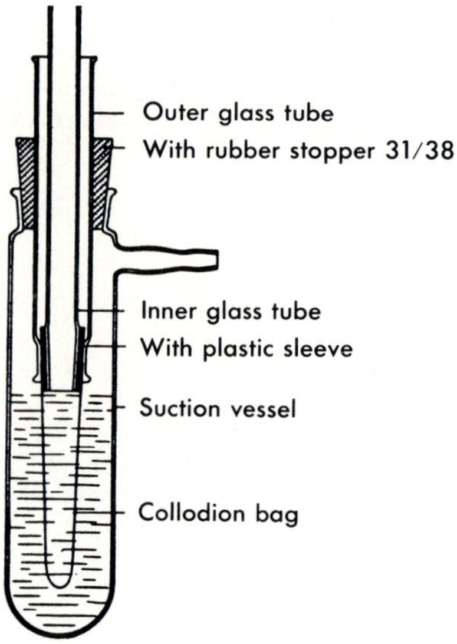

Fig. 20-12. Ultrafiltration apparatus for concentration of urine and CSF. (Courtesy Schleicher & Schuell, Inc., Keene, N. H.)

Method for concentrating urine and cerebrospinal fluid for electrophoresis[25]

Concentration of fluids of low protein content such as CSF and urine is required before electrophoresis can be performed. Molecular sieving methods such as starch and polyacrylamide gel do not require any or as much concentration of these substances as does ordinary zone electrophoresis on paper, cellulose acetate, or agar gel.

We have found the collodion bag ultrafiltration apparatus manufactured by Schleicher & Schuell Co. to meet our needs in this regard. The collodion bag retains proteins of a molecular weight of approximately 70,000 to 100,000 but may pass small compounds of 30,000 M.W. and below.

The bag is removed from the 20% alcohol storage container and is washed with distilled water. It is forced over the end of the plastic sleeve of the inner glass tube for about 1 cm. This is fastened tightly by the aid of the outer tube. (See Fig. 20-12.)

The assembled parts are fitted into the suction tube, which contains 0.9% NaCl. Take care that the water level inside the suction tube does not exceed the lower level of the inner and outer tubes. Attach the unit to a clamp on a ring stand.

Fill the bag with the fluid to be concentrated, via pipette through the inner glass tube. Using 10 ml of centrifuged urine or 5 ml of CSF is satisfactory.

Apply vacuum to the suction flask by means of a suitable vacuum pump. The specimen is concentrated to a volume of 0.1 ml or less and if desired may be somewhat reconstituted with a small amount of buffer, which is used for electrophoresis. Remove the concentrated fluid with the aid of a pipette. The bag may be reused if it is washed and replaced in the saline-containing suction tube.

The concentrated urine and CSF may be used in the same way as serum or plasma in the preceding electrophoretic methods.

Quality control in electrophoresis

In our laboratories we have used a number of commercially available lyophylized sera in the past for controls in electrophoretic methods. Careful attention must be made to the method by which these controls are run to establish the values given. In recent years we have found that freezing aliquots from a large serum pool and using them as needed works quite well. The value of the pool is determined, using numerous controls to establish a mean value. These controls are used after thawing for up to several days. All electrophoretic runs include at least one control serum. An excellent discussion in Nerenberg[9] deals with the problems encountered with quality control in electrophoresis.

Significance

Serum protein electrophoresis

A number of excellent references are available to the reader with regard to values found in the various disease states.[7,26-31]

Table 20-1 is a summary of various investigators' values (normal) that are established for the various types of serum protein electrophoresis. It must be emphasized, however, that the laboratory should establish its own values for its particular procedure. It must also be pointed out that values obtained on children, as shown in Table 20-2, vary from the adult levels.

Table 20-3 is included to exhibit the changes evidenced in the various pro-

Table 20-1. A summary of various investigators' values found for the protein components of human serum

	Free[69] (% total proteins)	Paper[28] (% total proteins ± 2 SD)	Cellulose acetate[70] (% total proteins ± 1 SD)	Agar gel[71] (% total proteins)
Albumin	55-64	42.5-59.9	56.5-71.1	45-68
Alpha$_1$ globulin	3.6- 7.2	3.1- 8.7	3.1- 5.6	2.8- 7.6
Alpha$_2$ globulin	7.6-10.1	6.1-12.7	6.7-10.3	6.0-17.1
Beta globulin	11.6-15.2	9.3-16.5	6.8-11.4	8.3-14.3
Gamma globulin	10.1-17.2	16.4-26.6	11.8-19.3	12.2-21.7

Table 20-2. Serum proteins in g/100 ml found in children at various age levels by cellulose acetate electrophoresis

	1 week	6 months	1 year	5 years
Albumin	3.4-5.5	2.62-5.10	3.98-5.20	3.80-5.78
Alpha$_1$ globulin	0.10-0.30	0.02-0.50	0.13-0.41	0.11-0.33
Alpha$_2$ globulin	0.31-0.52	0.42-0.91	0.35-0.69	0.41-1.02
Beta globulin	0.2-0.65	0.38-1.03	0.50-0.90	0.51-1.02
Gamma globulin	0.35-1.50	0.10-0.72	0.42-0.70	0.49-1.18

tein-disturbed diseases. Figs. 20-8 to 20-10 depict some of the electrophoretic patterns observed in disease states.

Cerebrospinal fluid protein electrophoresis

Normal values for cellulose acetate electrophoresis are given in Table 20-4. The clinical application of CSF electrophoresis may be found in the references.[7,32-38]

Urine protein electrophoresis

For a review of the urinary proteins and electrophoretic interpretation of disease states consult the references.[7,39-44]

The hemoglobins

The hypothesis of Pauling and co-workers[45] that the sickling phenomenon of red blood cells might be due to a molecular disease led to the discovery that patients with sickle cell anemia do, in fact, have hemoglobin with a different electrophoretic mobility than that of normal hemoglobin. Today the various molecular forms of hemoglobin known are vast in number.

The hemoglobin A molecule consists of four polypeptide chains. Two chains

Table 20-3. Electrophoretic fractionations of serum proteins in a variety of diseases

Category	No.	Protein concentrations (g/100 ml; mean ± standard deviation)					
		Total protein	Albumin	α_1-globulin	α_2-globulin	β-globulin	γ-globulin
Normal controls	26	7.14 ± 0.33	3.65 ± 0.31	0.42 ± 0.10	0.67 ± 0.12	0.91 ± 0.13	1.62 ± 0.18
Multiple myeloma	67	7.96 ± 1.15**	2.60 ± 0.73**	(total globulins = 5.35 ± 1.30**)			
Macroglobulinemia	13	7.84 ± 0.62**	2.59 ± 0.51**	0.32 ± 0.16	0.58 ± 0.21	0.82 ± 0.25	3.54 ± 0.52
Hodgkin's disease	14	6.38 ± 0.77**	2.45 ± 0.74**	0.59 ± 0.16**	0.85 ± 0.16**	0.81 ± 0.15	1.74 ± 0.55
Lymphatic leukemia and lymphomas	27	6.04 ± 0.82**	3.24 ± 0.71**	0.40 ± 0.13	0.73 ± 0.17	0.77 ± 0.15**	0.88 ± 0.49**
Myelogenous and monocytic leukemias	12	6.78 ± 0.70	2.64 ± 0.68**	0.44 ± 0.10	0.82 ± 0.25**	0.73 ± 0.13**	2.15 ± 0.58**
Hypogammaglobulinemia	18	5.86 ± 0.87**	3.67 ± 0.56	0.38 ± 0.19	0.74 ± 0.26	0.84 ± 0.23	0.17 ± 0.15**
Analbuminemia	1	4.8	0.1	0.4	1.0	1.5	1.8
Peptic ulcer	9	6.31 ± 0.57**	2.78 ± 0.42**	0.41 ± 0.10	0.79 ± 0.17	0.88 ± 0.17	1.43 ± 0.44
Ulcerative colitis	8	6.58 ± 0.32**	3.22 ± 0.57*	0.45 ± 0.17	0.80 ± 0.16	0.73 ± 0.16**	1.35 ± 0.33
Exudative enteropathy	3	5.07 ± 1.23**	1.76 ± 0.10**	0.57 ± 0.10*	1.20 ± 0.30**	0.80 ± 0.36	0.73 ± 0.50**
Acute cholecystitis	8	6.68 ± 0.42**	2.90 ± 0.47**	0.45 ± 0.16	0.84 ± 0.33	0.95 ± 0.21	1.53 ± 0.21
Nephrosis	24	4.54 ± 2.48**	1.44 ± 0.59**	0.40 ± 0.17	1.20 ± 0.46**	0.79 ± 0.22*	0.72 ± 0.35**
Glomerulonephritis	24	5.84 ± 0.64**	2.08 ± 0.56**	0.41 ± 0.14	1.02 ± 0.29**	0.90 ± 0.26	1.48 ± 0.35
Hepatic cirrhosis	95	6.54 ± 1.71**	2.31 ± 0.67**	0.38 ± 0.14	0.62 ± 0.20	0.85 ± 0.22	2.42 ± 0.70**
Viral hepatitis	18	6.30 ± 2.32	2.93 ± 1.19*	0.29 ± 0.17**	0.52 ± 0.24	0.78 ± 0.35	1.74 ± 0.92
Sarcoidosis	32	7.54 ± 2.07	3.14 ± 0.58**	0.42 ± 0.14	0.84 ± 0.20**	1.07 ± 0.20**	2.24 ± 0.41**
Alpha-3-globulinemia	7	6.10 ± 1.37	2.07 ± 0.54**	0.50 ± 0.20	0.53 ± 0.14*	0.80 ± 0.26	1.41 ± 0.54
				(alpha-3 globulin = 0.64 ± 0.43)			
Lupus erythematosus	37	6.78 ± 0.83*	2.48 ± 0.55**	0.41 ± 0.12	0.84 ± 0.30**	0.83 ± 0.19	2.18 ± 0.80**
Polyarteritis nodosa	9	6.22 ± 0.79**	2.61 ± 0.78**	0.49 ± 0.17	0.87 ± 0.19**	0.78 ± 0.14	1.42 ± 0.42
Rheumatoid arthritis	27	6.98 ± 0.55	2.81 ± 0.47**	0.45 ± 0.13	0.95 ± 0.22**	0.89 ± 0.13	1.85 ± 0.66*
Scleroderma	13	6.84 ± 1.11	3.54 ± 0.99	0.33 ± 0.17	0.62 ± 0.17	0.83 ± 0.13	1.57 ± 0.75
Rheumatic fever	23	6.99 ± 0.42	3.16 ± 0.69**	0.45 ± 0.15	0.84 ± 0.22**	0.87 ± 0.15	1.64 ± 0.32
Essential hypertension	17	6.92 ± 0.45	3.55 ± 0.70	0.41 ± 0.15	0.73 ± 0.18	0.84 ± 0.17	1.36 ± 0.32*
Congestive heart failure	39	6.50 ± 0.71**	2.92 ± 0.56**	0.41 ± 0.11	0.71 ± 0.30	0.92 ± 0.19	1.49 ± 0.36
Carcinomatosis	37	6.69 ± 0.62*	2.81 ± 0.55**	0.50 ± 0.16*	0.86 ± 0.22	0.89 ± 0.20	1.67 ± 0.54
Bacterial pneumonia	16	6.60 ± 0.62**	2.60 ± 0.62**	0.51 ± 0.15*	0.94 ± 0.23**	0.92 ± 0.12	1.65 ± 0.39
Bacterial meningitis	7	6.23 ± 0.58**	2.74 ± 0.80**	0.44 ± 0.15	0.87 ± 0.24*	0.79 ± 0.15	1.39 ± 0.37
Bronchiectasis	10	6.97 ± 0.50	3.20 ± 0.57*	0.45 ± 0.16	0.74 ± 0.20	0.87 ± 0.16	1.76 ± 0.50
Osteomyelitis	11	6.68 ± 0.86	2.73 ± 0.84**	0.48 ± 0.14	0.89 ± 0.23**	0.94 ± 0.24	1.65 ± 0.36
Myxedema	11	6.44 ± 0.68**	3.29 ± 0.78*	0.32 ± 0.12*	0.65 ± 0.16	0.81 ± 0.18	1.35 ± 0.24*
Thyrotoxicosis	14	6.67 ± 0.82*	3.07 ± 0.74**	0.43 ± 0.10	0.68 ± 0.15	0.80 ± 0.13	1.61 ± 0.27
Diabetes mellitus	36	6.40 ± 1.65**	2.96 ± 0.97**	0.34 ± 0.16	0.81 ± 0.27**	0.87 ± 0.28	1.41 ± 0.51

Reproduced with kind permission of the author and publisher from Sunderman, F. W., and Sunderman, F. W., Jr.: Serum proteins and the dysproteinemias, Philadelphia, 1964, J. B. Lippincott Co., p. 324.

*p = < 0.05 **p = < 0.01

Table 20-4. Normal values for cerebrospinal fluid electrophoresis utilizing vacuum ultrafiltration, microzone cellulose acetate, and ponceau S stain*

	Percent of total
Prealbumin	4.9 ± 1.2
Albumin	61.5 ± 5.3
Alpha$_1$ globulin	4.5 ± 1.4
Alpha$_2$ globulin	6.7 ± 1.8
Beta globulin	13.7 ± 3.6
Gamma globulin	8.8 ± 2.6

*Based on data from Kaplan, A.: Electrophoresis of cerebrospinal fluid proteins, Amer. J. Med. Sci. **253**:549, 1967.

are designated as the alpha (α) type and two as the beta (β) type ($\alpha_2\beta_2$). The alpha chain consists of 141 amino acids and the beta chain contains 146 amino acids. Four different chains have been found in the three normal hemoglobins—A, F, and A_2—and are referred to as alpha, beta, gamma, and delta chains. If one amino acid is substituted for another in either chain, an abnormal hemoglobin is obtained.

Singer and co-workers[20] in 1951 used an alkali denaturation method (see procedure) to establish the resistance of hemoglobin F (fetal) to denaturation. Hemoglobin F is the most common hemoglobin in the fetus and is gradually replaced by hemoglobin A in the normal developing infant. If an individual continues to produce Hb F beyond the 1% level, the clinician should be alert to a possible thalassemia or acquired anemia.

Hemoglobin A_2 is found in normal individuals at a mean level of 2.6%, a study of twelve patients with thalassemia minor revealed Hb A_2 to be in a concentration of 5% to 10% with no overlap between normal values.[46]

In the thalassemias a defect in the synthesis of the beta chains results in the doubling of the A_2 component. The difference between hemoglobins A and A_2 resides in the non-alpha chains—that is, in 34 amino acids or 6% of the molecule. This explains the possibility of their differences exhibited by immunologic methods.

In summary, the normal hemoglobins possess the following amino acid chain combinations:

$$\text{Hb A } \alpha_2\beta_2$$
$$\text{Hb F } \alpha_2\gamma_2$$
$$\text{Hb A}_2\ \alpha_2\delta_2$$

Ingram and Hunt[47,48] found that in the abnormal hemoglobins—S, C, and E—the difference from the normal hemoglobin molecule was simply limited to a single amino acid in the beta chain, 0.34% of the entire molecule. These molecules cannot be differentiated immunologically, however, as can the A_2 from the A. If the abnormality exists in the alpha chain, such hemoglobins as $I_{Norfolk}$, M_{Boston}, and $G_{Philadelphia}$, to mention a few, occur.

When a child receives genes from both parents for the hemoglobin abnormality, a homozygous state exists. If the sickle cell gene is involved, the condition resulting is termed sickle cell anemia. If the thalassemia gene is involved, the

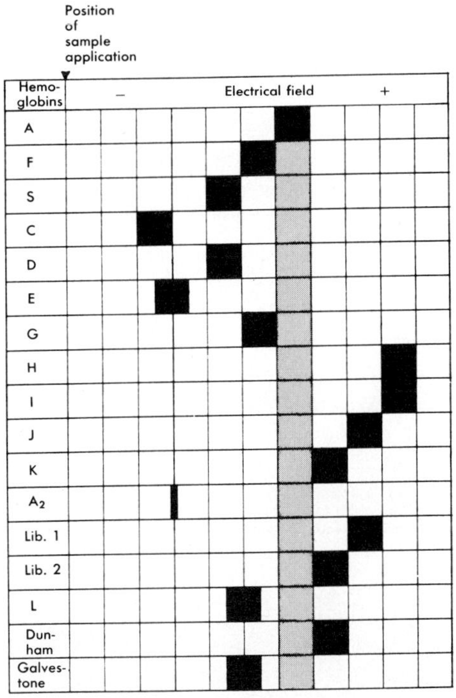

Fig. 20-13. Relative positions of hemoglobin variants after electrophoresis on paper with barbital buffer, pH 8.6. (Courtesy Hyland Division, Travenol Laboratories, Inc., Costa Mesa, Calif.)

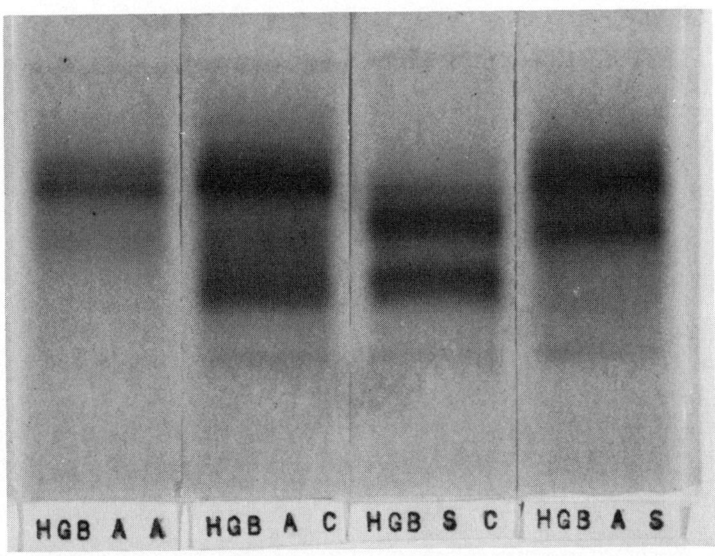

Fig. 20-14. Separation of hemoglobin variants with starch gel procedure in text. (From Marsh, C. L., and Jolliff, C. R.: Micro starch gel electrophoresis. In Smith, Ivor, editor: Chromatographic and electrophoretic technique, vol. 2, Zone electrophoresis, ed. 2, London, 1968, Wm. Heinemann Medical Books, Ltd.)

Table 20-5. Laboratory findings in the hemoglobinopathies*

Disease	Hemoglobin type	Fetal hemoglobin (% of total)	Sickling	Severity of anemia
Normal	A	Up to 2	0	0
Thalassemia major	A	10 to 100	0	Severe
Thalassemia minor	A	Up to 10	0	Mild
Sickle cell anemia	S	Up to 20	40 to 100	Severe
Sickle cell trait	A and S	Up to 2	5 to 95	0
Sickle cell–thalassemia	S and A	Up to 40	+ to +++	Severe
Sickle cell–hereditary spherocytosis	A and S	Up to 2	+	Mild to severe
Sickle cell–Hb C	S and C	Up to 40	+	Mild to severe
Sickle cell–Hb D	S and D	Up to 12	+	Moderate to severe
Sickle cell–Hb G	S and G	Up to 2	+	0
Hb C disease	C	Up to 7	0	Mild
Hb C trait	A and C	Up to 2	0	0
Hb C thalassemia	C and A	Up to 20	0	0 to mild
Hb C elliptocytosis	A and C	Up to 2	0	0 to mild
Hb D	A and D	Up to 2	0	0
Hb E disease	E	Up to 9	0	0 to mild
Hb E trait	A and E	Up to 2	0	0
Hb E thalassemia	E and A	20 to 40	0	Moderate to severe
Hb G disease	G	Up to 2	0	0
Hb G trait	A and G	Up to 2	0	0
Hb G thalassemia	G and A	Up to 2	0	Mild
Hb H thalassemia	H and A	Up to 2	0	Moderate
Hb I (trait?)	A and I	Up to 2	0	0

*From Miale, John B.: Laboratory medicine—hematology, ed. 3, St. Louis, 1967, The C. V. Mosby Co.

child suffers from thalassemia major. If, on the other hand, only one parent transmits the abnormal gene, a heterozygous state evolves and the above states are then referred to as sickle cell trait and thalassemia minor.

Death usually results in the early life of the child with sickle cell disease or thalassemia major, in contrast to those children with the heterozygous condition. Patients homozygous for the other hemoglobin types do not appear to be so critically affected.

Fig. 20-13 depicts the relative migrations of some of the various hemoglobins, by paper electrophoresis. Fig. 20-14 shows the migration pattern and differentiation by starch gel electrophoresis, which in our hands is the preferred method. Table 20-5 lists for reference some of the known diseases due to hemoglobin abnormalities.

A considerable number of references are available to those more interested in the structure, genetics, and biochemistry of the hemoglobin molecules.[49-55]

The lipoproteins

For fuller understanding of the implications of lipoprotein electrophoresis the reader is referred to the references.[56-60] Fig. 20-15 shows the five phenotypes of hyperlipoproteinemia as observed with paper electrophoresis using the method in this chapter. Fig. 20-16 reveals the phenotypes as observed with disc electro-

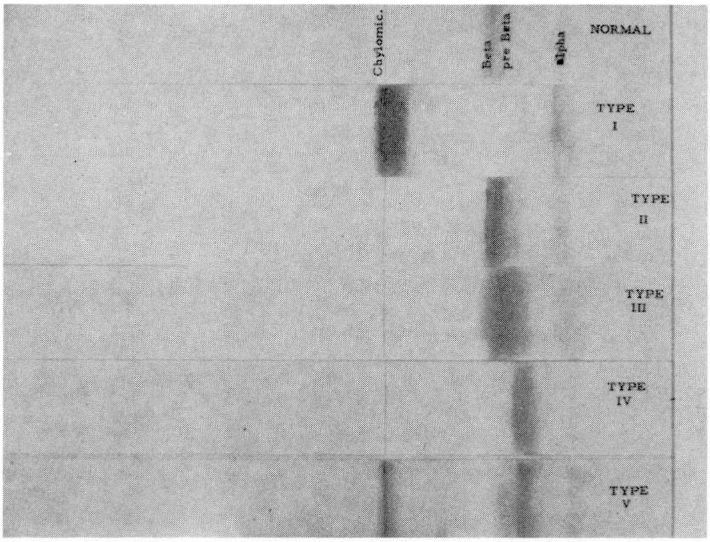

Fig. 20-15. Phenotypes of lipoprotein as obtained with paper electrophoresis.

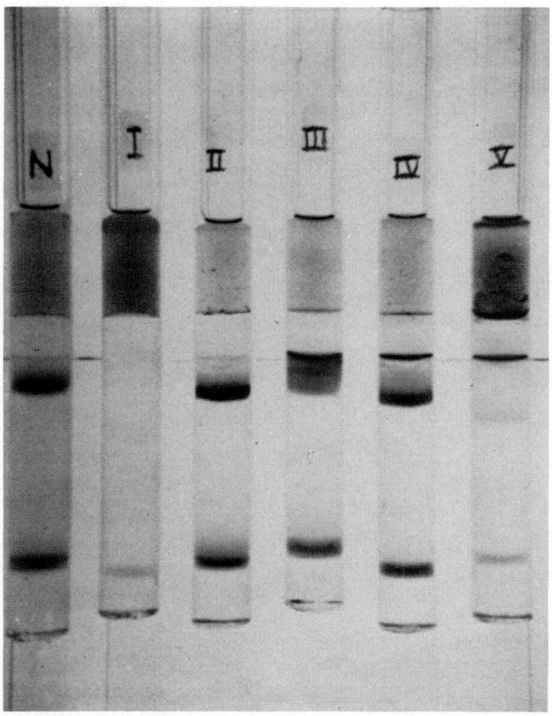

Fig. 20-16. Phenotypes of lipoprotein as obtained with polyacrylamide gel Quick-Disc method. (Courtesy Canalco Inc., Rockville, Md.)

Table 20-6. Lipoprotein phenotypes[56]

	Essential hypertriglyceridemia		Familial hypercholesterolemia		
	Fat-induced form Type I	Carbohydrate-induced form Type IV	Caloric (fat and carbohydrate)—induced form Type V	Without accompanying triglyceridemia Type II	With accompanying triglyceridemia Type III
Plasma	Opalescent, turbid	Clear or turbid	Opalescent, turbid	Clear	Clear or turbid
Cholesterol	Slightly elevated	Normal or slightly elevated	Slightly elevated	Elevated	Elevated
Triglycerides	Highly elevated	Elevated	Elevated	Normal	Elevated
Chylomicrons	Highly elevated	Normal	Highly elevated	Normal	Normal
β liproproteins	Normal	Normal	Normal	Increased	Increased
α_2 lipoproteins (pre-β lipoproteins)	Normal	Increased	Increased	Normal	Increased

phoresis referred to in the procedure section. Table 20-6 may be helpful in the classification of the phenotypes. In our laboratories, visual comparison of the phenotypes with a known normal control that is included with each electrophoretic run has sufficed to classify the lipoprotein abnormalities.

IMMUNOELECTROPHORESIS

Grabar and Williams[61] described in 1953 a combination of the gel immunodiffusion and gel electrophoresis method which they called immunoelectrophoresis. This method has become a tremendous asset to the clinical chemist in the study of serum and other body fluid protein abnormalities.

Double gel diffusion described by Ouchterlony[62] *implies* that both antigen and antibody are diffusing through the gel. If antigen to a particular antibody, also diffusing in the gel, is present, an immunoprecipitin band will form. (See Fig. 20-17.)

In 1955 Scheidegger[63] developed a technique called micro immunoelectrophoresis which is the basis, for the most part, of techniques used in the clinical laboratories today. It has been found that many of the supporting media such as agarose and cellulose acetate will allow for immunoelectrophoresis patterns. In our laboratories we have used with success for a number of years the agar plate method which is offered here. Given proper training and experience, the analyst can perform the method quite well in a manner of days.

Routine method for immunoelectrophoresis

Materials needed

Lantern slide, 3¼ × 4 inch
Immunoelectrophoresis pattern cutter (14 P.D.)
Deluxe pattern cutter (C. L. Davis) (Figs. 20-18 and 20-19)
Immunoelectrophoresis cell (C. L. Davis 117 A)

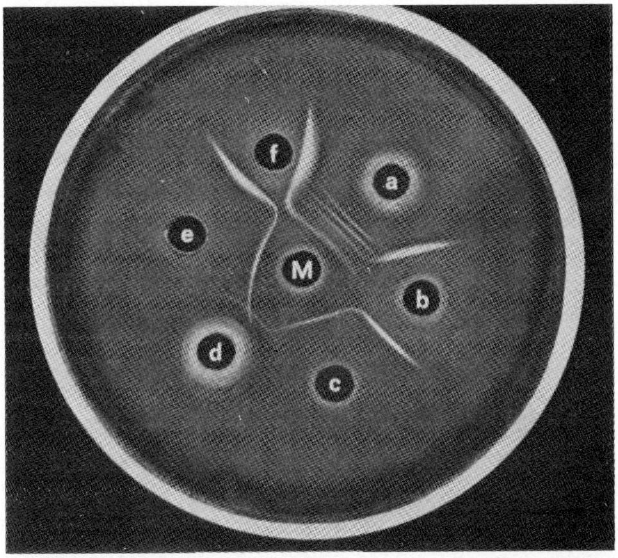

Fig. 20-17. Ouchterlony agar immunodiffusion plate: **M,** horse antiserum to human serum; **a,** human serum; **b,** rabbit antiserum to G globulin; **c,** G globulin; **d,** M globulin; **e,** albumin; **f,** rabbit antiserum to albumin. (Courtesy Behring Diagnostics, Inc., Woodbury, N.Y.)

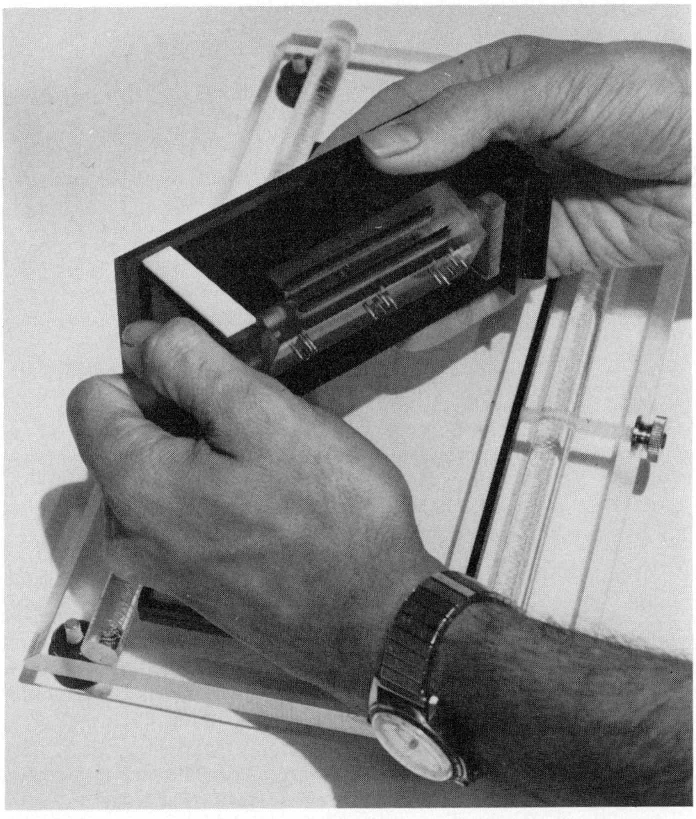

Fig. 20-18. Immunoelectrophoresis and immunodiffusion pattern cutter. (Courtesy C. L. Davis Co., Lincoln, Neb.)

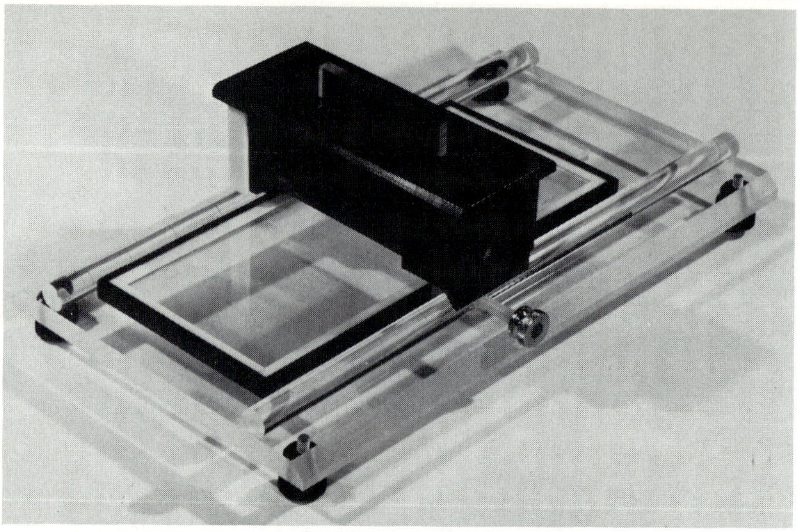

Fig. 20-19. Tray and jig for holding IEP plates in place during cutting. (Courtesy C. L. Davis Co., Lincoln, Neb.)

Power supply (ISCO 490) (Instrumentation Specialties Co.)
Slide washer (C. L. Davis 130 S.W.)
Destainer (C. L. Davis 12 D)
Humidity cabinet (C. L. Davis 113 H)
Pattern viewer, dark-field IEP box (C. L. Davis)
Polaroid MP-3 camera assembly (Polaroid)
Petri dishes, large, for staining
Meat keeper, plastic (Tupperware)
Diamond marking pencil
Bibulous paper booklet

Preparation of solutions and immunoelectrophoresis plates

Solution 1: buffer for immunoelectrophoresis chamber

Veronal	1.38 g
Sodium Veronal	8.70 g
Calcium lactate	0.38 g
Distilled water to	1000 ml

Solution 2: solution for preparation of Ionagar

Veronal	1.66 g
Sodium Veronal	10.50 g
Calcium lactate	1.54 g
Distilled water to	1000 ml

Add aqueous Merthiolate to a concentration of 1:10,000 to prevent bacterial and mold growth.

Preparation of Ionagar slides

Two parts solution 2 are mixed with 1 part distilled water. We usually use 66 ml of solution 2 and add distilled water to the 100 ml mark on a cylinder graduate. This is poured into a 250 ml flask to which has been added 2 g of Ionagar 2 (Colab). The agar is partially dissolved and then heated to boiling with constant swirling. Heat until the mixture foams almost to the top of the flask three times to ensure proper dissolution of the Ionagar. With a 10 ml pipette, transfer 10 ml of the hot mixture to a $3\frac{1}{4} \times 4$ inch lantern slide, previously cleansed so that it is free of all grease, and lying on a level table.

Carefully spread the agar on the slide so that the entire surface is covered. The slide will easily hold as much as 12 ml of agar and we usually add this amount instead of 10 ml.

After the slides have solidified, transfer them to a plastic (airtight) container that contains several paper towels soaked in water. Seal the container and keep slides under refrigeration until ready to use.

Antisera

Antisera to be used in immunoelectrophoresis may be prepared by the individual. However, for clinical purposes and for use in laboratories where the needed equipment and animals are not available, commercially available antisera have been found to satisfy requirements. Antisera are available from Hyland, Behringwerke A.G. (Behring Diagnostics), Pentex, Mann Research Laboratories, Melpar, and Kallestad Laboratories.

It is important that the antisera be kept at 4° C when not in use, or between uses. If any storage is intended, place in freezer and thaw before use. Do not repeat freezing and thawing but, rather, thaw and keep at 4° C when using.

For routine immunoelectrophoresis we maintain antisera to the following human serum components:

Human serum	Beta lipoprotein
Gamma G immunoglobulin	Haptoglobin
Gamma A immunoglobulin	Beta 1C/1A globulin
Gamma M immunoglobulin	$Alpha_2$ macroglobulin
Gamma D immunoglobulin	Transferrin
Gamma GK (I) kappa light chain	Ceruloplasmin
Gamma GL (II) lambda light chain	Albumin

Procedure

The previously prepared Ionagar 3¼ × 4 inch plates are removed from the humidity box in the refrigerator and allowed to come to room temperature. At this time, etch the number of the test on the non–agar-containing side of the plate. Place the plate in the cutting jig and carefully press the cutting tool into the agar at the desired location on the plate. Fig. 20-20 shows the position of the patterns on the plate.

It is important that the end pattern be far enough in from the edge so that distortion of the developing arcs does not occur because of the smaller volume of agar at the edges of the slide. If the jig is followed correctly, this will not result.

The pattern having been cut on the agar, the plate is removed from the jig. A cutoff 15-gauge needle (Fig. 20-21) is used to suck out the agar plug in the antigen (serum) well. The antibody trough agar is not removed at this time.

After the eight antigen wells are freed of agar, the antigen is introduced into the wells with the aid of a microliter syringe or capillary tube. Fig. 20-22 shows the placement of antigen.

A pooled control serum that approximates normal level for the age of the patient being investigated is used. To this serum is added bromphenol blue until the serum is colored sufficiently to allow for indicating migration of the albumin portion of the patterns.

The slide is now placed in the electrophoresis chamber as illustrated in Fig. 20-23.

Filter paper wicks (two thicknesses of bibulous paper, as used in microbiology) are cut in half and work quite well. The wicks are placed approximately ⅜ to ½ inch on the agar on each side and allowed to dip into the chamber buffer. Particular caution must be observed to have the filter paper wicks placed in a straight position on the slide as follows (Fig. 20-24).

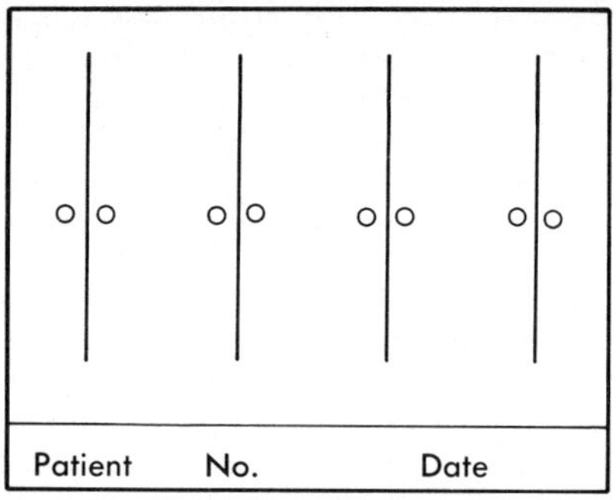

Fig. 20-20. Position of IEP patterns on 3¼ × 4 inch slide.

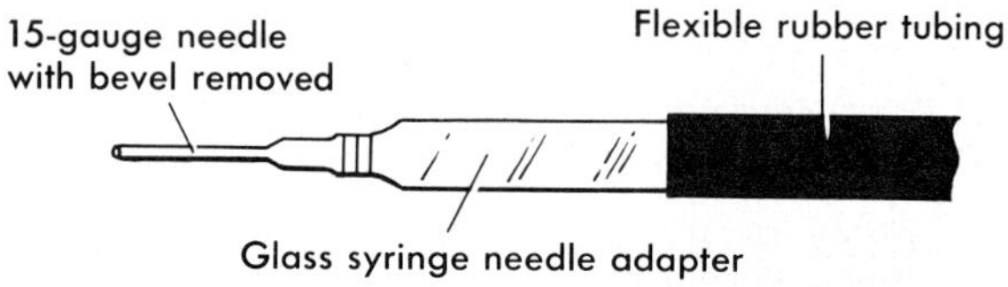

Fig. 20-21. Drawing of needle with adapter for use in withdrawing antigen well agar after cutting.

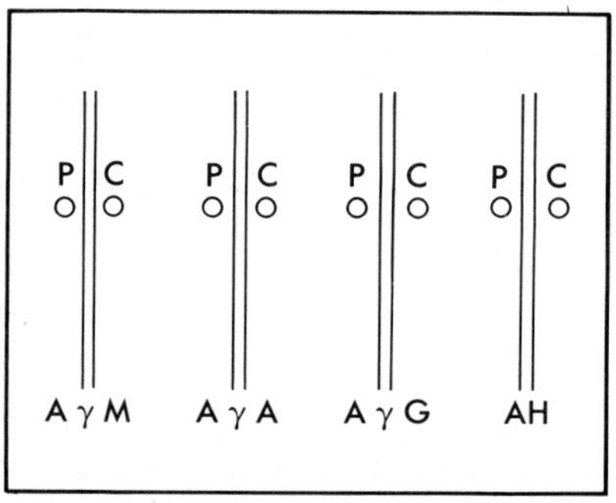

Fig. 20-22. Schematic for positions of antigen and antibody on an IEP run. **P,** Patient sera; **C,** control sera; **AH,** antihuman serum; **AγG,** anti–gamma G serum; **AγA,** anti–gamma A serum; **AγM,** anti–gamma M serum.

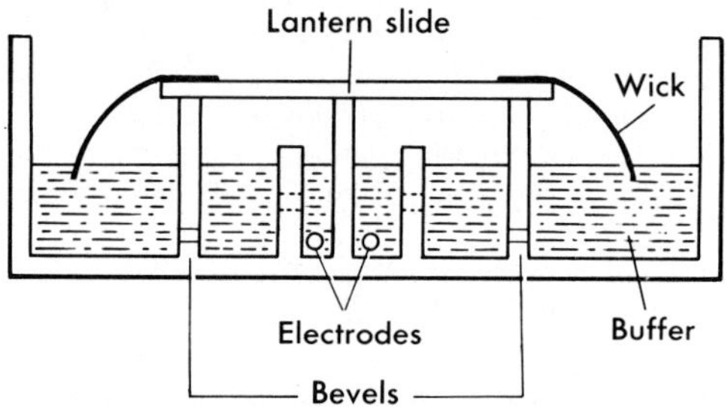

Fig. 20-23. Drawing of position of IEP slide, wicks, and buffer level in cell.

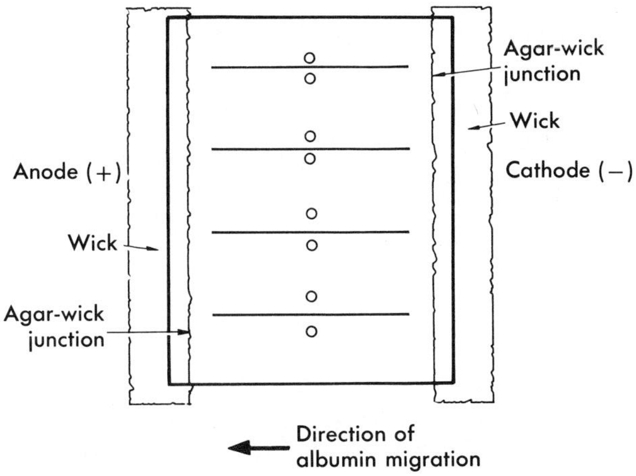

Fig. 20-24. Wick position preparatory to IEP run.

Be certain the wells of antigen are nearest the cathode (−), as migration of albumin is toward the anode (+).

Close the cover of the electrophoresis chamber and set voltage at 175 volts, which is about 40 ma. We have used constant voltage and have had excellent results.

When the albumin marker in the control has moved about 35 to 40 mm, shut off current and remove the plate from the chamber. With a scalpel cut the agar at the ends of each trough. Holding the plate agar upside down, carefully remove the agar in each of the four troughs. Holding it upside down lets the agar fall away from the slide and will prevent damage to the agar surface if the trough gel should have to be removed from the surface of the agar.

Fill the troughs with the appropriate antisera (see Fig. 20-22), being careful not to damage the troughs or wells in doing so.

Carefully place the slide in the humidity cabinet for 12 to 24 hours at room temperature, being certain that sufficient water is in the cabinet.

Photography of plates

After adequate time has elapsed (12 to 18 hours) for the arcs to develop, the plate is photographed for a permanent record of the test. This is accomplished by placing the plate on a view box so constructed as to allow a darkfield background with 45-degree angle lighting to the side. See Fig. 20-25.

We have found the Polaroid MP-3 camera assembly to be the most adaptable for this purpose. A Polaroid photograph is taken of the entire plate. If special attention is necessary regarding one or more of the arcs, an enlargement of the arc can be made by simply using the proper lens combination.

Generally two photographs of each plate are taken—one for the laboratory record and the second for the physician's record. All photographs are numbered with the patient or experiment number, and antisera troughs as well as antigen wells are identified. A marking pencil (Scientific Products P1226) works very well for this purpose.

Staining of patterns

After the photographs of the IEPs have been taken, the slide is washed under the cold water tap carefully so that the agar is not washed off. This is accomplished in about 10 seconds. The slide is then placed, agar side up, on a paper towel. Two pieces of bibulous paper are wetted

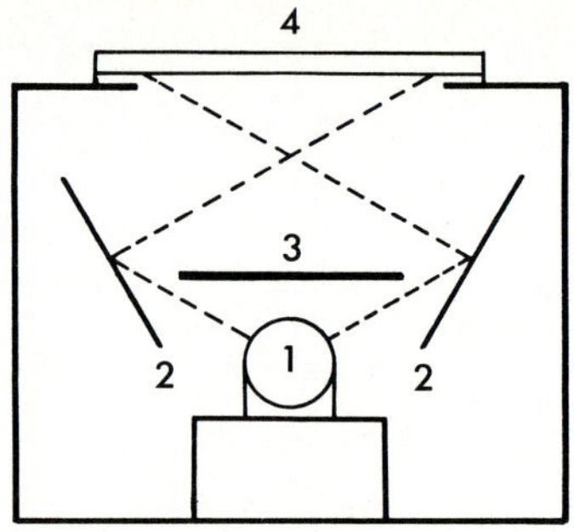

Fig. 20-25. Schematic of view box for use in photographing IEP and immunodiffusion patterns: **1,** fluorescent light; **2,** mirrors (adjustable to proper angle); **3,** metal plate, dark background; **4,** immunoelectrophoresis pattern.

with tap water, the two pieces are laid on the slide, and the slide is allowed to stand until all of the moisture is removed from the agar and the pattern remains as a thin film on the glass slide. This usually takes about 8 hours at room temperature but may be quickened by allowing warm dry air, as from a fan or hair drier, to flow over the wet bibulous paper. Do not allow to dry too fast and do not place in an oven, as cracking of the agar and distortion of the pattern will occur.

Once the slide has dried, it is washed in 0.85% NaCl, with several changes for a period of about 18 hours. A washing apparatus (C. L. Davis) is quite suitable for this purpose. The slide is then washed for several hours with several changes of distilled water. Again the slide washer may be used. The slide may be washed in a large Petri dish if a slide washer is not available. Frequent changes of saline and water are imperative. After the water wash, the slide is again laid on a paper towel and two pieces of bibulous paper are placed on top of the agar surface and the paper allowed to dry.

After drying, the slide may be stained by any number of methods or stains depending on the components the investigator is most interested in. For routine immunoelectrophoresis we use ponceau S and amido black 10B. The formula for each of these solutions is given:

Amido black 10B

Amido black 10B	1 g
Glacial acetic acid	100 ml
Methanol	700 ml
Distilled water	200 ml

Filter through glass wool before using.

Ponceau S

Ponceau S	1 g
Glacial acetic acid	100 ml
Methanol	700 ml
Distilled water	200 ml

The slides are now placed in a slide staining tray or slide stainer (C. L. Davis) for 5 minutes. The slides are allowed to drain dry of the amido black 10B stain or the ponceau S stain and then placed in a destaining solution until all unbound stain is removed.

Destaining solution

Glacial acetic acid	100 ml
Methanol	700 ml
Distilled water	200 ml

The slides are allowed to dry at room temperature. When completely dry, the pattern may be covered with a protective coating by using the coater from a Polaroid black and white film pack. The slides are now permanent and may be labeled with an ink pen, filed, and photographed or used in a regular lantern slide projector for better visualization or for presentation to groups.

Immunoquantitation procedures

Immunoelectrophoresis offers a qualitative method by which the various serum proteins are assessed. A quantitative procedure was necessary for the establishment of the values for protein components in the normal as well as a diseased individual. Various methods have been reported such as Oudin's single immunodiffusion,[64] Ouchterlony's double diffusion,[62] the single radial immunodiffusion of Mancini,[65,66] and electroimmunodiffusion.[67] We have used several of these methods; however, only two are presented here. The single agar diffusion method of Claman and Merrill affords an excellent method but the procedure is time-consuming and does not adapt well to the busy clinical laboratory. The single radial immunodiffusion method has found favor, in that commercially available materials are now abundantly produced for this method.

Methods for the quantitative determination of immunoglobulins

Preparation of single agar diffusion tubes and method of immunoglobulin quantitation (Claman and Merrill[68])

Preparation of reaction tubes

1. Purchase 3 feet of glass tubing, 1.7 mm I.D.
2. Cut tubing into 5 cm lengths.
3. Wash three times with distilled water and one time with 1:100 solution of Siliclad (Clay Adams). Leave in the solution for at least 5 minutes, being certain that tubes are completely submerged and no air pockets are in the tubes.
4. Wash again with distilled water several times.
5. Dry in oven at 100° C overnight and keep in clean container.

Preparation of agar tubes

1. Ionagar is dissolved in buffer used for immunoelectrophoresis (1% Ionagar). Be certain that buffer contains Merthiolate as specified in the immunoelectrophoretic procedure.
2. Dissolve by heating the Ionagar, cool to about 50° C, and add 1 ml of the Ionagar to 2 ml of the specific antisera (Hyland IgG, IgA, and IgM).
3. Place in a prewarmed container, being certain to maintain the temperature at 50° C throughout the entire procedure. With a blood cell pipette rubber tubing suck the antisera-Ionagar mixture into about half of the reaction tube.
4. Seal one end (end with the agar) of the reaction tube with hematocrit tube sealing compound.
5. Place tubes in a sealed container in the refrigerator and store at 4° C.

Preparation of the immunoglobulin standard

From pooled normal human serum make the following dilutions with 4% ovalbumin (Nutritional Biochemicals), three times crystallized.

Tube 1	Straight pooled serum
Tube 2	1:1 of tube 1
Tube 3	1:1 of tube 2
Tube 4	1:1 of tube 3

Freeze these standards in 0.2 ml aliquots until ready for use.

Procedure

1. In an appropriate rack, standard, control, and unknown test tubes are set up as follows: four rows of 3 tubes each will be used for the four

standards and one row of 3 tubes will be used for the control serum. An additional row of 3 tubes will be used for each unknown. The columns of the 3 tubes contain gamma G, gamma A, and gamma M, respectively.

	Column		
Test tube row	*1*	2	3
	Gamma G	Gamma A	Gamma M
1-4	Standard	Standard	Standard
5	Control	Control	Control
6-∞	Unknown	Unknown	Unknown

2. After the tubes have been placed in the racks in proper sequence, add the standard sera, control sera, and unknown sera to the reaction tubes with a 2½-inch 25-gauge needle on a tuberculin syringe. Add sera to the tubes, being careful to overlay the sera onto the agar surface. Fill the tubes with sera so that approximately 1 to 1.5 cm of tube contains sera above the agar-sera junction. Amount of sera over this amount is not too critical, but not less should be used.
3. Place the tubes in an upright position in a rack. We have found it desirable to carefully cover the top of each tube with a little piece of hematocrit tube sealer to prevent evaporation of the sera. Leave the tubes at room temperature for the following times before reading results:
   Gamma G tubes—24 hours
   Gamma A and gamma M tubes—72 hours
4. At the end of these times read the distance of the precipitation band or zone from the point of the sera-agar junction and record the distance in $\frac{1}{100}$-inch units.
5. A convenient ruler for this purpose is one manufactured by Rabone Chesterman in England.*
6. The values obtained from the standard tubes are then plotted on semilog paper K and E 385-50, with the ordinate being the mg/100 ml concentration of the immunoglobulin, and the abscissa being the $\frac{1}{100}$-inch value. A straight line is drawn or a slightly curved line, as sometimes may be the case, and the values for the unknowns and controls are then calculated.
7. The lowest limits of detection by this method have been found to be as follows:
   Gamma G—20 to 40 mg/100 ml
   Gamma A—23 to 38 mg/100 ml
   Gamma M— 7 to 15 mg/100 ml
8. Normal human values as determined by this method and as found by Claman and Merrill are as follows:
   Gamma G—760 to 2000; mean 1194
   Gamma A—118 to 1065; mean 395
   Gamma M— 64 to 380; mean 191

We have found these values to be high, especially for gamma A and gamma M, and suggest that you establish your own normal values.

Radial immunodiffusion method[65,66]

It is suggested that for the routine determination of serum immunoglobulins in the laboratory commercial plates be obtained. These plates have been found to be quite adequate and are now maintained under a system of quality control. We would suggest three types of plates:

Hyland Laboratories: 085-024 Anti–gamma G immunoplate, 085-130 Anti–gamma A immunoplate, and 085-026 Anti–gamma M immunoplate

Behring Diagnostics (Behringwerke): Partigen plates for IgA, IgG, and IgM quantitation (Separate instructions are available with these plates.)

---

*Distributed by Arthur H. Thomas Co. (42R).

Melpar: Immunodiffusion plates for IgG, IgA, IgM, and IgD
Kallestad: Quanti-plate for IgE, IgA, IgM, and IgD

Preparation of the plates

These plates may be prepared by using Ionagar Difco special Noble agar 2% with glycine 7.5%, NaCl 1%, and sodium azide 0.1%, pH 7.0 to 7.2. The antisera are added to the melted (50° C) agar in a concentration that depends on the strength of the antisera, which must be predetermined by using small amounts and thus finding the proper concentration.

Holes for the antigen are punched out by using a special template prepared for such a purpose.

Preparation of standards

Pooled normal human serum is checked for immunoglobulin concentration against an international standard.*

Procedure

Depending on the type of immunoplate being used, the time will vary, as well as incubation. It is suggested that instructions of the company producing the plates be followed. We have found that incubation at room temperature in a moist chamber such as the chamber used for the immunoelectrophoresis slides works very well.

It is imperative that sera having values over those of the highest standards be diluted with 4% albumin (p. 595). Under no circumstances make dilutions with saline or distilled water. Standards are prepared in accordance with method I (p. 595).

Readings are made with a magnifying lens and either a mm or a 1/100-inch grid against a darkfield background. Results are plotted the same as in method I. The diameter, rather than the precipitin line in the tube, is measured and plotted on semilog paper (p. 596).

A recent modification has been to photograph the immunoplates with the Polaroid MP-3 camera with the same view box that is used to photograph the immunoelectrophoresis slides (Fig. 20-26). The precipitin

*Send pooled serum to George B. Olson, Ph.D., NCI Immunoglobulin Reference Center, 6715 Electronic Dr., Springfield, Va. 22151. Send 5 ml of pooled serum. It is advisable to send two different pools—one for a standard and another for a control.

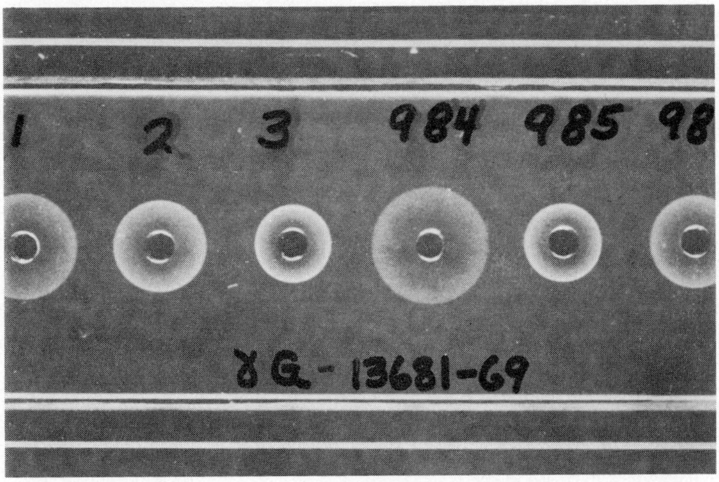

Fig. 20-26. Hyland immunoplate immunoglobulin quantitation. Numbers **1**, **2**, and **3** are standards. γG quantitation.

Table 20-7. Serum immunoglobulin levels of adults and of children from birth to 13 years*

Age	No.	Level of IgG† mg/100 ml (± 1 SD)	Significance of difference from adult level‡	Percent of adult level	Level of IgA† mg/100 ml (± 1 SD)	Significance of difference from adult level‡	Percent of adult level	Level of IgM† mg/100 ml (± 1 SD)	Significance of difference from adult level‡	Percent of adult level
Newborn	20	1145 ± 252 (750-1500)	N.S.	100%	N.D.	N.D.	1%	11.8 ± 2.6 (11-35)	P < 0.001	10%
1-3 mo.	7	371 ± 132 (270-780)	P < 0.001	31%	28 ± 21 (6-68)	P < 0.001	14%	66 ± 34 (12-87)	P < 0.01	65%
4-6 mo.	7	556 ± 111 (190-860)	P < 0.001	33%	56 ± 14 (10-96)	P < 0.001	28%	70 ± 23 (25-120)	P < 0.01	69%
7-12 mo.	7	721 ± 208 (350-1180)	P < 0.001	63%	81 ± 67 (35-165)	P < 0.001	40%	76 ± 16 (36-104)	P < 0.01	75%
2 yr.	26	705 ± 106 (520-1080)	P < 0.001	62%	69 ± 33 (36-165)	P < 0.001	35%	85 ± 28 (72-160)	P < 0.01	83%
3 yr.	21	866 ± 213 (500-1360)	P < 0.001	76%	81 ± 21 (45-135)	P < 0.001	40%	93 ± 37 (46-190)	P < 0.1	91%
4 yr.	21	1026 ± 208 (540-1440)	P < 0.01	92%	108 ± 42 (52-210)	P < 0.001	54%	121 ± 63 (52-200)	N.S.	118%
5 yr.	38	924 ± 234 (640-1420)	P < 0.001	81%	111 ± 41 (52-220)	P < 0.001	56%	112 ± 36 (40-180)	N.S.	110%
6 yr.	26	1031 ± 213 (650-1410)	P < 0.01	90%	126 ± 39 (83-217)	P < 0.001	63%	108 ± 33 (55-210)	N.S.	106%
7 yr.	19	1000 ± 203 (570-1320)	P < 0.05	87%	117 ± 49 (65-240)	P < 0.001	58%	121 ± 29 (60-175)	N.S.	118%
8 yr.	18	1040 ± 244 (730-1410)	N.S.	91%	136 ± 56 (74-260)	P < 0.001	68%	118 ± 33 (68-175)	N.S.	116%
9 yr.	20	975 ± 131 (760-1330)	P < 0.001	85%	144 ± 34 (108-200)	P < 0.001	72%	101 ± 38 (55-160)	N.S.	99%
10 yr.	22	1073 ± 181 (730-1350)	N.S.	94%	139 ± 34 (70-222)	P < 0.001	69%	109 ± 30 (80-150)	N.S.	106%
11 yr.	19	1082 ± 147 (850-1300)	N.S.	95%	141 ± 51 (91-255)	P < 0.001	70%	110 ± 23 (66-155)	N.S.	107%
12-13 yr.	26	1143 ± 228 (770-1510)	N.S.	100%	188 ± 46 (108-325)	P < 0.01	94%	114 ± 32 (70-150)	N.S.	112%
Adult	72	1142 ± 195 (770-1510)			201 ± 44 (134-297)			102 ± 30 (67-208)		

*From Uffelman, J. A., Engelhard, W. E., and Jolliff, C. R.: Quantitation of immunoglobulins in normal children, Clin. Chim. Acta **28:**185, 1970.

†Geometric mean.

‡Student 2 sample t test.

N.D.—Only 2 samples with detectable IgA (22 and 60 mg/100 ml).

rings are thereby enlarged and may be measured with the aid of a magnifying glass and a 1/100-inch ruler. The photographs may be kept as a visual record of the immunodiffusion quantitative procedure.

It has been our experience that the results of the total immunoglobulins do not always agree with the results of total gamma globulin as obtained with electrophoresis. There is gamma globulin in the beta as well as the $alpha_2$ region of the electropherogram, and this could well account for the discrepancy.

Normal values for the immunoglobulins that we have obtained on adults and children are found in Table 20-7. It is suggested, however, that each laboratory should establish its normal values for the various age groups.

Immunoquantitation of other proteins

A number of products have become available for the immunoquantitation of proteins other than the immunoglobulins. Some of these are listed here. For list of manufacturers see Appendix E.

Protein component	*Hyland immunoplate**	*Behring partigen plate**
Prealbumin		+
Albumin		+
$Alpha_1$ acid glycoprotein		+
$Alpha_1$ antitrypsin		+
Ceruloplasmin	+	+
Haptoglobin		+
$Alpha_2$ macroglobulin	+	+
$Alpha_2$ HS-glycoprotein		+
Beta lipoprotein		+
Transferrin	+	+
Fibrinogen		+
$Beta_2$ glycoprotein I		+
Complement C3 (B_{1c}/B_{1a})	+	+
C-reactive protein		+

The immunoglobulins

Of the serum proteins the immunoglobulins have become a focus point for the chemist and clinician alike. The gamma globulin fraction as seen in ordinary zone electrophoresis exhibits numerous changes in the various disease states such as infection, myeloma, hepatic disease, gastrointestinal enteropathies, urinary tract disease, and autoimmune mechanisms. With the discovery of the immunoglobulins and the establishment of the myeloma proteins or M proteins as abnormal immunoglobulins, a new vista was opened to the student of immunochemistry. Today the field is moving so rapidly that by the time books are published many new findings have been reported; thus it behooves authors to suggest to the reader numerous reviews on the subject, which may be consulted.[6,7,9,14,15,26,27,72-83]

An understanding of the immunoglobulin molecule is necessary, however, for the analyst or clinical chemist to talk intelligently about the subject. The World Health Organization has established an International Nomenclature for the Immunoglobulins[84] (Table 20-8) which has to some degree simplified the confusion about these molecules and their structure which previously existed.

Structure of the immunoglobulins. Fig. 20-27 shows the structure of IgG as

*These plates may be used with appropriate standards that are supplied by both companies.

Table 20-8. Human serum immunoglobulins*

Terminology†					
Proposed symbols	IgG	IgA	IgM	IgD	IgE§
	γG	γA	γM	γD	γE
Other notations	γ_2	β_2A	γ_1M	γ_1J	
	γSS	γ_1A	β_2M		
	6.6Sγ	(6.6S-13S)	18S		
	7Sγ				
Polypeptide chain					
Composition					
Type K (type I)‡	$\gamma_2\kappa_2$	$(\alpha_2\lambda_2)n$	$(\mu_2\kappa_2)5$	$\delta_2\kappa_2$	$\varepsilon_2\kappa_2$
Type L (type II)‡	$\gamma_2\lambda_2$	$(\alpha_2\lambda_2)n$	$(\mu_2\lambda_2)5$	$\delta_2\lambda_2$	$\varepsilon_2\lambda_2$

*From Fahey, John L.: Antibodies and immunoglobulins. I. Structure and function, J.A.M.A. 194:183, 1965.
†Terminology based on heavy polypeptide chain content of the immunoglobulin (established by a committee on nomenclature).[84]
‡Terminology types K and L (types I and II) based on light chain content.
§Author's addition.

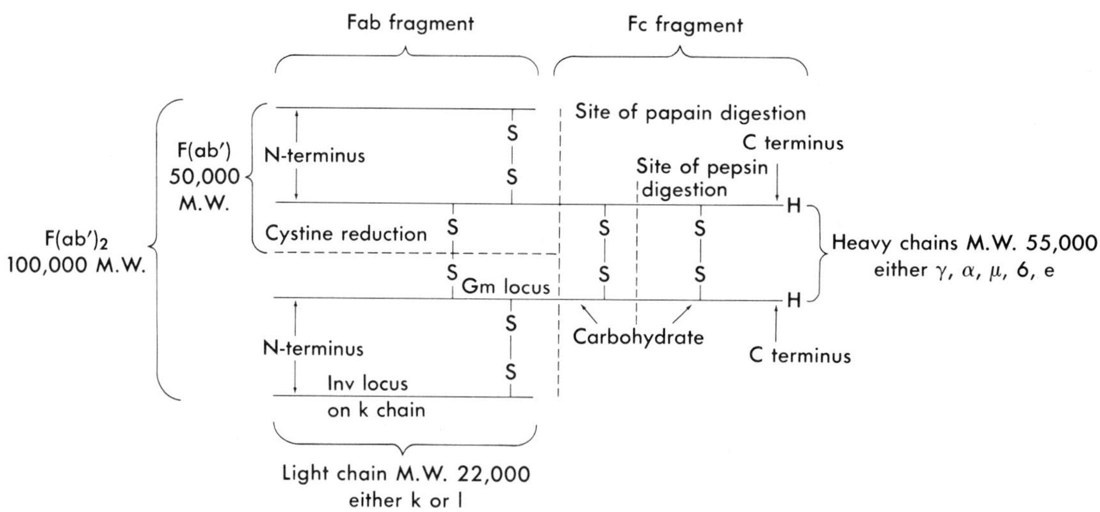

Fig. 20-27. Schematic representation of IgG molecule proposed by Porter.[28]

tentatively proposed by Porter.[85] One notes that the molecule is composed of two heavy (H) polypeptide chains (M.W. 55,000) and two light (L) polypeptide chains (M.W. 22,000), which are held together by interchain disulfide bonds and noncovalent bonds. When the interchain disulfide bonds are cleaved and then denatured, so as to break the existing noncovalent bonds between the chains, the heavy and light chains may be separated and the heavy chains characterized.

Porter[86] in 1959 found that papain hydrolysis followed by reduction allows splitting of the IgG molecule into three fragments, which were designated as two F(ab) (antibody) fragments and one Fc (crystallizable) fragment. Edelman and Poulik[87] showed that reduction by mercaptoethanol of the IgG mole-

Table 20-9. Immunoglobulin polypeptide chains*

	Heavy†	Light
Molecular weight	55,000	22,000
Electrophoretic mobility	Heterogeneous	Heterogeneous
Carbohydrate attached	+	0
Antigenic determinants		
Specific for IgG, IgA, IgM, IgD, IgE	+	0
Specific for type K and L (I and II)	0	+
Genetic factors		
Gm	+	0
Inv	0	+
Antibody specificity	+	0
Needed for full antibody activity	+	+
Placental transport	+	0
Regulation of catabolism	+	0
Complement binding	+	0
Skin sensitization	+	0
Reacts with rheumatoid factor	+	

*From Fahey, J. L.: Antibodies and immunoglobulins. I. Structure and function, J.A.M.A. 194:183, 1965.
†Most of these properties have been identified only for the γ heavy chains from IgG molecules.

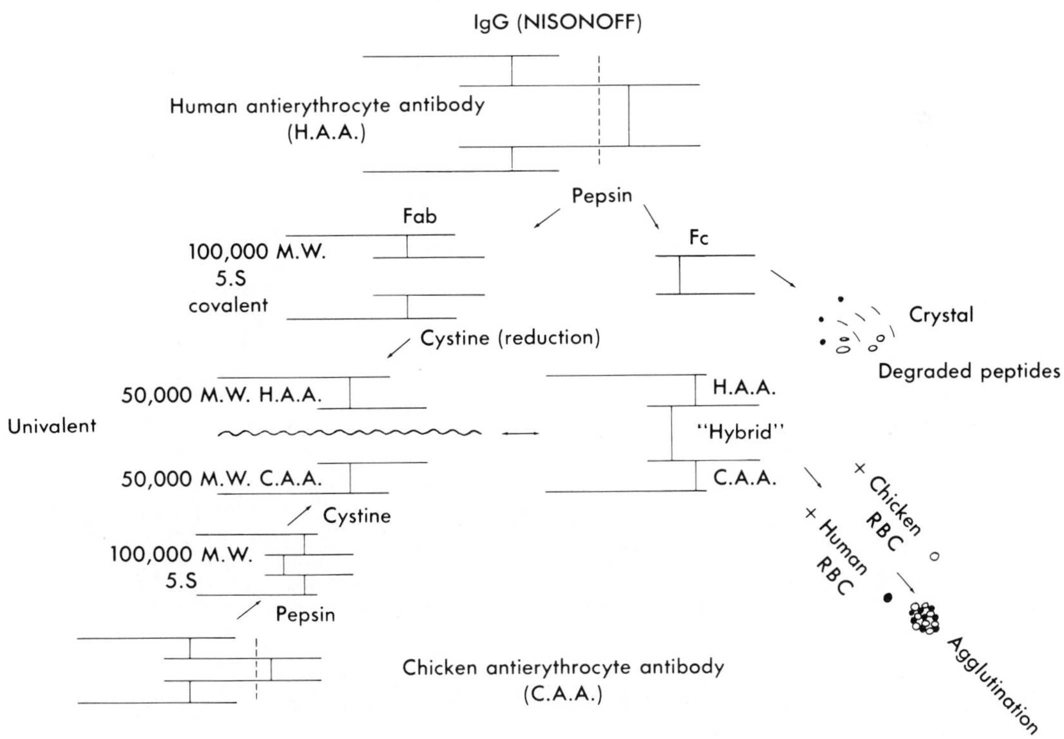

Fig. 20-28. Schematic representation of the production of F (ab′) fragment "hybrids." (Based on data from Nisonoff and associates.[88])

Table 20-10. Possible molecular structures of immunoglobulins based on heavy and light chains as well as heavy chain subclasses known at this time

IgGK	$(\gamma_2)_2\kappa_2$	IgAK	$\alpha_2\kappa_2$
	$(\gamma_1)_2\kappa_2$	IgAL	$\alpha_2 \ell_2$
	$(\gamma_3)_2\kappa_2$	IgMK	$\mu_2\kappa_2$
	$(\gamma_4)_2\kappa_2$	IgML	$\mu_2 \ell_2$
IgGL	$(\gamma_2)_2 \ell_2$	IgDK	$\delta_2\kappa_2$
	$(\gamma_1)_2 \ell_2$	IgDL	$\delta_2 \ell_2$
	$(\gamma_3)_2 \ell_2$	IgEK	$\epsilon_2\kappa_2$
	$(\gamma_4)_2 \ell_2$	IgEL	$\epsilon_2 \ell_2$

Table 20-11. Inv (genetic) factors known that reside on K chains of light chain in IgG immunoglobulin

Old terminology	New terminology
1	1
a	2
b	3

There are no Inv factors known as yet for the IgA or IgM kappa chains or the IgG, IgA, or IgM lambda chains.
It is known that the kappa chain (IgG) amino acid 189 Inv (2) is valine.
The amino acid on the non-Inv (2) kappa light chain of IgG at position 189 is leucine.

Table 20-12. IgG molecule heavy (H) chain subclasses

Old terminology	New terminology
γ_2a (Ne)	IgG2 or γ_2
γ_2b (We)	IgG1 or γ_1
γ_2c (Vi)	IgG3 or γ_3
γ_2d (Ge)	IgG4 or γ_4

Ratio found in normal subjects of $\gamma_2, \gamma_1, \gamma_3, \gamma_4 = 11{:}77{:}9{:}3$.

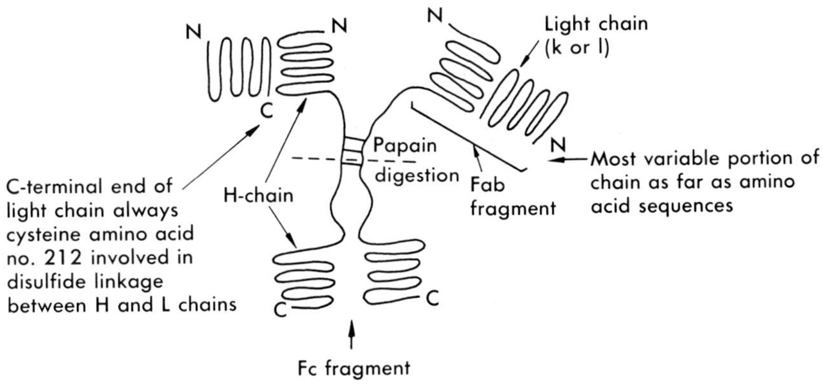

Fig. 20-29. Rabbit gamma G molecule, as proposed by Tanford,[97] showing possible N-terminal combining sites.

Table 20-13. Gm (genetic) factors known that reside on H chain of IgG immunoglobulin*

Old terminology	New terminology
a	1
x	2
$b^w b^2$	3 } Probably same
$f=b^2$	4 } Probably same
b & b'	5
c	6
r	7
e	8
p	9
b^α	10
b^B	11
b^γ	12
b^3	13
b^4	14
s	15
t	16
z	17
	18
	19
fud	20
g	21
y	22
n	23

*(Gm) locus controls the structural sequence on the Fc fragment (γ) H chain of IgG immunoglobulin. At present there are 20 (Gm) factors known to occur at this locus. All on γ_1 and γ_3 chains.

cule in the presence of urea and followed by separation in N-acetic or N-propionic acids with gel filtration resulted in the dissociation to heavy and light chains. For the properties of the H and L chains see Table 20-9.

Reduction and alkylation of the F(ab) fragments results in the separation of these units into the complete L chain and the Fd portion of the heavy (H) chain. Nisonoff and associates[88] found that treatment with pepsin resulted in the production of a 5 S fragment of 100,000 M.W. consisting of two F(ab) fragments F(ab).[2] These fragments may be reduced with cystine, resulting in two univalent 50,000 M.W. fragments, each possessing antibody activity. It is possible to stop the reduction and to recombine two different antibody IgG F(ab) molecules to produce a so-called "hybrid" (Fig. 20-28).

Classes of immunoglobulins. Each immunoglobulin molecule must possess either kappa or lambda L chains. The same molecule will not have one of each. Each class of immunoglobulin shares either kappa or lambda light chains, the light chains being the same for each class. The H chains determine the class specificity. The H chains are classified as either γ (IgG), α (IgA), μ (IgM), δ (IgD), or ϵ (IgE). Therefore it is possible to propose molecular formulas for the five known classes of immunoglobulins (Table 20-10).

Antigenic subclasses. The L chains possess Inv determinants on the kappa chains only. Three major Inv factors may be present: Inv 1, Inv 2, and Inv 3 (Table 20-11).

The H chain antigenic subclasses are mainly described for the γH chains. At least four subclasses exist: γ_1, γ_2, γ_3, and γ_4 (Table 20-12). The ratio of these chains among γG myelomas is 77:11:9:3.[89-91] The Gm types (Table 20-13) are found on possibly three loci. It was observed[92,93] that only two of the γH chain subclasses possess the three Gm tested for. The two are γ_1 and γ_2.

Combining sites of the immunoglobulins. The immunoglobulin molecule appears to have two binding sites at the N terminal ends of the L chains (Fig. 20-29). Great heterogeneity of the amino acid sequence occurs on the L polypeptide chain in the region of the combining site, which is only 5% of the molecule. In 1959 Porter[86] found the F(ab) fragments possessed the antibody activity, and both the H and L chains are involved. No reconstitution of antibody occurs when specific L chains are recombined with nonspecific H chains; therefore the H chains are a most important part of the F(ab) fragment and are required along with the L chain for optimal steric configuration of the combining site.

Diseases affecting the immunoglobulins

Polyclonal gammopathies. This term polyclonal gammopathies implies the production of immunoglobulins from a number of different clones of immunoglobulin-producing cells. In the polyclonal gammopathies the increase is in general associated with increased production of all classes of immunoglobulins (γG, γA, and γM). In some instances, however, this is not the case, and only one or two of the immunoglobulins may be increased, with the others remaining normal or somewhat depleted. A summary of disease entities that elicit a polyclonal increase is found in Table 20-14.

Monoclonal gammopathies. Osserman[94] has classified plasmacytic disorders as possessing the following: (1) the production of plasma or immunologically competent cells in the absence of an antigenic stimulus, (2) marked production of homogenous or myeloma M type proteins or polypeptide fragments (Bence Jones

Table 20-14. Serum immunoglobulin concentrations in diseases associated with polyclonal gammopathies*

Diseases	Serum immunoglobulin concentration		
	γG globulins	γA globulins	γM globulins
I. Liver diseases			
1. Infectious hepatitis	↑↔↑↑	N↔↑	N↔↑↑
2. Laennec's cirrhosis	↑↔↑↑↑	↑↔↑↑↑	N↔↑↑
3. Biliary cirrhosis	N	N	↑↔↑↑
4. Lupoid hepatitis	↑↑↑	↑	N↔↑↑
II. Collagen disorders			
1. Lupus erythematosus	↑↔↑↑	N↔↑	N↔↑↑
2. Rheumatoid arthritis	N↔↑	↑↔↑↑↑	N↔↑
3. Sjögren's syndrome	N↔↑	N↔↑	↓↔↑↑
4. Scleroderma	N↔↑	N	↑
III. Infections			
1. Tuberculosis	↑↔↑↑	N⇄↑↑↑	↓↔N
2. Subacute bacterial endocarditis	↑↔↑↑	↓↔N	↑↔↑↑
3. Leprosy	↑↔↑↑	N	↑
4. Trypanosomiasis	N↔↑	N↔↑	↑↑↔↑↑↑
5. Malaria	↓↔↑	N	↑↔↑↑
6. Kala-azar	↑↑	N	N
7. Infectious mononucleosis	↑↔↑↑	N↔↑	↑↔↑↑
8. Fungus diseases	N	N↔↑	N
9. Bartonellosis	↑	↓↔N	↑↑↔↑↑↑
10. Lymphogranuloma venereum	↑↑	N	N
11. Actinomycosis	↑↑↑	↑↑	↑↑↑
IV. Sarcoidosis	N↔↑↑	N↔↑↑	N↔↑
V. Miscellaneous			
1. Hodgkin's disease	↓↔↑↑	↓↔↑	↓↔↑↑
2. Monocytic leukemia	↑	↑	↑↑
3. Cystic fibrosis	↑↔↑↑	↑↔↑↑	N↔↑↑

*From Ritzmann, S. E., and Levin, W. C.: Polyclonal and monoclonal gammopathies, Lab Synopsis 2:212, 1967.
N = Essentially normal serum immunoglobulin levels.
↓ = Decreased immunoglobulin concentrations.
↑, ↑↑, ↑↑↑ = Slight, moderate or marked increases in serum immunoglobulin concentrations.
↔ = Range of immunoglobulin levels.

proteins) of these proteins, and (3) associated deficiency of the normal immunoglobulins.

The monoclonal gammopathies may also be found in instances in which a clear clinical picture of myeloma or macroglobulinemia is not apparent. In such instances these must be classified as a monoclonal gammopathy of undetermined etiology. An excellent review of selected cases may be found in Cawley[7] and Michaux and Heremans.[95] It must be stressed that the definitive diagnosis of multiple myeloma or macroglobulinemia must not be hurriedly made from the electrophoretic pattern alone, but confirmatory evidence from bone marrow aspirations, radiologic finding of skeletal lesions and immunoelectrophoretic and quantitative procedures must be available.

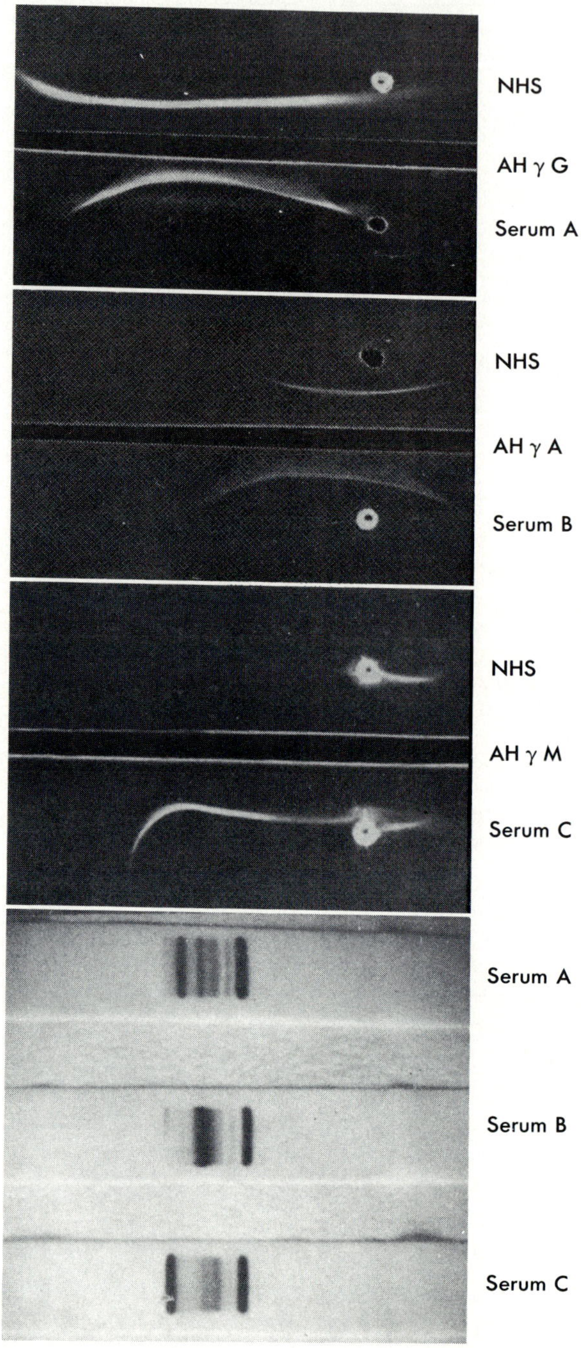

Fig. 20-30. Three types of monoclonal gammopathy as exhibited by cellulose acetate electrophoresis and immunoelectrophoresis. NHS (normal human serum), AHγG (anti–human γG), AHγA (anti–human γA), AHγM (anti–human γM), serum A (gamma G myeloma), serum B (gamma A myeloma), serum C (gamma M Waldenström's macroglobulinemia).

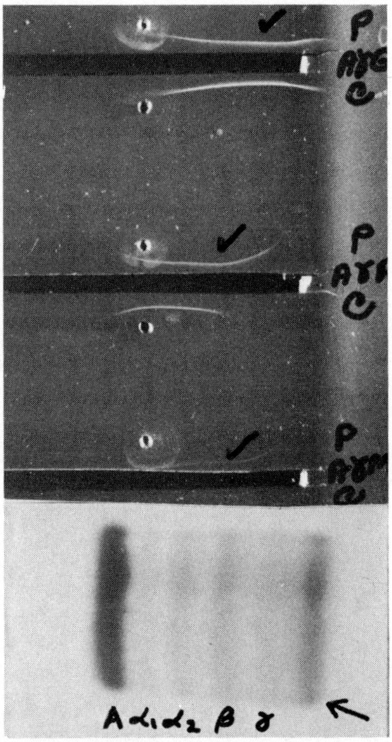
Monoclonal peak

Fig. 20-31. Cellulose acetate and immunoelectrophoretic pattern of a monoclonal gammopathy of undetermined origin. Note adequacy of all immunoglobulin fractions.

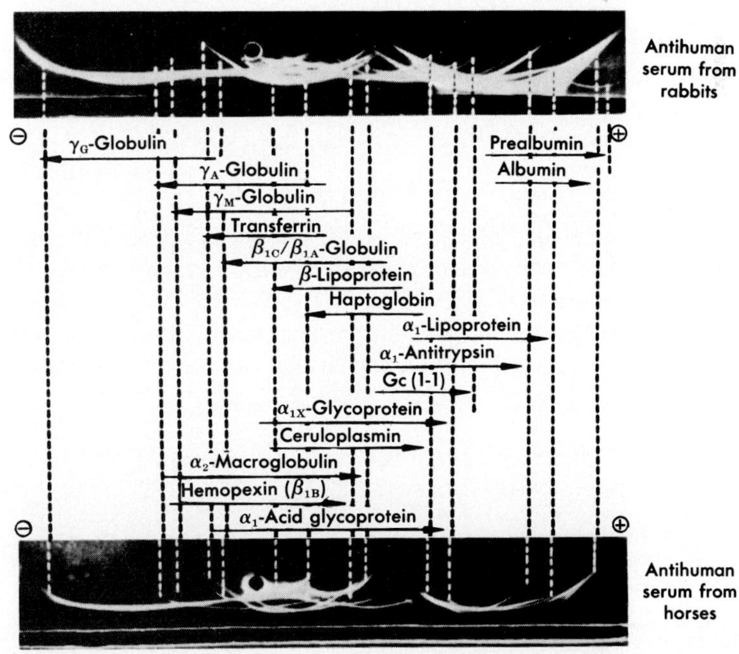

Fig. 20-32. Immunoelectrophoretic analysis of normal human serum showing labeled arcs. Note difference in arcs produced with the antiserum to human serum in different species. (Courtesy Behring Diagnostics, Inc., Woodbury, N. Y.)

Table 20-15. Classification of dysclonal or aclonal gammopathies*

Antibody-deficient syndrome	Findings
Transient hypogammaglobulinemia	Infants, very low γG, γA, γM, for first 9-15 months
Congenital agammaglobulinemia	
Sex-linked	Infants, extremely low γG, γA, γM, male only
Autosomal	Infants, very low γG, γA, γM, males and females
Acquired agammaglobulinemia	
Primary	Adults, sudden depression of γG, γA, and γM
Secondary	Adults, children, depression of γG, γA, and γM by systemic disease
Swiss-type agammaglobulinemia†	Infants, lymphopenia, thymic aplasia, autosomal
Dysgammaglobulinemia (congenital and acquired)	
Type I	Low γG and γA, increased γM, sex-linked
Type II	Low γA and γM, increased γG
Type III	Absence of γA

*From Cawley, L. P.: Electrophoresis and immunoelectrophoresis, Boston, 1969, Little, Brown & Co.
†Hereditary thymic aplasia.

Figs. 20-30 and 20-31 depict the electrophoretc and immunoelectrophoretic patterns in the various monoclonal gammopathies related to myeloma and macroglobulinemia as well as a monoclonal gammopathy of undetermined etiology. A normal immunoelectrophoretic pattern is shown to depict the arcs of the various serum proteins (Fig. 20-32).

Bence Jones protein may be detected by the use of specific antisera developed for the k or l light chains. H chain disease of Franklin and associates[96] react with H chain specific antiserum to γG globulin and not with antiserum to k or l light chains.

Dysclonal or aclonal gammopathies. The terms dysclonal and aclonal gammopathies have been suggested by Cawley[7] and reflect a quite acceptable nomenclature and method of classification for the conditions in which the immunoglobulins may be lower than normal or absent as far as our present-day methods are able to quantitate. Table 20-15 from Cawley[7] lists the conditions falling under these classifications. It must be kept in mind that the immunoglobulin levels of children vary from the adult until a certain age level is reached (Table 20-15).

REFERENCES

1. Strickland, R. D.: Electrophoresis, Anal. Chem. **38**:99R, 1966.
2. McDougall, E. I., and Synge, R. L.: Recent advances in electrophoretic techniques, Brit. Med. Bull. **22**:115, 1966.
3. Gray, G. W.: Electrophoresis, Sci. Amer. **185**: 45, 1951.
4. Moore, D. H., and Abramson, H. A.: Medical physics. In Glasser, Otto, editor: New medical physics, vol. 1, Chicago, 1955, Year Book Medical Publishers, Inc., pp. 403-407.
5. McDonald, H. J.: Ionography, electrophoresis in stabilized media, Chicago, 1955, Year Book Medical Publishers, Inc.
6. Wieme, R. J.: Agar gel electrophoresis, Amsterdam, 1965, Elsevier Publishing Co.
7. Cawley, L. P.: Electrophoresis and immunoelectrophoresis, Boston, 1969, Little, Brown & Co.
8. Henry, R. J.: Clinical chemistry. Principles and techniques, New York, 1967, Paul B. Hoeber, Inc., Medical Book Department of Harper & Row, Publishers, pp. 93-109, 211-226.
9. Nerenberg, S. T.: Electrophoresis—a labora-

tory manual, Philadelphia, 1966, F. A. Davis Co.
10. Bier, Milan, editor: Electrophoresis—theory, methods, and applications, vol. 2, New York, 1967, Academic Press, Inc.
11. Block, R. J. Durrum, E. L., and Zweig, G.: A manual of paper chromatography and paper electrophoresis, New York, 1955, Academic Press, Inc.
12. Lederer, M.: Paper electrophoresis, Amsterdam, 1955, Elsevier Publishing Co.
13. Audubert, R., and de Mende, S.: The principles of electrophoresis, New York, 1960, The Macmillan Co.
14. Smith, Ivor: Chromatographic and electrophoretic techniques, vol. 2, Zone electrophoresis, ed. 2, London, 1968, William Heinemann, Ltd.
15. Cawley, L. P.: Workshop manual on electrophoresis and immunoelectrophoresis, Chicago, 1964, American Society of Clinical Pathologists.
16. Tiselius, A.: A new apparatus for electrophoresis: analysis of colloidal mixtures, Trans. Faraday Soc. **33**:524, 1937.
17. Marsh, C. L., Jolliff, C. R., and Payne, L. C.: A rapid micromethod for starch-gel electrophoresis, Amer. J. Clin. Path. **41**:217, 1964.
18. Smithes, O.: Zone electrophoresis in starch gels. Grouped variations in the serum proteins of normal human adults, Biochem. J. **61**:629, 1955.
19. Smithes, O.: Zone electrophoresis in starch gels and its application to studies of serum proteins, Advances Protein Chem. **14**:65, 1959.
20. Singer, K., Chernoff, A. I., and Singer, L.: Studies on abnormal hemoglobins. I. Their demonstration in sickle cell anemia and other hematologic disorders by alkali denaturation, Blood **6**:413, 1951.
21. Beckman, L.: Isoenzyme variations in man. Monographs in human genetics, Philadelphia, 1966, S. Karger.
22. Vessell, E. S.: Multiple molecular forms of enzymes, Ann. N. Y. Acad. Sci. **151**:1968.
23. Kelsey, R. L., de Graffenried, T. P., and Donaldson, R. C.: Electrophoretic fractionation of serum glycoprotein on cellulose acetate, Clin. Chem. **11**:1058, 1965.
24. Wieme, R. J.: Agar gel electrophoresis, Amsterdam, 1965, Elsevier Publishing Co.
25. Burrows, S.: Simple method for concentration of cerebrospinal fluid for protein electrophoresis, Clin. Chem. **11**:1068, 1965.
26. Korngold, L.: Plasma proteins: methods of study and changes in disease. In Stefanini, Mario, editor: Progress in clinical pathology, vol. 1, New York, 1966, Grune & Stratton, Inc.
27. Putnam, F. W.: The plasma proteins, vol. 2, New York, 1960, Academic Press, Inc.
28. Sunderman, F. W., and Sunderman, F. W., Jr.: Serum proteins and the dysproteinemias, Philadelphia, 1964, J. B. Lippincott Co.
29. Dimopoullos, G. T.: Plasma proteins in health and disease, Ann. N. Y. Acad. Sci. **94**:1, 1961.
30. Lewis, L. A.: Electrophoresis in physiology, Springfield, Ill., 1960, Charles C Thomas, Publisher.
31. Sandor, G.: Serum proteins in health and disease, Baltimore, 1966, The Williams & Wilkins Co.
32. Ivers, R. R., McKenzie, B. F., McGuckin, W. F., and Goldstein, N. P.: Spinal fluid gamma globulin in multiple sclerosis and other neurologic disease. Electrophoretic patterns in 606 patients, J.A.M.A. **176**:515, 1961.
33. Goa, J., and Tueten, L.: Electrophoresis of CSF proteins in certain neurological diseases, Scand. J. Clin. Lab. Invest. **15**:152, 1963.
34. Brockenridge, C. J.: Cerebrospinal fluid protein fractions in health and disease, J. Clin. Path. **15**:206, 1962.
35. Clausen, J., Matzke, J., and Gerhardt, W.: Agar-gel micro electrophoresis of proteins in the cerebrospinal fluid. Normal and pathological findings, Acta Neurol. Scand. **40** (supp. 10):49, 1964.
36. Evans, J. H., and Quick, D. T.: Polyacrylamide gel electrophoresis of spinal-fluid proteins, Clin. Chem. **12**:28, 1966.
37. Greenhouse, A. H., and Speck, L. B.: The electrophoresis of spinal fluid proteins, Amer. J. Med. Sci. **248**:333, 1964.
38. Kaplan, A.: Electrophoresis of cerebrospinal fluid proteins, Amer. J. Med. Sci. **253**:549, 1967.
39. Berggard, I.: The plasma proteins in normal urine, Nature **187**:776, 1960.
40. Berggard, I.: Studies on the plasma proteins in normal human urine, Clin. Chem. Acta **6**: 413, 1961.
41. Berggard, I.: Identification and isolation of urinary proteins. In Peeters, H., editor: Colloquium on the protides of the biological fluids. Proceedings of the 12th Colloquium, 1964, Amsterdam, 1965, Elsevier Publishing Co., pp. 285-291.
42. Grant, G. H.: The proteins of normal urine, J. Clin. Path. **10**:360, 1957.
43. Schultze, H. E., and Heremans, J. F.: Molecular biology of human proteins. In Nature and metabolism of extracellular proteins, vol. 1, New York, 1966, American Elsevier Publishing Co., Inc., pp. 670-731.
44. Tidstrom, B.: Paper electrophoresis in the diagnosis of renal disease. In Peeters, H., editor: Colloquium on the protides of the biological fluids. Proceedings of the 12th Collo-

quium, 1964, Amsterdam, 1965, Elsevier Publishing Co.
45. Pauling, L., Itano, H. A., Singer, S. J., and Wells, I. C.: Sickle-cell anemia, a molecular disease, Science **110**:543, 1949.
46. Pinfield, A. S., and Rodgerson, D.: Quantitation of A_2 hemoglobin by electrophoresis on cellulose acetate, Clin. Chem. **12**:883, 1966.
47. Ingram, V. M.: Chemistry of the abnormal human hemoglobins in haematology and blood groups, papers from the British Medical Bulletin, Chicago, 1961, University of Chicago Press, pp. 26-31.
48. Hunt, J. A., and Ingram, V. M.: Allelomorphism and the chemical differences of the hemoglobins A, S, and C, Nature **181**:1062, 1958.
49. Heller, P.: The molecular basis of the pathogenecity of abnormal hemoglobins—some recent developments, Blood **25**:110, 1965.
50. Heller, P.: Hemoglobins and heredity. I & II. Postgrad. Med. **43**:63, 91, 1968.
51. Perutz, M. F.: The hemoglobin molecule, Sci. Amer. **211**:64, 1964.
52. Ingram, V. M.: Hemoglobin and its abnormalities, Springfield, Ill., 1961, Charles C Thomas, Publisher.
53. Jonxis, J. H. P., and Delafresnaye, J. F.: Abnormal hemoglobins, Springfield, Ill., 1959, Charles C Thomas, Publisher.
54. Jonxis, J. H. P., and Huisman, T. H. J.: A laboratory manual on abnormal haemoglobins, Oxford, England, 1958, Blackwell Scientific Publications.
55. Miale, J. B.: Laboratory medicine—hematology, St. Louis, 1967, The C. V. Mosby Co.
56. Fredrickson, D. S., and Lees, K. S.: A system for phenotyping hyperlipoproteinemia, Circulation **31**:321, 1965.
57. Frederickson, D. S., Lees, R. S., and Levy, R. I.: Genetically determined abnormalities in lipid transportation. In Paoletti, R., Kritchevsky, D., and Steinberg, D., editors: Progress in biochemical pharmacology, vol. 2 Basel, Switzerland, 1966, S. Karger, p. 343.
58. Levy, R. I., Lees, R. S., and Fredrickson, D. S.: The clinical and biochemical forms of two forms of familial hyperbetalipoproteinemia, J. Clin. Invest. **46**:1086, 1967.
59. Fredrickson, D. S., Levy, R. I., and Lees, K. S.: Fat transport in lipoproteins—an integrated approach to mechanism and disorders, New Eng. J. Med. **276**:32, 1967.
60. Lees, R. S., and Hatch, F. T.: Sharper separation of lipoprotein species by paper electrophoresis in albumin containing buffer, J. Lab. Clin. Med. **61**:318, 1963.
61. Garbar, P., and Williams, C. A., Jr.: Methode permettant l'etude conjuguee des proprietes electrophoretiques et immunochimiques d'un melonge des proteines: applications au serum sanguin, Biochim. Biophys. Acta **10**:193, 1953.
62. Ouchterlony, O.: Antigen-antibody reactions in gels, Acta Path. Microbiol. Scand. **26**:507, 1949.
63. Scheidegger, J. L.: A microtechnique of immunoelectrophoresis, Int. Arch: Allerg. **7**:103, 1955.
64. Oudin, J.: Immunologie, methode d'analyse immunochimique par precipitation specifique en milieu gelifie, C. R. Acad. Sci. **222**:115, 1946.
65. Mancini, G., Carbonara, A. O., and Heremons, J. F.: Immunochemical quantitation of antigens by single radial immunodiffusion, Immunochemistry **2**:235, 1965.
66. Fahey, J. L., and McKelvey, E. M.: Quantitative determination of serum immunoglobulins in antibody agar plates, J. Immun. **94**:84, 1965.
67. Merrill, D., Hartley, T. Z., and Claman, N. N.: Electroimmunodiffusion (E.I.D.): a simple, rapid method for quantitation of immunoglobulins in dilute biological fluids, J. Lab. Clin. Med. **69**:151, 1967.
68. Claman, H. N., and Merrill, D.: Quantitative measurement of human gamma—2, beta—2A and beta—2M serum immunoglobulins, J. Lab. Clin. Med. **64**:685, 1964.
69. Ehrmantrant, H. C.. Presentation, Joint meeting Northwest Society of Pathologists and Northwest Region College of American Pathologists, May, 1968.
70. Briere, R. O., and Mull, J. D.: Electrophoresis of serum protein with cellulose acetate, Amer. J. Clin. Path. **42**:547, 1964.
71. Camley, L. P., and Eberhardt, L.: Simplified gel electrophoresis. I. Rapid technique applicable to the clinical laboratory, Amer. J. Clin. Path. **38**:539, 1968.
72. Ouchterlony, O.: Handbook of immunodiffusion and immunoelectrophoresis, with appendices by Hirshfield, J., Clousen, J., and Sewell, M. M. H., Ann Arbor, Mich., 1968, Ann Arbor Science Publishers.
73. McKay, C. G.: Practical application of immunoelectrophoresis, Lab. Synopsis, vol. 2 (Behring Diagnostics), 1967.
74. Ritzman, S. E., and Levin, W. C.: Polyclonal and monoclonal gammopathies, Lab. Synopsis, vol. 2 (Behring Diagnostics), 1967.
75. Grabar, P., and Burtin, P.: Immuno-electrophoretic analysis. Application to human biological fluids, Amsterdam, 1964, Elsevier Publishing Co.
76. Waldenström, J. C.: Monoclonal and polyclonal hypergammaglobulinemia—clinical and biological significance, Nashville, Tenn., 1968, Vanderbilt University Press.

77. Janeway, C. A., Rosen, F. S., Merler, E., and Alper, C. A.: The gamma globulins, Boston, 1967, Little, Brown & Co.
78. Engle, R. L., Jr., and Wallis, L. A.: Immunoglobulinopathies, Springfield, Ill., 1969, Charles C Thomas, Publisher.
79. Samter, Max: Immunological diseases, Boston, 1965, Little, Brown & Co.
80. Wunderly, C.: Immunoelectrophoresis: methods, interpretation, results in advances in clinical chemistry, vol. 4, New York, 1961, Academic Press, Inc., p. 207.
81. Crowle, A. J.: Immunodiffusion, New York, 1961, Academic Press, Inc.
82. Peetoom, F.: The agar precipitation technique and its application as a diagnostic and analytical method, Springfield, Ill., 1963, Charles C Thomas, Publisher.
83. Schwick, G., and Storiko, K.: Qualitative and quantitative determination of plasma proteins by immunoprecipitation, Lab. Synopsis, vol. 1, (Behring Diagnostics), 1964.
84. Nomenclature for human immunoglobulins, Bull. WHO **30**:447, 1964.
85. Porter, R. R., and Press, E. M.: Immunochemistry, Ann. Rev. Biochem. **31**:365, 1962.
86. Porter, R. R.: Hydrolysis of rabbit gamma-globulin and antibodies with crystalline papain, Biochem. J. **73**:119, 1959.
87. Edelman, G. M., and Poulik, M. D.: Studies on structural units of the gamma-globulins, J. Exp. Med. **113**:861, 1961.
88. Nisonoff, A., Wissler, F. C., and Lipman, L. N.: Properties of major component of peptic digest of rabbit antibody, Science **132**:1770, 1960.
89. Grey, H. M., and Kunkel, H. G.: H chain subgroups of myeloma proteins and normal 7S gamma-globulin, J. Exp. Med. **120**:253, 1964.
90. Terry, W. D., and Fahey, J. L.: Subclasses of human gamma 2 globulin based on differences in polypeptide chains, Science **146**:400, 1960.
91. Terry, W. D.: Skin-sensitizing activity related to gamma polypeptide chain characteristics of human IgG, J. Immunol. **95**:1041, 1965.
92. Harboe, M., Osterland, C. K., Mannik, M. and Kunkel, H. G.: Genetic characteristics of human gamma globulins in myeloma proteins, J. Exp. Med. **116**:719, 1962.
93. Franklin, E. C., Fudenberg, H., Meltzer, M., and Stanworth, D. R.: The structural basis for genetic variations of normal human gamma-globulins, Proc. Nat. Acad. Sci. U. S. A. **48**: 914, 1962.
94. Osserman, E. F.: Plasma cells dyscrasia, general considerations. In Beeson, P. B., and McDermott, W., editors: Cecil-Loeb textbook of medicine, ed. 12, Philadelphia, 1967, W. B. Saunders Co.
95. Michaux, J. C., and Heremans, J.: Thirty cases of monoclonal immunoglobulin disorders other than myeloma or macroglobulinemia. A classification of diseases associated with the production of monoclonal type immunoglobulins, Amer. J. Med. **46**:562, 1969.
96. Franklin, E. C., Lowenstein, J., Bigelow, B., and Meltzer, M.: Heavy chain disease—a new disorder of serum gamma globulins. Report of the first case, Amer. J. Med. **37**:332, 1964.
97. Noelken, M. E., Nelson, C. A., Buckley, C. E. III, and Tanford, C.: Gross conformation of rabbit 7Sγ–immunoglobulin and its papain-cleaved fragments, J. Biol. Chem. **240**: 218, 1965.

• Appendixes

A Tables

B Fire precautions, chemical hazards, and antidotes of poisons

C Basic information for the student

D Reagents

E Company directory

• A

• Tables

Table A-1. 1968 table of relative atomic weights ($^{12}C = 12$)*

Element	Symbol	Atomic No.	Atomic Weight	Element	Symbol	Atomic No.	Atomic Weight
Actinium	Ac	89	[227]†	Fermium	Fm	100	[257]†
Aluminum	Al	13	26.9815	Fluorine	F	9	18.9984
Americium	Am	95	[243]†	Francium	Fr	87	[223]†
Antimony	Sb	51	121.75	Gadolinium	Gd	64	157.25
Argon	Ar	18	39.948	Gallium	Ga	31	69.72
Arsenic	As	33	74.9216	Germanium	Ge	32	72.59
Astatine	At	85	[210]†	Gold	Au	79	196.967
Barium	Ba	56	137.34	Hafnium	Hf	72	178.49
Berkelium	Bk	97	[247]†	Helium	He	2	4.0026
Beryllium	Be	4	9.0122	Holmium	Ho	67	164.930
Bismuth	Bi	83	208.980	Hydrogen	H	1	1.00797[a]
Boron	B	5	10.811[a]	Indium	In	49	114.82
Bromine	Br	35	79.904[b]	Iodine	I	53	126.9044
Cadmium	Cd	48	112.40	Iridium	Ir	77	192.2
Calcium	Ca	20	40.08	Iron	Fe	26	55.847[b]
Californium	Cf	98	[252]†	Krypton	Kr	36	83.80
Carbon	C	6	12.01115[a]	Lanthanum	La	57	138.91
Cerium	Ce	58	140.12	Laurentium	Lr	103	[256]†
Cesium	Cs	55	132.905	Lead	Pb	82	207.19
Chlorine	Cl	17	35.453[b]	Lithium	Li	3	6.939
Chromium	Cr	24	51.996	Lutetium	Lu	71	174.97
Cobalt	Co	27	58.9332	Magnesium	Mg	12	24.305
Copper	Cu	29	63.546[b]	Manganese	Mn	25	54.9380
Curium	Cm	96	[247]†	Mendelevium	Md	101	[257]†
Dysprosium	Dy	66	162.50	Mercury	Hg	80	200.59
Einsteinium	Es	99	[254]†	Molybdenum	Mo	42	95.94
Erbium	Er	68	167.26	Neodymium	Nd	60	144.24
Europium	Eu	63	151.96	Neon	Ne	10	20.179[b]

*By permission of the International Union of Pure and Applied Chemistry and Butterworths Scientific Publications.

†Value in brackets denotes the mass number of the isotope of longest known half-life (or a better known one for Cf, Po, Pm, and Tc).

[a]Atomic weight varies because of natural variation in isotopic composition: B, ±0.003; C, ±0.00005; H, ±0.00001; O, ±0.00001; Si, ±0.001; S, ±0.003.

[b]Atomic weight is believed to have following experimental uncertainty: Br, ±0.001; Cl, ±0.01; Cu, ±0.001; Fe, ±0.003; Ne, ±0.003; Ag, ±0.001. (For other elements, the last digit given for the atomic weight is believed reliable to ±0.5.)

Continued.

Table A-1. 1968 table of relative atomic weights (^{12}C = 12)—cont'd

Element	Symbol	Atomic No.	Atomic Weight	Element	Symbol	Atomic No.	Atomic Weight
Neptunium	Np	93	[237]†	Selenium	Se	34	78.96
Nickel	Ni	28	58.71	Silicon	Si	14	28.086[a]
Niobium	Nb	41	92.906	Silver	Ag	47	107.868[b]
Nitrogen	N	7	14.0067	Sodium	Na	11	22.9898
Nobelium	No	102	[255]†	Strontium	Sr	38	87.62
Osmium	Os	76	190.2	Sulfur	S	16	32.064[a]
Oxygen	O	8	15.9994[a]	Tantalum	Ta	73	180.948
Palladium	Pd	46	106.4	Technetium	Tc	43	[99]†
Phosphorus	P	15	30.9738	Tellurium	Te	52	127.60
Platinum	Pt	78	195.09	Terbium	Tb	65	158.924
Plutonium	Pu	94	[244]†	Thallium	Tl	81	204.37
Polonium	Po	84	[210]†	Thorium	Th	90	232.038
Potassium	K	19	39.102	Thulium	Tm	69	168.934
Praseodymium	Pr	59	140.907	Tin	Sn	50	118.69
Promethium	Pm	61	[147]†	Titanium	Ti	22	47.90
Protactinium	Pa	91	[231]†	Tungsten	W	74	183.85
Radium	Ra	88	[226]†	Uranium	U	92	238.03
Radon	Rn	86	[222]†	Vanadium	V	23	50.942
Rhenium	Re	75	186.2	Xenon	Xe	54	131.30
Rhodium	Rh	45	102.905	Ytterbium	Yb	70	173.04
Rubidium	Rb	37	85.47	Yttrium	Y	39	88.905
Ruthenium	Ru	44	101.07	Zinc	Zn	30	65.37
Samarium	Sm	62	150.35	Zirconium	Zr	40	91.22
Scandium	Sc	21	44.956				

Table A-2. Temperature conversion—centigrade to Fahrenheit

The values in the body of the table give, in degrees Fahrenheit, the temperatures indicated in degrees centigrade at the top and side. 1° C = 1.8° F; −273.16° C = −459.72° F = Absolute zero.

For temperatures below 0° C

Temp. °C	0	1	2	3	4	5	6	7	8	9
0	+ 32.0	30.2	28.4	26.6	24.8	23.0	21.2	19.4	17.6	15.8
− 10	+ 14.0	12.2	10.4	8.6	6.8	5.0	3.2	+1.4	−0.4	−2.2
− 20	− 4.0	5.8	7.6	9.4	11.2	13.0	14.8	16.6	18.4	20.2
− 30	− 22.0	23.8	25.6	27.4	29.2	31.0	32.8	34.6	36.4	38.2
− 40	− 40.0	41.8	43.6	45.4	47.2	49.0	50.8	52.6	54.4	56.2
− 50	− 58.0	59.8	61.6	63.4	65.2	67.0	68.8	70.6	72.4	74.2
− 60	− 76.0	77.8	79.6	81.4	83.2	85.0	86.8	88.6	90.4	92.2
− 70	− 94.0	95.8	97.6	99.4	101.2	103.0	104.8	106.6	108.4	110.2
− 80	−112.0	113.8	115.6	117.4	119.2	121.0	122.8	124.6	126.4	128.2

Table A-2. Temperature conversion—centigrade to Fahrenheit—cont'd

The values in the body of the table give, in degrees Fahrenheit, the temperatures indicated in degrees centigrade at the top and side. 1° C = 1.8° F; −273.16° C = −459.72° F = Absolute zero.

For temperatures below 0° C

Temp. °C	0	1	2	3	4	5	6	7	8	9
− 90	−130.0	131.8	133.6	135.4	137.2	139.0	140.8	142.6	144.4	146.2
−100	−148.0	149.8	151.6	153.4	155.2	157.0	158.8	160.6	162.4	164.2
−110	−166.0	167.8	169.6	171.4	173.2	175.0	176.8	178.6	180.4	182.2
−120	−184.0	185.8	187.6	189.4	191.2	193.0	194.8	196.6	198.4	200.2
−130	−202.0	203.8	205.6	207.4	209.2	211.0	212.8	214.6	216.4	218.2
−140	−220.0	221.8	223.6	225.4	227.2	229.0	230.8	232.6	234.4	236.2
−150	−238.0	239.8	241.6	243.4	245.2	247.0	248.8	250.6	252.4	254.2
−160	−256.0	257.8	259.6	261.4	263.2	265.0	266.8	268.6	270.4	272.2
−170	−274.0	275.8	277.6	279.4	281.2	283.0	284.8	286.6	288.4	290.2
−180	−292.0	293.8	295.6	297.4	299.2	301.0	302.8	304.6	306.4	308.2
−190	−310.0	311.8	313.6	315.4	317.2	319.0	320.8	322.6	324.4	326.2
−200	−328.0	329.8	331.6	333.4	335.2	337.0	338.8	340.6	342.4	344.2
−210	−346.0	347.8	349.6	351.4	353.2	355.0	356.8	358.6	360.4	362.2
−220	−364.0	365.8	367.6	369.4	371.2	373.0	374.8	376.6	378.4	380.2
−230	−382.0	383.8	385.6	387.4	389.2	391.0	392.8	394.6	396.4	398.2
−240	−400.0	401.8	403.6	405.4	407.2	409.0	410.8	412.6	414.4	416.2
−250	−418.0	419.8	421.6	423.4	425.2	427.0	428.8	430.6	432.4	434.2
−260	−436.0	437.8	439.6	441.4	443.2	445.0	446.8	448.6	450.4	452.2
−270	−454.0	455.8	457.6	459.4	..	..	..	..	..	..

For temperatures above 0° C

Temp. °C	0	1	2	3	4	5	6	7	8	9
0	32.0	33.8	35.6	37.4	39.2	41.0	42.8	44.6	46.4	48.2
10	50.0	51.8	53.6	55.4	57.2	59.0	60.8	62.6	64.4	66.2
20	68.0	69.8	71.6	73.4	75.2	77.0	78.8	80.6	82.4	84.2
30	86.0	87.8	89.6	91.4	93.2	95.0	96.8	98.6	100.4	102.2
40	104.0	105.8	107.6	109.4	111.2	113.0	114.8	116.6	118.4	120.2
50	122.0	123.8	125.6	127.4	129.2	131.0	132.8	134.6	136.4	138.2
60	140.0	141.8	143.6	145.4	147.2	149.0	150.8	152.6	154.4	156.2
70	158.0	159.8	161.6	163.4	165.2	167.0	168.8	170.6	172.4	174.2
80	176.0	177.8	179.6	181.4	183.2	185.0	186.8	188.6	190.4	192.2
90	194.0	195.8	197.6	199.4	201.2	203.0	204.8	206.6	208.4	210.2
100	212.0	213.8	215.6	217.4	219.2	221.0	222.8	224.6	226.4	228.2
110	230.0	231.8	233.6	235.4	237.2	239.0	240.8	242.6	244.4	246.2
120	248.0	249.8	251.6	253.4	255.2	257.0	258.8	260.6	262.4	264.2
130	266.0	267.8	269.6	271.4	273.2	275.0	276.8	278.6	280.4	282.2
140	284.0	285.8	287.6	289.4	291.2	293.0	294.8	296.6	298.4	300.2
150	302.0	303.8	305.6	307.4	309.2	311.0	312.8	314.6	316.4	318.2
160	320.0	321.8	323.6	325.4	327.2	329.0	330.8	332.6	334.4	336.2
170	338.0	339.8	341.6	343.4	345.2	347.0	348.8	350.6	352.4	354.2
180	356.0	357.8	359.6	361.4	363.2	365.0	366.8	368.6	370.4	372.2
190	374.0	375.8	377.6	379.4	381.2	383.0	384.8	386.6	388.4	390.2

Table A-3. Metric units

Quantity	Unit	Symbol	Relationship of units
Length	millimeter	mm	1 mm = 0.001 m
	centimeter	cm	1 cm = 10 mm
	decimeter	dm	1 dm = 10 cm
	meter	m	1 m = 100 cm
	kilometer	km	1 km = 1000 m
Area	square centimeter	cm^2	1 cm^2 = 100 mm^2
	square decimeter	dm^2	1 dm^2 = 100 cm^2
	square meter	m^2	1 m^2 = 100 dm^2
	are	a	1 a = 100 m^2
	hectare	ha	1 ha = 100 a
	square kilometer	km^2	1 km^2 = 100 ha
Volume	cubic centimeter	cm^3	1 cm^3 } = 0.001 L
	milliliter	ml	1 ml } = 0.001 L
	cubic decimeter	dm^3	1 dm^3 } = 1000 ml
	liter	L	1 L } = 1000 ml
	cubic meter	m^3	1 m^3 = 1000 L
Mass	milligram	mg	1 mg = 0.001 g
	gram	g	1 g = 1000 mg
	kilogram	kg	1 kg = 1000 g
	metric ton	t	1 t = 1000 kg

Table A-4. Metric decimal prefixes

Multiplication factors	Prefix	Symbol
1 000 000 000 000 = 10^{12}	tera	T
1 000 000 000 = 10^{9}	giga	G
1 000 000 = 10^{6}	mega	M
1 000 = 10^{3}	kilo	k
100 = 10^{2}	hecto	h
10 = 10^{1}	deka	da
1		
0.1 = 10^{-1}	deci	d
0.01 = 10^{-2}	centi	c
0.001 = 10^{-3}	milli	m
0.000 001 = 10^{-6}	micro	μ
0.000 000 001 = 10^{-9}	nano	n
0.000 000 000 001 = 10^{-12}	pico	p
0.000 000 000 000 001 = 10^{-15}	femto	f
0.000 000 000 000 000 001 = 10^{-18}	atto	a

Table A-5. Conversion factors

U. S. and metric units

Each unit in bold face type is followed by its equivalent in other units of the same quantity.

Acre—0.0015625 square mile (statute); 4.3560 × 10^4 square feet; 0.40468564 hectare

Bushel—(U. S.)—1.244456 cubic feet; 2150.42 cubic inches; 0.035239 cubic meter; 35.23808 liters

Centimeter—0.0328084 foot; 0.393701 inch

Circular mil—7.853982 × 10^{-7} square inches; 5.067075 × 10^{-6} square centimeters

Cubic centimeter—0.061024 cubic inch; 0.270512 dram (U. S. fluid); 16.230664 minims (U. S.); 0.999972 milliliter

Cubic foot—0.803564 bushel (U. S.); 7.480520 gallons (U. S. liquid); 0.028317 cubic meter; 28.31605 liters

Cubic inch—16.387064 cubic centimeters

Cubic meter—35.314667 cubic feet; 264.17205 gallons (U. S. liquid)

Foot—0.3048 meter

Gallon (U. S. liquid)—0.1336816 cubic foot; 0.832675 gallon (British); 231 cubic inches; 0.0037854 cubic meter; 3.785306 liters

Grain—0.06479891 gram

Gram—0.00220462 pound (avoirdupois); 0.035274 ounce (avoirdupois); 15.432358 grains

Hectare—2.471054 acres; 1.07639 × 10^5 square feet

Inch—2.54 centimeters

Kilogram—2.204623 pounds (avoirdupois)

Kilometer—0.621371 mile (statute)

Liter—0.264179 gallon (U. S. liquid); 0.0353157 cubic foot; 1.056718 quarts (U. S. liquid)

Meter—1.093613 yards; 3.280840 feet; 39.37008 inches

Mile (statute)—1.609344 kilometers

Ounce (U. S. fluid)—1.804688 cubic inches; 29.573730 cubic centimeters

Ounce (avoirdupois)—28.349523 grams

Ounce (apothecary or troy)—31.103486 grams

Pint (U. S. liquid)—0.473163 liter; 473.17647 cubic centimeters

Pound (avoirdupois)—0.453592 kilogram; 453.59237 grams

Pound (apothecary or troy)—0.3732417 kilogram, 373.24172 grams

Quart (U. S. dry)—1.10119 liters

Quart (liquid)—0.946326 liter

Radian—57.295779 degrees

Rod—5.0292 meters

Square centimeter—0.155000 square inch

Square foot—0.09290304 square meter

Square inch—645.16 square millimeters

Square meter—10.763910 square feet

Square yard—0.836127 square meter

Ton (short)—907.18474 kilograms

Yard—0.9144 meter

Table A-6. Conversion of percent transmittance, absorbance, and Klett units

% T	A	Klett units	% T	A	Klett units
100	0.000	0.0	90	0.046	23.0
99	0.004	2.0	89	0.051	25.5
98	0.009	4.5	88	0.056	28.0
97	0.013	6.5	87	0.061	30.5
96	0.018	9.0	86	0.066	33.0
95	0.022	11.0	85	0.071	35.5
94	0.027	13.5	84	0.076	38.0
93	0.032	16.0	83	0.081	40.5
92	0.036	18.0	82	0.086	43.0
91	0.041	20.5	81	0.092	46.0

$$A = \log \frac{100}{\%\ T} = \frac{\text{Klett units}}{500}.$$

Continued.

Table A-6. Conversion of percent transmittance, absorbance, and Klett units—cont'd

% T	A	Klett units	% T	A	Klett units
80	0.097	48.5	40	0.398	199.0
79	0.102	51.0	39	0.409	204.5
78	0.108	54.0	38	0.420	210.0
77	0.114	57.0	37	0.432	216.0
76	0.119	59.5	36	0.444	222.0
75	0.125	62.5	35	0.456	228.0
74	0.131	65.5	34	0.469	234.5
73	0.137	68.5	33	0.482	241.0
72	0.143	71.5	32	0.495	247.5
71	0.149	74.5	31	0.509	254.5
70	0.155	77.5	30	0.523	261.5
69	0.161	80.5	29	0.538	269.0
68	0.168	84.0	28	0.553	276.5
67	0.174	87.0	27	0.569	284.5
66	0.181	90.5	26	0.585	292.5
65	0.187	93.5	25	0.602	301.0
64	0.194	97.0	24	0.620	310.0
63	0.201	100.5	23	0.638	319.0
62	0.208	104.0	22	0.658	329.0
61	0.215	107.5	21	0.678	339.0
60	0.222	111.0	20	0.699	349.5
59	0.229	114.5	19	0.721	360.5
58	0.237	118.5	18	0.745	372.5
57	0.244	122.0	17	0.770	385.0
56	0.252	126.0	16	0.796	398.0
55	0.260	130.0	15	0.824	412.0
54	0.268	134.0	14	0.854	427.0
53	0.276	138.0	13	0.886	443.0
52	0.284	142.0	12	0.921	460.5
51	0.292	146.0	11	0.959	479.5
50	0.301	150.5	10	1.000	500.0
49	0.310	155.0	9	1.046	523.0
48	0.319	159.5	8	1.097	548.5
47	0.328	164.0	7	1.155	577.5
46	0.337	168.5	6	1.222	611.0
45	0.347	173.5	5	1.301	650.5
44	0.357	178.5	4	1.398	699.0
43	0.367	183.5	3	1.523	761.5
42	0.377	188.5	2	1.699	849.5
41	0.387	193.5	1	2.000	1000.0

Table A-7. Percent transmittance—absorbance conversion table

% T	(.00)	(.25)	(.50)	(.75)	% T	(.00)	(.25)	(.50)	(.75)
1	2.000	1.903	1.824	1.757	51	.292	.290	.288	.286
2	1.699	1.648	1.602	1.561	52	.284	.282	.280	.278
3	1.523	1.488	1.456	1.426	53	.276	.274	.272	.270
4	1.398	1.372	1.347	1.323	54	.268	.266	.264	.262
5	1.301	1.280	1.260	1.240	55	.260	.258	.256	.254
6	1.222	1.204	1.187	1.171	56	.252	.250	.248	.246
7	1.155	1.140	1.126	1.112	57	.244	.242	.240	.238
8	1.097	1.083	1.071	1.059	58	.237	.235	.233	.231
9	1.046	1.034	1.022	1.011	59	.229	.227	.226	.224
10	1.000	.989	.979	.969	60	.222	.220	.218	.216
11	.959	.949	.939	.930	61	.215	.213	.211	.209
12	.921	.912	.903	.894	62	.208	.206	.204	.202
13	.886	.878	.870	.862	63	.201	.199	.197	.196
14	.854	.846	.838	.831	64	.194	.192	.191	.189
15	.824	.817	.810	.803	65	.187	.186	.184	.182
16	.796	.789	.782	.776	66	.180	.179	.177	.176
17	.770	.763	.757	.751	67	.174	.172	.171	.169
18	.745	.739	.733	.727	68	.168	.166	.164	.163
19	.721	.716	.710	.703	69	.161	.160	.158	.157
20	.699	.694	.688	.683	70	.155	.153	.152	.150
21	.678	.673	.668	.663	71	.149	.147	.146	.144
22	.658	.653	.648	.643	72	.143	.141	.140	.138
23	.638	.634	.629	.624	73	.137	.135	.134	.132
24	.620	.615	.611	.606	74	.131	.129	.128	.126
25	.602	.598	.594	.589	75	.125	.124	.122	.121
26	.585	.581	.577	.573	76	.119	.118	.116	.115
27	.569	.565	.561	.557	77	.114	.112	.111	.109
28	.553	.549	.545	.542	78	.108	.107	.105	.104
29	.538	.534	.530	.527	79	.102	.101	.100	.098
30	.523	.520	.516	.512	80	.097	.096	.094	.093
31	.509	.503	.502	.498	81	.092	.090	.089	.088
32	.495	.491	.488	.485	82	.086	.085	.084	.082
33	.482	.478	.475	.472	83	.081	.080	.078	.077
34	.469	.465	.462	.459	84	.076	.074	.073	.072
35	.456	.453	.450	.447	85	.071	.069	.068	.067
36	.444	.441	.438	.435	86	.066	.064	.063	.062
37	.432	.429	.426	.423	87	.061	.059	.058	.057
38	.420	.417	.414	.412	88	.056	.054	.053	.052
39	.409	.406	.403	.401	89	.051	.049	.048	.047
40	.398	.395	.392	.390	90	.046	.045	.043	.042
41	.387	.385	.382	.380	91	.041	.040	.039	.037
42	.377	.374	.372	.369	92	.036	.035	.034	.033
43	.367	.364	.362	.359	93	.032	.030	.029	.028
44	.357	.354	.352	.349	94	.027	.026	.025	.024
45	.347	.344	.342	.340	95	.022	.021	.020	.019
46	.337	.335	.332	.330	96	.018	.017	.016	.014
47	.328	.325	.323	.321	97	.013	.012	.011	.010
48	.319	.317	.314	.312	98	.009	.008	.007	.006
49	.310	.308	.305	.303	99	.004	.003	.002	.001
50	.301	.299	.297	.295	100	.000	.000	.000	.000

Table A-8. Square roots of numbers from 1 to 100 in steps of 0.2

n	√n	n	√n	n	√n	n	√n	n	√n	n	√n	n	√n	n	√n	n	√n	n	√n
1.0	1.00	6.0	2.45	11.0	3.32	16.0	4.00	21.0	4.58	26.0	5.10	31.0	5.57	36.0	6.00	41.0	6.40	46.0	6.78
1.2	1.09	6.2	2.49	11.2	3.35	16.2	4.02	21.2	4.60	26.2	5.12	31.2	5.59	36.2	6.02	41.2	6.42	46.2	6.80
1.4	1.18	6.4	2.53	11.4	3.38	16.4	4.05	21.4	4.63	26.4	5.14	31.4	5.60	36.4	6.03	41.4	6.43	46.4	6.81
1.6	1.26	6.6	2.57	11.6	3.41	16.6	4.07	21.6	4.65	26.6	5.16	31.6	5.62	36.6	6.05	41.6	6.45	46.6	6.83
1.8	1.34	6.8	2.61	11.8	3.44	16.8	4.10	21.8	4.67	26.8	5.18	31.8	5.64	36.8	6.07	41.8	6.47	46.8	6.84
2.0	1.41	7.0	2.65	12.0	3.46	17.0	4.12	22.0	4.69	27.0	5.20	32.0	5.66	37.0	6.08	42.0	6.48	47.0	6.86
2.2	1.48	7.2	2.68	12.2	3.49	17.2	4.15	22.2	4.71	27.2	5.22	32.2	5.67	37.2	6.10	42.2	6.50	47.2	6.87
2.4	1.55	7.4	2.72	12.4	3.52	17.4	4.17	22.4	4.73	27.4	5.23	32.4	5.69	37.4	6.12	42.4	6.51	47.4	6.88
2.6	1.61	7.6	2.76	12.6	3.55	17.6	4.20	22.6	4.75	27.6	5.25	32.6	5.71	37.6	6.13	42.6	6.53	47.6	6.90
2.8	1.67	7.8	2.79	12.8	3.58	17.8	4.22	22.8	4.77	27.8	5.27	32.8	5.73	37.8	6.15	42.8	6.54	47.8	6.91
3.0	1.73	8.0	2.83	13.0	3.61	18.0	4.24	23.0	4.80	28.0	5.29	33.0	5.74	38.0	6.16	43.0	6.56	48.0	6.93
3.2	1.79	8.2	2.86	13.2	3.63	18.2	4.27	23.2	4.82	28.2	5.31	33.2	5.76	38.2	6.18	43.2	6.57	48.2	6.94
3.4	1.84	8.4	2.90	13.4	3.66	18.4	4.29	23.4	4.84	28.4	5.33	33.4	5.78	38.4	6.20	43.4	6.59	48.4	6.96
3.6	1.90	8.6	2.93	13.6	3.69	18.6	4.31	23.6	4.86	28.6	5.35	33.6	5.80	38.6	6.21	43.6	6.60	48.6	6.97
3.8	1.95	8.8	2.97	13.8	3.71	18.8	4.34	23.8	4.88	28.8	5.37	33.8	5.81	38.8	6.23	43.8	6.62	48.8	6.99
4.0	2.0	9.0	3.0	14.0	3.74	19.0	4.36	24.0	4.90	29.0	5.39	34.0	5.83	39.0	6.24	44.0	6.63	49.0	7.00
4.2	2.05	9.2	3.03	14.2	3.77	19.2	4.38	24.2	4.92	29.2	5.40	34.2	5.85	39.2	6.26	44.2	6.65	49.2	7.01
4.4	2.10	9.4	3.07	14.4	3.79	19.4	4.40	24.4	4.94	29.4	5.42	34.4	5.87	39.4	6.28	44.4	6.66	49.4	7.03
4.6	2.14	9.6	3.10	14.6	3.82	19.6	4.43	24.6	4.96	29.6	5.44	34.6	5.88	39.6	6.29	44.6	6.68	49.6	7.04
4.8	2.19	9.8	3.13	14.8	3.85	19.8	4.45	24.8	4.98	29.8	5.46	34.8	5.90	39.8	6.31	44.8	6.69	49.8	7.06
5.0	2.24	10.0	3.16	15.0	3.87	20.0	4.47	25.0	5.00	30.0	5.48	35.0	5.92	40.0	6.32	45.0	6.71	50.0	7.07
5.2	2.28	10.2	3.19	15.2	3.90	20.2	4.49	25.2	5.02	30.2	5.50	35.2	5.93	40.2	6.34	45.2	6.72	50.2	7.09
5.4	2.32	10.4	3.22	15.4	3.92	20.4	4.52	25.4	5.04	30.4	5.51	35.4	5.95	40.4	6.36	45.4	6.74	50.4	7.10
5.6	2.37	10.6	3.26	15.6	3.95	20.6	4.54	25.6	5.06	30.6	5.53	35.6	5.97	40.6	6.37	45.6	6.75	50.6	7.11
5.8	2.41	10.8	3.29	15.8	3.97	20.8	4.56	25.8	5.08	30.8	5.55	35.8	5.98	40.8	6.39	45.8	6.77	50.8	7.13

n	$\sqrt{n}$	n	$\sqrt{n}$	n	$\sqrt{n}$	n	$\sqrt{n}$	n	$\sqrt{n}$	n	$\sqrt{n}$	n	$\sqrt{n}$	n	$\sqrt{n}$	n	$\sqrt{n}$	n	$\sqrt{n}$
51.0	7.14	56.0	7.48	61.0	7.81	66.0	8.12	71.0	8.43	76.0	8.72	81.0	9.00	86.0	9.27	91.0	9.54	96.0	9.80
51.2	7.16	56.2	7.50	61.2	7.82	66.2	8.14	71.2	8.44	76.2	8.73	81.2	9.01	86.2	9.28	91.2	9.55	96.2	9.81
51.4	7.17	56.4	7.51	61.4	7.84	66.4	8.15	71.4	8.45	76.4	8.74	81.4	9.02	86.4	9.30	91.4	9.56	96.4	9.82
51.6	7.18	56.6	7.52	61.6	7.85	66.6	8.16	71.6	8.46	76.6	8.75	81.6	9.03	86.6	9.31	91.6	9.57	96.6	9.83
51.8	7.20	56.8	7.54	61.8	7.86	66.8	8.17	71.8	8.47	76.8	8.76	81.8	9.04	86.8	9.32	91.8	9.58	96.8	9.84
52.0	7.21	57.0	7.55	62.0	7.87	67.0	8.19	72.0	8.49	77.0	8.77	82.0	9.06	87.0	9.33	92.0	9.59	97.0	9.85
52.2	7.22	57.2	7.56	62.2	7.89	67.2	8.20	72.2	8.50	77.2	8.79	82.2	9.07	87.2	9.34	92.2	9.60	97.2	9.86
52.4	7.24	57.4	7.58	62.4	7.90	67.4	8.21	72.4	8.51	77.4	8.80	82.4	9.08	87.4	9.35	92.4	9.61	97.4	9.87
52.6	7.25	57.6	7.59	62.6	7.91	67.6	8.22	72.6	8.52	77.6	8.81	82.6	9.09	87.6	9.36	92.6	9.62	97.6	9.88
52.8	7.27	57.8	7.60	62.8	7.92	67.8	8.23	72.8	8.53	77.8	8.82	82.8	9.10	87.8	9.37	92.8	9.63	97.8	9.89
53.0	7.28	58.0	7.62	63.0	7.94	68.0	8.25	73.0	8.54	78.0	8.83	83.0	9.11	88.0	9.38	93.0	9.64	98.0	9.90
53.2	7.29	58.2	7.63	63.2	7.95	68.2	8.26	73.2	8.56	78.2	8.84	83.2	9.12	88.2	9.39	93.2	9.65	98.2	9.91
53.4	7.31	58.4	7.64	63.4	7.96	68.4	8.27	73.4	8.57	78.4	8.85	83.4	9.13	88.4	9.40	93.4	9.66	98.4	9.92
53.6	7.32	58.6	7.66	63.6	7.97	68.6	8.28	73.6	8.58	78.6	8.87	83.6	9.14	88.6	9.41	93.6	9.67	98.6	9.93
53.8	7.33	58.8	7.67	63.8	7.99	68.8	8.29	73.8	8.59	78.8	8.88	83.8	9.15	88.8	9.42	93.8	9.68	98.8	9.94
54.0	7.35	59.0	7.68	64.0	8.00	69.0	8.31	74.0	8.60	79.0	8.89	84.0	9.17	89.0	9.43	94.0	9.70	99.0	9.95
54.2	7.36	59.2	7.69	64.2	8.01	69.2	8.32	74.2	8.61	79.2	8.90	84.2	9.18	89.2	9.44	94.2	9.71	99.2	9.96
54.4	7.38	59.4	7.71	64.4	8.02	69.4	8.33	74.4	8.63	79.4	8.91	84.4	9.19	89.4	9.46	94.4	9.72	99.4	9.97
54.6	7.39	59.6	7.72	64.6	8.04	69.6	8.34	74.6	8.64	79.6	8.92	84.6	9.20	89.6	9.47	94.6	9.73	99.6	9.98
54.8	7.40	59.8	7.73	64.8	8.05	69.8	8.35	74.8	8.65	79.8	8.93	84.8	9.21	89.8	9.48	94.8	9.74	99.8	9.99
55.0	7.42	60.0	7.75	65.0	8.06	70.0	8.37	75.0	8.66	80.0	8.94	85.0	9.22	90.0	9.49	95.0	9.75	100.0	10.00
55.2	7.43	60.2	7.76	65.2	8.07	70.2	8.38	75.2	8.67	80.2	8.96	85.2	9.23	90.2	9.50	95.2	9.76		
55.4	7.44	60.4	7.77	65.4	8.09	70.4	8.39	75.4	8.68	80.4	8.97	85.4	9.24	90.4	9.51	95.4	9.77		
55.6	7.46	60.6	7.78	65.6	8.10	70.6	8.40	75.6	8.69	80.6	8.98	85.6	9.25	90.6	9.52	95.6	9.78		
55.8	7.47	60.8	7.80	65.8	8.11	70.8	8.41	75.8	8.71	80.8	8.99	85.8	9.26	90.8	9.53	95.8	9.79		

Table A-9. Interpretation of chemical resistance*

The chart should be taken as a general guide only. It states that, for most compounds of the class indicated, the specific material is rated either Excellent, Good, Fair, or Not recommended.

Because so many factors can affect the chemical resistance of a given product, we recommend you test under your own conditions if any doubt exists. The combination of compounds of two or more classes may cause an undesirable chemical effect. Other factors affecting chemical resistance include temperature, pressure and other stresses, length of exposure, and concentration of the chemical. As the maximum useful temperature of the plastic is approached, resistance to attack decreases.

Chemical resistance chart

Class of substances	CPE	LPE	PA	PP	TPX†	FEP	PC	NO	PS	SA	PVC
Acids, inorganic	E	E	E	E	E	E	E	G	N	E	G
Acids, organic	E	E	E	E	E	E	G	E	G	E	G
Alcohols	E	E	E	E	E	E	G	E	G	G	G
Aldehydes	G	G	G	G	G	E	F	G	N	F	F
Amines	G	G	G	G	G	E	N	F	G	G	N
Bases	E	E	E	E	E	E	N	E	G	E	E
Dimethyl sulfoxide (DMSO)	E	E	E	E	E	E	N	E	N	N	N
Esters	E	E	E	E	E	E	N	F	N	N	F
Ethers	G	G	G	G	G	E	F	N	F	N	F
Foods	E	E	E	E	E	E	E	G	E	G	G
Glycols	E	E	E	E	E	E	G	E	G	G	F
Hydrocarbons, aliphatic	G	G	G	G	G	E	F	G	N	E	F
Hydrocarbons, aromatic	G	G	G	G	F	E	N	N	N	N	N
Hydrocarbons, halogenated	G	G	G	G	F	E	N	N	N	N	N
Ketones	G	G	G	G	G	E	N	N	N	N	N
Mineral oil	E	E	E	E	E	E	E	E	G	G	E
Oils, essential	G	G	G	G	G	E	G	F	N	F	N
Oils, lubricating	G	E	E	E	E	E	G	E	G	G	E

Oils, vegetable	E	E	E	E	E	E	E	E	G	E	E
Proteins, unhydrolyzed	E	E	E	E	E	E	E	G	G	E	G
Salts	E	E	E	E	E	E	E	E	E	E	E
Silicones	G	E	E	E	E	E	E	E	G	G	G
Water	E	E	E	E	E	E	E	E	E	E	E

E = Excellent. Long exposures (up to one year) at room temperature have no effect.

G = Good. Short exposures (less than 24 hours) at room temperature cause no damge.

F = Fair. Short exposures at room temperature cause little or no damage under unstressed conditions.

N = Not recommended. Short exposures may cause permanent damage.

Resins code

CPE—Conventional polyethylene
LPE—Linear polyethylene
PA—Polyallomer
PP—Polypropylene
TPX—Polymethylpentene
FEP—Teflon FEP
PC—Polycarbonate
NO—Noryl
PS—General purpose polystyrene
SA—Styrene-acrylonitrile
PVC—Polyvinyl chloride

Physical properties

	CPE	LPE	PA	PP	TPX†	FEP	PC	NO	PS	SA	PVC
Temperature limit, °C	80	120	130	135	175	205	135	135	70	95	70
Specific gravity	0.92*	0.95	0.90	0.90	0.83	2.15	1.20	1.06	1.07	1.07	1.34
Tensile strength, psi	2000	4000	2900	5000	4000	3000	8000	9600	6000	11,000	6500
Brittleness temperature, °C	−100	−100	−40	0		−270	−135		‡	−25	−30
Water absorption, %	<0.01	<0.01	<0.02	<0.02	<0.01	<0.01	0.35	0.07	0.05	0.23	0.06
Flexibility	Excellent	Rigid	Slight	Rigid	Rigid	Excellent	Rigid	Rigid	Rigid	Rigid	Rigid
Transparency	Translucent	Opaque	Translucent	Translucent	Clear	Translucent	Clear	Opaque	Clear	Clear	Clear
Relative O_2 permeability	0.40	0.08	0.20	0.11	2.0	0.59	0.15		0.11	0.03	0.01
Autoclavable	No	With caution	Yes	Yes	Yes	Yes	Yes	Yes	No	No	No

*Courtesy Nalge Co.

†Trademark of Imperial Chemical Industries, Ltd., for their brand of methylpentene polymer.

‡Normally somewhat brittle at room temperatures.

Table A-10. Values of *t* for given degrees of freedom *(n)* and at specified levels of significance *(P)**

n	Level of significance (P) .90	.80	.70	.60	.50	.40	.30	.25	.20	.10	.05	.025	.02	.01	.005	.001	n
1	.158	.325	.510	.727	1.000	1.376	1.963	2.414	3.078	6.314	12.706	25.452	31.821	63.657	127.32	636.619	1
2	.142	.289	.445	.617	.816	1.061	1.386	1.604	1.886	2.920	4.303	6.205	6.965	9.925	14.089	31.598	2
3	.137	.277	.424	.584	.765	.978	1.250	1.423	1.638	2.353	3.182	4.176	4.541	5.841	7.453	12.941	3
4	.134	.271	.414	.569	.741	.941	1.190	1.344	1.533	2.132	2.776	3.495	3.747	4.604	5.598	8.610	4
5	.132	.267	.408	.559	.727	.920	1.156	1.301	1.476	2.015	2.571	3.163	3.365	4.032	4.773	6.859	5
6	.131	.265	.404	.553	.718	.906	1.134	1.273	1.440	1.943	2.447	2.969	3.143	3.707	4.317	5.959	6
7	.130	.263	.402	.549	.711	.896	1.119	1.254	1.415	1.895	2.365	2.841	2.998	3.499	4.029	5.405	7
8	.130	.262	.399	.546	.706	.889	1.108	1.240	1.397	1.860	2.306	2.752	2.896	3.355	3.832	5.041	8
9	.129	.261	.398	.543	.703	.883	1.100	1.230	1.383	1.833	2.262	2.685	2.821	3.250	3.690	4.781	9
10	.129	.260	.397	.542	.700	.879	1.093	1.221	1.372	1.812	2.228	2.634	2.764	3.169	3.581	4.587	10
11	.129	.260	.396	.540	.697	.876	1.088	1.214	1.363	1.796	2.201	2.593	2.718	3.106	3.497	4.437	11
12	.128	.259	.395	.539	.695	.873	1.083	1.209	1.356	1.786	2.179	2.560	2.681	3.055	3.428	4.318	12
13	.128	.259	.394	.538	.694	.870	1.079	1.204	1.350	1.771	2.160	2.533	2.650	3.012	3.372	4.221	13
14	.128	.258	.393	.537	.692	.868	1.076	1.200	1.345	1.761	2.145	2.510	2.624	2.977	3.326	4.140	14
15	.128	.258	.393	.536	.691	.866	1.074	1.197	1.341	1.753	2.131	2.490	2.602	2.947	3.286	4.073	15
16	.128	.258	.392	.535	.690	.865	1.071	1.194	1.337	1.746	2.120	2.473	2.583	2.921	3.252	4.015	16
17	.128	.257	.392	.534	.689	.863	1.069	1.191	1.333	1.740	2.110	2.458	2.567	2.898	3.222	3.965	17
18	.127	.257	.392	.534	.688	.862	1.067	1.189	1.330	1.734	2.101	2.445	2.552	2.878	3.197	3.922	18
19	.127	.257	.391	.533	.688	.861	1.066	1.187	1.328	1.729	2.093	2.433	2.539	2.861	3.174	3.883	19
20	.127	.257	.391	.533	.687	.860	1.064	1.185	1.325	1.725	2.086	2.423	2.528	2.845	3.153	3.850	20
21	.127	.257	.391	.532	.686	.859	1.063	1.183	1.323	1.721	2.080	2.414	2.518	2.831	3.135	3.819	21
22	.127	.256	.390	.532	.686	.858	1.061	1.182	1.321	1.717	2.074	2.406	2.508	2.819	3.119	3.792	22
23	.127	.256	.390	.532	.685	.858	1.060	1.180	1.319	1.714	2.069	2.398	2.500	2.807	3.104	3.767	23
24	.127	.256	.390	.531	.685	.857	1.059	1.179	1.318	1.711	2.064	2.391	2.492	2.797	3.090	3.745	24
25	.127	.256	.390	.531	.684	.856	1.058	1.178	1.316	1.708	2.060	2.385	2.485	2.787	3.078	3.725	25
26	.127	.256	.390	.531	.684	.856	1.058	1.177	1.315	1.706	2.056	2.379	2.479	2.779	3.067	3.707	26
27	.127	.256	.389	.531	.684	.855	1.057	1.176	1.314	1.703	2.052	2.373	2.473	2.771	3.056	3.690	27
28	.127	.256	.389	.530	.683	.855	1.056	1.175	1.313	1.701	2.048	2.368	2.467	2.763	3.047	3.674	28
29	.127	.256	.389	.530	.683	.854	1.055	1.174	1.311	1.699	2.045	2.364	2.462	2.756	3.038	3.659	29
30	.127	.256	.389	.530	.683	.854	1.055	1.173	1.310	1.697	2.042	2.360	2.457	2.750	3.030	3.646	30
40	.126	.255	.388	.529	.681	.851	1.050	1.167	1.303	1.684	2.021	2.329	2.423	2.704	2.971	3.551	40
60	.126	.254	.387	.527	.679	.848	1.046	1.162	1.296	1.671	2.000	2.299	2.390	2.660	2.915	3.460	60
120	.126	.254	.386	.526	.677	.845	1.041	1.156	1.289	1.658	1.980	2.270	2.358	2.617	2.860	3.373	120
∞	.126	.253	.385	.524	.674	.842	1.036	1.150	1.282	1.645	1.960	2.241	2.326	2.576	2.807	3.291	∞

*From Fisher, R. A., and Yates, F.: Statistical tables for biological, agricultural, and medical research, ed. 6, Edinburgh, 1963, Oliver & Boyd; and from Merrington, M.: Table of percentage points of the *t*-distribution, Biometrika 32:300, 1942.

Table A-11. Greek alphabet

Greek letter	Greek name	Greek letter	Greek name
Α α	Alpha	Ν ν	Nu
Β β	Beta	Ξ ξ	Xi
Γ γ	Gamma	Ο ο	Omicron
Δ δ	Delta	Π π	Pi
Ε ε	Epsilon	Ρ ρ	Rho
Ζ ζ	Zeta	Σ σ ς	Sigma
Η η	Eta	Τ τ	Tau
Θ θ	Theta	Υ υ	Upsilon
Ι ι	Iota	Φ φ	Phi
Κ κ	Kappa	Χ χ	Chi
Λ λ	Lambda	Ψ ψ	Psi
Μ μ	Mu	Ω ω	Omega

• B

• Fire precautions, chemical hazards, and antidotes of poisons

FIRE PRECAUTIONS AND CHEMICAL HAZARDS

acetone. Dilute with a spray of water to avoid spread of burning liquid. Use suitable gas mask.

alcohol. See acetone.

ammonia. Use water and dilute acid. Use a suitable gas mask.

benzol or benzene. Use water to cool containers that are endangered; extinguish flame with sand, earth, fire-foam or carbon tetrachloride fire extinguishers. Use a suitable gas mask.

calcium carbide. Do not use water as this generates acetylene, an inflammable and explosive gas; cut off electric current to avoid ignition of gas. Remove containers to a dry place. Use a gas mask.

carbon disulfide. Use water to cool containers that are endangered; extinguish blaze with sand, earth, fire-foam or carbon tetrachloride fire extinguishers. Use a suitable gas mask.

carbon tetrachloride. When a carbon tetrachloride type extinguisher is used on a fire in a confined space, the fire should be attacked from outside the enclosure, if possible, or the area should be vacated immediately after the fire is out. No one should return to the enclosure until the air is cleared of smoke and fumes. These precautions should be observed regardless of the means by which the fire is extinguished, however, since fire in a confined space rapidly produces a toxic atmosphere.

Do not put carbon tetrachloride on a sodium fire; violent explosions may be caused.

celluloid. Use large volumes of water and sand. The smoke contains oxides of nitrogen which are injurious. Use a suitable gas mask.

chlorine. Spray with water. The pungent nature of the gas makes the use of a gas mask imperative.

collodion. See carbon disulfide.

ether. See carbon disulfide.

gasoline See carbon disulfide.

hydrochloric acid. Use large volumes of water also chalk or soda. Use a gas mask.

hydrocyanic or prussic acid. A suitable gas mask is essential because of the extremely poisonous nature of the vapors. Provide ventilation.

lacquer solvents. See carbon disulfide.

magnesium. Do not use water. Use sand or earth to extinguish flames. Remove containers to a dry place.

nitric acid and oxides of nitrogen. Use large volumes of water. Do not use sand or earth. Use a gas mask.

potassium. Do not use water. Remove containers to dry place. Extinguish flames with sand or earth. For storage, potassium is kept immersed in petroleum.

potassium hydroxide. Use large volumes of water or dilute acids.

phosphorus. Use water and wet sand. Use gas mask. For storage, white phosphorus must be kept immersed in water. Red phosphorus is less dangerous.

sodium. See potassium

sodium hydroxide. See potassium hydroxide.

sulfur. Extinguish with water or sand. Use a gas mask.

sulfuric acid. See hydrochloric acid.

turpentine. See acetone.

ANTIDOTES OF POISONS

acetic acid. Emetics, magnesia, chalk, soap, oil.

acetylene. Same as for carbon monoxide.

arsenic, rat poison, paris green. Milk, raw egg, sweet oil, lime water, flour and water.

carbolic acid. Any soluble nontoxic sulfate, after provoking vomiting with zinc sulfate; uncooked white of egg in abundance, milk of lime, olive or castor oil with magnesia in suspension, ice, washing the stomach with equal parts water and vinegar; give alcohol or whiskey or about 4 fluid ounces of camphorated oil at one dose.

carbon monoxide. Remove to fresh air immediately and call for pulmotor; apply artificial respiration for at least 1 hour or until the pulmotor arrives. Administration of oxygen containing 5% of carbon dioxide is beneficial; inhalation of ammonia or amyl nitrite is often of value.

chloroform, chloral, ether. Dash cold water on head and chest, artificial respiration.

ethylene. Same as for carbon monoxide.

gas (illuminating). Same as for carbon monoxide.

hydrochloric acid. Magnesia, alkali carbonates, albumin, ice.

hydrocyanic or prussic acid. Hydrogen peroxide internally, and artificial respiration, breathing ammonia or chlorine from chlorinated lime, ferrous sulfate followed by potassium carbonate, emetics, warmth.

iodine. Emetics, stomach siphon, starchy foods in abundance, sodium thiosulfate.

lead acetate. Emetics, stomach siphon, sodium, potassium or magnesium sulfates, milk, albumin.

mercuric chloride or corrosive sublimate. Zinc sulfate, emetics, stomach siphon, raw white of egg, milk, chalk, castor oil, table salt, reduced iron.

nitrate of silver. Salt and water.

nitric acid. Same as for hydrochloric acid.

opium, morphine, laudanum, paregoric, etc. Strong coffee, hot bath. Keep awake and moving at any cost.

phosphoric acid. Same as for hydrochloric acid.

sodium hydroxide or potassium hydroxide. Vinegar, lemon juice, orange juice, oil, milk.

sulfuric acid. Same as for hydrochloric acid with the addition of soap or oil.

sulfurous acid or sulfur dioxide. Mustard plaster on chest, narcotics, expectorants.

wood alcohol (methyl alcohol or methanol). Emetic or wash out stomach (stomach tube) with a solution of 10 grains of sodium citrate per ounce of water. Give milk, raw white of egg, or flour in water; purgative of magnesium sulfate (15 Gm); stimulate, and combat collapse. In case of cardiac or pulmonary failure use artificial respiration. Physicians may administer atropine, digitalis, or strychnine as stimulants; to cause perspiration and elimination of the poison use 0.1 grain of pilocarpine hydrochloride.

• C

• Basic information for the student

ANALYTICAL BALANCE

An analytical balance is a highly sensitive instrument. With each analytical balance there is a set of weights. Both the balance and the weights should be handled with the utmost care.

A typical set of weights is illustrated in Fig. C-1. There are both gram weights and milligram weights. These weights are placed on the right-hand pan of the balance.

In the box of weights, there is also a small piece of bent wire that is known as a rider. When the rider is properly adjusted on the beam of the balance, it enables the analyst to weigh to the fourth decimal place. For example, it is thus possible to weigh 5.1001 g of substance. The chainomatic balances and the new automatic balances enable an analyst to weigh easily to the fourth decimal place. However, the analytical balance in Fig. C-1 does just as well. The following discussion concerns mainly the balance shown in Fig. C-1.

In any given weighing, such as 5.1001 g, the numbers on the left side of the decimal point represent grams, whereas the numbers on the right-hand side of the decimal point represent milligrams. Thus, in weighing 5.1001 g, we would use a 5 g weight and a 100 mg weight. Step 5 explains how the fourth decimal place is obtained.

In weighing 6.155 g, we would use the following weights:

Gram weights	*Milligram weights*
1 g	100 mg
5 g	50 mg
	5 mg
6 g	155 mg

An analytical balance is so sensitive that a 1 mg weight will cause a deflection of the pointer. The degree of deflection produced by the 1 mg weight is called the sensitivity of the balance. For example, if the pointer moves 3 units on the scale when a 1 mg weight is added to either pan, the sensitivity of the balance is 3.

This instrument should be placed where there is controlled temperature and there are no drafts, corroding fumes, dust, or vibrations. Be sure to set up away from ovens, radiators, or cold, damp areas, for such temperature changes may set up a convection current within the balance, affecting the swings of the beam and giving what is called an irregular falling off. Some laboratories have temperature and humidity controlled rooms as "weighing rooms." Although the "weighing room" is ideal, it is not practical for the average hospital laboratory or doctor's office. Place the balance in the cleanest and most draft-free area of the room, the part having the least vibration and temperature change. Extreme caution and observation should be made to eliminate errors caused by the above factors.

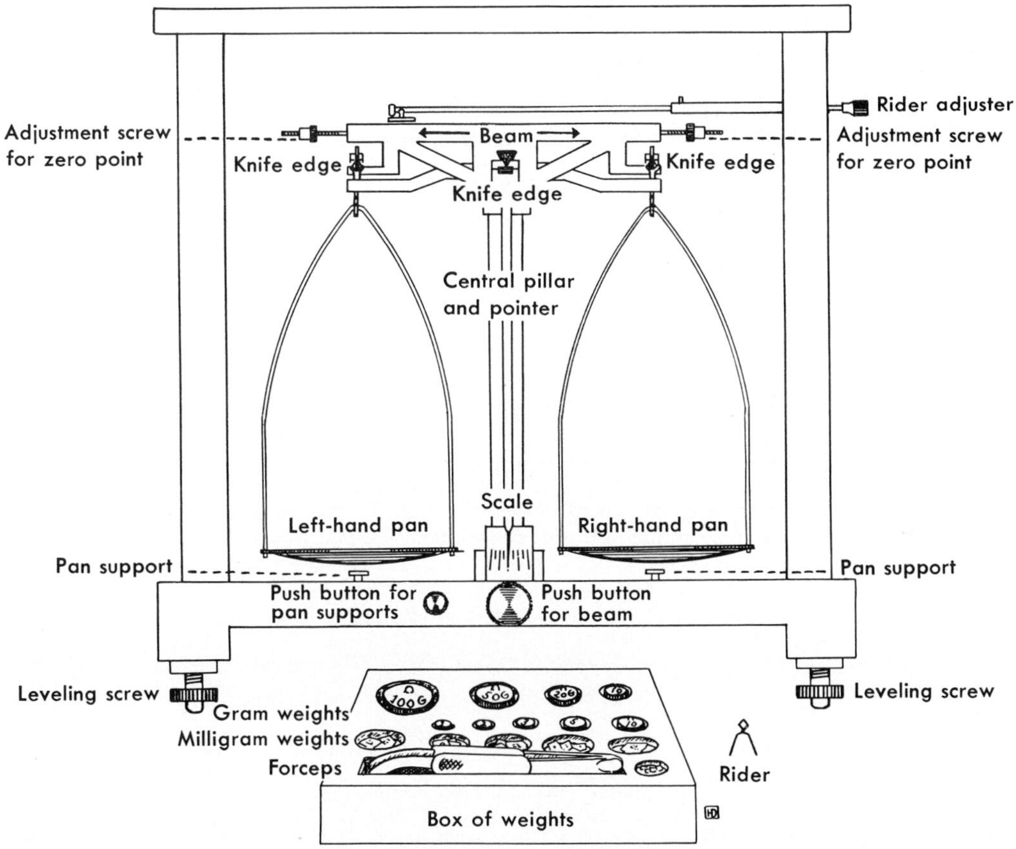

Fig. C-1. Analytical balance.

Before operating an analytical balance, the student should study the parts of the balance as illustrated in Fig. C-1.

During all weighing procedures, keep the following rules in mind.

1. *Always* handle the weights with a pair of forceps. This keeps them free from moisture and dirt.
2. *Never* put anything on or take anything off the pans unless they are resting on the pan supports. This preserves the knife edges on which the beam and stirrups rest and thus helps to maintain the accuracy of the balance.
3. Close the glass door of the balance before the actual weighing. This prevents drafts from moving the pans of the balance.

The steps outlined below for the operation of an analytical balance have been followed with good results by many analysts. However, they are by no means the final word, as the technique of weighing varies considerably. The student should get firsthand instruction from the chemist or supervisor in charge of the department.

1. Check to see if the balance is resting in a level position. The "level indicator" is usually located near the central pillar. If the balance is not level, adjust the leveling screws. (This adjustment of the level indicator rarely needs to be made.)
2. Carefully note which push button is used to lower the beam and which

push button is used to lower the pan supports. Now check the zero point by *carefully* lowering the beam onto the the knife edges and then *slowly* lowering the pan supports. The pointer should come to rest at the midpoint (zero point) on the scale. If it does not, ask the chemist to adjust the screw at the end of the beam. Now raise the pan supports and beam. (This adjustment of the zero point also rarely needs to be made.)

3. Many people have a problem at first with parallax in reading the indicator on the index scale. It is difficult to find the exact midpoint of the scale. The student will look at the indicator while squinting with one eye. Then he will "take bead" on the indicator with the other eye, and of course the exact midpoint moves. Next comes squirming in the chair, and the line of sight is definitely off. To obtain a consistent technique in weighing, place a permanent mark or dot on the balance door window, and always take your readings with the eye level so located that the "midpoint" appears directly behind this mark. This is a suggestion once made by Christian Becker, the maker of torsion balances.
4. At some time in your work, you may have to calibrate a set of weights for your own use. The simplest method is to compare the test set against a standard set (certified from the Bureau of Standards). Make a chart and record plus or minus weights of the test set against those of the control set. The tolerances of the weights shown are for Class S_1 weights, which are used for general routine laboratory work. These are tolerances allowed by the National Bureau of Standards.

Weight	*Tolerance*
100 g	1.0 mg
50	0.6
20	0.35
10	0.25
5	0.18
2	0.13
1	0.10
500 mg	0.080
200	0.060
100	0.050
50	0.042
20	0.035
10	0.030
5	0.028
2	0.025
1	0.025

5. When a student is learning to weigh a sample, the fourth decimal place is sometimes avoided because of a lack of confidence. If the balance has a calibrated beam and rider, as previously mentioned, and if this is a 5 mg rider, the third and fourth decimal places may be obtained. With the 5.1001 g quantity the rider would be placed as in Fig. C-2. (NOTE: Detach the rider adjuster from the rider before releasing the beam and pan supports to obtain the weight of specimen.) For 5.1021 g you could either use a 2 mg weight for the 2 and set the rider as illustrated in Fig. C-2 or set the rider at 2.1, giving the

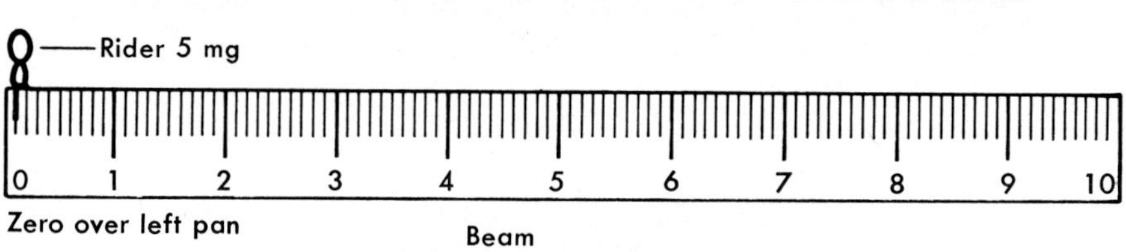

Fig. C-2. Beam of the analytical balance.

0.0021 weight needed. If no rider is available, a method is recommended whereby the fourth decimal place is found by calculation from the known sensitivity of the balance.

6. The sensitivity must first be determined. This is a more or less stable characteristic so the sensitivities are done only at intervals or when the balance is moved to a new location. The zero point in practical weighing need not be the exact zero on the scale, but it should be determined each day when beginning the balance work for the day.
7. Sensitivity means the number of points of swing equivalent to 1 mg. A scale for the pointer numbered 0 to 20 from left to right will simplify calculations. Those that are not numbered may be marked for the individual who has difficulty in recording swings. (Fig. C-3.) A beginner should determine the "rest-point" several times until the final point is reproducible. Due to friction, each swing is somewhat shorter than the preceding one. With the pans empty set the beam in motion by slowly releasing the beam arrest and the pan arrest to get the desired amplitude of swing. The beam may be set into motion by placing the rider on the beam momentarily and then removing it. Do not start the oscillations by tapping the pans or by blowing air into the balance. The first few swings will be erratic due to vibrations caused in releasing the beam and pan arrests. The pointer should swing about ten divisions on the scale. Average the readings of the three consecutive swings on the left and the readings of the two consecutive swings on the right. Average these two results to obtain the midpoint. This is called the equilibrium point of the balance under zero load. Example:

	Left ← Swings	→ Right
	5	15
	6	14
	7	
Totals	18	29
Average	6 (18 ÷ 3)	14.5 (29 ÷ 2)
Total	20.5 (6 + 14.5)	
Midpoint or equilibrium	10.2 (20.5 ÷ 2)	

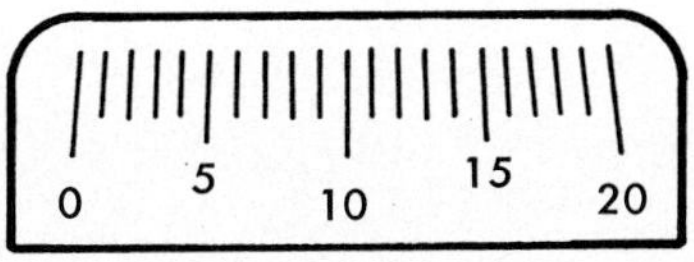

Fig. C-3. Scale of the analytical balance.

8. Now place the rider on 1 mg notch of the beam. Again find the equilibrium or midpoint in the same way. Of course, it will be less than 10.2. Say, for illustration, the midpoint calculates 6.2. Subtracting the 6.2 from the 10.2 gives a sensitivity of 4.0 for the empty balance. Therefore, four points of the swing will equal 1 mg. The sensitivity of the balance is the tangent of the angle through which the equilibrium position of the pointer is displaced by an excess load (usually, 1 mg) on one of the pans. The angle is so small the sensitivity may be taken as the number of scale divisions displaced by the overload of the 1 mg on the pan. The analyst should determine the sensitivity of the balance at 0, 1, 5, 10, 20, and 50 g. Plot the sensitivity in scale divisions against the load and draw a smooth curve through the points. This will give a chart in which sensitivity may be read directly for any given load. Each of the midpoints is obtained as explained in the directions given above.

9. Practice weighing a small beaker and a small rubber stopper together and then separately until you can check with accuracy to the fourth place. For example, if the actual weight of the two articles is 5.1001 g and you have on the pan a 5 g weight and a 100 mg weight, the swing of the pointer is light. Now determine the equilibrium or midpoint. It should average 10.6. The midpoint of the empty balance was 10.2. Using the sensitivity of 4.0, you found that 4 points of the swing = 1 mg (step 8). Therefore, 0.4 of a point of swing = 0.1 mg or 0.0001 g. The total weight would then be 5.1001 g.
10. Watchglasses, weighing bottles, or 3 × 5 inch glassine weighing paper may be used as containers for reagents. *Never* put any reagent directly on the pans. Now determine which is the heavier by lowering the beam and very *gently* partially lowering the pan supports. The heavier container, of course, will be the one that drops lower. After raising the supports and the beam, transfer the lighter containers to the right-hand pan of the balance.
11. Balance the pans by adding weights to the right-hand pan of the balance. The pointer will come to rest at *exactly* midpoint of the scale. Using the forceps, place the desired weights on the right-hand pan.
12. With a clean dry spatula add small portions of reagent to the left-hand pan of the balance. When it appears to the naked eye that an amount equal to the weights has been added, lower the beam and very gently *partially* lower the pan supports. Note whether too much or too little has been added and then raise the pan supports and beam. Using a clean dry spatula, add or remove the reagent from the pan until the pointer can be made to come to rest *exactly* at the midpoint of the scale. When you add or remove the reagent, *make certain* that the pan supports are raised so that the "shock" is taken off the knife edges of the beam. (NOTE: *Never* place excess reagent back into the bottle.)
13. Check the weights to see if you have weighed the amount desired. Check the reagent to see if you have weighed the reagent desired.
14. Place a clean dry funnel in a volumetric flask of the desired volume. Carefully transfer the reagent into the funnel, making certain not to lose any. Wash the reagent down the funnel and into the flask with the solvent.
15. Using forceps, place the weights back in the box.
16. Clean the balance, using a small camel's-hair brush for the pans. Close the door of the balance.

TYPES OF SOLUTIONS

Saturated solutions

When a solid is dissolved in a liquid, the solid is known as the solute and the liquid is known as the solvent. For example, if you add sugar to coffee, the sugar is the solute and the coffee is the solvent.

A saturated solution is one that contains as much solute as it can hold in solution. To illustrate, if you were to add 25 teaspoons of salt to a glass of water, some salt would remain undissolved. The supernatant would then be a saturated solution.

Percent solutions

A percent solution contains a measured amount of the solute in a specified volume of solution. The solute may be either a solid or a liquid. For example, sodium chloride is a solute that is a solid, whereas acetic acid is a solute that is a liquid.

In most percent solutions, distilled water is used as the solvent. However, in some instances, alcohol or another liquid may be used.

Molar solutions

A molar (M) solution contains 1 gram-molecular weight per liter of solution.

Molal solutions

Whereas the *molar* solution contains 1 mole (molecular weight in grams) of solute in 1000 ml of solution, the *molal* solution contains 1 mole of solute in 1000 g (1 kg) of solvent. A more technical definition of molal solution states that it contains 1 gram-molecular weight of solute dissolved in 1 kg of solvent.

An easy way to remember the difference is to note that the molar solution is always on a solute per solution basis and the molal solution is always on a solute per solvent basis.

Normal solutions

Normal solutions are used extensively in the laboratory. Therefore, the student should thoroughly master the material of this section. The discussion covers the following: composition of normal solutions, equivalent weights, and standardi-ization.

Composition of normal solutions

The word "normal" is abbreviated with the capital letter N.

The normality of a solution is expressed with numbers. For example, 0.1 N NaOH is one tenth normal sodium hydroxide.

Unfortunately for the student, some chemists write the numbers as fractions. Thus they may write 0.1 as N/10. Or they may write 0.01 N as N/100. In order to avoid confusion, the student should note the type of labeling used in his particular laboratory.

A replaceable hydrogen atom is one that is capable of ionizing. Hydrochloric acid, HCl, has 1 replaceable hydrogen atom. Sulfuric acid, H_2SO_4, has 2 replaceable hydrogen atoms. Acetic acid, $HC_2H_3O_2$, has 4 hydrogen atoms but only 1 is capable of ionizing. Therefore, acetic acid has only 1 replaceable hydrogen atom. The number of replaceable hydrogen atoms in an acid is an inherent property, which the chemist has found through experimentation and a study of molecular structure.

An equivalent weight of an *acid* is the molecular weight divided by the number of replaceable hydrogen atoms. For example, hydrochloric acid (HCl) has 1 replaceable hydrogen atom. Its equivalent weight is, therefore, the molecular weight divided by 1:

$$\text{Equivalent weight HCl} = \frac{\text{Molecular weight}}{\text{Number of replaceable hydrogen atoms}}$$

$$= \frac{36.461\text{ g}}{1} = 36.461\text{ g}$$

Sulfuric acid (H_2SO_4) has 2 replaceable atoms, and, consequently, its equivalent weight is the molecular weight divided by 2:

$$\text{Equivalent weight } H_2SO_4 = \frac{\text{Molecular weight}}{\text{Number of replaceable hydrogen atoms}}$$

$$= \frac{98.078\text{ g}}{2} = 49.039\text{ g}$$

The equivalent weight of a *base* is the molecular weight divided by the number of replaceable hydroxyl ions. For example, sodium hydroxide (NaOH) has 1 replaceable hydroxyl ion. Therefore, its equivalent weight is the molecular weight divided by 1:

$$\text{Equivalent weight NaOH} = \frac{\text{Molecular weight}}{\text{Number of replaceable hydroxyl ions}}$$

$$= \frac{39.997 \text{ g}}{1} = 39.997 \text{ g}$$

Calcium hydroxide, $Ca(OH)_2$, has 2 replaceable hydroxyl ions. Therefore, its equivalent weight is the molecular weight divided by 2:

$$\text{Equivalent weight Ca(OH)}_2 = \frac{\text{Molecular weight}}{\text{Number of replaceable hydroxyl ions}}$$

$$= \frac{74.0947 \text{ g}}{2} = 37.047 \text{ g}$$

The equivalent weights of *salts,* such as silver nitrate and potassium permanganate, are determined by the number of electrons which the salts give or take during a reaction. For example, silver nitrate, $AgNO_3$, gives 1 electron. Therefore, the equivalent weight is the molecular weight divided by 1. Potassium permanganate, $KMnO_4$, takes 5 electrons. Therefore the equivalent weight is the molecular weight divided by 5.

The equivalent weights of the compounds that the student is likely to meet are given in Table C-1.

An equivalent weight is contained in 1 L of a 1 normal (1 N) solution. For example, 49.039 g of sulfuric acid is present in a liter of 1 N H_2SO_4, and 39.997 g of sodium hydroxide is present in a liter of 1 N NaOH.

It follows that 2 equivalent weights will be present in 1 L of a 2 N solution, and 0.1 equivalent weight in 1 L of an 0.1 N solution.

Some chemists prefer to discuss equivalent weights in terms of milliequivalent weights. The prefix "milli" means 1/1000. Consequently, a milliequivalent weight is 1/1000 of an equivalent weight. The word "millequivalent" is often abbreviated mEq (or meq). Milliliter is abbreviated ml. There is 1 mEq in 1 ml of every 1 N solution. This may be expressed by the following equation:

$$\text{Volume (in ml)} \times \text{Normality} = \text{Milliequivalent}$$

or

$$V \times N = \text{mEq}$$

Table C-1. Equivalent weights of compounds

Compound	Formula	Molecular weight	Equivalent weight
Hydrochloric acid	HCl	36.461	36.461
Sulfuric acid	H_2SO_4	98.078	49.039
Oxalic acid	$H_2C_2O_4 \cdot 2H_2O$	126.067	63.033
Sodium hydroxide	NaOH	39.997	39.997
Potassium hydroxide	KOH	56.109	56.109
Silver nitrate	$AgNO_3$	169.874	169.874
Potassium permanganate	$KMnO_4$	158.038	31.607

Equivalent weights

The system of using equivalent weights is a useful tool in chemistry calculations because the differences in the weights (or masses) of various atoms need not be considered. For example, 3 gram atoms of sodium have a combined weight of 69 g, whereas 2 gram atoms of chlorine have a weight of 71 g. Even though their weights are nearly the same, it is known that only 2 atoms of sodium can combine with the 2 atoms of chlorine. This, then, leaves an excess of sodium even though there are fewer grams of sodium than there are of chlorine.

On the other hand, if it were stated that there are 3 equivalents of sodium and only 2 equivalents of chlorine it would be readily apparent that there is an excess of sodium.

Another example of the usefulness of the equivalent system is in the neutralization of acids and bases. Keep in mind, however, that the same scheme may be used in any situation where chemical reaction occurs.

By definition 1 ml of 1 N solution contains 1 mEq of material. Therefore, 2 ml contains 2 mEq, and 10 ml contains 10 mEq. Generally then, volume in milliliters times normality equals milliequivalents. Stated mathematically, $V \times N =$ mEq. If one has, for example, 10 ml of a 1 N acid, it will contain 10 mEq of acid. This amount of acid will exactly neutralize 10 mEq of any alkali. If one wants to calculate the normality of an alkali and it is found that 5 ml of the alkali is required to neutralize the 10 ml of 1 N acid, it is known that there must be 10 mEq of alkali in the 5 ml volume. The normality of the alkali must then be 2, since 5 ml of a 2 N alkali would contain 10 mEq. Examples that follow will illustrate this principle of $V \times N =$ mEq.

Since the equivalent weight of acid is the same as that of alkali, volume of acid × normality of acid = volume of base × normality of base.

This same system applies to many other types of calculation. For example, for a mixture of 10 ml of 1 N HCl with 5 ml of 2 N NaOH the following calculations may be made:

$$10 \times 1 = 10 \text{ mEq of HCl}$$

and

$$5 \times 2 = 10 \text{ mEq of NaOH}$$

Therefore, the alkali exactly neutralizes the acid.

In addition, the amount of NaCl that could be produced in such a combination is easily determined. Since 1 mEq of HCl and 1 mEq of NaOH produce 1 mEq of NaCl, the mixture just described would yield 10 mEq of NaCl. Since 1 mEq requires 58.5 mg, 10 mEq would represent 585 mg of NaCl.

At this point it might be brought out that all quantities need not necessarily be discussed in terms of specific volumes. For example, if one had a flask that contained exactly 40 mg of NaOH, the amount of water in which it was dissolved would not alter the absolute quantity. When titrating this amount of NaOH with a standard acid it would not matter whether the NaOH was dissolved in 5 ml or 500 ml of water. Consider the following examples:

1. Assume that 40 mg of NaOH is dissolved in 5 ml of H_2O. Therefore there is 1 mEq in a volume of 5 ml. Since $V \times N =$ mEq, you can calculate that the value of N will be 0.2. It would require 1 mEq of acid to neutralize the NaOH. Suppose that 0.1 N HCl is used. Then:

$$V_A \times N_A = V_B \times N_B$$

$$V \times 0.1 = 5 \times 0.2$$

$$V = 10 \text{ ml}$$

Therefore 10 ml of 0.1 N HCl would be required to neutralize 40 mg of NaOH dissolved in 5 ml of H_2O.

2. If 40 mg of NaOH is dissolved in 10 ml of H_2O, $10 \times N = 1$ or $N = 0.1$. Suppose the same HCl solution is used. Then:

$$V_A \times 0.1 = 10 \times 0.1$$

$$V = 10 \text{ ml}$$

Note that in both cases 10 ml of the 0.1 N HCl is required. This should be evident, since 10 ml of 0.1 N HCl constitutes 1 mEq. In other words, the 1 mEq of HCl neutralizes the 1 mEq of NaOH regardless of the amount of H_2O used to dissolve the NaOH.

The following problem demonstrates the usefulness of the equivalent system. Suppose 20 ml of 0.2 N HCl is added to 10 ml of 0.3 N NaOH: (1) Is the resulting solution acid or alkaline? (2) What is the final normality? (3) How much NaCl is produced?

The amounts of the acid and alkali are calculated with the equation $V \times N =$ mEq as follows:

$$20 \times 0.2 = 4 \text{ mEq of HCl}$$

$$10 \times 0.3 = 3 \text{ mEq of NaOH}$$

Since there is more HCl than required to neutralize the NaOH, the resulting solution will be acid. In this situation, 3 mEq of HCl will be utilized to neutralize the 3 mEq of NaOH. Therefore, there will be 1 mEq of HCl in the final solution, which has a volume of 30 ml. There is, therefore, 1 mEq of HCl in a volume of 30 ml, or $30 \times N = 1$. The final N of the HCl is then $\frac{1}{30}$. Further, 3 mEq of NaOH with 3 mEq of HCl will produce 3 mEq of NaCl. Since 1 mEq of NaCl weighs 58.5 mg there would be 3×58.5 or 175.5) mg of NaCl produced.

Many other uses of the mEq system, too numerous to itemize, make this an essential tool for use in quantitative chemistry.

By way of summary—an equivalent weight of an acid is the molecular weight divided by the number of replaceable hydrogen atoms, whereas the equivalent weight of a base is the molecular weight divided by the number of replaceable hydroxyl ions. A 1 normal solution of an acid or base contains 1 equivalent weight per liter of solution.

Standardization

Suppose we have a hydrochloric acid solution of unknown normality, and the problem is to determine the exact normality of this solution. The following two steps are necessary:

Step 1: Titration or standardization
Step 2: Calculation

Step 1: Titration. Titration is a volumetric operation measuring the concentration of a solution or a constituent within a solution by measuring the reaction of this unknown against a standard of known concentration. The proper indicator is used to aid in determining the end point. In this discussion hydrochloric acid is the solution to be standardized. First, 10 ml of the HCl are pipetted into a flask or beaker. A few drops of the indicator phenolphthalein are then added. Next, 0.1 N NaOH is slowly added from a burette until the indicator changes from colorless to a faint pink, thus signifying the completion of neutralization. The number of milliliters required to reach this end point is recorded and used in the calculation given below. Let us assume, for the sake of illustration, that 20 ml of the 0.1 N NaOH was used. Consequently, from the above titration, we would have the following data:

Volume of acid (HCl) used:	10 ml
Normality of base (NaOH) used:	0.1 N
Volume of base (NaOH) used:	20 ml

Step 2: Calculation. The normality of the hydrochloric acid is found by writing down the following equation, filling in the blanks with the available data, and then solving the equation.

$$\begin{array}{ccccccc} \text{Volume of acid} & \times & \text{Normality of acid} & = & \text{Volume of base} & \times & \text{Normality of base} \\ (\quad) & \times & (\quad) & = & (\quad) & \times & (\quad) \end{array}$$

For example, we first write down the equation as indicated above and then proceed to fill in the blanks in the following manner. Since we are looking for the normality of the acid, a question mark is put under the Normality of acid label. Because the volume of acid used was 10 ml, a 10 is put below Volume of acid. Since the normality of the base was 0.1 N, an 0.1 is put below the Normality of base. Then, as we are assuming the volume of the base was 20 ml, a 20 is placed below the Volume of base. Thus, we have:

$$\begin{array}{ccccccc} \text{Volume of acid} & \times & \text{Normality of acid} & = & \text{Volume of base} & \times & \text{Normality of base} \\ 10 & \times & ? & = & 20 & \times & 0.1 \end{array}$$

To solve the equation, we must first eliminate the number 10 on the side of the unknown. Since there is a rule in algebra which states that both sides of an equation can be divided by the same number, we divide both sides by 10:

$$\frac{10 \times \text{Normality of acid}}{10} = \frac{20 \times 0.1}{10}$$

Canceling the 10's, we have:

$$\text{Normality of acid} = 2 \times 0.1$$

Multiplying, we have:

$$\text{Normality of acid} = 0.2$$

Consequently, the exact normality of the HCl is 0.2 N.

Another way to calculate the normality is shown below:

$$\begin{array}{ccccccc} \text{Volume of acid} & \times & \text{Normality of acid} & = & \text{Volume of base} & \times & \text{Normality of base} \\ 10 & \times & x & = & 20 & \times & 0.1 \end{array}$$

$$10 \times x = 20 \times 0.1$$

$$10\ x = 2.0$$

$$x = 0.2$$

IONIZATION CONSTANTS

According to the law of mass action, the rate of a chemical reaction is proportional to the molecular concentration of the reactants. As the concentration of any weak electrolyte becomes less, the percent of ionization becomes greater. These changes react proportionally to each other, or in other words, offset each other, and the ionization constant remains the same. These weak electrolyte concentrations are expressed in moles per liter; when applied to an ionization, equilibrium is usually defined as the ionization constant or K_i. The law of mass action does not hold for strong electrolytes. An example would be HCl and H_2SO_4 that have no usable ionization constant.

In solving problems concerning these constants it must be remembered that the constant involved is not always known to a precision greater than two or three significant figures. Therefore, in various books you may see examples with the word "approximate" written to the side.

The concentration value of an aqueous solution needed for the calculation of the ionization constant may be obtained by measuring the concentration of ions in the solution, by means of electric current; the more ions there are in a solution, the more current flows through the solution.

The following examples are given to help you understand various reactions that may take place with the weak acids and bases. Acetic acid will be the main example used for calculating various problems. Remember that opposing arrows in the equations indicate reversible reactions—meaning that the reaction will go in both directions, depending upon the conditions of the reaction, which will then reach a point of equilibrium where no further change is apparent. The ions are still in constant motion, separating as rapidly as they are bound together, only there is no change in the balance or equilibrium. The brackets [ ] enclosing the constant factor equations are used as a symbol of molecular concentration.

In relation to their ionization, weak acids may be classified as monobasic (monoprotic), dibasic (diprotic), or polybasic (triprotic, etc.). These are classifications according to the number of positive hydrogen ions (protons) that can be replaced per molecule.

Monobasic acid

A monobasic acid is an acid having 1 replaceable hydrogen atom per molecule. Example—acetic acid:

$$HC_2H_3O_2 \rightleftarrows H^+ + C_2H_3O_2^-$$

$$\frac{[H^+] \times [C_2H_3O_2^-]}{[HC_2H_3O_2]} = K_1$$

Dibasic acid

A dibasic acid is an acid having 2 replaceable hydrogen atoms per molecule. The ionization takes place in two steps and each reaction has its own equilibrium constant (K_i), designated K_1 or K_2. Example—carbonic acid:

1. $H_2CO_3 \rightleftarrows H^+ + HCO_3^{-1}$

$$\frac{[H^+] \times [HCO_3^-]}{[H_2CO_3]} = K_1$$

2. $HCO_3^{-1} \rightleftarrows H^+ + CO_3^{-2}$

$$\frac{[H^+] \times [CO_3^{-2}]}{[HCO_3^-]} = K_2$$

Polybasic acid

An acid having more than 2 replaceable hydrogen atoms per molecule, thus requiring more than two steps in order to obtain an ionization equilibrium, is termed a polybasic acid. Each step has its K value. Example—phosphoric acid:

1. $H_3PO_4 \rightleftarrows H^+ + H_2PO_4^{-1}$

$$\frac{[H^+] \times [H_2PO_4^-]}{[H_3PO_4]} = K_1$$

2. $H_2PO_4^{-1} \rightleftarrows H^+ + HPO_4^{-2}$

$$\frac{[H^+] \times [HPO_4^{-2}]}{[H_2PO_4^-]} = K_2$$

3. $HPO_4^{-2} \rightleftarrows H^+ + PO_4^{-3}$

$$\frac{[H^+] \times [PO_4^{-3}]}{[HPO_4^{-2}]} = K_3$$

Examples

A 0.1 molar (M) solution of acetic acid is 1.34% ionized. This has been established by the measurement, mentioned earlier, of electric conductance. Since 1.34% of the hydrogen acetate is ionized, then the remaining covalent molecules would be 98.66%. As has been discussed, the ion and molecule concentrations remain constant although there is a continual reaction taking place. Water is not included in the equilibrium constant equation because water concentration is practically unchanged in a dilute aqueous solution. However, because of the importance of water and its ions, the numerical value of that K_i (constant) should be remembered. (Table C-2.)

1. Calculate the ionization constant (K_i) for 0.1 M acetic acid at 25° C:

$HC_2H_3O_2$	$\rightleftarrows$	H^+	+	$C_2H_3O_2^-$
98.66% of 0.1 M		1.34% of 0.1 M		1.34% of 0.1 M
or		or		or
0.09866 M		0.00134 M		0.00134 M

$$\frac{[H^+] \times [C_2H_3O_2^-]}{[HC_2H_3O_2]} = K_i$$

$$\frac{0.00134 \times 0.00134}{0.1^* - 0.00134} = \frac{0.00000179}{0.09866}$$

$$\frac{0.00000179}{0.09866} = 0.0000182 \text{ or } 1.8 \times 10^{-5} = K_i$$

2. Calculate the percent ionization in a 0.2 M solution of acetic acid at 25° C:

$$K_i \text{ constant} = 1.8 \times 10^{-5}$$

Let x = hydrogen concentration in a 0.2 M solution

$$\frac{[H^+] \times [C_2H_3O_2^-]}{[HC_2H_3O_2]} = 1.8 \times 10^{-5}$$

$$\frac{x \cdot x}{0.2 - x} = 1.8 \times 10^{-5}$$

$$x^2 = 0.000018 \times 0.2$$

or

$$(1.8 \times 10^{-5}) \times 0.2, \text{ since } 0.2 - x = \text{approximately } 0.2$$

$$x^2 = 0.0000036 \text{ or } 3.6 \times 10^{-6}$$

$$x = 0.0019 \text{ or } 1.9 \times 10^{-3}$$

$$\frac{0.0019}{0.2} \times 100 = 0.95\% \text{ ionization of a 0.2 M solution of acetic acid}$$

*0.1 M acetic acid.

Table C-2. Ionization constants at 25° C (other temperatures noted in parentheses)

Name	Ionization reaction	K_i
Acetic acid	$HC_2H_3O_2 \rightleftarrows H^+ + C_2H_3O_2^-$	1.8×10^{-5}
Ammonium hydroxide	$NH_3 + H_2O \rightleftarrows NH_4^+ + OH^-$	1.8×10^{-5}
		(0°) 1.4×10^{-5}
Arsenic	$H_3AsO_4 \rightleftarrows H^+ + H_2AsO_4^{-1}$	K_1 5×10^{-3}
	$H_2AsO_4^{-1} \rightleftarrows H^+ + HAsO_4^{-2}$	K_2 8.3×10^{-8}
	$HAsO_4^{-2} \rightleftarrows H^+ + AsO_4^{-3}$	K_3 6×10^{-10}
Arsenious acid	$H_3AsO_3 \rightleftarrows H^+ + H_2AsO_3^-$	6×10^{-10}
Benzoic acid	$HC_7H_5O_2 \rightleftarrows H^+ + C_7H_5O_2^-$	6.3×10^{-5}
Boric acid	$H_3BO_3 \rightleftarrows H^+ + H_2BO_3^-$	5.8×10^{-10}
n-Caprylic alcohol	$HC_8H_{15}O_2 \rightleftarrows H^+ + C_8H_{15}O_2^-$	1.27×10^{-5}
Carbonic acid	$H_2CO_3 \rightleftarrows H^+ + HCO_3^{-1}$	K_1 4.5×10^{-7}
	$HCO_3^{-1} \rightleftarrows H^+ + CO_3^{-2}$	K_2 5.6×10^{-11}
Creatine (anhydrous)	$C_4H_9O_2N_3 \rightleftarrows H^+ + C_4H_8O_2N_3^-$	(40°) 1.9×10^{-11}
Creatinine	$C_4H_7ON_3 \rightleftarrows H^+ + C_4H_6ON_3^-$	(40°) 3.7×10^{-11}
Formic acid	$HCHO_2 \rightleftarrows H^+ + CHO_2^-$	1.8×10^{-4}
Hydrochloric acid	$HCl \rightleftarrows H^+ + Cl^-$	No constant
Lactic acid	$HC_3H_5O_3 \rightleftarrows H^+ + C_3H_5O_3^-$	1.387×10^{-4}
Phosphoric acid	$H_3PO_4 \rightleftarrows H^+ + H_2PO_4^{-1}$	K_1 7.5×10^{-3}
	$H_2PO_4^{-1} \rightleftarrows H^+ + HPO_4^{-2}$	K_2 6.2×10^{-8}
Phosphorous acid	$H_3PO_3 \rightleftarrows H^+ + H_2PO_3^{-1}$	K_1 1.6×10^{-2}
	$H_2PO_3^{-1} \rightleftarrows H^+ + HPO_3^{-2}$	K_2 7×10^{-7}
Picric acid	$HC_6H_2O_7N_3 \rightleftarrows H^+ + C_6H_2O_7N_3$	4.2×10^{-1}
Sulfuric acid	$H_2SO_4 \rightleftarrows H^+ + HSO_4^{-1}$	No constant
	$HSO_4^{-1} \rightleftarrows H^+ + SO_4^{-2}$	K_2 1.2×10^{-2}
Sulfurous acid	$H_2SO_3 \rightleftarrows H^+ + HSO_3^{-1}$	K_1 1.7×10^{-2}
	$HSO_3^{-1} \rightleftarrows H^+ + SO_3^{-2}$	K_2 5×10^{-6}
Trichloroacetic acid	$HC_2O_2Cl_3 \rightleftarrows H^+ + C_2O_2Cl_3^-$	1.3×10^{-1}
Urea	$CH_4ON_2 \rightleftarrows H^+ + CH_3ON_2^-$	1.5×10^{-14}
Uric acid	$C_5H_4O_3N_4 \rightleftarrows H^+ + C_5H_3O_3N_4$	1.3×10^{-4}
Water	$H_2O \rightleftarrows H^+ + OH^-$	1.0×10^{-14}

3. A normal solution of acetic acid is 0.42% ionized. What is the ionization constant? Referring to problem 1, solve as follows:

$$\frac{0.0042 \times 0.0042}{1 - 0.0042} = 0.0000177 \text{ or } 1.8 \times 10^{-5} = K_i$$

4. Sufficient sodium acetate was dissolved in a 0.1 M solution of acetic acid to bring the molarity to a 1.5 solution. What is the hydrogen ion concentration of this 1.5 M solution?

$$\frac{[H^+] \times 1.5}{0.1} = 1.8 \times 10^{-5} = K_I$$

$$15\ [H^+] = 0.000018 \text{ or } 1.8 \times 10^{-5}$$

$$[H^+] = 0.0000012 \text{ or } 1.2 \times 10^{-6}\ M$$

Problems

1. Calculate the ionization constant of formic acid, $HCHO_2$, which ionizes 4.2% in a 0.10 M solution.

   Answer: 1.8×10^{-4} or 0.00018.

2. Refer to Table C-2 for the ionization constant of ammonium hydroxide at 25° C. (a) Determine the degree of ionization. (b) Find the OH^- concentration of a 0.08 M solution of NH_4OH.

   Answers: (a) 1.5%. (b) 1.2×10^{-3} mole/L.

3. A solution of acetic acid is 1.0% ionized. (a) What is the molarity of the solution? (b) What is the hydrogen concentration of the solution?

   Answers: (a) 0.18 M. (b) 1.8×10^{-3} or 0.0018 mole/L.

CHEMICAL INDICATORS

Indicators show color changes according to the concentration of the hydrogen (hydroxyl) ion. These are used to determine the strength of solutions and the end points of reactions. A few types of indicators are:

1. Achromatic—a mixture of indicators chosen for their complementary colors so that at the end of the reaction the end point is colorless
2. Absorption—an indicator that shows the end point of a titration by changing the indicator's colloidal condition, so that the precipitate, which has been colored by the indicator, will either lose or gain the color
3. External (outside)—an indicator that is used outside the solution being titrated, with the indicator placed in a small beaker while the solution being titrated is added dropwise into the beaker to obtain the end point
4. Fluorescent—an indicator that shows the end point of a reaction by either an increase or a decrease in fluorescence
5. Internal—an indicator that is added to the solution being analyzed
6. Neutralization—the indicator that undergoes a distinct color change at the equivalence point where the acid and base solutions are neutralized in titrations
7. Oxidation-reduction—an indicator that undergoes a distinct color change at a definite state of oxidation of a solution
8. Turbidity—indicators of a type that shows the end point by an increase in turbidity or precipitation; usually colloidal solutions
9. Universal—a selected mixture of indicators giving a wide range of pH values

Five common indicators

Litmus paper. Litmus paper is a strip of chemically treated paper. It turns red in an acid solution and blue in a basic solution.

Nitrazine paper. Nitrazine paper is also a strip of chemically treated paper. This indicator turns from green to yellow in an acid solution and blue in a basic solution. It is used to test urine for acid or basic properties.

Töpfer's reagent. Töpfer's reagent is a dye dissolved in alcohol. It is red up to a pH of 2.8. As the pH value increases, it turns light orange and finally yellow at pH 4.0. Töpfer's reagent is used to test gastric juices for hydrochloric acid and to determine the strength of weak acid and basic solutions.

Phenolphthalein. Phenolphthalein is an organic salt dissolved in alcohol. It is colorless up to pH 8.3. As the pH value increases beyond 8.3, it turns pink and finally bright red at pH 10.0. Phenolphthalein is used to test the strength of weak acid and basic solutions.

pHydrion. pHydrion is a chemically treated paper strip that gives specific pH values. The color of the paper is compared to the color chart in the container.

Table C-3. Indicators

Indicator	Color change		pH range	Preparation	
	Acid	Alkaline			
Thymol blue					
Acid	Red	Yellow	1.2-2.8	Thymol blue (thymolsulfonphthalein)	40.0 mg
Alkaline	Yellow	Blue	8.2-9.6	Absolute ethyl alcohol	100.0 ml
Bromphenol blue	Violet	Blue	3.0-3.6	Tetrabromophenolsulfonphthalein	0.1 g
				Ethyl alcohol (20%)	100.0 ml
Töpfer's reagent	Red	Yellow	3.0-4.0	p-Dimethylaminoazobenzene	0.5 g
				Ethyl alcohol (95%)	100.0 ml
Congo red	Blue	Red	3.0-5.0	Congo red	0.5 g
				Distilled water	90.0 ml
				Ethyl alcohol (95%)	10.0 ml
Methyl orange	Orange-red	Yellow	3.1-4.4	Methyl orange	100.0 mg
				Distilled water	100.0 ml
Bromcresol green	Yellow	Blue	3.8-5.4	Tetrabromometacresolsulfonphthalein	40.0 mg
				Absolute ethyl alcohol	100.0 ml
Methyl red	Red	Yellow	4.2-6.3	Methyl red	1.0 g
				Ethyl alcohol (95%)	300.0 ml
				Dilute to 500 ml with distilled water	
Chlorphenol red	Yellow	Red	4.8-6.4	Chlorphenol red	40.0 mg
				Absolute ethyl alcohol	100.0 ml
Alizarin red	Yellow	Red	5.5-6.8	Sodium alizarin monosulfonate	1.0 g
				Distilled water	100.0 ml
Bromthymol blue	Yellow	Blue	6.0-7.6	Dibromthymolsulfonphthalein	40.0 mg
				Absolute ethyl alcohol	100.0 ml
Neutral red	Red	Yellow	6.8-8.0	Neutral red	0.5 g
				Ethyl alcohol (95%)	300.0 ml
				Dilute to 500 ml with distilled water	
Phenol red	Yellow	Red	6.8-8.4	Phenol red (phenolsulfonphthalein)	20.0 mg
				Ethyl alcohol (95%)	100.0 ml
Phenolphthalein	Colorless	Red	8.3-10.0	Phenolphthalein	1.0 g
				Ethyl alcohol (95%)	100.0 ml

Reason for the color changes of indicators

Indicators are usually organic dyes whose color in a solution changes according to the degree of ionization of the indicator. In a basic indicator there will be more negative ions than in a solution that is neutral. The weak acid indicator has even fewer ions. In the case of the indicator methyl red, since the negative ion differs from the molecule, the acid color of red changes to the base color of yellow according to the ionization taking place. In water solution these indicator ions are in balance (equilibrium) with the molecule, symbolized as follows:

$$\underset{\text{Color A}}{\text{H Ind}} \rightleftarrows H^+ + \underset{\text{Color B}}{\text{Ind}^-}$$

If acid with more H^+ is added, the equilibrium will be pushed to the left; color A is dominant. If base is added, uniting with the H^+ ions, the equilibrium will move to the right; color B is dominant.

In such an equilibrium there is always a constant ratio between the product of the ions and the molecule:

$$\frac{[H^+] \times [\text{Ind}^-]}{[\text{H Ind}]} = K \text{ (constant)}$$

or

$$[H^+] \times \underset{\text{Color A}}{[\text{Ind}^-]} = K \underset{\text{Color B}}{[\text{H Ind}]}$$

At the end point of a reaction, the two color forms will be in equal concentrations, and the amount of $[H^+]$ will equal the ionization K of the indicator.

• • •

Since each indicator has its own K value, the indicator that is to be used must be selected to fit the ionization value of the mixture being measured. A list of indicators and their pH range is given in Table C-3.

• D

• Reagents

PREPARATION OF REAGENTS

In preparing the reagents in this section only the purest chemicals available should be used unless another grade is specifically designated. In some cases a specific supplier of a chemical is recommended. These recommendations are generally those of the person or persons who developed the individual method.

An analytical balance should always be used for weighing chemicals unless the amount to be weighed exceeds the load limit of the balance; then a torsion balance may be used. Material for standards and buffers should be weighed accurately to the fourth decimal place (on an analytical balance).

When measuring liquid materials, use class A volumetric pipettes unless otherwise specified in the directions. Do not pipette strong acids or alkalies by mouth. Attach a rubber bulb or vacuum line to the mouthpiece when drawing up these materials into a pipette. Hot and cold liquids should be allowed to come to room temperature before measurement.

Only distilled or deionized water is to be used in preparing any of the aqueous reagents listed. The source of the water supply should be checked periodically to be sure that water of the highest purity is being delivered.

When using a magnetic stirring bar be sure to remove the bar from a volumetric flask before diluting a solution to volume. After dilution to volume, a solution or standard should be mixed thoroughly. One of the most common errors in preparing standards, particularly, is not mixing thoroughly after dilution.

Some reagents require adjusting to a certain pH. This must be done by using a pH meter. The pH adjustment can be simplified somewhat by placing the solution in a beaker on a magnetic stirrer. With the stirring bar spinning and a combination electrode suspended in the solution, acid or alkali may be added dropwise and the change in pH noted almost immediately.

Some reagents are light-sensitive and it is recommended that these be stored in amber or brown bottles. Polyethylene bottles are recommended for storing most other reagents, mainly for reasons of safety. All bottles should be clearly labeled as to content and the date of preparation. If the reagent has an expiration date, this should be included on the label.

The following reagents are listed in alphabetic order with their identification numbers. Directions for preparing the reagents are then given in numerical order. For example, the first reagent mentioned in the text is listed as "(D-1) Sulfuric acid, 0.67 N."

Alphabetic list of reagents

Acetate buffer (D-162)
Acetic acid, 1.0% (D-245)
Acetic acid, 5% (D-260)
Acetic acid, 10% (v/v) (D-205)
Acetic acid, 20% (v/v) (D-234)
Acetic acid, 30% (D-276)
Acetic acid, 50% (D-262)
Acetic acid, 95% (D-99)
Acetic anhydride–dioxane (A-D) reagent (D-209)
Acetone-alcohol mixture (D-203)
Acetone-ether mixture (D-206)
Acid antifoam diluent (D-12)
Acid buffer, 0.09 M citric acid, pH 4.8 (D-231)
Acid buffered substrate, pH 4.8-4.9 (D-232)
Acid molybdate (D-51)
Acid reagent, 2 M NaH_2PO_4 (D-214)
Albumin stock standard, 10 g/100 ml (D-93)
Alcoholic KOH hydroxide wash solution, 2% (D-58)
Alkali-hydrochlorite reagent (D-96)
Alkaline buffer, pH 10.6-10.7 (D-213)
Alkaline buffer, 0.1 M glycine, pH 10.5 (D-227)
Alkaline buffer with magnesium (D-64)
Alkaline buffered substrate, pH 10.3-10.4 (D-229)
Alkaline copper reagent (Folin-Wu) (D-143)
Alkaline copper reagent (Nelson-Somogyi) (D-139)
Alkaline iodide (D-86)
Alkaline potassium ferricyanide (D-23)
2-Amino-2-methyl-1-propanol, 50% (D-103)
4-Aminoantipyrine (D-66)
4-Aminoantipyrine, 0.6% (D-126)
Ammonium chloride stock standard, 1 M (D-21)
Ammonium hydroxide, 5% (D-192)
Ammonium hydroxide, 10% (w/v) (D-291)
Ammonium hydroxide, dilute (0.5%) (D-183)
Ammonium oxalate, 0.1 M (D-293)
Ammonium oxalate, 4% (D-182)
Ammonium oxalate, 10% (D-190)
Ammonium pyrrolidine dithiocarbamate, 1% (D-222)
Ammonium sulfamate, 0.5% (D-331)
Ammonium sulfate, saturated (D-254)
AMP stock buffer, 50% (w/v) (D-59)
AMP working buffer (D-60)
Antifoam solution (D-309)
Arsenious acid, 0.2 N (D-313)
Arsenomolybdate color reagent (D-140)
Ascorbic acid, 4% (D-84)
Azoene fast red dye diluent (D-56)
Azoene fast red dye solution (D-57)

Barium chloride, 5% (D-294)
Barium chloride, 10% (D-246)
Barium hydroxide, 0.3 N (D-5)
Benedict's qualitative reagent (D-270a)
Benedict's quantitative reagent (D-270b)
Benzidine, saturated, in glacial acetic acid (D-239)
Benzoic acid, 0.2% (D-141)
Benzoic acid, saturated (D-30)
Biuret reagent (D-197)
Biuret reagent (Robot) (D-134)
Biuret stock solution (D-87)
Biuret working solution (D-88)
Bromide stock standard, 20 mg/ml (D-326)
Bromine water (D-285)
BSP standard, 5 mg/100 ml (D-215)
BUN acid, working (D-29)
BUN color reagent, working (D-26)

Caffeine diluent (D-79)
Calcium base, 3.75% diethylamine (D-46)
Calcium chloride, 2.5% (D-200)
Calcium chloride, 10% (D-283)
Calcium standard (1 ml = μg Ca) (D-189)
Calcium stock standard, 50 mg Ca/100 ml (D-47)
Cal-red indicator (D-187)
Carbonate buffer, pH 10.0 (D-123)
Carbonate-bicarbonate buffer (D-17)
Ceric ammonium sulfate, 0.02 N (D-307)
Ceric ammonium sulfate, 0.04 N (D-314)
Chloric acid reagent (D-311)
Chloride color reagent (D-11)
Cholesterol color reagent (D-98)
Cholesterol standard, 0.2 mg/ml (D-210)
Cholesterol stock standard, 1.25 mg/ml (D-343)
Chromic acid (D-312)
Chromotropic acid, 0.5% (D-318)
Citrate buffer (D-55)
CO_2 color reagent (D-18)
Color reagent (ferric-mercuric reagent) (D-327)
Copper sulfate, 4% (D-323)
Coupling reagent (naphthyl ethylenediamine solution) (D-332)
CPK blank substrate (D-114)
Creatine substrate (D-115)
Creatinine stock standard, 1 mg/ml (D-36)
Cresolphthalein complexone, 0.01%, with 8-hydroxyquinoline (D-45)
Cupric sulfate, 10% (D-102)
Cupric sulfate solution, buffered (D-105)
Cyanide-urea working solution (D-42)

Diacetyl monoxime stock, 2.5% (D-24)
Diazo blank solution (D-133)
Diazo reagent (D-132)
Diazo reagent, combined (D-82)
Digitonin solution (D-204)
2,4-Dinitrophenylhydrazine, 0.001 M (D-116)
Diphenylcarbazone indicator (D-175)
2,2′-Dipyridyl, 0.4% (D-340)
Disodium phenyl phosphate substrate (D-124)

EDTA solution (D-186)
Ehrlich's diazo reagent (D-282)

Reagents—directions

(D-1) Sulfuric acid, 0.67 N

Transfer exactly 19 ml of concentrated sulfuric acid, H_2SO_4, to a 1 L volumetric flask containing about 700 ml of water. Dilute to volume and mix. Standardize.

NOTE: With phenolphthalein as an indicator, 20 ml of the above solution should require 13.33 ml of 1 N NaOH for neutralization. If the solution is too acid, adjust the normality to 0.67 N by dilution.

(D-2) Sodium tungstate, 10%

Weigh 100 g of sodium tungstate, Na_2WO_4, and transfer to a 1 L volumetric flask. Dissolve and dilute to volume with water. Stopper and mix thoroughly. Store in a Pyrex or polyethylene bottle.

(D-3) Sulfuric acid, 0.08 N

Pipette exactly 2.3 ml of concentrated sulfuric acid into a 1 L volumetric flask containing about 700 ml of water. Dilute to volume and mix. Standardize.

NOTE: With phenolphthalein as an indicator, 20 ml of the above solution should require 16.67 ml of 0.1 N NaOH for neutralization. If the solution is too acid, adjust the normality to 0.08 N by dilution.

(D-4) Zinc sulfate, 5%

Weigh 50 g of zinc sulfate, $ZnSO_4{\cdot}7H_2O$, and transfer to a 1 L volumetric flask. Dissolve and dilute to volume with water. Mix thoroughly.

(D-5) Barium hydroxide, 0.3 N

Weigh 47 g of barium hydroxide, $Ba(OH)_2{\cdot}8H_2O$, and transfer to a 1 L volumetric flask. Dissolve and dilute to volume with water. Mix. Filter if cloudy. Store in a well-closed container with a minimum of air space.

NOTE: Reagents D-4 and D-5 must neutralize each other. Pipette 10 ml of $ZnSO_4$ solution into a 250 ml Erlenmeyer flask. Use a volumetric pipette. Add 50 ml of distilled water and 4 drops of phenolphthalein indicator. Titrate with the barium hydroxide very slowly until 1 drop turns the solution to a faint pink. If this is done too rapidly a false end point will be obtained. It should require 10 ml (± 0.05 ml) of barium chloride to reach the end point. If either solution is too strong, adjust with water and titrate. The adjusted or "matched" solutions are now placed into aspirator bottles with the barium hydroxide bottle containing a soda-lime tube in the stopper. The soda-lime tube is to prevent the solution from absorbing atmospheric CO_2. Soda lime is a mixture of CaO and NaOH.

(D-6) Lithium nitrate stock, 1000 mEq Li/L in 2 N H_2SO_4

Weigh 69.0 g of lithium nitrate, $LiNO_3$, and transfer to a 1 L volumetric flask containing about 700 ml of water. Mix until dissolved. Slowly add 98.08 g (53.5 ml) of concentrated sulfuric acid with mixing. Allow to cool to room temperature. Dilute to volume and mix. Store in a polyethylene bottle.

(D-7) Lithium nitrate working solution, 125 mEq Li/L in 0.25 N H_2SO_4

Dilute 125 ml of lithium nitrate stock solution to 1000 ml with water. Add 0.5 ml of Brij 35 and mix.

(D-8) Mercuric thiocyanate, saturated

Place a 2 L Erlenmeyer flask on a magnetic stirrer. Add 1 L of water and about 2 g of mercuric thiocyanate, $Hg(SCN)_2$. Stir for several hours and then allow to sit overnight. Filter through a good grade of filter paper and store in an amber reagent bottle.

(D-9) Ferric nitrate, 0.5 M

Weigh 202 g of ferric nitrate, $Fe(NO_3)_3 \cdot 9H_2O$, and transfer to a 1 L volumetric flask. Add about 700 ml of water and mix until all the ferric nitrate is dissolved. Slowly add 31.5 g (21 ml) of concentrated nitric acid and mix. Dilute to volume and mix thoroughly. Filter through a good grade of filter paper and store in an amber reagent bottle.

(D-10) Mercuric nitrate, 0.2 M in 0.2 N HNO_3

Place about 700 ml of water in a 1 L volumetric flask. Slowly add 12.6 g (8.4 ml) of concentrated nitric acid. Swirl to mix thoroughly. Weigh 68.5 g of mercuric nitrate, $Hg(NO_3)_2 \cdot H_2O$, and transfer to the flask. Mix until completely dissolved. Dilute to volume and mix again. Filter through a good grade of filter paper and store in a polyethylene bottle.

(D-11) Chloride color reagent

Combine 700 ml of saturated mercuric thiocyanate, 300 ml of ferric nitrate solution, and 3.5 ml (see step 11 under operating sequence, electrolytes) of mercuric nitrate solution. Mix thoroughly, filter, and store in an amber reagent bottle.

(D-12) Acid antifoam diluent

To about 700 ml of water in a 1 L volumetric flask, slowly add 2.8 ml of concentrated sulfuric acid and mix. Add 1 ml of Dow Corning Antifoam B, dilute to volume with water and mix thoroughly. Store in a glass or polyethylene bottle. Shake well each day before use. If bubbling in the CO_2 trap still occurs, add another ml/L of Antifoam B.

(D-13) Sodium hydroxide wash solution

Weigh 40 g of sodium hydroxide and transfer to a 1 L volumetric flask containing about 700 ml of water. Mix until all sodium hydroxide is dissolved. Dilute to volume and mix. Store in a polyethylene bottle.

(D-14) Phenolphthalein indicator, 1%

Weigh 1.0 g of phenolphthalein powder and transfer to a 100 ml volumetric flask. Dissolve and dilute to volume with methyl alcohol. Store in a 4-ounce amber bottle.

(D-15) Sodium carbonate, 1 M

Very carefully weigh 106.0 g of anhydrous sodium carbonate and transfer to a 1 L volumetric flask. Add about 600 ml of water and mix until the sodium carbonate is completely dissolved. Dilute to volume and mix thoroughly. Check the molarity of the sodium carbonate by the following procedure: Measure 20 ml of 1 N HCl with a volumetric pipette and transfer to a 125 ml Erlenmeyer flask. Add 3 drops of 1% methyl orange indicator. Titrate the HCl with the sodium carbonate until the indicator changes from red to yellow. Note the volume of sodium carbonate required to reach this end point. It should require 10.0 ml if the sodium carbonate is 2 N (1 M). Adjust if necessary. This solution may be divided into two portions—one to be used for the carbonate-bicarbonate buffer and the other for the sodium carbonate stock standard. Store in tightly capped 500 ml polyethylene bottles. Do not interchange bottles and do not use reagent from one bottle for both purposes. There is the danger of contaminating the stock standard.

(D-16) Sodium bicarbonate, 1 M

Weigh 84.0 g of sodium bicarbonate and transfer to a 1 L volumetric flask. Add about 600 ml of water and mix until all the bicarbonate is dissolved. Dilute to volume with water and mix thoroughly. Check the molarity by the same procedure as that given for the sodium carbonate. This time, however, it should require 20 ml to reach an end point if the sodium bicarbonate solution is 1 N (1 M). Adjust if necessary. Store in a 1 L polyethylene bottle.

(D-17) Carbonate-bicarbonate buffer

To 35 ml of 1 M sodium carbonate solution, add 70 ml of 1 M sodium bicarbonate solution. Mix well and store in a 4-ounce polyethylene bottle.

(D-18) CO_2 color reagent

Prepare this reagent fresh on the day of use. Pipette 1.5 ml of carbonate-bicarbonate buffer* and 0.9 ml of phenolphthalein indicator* and transfer to a 1 L graduated cylinder. Dilute to 1 L with water, add 0.5 ml of Brij 35 and transfer to a 1 L bottle. Stopper and mix well. Avoid prolonged exposure to air.

(D-19) Sodium chloride stock standard, 1 M

Weigh 58.448 g of sodium chloride and transfer to a 1 L volumetric flask. Dissolve and dilute to volume with water. Store in a tightly capped polyethylene bottle. This standard contains 1000 mEq Na/L and 1000 mEq Cl/L.

---

*These amounts may need to be adjusted as directed in operating sequence (electrolytes), steps 7 and 10.

(D-20) Potassium chloride stock standard, 1 M

Weigh 7.4557 g of potassium chloride and transfer to a 1 L volumetric flask. Dissolve and dilute to volume with water. Store in a tightly capped polyethylene bottle. This standard contains 100 mEq K/L and 100 mEq Cl/L.

(D-21) Ammonium chloride stock standard, 1 M

Weigh 53.5 g of ammonium chloride, NH_4Cl, and transfer to a 1 L volumetric flask. Dissolve and dilute to volume with water. Store in a tightly capped polyethylene bottle. This standard contains 1000 mEq Cl/L.

(D-22) Saline, 0.9%

Weigh 9.0 g of sodium chloride and transfer to a 1 L volumetric flask. Dissolve and dilute to volume with water. Store in a Pyrex or polyethylene bottle.

(D-23) Alkaline potassium ferricyanide

Weigh 9.0 g of sodium chloride and transfer to a 1 L volumetric flask containing about 600 ml of water. Mix until dissolved. Add 0.25 g of potassium ferricyanide, $K_3Fe(CN)_6$, and mix until dissolved. Slowly and with mixing (e.g., magnetic stirrer) add 20.0 g of sodium carbonate, Na_2CO_3, and mix until dissolved. Dilute to volume, mix thoroughly, and filter. Add 2 ml of Brij 35 and mix. Store in a brown glass or polyethylene bottle. If the solution has been stored for a period of time, filter just prior to use.

(D-24) Diacetyl monoxime stock, 2.5%

Weigh 25.0 g of diacetyl monoxime (2,3-butanedione-2-oxime) and transfer to a 1 L volumetric flask. Add about 500 ml of water and mix until dissolved. Dilute to volume with water and mix thoroughly. Filter and store in a brown bottle.

(D-25) Thiosemicarbazide stock, 0.5%

Weigh 5.0 g of thiosemicarbazide and transfer to a 1 L volumetric flask containing about 500 ml of water. Mix until dissolved. Dilute to volume with water and mix thoroughly. Store in a brown bottle.

(D-26) Working BUN color reagent

Measure 67 ml of stock diacetyl monoxime and transfer to a 1 L volumetric flask containing about 500 ml of water. Add 67 ml of stock thiosemicarbazide and dilute to volume with water. Add 0.5 ml of Brij 35 and mix thoroughly. Store in a polyethylene or brown glass bottle.

(D-27) Ferric chloride–phosphoric acid stock

Dissolve 15.0 of ferric chloride, $FeCl_3 \cdot 6H_2O$, in about 30 ml of water and transfer to a 500 ml glass-stoppered graduated mixing cylinder. Slowly add 300 ml of 85% phosphoric acid. Dilute to the 450 ml mark with water and mix. Store in a brown bottle.

(D-28) Sulfuric acid stock, 20%

Place a 1 L Erlenmeyer flask containing about 600 ml of water into a pan of ice water for about 10 minutes. Slowly add 200 ml of concentrated sulfuric acid to the flask and mix by swirling. Allow the solution to come to room temperature and then transfer to a 1 L volumetric flask. Rinse the Erlenmeyer flask three times with 30 ml of water and transfer washings to the volumetric flask. Dilute to volume with water and mix.

(D-29) Working BUN acid

Pipette 1.0 ml of stock ferric chloride–phosphoric acid into a 1 L volumetric flask and dilute to volume with 20% sulfuric acid. Mix thoroughly and store in a polyethylene bottle.

(D-30) Benzoic acid, saturated

Place a beaker or flask containing 2 L of water on a magnetic stirrer. Add about 8 g of benzoic acid and allow to stir overnight. Filter and store in a polyethylene bottle.

(D-31) Glucose stock standard, 10 mg/ml

Weigh exactly 10.000 g of dextrose on an analytical balance and transfer to a 1 L volumetric flask. Dissolve and dilute to volume with saturated benzoic acid. Mix thoroughly and store in a tightly closed polyethylene or Pyrex bottle.

(D-32) Standard diluent (0.01 N H_2SO_4 containing 40 mg/L PMA)

Weigh 0.2 g of phenylmercuric acetate (PMA) (Eastman P-4267) and transfer to a 250 ml beaker containing 100 ml of water. Heat on a steam bath until completely dissolved. Cool to room temperature and transfer to a 5 L volumetric flask (Corning 94450). Rinse the beaker several times with water and transfer the washings to the flask. Add 1.4 ml of concentrated sulfuric acid and mix. Dilute to volume with water and mix.

(D-33) Urea nitrogen stock standard, 10 mg urea N/ml

Weigh exactly 21.4383 g of urea and transfer to a 1 L volumetric flask. Dissolve and dilute to volume with Standard diluent (D-32). Mix thoroughly and store in a tightly closed polyethylene or Pyrex bottle.

(D-34) Sodium hydroxide, 0.5 N

Prepare by using a concentrated solution of NaOH that has been standardized.

$$\text{ml} \times \text{N} = \text{ml} \times \text{N}$$

(D-35) Picric acid, saturated

Heat 1 L of distilled water to boiling in a 2 L flask or beaker. Remove from heat and add 11.75 g of picric acid. (Keep away from an open flame). Allow to cool to room temperature and filter. Store in a brown, glass-stoppered bottle.
NOTE: Reagent grade picric acid will have 10% to 12% water added.

(D-36) Creatinine stock standard, 1 mg/ml

Weigh 1.00 g of creatinine and transfer to a 1 L volumetric flask. Dissolve and dilute to volume with 0.1 N HCl. Mix and store in a polyethylene bottle.

(D-37) Hydrochloric acid, 0.1 N

Using a 10 ml serologic pipette attached to a rubber bulb or vacuum line, transfer 8.4 ml of concentrated, reagent grade hydrochloric acid to a 1 L volumetric flask containing about 700 ml of water. Dilute to volume and mix. Standardize. Be sure that neither the concentrated nor the dilute acid is exposed to ammonia fumes. *Do not pipette strong acids by mouth.*

(D-38) Hydrochloric acid, 0.02 N

Measure 1.67 ml of concentrated hydrochloric acid into a 1 L volumetric flask containing about 700 ml of water. Dilute to volume with water and mix. Store in a glass or polyethylene bottle.

(D-39) Phosphotungstic acid

Measure 300 ml of water and transfer to a 1 L boiling flask with a ground glass joint top. Add 40 g of sodium tungstate and swirl until dissolved. Add several glass beads to the flask. Add 32 ml of 85% *o*-phosphoric acid and mix. Attach a reflux condenser to the flask and boil gently for 2 hours. Cool to room temperature. While the reflux condenser is still attached to the boiling flask, rinse the inside of the condenser with about 50 ml of water and allow the washings to run down into the flask. Transfer the contents of the flask to a 1 L volumetric flask. Rinse the boiling flask three times with about 30 ml of water and transfer washings to the volumetric flask. Dilute to volume with water. Add 32 g of lithium sulfate, Li_2SO_4, and mix until completely dissolved. Filter and store in an amber bottle. This reagent is stable if kept refrigerated when not in use.

(D-40) Sodium cyanide stock, 10%

Weigh 100 g of sodium cyanide, NaCN, and transfer to a 1 L volumetric flask. Add about 800 ml of water and mix until all of the cyanide is dissolved. Add 2 ml of concentrated ammonium hydroxide and dilute to volume with water. Filter and store in a polyethylene bottle in a refrigerator.
CAUTION: This is extremely poisonous. *Do not pipette by mouth.*

(D-41) Urea stock, 20%

Weigh 200 g of urea, NH_2CONH_2, and transfer to a 1 L volumetric flask. Dissolve and dilute to volume with water. Filter and store in a polyethylene bottle.

(D-42) Cyanide-urea working solution

This reagent should be freshly prepared on the day of use. Allow about 150 ml for 1 hour's running time. Mix one part sodium cyanide stock with one part urea stock. *Do not pipette by mouth.*

(D-43) Uric acid stock standard, 1 mg uric acid/ml

Weigh 1.0 g of uric acid and transfer to a 1 L volumetric flask. Weigh 0.6 g of lithium carbonate and transfer to a 250 ml beaker containing 150 ml of water. Mix until completely dissolved. Filter off any insoluble material and heat the solution to 60° C. Add this solution to the volumetric flask and swirl the flask under running hot water until the uric acid is dissolved (about 5 minutes). Then cool the flask under cold running water. Add 20 ml of 37% formalin and about 400 ml of water. With mixing add 25 ml of 1 N sulfuric acid. Dilute to volume with water and mix thoroughly. Store in an amber bottle.

(D-44) Hydrochloric acid, 0.25 N

Place about 500 ml of water in a 1 L volumetric flask. Carefully measure 21 ml of concentrated hydrochloric acid in a 25 ml graduated cylinder and add to the flask. Rinse the cylinder with water and add the washings to the flask. Swirl to mix. Dilute to volume with water and mix. Add 0.5 ml of Brij 35 and mix thoroughly. Store in a polyethylene bottle.

(D-45) Cresolphthalein complexone, 0.01%, with 8-hydroxyquinoline

Weigh 0.100 g of cresolphthalein complexone and transfer to a 1 L volumetric flask containing about 700 ml of 0.25 N hydrochloric acid. Place the flask on a magnetic stirrer and allow to stir until all the dye is dissolved. Then add 2.5 g of 8-hydroxyquinoline and mix until dissolved. Dilute to volume with 0.25 N hydrochloric acid and mix. Filter and store in a polyethylene bottle. *Do not* add any wetting agent to this reagent.

(D-46) Calcium base, 3.75% diethylamine

It is recommended that this reagent be prepared in a hood. Place a 1 L volumetric flask containing about 500 ml of water on a magnetic stirrer. Add 37.5 ml of diethylamine. Dilute to volume and mix. Store in a polyethylene bottle.

(D-47) Calcium stock standard, 50 mg Ca/100 ml

Weigh 1.250 g of calcium carbonate and transfer to a 1 L volumetric flask. Add about 100 ml of water. *Very slowly* add 7 ml of concentrated hydrochloric acid. Swirl the flask gently until all of the calcium carbonate is dissolved. Dilute to volume with water and mix thoroughly. Store in a polyethylene bottle.

(D-48) Magnesium stock standard, 100 mg Mg/100 ml

Weigh 8.36 g of magnesium chloride, $MgCl_2 \cdot 6H_2O$, and transfer to a 1 L volumetric flask containing about 500 ml of water. Dissolve and dilute to volume with water. Mix thoroughly and store in a polyethylene bottle.

(D-49) Hydrazine sulfate, 0.2%, in 1 N H_2SO_4

Place a 2 L volumetric flask containing about 1500 ml of water on a magnetic stirrer. Slowly add 56 ml of concentrated sulfuric acid and allow to stir until thoroughly mixed. Add 4.0 g of hydrazine sulfate and stir until dissolved. Dilute to volume with water and mix. Store in a polyethylene or amber glass bottle.

(D-50) Stannous chloride–hydrazine sulfate working solution

Weigh 0.200 g of stannous chloride, $SnCl_2 \cdot 2H_2O$, and transfer to a 1 L volumetric flask containing about 750 ml of 0.2% hydrazine sulfate. Mix until dissolved. Dilute to volume with 0.2% hydrazine sulfate and mix thoroughly. Store in a polyethylene or amber glass bottle. This reagent is stable for at least 2 weeks at room temperature.

(D-51) Acid molybdate

Place a 1 L volumetric flask containing about 800 ml of water into a pan of ice water for about 10 minutes. Then slowly add 35 ml of concentrated sulfuric acid and swirl to mix. Add 10 g of ammonium molybdate, $(NH_4)_6Mo_7O_{24} \cdot 4H_2O$, and mix until dissolved. Allow the solution to come to room temperature and dilute to volume with water. Mix thoroughly and store in a polyethylene bottle.

(D-52) Phosphorus stock standard, 1 mg P/ml

Weigh exactly 4.3937 g of monobasic potassium phosphate, KH_2PO_4, on an analytical balance and transfer to a 1 L volumetric flask. Dissolve and dilute to volume with water. Mix thoroughly and store in a polyethylene bottle.

(D-53) SGOT substrate

Place a 1 L beaker containing about 800 ml of water on a magnetic stirrer. Slowly add 33.5 g of dibasic potassium phosphate, K_2HPO_4, and mix until completely dissolved. Add 1.0 g of monobasic potassium phosphate, KH_2PO_4, and mix until dissolved. Add 2.66 g of *l*-aspartic acid and mix until dissolved. Add 1.462 g α-ketoglutaric acid and mix until dissolved. Add 1.0 g of tetrasodium EDTA (tetrasodium ethylenediamine tetra-acetate) and mix until dissolved. Check the pH with a pH meter. The substrate should be pH 7.40. If necessary, adjust the solution to pH 7.40 with either 0.1 N HCl or 0.1 N NaOH. Transfer the solution to a 1 L volumetric flask. Rinse the beaker three times with about 20 ml of water and add washings to the flask. Dilute to volume with water and mix thoroughly. Filter and transfer to a polyethylene bottle. Add 0.5 ml of chloroform and 1.0 ml of Triton X-405. Mix and store in a refrigerator.

(D-54) Phenyl mercuric acetate, 0.04%

Weigh 0.4 g of phenyl mercuric acetate and transfer to a 1 L volumetric flask containing about 850 ml of hot water. Mix until dissolved. Allow to cool to room temperature, dilute to volume, and mix. Filter and store in a polyethylene bottle.

(D-55) Citrate buffer

Weigh 32.4 g of sodium citrate, $Na_3C_6H_5O_7 \cdot 2H_2O$, and transfer to a 1 L volumetric flask containing about 500 ml of water and 50 ml of 0.04% phenyl mercuric acetate. Mix until dissolved. Add 18.94 g of citric acid, $H_3C_6H_5O_7 \cdot H_2O$, and mix until dissolved. Dilute to volume with water and mix. Add 1 ml of Triton X-405 and mix thoroughly. Check the pH with a pH meter and adjust to pH 4.5, if necessary, using either 0.1 N NaOH or 0.1 N HCl. Store in a poylethylene bottle.

(D-56) Azoene fast red dye diluent

Measure 150 ml of 95% ethyl alcohol and 5 ml of 1 N hydrochloric acid and transfer to a 1 L volumetric flask containing about 600 ml of water. Dilute to volume and mix. Store in a polyethylene bottle.

(D-57) Azoene fast red dye solution (prepare fresh daily)

Weigh 0.400 g of azoene fast red, PDC (fast ponceau L), and transfer to a 200 ml volumetric flask. Dissolve and dilute to volume with azoene fast red dye diluent. Mix, with as little exposure to air as possible, and filter in a refrigerator. Keep this reagent in an amber bottle in ice during use. This amount will be sufficient for 2½ hours of running time.

(D-58) Alcoholic KOH wash solution, 2%

Dissolve 2 g of potassium hydroxide pellets in 100 ml of 95% ethyl alcohol.

(D-59) AMP stock buffer, 50% (w/v)

Weigh 500 g of 2-amino-2-methyl-1-propanol* in a beaker or Erlenmeyer flask. Transfer to a 1 L volumetric flask. Rinse the beaker or flask three times with 100 ml quantities of water. Add the washings to the volumetric flask. Dilute to volume with water and mix thoroughly. Store in a Pyrex or polyethylene bottle.

(D-60) AMP working buffer

Place a beaker containing about 700 ml of water on a magnetic stirrer. Add 118 ml of AMP stock buffer. Using a pH meter, adjust the pH to 10.25 ± 0.05 with 5.0 N hydrochloric acid. About 25 to 27 ml is required. Transfer to a 1 L volumetric flask and rinse the beaker two times with 50 ml quantities of water. Transfer washings to the volumetric flask. Dilute to volume with water. Add 1 ml of Triton X-405 and mix thoroughly. Store in a polyethylene bottle.

(D-61) Magnesium chloride, 1.0 M

Weigh 9.532 g of magnesium chloride, $MgCl_2 \cdot 6H_2O$, and transfer to a 100 ml volumetric flask containing about 80 ml of water. Swirl until dissolved. Add two drops of concentrated hydrochloric acid. Dilute to volume with water and mix. Store in a Pyrex or polyethylene bottle.

(D-62) *p*-Nitrophenyl phosphate substrate

Measure 800 ml of AMP working buffer and transfer to a 1 L volumetric flask. Add 2.0 g of *p*-nitrophenyl phosphate and swirl to dissolve. Add 2 ml of 1 M magnesium chloride. Dilute to volume with water and mix thoroughly. This substrate is stable for 2 days when stored in the refrigerator. During use, the bottle should be kept in an ice bath.

(D-63) Magnesium chloride, 1% Mg^{++}

Weigh 8.36 g of magnesium chloride, $MgCl_2 \cdot 6H_2O$, and transfer to a 100 ml volumetric flask. Dissolve and dilute to volume with water. Mix thoroughly and store in a polyethylene bottle.

(D-64) Alkaline buffer with magnesium

Weigh 1.58 g of anhydrous sodium carbonate and transfer to a 1 L volumetric flask containing about 500 ml of water. Mix until all the carbonate is dissolved. Then add 0.84 g of sodium bicarbonate and mix until it is dissolved. Add 2 ml of magnesium chloride, 1% Mg^{++}, and 1 ml of Triton X-405. Dilute to volume with water and mix thoroughly. Filter through a good grade of filter paper. Check the pH with a pH meter. Adjust if necessary to pH 10.00 ± 0.05. Store in a polyethylene bottle.

(D-65) Phenyl disodium phosphate substrate

Prepare fresh daily. Weigh 2.0 g of phenyl disodium phosphate (phenol-free), $Na_2C_6H_5PO_4$, and transfer to a 1 L volumetric flask. Dissolve and dilute to volume with alkaline buffer with magnesium. Mix thoroughly and store in a polyethylene bottle in a refrigerator until needed.

(D-66) 4-Aminoantipyrine

Prepare fresh daily. Weigh 1.0 g of 4-aminoantipyrine and transfer to a 1 L volumetric flask. Dissolve and dilute to volume with water. Store in an amber bottle in a refrigerator until needed.

(D-67) Potassium ferricyanide, 0.5%

Weigh 5.0 g of potassium ferricyanide, $K_3Fe(CN)_6$, and transfer to a 1 L volumetric flask. Dissolve and dilute to volume with water. Filter through a good grade of filter paper and store in an amber bottle.

(D-68) LDH substrate

Place a 1 L volumetric flask on a magnetic mixer and add about 50 ml of water. Begin mixing gently and slowly add 6 g of sodium hydroxide. When it has dissolved, slowly add 15 ml of 85% *dl*-lactic acid to the warm solution and continue mixing for at least 30 minutes. Add about 500 ml of water to the flask and then add 12.1 g of Tris [Tris(hydroxymethyl)aminomethane] to the mixing solution. When the Tris is dissolved, add water to bring the volume to about 950 ml. Using a pH meter, adjust the pH of the substrate to 8.8 ± 0.05 with 1 N HCl. This will require about 7.5 ml. Dilute to volume with water, add 1.5 ml of Triton X-100, and mix well. Filter and store in a polyethylene bottle. Although this reagent is stable for at least 4 weeks at room temperature, it should be stored in a refrigerator for maximum stability.

(D-69) Phosphate buffer, 0.067 M, pH 7.4

Place a 1 L beaker containing about 800 ml of water on a magnetic mixer. Add 9.1 g of mono-

*Eastman P-4780.

basic potassium phosphate, KH_2PO_4, to the mixing water. When the phosphate is dissolved, add 2.2 g of sodium hydroxide and continue mixing until it is dissolved. Allow to cool to room temperature. Using a pH meter, check the pH of the buffer. Adjust to pH 7.4, if necessary, with 0.01 N NaOH or 0.01 N HCl. Transfer the solution to a 1 L volumetric flask. Rinse the beaker with water and add washings to the flask. Dilute to volume with water and mix well. Store in a polyethylene bottle.

(D-70) NAD-diaphorase

Weigh 0.30 g of bovine albumin fraction V powder and transfer to a 250 ml beaker containing 175 ml of 0.067 M phosphate buffer, pH 7.4. Place the beaker on a magnetic mixer and mix until dissolved. Using a pH meter, check pH and adjust to 7.4 if necessary. Add 605 Worthington units of diaphorase and mix. Add 1.2 g of nicotinamide adenine dinucleotide (NAD) and mix until dissolved. Transfer to a 200 ml volumetric flask. Rinse beaker with three 5 ml portions of phosphate buffer and add washings to the flask. Dilute to volume with phosphate buffer and mix thoroughly. This solution should be divided into quantities sufficient for daily use and stored in a freezer. About 15 ml/hr is used. Thaw the reagent at room temperature and place in an ice bath during use. It is stable for 2 days when kept refrigerated or iced. This material is available from Technicon Corp. in a lyophilized form. It can be diluted with distilled water in quantities sufficient for daily use.

(D-71) INT dye

Weigh 0.625 g of INT, 3-*p*-nitrophenyl-1-2-*p*-iodophenyl-5-phenyl tetrazolium chloride, and transfer to a 1 L volumetric flask. Dissolve and dilute to volume with water. Store in an amber glass bottle. The dye is stable for at least 3 months at room temperature if it is protected from light.

(D-72a) LDH color reagent

Weigh 0.20 g of 2-*p*-iodophenyl-3-*p*-nitrophenyl-5-phenyl tetrazolium chloride (INT)* and transfer to a 100 ml beaker containing about 75 ml of water. Place on a magnetic stirrer and mix until dissolved. Add 0.50 g of nicotinamide adenine dinucleotide (NAD) and mix until dissolved. Add 0.05 g of phenazine methosulfate (PMS) and mix until dissolved. Transfer immediately to a low-actinic 100 ml volumetric flask. Rinse the beaker with water and transfer washings to the flask. Dilute to volume with water and mix. This reagent should be protected from light at all times as it is very light-sensitive. Store in a low-actinic glass reagent bottle in a refrigerator. It is stable for several weeks.

*Dajac Laboratories.

(D-72b) LDH color reagent

Weigh 0.100 g of 2-*p*-iodophenyl-3-*p*-nitrophenyl-5-phenyl tetrazolium chloride (INT)* and transfer to a 100 ml beaker containing about 75 ml of water. Place on a magnetic stirrer and mix until dissolved. Add 0.250 g of nicotinamide adenine dinucleotide (NAD) and mix until dissolved. Add 0.025 g of phenazine methosulfate (PMS) and mix until dissolved. Transfer immediately to a low-actinic 100 ml volumetric flask. Rinse the beaker with water and transfer washings to the flask. Dilute to volume with water and mix. This reagent should be protected from light at all times as it is very light-sensitive. Store in a low-actinic glass reagent bottle in a refrigerator. It is stable for several weeks.

NOTE: This reagent is one half the concentration of the color reagent used in the manual and Robot Chemist methods.

(D-73) LDH buffer reagent

Dissolve 1.0 g of ethoxylated oleyl alcohol (Lipal 10-OA)† in 10 ml of water by heating to about 95° C. Add about 50 ml of water and add 12.1 g of tris(hydroxymethyl)aminomethane. Mix to dissolve. Adjust to pH 8.2 with 3 N HCl. Transfer to a 100 ml volumetric flask. Rinse the beaker with water and add washings to the flask. Dilute to volume with water and mix. Store in a refrigerator and discard if any visible growth occurs. Lipal 10-OA may be omitted if only serum is to be assayed.

(D-74) LDH substrate, 0.1 M *l*(+) lactate

Add 5.0 ml of *l*(+) lactic acid, 20% solution, to about 50 ml of water. Adjust to pH 5.5 with 1 N NaOH. Dilute to 120 ml with water and mix. Add a few drops of chloroform and store in a refrigerator.

(D-75) LDH control reagent

Weigh 0.2 g of potassium oxalate and 0.2 g of ethylenediamine tetra-acetic acid, disodium dihydrate, and transfer to a 100 ml volumetric flask. Dissolve and dilute to volume with water. Store in a polyethylene bottle.

(D-76) *l*(+) Lactate, 2 M

Dilute 60 ml of 30% *l*(+) lactic acid (Sigma L1875) to 100 ml with water. Adjust to pH 5.5. Store in a refrigerator.

(D-77) *l*(+) Lactate, 0.02 M, in 2 M urea

Weigh 12.012 g of urea and transfer to a 100 ml volumetric flask containing about 80 ml of water.

*Dajac Laboratories.
†Drew Chemical Co.

Mix to dissolve. Add 1 ml of 2 M *l*(+) lactate and dilute to volume with water. Mix well. Adjust to pH 5.5. Transfer to a polyethylene bottle and store in a refrigerator.

(D-78) Lieberman-Burchard color reagent

Measure 350 ml of glacial acetic acid and transfer to a 2 L Erlenmeyer flask. Slowly add 550 ml of acetic anhydride. Seal to prevent absorption of atmospheric moisture. Cool to 5° C by overnight refrigeration or by placing in an ice bath. Cool 100 ml of concentrated sulfuric acid to 5° C. When both reagents are chilled, place the Erlenmeyer flask on a magnetic mixer. *Slowly* add the 100 ml of sulfuric acid to the stirring mixture. Add 10 g of anhydrous sodium sulfate and continue stirring until dissolved.

CAUTION: Reagents must be precooled to prevent charring when the sulfuric acid is added to the acetic acid–anhydride mixture. Divide into convenient aliquots and store in glass bottles in the refrigerator.

(D-79) Caffeine diluent

Place about 500 ml of moderately hot (50° to 60° C) distilled water in a 1 L volumetric flask. Add 37.5 g of caffeine, purified alkaloid, $C_{18}H_{10}N_4O_2$, 57.0 g of sodium benzoate, USP, C_6H_5COONa, and 94.5 g of sodium acetate, $CH_3COONa \cdot 3H_2O$. Mix until dissolved. Allow to cool to room temperature and dilute to volume with water. Mix thoroughly. Store in a polyethylene bottle. This reagent is stable for at least 6 months at room temperature. The formula has been increased by 50% over the formulation in previously published AutoAnalyzer methods. This was done to solve the problem of an occasional false low bilirubin due to insufficient caffeine concentration (Technicon Laboratory method file N-51a).

(D-80) Sulfanilic acid (diazo I)

Weigh 10 g of sulfanilic acid and transfer to a 1 L volumetric flask containing about 600 ml of water. Mix until dissolved and add 15 ml of concentrated hydrochloric acid. Dilute to volume with water and mix thoroughly. Store in a polyethylene bottle.

(D-81) Sodium nitrite, 0.5% (diazo II)

Weigh 0.5 g of sodium nitrite, $NaNO_2$, and transfer to a 100 ml volumetric flask. Dissolve and dilute to volume with water. Mix thoroughly and store in an amber glass bottle in a refrigerator. This reagent is stable for about 3 weeks.

(D-82) Diazo reagent, combined

Prepare fresh daily. Mix 2.5 ml of diazo II reagent with every 100 ml of diazo I.

(D-83) Tartrate buffer

Weigh 50 g of sodium hydroxide and 175 g of potassium sodium tartrate and transfer to a 1 L volumetric flask containing about 500 ml of water. Dissolve and dilute to volume with water. Mix thoroughly and store in a polyethylene bottle. This buffer is stable for at least 6 months at room temperature.

(D-84) Ascorbic acid, 4%

Prepare fresh daily. Weigh 1 g of ascorbic acid and transfer to a 25 ml volumetric flask. Dissolve and dilute to volume with water. Mix thoroughly.

(D-85) Hydrochloric acid, 0.05 N

Place about 900 ml of water in a 1 L volumetric flask and add 4.2 ml of concentrated hydrochloric acid. Dilute to volume and mix well. Store in a polyethylene or amber glass bottle.

(D-86) Alkaline iodide

Weigh 8.0 g of sodium hydroxide and transfer to a 1 L volumetric flask containing about 800 ml of water. Mix until dissolved. Add 5.0 g of potassium iodide and mix until dissolved. Dilute to volume with water and mix thoroughly. Store in a polyethylene bottle.

(D-87) Biuret stock solution

Weigh 45.0 g of potassium sodium tartrate, $KNaC_4H_4O_6 \cdot 4H_2O$, and transfer to a 1 L volumetric flask. Add 400 ml of 0.2 N sodium hydroxide and mix until the tartrate is dissolved. Add 15.0 g of copper sulfate, $CuSO_4 \cdot 5H_2O$, and mix until dissolved. Dilute to volume with 0.2 N sodium hydroxide and mix thoroughly. Filter and store in a Pyrex or polyethylene bottle.

(D-88) Biuret working solution

Thoroughly mix 200 ml of stock biuret with 800 ml of alkaline iodide solution. Store in a polyethylene bottle.

(D-89) Protein blank solution

Weigh 9.0 g of potassium sodium tartrate and transfer to a 1 L volumetric flask. Dissolve and dilute to volume with alkaline iodide solution. Store in a polyethylene bottle.

(D-90) HABA dye stock solution, 6×10^{-3} M

Dissolve 1.452 g of 2-(4′-hydroxyazobenzene)benzoic acid (HABA)* in 10 ml of 1 N sodium hydroxide in a 1 L volumetric flask. Dilute to volume with water and mix thoroughly. Store in an amber bottle.

*Dajac Laboratories.

(D-91) Phosphate buffer

Weigh 4.662 g of dibasic potassium phosphate, K_2HPO_4, and transfer to a 2 L volumetric flask containing about 800 ml of water. Mix until dissolved. Add 16.595 g of monobasic potassium phosphate and mix until dissolved. Check pH with a pH meter. Adjust to pH 6.05 ± 0.05 if necessary. Dilute to volume and mix.

(D-92) HABA dye working solution, 6×10^{-4} M

Measure 100 ml of stock HABA dye and transfer to a 1 L volumetric flask. Dilute to volume with phosphate buffer and mix thoroughly. Check pH with a pH meter and adjust to pH 6.2 to 6.4 if necessary. Store in a brown bottle. This reagent is stable for 3 days if it is kept refrigerated between use. Bring to room temperature before using.

(D-93) Albumin stock standard, 10 g/100 ml

To prepare a standard containing 10 g of protein/100 ml it is necessary to know the Kjeldahl nitrogen factor for the lot of bovine albumin used. Crystalline bovine albumin fraction V, may be obtained from Armour Pharmaceutical Co. To calculate the amount of bovine albumin needed use the following formula:

$$10 \text{ g} \times \frac{160}{\text{N factor}} = \text{grams of albumin needed}$$

Weigh this amount and very carefully transfer it to a 250 ml beaker. The albumin is very light and fluffy and, unless care is taken, some of the material may be lost in the transfer. Add about 80 ml of saline and carefully stir until all the albumin is in solution. Transfer the solution to a 100 ml volumetric flask. Rinse the beaker with small quantities of saline and add washings to the flask. Dilute to volume with saline, add 0.05 g of sodium azide, and mix thoroughly. Duplicate Kjeldahl nitrogen determinations should be performed on this solution to determine the actual protein concentration.

(D-94) Buffered urease solution

Weigh 150 mg of urease* and transfer to a 100 ml volumetric flask. Add 1.0 g of ethylenediamine tetraacetic acid (EDTA). Dissolve and dilute to volume with water. This solution is stable for 1 month when stored in a refrigerator.

(D-95) Phenol color reagent

Weigh 50.0 g of reagent grade phenol and 0.25 g of reagent grade sodium nitroprusside and transfer to a 1 L volumetric flask. Dissolve and dilute to volume with water. This solution is stable for at least 2 months if stored in a brown bottle in a cool place.

(D-96) Alkali-hypochlorite reagent

Weigh 25.0 g of reagent grade NaOH and transfer to a 1 L volumetric flask. Add 40.0 ml of Clorox (commercial bleach). Dilute to volume with water and mix. Store in an amber bottle protected from light and heat. This solution is stable for at least 2 months.

(D-97) Urea stock standard, 1 mg urea N/ml

Weigh 2.145 g of urea and transfer to a 1 L volumetric flask. Dissolve and dilute to volume with water. Add a few drops of chloroform. Store in a refrigerator in a glass bottle.

(D-98) Cholesterol color reagent

Measure 600 ml of acetic anhydride and transfer to a 2 L Erlenmeyer flask. Cool to 0° to 4° C in an ice and salt water bath. In a separate container, cool 100 ml of concentrated sulfuric acid to the same temperature. Add the cold sulfuric acid to the cold acetic anhydride *slowly with mixing*. Do not allow the temperature of the mixture to exceed 10° C. Add 300 ml of glacial acetic acid and mix well. Add 20 g of anhydrous sodium sulfate and mix until dissolved. If the temperature is controlled properly during mixing, the solution should be colorless. Store in a refrigerator. Discard if a yellow color develops.

(D-99) Acetic acid, 95%

Measure 50 ml of water and transfer to a 2 L Erlenmeyer flask. Add 950 ml of glacial acetic acid with mixing. Store in a glass bottle at room temperature.

(D-100) Glycine buffer

Weigh 7.5 g of glycine and transfer to a 2 L beaker containing about 900 ml of water. Place on a magnetic stirrer and mix. Adjust to pH 9.0, using 10 N sodium hydroxide followed by 1 N sodium hydroxide. Transfer to a 1 L volumetric flask. Rinse beaker, adding washings to the volumetric flask. Dilute to volume with water. Mix and store in a polyethylene bottle.

(D-101) Uricase solution, buffered

Prepare just prior to use. Dissolve 120 mg of high-activity uricase (45 units/mg) in 100 ml of glycine buffer solution.

---

*Urease, Type II, about 800 to 1000 Sumner units/g, Sigma Chemical Co.

(D-102) Cupric sulfate, 10%

Weigh 10 g of cupric sulfate and transfer to a 100 ml volumetric flask. Dissolve and dilute to volume with water. Store in a polyethylene bottle.

(D-103) 2-Amino-2-methyl-1-propanol, 50%

Dissolve 50 g of 2-amino-2-methyl-1-propanol in 50 ml of water. Store in an amber glass bottle.

(D-104) Neocuproine solution

Weigh 1 g of neocuproine hemihydrate and transfer to a 250 ml beaker containing about 200 ml of water. Place on a magnetic stirrer and mix until dissolved. Adjust to pH 2.0 using 5 N hydrochloric acid. Readjust to pH 4.5 with 1 N sodium hydroxide. Transfer the solution to a 250 ml volumetric flask. Rinse the beaker and add the washings to the flask. Dilute to volume with water. Store in a Pyrex or polyethylene bottle.

(D-105) Cupric sulfate solution, buffered

Add 19 ml of 50% 2-amino-2-methyl-1-propanol solution and 3.9 ml of 10% cupric sulfate solution to about 800 ml of water in a 1 L beaker. Adjust to pH 9.0 with 5 N hydrochloric acid. A cupric-amine complex precipitates at first but redissolves as the pH is lowered. Transfer to a 1 L volumetric flask. Rinse the beaker and add washings to the flask. Dilute to volume with water and mix. Store in a polyethylene bottle. This solution is stable indefinitely.

(D-106) Uric acid color reagent, combined

Mix 3 volumes of the neocuproine solution with 20 volumes of buffered cupric sulfate solution. Adjust to pH 9.0 if necessary. Stable for about 2 days if refrigerated.

(D-107) Uric acid stock standard, 2 mg/ml

Weigh 0.200 g of uric acid and transfer to a 100 ml volumetric flask containing about 90 ml of 0.1 M lithium carbonate. Place on a magnetic stirrer and mix until dissolved. Remove stirring bar and dilute to volume with 0.1 M lithium carbonate. Mix thoroughly. This standard is stable for about 1 week when stored in a refrigerator.

(D-108) Starch substrate

Weigh 13.3 g of anhydrous disodium phosphate and 4.3 g of benzoic acid and transfer to a 500 ml beaker. Add about 250 ml of water and mix to dissolve. Heat to boiling. In a separate beaker mix 0.200 g of Merck's soluble starch (Lintner) in 5 ml of cold water and add it to the boiling mixture. Rinse the beaker with additional cold water and add the washings to the boiling solution. Continue boiling for 1 minute. Cool to room temperature. Transfer to a 500 ml volumetric flask. Rinse the beaker with water and add the washings to the flask. Dilute to volume with water and mix. This solution is stable at room temperature and should remain water-clear.

(D-109) Iodine reagent

a. Stock iodine solution, 0.1 N
   Weigh 3.567 g of potassium iodate, KIO_3, and 45 g of potassium iodide and transfer to a 1 L volumetric flask containing about 800 ml of water. Mix to dissolve. Slowly and with mixing add 9 ml of concentrated hydrochloric acid. Dilute to volume with water and mix.
b. Working iodine solution, 0.01 N
   Weigh 25 g of potassium fluoride and transfer to a 500 ml volumetric flask containing about 350 ml of water. Mix to dissolve. Add 50 ml of the stock iodine solution and dilute to volume with water. This solution is stable for 1 to 2 months when stored in an amber bottle in a refrigerator.

(D-110) Sodium phosphate, dibasic, 0.1 M

Weigh 14.1959 g of dibasic sodium phosphate, Na_2HPO_4, and transfer to a 1 L volumetric flask. Dissolve and dilute to volume with water.

(D-111) Potassium phosphate, monobasic, 0.1 M

Weigh 13.609 g of monobasic potassium phosphate, KH_2PO_4, and transfer to a 1 L volumetric flask. Dissolve and dilute to volume with water.

(D-112) Phosphate buffer, pH 7.4

Mix 840 ml of 0.1 M Na_2HPO_4 with 160 ml of 0.1 M KH_2PO_4. Adjust to pH 7.4 if necessary. Store in a polyethylene bottle.

(D-113) Pyruvic acid standard, 0.5 μM/ml

Weigh 0.055 g of pyruvic acid, sodium salt, and transfer to a 1 L volumetric flask. Dissolve and dilute to volume with phosphate buffer. Store in an amber bottle in a refrigerator.

(D-114) CPK blank substrate

Weigh 0.73 g of adenosine-5′-triphosphate (ATP), disodium $3H_2O$, crystalline (Sigma grade), 0.1864 g of phospho(enol)pyruvic acid (PEP), tricyclohexylamine salt, crystalline, 0.2964 g of magnesium sulfate, $MgSO_4{\cdot}7H_2O$, and 3.64 g of tris(hydroxymethyl)aminomethane (Sigma Trizma base) and transfer to a 200 ml volumetric flask. Add about 160 ml of water and mix to dissolve. Add 0.6 ml of pyruvate kinase, type II, crystalline (from rabbit skeletal muscle); ammonium sulfate suspension, 10 mg/ml. Mix and dilute to volume with water. Adjust to pH 9.0 if necessary, using 10% HCl or 10% NaOH.

(D-115) Creatine substrate

Weigh 0.9956 g of creatine·H_2O and transfer to a 100 ml volumetric flask. Dissolve and dilute to volume with CPK blank substrate. These two substrates are stable for up to 1 month when stored in a refrigerator.

(D-116) 2,4-Dinitrophenylhydrazine, 0.001 M

Weigh 0.1981 g of 2,4-dinitrophenylhydrazine and transfer to a 1 L volumetric flask. Dissolve and dilute to volume with 1 N HCl. Store in a polyethylene bottle in a refrigerator.

(D-117) Mercaptoethanol, 0.25 M

Pipette 17.1 ml of mercaptoethanol into a 1 L volumetric flask. Dilute to volume with water and mix. Store in an amber bottle in the refrigerator. This solution is stable for about 1 month.

(D-118) Glycine buffer

Weigh 15.0 g of glycine and 0.190 g of magnesium chloride, $MgCl_2$, and transfer to a 1 L volumetric flask containing about 750 ml of water. Mix until dissolved. Add 170 ml of 1 N sodium hydroxide, mix, and dilute to volume with water. Mix well and store in a polyethylene bottle in a refrigerator.

(D-119) *p*-Nitrophenyl phosphate stock substrate

Weigh 0.8 g of disodium *p*-nitrophenyl phosphate and transfer to a 100 ml volumetric flask. Dissolve and dilute to volume with 0.002 N hydrochloric acid. Adjust to pH 6.5 to 8.0 if necessary. Store in a polyethylene bottle in a refrigerator.

(D-120) *p*-Nitrophenyl phosphate working substrate

Mix equal parts of glycine buffer and *p*-nitrophenyl phosphate stock substrate. Adjust to pH 10.3 to 10.4 if necessary. Mix just prior to use or store in a freezer.

(D-121) Hydrochloric acid, 6 N

Slowly and with mixing add 500 ml of concentrated hydrochloric acid to 500 ml of water. Mix well and store in a Pyrex bottle.

(D-122) Sodium hydroxide, 0.0075 N

Pipette 7.5 ml of 1 N sodium hydroxide into a 1 L volumetric flask containing about 800 ml of water. Mix, dilute to volume with water, and mix again. Store in a polyethylene bottle.

(D-123) Carbonate buffer, pH 10.0

Weigh 6.36 g of anhydrous sodium carbonate and 3.36 g of sodium bicarbonate and transfer to a 1 L volumetric flask containing about 700 ml of water. Mix to dissolve. Dilute to volume with water and mix well. Store in a polyethylene bottle.

(D-124) Disodium phenyl phosphate substrate

Weigh 2.18 g of disodium phenyl phosphate and dissolve and dilute to 1 L. Quickly bring to a boil to "sterilize" and cool immediately in a water bath. Add a few drops of toluol as a preservative and store in a refrigerator.

(D-125) Sodium bicarbonate, 0.5 M

Weigh 42 g of sodium bicarbonate and transfer to a 1 L volumetric flask. Dissolve and dilute to volume with water. Store in a polyethylene bottle.

(D-126) 4-Aminoantipyrine, 0.6%

Weigh 3.0 g of 4-aminoantipyrine and transfer to a 500 ml volumetric flask. Dissolve and dilute to volume with water. Store in a polyethylene bottle in a refrigerator.

(D-127) Potassium ferricyanide, 2.4%

Weigh 12.0 g of potassium ferricyanide, $K_3Fe(CN)_6$, and transfer to a 500 ml volumetric flask. Dissolve and dilute to volume with water. Store in a polyethylene bottle.

(D-128) SGOT substrate, α-ketoglutarate, 2 mM/L; *dl*-aspartate, 200 mM/L

Weigh 0.0292 g of α-ketoglutaric acid and 2.66 g of *dl*-aspartic acid and transfer to a small beaker. Add 20 ml of 1 N sodium hydroxide and stir until the acids are dissolved. Using a pH meter, adjust the solution to pH 7.4 by adding 1 N sodium hydroxide dropwise. Transfer quantitatively to a 100 ml volumetric flask, using buffer D-342 to rinse the beaker. Dilute to volume with buffer and mix. Add 1 ml of chloroform as a preservative and store in a refrigerator.

(D-129) Sodium hydroxide, 0.4 N

This may be prepared from a concentrated solution of sodium hydroxide that has been standardized.

Or: Weigh 16 g of sodium hydroxide pellets and transfer to a 1 L volumetric flask. Dissolve and dilute to volume with water. Mix well and store in a polyethylene bottle.

(D-130) SGPT substrate, α-ketoglutarate, 2 mM/L; *dl*-alanine, 200 mM/L

Weigh 0.0292 g of α-ketoglutaric acid and 1.78 g of *dl*-alanine and transfer to a 100 ml beaker. Add 20 ml of water and stir to dissolve. Using a pH meter, adjust the solution to pH 7.4 by adding 1 N sodium hydroxide dropwise. This will require about 10 drops. Transfer quantitatively to a 100 ml volumetric flask, using buffer D-342 to rinse the beaker. Dilute to volume with buffer and mix. Add 1 ml of chloroform as a preservative and store in a refrigerator.

(D-131) Sulfanilic acid, solution A, 0.1%

Weigh 100 mg of sulfanilic acid and transfer to a 100 ml volumetric flask. Add 1.5 ml of concentrated hydrochloric acid. Dilute to volume with water.

(D-132) Diazo reagent

Prepare just prior to use. Mix 10 ml of 0.1% sulfanilic acid with 0.3 ml of 0.5% sodium nitrite.

(D-133) Diazo blank solution

Transfer 1.5 ml of concentrated hydrochloric acid to a 100 ml volumetric flask. Dilute to volume with water and mix.

(D-134) Biuret reagent

a. Dissolve 17.3 g of copper sulfate, $CuSO_4 \cdot 5H_2O$, in 100 ml of hot water.
b. Dissolve 173 g of sodium citrate and 100 g of anhydrous sodium carbonate in 800 ml of water with heating.

Cool both solutions to room temperature and transfer both to a 1 L volumetric flask. Rinse containers and transfer washings to the flask. Mix and dilute to volume with water. Mix again and store in a polyethylene bottle.

(D-135) Sodium hydroxide, 15%

Weigh 150 g of sodium hydroxide and transfer to a 1 L volumetric flask containing about 800 ml of water. Mix to dissolve. Cool to room temperature and dilute to volume with water. Store in a polyethylene bottle.

(D-136) Phosphate buffer, pH 6.2-6.4

Weigh 11.75 g of dibasic potassium phosphate, K_2HPO_4, and 41.85 g of monobasic potassium phosphate, KH_2PO_4, and transfer to a 1 L volumetric flask containing about 700 ml of water. Mix to dissolve. Add 50 ml of 37% formaldehyde and mix. Adjust to pH 6.2 to 6.4 if necessary. Dilute to volume with water and mix. Store in a polyethylene bottle. This reagent is stable for at least 2 years at room temperature.

(D-137) HABA dye stock solution

Weigh 1.815 g of HABA dye, 2,4′-hydroxyazobenzene benzoic acid, and transfer to a 250 ml volumetric flask containing 12.5 ml of 1 N sodium hydroxide. Mix to dissolve. Dilute to volume with water. Mix and store in a polyethylene bottle.

(D-138) HABA dye working solution

Dilute 100 ml of stock HABA dye solution to 1 L with phosphate buffer. Allow to age at least 24 hours before use.

(D-139) Alkaline copper reagent (Nelson-Somogyi)

Weigh 24.0 g of sodium carbonate and 12.0 g of sodium potassium tartrate (Rochelle salt) and transfer to a 1 L volumetric flask. Add about 250 ml of water and swirl until all the material is dissolved. Weigh 4.0 g of copper sulfate ($CuSO_4 \cdot 5H_2O$) and transfer to a 250 ml beaker containing about 50 ml of water. Stir with a glass stirring rod until all crystals are dissolved. Slowly add this to the carbonate-tartrate solution and mix well. Weigh 16.0 g of sodium bicarbonate and transfer to the flask (use a powder funnel). Weigh 180 g of sodium sulfate and transfer to a 1000 ml beaker containing about 500 ml of hot water. Allow this solution to boil for 2 or 3 minutes to expel air; then cool. Add this to the flask containing the other reagents and dilute to the mark with water. Mix thoroughly. If a sediment forms after several days of standing the solution may be filtered.

(D-140) Arsenomolybdate color reagent

Weigh 50.0 g of ammonium molybdate and transfer to a 2000 ml beaker or flask containing 900 ml of water. Mix until all the molybdate goes into solution. Slowly add 42.0 ml of concentrated sulfuric acid, with constant stirring. Add 6.0 g of disodium arsenate ($Na_2HASO_4 \cdot 7H_2O$) and mix. Transfer to a brown, glass-stoppered bottle and place in an incubator at 37° C for 24 to 48 hours.

(D-141) Benzoic acid, 0.2%

Weigh 2.0 g of benzoic acid and transfer to a 1 L volumetric flask. Dilute to the mark with water and mix. Benzoic acid dissolves very slowly, so it is best to let this solution stand in a warm place for several hours before using.

(D-142) Glucose stock standard, 1.0 mg/ml

Carefully weigh exactly 1.000 g of glucose on an analytical balance and transfer to a 1 L volumetric flask. Dissolve and dilute to the mark with 0.2% benzoic acid solution. Mix well and store in a Pyrex or polyethylene bottle.

(D-143) Alkaline copper reagent (Folin-Wu)

Weigh 40.0 g of sodium carbonate and transfer to a 1 L volumetric flask containing about 400 ml of water. Swirl until all material is dissolved. Add 7.5 g of tartaric acid and 4.5 g of copper sulfate ($CuSO_4 \cdot 5H_2O$). When this is dissolved, dilute to the mark and mix thoroughly. If a sediment forms, the solution may be filtered.

(D-144) Phosphomolybdic acid

Place 400 ml of water and 400 ml of 10% sodium hydroxide in a 2 L Erlenmeyer flask. Add 70.0 g of molybdic acid and 10.0 g of sodium tungstate. Boil for 30 to 40 minutes to remove any ammonia. Cool to room temperature and add 250 ml of *o*-phosphoric acid (85%). Mix and transfer to a 1 L volumetric flask. Rinse the Erlenmeyer flask several times with about 10 ml of water and add these rinsings to the volumetric flask. Dilute to the mark with water and mix well. Store in a Pyrex or polyethylene bottle.

(D-145) Phosphomolybdic acid, alternate

Mix 500 ml of reagent D-144 with 35 ml of water and 215 ml of *o*-phosphoric acid. When using this reagent instead of D-144, 3 ml should be used instead of the usual 2 ml.

(D-146) Sodium hydroxide, 10% (2.5 N)

Dilute solutions of sodium hydroxide should be prepared by using a nearly saturated solution that has been standardized. If the normality of the concentrated solution is known, then the volume needed to make a dilute solution may be determined with the following formula:

$$\text{ml} \times \text{N} = \text{ml} \times \text{N}$$

Or: Weigh 100 g of sodium hydroxide and transfer to a 1 L volumetric flask containing 800 ml of water. Mix to dissolve, cool to room temperature, dilute to volume and mix.

(D-147) Picric acid, 0.040 M

Weigh 9.16 g of anhydrous or 10.17 g of reagent grade picric acid (containing 10% to 12% added water) and transfer to a 1 L volumetric flask. Dissolve and dilute to volume with water. Mix well and store in an amber bottle.

(D-148) Sodium hydroxide, 0.75 N

Weigh 30 g of sodium hydroxide and transfer to a 1 L volumetric flask containing about 800 ml of water. Mix to dissolve. Dilute to volume with water and mix thoroughly. Store in a polyethylene bottle.

(D-149) Uric acid stock standard, 1 mg/ml

Weigh 1.000 g of uric acid and transfer to a 1 L volumetric flask (Pyrex). Weigh 0.5 g of lithium carbonate and transfer to a 250 ml Erlenmeyer flask containing 150 ml of hot water. Add this warm solution of lithium carbonate to the flask containing the uric acid. Shake until all the uric acid is dissolved. Add about 300 ml of water and 25 ml of 40% formaldehyde; mix. Add 3 ml of glacial acetic acid and mix. After the liberated CO_2 has been removed from the solution by shaking, dilute to volume with water. Mix well. Store in a brown bottle.

(D-150) Polyanethol sodium sulfonate (Liquoid)

Weigh 2.0 g of polyanethol sodium sulfonate (Liquoid)* and carefully transfer to a 100 ml volumetric flask. Use a funnel. This material is very light and fluffy, so care must be taken to avoid losing any. Wash any material remaining in the funnel down into the flask with about 80 ml of distilled water. When all the Liquoid is dissolved, dilute to the 100 ml mark and mix thoroughly. Transfer to a brown bottle and store in a refrigerator.

(D-151) Glycerin-silicate reagent

Weigh 50.0 g of crystalline sodium silicate "soluble" and transfer to a 1 L beaker containing 500 ml of hot water. Stir until dissolved. Add 100 ml of glycerin. Stir until thoroughly mixed. Allow to come to room temperature. If the solution is cloudy, filter through hardened paper. Store in a Pyrex or polyethylene bottle.

(D-152) Phosphotungstic acid

Weigh 50.0 g of sodium tungstate, $Na_2WO_4 \cdot 2H_2O$, and transfer to a 1 L round-bottom flask. Add 400 ml of water and 40 ml of 85% *o*-phosphoric acid. Boil gently under a reflux condenser for 4 hours. Allow to cool to room temperature. Transfer to a 500 ml volumetric flask. Rinse the boiling flask twice with 10 ml of water and transfer rinsings to the volumetric flask. Dilute to the 500 ml mark with water. Mix thoroughly and store in a brown bottle.

(D-153) Uric acid special reagent

To 180 ml of water, add 80 ml of phosphotungstic acid (D-152) and 100 ml of Liquoid solution (D-150). Mix well and store in a brown bottle.

---

*May be obtained from Hoffman-La Roche, Inc.

(D-154) Sodium carbonate, 14%

Weigh 140.0 g of anhydrous sodium carbonate and transfer to a 1 L volumetric flask. Add about 800 ml of water and mix until all the carbonate is dissolved. Dilute to volume and mix again. Store in a Pyrex or polyethylene bottle.

(D-155) Sulfuric acid (digestion mixture), 30%

Measure 340 ml of water and transfer to a 1 L beaker or Erlenmeyer flask. Place the beaker or flask in a large pan of ice for 10 or 15 minutes. *Very carefully* add 150 ml of concentrated sulfuric acid. This mixture will be very hot! Carefully mix and allow to stand in the ice water until cool. Store in a Pyrex or polyethylene bottle.

(D-156) Hydrogen peroxide, 30% (Superoxol)

This may be obtained from any Merck supplier. It is a strong oxidant and should be handled with care. Store in a refrigerator when not in use.

(D-157) Nessler's reagent

Weigh 1000 g of sodium hydroxide pellets and dissolve in about 1500 ml of water. Cool and dilute to 2 L. Transfer this to a 5-gallon Pyrex bottle and add 8 L of water. Dissolve 150 g of potassium iodide (KI) in about 200 ml of water. Gradually add 200 g of mercuric iodide to the KI solution, with vigorous stirring. Dilute this solution to 2 L and add to the NaOH solution. Add 2 more liters of water. Stopper and mix thoroughly. Allow to settle for at least 3 weeks. The clear supernatant is siphoned off as needed.
NOTE: Do *not* make this reagent near the protein-bound iodine area if you are doing PBIs.

(D-158) Stock NPN or BUN standard, 0.5 mg N/ml

Weigh 2.3585 g of anhydrous ammonium sulfate, $(NH_4)_2SO_4$, and transfer to a 1 L volumetric flask. Dissolve and dilute to the mark with 0.1 N hydrochloric acid (ammonia-free).

(D-159) Urease solution

Weigh 15.0 g of jack bean meal and transfer to a 250 ml Erlenmeyer flask. Add 10.0 g of dry Permutit and 100 ml of approximately 20% ethyl alcohol (21 ml of 95% alcohol diluted to 100 ml with water). Stopper and shake for 30 minutes. Filter in refrigerator. Cover funnel with watchglass. This solution is stable for about 2 weeks when stored in the refrigerator.

(D-160) Potassium persulfate, 2.5%

Weigh 2.5 g of reagent grade, nitrogen-free potassium persulfate and transfer to a 100 ml volumetric flask. Dissolve and dilute to volume with water. Prepare a fresh solution weekly. Store in a refrigerator and remove only long enough to obtain the amount needed, as decomposition is rapid at higher temperatures.

(D-161) Potassium gluconate, 1%

Weigh 1.0 g of potassium gluconate and transfer to a 100 ml volumetric flask. Dissolve and dilute to volume with water. Prepare a fresh solution weekly and store in a refrigerator.

(D-162) Acetate buffer

Weigh 15.0 g of sodium acetate and transfer to a 100 ml volumetric flask. Add 1.0 ml of glacial acetic acid. Dilute to volume with water and mix.

(D-163) Uranyl zinc acetate reagent

Weigh 10.0 g of uranyl acetate and dissolve in a beaker containing 50 ml of boiling water and 2.0 ml of glacial acetic acid. In another beaker, dissolve 30.0 g of zinc acetate in 50 ml of boiling water and 1.0 ml of glacial acetic acid. Mix the two solutions and heat again to the boiling point. Let the beaker stand overnight at room temperature. Filter and then mix with an equal volume of 95% ethyl alcohol. Place in a refrigerator for 2 days and then filter. This reagent is stable at room temperature.

(D-164) Trichloroacetic acid, 10%

Trichloroacetic acid (TCA) is extremely hygroscopic (capable of absorbing water from the atmosphere) and is therefore difficult to weigh accurately. It is recommended that reagent grade TCA be purchased in ¼-pound (113.4 g) bottles and that an entire bottle be dissolved and diluted to the desired concentration. To prepare 10% TCA, dissolve and dilute a ¼-pound or 113.4 g bottle of TCA (check weight on bottle) to 1134 ml. Store in a Pyrex bottle.

(D-165) Sodium standard, 0.64 mg Na/ml

Weigh 0.162 g of sodium chloride and transfer to a 100 ml volumetric flask. Dissolve and dilute to volume with water.

(D-166) Sodium cobaltinitrite reagent

a. Weigh 25.0 g of cobalt nitrate, $Co(NO_3)_2$, and transfer to a 400 ml beaker containing 50 ml of water. Stir to dissolve and then add 12.5 ml of glacial acetic acid.
b. Weigh 120.0 g of sodium nitrite, $NaNO_2$, and dissolve in 180 ml of water.

Place the beaker containing *solution a* in a fume hood and to it, add 210 ml of *solution b*. Bubble air through this solution until all fumes of nitrous oxide are driven off. Filter. This reagent is stable for about 1 month when stored in a refrigerator.

(D-167) Sodium acetate, half-saturated

Dilute saturated sodium acetate (D-281) with an equal volume of water.

(D-168) Wash solution, saturated with potassium sodium cobaltinitrite

Weigh 0.070 g of potassium sulfate, K_2SO_4, and dissolve in 100 ml of water. Slowly add 25 ml of half-saturated sodium acetate and 12.5 ml of sodium cobaltinitrite reagent. Allow to stand for about 30 minutes and then filter through a sintered glass filter. Wash the precipitate twice with 70% ethyl alcohol and once with 95% ethyl alcohol. Transfer the precipitate to an Erlenmeyer flask and prepare a saturated solution with 10% ethyl alcohol. Allow any excess of the precipitate to remain in the bottom of the flask. Remove and filter a portion of the supernatant on the day of use.

(D-169) Glycine, 7.5%

Weigh 7.5 g of glycine and transfer to a 100 ml volumetric flask. Dissolve and dilute to volume with water. Add 5 drops of chloroform as a preservative.

(D-170) Sodium carbonate, 25%

Weigh 25.0 g of anhydrous sodium carbonate, Na_2CO_3, and transfer to a 100 ml volumetric flask containing about 50 ml of water. Dissolve and dilute to volume with water and mix. It may be necessary to warm the solution slightly to aid in dissolving the carbonate. Store in a warm place in a polyethylene bottle.

(D-171) Phenol reagent of Folin-Ciocalteu stock

Weigh 100 g of sodium tungstate, $Na_2WO_4{\cdot}2H_2O$, and 25.0 g of sodium molybdate, $Na_2MoO_4{\cdot}2H_2O$, and transfer to a 2 L boiling flask containing 700 ml of water. Swirl to dissolve and add 50 ml of 85% *o*-phosphoric acid and 100 ml of concentrated hydrocholoric acid. Attach a reflux condenser to the flask (apparatus must have ground glass joints) and reflux gently for 10 hours. Add 150 g of lithium sulfate, Li_2SO_4, 50 ml of water, and 2 or 3 drops of bromine. Place the flask in a fume hood without the condenser and boil the mixture for 15 minutes to remove excess bromine. Cool to room temperature and dilute to 1 L with water. This reagent should be yellow without any trace of a green tint. If it turns green during storage it can be restored by adding 2 or 3 drops more of bromine and boiling again. Store in a refrigerator.

(D-172) Potassium standard, 0.2 mg K/ml

Weigh 0.446 g of potassium sulfate, K_2SO_4, and transfer to a 1 L volumetric flask. Dissolve and dilute to volume with water. Mix thoroughly.

(D-173) Standard silver nitrate solution

Weigh 2.907 g of silver nitrate and transfer to a 1 L volumetric flask. Dissolve and dilute to volume. Mix thoroughly. One milliliter is equivalent to 1 mg of sodium chloride. Standardize against sodium chloride and adjust if necessary. Store in a brown bottle.

(D-174) Thiocyanate solution, standard

Weigh 1.7 g of potassium thiocyanate, KCNS, or 1.4 g of ammonium thiocyanate, NH_4CNS, and transfer to a 1 L volumetric flask. Dissolve and dilute to volume with water. Mix thoroughly. Standardize against the standard silver nitrate solution under the conditions called for in the chloride procedure. Adjust if necessary so that 5 ml of the standard thiocyanate solution is equivalent to 5 ml of the standard silver nitrate solution.

(D-175) Diphenylcarbazone indicator

Weigh 0.100 g of diphenylcarbazone and transfer to a 100 ml volumetric flask. Dissolve and dilute to volume with 95% ethyl alcohol. Mix thoroughly. Store in a refrigerator in a brown bottle.

(D-176) Standard sodium chloride, 10 mEq/L or 58.5 mg/100 ml

Weigh 0.585 g of dry sodium chloride and transfer to a 1 L volumetric flask. Dissolve and dilute to volume. Mix thoroughly.

(D-177) Standard mercuric nitrate solution

Weigh 3.0 g of mercuric nitrate and transfer to a 1 L volumetric flask containing about 500 ml of water and 20 ml of 2 N nitric acid. Swirl to dissolve and dilute to volume with water. Mix thoroughly. Standardize this solution as follows: pipette 2 ml of standard sodium chloride solution into a small flask, add 4 drops of diphenylcarbazone indicator, and titrate with the mercuric nitrate solution. The number of milliliters of the mercuric nitrate solution used equals the value E used in the calculations of the chloride procedure. The standard mercuric nitrate solution is stable indefinitely.

(D-178) Lactic acid, 0.1 N

Pipette 4.4 ml of concentrated lactic acid (85%) into a 500 ml volumetric flask. Dilute to volume with water. This solution is stable for about 2 weeks.

(D-179) Sodium hydroxide, 5 N (carbonate-free)

Prepare by using a nearly saturated solution of sodium hydroxide that has been standardized:

$$ml \times N = ml \times N$$

(D-180) Sodium oxalate, 0.100 N

Dry about 10 g of anhydrous sodium oxalate, $Na_2C_2O_4$, in a 110° C oven for 12 hours or overnight. Cool in a desiccator; then weigh 6.7000 g and transfer to a 1 L volumetric flask. Add about 200 ml of water and 5 ml of concentrated sulfuric acid. Swirl to dissolve and dilute to volume with water. Mix thoroughly.

(D-181) Potassium permanganate, 0.100 N

Weigh 3.2 g of potassium permanganate, $KMnO_4$, and transfer to a 1 L volumetric flask. Dissolve and dilute to volume with water. Mix and allow to stand in a dark place at room temperature for about a week. A sediment of MnO_2 will form. Being careful not to disturb the sediment, slowly pour about 700 ml of the solution into a brown, glass-stoppered bottle. Discard the remainder. This solution will be diluted and standardized before it is used in the determination of calcium. Store in a refrigerator.

(D-182) Ammonium oxalate, 4%

Weigh 40.0 g of ammonium oxalate, $(NH_4)_2C_2O_4$, and transfer to a 1 L volumetric flask. Dissolve and dilute to volume with water. Mix.

(D-183) Ammonium hydroxide, dilute (0.5%)

Transfer 2 ml of concentrated (28%) ammonium hydroxide to a flask and add 98 ml of water. This solution is stable for about a month.

(D-184) Sulfuric acid, dilute (1 N)

Transfer 28.0 ml of concentrated sulfuric acid to a 1 L Pyrex volumetric flask containing about 600 ml of water. Swirl to mix and allow to cool to room temperature. Dilute to volume with water and mix.

(D-185) Potassium hydroxide, 1.25 N

Weigh 70.0 g of potassium hydroxide pellets and transfer to a 1 L volumetric flask containing about 600 ml of water. Swirl to dissolve. Cool to room temperature and dilute to volume with water. Stopper and mix well. Store in a polyethylene bottle.

(D-186) EDTA solution

Weigh 0.395 g of disodium ethylenediamine tetraacetate dihydrate and transfer to a 1 L volumetric flask. Dissolve and dilute to volume with water. Mix well. Store in a Pyrex or polyethylene bottle.

(D-187) Cal-red indicator

Weigh 1.0 g of "cal-red dilute"* and transfer to a mortar of about 50 ml capacity. Add 10 ml of distilled water and grind with the pestle until a homogeneous suspension is obtained. (Not all of the material will dissolve.) This suspension should be stored in a refrigerator and it is usually stable for about 2 weeks. If, however, it does not give sharp end points, then it must be discarded. With some batches of indicator it may be necessary to prepare a new suspension daily.

(D-188) Sodium citrate, 0.05 M

Weigh 14.7 g of sodium citrate, $Na_3C_6H_5O_7 \cdot 2H_2O$, and transfer to a 1 L volumetric flask. Dissolve and dilute to volume with water. Store in a Pyrex or polyethylene bottle.

(D-189) Calcium standard (1 ml = 100 μg Ca)

Using an analytical balance, weigh 0.2472 g of calcium carbonate, $CaCO_3$, and transfer to a 1 L volumetric flask. Slowly add 7 ml of dilute hydrochloric acid (1 part concentrated HCl + 9 parts distilled H_2O). Allow to stand until all $CaCO_3$ is dissolved and then add about 900 ml of distilled water. Adjust the pH to 6.0 with 50% ammonium acetate. Dilute to volume with water and mix thoroughly. Store in a Pyrex or polyethylene bottle.

(D-190) Ammonium oxalate, 10%

Weigh 10.0 g of ammonium oxalate, $(NH_4)_2C_2O_4$, and transfer to a 100 ml volumetric flask. Dissolve and dilute to volume with water. Mix well and store in a polyethylene bottle.

(D-191) Hydrochloric acid, 1 N

Using a 10 ml serologic pipette attached to a rubber ball or vacuum line, transfer 8.4 ml of concentrated hydrochloric acid to a 100 ml volumetric flask containing about 70 ml of water. Dilute to volume and mix. *Do not pipette strong acids by mouth.*

---

*Scientific Service Laboratories, Inc.

(D-192) Ammonium hydroxide, 5%

Mix 20 ml of concentrated ammonium hydroxide, NH_4OH (28%), with 80 ml of water. Store in a glass-stoppered bottle.

(D-193) Molybdate reagent

Weigh 25.0 of ammonium molybdate, $(NH_4)_2MoO_4$, and transfer to a 1 L volumetric flask containing about 300 ml of water. Swirl to dissolve. Add 300 ml of 10 N sulfuric acid (D-194). Dilute to volume with water and mix. Store in a glass-stoppered bottle.

(D-194) Sulfuric acid, 10 N

Measure 282 ml of concentrated sulfuric acid and add slowly to a 1 L Pyrex volumetric flask containing about 600 ml of water. Carefully swirl to mix. Cool to room temperature and dilute to volume with water. Mix.

(D-195) Reducing reagent (amino-naphtholsulfonic acid)

Weigh 14.64 g of sodium bisulfite, $NaHSO_3$, and 0.25 g of 1-amino-2-naphthol-4-sulfonic acid* and transfer to a 100 ml volumetric flask containing about 75 ml of water. Swirl to dissolve. Add 0.5 g of sodium sulfite, Na_2SO_3. Dissolve and dilute to volume. Transfer to a brown bottle and store in a refrigerator. If a precipitate forms, filter in the refrigerator so the reagent does not become warm.

(D-196) Phosphorus, stock standard, 0.1 mg P/ml

Weigh 0.4394 g of monobasic potassium phosphate, KH_2PO_4, and transfer to a 1 L volumetric flask. Add 10 ml of 10 N sulfuric acid and dilute to volume with water. Mix thoroughly.

(D-197) Biuret reagent

Weigh 45.0 g of sodium potassium tartrate and transfer to a 1 L volumetric flask containing about 600 ml of 0.2 N NaOH (prepare from a concentrated solution of NaOH that has been standardized). Mix until dissolved. Add 15.0 g of copper sulfate, $CuSO_4 \cdot 5H_2O$, and dissolve completely. Add 5.0 g of potassium iodide and mix. Dilute to volume with 0.2 N NaOH and mix thoroughly. Store in a Pyrex or polyethylene bottle.

(D-198) Sodium sulfate, 24.5%

Weigh 245.0 g of reagent grade sodium sulfate and transfer to a 1 L volumetric flask containing about 700 ml of water. Dissolve and dilute to volume. This solution should be kept in a warm place (26° to 30° C).

*Eastman, purified.

(D-199) Thymol-barbital buffer, pH 7.55

Boil 1100 ml of distilled water for about 5 minutes to drive off CO_2. Allow to cool to about 95° C. Weigh 6.0 g of thymol crystals and transfer to a 2 L Erlenmeyer flask. Add 300 ml of the hot water. Add 3.09 g of barbital, 1.69 g of sodium barbital, and 720 ml of hot water. Stopper and shake flask for 5 minutes. (Vigorous stirring with a magnetic stirrer works very well.) Release pressure from time to time by removing the stopper. Allow to cool to 25° C and add 20 ml of water to compensate for evaporation. Seed the solution with about 1 g of thymol crystals. Shake the flask vigorously until the supernatant solution becomes clear. Allow to stand at room temperature (25° C) until the next day. Then filter through Whatman No. 1 paper. Check pH. It must be between 7.50 and 7.60 at 25° C. If the pH is outside these limits, adjustments must be made by altering the ratio of barbital to sodium barbital in a new preparation. Store in a tightly stoppered container to protect it against changes in pH from uptake of CO_2.

(D-200) Calcium chloride, 2.5%

Weigh 2.5 g of calcium chloride, $CaCl_2$, and transfer to a 100 ml volumetric flask. Dissolve and dilute to volume with water. Mix.

(D-201) Sodium hydroxide, 0.2 N

Dilute 20.0 ml of 1 N sodium hydroxide (D-251) with 80.0 ml of water. Mix well. Store in a polyethylene bottle.

(D-202) Protein standard, 3 mg/ml

Weigh 0.300 g of pure albumin* and transfer to a 100 ml volumetric flask containing about 50 ml of saline (D-22). Swirl flask to dissolve and dilute to volume with saline. Stopper and mix. Pour about 3 ml aliquots of the standard into Pyrex test tubes and freeze for future use. When thawing protein standards, place the tubes in a beaker of water at room temperature. *Never use hot water.* Once a standard has been thawed, do not refreeze. Discard the unused portion.

(D-203) Acetone-alcohol mixture

Mix 500 ml reagent grade acetone with 500 ml of 95% ethyl alcohol.

(D-204) Digitonin solution

Mix 525 ml of 95% ethyl alcohol with 475 ml of water. Add 5.0 g of digitonin and stir to dissolve.

(D-205) Acetic acid, 10% (v/v)

Measure 100 ml of glacial acetic acid and transfer to a 1 L volumetric flask containing about 600 ml

*Sigma Chemical Co.

of water. Mix, dilute to volume with water, and mix again.

(D-206) Acetone-ether mixture

Mix 200 ml of reagent grade ether with 100 ml of reagent grade acetone.

(D-207) Potassium hydroxide solution

Weigh 20.0 g of reagent grade potassium hydroxide and transfer to a 100 ml beaker containing 40 ml of distilled water. Stir until dissolved. Store in a polyethylene bottle.

(D-208) Phenolphthalein, 1%

Weigh 1.0 g of phenolphthalein and transfer to a 100 ml volumetric flask. Dissolve and dilute to volume with 95% ethyl alcohol. Mix.

(D-209) Acetic anhydride–dioxane (A-D) reagent

Mix 600 ml of reagent grade acetic anhydride with 400 ml of purified dioxane. This should be done *under a hood.* Store in a brown, glass-stoppered bottle.

(D-210) Cholesterol standard, 0.2 mg/ml

Weigh 0.025 g of pure, dry cholesterol and 0.0277 g of cholesteryl acetate and transfer to a 500 ml Erlenmeyer flask. Add 240 ml of acetic anhydride–dioxane mixture (D-209) and heat 15 minutes in a boiling water bath. Cool and wipe outside of flask to remove moisture. Transfer to a 250 ml volumetric flask. Rinse the Erlenmeyer flask with 5 ml of A-D solution and add this to the volumetric flask. Dilute to volume wtih A-D solution and mix thoroughly. Store in a glass-stoppered, brown bottle.

(D-211) Sodium nitrite stock solution B, 5%

Weigh 5.0 g of sodium nitrite and transfer to a 100 ml volumetric flask. Dissolve and dilute to volume with water. Store in a refrigerator.

(D-212) Sodium nitrite, working solution, B, 0.5%

Prepare this solution just before use. Mix 1.0 ml of stock solution B (D-211) with 9.0 ml of water.

(D-213) Alkaline buffer, pH 10.6-10.7

Weigh 24.4 g of dibasic sodium phosphate, $Na_2HPO_4 \cdot 7H_2O$, (or 12.92 g of Na_2HPO_4) and 3.54 g of $Na_3PO_4 \cdot 12H_2O$ and 6.4 g of sodium *p*-toluenesulfonate and transfer to a 1 L volumetric flask. Dissolve and dilute to volume with water. Check pH and adjust if necessary.

(D-214) Acid reagent, 2 M NaH_2PO_4

Weigh 27.6 g of $NaH_2PO_4 \cdot H_2O$ and transfer to a 100 ml volumetric flask. Dissolve and dilute to volume with water.

(D-215) BSP standard, 5 mg/100 ml

Pipette 1.0 ml of the intravenous test solution (50 mg/ml)* into a 1 L volumetric flask. Dilute to volume with water. This diluted standard is stable for 1 week.

(D-216) Lanthanum stock solution, 5% in 25% (v/v) HCl

Weigh 58.65 g of lanthanum oxide, La_2O_3, and transfer to a 1 L volumetric flask. Add about 50 ml of water. *Very slowly* add 250 ml of concentrated hydrochloric acid and mix until dissolved. Dilute to volume with water and mix well. Store in a polyethylene bottle.

(D-217) Lanthanum working solution, 0.5% in 2.5% (v/v) HCl

Transfer 100 ml of lanthanum stock solution to a 1 L volumetric flask. Dilute to volume with water, add 1 ml of Brij 35, and mix. Store in a polyethylene bottle.

(D-218) Magnesium stock standard, 100 mEq/L

Weigh 1.216 g of pure magnesium metal and transfer to a 1 L volumetric flask. Dissolve in 9 to 10 ml of 12 N HCl. Dilute to volume with water and mix thoroughly. Store in a polyethylene bottle.

(D-219) Trichloroacetic acid, 5%

Dilute 50 ml of 10% trichloroacetic acid (D-164) with 50 ml of water. Mix.

(D-220) Sodium diethyldithiocarbamate, 1%

Weigh 1.0 g of sodium diethyldithiocarbamate and transfer to a 100 ml volumetric flask. Dissolve and dilute to volume with water. Mix well and store in a polyethylene bottle.

(D-221) Sodium ethylenediamine tetraacetate (EDTA), 1%

Weigh 1.0 g of sodium ethylenediamine tetraacetate and transfer to a 100 ml volumetric flask. Dissolve and dilute to volume with water. Mix well and store in a polyethylene bottle.

---

*Hynson, Westcott & Dunning.

(D-222) Ammonium pyrrolidine dithiocarbamate, 1%

Weigh 1.0 g of ammonium pyrrolidine dithiocarbamate and transfer to a 100 ml volumetric flask. Dissolve and dilute to volume with water. Mix well and store in a polyethylene bottle.

(D-223) Iodine indicator

a. Stock, 0.1 N iodine in 30% potassium iodide
   Weigh 13.0 g of elemental iodine and 300 g of potassium iodide and transfer to a 1 L volumetric flask. Dissolve and dilute to volume with water. Mix thoroughly.
b. Working, 0.002 N iodine in 2% potassium iodide
   Pipette 20.0 ml of stock iodine solution into a 1 L volumetric flask containing about 600 ml of water. Add 14.0 g of potassium iodide. Dissolve and dilute to volume with water. Mix thoroughly. Store in a brown bottle.

(D-224) Sodium hydroxide, 0.05 N

Prepare from a concentrated solution of sodium hydroxide that has been standardized.

$$\text{ml} \times \text{N} = \text{ml} \times \text{N}$$

Standardize this dilute solution against a standard acid solution and adjust if necessary to exactly 0.05 N.

(D-225) Phosphate buffer, pH 7.0

Weigh 5.785 g of dibasic sodium phosphate, Na_2HPO_4, and 3.532 g of monobasic potassium phosphate, KH_2PO_4, and transfer to a 1 L volumetric flask. Dissolve and dilute to volume with water. Mix thoroughly.

(D-226) Olive oil substrate

Weigh 5.0 g of acacia and 0.4 g of sodium benzoate, C_6H_5COONa, and dissolve in 100 ml of water. Transfer this solution to a blendor and add 100 ml of pure olive oil. Blend for 10 minutes to obtain a homogeneous emulsion. Store in a refrigerator. Shake vigorously just before using.

(D-227) Alkaline buffer, 0.1 M glycine, pH 10.5

Weigh 7.5 g of glycine and 0.095 g of magnesium chloride and transfer to a 1 L volumetric flask containing about 750 ml of water. Add 85 ml of 1 N sodium hydroxide (D-251) and dilute to volume with water. Mix thoroughly. Store in a refrigerator.

(D-228) *p*-Nitrophenyl phosphate, stock substrate

Dissolve 0.100 g of *p*-nitrophenyl phosphate* in 25 ml of water.

*Available from Sigma Chemical Co.

(D-229) Alkaline buffered substrate, pH 10.3 to 10.4

Mix equal volumes of alkaline buffer (D-227) and stock substrate (D-228).

(D-230) Sodium hydroxide, 0.02 N

Pipette 20 ml of 1 N NaOH (D-251) into a 1 L volumetric flask and dilute to volume with water. Mix well.

(D-231) Acid buffer, 0.09 M citric acid, pH 4.8

Weigh 18.907 g of citric acid and transfer to a 1 L volumetric flask containing about 600 ml of water. Add 180 ml of 1 N sodium hydroxide (D-251) and 100 ml of 0.1 N hydrochloric acid (D-37). Dilute to volume with water and mix. Store in a refrigerator.

(D-232) Acid buffered substrate, pH 4.8 to 4.9

Mix equal volumes of acid buffer (D-231) and stock substrate (D-228).

(D-233) Sodium hydroxide, 0.1 N

Transfer 100 ml of 1 N sodium hydroxide to a 1 L volumetric flask and dilute to volume with water. Mix.

(D-234) Acetic acid, 20% (v/v)

Transfer 20 ml of glacial acetic acid to a 100 ml volumetric flask and dilute to volume with water. Mix.

(D-235) Tartrate acid buffer, pH 4.8

Weigh 18.9 g of citric acid, $C_6H_8O_7 \cdot H_2O$, and 6.0 g of *l*(+) tartaric acid and transfer to a beaker containing about 600 ml of water. Stir to dissolve. Add 100 ml of 0.1 N hydrochloric acid (D-37) and mix. Add enough 1.0 N sodium hydroxide (D-251) to bring the pH to 4.8. Use a pH meter. Transfer quantitatively to a 1 L volumetric flask and dilute to volume with water. Store in a refrigerator.

(D-236) *p*-Nitrophenol, 10 mM/L

Weigh 0.1391 g of *p*-nitrophenol and transfer to a 100 ml volumetric flask. Dissolve and dilute to volume with water.

(D-237) Pyruvate standard, 2 mM/L

Weigh 0.0200 g of pure sodium pyruvate and dissolve in 100 ml of phosphate buffer.

(D-238) Töpfer's reagent

Weigh 0.5 g of dimethylaminoazobenzene and dissolve in 100 ml of 95% ethyl alcohol.

(D-239) Benzidine, saturated, in glacial acetic acid

The benzidine powder must be labeled "for blood tests." Weigh 4 g of benzidine powder and transfer to a brown bottle. Add 50 ml of glacial acetic acid. Shake to dissolve as much as possible. This solution is stable for about 1 month.

(D-240) Sodium hydroxide, 0.01 N

Pipette 1.0 ml of 1 N sodium hydroxide (D-251) into a 100 ml volumetric flask. Dilute to volume with water and mix.

(D-241) Ferric chloride, 10%

Weigh 10 g of ferric chloride, $FeCl_3$, and transfer to a 100 ml volumetric flask. Dissolve and dilute to volume with water. Mix.

(D-242) Sodium carbonate, 1%

Weigh 1 g of sodium carbonate, Na_2CO_3, and transfer to a 100 ml volumetric flask. Dissolve and dilute to volume with water. Mix.

(D-243) Starch, 1%

Heat 100 ml of distilled water to boiling. Add a mixture of 1 g of soluble starch in 5 ml of distilled water, stirring constantly until the fluid again reaches the boiling point. Let cool. If the solution is to be kept more than a few days, add a few drops of toluol or chloroform and store in a refrigerator.

(D-244) Gram's iodine solution

Weigh 0.4 g of iodine and 0.8 g of potassium iodide and dissolve in 120 ml of water. Store in a brown bottle.

(D-245) Acetic acid, 1.0%

Pipette 1 ml of glacial acetic acid into a 100 ml volumetric flask. Dilute to volume with water and mix.

(D-246) Barium chloride, 10%

Weigh 10 g of barium chloride and transfer to a 100 ml volumetric flask. Dissolve and dilute to volume with water. Mix.

(D-247) Mercuric chloride, 10%

Weigh 10 g of mercuric chloride, $HgCl_2$, and transfer to a 100 ml volumetric flask. Dissolve and dilute to volume with water. Mix.

(D-248) Ethyl alcohol, 95%, containing 0.4% capryl alcohol

Pipette 4 ml capryl alcohol into a 1 L volumetric flask and dilute to volume with 95% ethyl alcohol. Mix.

(D-249) Potassium hydroxide, 33%

Weigh 330 g of potassium hydroxide pellets and transfer to a 1 L Pyrex volumetric flask containing about 600 ml of water. Swirl to dissolve and allow to cool to room temperature. Dilute to volume and mix. Store in a polyethylene bottle.

(D-250) Hydrochloric acid, 25%

Measure 676 ml of concentrated hydrochloric acid (37%) and transfer to a 1 L volumetric flask containing 300 ml of water. Dilute to volume with water and mix.

(D-251) Sodium hydroxide, NaOH, 1.00 N

Prepare from a concentrated solution of sodium hydroxide that has been standardized. Standardize the diluted solution against 1 N acid. Adjust to 1.000 N if necessary.

(D-252) Thymol blue, 0.2% in 50% ethyl alcohol

Weigh 0.2 g of thymol blue and transfer to a 100 ml volumetric flask. Add 53 ml of 95% ethyl alcohol and dilute to volume with water and mix.

(D-253) Pandy's reagent

Weigh 10 g of phenol, C_6H_5OH, and dissolve in 100 ml of water. Place in an incubator at 37° C for 2 or 3 days. Use only the clear supernatant solution.

(D-254) Ammonium sulfate, saturated

Weigh 80 g of ammonium sulfate, $(NH_4)_2SO_4$, and dissolve in 100 ml of water.

(D-255) Trichloroacetic acid, 3%

Trichloroacetic acid crystals are difficult to weigh accurately because they absorb moisture from the air very rapidly. It is recommended that dilute solutions of TCA be made from more concentrated solutions. Measure 300 ml of 10% trichloroacetic acid (D-164) and transfer to a 1 L volumetric flask. Dilute to volume with water and mix.

(D-256) Formaldehyde, 2%

Mix 5 ml of formalin (40% formaldehyde) with 95 ml of water.

(D-257) Sodium nitrite, 0.06%

Weigh 0.060 g of sodium nitrite, $NaNO_2$, and transfer to a 100 ml volumetric flask. Dissolve and dilute to volume with water. Mix. Transfer to a brown bottle and store in a refrigerator.

(D-258) Indole-3-acetic acid reagent, 0.5% in 95% ethyl alcohol

Weigh 0.5 g of indole-3-acetic acid and transfer to a 100 ml volumetric flask. Dissolve and dilute to volume with 95% ethyl alcohol. Mix thoroughly.

(D-259) PSP stock standard, 120 mg/500 ml

Weigh 0.120 g of phenolsulfonphthalein (phenol red) and transfer to a 500 ml volumetric flask. Add 1 ml of 10% NaOH. Dilute to volume with water. Mix thoroughly.

(D-260) Acetic acid, 5%

Pipette 5 ml of glacial acetic acid into a 100 ml volumetric flask. Dilute to volume with water and mix.

(D-261) Sodium chloride, saturated

Weigh 40 g of sodium chloride, NaCl, and transfer to a flask or beaker containing 100 ml of hot distilled water. Stir to dissolve as much as possible. Allow any undissolved salt to remain on the bottom of the bottle in which the solution is stored.

(D-262) Acetic acid, 50%

Add 50 ml of glacial acetic acid to 50 ml of water. Mix.

(D-263) Sulfosalicylic acid, 20%

Weigh 20 g of sulfosalicylic acid and transfer to a 100 ml volumetric flask. Dissolve and dilute to volume with water. Mix.

(D-264) Exton's reagent

Weigh 200 g of sodium sulfate, $Na_2SO_4 \cdot 10H_2O$, and transfer to a 1 L volumetric flask containing 800 ml of water. Swirl to dissolve. Add 50 g of sulfosalicylic acid. Dilute to volume with water and mix thoroughly.

(D-265) Roberts' reagent

To 1 L of distilled water, add magnesium sulfate, with stirring, until no more dissolves. Add 200 ml of concentrated nitric acid. Mix.

(D-266) Hydrochloric acid, 5% (v/v)

Dilute 5 ml of concentrated hydrochloric acid with 95 ml of water. Mix.

(D-267) Esbach's reagent

Weigh 1 g of picric acid and 2 g of citric acid and dissolve in 100 ml of water.

(D-268) Tsuchiya's reagent

Weigh 15.0 g of phosphotungstic acid and transfer to a 2 L Erlenmeyer flask containing 1 L of 95% ethyl alcohol. Add 50.0 ml of concentrated hydrochloric acid and mix thoroughly.

(D-269) Sulfosalicylic acid, 3%

Weigh 3 g of sulfosalicylic acid and transfer to a 100 ml volumetric flask. Dissolve and dilute to volume with water. Mix.

(D-270a) Benedict's qualitative reagent

Weigh 100 g of anhydrous sodium carbonate, Na_2CO_3, *or* 200 g of hydrated sodium carbonate, $Na_2CO_3 \cdot 10H_2O$. Place in a Pyrex container. Add about 800 ml of distilled water. Add 173 g of sodium citrate. Heat to dissolve. Cool to room temperature. Weigh 17.3 g of copper sulfate, $CuSO_4 \cdot 5H_2O$. With the aid of heat, dissolve in 100 ml of distilled water. Cool to room temperature. With constant stirring, pour the copper sulfate solution slowly into the solution of carbonate and citrate salts. Make up to 1 L with distilled water. Mix. This preparation keeps indefinitely.

(D-270b) Benedict's quantitative reagent

Weigh out the first four reagents listed below; place in separate labeled containers; and then prepare 5 ml of the 5% potassium ferrocyanide solution.

Copper sulfate (crystallized) $CuSO_4 \cdot 5H_2O$	18.0 g
Sodium carbonate (anhydrous) Na_2CO_3 (or 200 g of crystalline $Na_2CO_3 \cdot 10H_2O$)	100.0 g
Sodium citrate (or potassium citrate)	200.0 g
Potassium thiocyanate (sulfocyanate)	125.0 g
Potassium ferrocyanide, 5%	5.0 ml

Pour about 600 ml of distilled water into a 1 L Pyrex beaker. Heat to about 70° C. With stirring, transfer the above weights of sodium carbonate, sodium citrate, and potassium thiocyanate to the beaker. Stir until dissolved. Filter. Dissolve the copper sulfate in about 100 ml of distilled water. With stirring, slowly add this copper solution to the 1 L beaker. Now add the 5.0 ml of the 5% potassium ferrocyanide solution. Mix. Allow the solution to cool. Transfer all of the solution to a 1 L volumetric flask. Add distilled water to the 1 L mark. Mix. Five milliliters of this reagent should be reduced by 1.0 ml of a 1% glucose solution.

(D-271) Sodium nitroprusside, saturated

Add 40 g of sodium nitroprusside (sodium nitroferricyanide) to 100 ml of water in a brown bottle. Shake to dissolve as much as possible. Allow any undissolved salt to remain in the bottom of the bottle.

(D-272) Sodium nitroprusside reagent

Weigh 10 g of sodium nitroprusside (sodium nitroferricyanide) and transfer to a flask containing

95 ml of water and 2 ml of concentrated sulfuric acid. Mix and store in a brown bottle.

(D-273) Rantzman's reagent

Weigh 37.5 g of ammonium nitrate and 2.5 g of sodium nitroprusside (sodium nitroferricyanide) and transfer to a 100 ml volumetric flask. Dissolve and dilute to volume with water. Mix. Store in a brown bottle. This reagent is stable for about 2 months.

(D-274) Sodium hydroxide, 40%

Weigh 40 g of sodium hydroxide pellets and transfer to a Pyrex 100 ml volumetric flask containing about 50 ml of water. Swirl to dissolve. Cool to room temperature, dilute to volume, and mix.

(D-275) Salicylaldehyde, 10%

Place 10 ml of salicylaldehyde (salicylous acid) in a container. Add 90 ml of 95% ethyl alcohol. Mix.

(D-276) Acetic acid, 30%

Add 30 ml of glacial acetic acid to 70 ml of water and mix.

(D-277) Lugol's solution

Weigh 5 g of iodine and 10 g of potassium iodide and transfer to a brown bottle. Add 100 ml of water and mix.

(D-278) Fouchet's reagent

Weigh 25 g of trichloroacetic acid and transfer to a flask containing 100 ml of water. Add 10 ml of 10% ferric chloride solution (D-241) and mix.

(D-279) Iodine, ethyl alcohol solution

Weigh 0.7 g of iodine and 0.5 g of potassium iodide and transfer to a 125 ml Erlenmeyer flask. Add 5 ml of water and mix. Add 95 ml of 95% ethyl alcohol and mix.

(D-280) Ehrlich's reagent

Pour 100 ml of distilled water into a brown bottle. Add 0.7 g of *p*-dimethylaminobenzaldehyde. Carefully add 150 ml of concentrated hydrochloric acid. Mix.

(D-281) Sodium acetate, saturated

Heat 100 ml water to 45° C and add 130 g of sodium acetate, $NaC_2H_3O_2 \cdot 3H_2O$. Mix well and place in a 37° C water bath overnight. Cool to room temperature and allow excess sodium acetate crystals to settle out.

(D-282) Ehrlich's diazo reagent

A. Add about 500 ml of distilled water to a 1 L volumetric flask. Add 50 ml of concentrated hydrochloric acid. Weigh 5.0 g of sulfanilic acid and transfer to the volumetric flask. Add distilled water to the 1 L mark. Mix.
B. Weigh 0.5 g of sodium nitrite. Dissolve in 100 ml of distilled water. Pour into a brown bottle. Mix. Store in a refrigerator.

Place 25 ml of solution A in a flask. Add 0.5 ml in solution B. Mix. This reagent must be prepared just before use, as it does not keep.

(D-283) Calcium chloride, 10%

Weigh 10 g of anhydrous calcium chloride, $CaCl_2$, and transfer to a 100 ml volumetric flask. Dissolve and dilute to volume with water. Mix.

(D-284) Zinc acetate, saturated alcoholic solution

Place 100 ml of ethyl alcohol in a beaker. Add zinc acetate, with stirring, until no more dissolves.

(D-285) Bromine water

Cautiously add a few drops of liquid bromine to 100 ml of distilled water. Mix. Store in a brown bottle. When the solution loses its color, prepare a fresh solution.

(D-286) Orthotolidine reagent

Dissolve 4 g of orthotolidine in 100 ml of glacial acetic acid. Keep in a brown bottle and store in a refrigerator. Prepare fresh every month.

(D-287) Guaiac reagent

Weigh 1 g of powdered guaiac and transfer to a brown bottle. Add 10 ml of 95% ethyl alcohol and mix. Add 10 ml of 30% hydrogen peroxide and mix. This reagent is stable for about a month.

(D-288) Sulkowitch reagent

Weigh 2.5 g of ammonium oxalate and 2.5 g of oxalic acid and transfer to a brown bottle. Add 145 ml of water and 5 ml of glacial acetic acid. Mix.

(D-289) Trichloroacetic acid, 20%

Dissolve and dilute a ¼-pound bottle (see D-164) of trichloroacetic acid to 567 ml. Mallinckrodt trichloroacetic acid AR is recommended because of its extremely low iron content.

(D-290) Sodium acetate, 20%

Weigh 33.0 g of sodium acetate, $CH_3COONa \cdot 3H_2O$, and transfer to a 100 ml volumetric flask. Dissolve and dilute to volume with water. Mix.

(D-291) Ammonium hydroxide, 10% (w/v)

Measure 36 ml of concentrated ammonium hydroxide (28%) and transfer to a 100 ml volumetric flask. Dilute to volume with water and mix.

(D-292) Sulfuric acid, 2 N

Measure 56 ml of concentrated sulfuric acid in a 100 ml graduated cylinder and transfer to a 1 L volumetric flask containing about 600 ml of water. Swirl to mix. Cool to room temperature. Dilute to volume with water and mix.

(D-293) Ammonium oxalate, 0.1 M

Weigh 1.42 g of hydrated ammonium oxalate, $(NH_4)_2C_2O_4 \cdot H_2O$, and transfer to a 100 ml volumetric flask. Dissolve and dilute to volume with water. Mix. Store in a refrigerator.

(D-294) Barium chloride, 5%

Weigh 5.0 g of barium chloride and transfer to a 100 ml volumetric flask. Dissolve and dilute to the mark with distilled water. Store in a Pyrex or polyethylene bottle.

(D-295) Nitrosonaphthol reagent, 0.1%

Weigh 0.100 g of 1-nitroso-2-naphthol and transfer to a 100 ml volumetric flask. Dissolve and dilute to volume with 95% ethyl alcohol. Mix. Transfer to a brown bottle and store in a refrigerator.

(D-296) Sodium nitrite, 2.5%

Weigh 2.5 g of sodium nitrite, $NaNO_2$, and transfer to a 100 ml volumetric flask. Dissolve and dilute to volume with water. Mix. Store in a refrigerator.

(D-297) Sodium nitrite, 0.5%

Pipette 20 ml of 2.5% sodium nitrite into a 100 ml volumetric flask. Dilute to volume with water and mix. Store in a refrigerator.

(D-298) 5-HIAA standard, 10 mg/100 ml

Weigh 0.0200 g of the mono-dicyclohexylammonium salt of 5-hydroxyindole-3-acetic acid* and transfer to a 100 ml volumetric flask. Dissolve and dilute to volume with 0.1 N sulfuric acid. Mix thoroughly and transfer to a brown bottle. Store in a refrigerator.

(D-299) Ferric chloride, 5%

Weigh 5.0 g of $FeCl_3$ or 8.3 g of $FeCl_3 \cdot 6H_2O$ and transfer to a 100 ml volumetric flask. Dissolve and dilute to volume with water. Mix.

(D-300) Zinc sulfate, 10%

Weigh 100 g of zinc sulfate, $ZnSO_4 \cdot 7H_2O$, and transfer to a 1 L volumetric flask containing about 600 ml of water. Swirl to dissolve and dilute to volume with water. The 10% zinc sulfate and 0.5 N sodium hydroxide solutions should be so related that when 10 ml of the zinc sulfate is diluted with 50 ml of water and titrated with the sodium hydroxide, 10.8 to 11.2 ml of the 0.5 N NaOH is required to produce a permanent pink with phenolphthalein.

(D-301) Zinc sulfate working solution

Mix 100 ml of 10% zinc sulfate with 700 ml of water.

(D-302) Sodium carbonate, 4.0 N

Weigh 212 g of anhydrous sodium carbonate, Na_2CO_3, and transfer to a 1 L volumetric flask. Dissolve and dilute to volume with water. Mix thoroughly.

(D-303) Hydrochloric acid, 2.0 N

Measure 167 ml of concentrated hydrochloric acid, HCl, and transfer to a 1 L volumetric flask containing about 600 ml of water. Dilute to volume and mix.

(D-304) Sulfuric acid, 7.0 N

Measure 195 ml of concentrated sulfuric acid, H_2SO_4, and transfer slowly to a Pyrex 1 L volumetric flask containing about 700 ml of cold distilled water. Place flask in an ice bath. When the contents of the flask cool to the temperature of the bath, swirl the flask to mix and return it to the ice bath for about 10 minutes. Then remove the flask from the ice bath and allow the solution to come to room temperature. Dilute to volume with water and mix.

(D-305) Sulfuric acid, dilute

Dilute 800 ml of 7 N sulfuric acid to 2 L with water.

(D-306) Iodine standards

a. Concentrated stock standards, 100 μg/ml
   Weigh 0.1308 g of desiccated potassium iodide, KI, and transfer to a 1 L volumetric flask. Dissolve and dilute to volume with double-distilled water. Mix thoroughly.
b. Dilute stock standard, 0.2 μg/ml
   Pipette 2 ml of concentrated stock standard into a 1 L volumetric flask. Dilute to volume with double-distilled water and mix thoroughly.
c. Dilute working standard, 0.04 μg/ml
   Pipette 10 ml of dilute stock standard into a 50 ml volumetric flask and dilute to volume with double-distilled water. Mix thoroughly.

(D-307) Ceric ammonium sulfate, 0.02 N

Transfer 500 ml of water to a 1 L volumetric flask. Add 230 ml of 7 N sulfuric acid. Weigh 12.65 g

*May be obtained from Sigma Chemical Co.

of ceric ammonium sulfate, $(NH_4)_4Ce(SO_4)_4 \cdot 2H_2O$,* and add to the flask. Swirl to dissolve. Dilute to volume with water and mix.

(D-308) Sodium arsenite, 0.1 N

Weigh 12.99 g of sodium arsenite, $NaAsO_2$, and transfer to a 1 L volumetric flask. Dissolve and dilute to volume with water. Mix.

(D-309) Antifoam solution

Dissolve 35 g of Dow Corning Antifoam AF Emulsion in 500 ml of water. A Waring blendor will facilitate the mixing of the two ingredients.

(D-310) Perchloric acid, 10%

Place about 700 ml of double-distilled water in a 1 L Pyrex volumetric flask and *slowly* add 143 ml of 70% perchloric acid. Swirl to mix and cool to room temperature. Dilute to volume and mix again. Store in a brown glass bottle.

(D-311) Chloric acid reagent

Weigh 500 g of potassium chlorate, $KClO_3$, and transfer to a 2000 ml flask or beaker. Add 900 ml of double-distilled water and heat on an electric hot plate to dissolve. *Do not use an open flame.* While this solution is still hot, *slowly* add 400 ml of 70% perchloric acid with stirring. Cool to room temperature; then place in a freezing compartment for 24 hours. Decant and filter through glass wool while cold. Store the chloric acid in a brown glass bottle in a refrigerator. Flush the precipitated salts down the sink with plenty of water.

(D-312) Chromic acid

Weigh 3.03 g of chromium trioxide, CrO_3, and transfer to a 1 L volumetric flask. Dissolve and dilute to volume with double-distilled water. Store in a brown bottle.

(D-313) Arsenious acid, 0.2 N

Weigh 9.89 g of arsenic trioxide, As_2O_3, and 7.0 g of sodium hydroxide, NaOH, and dissolve in 100 ml of double-distilled water. When this is completely dissolved, add 300 ml of water and 3 drops of phenolphthalein indicator. Add concentrated sulfuric acid dropwise with stirring until the red color disappears. Then add 42 ml of concentrated sulfuric acid and dilute to 1000 ml with water. Store in a polyethylene bottle.

(D-314) Ceric ammonium sulfate, 0.04 N

Weigh 25.31 g of ceric ammonium sulfate, $(NH_4)_4Ce(SO_4)_4 \cdot 2H_2O$,* and transfer to a 1 L volumetric flask. Add 700 ml of water and 103 ml of concentrated sulfuric acid. Mix. Cool to room temperature and dilute to volume. Mix. Store in a polyethylene bottle.

*The ceric salt may be obtained from the G. Frederick Smith Chemical Co.

(D-315) Sulfuric acid, 10 N, with 100 mg NaCl/ml

Weigh 100 g of sodium chloride, NaCl, and transfer to a 1 L Pyrex volumetric flask. Add about 600 ml of water and swirl to dissolve. *Slowly* add 278 ml of concentrated sulfuric acid. Cool to room temperature and dilute to volume with water. Store in a Pyrex glass bottle.

(D-316) Iodate stock standard, 100 μg iodine/ml

Weigh 0.1685 g of potassium iodate, KIO_3, which has been dried in a vacuum desiccator, and transfer to a 1 L volumetric flask. Dissolve and dilute to volume with double-distilled water.

(D-317) Picric acid solution, 1%

Weigh 1.0 g of picric acid and transfer to a 100 ml volumetric flask. Dissolve and dilute to volume with water. Store in an amber glass bottle.

(D-318) Chromotropic acid, 0.5%

Weigh 0.5 g of chromotropic acid (4,5-dihydroxy-2,7-naphthalenedisulfonic acid) and transfer to a 100 ml volumetric flask. Dissolve and dilute to volume with water. Mix and store in an amber glass bottle.

(D-319) Sulfuric acid, 10%

Place about 60 ml of water in a Pyrex 100 ml volumetric flask. Place the flask in an ice bath for about 10 minutes. Add 10 ml of concentrated sulfuric acid to the flask and mix. Allow to come to room temperature. Dilute to volume with water and mix.

(D-320) Potassium permanganate, 5%

Weigh 5 g of potassium permanganate and transfer to a 100 ml volumetric flask. Dissolve and dilute to volume with water. Mix and store in an amber glass bottle.

(D-321) Sodium bisulfite, 0.15 M

Weigh 2.8525 g of sodium metabisulfite, $Na_2S_2O_5$, and transfer to a 100 ml volumetric flask. Dissolve and dilute to volume with water. Mix and store in a polyethylene bottle. (Sodium metabisulfite, $Na_2S_2O_5$, when dissolved in water yields sodium bisulfite, $NaHSO_3$.)

(D-322) *p*-Hydroxybiphenyl, 1% in 0.5 N NaOH

Weigh 1.0 g of *p*-hydroxybiphenyl and transfer to a 100 ml volumetric flask. Dissolve and dilute to volume with 0.5 N NaOH. Mix and store in an amber glass bottle.

(D-323) Copper sulfate, 4%

Weigh 4.0 g of copper sulfate and transfer to a 100 ml volumetric flask. Dissolve and dilute to volume with water. Store in a polyethylene bottle.

(D-324) Sodium hydroxide, 0.45 N

Weigh 18.0 g of sodium hydroxide pellets and transfer to a 1 L volumetric flask. Dissolve and dilute to volume with water. Mix thoroughly. Standardize and adjust if necessary. Store in a polyethylene bottle.

(D-325) Gold chloride, 0.5% in water

Weigh 0.5 g of gold chloride, $AuCl_3 \cdot HCl \cdot 3H_2O$, and transfer to a 100 ml volumetric flask. Dissolve and dilute to volume with water. Store in a refrigerator.

(D-326) Bromide stock standard, 20 mg/ml

Weigh 2.575 g of sodium bromide, NaBr, and transfer to a 100 ml volumetric flask. Dissolve and dilute to volume with water. Mix.

(D-327) Color reagent (ferric-mercuric reagent)

Weigh 40.0 g of mercuric chloride, $HgCl_2$, and transfer to a 1 L volumetric flask containing about 800 ml of water. Swirl to dissolve. Add 120 ml of 1 N hydrochloric acid (D-191) and 40.0 g of ferric nitrate, $Fe(NO_3)_3 \cdot 9H_2O$. When all the ferric nitrate is dissolved, dilute to volume with water and mix.

(D-328) Salicylate stock standard, 1 mg/ml

Weigh 1.160 g of sodium salicylate, $C_7H_5O_3 \cdot Na$, and transfer to a 1 L volumetric flask. Dissolve and dilute to volume with water. Add a few drops of chloroform as a preservative. Store in a refrigerator. This solution is stable for about 6 months.

(D-329) Salicylate standard solution, 0.05 mg/ml

Pipette 5.0 ml of stock salicylate solution (D-328) into a 100 ml volumetric flask and dilute to volume with water. Add a few drops of chloroform as a preservative. Store in a refrigerator. This solution is stable for about 6 months.

(D-330) Salicylate standard solution 0.20 mg/ml

Pipette 20.0 ml of stock salicylate solution (D-328) into a 100 ml volumetric flask and dilute to volume with water. Add a few drops of chloroform as a preservative. Store in a refrigerator. This solution is stable for about 6 months.

(D-331) Ammonium sulfamate, 0.5%

Weigh 0.50 g of ammonium sulfamate, $NH_2SO_3NH_4$, and transfer to a 100 ml volumetric flask. Dissolve and dilute to volume with water. Store in a refrigerator.

(D-332) Coupling reagent (naphthyl ethylenediamine solution)

Weigh 0.20 g of *n*-(1-naphthyl)-ethylenediamine dihydrochloride, $C_{10}H_7NHCH_2CH_2NH_2 \cdot 2HCl$, and transfer to a 100 ml volumetric flask. Dissolve and dilute to volume with 95% ethyl alcohol. Mix and transfer to a brown bottle. Store in a refrigerator.

(D-333) Sulfanilamide stock standard, 0.1 mg/ml

Weigh 0.100 g of pure sulfanilamide and transfer to a 1 L volumetric flask. Dissolve and dilute to volume with water. Store in a refrigerator.

(D-334) Sodium nitrate, saturated

Weigh 130 g of sodium nitrate, $NaNO_3$, and transfer to a 125 ml Erlenmeyer flask. Add 100 ml of of hot distilled water and mix. Cool to room temperature. There should be some undissolved crystals of sodium nitrate remaining in the bottom of the flask.

(D-335) Lead nitrate, saturated

Weigh 75 g of lead nitrate, $Pb(NO_3)_2$, and transfer to a 125 ml Erlenmeyer flask. Add 100 ml of hot distilled water and swirl to dissolve. Cool to room temperature. There should be some undissolved crystals of lead nitrate remaining in the bottom of the flask.

(D-336) Lithium stock standard, 10 mEq Li/L

Weigh 0.4240 g of lithium chloride which has been dried at 125 C for ½ hour and cooled in a dessicator. Dissolve and dilute to 1000 ml with distilled water and mix well. Store in a Pyrex or polyethylene bottle.

(D-337) Lithium working standard, 1 mEq Li/L

Dilute 10 ml of lithium stock standard to 100 ml with distilled water. Store in a Pyrex or polyethylene bottle.

(D-338) Sterox SE, 1%

Pipette 9.5 ml (10 g) of Sterox SE into a 1 L volumetric flask. Allow pipette to drain completely.

Slowly run water down the side of the flask and dilute to volume. If excessive foaming occurs, add one drop of caprylic alcohol. Mix well and store in a polyethylene bottle.

(D-339) Sterox SE, 0.02%

Pipette 20 ml of 1% Sterox SE into a 1 L volumetric flask. Dilute to volume with water and mix. Store in a polyethylene bottle.

(D-340) 2,2′-Dipyridyl, 0.4%

Weigh 0.40 g of 2,2′-dipyridyl and transfer to a 100 ml volumetric flask containing 5 ml of glacial acetic acid. Swirl to dissolve. Dilute to volume with iron-free water.

(D-341) Iron stock standard, 100 mg Fe/100 ml

Weigh exactly 0.100 g of pure iron wire and transfer to a 100 ml volumetric flask. Add 50 ml of 2 N HCl and allow the iron wire to dissolve. This may take several days. When all the wire is dissolved, dilute to volume with water.

(D-342) Phosphate buffer, 0.1 M, pH 7.4

a. Disodium phosphate, 0.1 M. Weigh 14.2 g of disodium phosphate, Na_2HPO_4, and transfer to a 1 L volumetric flask. Dissolve and dilute to volume with water. Mix.
b. Potassium dihydrogen phosphate, 0.1 M. Weigh 13.609 g of potassium dihydrogen phosphate, KH_2PO_4, and transfer to a 1 L volumetric flask. Dissolve and dilute to volume with water. Mix.

Mix 420 ml of 0.1 M disodium phosphate with 80 ml of 0.1 M potassium dihydrogen phosphate.

(D-343) Cholesterol stock standard, 1.25 mg/ml

Dissolve 125 mg of pure anhydrous cholesterol in glacial acetic acid and dilute to 100 ml.

• E

• Company directory

Abbott Laboratories
14th St. and Sheridan Rd.
North Chicago, Ill. 60064

Advanced Instruments, Inc.
45 Kenneth St.
Newton Highlands, Mass. 02161

Aldrich Chemical Co., Inc.
2371 N. 30th St.
Milwaukee, Wis. 53210

Allied Chemical Corp.
Industrial Chemicals Division
P. O. Box 70
Morristown, N. J. 07960

Aluminum Company of America
1501 Alcoa Bldg.
Pittsburgh, Pa. 15219

Alupharm Chemicals
610-612 Commercial Pl.
P. O. Box 30628
New Orleans, La. 70130

American Cyanamid Co.
Process Chemicals Department
Wayne, N. J. 07470

American Instrument Co., Inc.
8030 Georgia Ave.
Silver Spring, Md. 20910

Ames Co., Division of Miles Laboratories, Inc.
1127 Myrtle St.
Elkhart, Ind. 46514

Analtech Inc.
100 S. Justison St.
Wilmington, Del. 19801

H. Reeve Angel & Co., Inc.
9 Bridewell Pl.
Clifton, N. J. 07014

Arco Chemical Co.,
Division of AtlanticRichfieldCo.
260 S. Broad St.
Philadelphia, Pa. 19101

Armour Pharmaceutical Co.
P. O. Box 511
Kankakee, Ill.

Atlas Chemical Industries, Inc.
Wilmington, Del. 10899

Bausch & Lomb, Inc.
635 St. Paul St.
Rochester, N. Y. 14602

Beckman Instruments, Inc.
2500 Harbor Blvd.
Fullerton, Calif. 92634

Beckman Instruments, Inc.
Spinco Division
1117 California Ave.
Palo Alto, Calif. 94304

Behring Diagnostics, Inc., Subsidiary of American Hoechst Corp.
Behringwerke A. G.
400 Crossways Park Dr.
Woodbury, N. Y. 11797

Bio-Rad Laboratories
32nd and Griffin Ave.
Richmond, Calif. 94804

Blaisdell Publishing Co.
275 Wyman St.
Waltham, Mass. 02154

Brinkmann Instruments, Inc.
Cantiaque Rd.
Westbury, N. Y. 11590

Buchler Instruments, Inc.
1327 16th St.
Fort Lee, N. J. 07024

Calbiochem
3625 Medford St.
Los Angeles, Calif. 90063

Canalco Inc.
Canal Industrial Corp.
5635 Fisher Lane
Rockville, Md. 20852

Clay Adams, Division of
Becton, Dickinson & Co.
299 Webro Rd.
Parsippany, N. J. 07054

Cobe Laboratories, Inc.
1201 Oak St.
Denver, Colo. 80215

Coffman & Ladimer
Columbus, Ohio

Colab Laboratories, Inc.
1526 Halsted St.
Chicago Heights, Ill. 60411

Coleman Instruments,
Division of The Perkin-Elmer Corp.
42 W. Madison St.
Maywood, Ill. 60153

Dade, Division of American Hospital Supply Corp.
P. O. Box 672
Miami, Fla. 33152

Dajac Laboratories,
Division of The Borden Chemical Co.
5000 Langdon St.
Philadelphia, Pa. 19124

The C. L. Davis Co.
6955 Y St.
Lincoln, Neb. 68505

F. A. Davis Co.
1914 Cherry St.
Philadelphia, Pa. 19103

Denver Chemical Manufacturing Co., Inc.
35 Commerce Rd.
Stamford, Conn.

Distillation Products Industries,
Division of Eastman Kodak Co.
755 Ridge Road W.
Rochester, N. Y. 14603

The Dow Corning Corp.
P.O. Box 1592
2030 Abbott Rd. Center
Midland, Mich. 48640

Drew Chemical Co.
Boonton, N. J.

E-C Apparatus Corp.
755 St. Marks St.
Philadelphia, Pa. 19104

Eastman Organic Chemicals,
Division of Eastman Kodak Co.
343 State St.
Rochester, N. Y. 14650

Fisher Scientific Co.
711 Forbes Ave.
Pittsburgh, Pa. 15219

Floridin Co.
Three Penn Center
Pittsburgh, Pa. 15235

Fluka, A. G.
Buchs, S. G.
Switzerland

Gelman Instrument Co.
600 S. Wagner Rd.
Ann Arbor, Mich. 48106

General Diagnostics
Division of Warner-Lambert Pharmaceutical Co.
Morris Plains, N. J. 07950

W. R. Grace & Co.,
Davison Chemical Division
P. O. Box 188
Pompton Plains, N. J. 07444

Hamilton Co.
P. O. Box 307
12440 E. Lambert Rd.
Whittier, Calif. 90608

Harleco, Division of
American Hospital Supply Corp.
60th and Woodland Ave.
Philadelphia, Pa. 19143

Helena Laboratories
16147 Philomene
Allen Park, Mich. 48101

Hoffman-LaRoche Inc.
Kingsland Rd. and Bloomfield Ave.
Nutley, N. J. 07110

Howe & French, Inc.
99 Broad St.
Boston, Mass. 02110

Hyland, Division of Travenol Laboratories, Inc.
P. O. Box 2214
Costa Mesa, Calif. 02626

Hynson, Westcott & Dunning, Inc.
Charles and Chase Sts.
Baltimore, Md. 21201

Instrumentation Laboratory, Inc.
113 Hartwell Ave.
Lexington, Mass. 02173

Instrumentation Specialties Co., Inc.
4700 Superior
Lincoln, Neb. 68504

Johns-Manville Products Corp.
Celite Division
22 E. 40th St.
New York, N. Y. 10016

Kallestad Laboratories, Inc.
4005 Vernon Ave.
Minneapolis, Minn. 55416

K & K Laboratories, Inc.
121 Express St.
Plainview, N. Y. 11803

Kensington Scientific Corp.
1819 67th St.
Oakland, Calif. 94608

E. Leitz, Inc.
468 Park Ave. South
New York, N. Y. 10016

Eli Lilly & Co.
740 S. Alabama St.
Indianapolis, Ind. 46206

The London Co.
811 Sharon Dr.
Cleveland, Ohio 44145

3M Company
3M Center
St. Paul, Minn. 55101

Macalaster Bicknell Co. of Conn. Inc.
181 Henry St.
New Haven, Conn. 06507

Mallinckrodt Chemical Works
2nd and Mallinckrodt Sts.
St. Louis, Mo. 63160

Mann Research Laboratories
136 Liberty St.
New York, N. Y. 10006

The Matheson Co.
932 Paterson Plank Rd.
East Rutherford, N. J. 07073

Matheson, Coleman & Bell
2909 Highland Ave.
Norwood, Ohio 45212

Melpar Division, American Standard Co.
7700 Arlington Blvd.
Falls Church, Va. 22046

Miles Chemical Co.
Clifton, N. J.

Millipore Corp.
Ashby Rd.
Bedford, Mass. 01730

The C. V. Mosby Co.
3207 Washington Blvd.
St. Louis, Mo. 63103

Nalgene Labware Division,
Nalge Co.
75 Panorama Creek Dr.
Rochester, N. Y. 14625

National Aniline Division,
Allied Chemical Corp.
P. O. Box 70
Morristown, N. J. 07960

National Instrument Laboratories, Inc.
12300 Parklawn Dr.
Rockville, Md. 20852

Nutritional Biochemicals Corp.
26201 Miles Rd.
Cleveland, Ohio 44128

Palo Laboratories
New York, N. Y.

Pentex, Inc.
P. O. Box 272
195 W. Birch St.
Kankakee, Ill. 60901

Perkin-Elmer Corp.
723G Main Ave.
Norwalk, Conn. 06852

Phipps & Bird, Inc., Manufacturing Division
303 S. Sixth St.
Richmond, Va. 23205

Photovolt Corp.
1115 Broadway
New York, N. Y. 10010

Quickfit, Inc.
7 Just Rd.
Fairfield, N. J. 07006

Rohm & Haas Co.
Independence Mall West
Philadelphia, Pa. 19105

Schering Diagnostics
40 Markley St.
Port Reading, N. J. 07064

Schleicher & Schuell Co.
543 Washington St.
Keene, N. H. 03431

Schoeffel Instrument Corp.
24 Booker St.
Westwood, N. J. 07675

Scientific Service Laboratories, Inc.
P. O. Box 175
Dallas, Texas

Scientific Products
Division of American Hospital Supply Corp.
1210 Leon Pl.
Evanston, Ill. 60201

Shandon Scientific Co.
515 Broad St.
Sewickley, Pa. 15143

Sharp & Dohme
Division of Merck & Co., Inc.
West Point, Pa. 19486

Sigma Chemical Co.
3500 DeKalb St.
St. Louis, Mo. 63118

Simmler & Son Inc.
3755 Forest Park Ave.
St. Louis, Mo. 63108

G. Frederick Smith Chemical Co.
867 McKinley Ave.
P. O. Box 23344
Columbus, Ohio 43223

E. R. Squibb & Sons
745 Fifth Ave.
New York, N. Y. 10022

Technicon Corp.
511 Benedict Ave.
Tarrytown, N. Y. 10591

Arthur H. Thomas Co.
P. O. Box 779
Vine St. at Third
Philadelphia, Pa. 19105

Travenol Laboratories, Inc.
Artificial Organs Division
6301 Lincoln Ave.
Morton Grove, Ill. 60053

G. K. Turner Associates, Inc.
2524 Pulgas Ave.
Palo Alto, Calif. 94303

Ultra-Violet Products, Inc.
5114 Walnut Grove Ave.
San Gabriel, Calif. 91778

UniChem Corp.
3897 Stephens Court
Tucker, Ga. 30084

Union Carbide Corp.
270 Park Ave.
New York, N. Y. 10017

George T. Walker
Minneapolis, Minn.

Warner-Chilcott, General Diagnostics Division
201 Tabor Rd.
Morris Plains, N. J. 07950

Waters Associates, Inc.
61 Fountain St.
Framingham, Mass. 01701

Wheaton Glass Co.
1000 N. Tenth St.
Millville, N. J. 08332

Worthington Biochemical Corp.
Hall Mills Rd.
Freehold, N. J. 07728

• INDEX

C

E

J

K

O

Q

T

U

V